Eighteenth International Seaweed Symposium

Advances in Applied Phycology 1

Series Editor:
Michael A. Borowitzka
School of Biological Sciences & Biotechnology
Murdoch University, Murdoch, Western Australia

Aims and Scope

Applied Phycology, the practical use of algae, encompasses a diverse range of fields including algal culture and seaweed farming, the use of algae to produce commercial products such as hydrocolloids, carotenoids and pharmaceuticals, algae as biofertilizers and soil conditioners, the application of algae in wastewater treatment, renewable energy production, algae as environmental indicators, environmental bioremediation and the management of algal blooms. The commercial production of seaweeds and microalgae and products derived therefrom is a large and well established industry and new algal species, products and processes are being continuously developed.

The aim of this book series, *Advances in Applied Phycology*, is to present state-of-the-art syntheses of research and development in the field. Volumes of the series will consist of reference books, subject-specific monographs, peer reviewed contributions from conferences, comprehensive evaluations of large-scale projects, and other book-length contributions to the science and practice of applied phycology.

Eighteenth International Seaweed Symposium

Proceedings of the Eighteenth International Seaweed Symposium,
held in Bergen, Norway, 20 – 25 June 2004

Edited by

Robert Anderson, Juliet Brodie, Edvar Onsøyen and Alan T. Critchley

Hosted by

Norwegian Institute for Water Research (NIVA)
Institute of Marine Research (IMR)
Norwegian University of Science and Technology (NTNU)

Reprinted from the Journal of Applied Phycology, volume 18, nos. 3–5 (2006)

A C.I.P. Catalogue record for this book is available from the Library of Congress.

ISBN-13 978-90-481-7420-1 e-ISBN-13 978-1-4020-5670-3

Published by Springer,
P.O. Box 17, 3300 AA Dordrecht, The Netherlands.

www.springer.com

Printed on acid-free paper

TABLE OF CONTENTS

The successful execution of the
International Seaweed Symposium XVIII
was made possible by the
International Seaweed Association

ISA is particularly grateful to the following for allowing it to use part or all of their contribution for publishing these Proceedings of ISS XVIII

Degussa Texturant Systems
Japan Seaweed Association
Marinalg International
Multiexport, SA

University of British Columbia (Graduate Student Awards)

International Seaweed Association Executive Council

Juan Correa (Chile) President 2001–2004
Harris "Pete" Bixler (USA) Treasurer and President – elect 2004–2007
Tor L. Bokn (Norway)
Alan T. Critchley (France) (Editor-in-Chief) appointed 2003
Michael Friedlander (Israel)
Guillermo Garcia Reina (Spain, Canary Isl.)
In Kyu Lee (South Korea)
Masao Ohno (Japan)
Marianne Pedersén (Sweden)
Peter Salling (Spain)
Adelaida Semesi (Tanzania) (deceased)
Dimitri Stancioff (USA)

Eurico Oliveira (Brazil) Secretary – appointed 2004–2007
Mike Guiry (Ireland) Webmaster – appointed 2004

Jack McLachlan (Canada) Honorary Life Member
Mark A. Ragan (Australia) Honorary Life Member

National Organising Committee

Chair: Tor L. Bokn (NIVA, Oslo)
Secretary: Kjersti Sjøtun (IMR, Bergen)
Treasurer: Arild Steinnes (FMC Biopolymer, Drammen)
Officers: Jens Borum (University of Copenhagen, Copenhagen)
Kurt Ingar Draget (NTNU, Trondheim)
Stein Fredriksen (University of Oslo, Oslo)
Magne Gilje (FMC Biopolymer (Emeritus), Haugesund)
Marianne Pedersén (Stockholm University, Uppsala)
Olav Smidsrød (NTNU, Trondheim)
PCO: Kari Holmedal (PLUS Convention Norway AS, Bergen)

Acknowledgements

The organising committee of the XVIII International Seaweed Symposium also gratefully acknowledges the generous support of these sponsors:

Norwegian Research Council
Ministry of Foreign Affairs
Mayor of Bergen City
FMC Biopolymer
Algea as
The University of British Columbia
Sunniva and Egil Baardseth's Legacy
The Japan Seaweed Association
Marinalg International
MSC Co., Ltd., Korea
Hotel Augustin, Bergen
Rieber AS
Norwegian Institute for Water Research

Reviewers

The Editors of the Proceedings of the XVIIIth International Seaweed Symposium are grateful to the following people for reviewing manuscripts (in some cases more than one):

Anderson, R. J.; Ang, P. O.; Ask, E.; Avila, M.; Bartsch, I.; Bolton, J. J.; Buschmann, A.; Christensen, B. E.; Christie, H.; Collen, J.; Correa, J.; DeClerck, O.; Deysher, L.; Draget, K. I.; Dring, M.; Druehl, L.; Dunton, K.; Edwards, M.; Ertesvåg, H.; Farnham, W.; Fredericq, S.; Freshwater, W.; Friedlander, M.; Fujita, D.; Gåserød, O.; Gericke, N.; Givernaud, T.; Graham, M. H.; Griffiths, C. L.; Gurgel, F.; Holmes, M.; Hultmann, L.; Hurtado, A.; Indergaard, M.; John, D.; Kain, J.; Korez, R.; Kraan, S.; Larsen, B.; Leander, B.; Levy, I.; Lin, S-M.; Lindstrom. S.; Luxton, D.; MacNeill, S.; Maggs, C.; Mathieson, A.; McIvor, L.; Millar, A.; Molloy, F.; Moy, F. E.; Myklestad, S.; Nelson, W.; Neori, A.; Ohno, M.; Oliveira, E.; Oliveira, M.; Onsøyen, E.; Palmer, J.; Pedersén, M.; Pickering, T.; Potin, P.; Probyn, T.; Prud'Homme van Reine, W.; Rustad, T.; Sahoo, D.; Schaffelke, B.; Scrosati, R.; Sivertsen, K.; Skjåk-Bræak, G.; Smidsrød, O.; Smit, A. J.; Stekoll, M.; Stirk, W.; Tittley, I.; Troell, M.; Tronchin, E.; Ugarte, R.; Vairappan, C. S.; Wallentinus, I.; Weinberger, F.; Zuccarello, G. C.

Opening Address by Professor Juan Correa (Monday, June 21, 2004)

Professor Correa took great pleasure in declaring the XVIIIth International Seaweed Symposium meeting open. A hearty welcome was given to all presenters and participants. Professor Correa thanked all those responsible for making this meeting possible in Bergen and in particular Harris "Pete" Bixler, the ISS President-Elect and Eurico Oliveira, the Secretary, for the tremendous time and effort they had given to the business of the International Seaweed Association and in particular help with the logistics for the meeting in Bergen. There was a very good scientific and social programme ahead and he looked forward eagerly to all oral and poster presentations. The podium was then passed to John Rasmussen (President) and Pierre Kirsch (Secretary) to make the prestigious Marinalg International Awards.

Marinalg International Awards

Awards for presentations made at the XVIIth International Seaweed Symposium held in Cape Town, South Africa, in January 2001 and honoured at the XVIIIth International Seaweed Symposium held in Bergen, Norway, in June 2004.

Good morning. My name is John Rasmussen and I have the pleasure of representing Marinalg International, a world Association of Seaweed Processors dedicated to production and sales of ingredients for food, pharmaceutical, medical, cosmetic and feed applications.

I am proud to join this International Seaweed Symposium and I am proud to be in Norway. Norway has a long tradition in the Seaweed Industry and the Norwegian universities and Institutes have contributed with a big amount of valuable R&D work. When I was introduced to the seaweed industry 36 years ago some of the first articles I was asked to read to be introduced to carrageenan structure and functionalities were written by scientists from the University of Trondheim.

Within the seaweed world the International Seaweed Symposium, held every 3 years, is a great event which attracts people from all parts interested in Seaweed. Indeed the ISS is covering basic research on algae as well as work on the functionalities of the ingredients produced from seaweeds. Scientists, university and academic experts, R&D people from the industries using seaweed as raw materials and even people from the business and regulatory side of the industry are attracted.

Why?

Simply because the International Seaweed Symposium presents the latest developments on all aspects from micro- to macroalgae. Furthermore this event creates a fantastic environment for creating new contacts and relationships across the different areas of seaweed interests.

Marinalg International is a world-wide association founded in 1976 and which counts members from all continents – members who are producing agar, alginates and carrageenan both refined and semi-refined, all produced from seaweeds.

The members of Marinalg International appreciate very much the work done in universities and institutes around the world and have through many years sponsored the International Seaweed Symposium by contributing to the publication of the proceedings presented during the symposium and we are doing this also this year. However, Marinalg International also has a long tradition for awarding some of the presentations made at the previous symposium which today mean presentations from the 17th International Seaweed Symposium held in Cape Town 2001.

How do we select the winning papers. I can share with you – it is a very difficult job to make the choice among so many great presentations on a variety of titles. Our R&D people put forth a determined effort to judge papers as fairly as we can. The papers presented at the symposium are evaluated only after their publication. Each member company of Marinalg International is given the task of independently to select the papers to award one presentation within each of the areas Agar, Alginate and Carrageenan. The results of their evaluations are forwarded to the Marinalg International headquarter and the summary ands conclusions of the evaluations are drawn neutrally by the General Secretary of the Association."

Now, let us move to the winners from the 17th International Seaweed Symposium.

AGAR:
Th. Givernaud, A Mouradi, A Hassani, R Akallal and J Riyahi (Morocco)
"Design of a new technique for the reseeding of over-harvested beds of *Gelidium sesquipedale* (Turn.) Thuret (Rhodophyta, Gelidiales) in Morocco"

ALGINATE:
Ø Skaugrud and M Dornish (Norway)
"Biostructures of ultra-pure alginate for tissue engineering, directed drug delivery and cell encapsulation applications."

CARRAGEENAN:
Marcela Avila, A Candia, H Romo, H Pavez and C Torrijos (Chile)
"Exploitation and cultivation of *Gigartina skottsbergii* in Southern Chile".

Awards for Presentations made at XVIII ISS, Bergen

Three awards were made based on presentations and were distributed in the Closing Ceremony.

Sunniva and Egil Baardseth Legacy Award

Best student poster:
AFLP fingerprints reveal more than one introduction of the red alga *Heterosiphonia japonica* to the Norwegian Coast
Marit R. Bjærke and J. Rueness
Section for Marine Biology and Limnology, Department of Biology, University of Oslo, PO Box 1069, Blindern, N-0316 Oslo, Norway.

Best student paper:
Protoplast generation in vitro extruded protoplasm of the marine green alga *Bryopsis plumosa.*
Kim, GH and Tatyana Y. Klotchkova
Harmful Algal Blooming Control Laboratory, Department of Biology, Kongku National University, Kongju 314–701, Korea

University of British Columbia
Graduate Student Paper Awards

First prize:
Inducible response in two brown macroalgae form the northern central coast of Chile.
Erasmo Macaya [1], Rothäusler, E.[1], Thiel, M.[1], Molis, M.[2] and Wahl, M.[2].

Facultad Ciencias del Mar, [1]Universidad Católica del Norte, Larrondo 1281, Coquimbo, Chile. [2]Institut für Meereskunde, Düsterbrooker Weg 20, 24 105 Kiel, Germany.

Second prize:
Physiological response in *Palmaria palmata* to light micro-conditions and upwelled nutrients in summer.
Brezo Martínez[1], Rico, JM[2], Viejo, R[1]
[1]Área de Biodiversidad y Conservación. Universidad Rey Juan Carlos, Tulipán s/n, E-28933, Móstoles, Spain. [2]Área de Ecología, Universidad de Oviedo, Catedrático Rodrigo Uría n/s, E-33071 Oviedo, Spain.

Third prize:
The diversity, biology and distribution of the Gelidiaceae (Rhodophyta) of South Africa.
Enrico Tronchin[1], Freshwater, DW[2], Bolton, JJ[1], Anderson, RJ[3] and De Clerck, O[4].
[1]Department of Botany, University of Cape Town, Private Bag, Rondebosch 7701, South Africa. [2]Centre for Marine Science, University of North Carolina-Wilmington, 5600 Marvin Moss lane, Wilmington, NC 28409, USA. [3]Seaweed Research Unit, Marine and Coastal Management, Private Bag X2, Roggebaai 8012, South Africa. [4]Research Group Phycology, Biology Department, Ghent University, Krijgslaan 281/S8, 9000 Ghent, Belgium.

Japan Seaweed Association Poster Awards

Office of the JSA: Usa Marine Biological Institute, Kochi University, Usa-cho, Kochi, 781-11, Japan (Secretary, Professor M Ohno).

The First Prize was jointly awarded to:

- Phase behaviour of fish gelatin/carrageenan system.
  IJ Haug, KI Draget and O Smidsrød
  Department of Biotechnology/ NOBIPOL-NTNU, Norway,

and

- Carrageenan yield and gel properties of *Eucheuma isiforme* (Rhodophyta, Gigartinales) from Yucatan coast.
  Y Freile-Pelegrín, D Robledo and MA Domínguez Perez
  CINVESTAV-IPN/Unidad Mérida, Mexico.

Closing Ceremony
Tor L Bokn

Dear Friends. Every nice event has to come to an end. We are at that point now. It is my duty, but also my great pleasure to thank all of you for your co-operation, your kind attitude to all of us in the local committee.

Yes, we are exhausted, but this is a good feeling. We will give our salute to you and to our brave students assisting us with all topics.

The organising committee of the XVIIIth International Seaweed Symposium gratefully acknowledges the generous support of these sponsors:

Norwegian Research Council
Ministry of Foreign Affairs
Mayor of Bergen City
FMC Biopolymer
Algea as
The University of British Columbia

Sunniva and Egil Baardseth's Legacy
The Japan Seaweed Association
Marinalg International
MSC Co., Ltd., Korea
Hotel Augustin, Bergen
Rieber AS
Norwegian Institute for Water Research

We are grateful to all of them. We also give our thanks to the International Seaweed Association Council – ISAC for their efforts.

I will call upon Kari Holmedal to the stage please! Here is the lady chasing you through the world for money.

Now I call upon the chairpersons for the awards committees for the presentations of the University of British Columbia, Best Student Presentation and the Japan Seaweed Association, Best Poster Award (Professors Bolton, Wallentinus, Aruga and Ohno).

Finally we call upon Professors Aruga and Ohno the organisers of the XIXth International Seaweed Symposium, Kobe, Japan, March 26–31, 2007. Theme: *SEAWEEDS: Science and Technology for Traditional and Modern Utilisation.*

New President's Closing Remarks – Incoming President of the International Seaweed Association: Harris "Pete" Bixler.

First of all let me bring the members up to date on matters of your governing council that provides continuity between symposia. It is with regret that our by-laws require that we loose Marianne Pedersén who has given us wise council for more years than she would probably like me to mention. I cannot leave reference to Marianne without adding my personal thanks for her beautiful dedication to Adelaida Semesi on Monday. Tor Bokn will all be going off the council since he was filling the unexpired term of Peter Gacesa. However, I am sure we will still be hearing from him at future symposia. Let us give our hands to Marianne and Tor for their years of service.

Replacing these councillors will be Rob Anderson from Marine and Coastal Management, South Africa, Thierry Chopin of the University of New Brunswick, Canada and Rhodora Azanza of the University of the Philippines at Diliman.

It is also my pleasure to thank Tor Bokn and his National Organising Committee for an excellent symposium. We were all impressed with the organisation of the Programme Book and the posters. It was easy to navigate form one session of interest to another. I understand we have Kurt Draget to thank for that and as Treasurer of ISA I have a warm spot in my heart for Arild Steinnes who stood as Treasurer of the LOC. Yes, Bergen was expensive but it is a beautiful city that we have all had a chance to enjoy through the imaginative outside and accompanying persons events or through our individual resources. Thanks Tor and Committee....and by the way, I think it was a stroke of greatly deserved national pride when you included pieces by Grieg in the Opening Ceremonies.

Concluding ISA matters I would like to acknowledge the new President-elect of ISA who will follow me as President after the ISS XIX in March, 2007. Thierry Chopin was elected unanimously to this post on Sunday. It is a great pleasure for me to have Thierry in this position since we are only a few hours apart by car which should facilitate good exchange of ideas for building ISA in the seaweed science community. Congratulations Thierry.

We now move on to the XIXth International Seaweed Symposium. It is my pleasure to inform the members of the International Seaweed Association that the XIXth ISS will be held March 26–31, 2007 in Kobe, Japan. The co-chairs of the LOC, Professor Yusho Aruga and Professor Masao Ohno, are here with us and Professor Ohno, a loyal and long time Council member of ISA, will give a brief introduction to ISS XIX. I should also point out that

Professor Aruga is now President of the prestigious Japan Seaweed Association. I can assure you that the cost of accommodations and meals will be less than here in Bergen making it easier for students to attend. There will be an exhibition of seaweed products in the Kobe Convention Centre running in parallel with ISS XIX. The exhibition will be open to all ISS XIX attendees and will be an interesting complement to our Technical Program.

And now, unless Eurico Oliveira, our very efficient Secretary, stops me short because I have neglected some important closing duty I declare the XVIIIth ISS adjourned!

XVIIIth International Seaweed Symposium, Bergen, Norway, June 21, 2004

In memory of Professor Adelaida Kleti Semesi, University of Dar es Salaam, Tanzania

Dear Friends and colleagues,

Professor Adelaida Semesi, Tanzania, a member of our executive council, passed away on February 6th 2001. She was an outstanding scholar in marine sciences. She pioneered research on seaweeds and seaweed farming in Tanzania. Mama Semesi, which she also was called, was a great botanist who also was trained in microbiology. She worked at the University of Dar es Salaam for over twenty years and became the Head of the Department of Botany and Associate Dean of the Faculty.

From the year 1996–2000 Adelaida Semesi worked as a professor at the Centre for International Environment and Development Studies, here in Norway at the Agricultural University at Ås teaching in tropical ecology. In 2000 she became the Director of the Marine Science Institute in Zanzibar, University of Dar es Salaam.

She worked as a scientist in various international laboratories, including the University of South Florida in Tampa, USA, the University of Nijmegen in the Netherlands and the University of Ibadan in Nigeria. She was a council member of many associations like the International Society of Mangrove Ecosystems and the Western Indian Ocean Marine Science Association (Wiomsa). A fund, the Professor Adelaida Kleti Semesi Memorial Trust Fund, has been launched in her memory.

I met Adelaida Semesi for the first time in 1988 when we wrote an application to Sida/Sarec, Sweden for a research programme in marine sciences between Tanzania and Sweden. This marine science programme has now been running successfully for more than ten years with many PhD students from both Tanzania and Sweden.

Adelaida was a very warm friend and an enthusiastic scientist. She was very quick in writing manuscripts and applications. I remember when she told me that in the morning she had to get up early to go out and cut grass for her seven cows before she went to the university. I do not think I would have managed that. When she started to study the structures of carrageenans in red algae as a PhD she visited many laboratories abroad to learn the analytical procedures e.g. the Unilever Research Company in UK.

I once asked her how she had managed, as a woman in Africa with four children, to become the Head of the Department of Botany at the University of Dar es Salaam. Her modest answer to me was that she had a very understanding husband and a supportive family.

Adelaida gave us many happy moments with good laughter. We shared a room at the Seaweed Symposium in Valdivia. I became very ill with a stomach infection and she offered me all sorts of cures, for example bananas. She was such an adorable person and full of energy and smiles all the time. It is just unbelievable that she is not with us any more. Today we will remember her as the loveable person she was. She had so many ideas for how to improve the world. We will always remember her for her contributions to marine science development in Tanzania and the Western Indian Ocean. It is a great loss for Science, for Africa and for us. In the executive council of the International Seaweed Association she raised many questions and came up with many new ideas. We miss a great friend and a fantastic person. May you rest in peace Adelaida and thanks for all.

Marianne Pedersén
Stockholm University
Sweden

Professor Adelaida Kleti Semesi,
University of Dar es Salaam, Tanzania
Passed away February 6, 2001

List of registrants

Alekseeva, Svetlana
Pacific Institute of Bioorganic Chemistry
RUSSIA
mik@piboc.dvo.ru

Aminina, Natalia
Pacific Scientific Research Fisheries Centre
RUSSIA
aminina@tinro.ru

Andersen, Sanne Hjorth
National Environmental Research Institute
DENMARK
sha@dmu.dk

Andersen, Morten Birket
Cambrex Aps
DENMARK
morten.andersen@cambrex.com

Anderson, Robert
Marine and Coastal Management
SOUTH AFRICA
anderson@botzoo.uct.ac.za

Andersson, Markus
University of Uppsala
SWEDEN
markus.andersson@ebc.uu.se

Ang, Put Jr.
The Chinese University of Hong Kong
CHINA
put-ang@cuhk.edu.hk

Angelfoss, Helle
University of Bergen
NORWAY
helle.angelfoss@student.uib.no

Armenta Gonzalez, Andres
Productos del Pacifico S.A De C.V.
MEXICO
andargon1@starband.net

Aruga, Yusho
Tokyo University of Agriculture
JAPAN
aruga@nodai.ac.jp

Ask, Erick
FMC BioPolymer
USA
erick_ask@fmc.com

Avila, Marcela
Instituto Fomento Pesquero
CHILE
mavila@ifop.cl

Azanza, Rhodora V.
University of the Philippines
PHILIPPINES
rhod@upmsi.ph

Azis, Muhammed
Pt. Batara Laut Celebes
INDONESIA
bataracelebes@yahoo.com

Baek, Jae Min
Seaweed Research Center, NFRDI
SOUTH KOREA
jmbaek@momaf.go.kr

Baricuatro, Farley
FMC Marine Colloids Philippines Inc.
PHILIPPINES
farley_baricuatro@fmc.com

Barrett, Tony
Kilkieran
IRELAND
tbarrett@arramara.ie

Bartsch, Inka
Alfred-Wegener-Institut für Polar-und Meeresforschung
GERMANY
ibartsch@awi-bremerhaven.de

Batista De Vega, Gloria
University of Panama
PANAMA
gloriabv@cwpanama.net

Baweja, Pooja
University of Delhi
INDIA
poojabaweja2003@yahoomail.com

Bengtsson, Mia
Stockholm University
SWEDEN
mia.bengtsson@botan.su.se

Bhattacharya, Debashish
University of Iowa
USA
dbhattac@blue.weeg.uiowa.edu

Billard, Emmanuelle
Station Biologique de Roscoff UMR CNRS 7127
FRANCE
billard@sb-roscoff.fr

Birkeland, Gunnar
FMC Biopolymer
NORWAY
gunnar_birkeland@fmc.com

Bixler, Harris
Ingredients Solutions, Inc.
USA
pbixler@isinc.to

Bjærke, Marit Ruge
University of Oslo
NORWAY
m.r.bjarke@bio.uio.no

Bleicher-Lhonneur, Genevieve
Degussa Texturant Systems France SAS
FRANCE
genevieve.bleicher-lhonneur@degussa.com

Blouin, Nicolas Achille
University of Maine
USA
nicolas_blouin@umit.maine.edu

Blümel, Christian
University of Rostock
GERMANY
christian.bluemel@biologie.uni-rostock.de

Bodeau, Christine
Science Et Mer Laboratories
FRANCE
c.bodeau@wanadoo.fr

Bokn, Tor L.
NIVA
NORWAY
tor.bokn@niva.no

Bolton, John
Universtity of Cape Town
SOUTH AFRICA
bolton@botzoo.uct.ac.za

Boo, Sung Min
Chungnam National University
KOREA
smboo@cnu.ac.kr

Boonprab, Kangsadan
Kasetsart University
THAILAND
ffisksb@nontri.ku.ac.th

Borum, Jens
University of Copenhagen
DENMARK
jborum@zi.ku.dk

Brault, Dominique
CEVA
FRANCE
dominique.brault@ceva.fr

Brawley, Susan
University of Maine
USA
brawley@maine.edu

Brock, Elisabet
Göteborg University
SWEDEN
elisabet.brock@marbot.gu.se

Brodie, Juliet
Bath Spa University College
UNITED KINGDOM
j.brodie@bathspa.ac.uk

Bruns, Svenja
Queen's University of Belfast
UNITED KINGDOM
svenja.bruns@gmx.de

Buschmann, Alejandro
Universidad De Los Lagos
CHILE
abuschma@ulagos.cl

Carnachan, Susie
Gracefield Research Centre
NEW ZEALAND
s.carnachan@irl.cri.nz

Cecere, Ester
Iamc Cnr
ITALY
cecere@istta.le.cnr.it

Chapman, Anthony
Dalhousie University
CANADA
achapman@dal.ca

Chirapart, Anong
Kasetsart University
THAILAND
ffisanc@nontri.ku.ac.th

Choi, Han-Gil
Wonkwang University
SOUTH KOREA
hgchoi@wonkwang.ac.kr

Chopin, Thierry
University of New Brunswick
CANADA
tchopin@unbsj.ca

Chow Ho, Fungyi
University of Sao Paulo
BRAZIL
fchow@ib.usp.br

Christie, Hartvig
Norwegian Institute for Nature Research
NORWAY
hartvig.christie@nina.no

Ciancia, Marina
University of Buenos Aires
ARGENTINA
labcer@qo.fcen.uba.ar

Colin, Sebastien
UMR 7139 CNRS-Goëmar-UPMC
FRANCE
colin@sb-roscoff.fr

Collen, Jonas
UMR 7139 CNRS-Goëmar-UPMC
FRANCE
nyvall@sb-roscoff.fr

Connan, Solene
Lebham-Iuem
FRANCE
solene.connan@univ-brest.fr

Correa, Juan
Pontifica Universidad Catolica De Chile
CHILE
jcorrea@bio.puc.cl

Coury, Dan
Mie University
JAPAN

Coyer, James
University of Groningen
NETHERLANDS
coyeja@biol.rug.nl

Critchley, Alan T.
Degussa Texturant Systems France Sas
FRANCE
alan.critchley@degussa.com

De Clerck, Olivier
Research Group Phycology, Biology Dept., Ghent University
BELGIUM
olivier.declerck@ugent.be

De Vries, Joop
Danisco AS
DENMARK
g8pbr@danisco.com

Delara-Isassi, Graciela
Universidad Autonoma Metropolitana-Iztapalapa
MEXICO
grace@xanum.uam.mx

Delaroque, Nicolas
Max-Planck-Institute for Chemical Ecology
GERMANY
ndelaroque@ice.mpg.de

Destombe, Christophe
Station Biologique de Roscoff UMR CNRS 7127
FRANCE
destombe@sb-roscoff.fr

Deveau, Jean-Paul
Acadian Seaplants Limited
CANADA
kbutler@acadian.ca

Deveau, Louis
Acadian Seaplants Limited
CANADA
info@acadian.ca

Deysher, Larry
Ocean Imaging
USA
deysher@oceani.com

Dion, Patrick
CEVA
FRANCE
laurence.gautierbillon@ceva.fr

Draget, Kurt Ingar
NTNU
NORWAY
kurt.i.draget@biotech.ntnu.no

Dring, Matthew
Queen's University Belfast
IRELAND
m.dring@qub.ac.uk

Druehl, Louis
Bamfield Marine Sciences Centre
CANADA
LDRUEHL@ISLAND.NET

Dunton, Kenneth
University of Texas at Austin
USA
dunton@utmsi.utexas.edu

Edwards, Matthew
San Diego State University
USA
edwards@sciences.sdsu.edu

Eggereide, Sarah Fagertun
University of Bergen
NORWAY
sarah.eggereide@student.uib.no

Ekelund, Nils
Mid Sweden University
SWEDEN
nils.ekelund@mh.se

Endo, Terumasa
Nihon Fuji Industries Corp
PHILIPPINES
nfmulti@skyinet.net

Engel, Carolyn R.
Station Biologique de Roscoff UMR CNRS 7127
FRANCE
engel@sb-roscoff.fr

Eriksen, Mikael
Eurogum AS
DENMARK
me@eurogum.dk

Falduto, Daniela
University of Messina
ITALY
d.falduto@tiscali.it

Faugeron, Sylvain
Pontifica Universidad Catolica De Chile
CHILE
sfaugero@bio.puc.cl

Fazal, Murtaza
C-Weed Corporation
TANZANIA
morty@fazal.com

Fei, Xiugeng
IOCAS Chinese Academy of Sciences
CHINA
fei@ms.qdio.ac.cn

Floc'h, Jean Yves
Lebham-Iuem
FRANCE
j-y.floch@vanadoo.fr

Fredericq, Suzanne
University of Louisiana at Lafayette
USA
slf9209@louisiana.edu

Fredriksen, Stein
Universitetet I Oslo
NORWAY
stein.fredriksen@bio.uio.no

French, Rosabelle
Helen S. Dunn School
USA
rfrench@midmaine.com

Friedlander, Michael
Israel Oceanographic and Limnological Research
ISRAEL
michael@ocean.org.il

Fujita, Daisuke
Tokyo University of Marine Science and Technology
JAPAN
d-fujita@s.kaiyodai.ac.jp

Fujiyoshi, Eiji
Seikai National Fisheries Research Institute
JAPAN
fujieiji@affrc.go.jp

Furnari, Giovanni
Dipartimento Di Botanica Dell Universita
ITALY
g.furnari@mbax.dipbot.unict.it

Gabrielsen, Bjørn Olav
Alegea As
NORGE
bjorn.olav.gabrielsen@algea.com

Gargiulo, Gaetano Maurizio
University of Messina
ITALY
ggargiulo@unime.it

Genovese, Giuseppa
University of Messina
ITALY
ggenovese@unime.it

Gilje, Magne
NORWAY
m-gilje@frisurf.no

Graham, Michael H.
Moss Landing Marine Laboratories
USA
mgraham@mimi.calstate.edu

Guanpin, Yang
CHINA
yguanpin@mail.ouc.edu.cn

Gurgel, Frederico
University of Louisiana at Lafayette
USA

Gåserød, Olav
FMC Biopolymer
NORWAY
olav_gaserod@fmc.com

Ha, Jin Hwan
Cheju National University
SOUTH KOREA
jinhha@cheju.ac.kr

Haetta, Per
Biosign ASP
DENMARK

Hafting, Jeff
Big Island Abalone Corporation
USA
jhafting@bigislandabalone.com

Hagen, Nils T
Bodø Regional University
NORWAY
nils.hagen@hibo.no

Haji Gapor, Razaili
Fisheries Development Authority of Malaysia (LKIM)
MALAYSIA
iliazar@yahoo.com

Hara, Yoshiaki
Yamagata University
JAPAN
hara@sci.kj.yamagata-u.ac.jp

Haug, Ingvild J
Nobipol - NTNU
NORWAY
ingvild.haug@biotech.ntnu.no

Hee, Torben
FMC
DENMARK
toreben_hee@fmc.com

Hernandez, Gustavo
Centro Interdisciplinario De Ciencias Marinas – IPN
USA
gearmona@ipn.mx

Hertz, Ole
Arctic Ecological Research
DENMARK
olehertz@post.tdcadsl.dk

Holte, Øyvind
University of Oslo
NORWAY
oyvind.holte@farmasi.uio.no

Hommersand, Max
University of North Carolina
USA
hommersand@bio.unc.edu

Honda, Masaki
Abiko Research Laboratory
JAPAN
m-honda@criepi.denken.or.jp

Hong, Yong-Ki
Pukyong National University
KOREA
ykhong@pknu.ac.kr

Husa, Vivian
Havforskningsintstituttet
NORWAY
vivianh@imr.no

Huusfeldt, Trine
Biosign ASP
DENMARK
th@biosign.dk

Hwang, Eun Kyoung
Seaweed Research Centre, NFRDI
KOREA
ekhwang@hotmail.com

Hwang, Mi Sook
Seaweed Research Center
KOREA
mishwang@nfrdi.re.kr

Iri, Tadao
Proagar S.A
CHILE
proaiti@sunnet.cl

Israel, Alvaro
Israel Oceanographic and Limnological Research
ISRAEL
alvaro@ocean.org.il

Isæus, Martin
University of Stockholm
SWEDEN
martin.isaeus@botan.su.se

Iwamoto, Katsuaki
JAPAN
iwamoto@marine-science.co.jp

Jedrzejczak, Marcin Filip
Polish Academy of Science
POLAND
humbak@praeter.pl

Jeon, You Jin
Cheju National University
SOUTH KOREA
youjinj@cheju.ac.kr

Jones, Joanna M.
Australian National University
AUSTRALIA
jo@jokain.net

Kadoya, Kiyoshi
Kadoya & Co Ltd
JAPAN
agarcokk@gold.ccn.ne.jp

Kajiwara, Tadahiko
Yamaguchi University
JAPAN
kajiwara@agr.yamaguchi-u.ac.jp

Kakinuma, Makoto
Mie University
JAPAN
kakinuma@bio.mie-u.ac.jp

Kakita, Hirotaka
Institute of Marine Resources and Environment
JAPAN
h-kakita@aist.go.jp

Karez, Rolf
Landesamt für Natur und Umwelt (LANU)
GERMANY
rkarez@lanu.landsh.de

Karsten, Ulf
University of Rostock
GERMANY
ulf.karsten@biologie.uni-rostock.de

Kattan, Daniel
Peruvian Seaweeds Srl
PERU
danny@peruvianseaweeds.com

Kawai, Hiroshi
Kobe University
JAPAN
kawai@kobe-u.ac.jp

Kim, Nam-Gil
Gyeonsang National University
KOREA
ngkim@gaecjil.gsmu.ac.kr

Kim, Kil Jae
MSC Co Ltd
KOREA
msc@msckorea.com

Kim, Soo Hyun
Cheju National University
SOUTH KOREA
kshyun@cheju.ac.kr

Kim, Myung Sook
Pusan National University
KOREA
myungskim@pusan.ac.kr

Kim, Gwang Hoon
Kongju National University
SOUTH KOREA
ghkim@kongju.ac.kr

Kim, Hyung-Geun
Kangnung National University
CHINA
kim@kangnung.ac.kr

Kimura, Hajime
Wakayama Research Center of Agriculture
JAPAN
kimura_h0005@pref.wakayama.lg.jp

Kirihara, Shinji
Aomori Prefectural Fisheries Research Center
JAPAN
shinji_kirihara@ags.pref.aomori.jp

Kirsch, Pierre
Marinalg International
FRANCE
kirsch@khlaw.be

Kitade, Yukihiro
Graduate School of Fisheries Sciences, Hokkaido University
JAPAN
ykitade@msb.biglobe.ne.jp

Kloareg, Bernard
UMR 713G Station Biologique
FRANCE
kloareg@sb-roscoff.fr

Klochkova, Tatyana A.
Kongju National University
SOUTH KOREA

Koivikko, Riitta
University of Turku
FINLAND
riliko@utu.fi

Koo, Jae-Geun
Kunsna National University
KOREA
kseaweed@kunsan.ac.kr

Korolyova, Tatyana N
Kamchat NIRO
RUSSIA
tatyana@port.kamchatka.ru

Kraan, Stefan
National University of Ireland
IRELAND
kraanska@hotmail.com

Kreag, John
Acdi/voca Seegaad Project
TANZANIA
acdivoca@kaributanga.com

Kristiansen, Aase
University of Copenhagen
DENMARK
aasek@bi.ku.dk

Krupnova, Tatiana
Pacific Research Centre TINRO
RUSSIA
tatkr51-tinro@yandex.ru

Kudo, Hajime
Yamagata University
JAPAN
hajime@sbiol.kj.yamagata-u.ac.jp

Lago-Leston, M.Asuncion
Universidade do Algarve
PORTUGAL
alago@ualg.pt

Lee, Wook Jae
Chungnam National University
KOREA
algae@cnu.ac.kr

Lee, Ki Wan
Cheju National University
SOUTH KOREA
kiwanee@cheju.ac.kr

Lee, Joon-Back
College of Ocean Sciences
KOREA
jblee@cheju.ac.kr

Lee, Antonio
Solomon Seaweed
SOLOMON ISLANDS
lkk@business.com.sb

Leonardi, Patricia
Universidad Nacional Del Sur
ARGENTINA
leonardi@criba.edu.ar

Levy, Israel
Noritech-Seaweed Biotechnologies Ltd.
ISRAEL
israel@noritech.co.il

Lewin, Ralph
University ff California, San Diego
USA
rlewin@ucsd.edu

Lim, Tae IL
Taerim Trading Co, Ltd
SOUTH KOREA
tr0316@naver.com

Lin, Showe-Mei
National Taitung University
TAIWAN
smlin@nttu.edu.tw

Lindén, Åsa
Södertörns University College
SWEDEN
asa.linden@sh.se

Lindstrom, Sandra
University of British Columbia
CANADA
sandracl@interchange.ubc.ca

Lion, Ulrich
Max-Planck-Institute for Chemical Ecology
GERMANY
ulion@ice.mpg.de

Listak, Madis
Tallinn Technical University
ESTONIA
madis.listak@mail.ee

Lourenco, Sergio O.
Universidade Federal Fluminense
BRAZIL
solourenco@yahoo.com

Lugazo, Zuberi
Acdi/voca Seegaad Project
TANZANIA
seegaad@kaributana.com

Luhan, Maria Rovilla
Southeast Asian Fisheries Development Center
PHILIPPINES
mrhluhan@aqd.seafdec.org.ph

Lundsør, Elisabeth
University of Bergen
NORWAY
elisabeth.lindsor@imr.no

Luong-Van, Jim Thinh
Charles Darwin University
AUSTRALIA
jim.luong-van@ntu.edi.au

Lurton, Luc
Ceva
FRANCE
laurence.gautierbillon@ceva.fr

Luxton, David
D.Luxton & Associates Ltd.
NEW ZEALAND
dlatganz@enternet.co.nz

Lüder, Ulrike
Alfred Wegner Institute for Polar and Marine Research
GERMANY
ulueder@awi-bremerhaven.de

Macaya, Erasmo
Universidad Catolica Del Norte
CHILE

Maly, Ritha
Government
TANZANIA
fisheries@accensstranzania.com

Manevelt, Gavin
University of the Western Cape
SOUTH AFRICA
gmaneveldt@uwc.ac.za

Mansilla, Andres
Universidad De Magallanes
CHILE
andreas.mansilla@umag.cl

Marcos Ramirez, Roberto
Productos Del Pacifico S.A De C.V.
MEXICO
kelpropac@yahoo.com.mx

Martin, Georg
Estonian Marine Institute
ESTONIA
georg@klab.envir.ee

Martinez, Maria Brezo
Universidad Rey Juan Carlos
SPAIN
bmartinez@escet.urjc.es

Martinez, Enrique A
Centro De Estodios Avanzados En Zonas Aridas
CHILE
emartine@userena.cl

Mazloomi Arjagh, Mohammad
Iranian Fisheries Research Organization
IRAN
arjagh@yahoo.com

McNeill, Sally
Gracefield Research Centre
NEW ZEALAND
s.mcneill@irl.cri.nz

Michel, Gurvan
UMR 7139 CNRS-Goëmar-UPMC
FRANCE
gurvan@sb-roscoff.fr

Mikhaylova, Tatiana
SevPINRO
RUSSIA
tania@sevpinro@ru

Miravalles, Alicia
Universidad Nacional Del Sur
ARGENTINA
abmiraua@criba.edu.ar

Mohandoss, Sidharthan
Soonchunhyang University
SOUTH KOREA
akmsiddhu@hotmail.com

Morabito, Marina
University of Messina
ITALY
marina.morabito@tiscali.it

Mortensen, Agnes Mols
University of Copenhagen
DENMARK
agnes@nordlysid.com

Moujahid, Abderrahman
University of Hasjan
MOROCCO
moujahiolo@yahoo.fr

Moy, Frithjof E.
Norwegian Institute for Water Research Niva
NORWAY
frithjof.moy@niva.no

Munda, Ivka Maria
The Slovene Academy of Science
SLOVENIA
spela@zrc-sazu.si

Myklestad, Sverre M
Norwegian University of Science and Technology
NORWAY
sverre.myklestad@chembio.ntnu.no

Nagahisa, Eizo
Kitasato University, School of Fisheries Sciences
JAPAN
nagahisa@kitasato-u.ac.jp

Nam, Ki Wan
Pukyong National University
KOREA
kwnam@pknu.ac.kr

Namudu, Merekeleni
University of the South Pacific
FIJI
mere_namudu@hotmail.com

Nanba, Nobuyoshi
Kitasato University
JAPAN
nanba@kitasato-u.ac.jp

Nelson, Wendy
National Institute for Water & Atmospheric Research
NEW ZEALAND
w.nelson@niwa.co.nz

Neori, Amir
Israel Oceanographic and Limnological Research
ISRAEL
aneori@shani.net

Nishide, Eiichi
JAPAN
west620@yhb.att.ne.jp

Notoya, Masahiro
Tokyo University of Marine Science and Technology
JAPAN
notoya@kaiyodai.ac.jp

Nyberg, Cecilia
University of Göteborg
SWEDEN
cecilia.nyberg@marbot.gu.se

Nyberg, Maria
Mid Sweden University
SWEDEN
maria.nyberg@mh.se

Nygård, Charlotta
Mid Sweden University
SWEDEN
lotta.nygard@mh.se

Nygaard, Kari
Norwegian Institute of Water Research
NORWAY
kari.nygaard@niva.no

Nyvall Collen, Pi
UMR 7139 CNRS-Goëmar-UPMC
FRANCE
nyvall@sb-roscoff.fr

Ogawa, Hisao
Kitasato University
JAPAN
seaweeds@kitasato-u.ac.jp

Ohno, Masao
Kochi University
JAPAN
mohno@cc.kochi-u.ac.jp

Olivera, Mariana C.
University of Sao Paulo
BRAZIL
mcdolive@usp.br

Olivera, Eurico
University of Sao Paulo
BRAZIL
euricodo@usp.br

Olsen, Bernt
University of Bergen
NORWAY
bernt.olsen@student.uib.no

Paes De Barros, Marcelo
Universidade Cruzeiro Do Sul
BRAZIL
marcelo.barros@unicsul.br

Paoletti, Sergio
University of Trieste- Biochemistry Dept.
ITALY
paolese@bbcm.units.it

Parente, Manuela
University of Portsmouth
UNITED KINGDOM
mparente@notes.uae.pt

Park, Chan Sun
Mokpo National University
SOUTH KOREA
espark85@mokpo.ac.kr

Pedersen, Are
University of Connecticut
USA
areped@attglobal.net

Pedersen, Morten Foldager
University of Roskilde
DENMARK
mfp@ruc.dk

Pedersén, Marianne
University of Stockholm
SWEDEN
marianne.pedersen@botan.su.se

Pelegrin, Yolanda Freile
Cinvestav – IPN Unidad Merida
MEXICO
freile@mda.cinvestav.mx

Pena Freire, Viviana
University of A Coruna
SPAIN
viviana@udc.es

Pendle, Derrick
Atoll Seaweed Company Limited
KIRIBATI

Petrocelli, Antonella
Instituto Ambiente Marino Costiero IAMC
ITALY
antonella.petrocelli@istta.le.cnr.it

Phillips, Julie
University of Queensland
AUSTRALIA
j.phillips@mailbox.uq.edu.au

Piantini, Rene
Prodalmar Ltda
CHILE
rpiantini@prodalmar.cl

Pickering, Timothy
The University of the South Pacific
FIJI
pickering_t@usp.ac.fj

Pino, Hugo
Alimentos Multiexport S.A.
CHILE
hpino@mulitexport.cl

Podkorytova, Antonina
Russian Federal Research Institute of Fisheries & Oceanography
RUSSIA
podkor@vniro.ru

Pohnert, Georg
Max Planck Institute for Chemical Ecology
GERMANY
pohnert@ile.mpg.de

Porse, Hans
Cp Kelco Aps
DENMARK
hans.porse@cpkelco.com

Potin, Philippe
UMR 7139 CNRS-Goëmar-UPMC
FRANCE
potin@sb-roscoff.fr

Prud'homme Van Reine, Willem
National Herbarium Nederland
NETHERLANDS
prudhomme@nhn.leidenuniv.nl

Ramirez, Daniel Robledo
Cinvestav – IPN Unidad Merida
MEXICO
robledo@mda.cinvestav.mx

Ramirez, Alberto
Fundacion Chile
CHILE

Rasmussen, John
Danisco AS
DENMARK
john.rasmussen@danisco.com

Renoux, Aline
Universitè Antilles Guyane
GUADELOUPE

Repina, Olga
Northern Branch of The Polar Research Inst.
RUSSIA
repina@sevprino.ru

Rindi, Fabio
National University of Ireland
IRELAND
fabio.vindi@nuigalway.ir

Riosmena-Rodriguez, Rafael
Universidad Autonoma De Baja California Sur
MEXICO
riosmena@uacbs.mx

Robertson-Andersson, Deborah
University of Cape Town
SOUTH AFRICA
broberts@botzoo.uct.ac.za

Rodriguez, Ignacio Sanchez
Centro Interdisociplinario De Ciencias Marinas
MEXICO
isanchez@ipn.mx

Romo, Hector
Universidad De Concepcion
CHILE
hromo@udec.cl

Rothman, Mark
Marine and Coastal Management
SOUTH AFRICA
mrothman@botzoo.uct.ac.za

Rueness, Jan
University of Oslo
NORWAY
jan.rueness@bio.uio.no

Rønningen, Vera
University of British Columbia
CANADA
verar@interchange.ubc.ca

Sahoo, Debasish
University of Delhi
INDIA
debasishsahoo@yahoo.co.in

Salling, Peter
Hispanagar
SPAIN
psalling@afina.es

Sanderson, John Craig
Scottish Institute of Marine Sciences (UHI)
UK
craig.sanderson@sams.ac.uk

Santelices, Bernabe
Pontifica Universidad Catolica De Chile
CHILE
bsanteli@bio.puc.cl

Sato, Minoru
Tohoku University
JAPAN
msato@bios.tohoku.ac.jp

Schaffelke, Britta
CRC Reef Research Centre
AUSTRALIA
britta.schaffelke@crcreef.com

Schsching, Eleng
Murmansk State Technical University
RUSSIA

Schygula, Christof
University of Rostock
GERMANY
christof@schygula@biologie.uni-rostock.de

Searle, Richard
ISP Alginates Ltd.
SCOTLAND
rsearle@ispcerp.com

Segovia, Danilo
Fundacion Chile
CHILE

Seth, Abhiram
Pepsi Foods Private Limited
INDIA

Shin, Hyun-Woung
Soonchunhyang University
SOUTH KOREA
hwshin@sch.ac.kr

Sieber, Volker
Degussa Ag Projekthaus Biotechnologie
GERMANY
volker.sieber@degussa.com

Silva, Paul C.
University of California, Berkeley
USA
psilva@uclink4.berkeley.edu

Sivertsen, Knut
Finnmarik University College
NORWAY
knuts@hifm.no

Sjøtun, Kjersti
Havforskningsinstituttet
NORWAY
kjersti.sjoetun@imr.no

Skage, Morten
Zoologisk Institutt
NORWAY
morten.skage@zoo.uib.no

Skjåk-Bræk, Gudmund
NTNU
NORWAY
gudmund@biotech.ntnu.no

Smidsrød, Olav
NTNU
NORWAY
olav.smidsroed@chembio.ntnu.no

Smith, Carolyn
Old Town Elementary School
USA
carolynsmith@otsd.org

Snoeijs, Pauli
Uppsala University
SWEDEN
pauli.snoeijs@ebc.uu.se

Sobo, Fatma
Fisheries Division
TANZANIA
fsobo@hotmail.com

Soler, Anna
Martin Ryan Institute
IRELAND
anna.soler@nuigalway.ie

Soler-Onis, Emilio
Campus Universitario De Tafira
SPAIN
esoleronis@hotmail.com

Soriano, Eduardo
Soriano S. A.
ARGENTINA
jsoriano@interar.com.ar

Soriano, Gonzalo
Soriano Sacifio Y Dem
ARGENTINA
gonza@soriano-sa.com.ar

Sousa Pinto, Isabel
University of Porto
PORTUGAL
ispinto@cimar.org

Stancioff, Dimitri
USA
estaneioff@midmaine.com

Steinnes, Arild
FMC Biopolymer
NORWAY
arild_steinnes@fmc.com

Stekoll, Michael
University of Alaska
USA

Stengel, Dagmar
Martin Ryan Institute
IRELAND
dagmar.stengel@nuigalway.ie

Stirk, Wendy
University of Kwazulu Natal
SOUTH AFRICA
stirk@nu.ac.za

Sulu, Reuben
University of the South Pacific
FIJI
sulu_r@usp.ac.fj

Suzuki, Hisashi
Gifu Prefectural Research Institute of Industrial Product
JAPAN
suzuh@iri.rd.pref.gifu.jp

Teas, Jane
University of South Carolina
USA
teas@palmettohealth.org

Thomsen, Mads
University of Virginia
NEW ZEALAND
mads_thomsen@mailcity.com

Tiroba, Gideon
Dept of Fisheries and Marine Resources
SOLOMON ISLANDS
rfep@solomon.com.sb

Toledo, Maria- Isabel
Pontificia Universidad Catolica De Valparaiso
CHILE
itoledo@ucv.cl

Tomayao, Tita
Fmc Bio Polymer
PHILIPPINES
tita_tomayao@fmc.com

Tronchin, Enrico
University of Cape Town
SOUTH AFRICA
tronchin@botzoo.uct.ac.za

Uchida, Motoharu
National Research Institute of Fisheries and Environment of Inland Sea
JAPAN
uchida@affrc.go.jp

Ugarte, Raul
Acadian Seaplants LTD.
CANADA
rugarte@acadian.ca

Ursi, Suzana
Universidade De São Paulo
BRAZIL
suzursi@ib.usp.br

Urvantseva, Angela
Pacific Institute of Bioorganic Chemistry
RUSSIA
mik@piboc.dvo.ru

Usov, Anatolii
Russian Academy of Sciences
RUSSIA
usov@ioc.ac.ru

Vairappan, Charles S.
University of Malaysia Sabah
MALAYSIA
csv@ums.edu.my

Vasquez, Julio
Universidad Catolica Del Norte
CHILE
jvasquez@ucn.cl

Vea, Jostein
FMC Biopolymer
NORWAY
jostein_vea@fmc.com

Venkatesalu, Venugopalan
Annamalai University
INDIA
venkatesalu@yahoo.com

Veyret, Melanie
Station Biologique De Roscoff UMR CNRS 7127
FRANCE
veyret@sb-roscoff.fr

Viejo, Rosa
University of Rey Juan Carlos
SPAIN
rviejo@escet.urjc.es

Viera-Rodriguez, M Ascension
Campus Universitario De Tafira
SPAIN
mviera@dbio.ulpgc.es

Villena, Gunter
Peruvian Seaweeds Srl
PERU
gunter@peruvianseaweeds.com

Vinales, Jamie
Alimentos Multiexport S.A.
CHILE
jvinales@multiexport.cl

Voisin, Marie
Station Biologique de Roscoff UMR CNRS 7127
FRANCE
voisin@sb-roscoff.fr

Vroom, Peter
Coral Reef Ecosystem Division (RED)
USA
peter.vroom@noaa.gov

Wallentinus, Inger
University of Göteborg
SWEDEN
inger.wallentinus@marbot.gu.se

Watson, Duika Burges
University of Tasmania
AUSTRALIA
d_burges@utas.edu.au

Weinberger, Florian
UMR 7139 CNRS-Goëmar-UPMC
FRANCE
nyvall@sb-roscoff.fr

Wikström, Sofia A
Stockholm University
SWEDEN

Wong, Chak Ching
Hong Kong Sheli Ltd.
CHINA

Wright, Jeff
University of Wollongong
AUSTRALIA
jeffw@uow.edu.au

Yan, Xing-Hong
College of Science and Technology of Agua-Life
CHINA
xhyan@shfu.edu.cn

Yang, Eun Chan
Chungnam National University
KOREA

Yeo, Hwan-Goo
Hanseo University
SOUTH KOREA
yeohg@hanseo.ac.kr

Yermak, Irina
Pacific Institute of Bioorganic Chemistry
RUSSIA
iren@piboc.dvo.ru

Yokoya, Nair S.
Instituto De Botanica
BRAZIL
nyokoya@hotmail.com

Yokoyama, Takehiko
School of Fisheries Sciences Kitasato University
JAPAN
yokotake@kitasoto-u.ac.jp

Yoon, Ho-Dong
National Fisheries Research & Development Institute
KOREA
hdyoon@nfrda.re.kr

Yoon, Seung Je
Pukyong National University
KOREA
hoasungin@hotmail.com

Yoshimura, Cristalina Yoshie
University of Sao Paulo
BRAZIL
cristal@ib.usp.br

Zamorano, Jaime
Gelymar S.A
CHILE
jzamorano@gelymar.com

Zertuche, Jose
Universidad Autonoma De Baja California Sur
MEXICO
zertuche@uabc.mx

Zhang, Junzeng
Ocean Nutrition Canada LTD
CANADA
jzhang@ocean-nutrition.com

Zuniga, Elisa A.
Universidad De Santiago De Chile
CHILE
elisa.zuniga@usach.cl

Zvyagintseva, Tatyana
Pacific Institute of Bioorganic Chemistry
RUSSIA
zvyag@piboc.dvo.ru

Journal of Applied Phycology (2006) 18: 227–234
DOI: 10.1007/s10811-006-9022-1

Advances in seaweed aquaculture among Pacific Island countries

Timothy Pickering
Marine Studies Programme, The University of the South Pacific, Private Bag, Suva, Republic of Fiji
*Author for correspondence: e-mail: pickering_t@usp.ac.fj

Key words: seaweed, aquaculture, Pacific Island countries, *Kappaphycus*, *Cladosiphon*

Abstract

Recent developments in the seaweed aquaculture industries of Pacific islands are reviewed from the perspective of technical, production, geographic, marketing, species-diversification, socio-economic and institutional-support advances. Successful commercial aquaculture of seaweeds in the Pacific island region is presently based on two species, *Kappaphycus alvarezii* in Kiribati, Fiji and Solomon Islands, and *Cladosiphon* sp. in Tonga. It is possible that other candidate species could be considered for aquaculture for food (e.g. *Caulerpa racemosa* or *Meristotheca procumbens*) or extraction of agar (*Gracilaria*), although further research on the technical feasibility of aquaculture methods to produce sufficient tonnage, and particularly on their marketing, is needed. While the Pacific island region may be environmentally ideal for seaweed aquaculture, the limitations of distance from main centres and distance from markets, vulnerability to world price fluctuations, and socio-economic issues, make it unlikely that the Pacific Island region will ever rival the scale of Asian seaweed production. Regional seaweed farming can nevertheless make a useful contribution to supplement other sources of income, and can be an important economic boost for isolated outer islands where few alternative income-generating opportunities exist.

Introduction

The "Pacific Islands region" for the purposes of this paper comprises those countries and territories that are members of the Secretariat for the Pacific Community (SPC) and include the Federated States of Micronesia (Yap, Chuuk, Pohnpei and Kosrae), the Northern Mariana Islands, Marshall Islands, Nauru, Palau, Kiribati, Papua New Guinea (PNG), Solomon Islands, Vanuatu, New Caledonia, Fiji, American Samoa, Samoa, the Cook Islands, French Polynesia, Niue, Pitcairn, Tokelau, Tonga, Tuvalu, and Wallis and Futuna. Uwate et al. (1984) and Adams et al. (2001) have published reviews of aquaculture activities in the Pacific Islands region, and South and Pickering (2006) includes mention of the main seaweed aquaculture activities. Currently there are two species which provide a basis for commercial aquaculture; the red seaweed *Kappaphycus* Doty, and the brown seaweed *Cladosiphon* sp. The latter is known to occur naturally in Tonga and in New Caledonia; all cultured stocks of *Kappaphycus*, however, originated from outside the region.

Kappaphycus farming has been strongly promoted in the Pacific region because it requires a low level of technology and investment, can be operated at the family level, has relatively little environmental impact, does not require refrigeration or high-tech post-harvest processing within the country, and is normally compatible with traditional fishing and other subsistence uses of the inshore environment. It is a potential source of income and employment in rural areas with few other income-generating opportunities, and in particular is an activity that can provide income for women.

South (1993) reviewed the farming of *Kappaphycus* in the Pacific Islands up until the early 1990's, and reviews since then include Ask (2003), Ask et al. (2003c), Luxton and Luxton (1999), Luxton (2003), Pickering (2003) and South and Pickering (2006).

Various difficulties affecting the initial attempts to cultivate *Kappaphycus* in the 1980's (e.g. Tonga, Solomon Islands, Federated States of Micronesia and Tuvalu) led to the activity being abandoned in most of these countries (South, 1993).

Early results in Fiji proved encouraging, and this led to the establishment of an industry with total production of 684.4 t between 1985 and 1990 and highest yearly production of 277 t in 1987, however for a range of reasons reviewed by Ask et al., (2003c), Luxton (2003) and South and Pickering (2006) production ceased in 1993. A re-vitalization of the Fiji industry from 1997 onwards is also reviewed by Ask et al. (2003c). Annual seaweed production under this new initiative rose to 419 t by 2000, but levels have not reached the projections forecast by Ask et al. (2003c) and production is now declining. Feedback from farmers during 2003 indicated a loss of enthusiasm due to long delays in payments for seaweed produced; in some cases as long as 6 months (Pickering et al., 2003).

Seaweed production in the region has been both greatest and most consistent in Kiribati (Why, 1987; Uan, 1990; South, 1993; JICA, 1996). Commercial *Kappaphycus* cultivation commenced in the mid 1980's initially in the Gilbert Group, and later returned to the Line and Phoenix Groups where the early trials had taken place (JICA, 1996). In 1991 the commercial farming and marketing activities were handed over to the 100% government-owned Atoll Seaweed Co. Ltd. Production in recent years has been greatly dominated by a single atoll, Tabuaeran (Fanning Island) in the Line Islands, which has oceanographic conditions suited to rapid plant growth, and no copra industry owing to aged palms.

This paper reviews recent developments that might be considered "advances" in seaweed aquaculture among the countries and territories of the Pacific Islands region, including not only "technical advances" but also "production", geographic", "marketing", "species-diversification", "socio-economic" and "institutional support" advances.

Technical advances

The Pacific Islands' *Kappaphycus* industry has its origins in the Philippines and uses similar methods (McHugh & Philipson, 1989; Adams & Foscarini, 1990), so there is little to report from the Pacific region that could be considered as substantive "advances" in cultivation technology compared with Asian practices. Three principal farming methods have been tried in the South Pacific: off-bottom (fixed monofilament lines between posts driven into the substratum); floating rafts; and floating long-lines (Prakash & Foscarini, 1990; Ask 1999). Commercial cultivation in Fiji, Kiribati and Solomon Island is nowadays almost entirely by the off-bottom method. Kiribati also uses net cages for seed-stock farms, to protect plants from fish grazing.

Recently there has been a resurgence of interest in the raft method of cultivation in Solomon Islands, to reduce grazing by fish. Rafts are relatively easy to move around, to find locations where fish grazers are less abundant; placing rafts in depths of at least 5–10 m often gives good results (Alex Meloty pers. comm.). Another theoretical advantage of rafts is that a wider choice of farm sites becomes possible, because seabed type and water depth are no longer site-selection issues. In Kiribati trials of PVC-pipe rafts are now being carried out in the Gilbert Group on Abaiang and Nuotaea atolls, especially in areas where water flow is lacking (Ienimoa Kiatoa, pers. comm.). The disadvantage of rafts is that they require more labour and materials to set up (in Kiribati, even bamboo would need to be imported). In terms of their advantages, grazing by fishes on Pacific Island farms occurs at levels that can be tolerated for the most part, and there is currently no shortage of suitable reef space for off-bottom culture. The ultimate test of whether or not rafts are an improvement over off-bottom cultivation will lie in the proportion of farmers that willingly adopt this method of cultivation.

A recent advance in Asian cultivation technology is the Made Loop, described by Ask et al. (2003a) as a simple, low-tech and rapid way of attaching and harvesting seaweed plants on lines. These lines take longer to make than raffia tie-ties but are said to last longer. Furthermore, harvest is quicker, material cost is no higher, and seaweed quality is higher since less stray tie-tie material tangles in extraction machinery. Information about the Made Loop was incorporated into Pacific Island regional training materials, and most seaweed project officers in Fiji and Solomon Islands now know about it. Farmers in Fiji, however, are almost entirely using the raffia tie-tie method because Government provides farm materials to farmers and has made a multi-thousand-dollar investment in a special imported UV-resistant raffia tie-tie material which is still available in bulk quantities. Solomon Islands farmers tried out the Made Loop method after demonstration at SPC-funded training workshops held in November 2002, and many in Rarumana and Waghena are now using this method. It has not yet been tried in Kiribati as they only recently heard about it.

The main technical advance current in the Pacific Islands region is policy and research to develop better methods for translocation and quarantine of *Kappaphycus* between countries. There are intentions to introduce present varieties in the region to new locations, and to introduce new varieties from outside the region, but few guidelines are in place to manage and reduce environmental risks. SPC's Regional Aquaculture Programme, which advocates that member countries adopt responsible aquaculture practices in line with FAO guidelines, is working to conduct Import Risk Assessments (IRA's) and develop regional templates about translocation procedures for a range of aquacultured commodities including *Kappaphycus*. This has raised some research requirements, for example on the efficacy of treatments to remove other "hitch-hiker" species from translocated *Kappaphycus*, which are presently being fulfilled by the Institute of Marine Resources at the University of the South Pacific (Sulu & Pickering, Pers. comm.).

Production advances

Only in the Solomon Islands is production of *Kappaphycus* presently "advancing" (Table 1). A Fisheries Department/European Union Rural Fisheries Enterprise Project (RFEP), initially aimed at commercialization of rural artisanal fisheries, changed its focus in 2001 from finfish to seaweed in order to better meet the project's aims of "poverty alleviation through income generation, and empowerment of women through fisheries" (Rory Stewart, pers. comm.). By the first half of 2004 there were 250 seaweed farmers (Alex Meloti, pers. comm.), mainly at Rarumana in the New Georgia Group (80 farmers) and Waghena in Choiseul (170 farmers), who in the first 5 months of 2004 alone produced 51 t of dried *Kappapycus* (Table 1). Grazing by fishes has been a persistent but tolerable problem, while more serious have been difficulties in inter-island shipping, and unrealistic expectations in price created by statements of some politicians. These difficulties aside, the combination of a tightly-run seaweed project able to resist political pressures in the allocation of its resources, careful selection of appropriate private-sector partners for buying, and selection of communities who see seaweed farming as a worthwhile livelihood, appear important factors contributing toward this good start in Solomon Islands seaweed production.

In Fiji and in Kiribati, *Kappaphycus* production has recently been declining (Table 1). In Fiji, momentum was lost owing to lack of clarity and disagreement between the Department of Fisheries and the company nominated as sole exporter of Fiji seaweed, REL Ltd, over the extent of each other's responsibilities to support the fledgling industry. REL Ltd has adopted the practice of regarding the prevailing FJ$ 0.50 (US$ 0.27) beach price as a "landed in Port of Suva" price, leaving either farmers or tax-payers to cover the freight charges from outer islands to the baling point in Suva. Payment to farmers for seaweed delivered has also been very late at times, sometimes by up to 6 months (Pickering et al., 2003), which makes farmers lose heart and cease production. Government support for seaweed has been spread very widely and at times diverted to communities or projects (such as road-building to Kiuva Village) selected more for political and electioneering reasons rather than focussed upon a select few communities where seaweed is most likely to succeed. Mainly for these reasons, only 20 t was produced in 2003. Indications are that seaweed buying and marketing for the current project in Fiji may have been privatised at too early a stage and to a company too lightly capitalised, before sufficient volume had been built up to make the industry viable. Government needs to again review the current industry, learn from mistakes, and decide if it is still worth using tax-payers money to build up production to a level where there can be a smoother transition to a stand-alone industry run by the private sector.

Production in Kiribati has fluctuated widely but has always remained in the hundreds of tonnes. The centre of production moved away from the Gilberts Group to the Line and Phoenix Groups, with Tabueran providing the bulk of national output. The seaweed industry thus

Table 1. Commercial production (t) of *Eucheuma/Kappaphycus* from four Pacific Island countries between January 1983 and June 2004 (adapted from South 1993c, JICA 1996, Derrick Pendle pers. comm., Ienimoa Kiatoa pers. comm., Sam Mario pers. comm., Rory Stewart pers. comm., Alex Meloty pers. comm.).

	1983	1984	1985	1986	1987	1988	1989	1990	1991	1992	1993	1994	1995	1996	1997	1998	1999	2000	2001	2002	2003	Mid-2004
Fiji	–	–	30.0	173.4	277.0	60.3	80.3	87.4	55.0	60.0	–	–	–	–	–	19.8	300.0	418.6	240.0	80.0	20	13.7
Kiribati	–	–	24.9	66.0	30.0	45.1	149.2	637.0	1019.7	434.0	205.0	396.0	654.9	1249.2	924.4	742.2	1170.3	1437.8	1158.2	530.8	488.0	500
Tonga	3.0	11.0	5.0	2.0	1.5	–	–	–	–	–	–	–	–	–	–	–	–	–	–	–	–	–
Solomon Islands	–	–	–	–	–	–	–	–	–	–	–	–	–	–	–	–	–	–	0.6	2.8	16.9	51

keenly felt the effects of alternative tourism-related economic benefits (described in Low, 2004) since Norwegian Cruise Lines (NCL) vessels began weekly calls to Tabuaeran in 2001. To counter this, Atoll Seaweed Company Ltd has embarked on socio-economic study of farmer motivations at Tabuaeran (Antoine Teitelbaum, pers. comm.) and on a strategy to re-vitalize former farming locations in the Gilberts Group closer to the main port at Tarawa (Derek Pendle, pers. comm.). There are also signs that, through reflagging of its Hawaii-based vessels to exempt them from the US Jones Act (Magin, 2003), during 2004 NLC ship visits to Tabuaeran will decline or even cease. Kiribati production is recovering and is likely to continue to fluctuate around 1000±500 t for the foreseeable future.

Production of *Cladosiphon* Kützing in Tonga, largely from a highly seasonal fishery on the island of Tongatapu, is said to be around 400 t (wet) per year (Anon, 2004). This is salted and sent by shipping container to Japan where it is sold as the edible seaweed known to Japanese as *mozuku*. The fishery now appears fully developed, with three companies involved in export, each trading about 100–200 t annually. Aquaculture trials were conducted in Tonga in order to increase production from the resource, and to try and extend the growing season. The farming method involves placing nets similar to *Porphyra* (*nori*) nets on the seabed, which become seeded by natural spore-fall. Aquaculture of *Cladosiphon* sp. in Tonga is now said to be at an advanced trial stage (Silika Ngahe, pers. comm.).

Geographic advances

There are on-going efforts to extend the geographical range of active seaweed farming, plus a gradual shift in the locations where most farming occurs. The first phase of Fiji's industry in the mid-1980's focused on the north coast of the main island of Viti Levu (Ra province), an area offering alternative livelihoods like sugar cane, tourism and commercial fishing. By the early 1990's most production was from the isolated village of Kiuva in Tailevu province, or from Cakaudrove in Vanua Levu, where the main alternatives are copra, fishing or subsistence agriculture. From 1997 onwards the islands of the southern Lau group have featured strongly, in particular the isolated island of Ono-I-Lau (the "Tabuaeran" of Fiji). It is thus clear that communities may be ripe for the introduction of seaweed farming activity if alternative livelihoods are limited to copra or subsistence agriculture and fishing.

Similarly, in Kiribati the industry began in the Gilbert Group of islands but in recent years 90–99% of national production has come from Tabuaeran in the Line Islands (Atoll Seaweed Company, pers. comm.). Efforts are now underway to re-establish farming in the Gilberts Group. In the Solomon islands production is presently centred on two places in Western Province (Rarumana, and Waghena), but cultivation in northern Malaita Island is well underway and the RFEP has mapped out a strategy for progressive introduction of farming to communities in Central District (Guadalcanal, Gela, Savo and Russell Islands) and Makira Island. Other countries like Vanuatu, Republic of Marshall Islands and PNG now have seaweed projects in their planning stages.

Species-diversification advances

The mainstay of Pacific Island seaweed production has been the *tambalang* variety of *Kappaphycus alvarezii* (Doty) Doty ex P.C. Silva. However, its thick thalli mean that it can take 3–5 days to sun-dry and this can be problematic in districts with higher average rainfall. To counter this, a variety of *K. striatum* (Schmitz) Doty ex. P. Silva *(sacol)* was introduced to Fiji during its second phase of industry development because the thalli are thinner and dry faster (in as little as 24 h). In 2002 some of this Fiji *sacol* material was taken to Solomon Islands and tested there. The *sacol* variety is more popular with buyers because of its lower moisture content and less probability of spoilage from rain while drying. The *tambalang* variety is much preferred by farmers, however, because plants are thicker and heavier, thus yielding more weight of seaweed per line (a unit of effort) than *sacol*.

As part of efforts to re-vitalize seaweed farming in the Gilberts Group of Kiribati, Atoll Seaweed Company has been looking for a seaweed variety better suited to environmental conditions there. With assistance from their overseas buyer CP Kelco, they have plans to introduce a new variety of *K. alvarezii* during 2004 (Antoine Teitelbaum, pers. comm.).

The development of a *Cladosiphon* sp. export industry in Tonga now brings the total number of commercially-aquacultured seaweeds in Pacific Island countries to two. Other edible species, such as species of *Caulerpa*, *Codium* (Chlorophyceae), and *Gracilaria*, *Hypnea* and *Halymenia* (Rhodophyceae), are already commercialized as artisanal fisheries in various places (South and Pickering, 2006). One edible species found in Rotuma Island (Fiji) is the red seaweed *Meristotheca procumbens* P. Gabrielson et Kraft (N'Yeurt, 1996: 416). These currently have draw-backs either for aquaculture (difficult to culture) or as exports (either too perishable, or little market demand).

Marketing advances

A feature of external marketing arrangements in the Pacific region to date has been for a seaweed-producing country to secure a long-term contract with an overseas buyer. This provides producers with some degree of certainty about the range of export prices they can expect over the medium-term, and helps smooth any fluctuations in world prices for dried *Kappaphycus*. In Fiji the buyer is FMC Corporation, who set a price by Memorandum of Understanding (MOU) with the Fiji Government at US$ 0.55 per kg FOB. The local company nominated by FMC under the MOU to have exclusive right to export dried seaweed from Fiji is a private company, REL Ltd (newly set up for this purpose), who now pay USD 0.27 for seaweed once it has been brought to the main island of Viti Levu (it is no longer a "beach price"). In Kiribati the company that exports seaweed is the Government-owned Atoll Seaweed Company Ltd, which pays farmers a beach price of AU$ 0.60 (USD 0.42) per kg (of which AU$ 0.15 is a government subsidy), and is contracted to sell to CP Kelco.

A new development, then, is the situation in Solomon Islands where presently there are two local private companies (long-established in other local enterprises) gearing up to purchase dried seaweed from farmers (beach price is currently US$ 0.26 per kg) and export it. These companies have not been placed under any obligation to sell to any particular international buyer so are free to strike their own deals. Currently, both are choosing to sell to Degussa. Under a seaweed marketing plan set up by the RFEP, there will be two export licences granted until production reaches 300 t per annum, when a further licence will be granted. The situation will be further reviewed if production exceeds 500 t (Rory Stewart, pers. comm.).

Both Kiribati and Solomon Islands have succeeded in negotiating export prices superior to that enjoyed by Fiji. Though these are contract prices which are commercially sensitive and not readily available, it is believed that both countries now receive prices in the range of US$ 0.68–0.73 per kg. In addition, they are selling the dried seaweed with its salt and at 35% water content, compared with the Fiji exporter REL Ltd who must remove loose salt and achieve 30% water content.

In contrast to the sometimes volatile nature of Asian marketing arrangements, Pacific Island producers have tended to prefer stable, long-term and trusting relationships with a particular buyer. There have been and always will be maverick marketeers who urge that spot-market prices be taken, but cooler heads see the benefits of price stability and value an on-going buyer-supplier relationship. Trust is seen as important, given the general difficulty in obtaining accurate market intelligence upon which any price adjustments might be negotiated.

Export of *Cladosiphon* sp. from Tonga is now well-established as a profitable but highly seasonal fishery employing several hundred people, with three private companies involved (Silika Ngahe, pers. comm.). The main product is salted plants bound for the edible-seaweed market in Japan. *Cladosiphon* sp. has also been used as an ingredient in very expensive cosmetics and face creams (Hideyuki Tanaka, pers. comm.). *Cladosiphon* sp. is also found in New Caledonia where a local businessman, inspired by Tonga's example, has been experimenting with aquaculture techniques.

The potential of other edible seaweeds as commercial products merits further study. The possible cultivation of *Meristotheca procumbens*, a highly desirable food item in Japan (H. Tanaka, FAO/UNDP Suva; pers. comm.) falls into this category. Another candidate is *Caulerpa racemosa* (Forsskål) J. Agardh, a favourite food item in most Pacific Island countries and a high-value edible species in Japan. There is a potential to develop the fishery for the export market, although supply and post harvest problems (e.g. spoilage during transport) need to be overcome. Niche markets also exist for seaweed products like soaps, cosmetics and medicines. Commercial soap manufacturers in Fiji like Sandollars Ltd and Mokosoi Ltd now offer a range of seaweed/coconut soaps which use farmed *Kappaphycus* as an ingredient, and these are sold in tourist hotels, airport duty-free stores and gift shops. Opportunities for seaweeds to provide the basis for household-level small businesses in the Pacific region making soaps, cosmetics and medicines are being explored (Novaczek, 2001a,b; Novaczek & Athy, 2001; Novaczek, 2003).

Socio-economic advances

The main advances in socio-economic aspects of seaweed aquaculture in Pacific Island countries have been firstly an increased appreciation of the importance of socio-economic research in addition to technical research, and secondly better understanding of the socio-economic conditions associated with successful seaweed farming projects. Initially, during the 1980's, communities were chosen for seaweed projects purely for biological/environmental reasons. For example in Fiji the areas chosen were in Ra Province where there are strongly competing livelihoods, and project managers expected farmers to spend several days away from home and far out at sea on platforms erected near favourable growing sites (Sam Mario, pers. comm.). Later, a wide range of communities was chosen for seaweed projects and supported through provision of fibreglass punts and 40 hp outboard motors. As soon as possible, many farmers abandoned seaweed farming and used the boats for other pursuits like fishing for beche-de-mer.

Such experiences across the Pacific, in Asia and in Africa, led FMC Corporation to support post-graduate research on seaweed socio-economics at the University of the South Pacific, to identify the critical factors (alternative livelihoods, population demographics, geography, local traditions, etc.) which would enable better success in allocating available project resources for seaweed farming development. The results of this research will be reported separately (Namudu & Pickering, in press). In the Solomon Islands RFEP funded a social and economic impact assessment of seaweed development on the Rarumana community to determine the scope for future expansion and review the needs for further donor support (Wale, 2003). The Atoll Seaweed Company in Kiribati has commissioned two recent studies, one for Tabuaeran to clarify the reasons for the decline in production and develop strategies to overcome threats to seaweed production from alternative livelihoods, the other to identify socio-economic factors limiting seaweed production in the Gilberts Group (Antoine Teitelbaum, pers. comm.).

Through such research, and by trial and error, a better picture is emerging of the requirements for successful seaweed aquaculture projects in the Pacific region. Selection criteria being applied by RFEP in Solomon Islands to identify further areas for farm development have three main steps: (1) to be selected, a community should firstly be near farm sites where environmental conditions are conducive to good seaweed growth, (2) demographically, the community should be large, and (3) the availability of alternative livelihoods should be considered, for example in rural Solomon Islands seaweed farming can easily compete with copra, fishing and agriculture but not with logging (Rory Stewart,pers. comm.).

Similarly, of the business models possible for seaweed farming ventures, one has emerged a clear winner. Three main possibilities exist; (1) community (e.g. church) or tribe/clan groups farming cooperatively, (2) company farms, or "contract farming" (COFA), owned by buyers and operated by labour hired on a daily wage basis, and (3) individual or household (nuclear family) operated farms. Compared with the first two, the third (household farms) has been by far the most successful and dominates the industries in all three countries that have industries. Efforts have been made to establish COFA farms in the Gilbert Islands of Kiribati to revive production there and counter the downturn at Tabuaeran. However, the success of the COFA model in achieving this goal has not yet been demonstrated (Antoine Teitelbaum, pers. comm.).

Institutional support advances

Over the last four years, the region has advanced in terms of institutional support for aquaculture at the regional level. SPC obtained Australian AUSAid funding to set up a Regional Aquaculture Programme, which coordinates the provision of technical support and training to member countries, acts as a clearing house for information on aquaculture, and coordinates regional mechanisms for priority-setting in terms of the types of aquaculture to be supported. A regional aquaculture strategy links SPC with long-term applied research on commercial feasibility by Worldfish Centre's regional office based in Noumea, New Caledonia, and with short-term applied research and post-graduate student research at the Marine Studies Programme and Institute of Marine Resources of the University of the South Pacific (USP). *Kappaphycus* has been identified as a priority commodity for regional institutional support under these arrangements. Additionally, USP has enjoyed support for aquaculture training activities under the Canada-South Pacific Ocean Development Program Phase II (C-SPODP-II).

The Food and Agricultural Organisation's (FAO) Technical Co-operation Programme (TCP) based in Apia, Samoa, responded to increasing interest in seaweed farming by funding a consultancy to assess the feasibility of seaweed farming in selected other Pacific Island countries (Luxton, 2002). This led to an FAO TCP project to establish *Kappaphycus* farming in Marshall Islands, and to prioritization of Milne

Bay Province (MBP) in Papua New Guinea for a seaweed project (Kinch et al., 2003). The United Nations Development Programme (UNDP) has, separately, contracted non-governmental organisation Conservation International (CI) to execute a Milne Bay Community-based Coastal and Marine Conservation Programme (CMCP) and this has an alternative income generation component, under which seaweed farming is now being considered .

Support, under the regional aquaculture strategy for seaweed farming, has come in the form of training workshops and production of training materials. A training video on the seaweed farming techniques typically used in Fiji was produced by the Marine Studies Programme at USP in 2003 with funding from C-SPODP-II, and is available in English, Fijian and Solomon Island Pidgin languages, and the much-in-demand FMC Cottonii and Spinosum Cultivation Handbook (Ask, 1999) has been reprinted. SPC produced its own video to complement the USP one, through its focus on raising awareness in communities about the potential socio-economic benefits of seaweed farming, and produced a farming manual in a comic-book format. A spreadsheet-based interactive economic model has been developed jointly by USP, SPC and Queensland Department of Primary Industry to predict the viability of different seaweed farming scenarios according to user data inputs, or to compare the benefits of seaweed farming with alternative livelihoods like copra, artisanal fishing or agriculture (Johnstone & Pickering, 2003).

Donor interest in seaweed farming projects under bi-lateral arrangements in the region is also increasing. Currently, the most active in their support are the European Union in both Kiribati and Solomon Islands, and AUSAid in Solomon Islands (Rory Stewart, pers. comm.).

Conclusion

Successful commercial aquaculture of seaweeds in the region is presently based on two species, *Kappaphycus alvarezii* in Kiribati, Fiji and Solomon Islands, and *Cladosiphon* sp. in Tonga. It is possible that other candidate species could be considered for aquaculture for food (e.g. *Caulerpa racemosa* or *Meristotheca procumbens*) or extraction of agar (*Gracilaria*), although further research on the technical feasibility of aquaculture methods to produce sufficient tonnage, and particularly on marketing, is needed.

While the Pacific Island region may be environmentally ideal for seaweed aquaculture, the limitations of distance from main centres and markets, vulnerability to world price fluctuations, and socio-economic problems, make it unlikely that the Pacific Island region will ever rival the scale of Asian seaweed producing nations. Seaweed farming from this region can nevertheless make a useful contribution to supplement other world sources, and it can be an important economic boost for the less-developed outer islands of Pacific nations where few alternative income-generating opportunities exist.

Acknowledgments

The author is grateful to Sam Mario, Gideon Tiroba, Silika Ngahe, Rory Stewart, Alex Meloty, Derek Pendle, Ienimoa Kiatoa, Antoine Teitelbaum, Merekeleni Namudu and Sompert Gereva for providing much information about recent developments in seaweed aquaculture in their respective countries.

References

Adams, T, Foscarini R (eds), (1990) Proceedings of the Regional Workshop on Seaweed Culture and Marketing, 14–17 November 1989. Pacific Regional Aquaculture Development Project, FAO, Suva. pp. 86.

Adams T, Bell J, Labrosse P (2001) Current status of aquaculture in the Pacific Islands. In Subasinghe RP, Bueno PB, Phillips MJ, Hough C, McGladdery SE, Arthur JR (eds), Aquaculture in the Third Millenium. Technical Proceedings of the Conference on Aquaculture in the Third Millennium. Bangkok, Thailand. 20–25 February 2000. NACA, Bangkok and FAO, Rome. pp. 295–301.

Anon. (2004) Surialink Seaplants Handbook. www.surialink.com/HANDBOOK/ Genera/browns/Cladosiphon/Cladosiphon.htm.

Ask EI (1999) Cottonii and spinosum cultivation handbook. Unpublished Report, FMC Corporation, Philadelphia PA, USA. 52 pp.

Ask EI (2003) Creating a sustainable commercial *Eucheuma* cultivation industry: The importance and necessity of the human factor. In Chapman ARO, Anderson RJ, Vreeland VJ, Davison IR (eds.), Proceedings of XVIIth International Seaweed Symposium, Oxford University Press, pp. 13–18.

Ask E, Azanza R, Simbik M, Cay-An R, Lagahid J (2003a) Technological improvements in commercial *Eucheuma* cultivation (a short communication). Science Diliman 15: 47–51.

Ask EI, Ledua E, Batibasaga A, Mario S (2003b) Developing the cottonii (*Kappaphycus alvarezii*) cultivation industry in the Fiji Islands. In Chapman ARO, Anderson RJ, Vreeland VJ, Davison IR (eds.), Proceedings of XVIIth International Seaweed Symposium, Oxford University Press, pp. 81–85.

JICA (Japanese International Cooperation Agency) (1996) Present and Future of Aquaculture Research and Development in the Pacific Island Countries. Proceedings of the International Workshop held from 20th November – 24th November, 1995 at Ministry of Fisheries, Tonga. Nuku'alofa, 423. pp.

Johnstone W, Pickering T (2003) The economics of aquaculture in comparison with other rural development opportunities in Pacific Island countries: Outcomes of a meeting held at the University of the South Pacific, 29 September – 3 October 2003. Marine Studies Programme Technical Report 2003/07, The University of the South Pacific, 25 pp.

Kinch J, Bagita J, Bate M (2003) Exploring the potential for seaweed farming in Milne Bay Province, Papua New Guinea. SPC Fisheries Newsletter #104, Secretariat for the Pacific Community, pp. 25–31.

Low L (2004). Heavenly Footprint. www.suite101. com/article. cfm/15912/107695.

Luxton DM, Luxton PM (1999) Development of commercial *Kappaphycus* production in the Line Islands, Central Pacific. Hydrobiologia 398/399: 477–486.

Luxton DM (2002) Development of commercial opportunities for cottonii seaweed (*Kappaphycus*) mariculture in the South Pacific. FAO Draft Regional Technical Co-operation Programme Proposal – SAPA/RAPI.

Luxton DM (2003) *Kappaphycus* agronomy in the Pacific Islands. In Chapman ARO, Anderson RJ, Vreeland VJ, Davison IR (eds.), Proceedings of XVIIth International Seaweed Symposium, Oxford University Press, pp. 41–47.

Magin JL (2003) Cruise exemption passes Congress – Norwegian Cruise Line will no longer have to make a stop in a foreign port. http://starbulletin.com/2003/02/14/business/story3.html.

McHugh DJ, Philipson PW (1989) Post-harvest technology and marketing of cultured *Eucheuma* Seaweeds. In P.W. Philipson (ed.), The Marketing of Marine Products in the South Pacific. University of the South Pacific, Institute of Pacific Studies, Suva: 143–163.

Novaczek I (2001a) Sea Plants. Community Fisheries Training Pacific Series 3. Fiji: The University of the South Pacific/Secretariat for the Pacific Community. pp. 31.

Novaczek I (2001b) A guide to the common edible and medicinal sea plants of the Pacific islands. Community Fisheries Training Pacific Series 3A, Supplementary Resource to Sea Plants: Pacific Series 3. Fiji: The University of the South Pacific/Secretariat for the Pacific Community, pp. 40.

Novaczek I, Athy A (2001) Sea vegetable recipes for the Pacific Islands. Community Fisheries Training Pacific Series 3B, Supplementary Resource to Sea Plants: Pacific Series 3. Fiji: The University of the South Pacific/Secretariat for the Pacific Community, pp. 36.

Novaczek I (2003) Seaweed: A promising option for women's small business development in the Pacific region. SPC Women in Fisheries Information Bulletin #13, December 2003, pp. 17–18.

N'Yeurt ADR (1996) A preliminary floristic survey of the benthic marine algae of Rotuma Island. Aust. Syst. Bot. 9: 361–490.

Pickering TD (2003) Seaweed. In: Profiles of high interest aquaculture commodities for Pacific Islands countries. SPC Aquaculture Technical Papers/Secretariat for the Pacific Community, 59–61.

Pickering TD, Mate F, Namudu M, Lasi F (2003) Report on the outcomes of the train-the-trainers workshop on *Kappaphycus* seaweed farming in Pacific Island countries. Marine Studies Programme Technical Report 2003/06, The University of the South Pacific, pp. 26.

Prakash J, Foscarini R (1990) Handbook on *Eucheuma* Seaweed Cultivation in Fiji. Fiji Ministry of Primary Industries, Fisheries Division, and South Pacific Aquaculture Development Project, FAO. pp. 42.

South GR (1993) Seaweeds. In: Wright A, Hill L. (eds), Nearshore Marine Resources of the South Pacific. Information for Fisheries Development and Management.Institute of Pacific Studies, Suva, Forum Fisheries Agency, Honiara, International Centre for Ocean Development Canada, pp. 683–710.

South GR, Pickering TD (2006) The seaweed resources of the Pacific Islands. In: Seaweed Resources. Critchley AT, Ohno M, Largo D (eds), Expert Centre for Taxonomic Identification (ETI), Univ. Amsterdam (CD-ROM series).

Uan J (1990) Kiribati. In Adams T, Foscarini R (eds). Proceedings of the Regional Workshop on Seaweed Culture and Marketing, Suva, Fiji, 14–17 November 1989. South Pacific Regional Aquaculture Development Project. FAO, pp.10–15.

Uwate R, Kunatuba P, Roberti B, Tekanai C (1984) A Review of Aquaculture Activities in the Pacific Islands Region. PIDP, East-West Center, Honolulu.

Wale R (2003) Social and economic impact assessment of the seaweed development project in Rarumana community, Parara Island, Western Province. Educe Consulting Network (unpublished report). pp. 49.

Why SJ (1987) *Eucheuma* seaweed farming development, 1985–1987. Fisheries Division, Kiribati.

Journal of Applied Phycology (2006) 18: 235–240
DOI: 10.1007/s10811-006-9024-z

Experimental tank cultivation of *Porphyra* in Israel

A. Israel[1,*], I. Levy[2] & M. Friedlander[1]
[1]*Israel Oceanographic & Limnological Research, Ltd., The National Institute of Oceanography, Tel Shikmona, P.O. Box 8030, Haifa 31080, Israel;* [2]*Noritech Seaweed Biotechnologies, Ltd., New Industrial Park, Bldg. 7, P.O. Box 620, Yoqneam 20692, Israel*

*Author for correspondence: e-mail: alvaro@ocean.org.il; fax: +972-4-851-1911

Key words: biomass yields, environment, *Porphyra*, tank cultivation

Abstract

Outdoor tank cultivation of several *Porphyra* (nori) species was carried out from late November 2002 through early May 2003 using 40 L (with a surface of 0.25 m^2), 600 L (1 m^2), and 24,000 L (30 m^2) fiberglass or PVC tanks provided with continuous aeration and seawater flow. Sexual and asexual spores produced from cultured conchocelis and frozen thalli in the laboratory, respectively, were subsequently grown to produce young fronds (ca. 5–10 cm) in an average time of 8 weeks. Growth in outdoor tanks and ponds was possible for a period of up to 20 weeks (i.e. growth season), with yields above 100 g FW m^{-2} d^{-1} occurring during 12–14 weeks from late December through late March, when seawater temperatures were below 20 °C. These yields correlated with the species and depended on the type of tanks in which the algae were cultivated, with the highest yields observed for *Porphyra* sp. and *Porphyra yezoensis* when fertilized twice a week with NH_4Cl and NaH_2PO_4 in 40 L tanks. Calculations of productivity for an entire growth season based on $\geq$100 g FW m^{-2} d^{-1} yields exceed the average productivities using seeded nets in open sea, for all *Porphyra* species tested (0.96–4.06 kg DW m^{-2} season^{-1} vs. 0.7–1.0 kg DW m^{-2} of net season^{-1}). Therefore, tank cultivation of *Porphyra* can offer an additional source of nori biomass to international markets. Land-based tank cultivation also offers an environmentally friendly practice that allows for the manipulation of growth conditions to enrich seaweeds with specific, valuable chemicals such as protein and minerals.

Introduction

Out of the approximately 130 identified species of *Porphyra* only a few have served as commercial, sea-vegetable foods (nori, purple laver). Several variants of these naturally occurring species have been produced to enhance yields and culinary parameters of nori production in Japan, Korea and China. Until the late 1980's, production of nori was almost equally balanced with consumption (Miura & Aruga, 1987). Rapidly expanding seaweed markets and degradation of marine environments have both led to steadily increasing demands for nori worldwide (McHugh, 2003; Merrill, 1993).

On-land seaweed tank cultivation has several advantages over traditional, open-sea aquaculture. In tanks, algal growth can to some extent be manipulated and seaweeds can be enriched with desired bio-chemicals. Abrupt pollution events may become detrimental for cultivation in the open sea. This and other environmental factors during the growth season make biomass yields unpredictable. Nevertheless, few studies have reported successful, sustainable cultivation of seaweeds in land based tanks or ponds, and even fewer have described tank cultivation for *Porphyra*. In fact it is likely that tank cultivation is more common than reported in the scientific literature, as seaweeds have gained crucial roles in, for example, developing integrated aquaculture (Chopin et al., 2001; Fei, 2004). Mencher et al. (1983) described the use of ocean thermal energy conversion effluents to cultivate *Porphyra* in 1 m^3 tank compartments. Yamamoto et al. (1991) tested outdoor raceways as an alternative cultivation approach to grow *P. yezoensis* in Japan. Hafting (1999a,b) demonstrated monospore production, after cutting and maceration of foliose thalli as seeds, to establish a tank cultivation technology for *Porphyra*, while Notoya (1999) also proposed seed production from tissue culture of both monoecious and dioecious species. In Israel, tank cultivation has been a common practice for at least a decade, both experimentally and commercially (Friedlander & Levy, 1995; Neori et al., 2000).

The current study describes a tank cultivation technology implemented for various *Porphyra* species, and presents their fresh weight and dry weight yields during a full growing season, while discussing the advantages of land-based cultivation over conventional, open-sea seaweed culture.

Materials and methods

Algal material

The species tested for tank cultivation were *Porphyra linearis* Greville, a winter annual species of uncertain taxonomic determination collected from a nearby shore, *Porphyra tenera* Kjellman and *Porphyra yezoensis* Ueda brought to our laboratory from commercial cultivars in Japan, and *Porphyra* sp. collected from East Taiwan in 1997, also of uncertain taxonomical status. These species are part of a seaweed culture collection maintained at Israel Oceanographic & Limnological Research, Ltd (IOLR), Haifa, Israel. They were maintained in a growth chamber at 15°C, 70 μmol photon $m^{-2}\,s^{-1}$ and 10 h photoperiod as seedlings or cultured conchocelis.

Indoor seedling production

Seedlings were obtained from sexual spores (conchospores, obtained from all species except *Porphyra* sp.) by manipulation of conchocelis filaments to obtain mature conchosporangia (Sidirelli-Wolff, 1992), or from vegetative spores (archeospores, obtained from all species) after foliose thalli were frozen at –20°C for 24 h and thawed in seawater. In the latter case spore release was considered terminated after 24–48 h. The spores obtained using either of the two sources were grown in 250 ml glass beakers ($n = 20-30$) filled with enriched (PEM-II; Provasoli, 1968) seawater medium at 15 °C, 70 μmol photon $m^{-2}\,s^{-1}$ and 10 h photoperiod in a growth chamber. The beakers were placed on a shaker set at low speed and the seawater media changed every 5–7 days. Seedlings were allowed to reach about 0.5 cm length before being transferred to outdoor cultivation settings.

Outdoor growth

Outdoor culture extended from 22 November 2002 to 28 April 2003, totaling 5 months. Few experiments were continued until mid May 2003. An average seedling biomass of 10–15 g FW was transferred to 5–8, 40 L (0.25 m^2), fiberglass tanks equipped with running seawater and aeration, similar to the system described by Israel et al. (1999). The tanks were covered with plastic nets to reduce irradiance to approximately one third of incident sunlight. Transfer from seedling incubators to outdoor tanks was carried out when ambient seawater temperature was 20 °C or below. Biomass yields were determined by weighing the algae every one or two weeks after carefully allowing excess seawater to drip off the algal material.

Porphyra thalli averaging 2–3 cm long were then moved to 2–3, 600 L (1 m^2) plastic tanks, also receiving continuous seawater flow, aeration and reduced sunlight. Next, when the young thalli reached ca. 5–12 cm long, 10–15 kg FW were used to inoculate 2, 24,000 L (30 m^2) PVC or concrete ponds supplied with running seawater and aeration. During all steps of outdoor cultivation the algae were pulse fed twice a week for 24 h with NH_4Cl and NaH_2PO_4 added to the medium to reach 1.0 and 0.1 mM final concentrations, respectively (Friedlander & Levy, 1995). The data were statistically analyzed with two-way ANOVA and Duncan's tests.

Results

Seawater temperatures and growth

Seawater temperatures below 20 °C were regarded as a prerequisite for outdoor culture in all four species. Temperatures in *Porphyra* culture tanks varied similarly for all three types of tanks, and they fell below 20 °C from late November 2002 to late April 2003, approximately 20 weeks (Figure 1). Average daily yields determined in 40 L tanks of *Porphyra* sp., *P. yezoensis* and *P. tenera* ranged from 126 to 305 g FW $m^{-2}\,d^{-1}$ and these yields decreased in a similar fashion for all species when seawater temperatures approached 18 °C (Figure 2). *Porphyra* sp. exhibited the highest yields at lower temperatures while *P. tenera* was the most sensitive to higher temperatures (Figure 2).

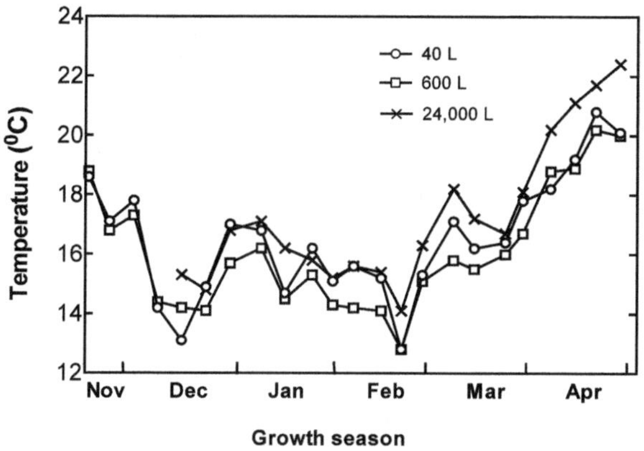

Figure 1. Fluctuations in seawater temperatures in outdoor tanks and ponds of different volumes ($n = 3$ each) containing *Porphyra* from 22 November 2002 to 28 April 2003.

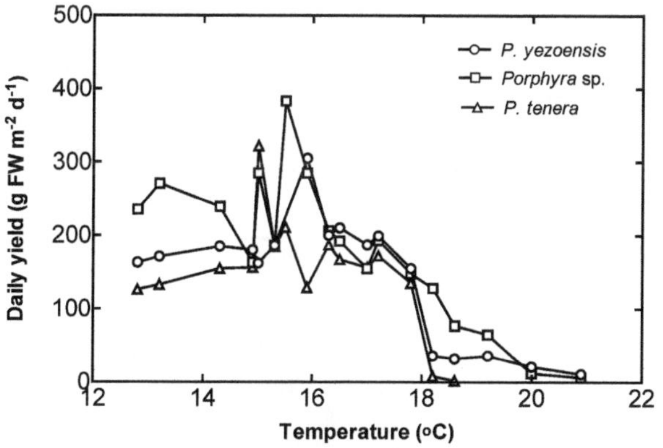

Figure 2. Seawater temperature and average daily biomass yields (g FW m^{-2} day^{-1}) for three *Porphyra* species cultivated in 40 L fiberglass tanks. Yields were calculated from determinations of fresh weight increases every week. (n = 5–8 tanks; S.Ds. were up to 27% of means).

Biomass yields

Daily fresh weight yields in 600 L tanks were generally high between mid December 2002 and late March 2003, reaching peaks of 500 and 545 g FW m^{-2} d^{-1} for *Porphyra* sp. and *P. tenera*, respectively (Figure 3). The lowest yields were observed in *P. linearis,* ranging from 17–25 g FW m^{-2} d^{-1} with a maximum of 120 g FW m^{-2} d^{-1} during mid February 2003 (Figure 3). The first innoculum of *Porphyra* was transferred from 600 L tanks to 30 m^2 ponds during late December 2002, about 4 weeks after outdoor cultivation was initiated. Daily fresh weight yields in these ponds increased progressively towards mid December 2002 and peaked during January-February 2003, decreasing

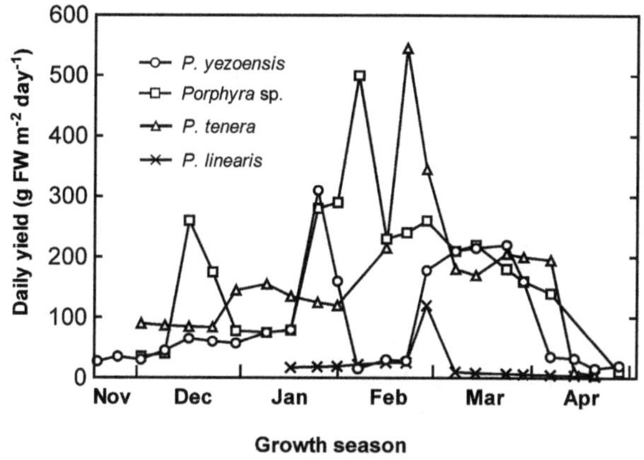

Figure 3. Average daily biomass yield (g FW m^{-2} day^{-1}) in four *Porphyra* species cultivated from 22 November 2002 to 28 April 2003 (155 d) in 600 L outdoor tanks. Yields were calculated from determinations of fresh weight increases every two weeks. (n = 2–3 tanks; S.Ds. were up to 33% of means).

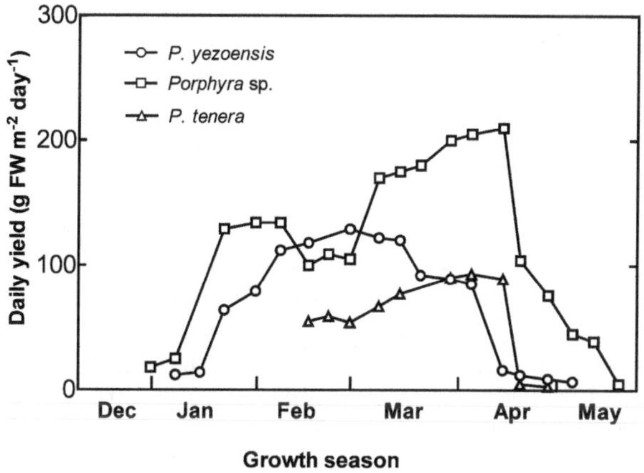

Figure 4. Average daily biomass yield (g FW m^{-2} day^{-1}) in three *Porphyra* species cultivated from 1 January 2003 to 13 May 2003 (134 d) in 24,000 L outdoor ponds. Yields were calculated from determinations of fresh weight increases every two weeks. (n = 2 ponds; S.Ds. were up to 37% of means).

abruptly for all species towards the end of March 2003 (Figure 4). The highest productivity occurred in *Porphyra* sp. with yields of 210 g FW m^{-2} d^{-1} and the lowest one in *P. tenera* with yields of 93 g FW m^{-2} d^{-1} (Figure 4).

Calculations of average seasonal yield per species as related to tank shapes and volumes are summarized in (Figure 5), and they were based on growth experiments presented in (Figures 2–4). From these calculations one can conclude that the highest yields were obtained in 40 L tanks, while the lowest yields occurred in 24,000 L ponds with 600 L tanks in the mid range of

Table 1. Weekly and 20-week biomass yields ± S.Ds produced on a m^2 basis, and maximal possible dry weight (DW) production estimated from the highest growth rates for *Porphyra* species grown in 600 L tanks and 24,000 L ponds

	Species			
Yields	*Porphyra* sp (Taiwan cultivar)	*Porphyra yezoensis* (Japan cultivar)	*Porphyra tenera* (Japan cultivar)	*Porphyra linearis* (Israel cultivar)
Average weekly DW (WDW) (g DW m^{-2} $week^{-1}$)	69.3 ± 20.8	41.0 ± 10.3	74.6 ± 22.7	2.4 ± 2.1
WDW – 20 (g DW m^{-2} 20 $weeks^{-1}$)	1386.0 ± 415.8	820.0 ± 246.1	1492.0 ± 447.6	48.0 ± 19.2
Maximal observed WDW (g DW m^{-2} $week^{-1}$)	202.9 ± 30.9	128.1 ± 18.4	179.8 ± 33.6	47.9 ± 18.8
Maximal seasonal calculated DW (kg m^{-2} $season^{-1}$)	4.06 ± 1.02	2.56 ± 0.83	3.60 ± 0.93	0.96 ± 0.39

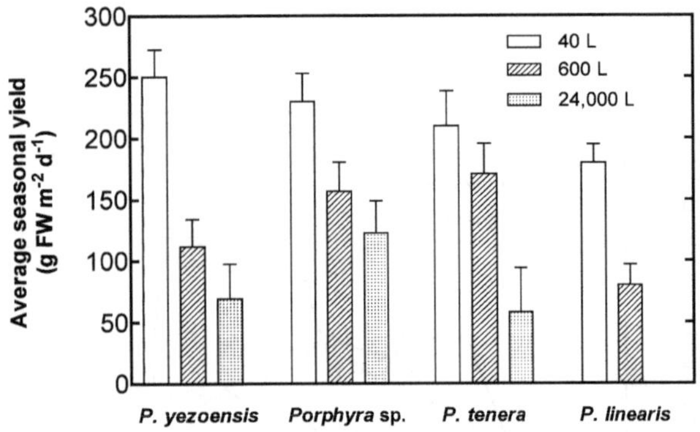

Figure 5. Average seasonal biomass yield (g FW m^{-2} day^{-1}) in four *Porphyra* species cultivated from 22 November 2002 to 13 May 2003 (170 d) in outdoor tanks and ponds of different volumes. Yields were calculated from the mean daily fresh weight increases in Figures 2–4 for the entire growth season.

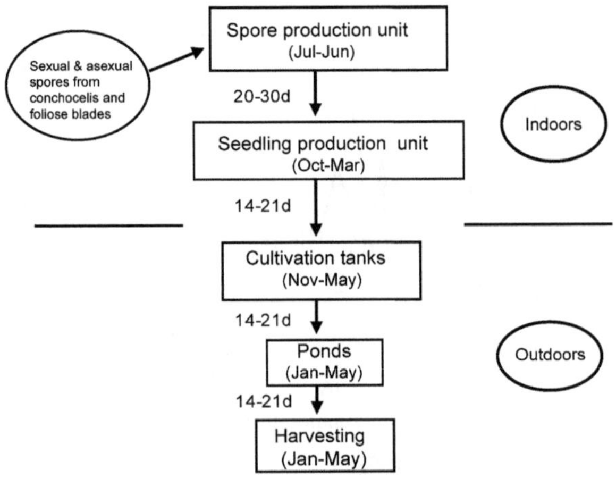

Figure 6. Major steps of indoor and outdoor *Porphyra* cultivation and their average time lengths (d). Most suitable months over a calendar year for each step are shown in parenthesis.

biomass yields ($P < 0.05$, Figure 5). Weekly yields for all species are also indicated on a dry weight (DW) basis and were estimated from yields in 600 L and 24,000 L tanks (Table 1). The highest DW production during the 20-week growth season occurred in *P. tenera* and *Porphyra* sp. followed by *P. yezoensis* ($P < 0.05$). Significantly lower productions were observed in *P. linearis* ($P < 0.01$). Assuming that high growth rates can be maintained during the whole growth season (i.e. maximal seasonal, Table 1), the calculated maximal biomass that can possibly be reached with the current technology ranged from 2.56–4.06 Kg DW m^{-2} $season^{-1}$ for the three more productive species ($P < 0.05$, Table 1).

The completion of the main steps involved in *Porphyra* cultivation and how they extend throughout a calendar year are described in (Figure 6). Generally, seedlings suitable for outdoor growth can be generated within 3–4 weeks after spore release under controlled indoor conditions. Outdoor cultivation to produce mature blades can take 30–60 days. Growth in ponds took on average another 2–3 weeks before they could be harvested (Figure 6). Consequently, spore production can be carried out all year around as long as their source (seedlings and/or free conchocelis) exists in sufficient amounts and their physiological conditions are optimal. Seedling production could expand from October to March to provide sufficient biomass for outdoor tanks when seawater temperatures are suitable (i.e. 20 °C). Ponds could likely be innoculated by early January with the first harvests anticipated by the end of January under local conditions (Figure 6).

Discussion

The optimal seawater temperatures as determined from maximal yields in 40 L tanks were between 13–17 °C for all species cultivated. Beyond approximately 18 °C all average daily yields substantially decreased until growth almost stopped above 20 °C, similar to optimal temperature responses for growth reported previously for *P. yezoensis* (Yamamoto et al., 1991) and *P. linearis* (Katz et al., 2000). Thus, as expected from the characteristic seasonal dynamics of *Porphyra* in nature, namely, active growth during cold months, temperature is a crucial limiting factor for outdoor cultivation in tanks and ponds. By comparison, *Porphyra* cultivation on nets, in Asian latitudes, begins when seawater temperature falls to 23–22 °C in the autumn. Seawater temperature then falls further and ranges from approximately 15 to 20 °C at the end of the period of making nursery nets, when some of the nets are stored in refrigerators as frozen seed nets, and others are cultivated in the sea for the first harvest (Yamamoto et al., 1991). In vitro optimal temperatures are around 15 °C yet for nori cultivation farms in the sea the optimum is considered to range from 8–10 °C (Yamamoto et al., 1991).

P. yezoensis, *P. tenera* and *Porphyra* sp. showed remarkable adaptation to free floating conditions in the new habitat created by tanks and ponds. Growth of *P. linearis*, on the other hand, was less successful even though we expected conditions to be more favorable for this local species. Therefore, tank cultivation may not fit all species but only those sustaining constant growth, similar to other seaweeds cultured previously (Friedlander & Levy, 1995; Israel et al., 1999). This may also account for an intrinsic resistance to epiphytes and/or bacterial infections that, although not quantified, were exceptionally high for *P. linearis* as compared to the other three species experimented.

The high productivities determined in this study can be explained by a number of factors. First, cultures of free-floating blades are kept in constant motion by agitation (aeration) of the medium, resulting in more efficient use of nutrients since agitation reduces boundary layers and prevents diffusion rates from limiting growth (Hafting, 1999a,b). Second, blades receive equivalent illumination with potentially higher production than in ocean-based net cultivation (Hafting, 1999a). Third, by regulating seawater exchange rates and seaweed density, tank cultivation may prevent the negative effects of epiphytes. Fourth, pond culture can be used for 6–8 cycles of 2–3 weeks each, during the growth season. Another significant advantage of pond cultivation is the ability to control mineral nutrition, resulting in the manipulation of organic and inorganic components of *Porphyra*.

Productivities of *Porphyra* species reported in this study were 50–300% higher than yields indicated for crops grown on seeded nets (Tanaka et al., 1997), even for the least productive species, *P. linearis*. These calculations were based on average biomass production of $\geq$100 g FW m^{-2} d^{-1} that is possible for at least 12 weeks out of the 20 weeks growth season. Tank and pond cultivation of the foliose phase of *Porphyra*, which is presented here for the first time, suggests a controlled high yield alternative for the traditional open sea net culture. Selection of an appropriate cultivation site can extend the growth season of *Porphyra*, and a choice of another commercial seaweed species may allow the use of the culture facilities for a whole year.

Acknowledgments

This study was partially supported by the Canada-Israel Industrial & Development Foundation (CIIRDF) and Noritech Seaweed Biotechnologies, Ltd. The authors are thankful to Mr. S. Piko for his technical assistance.

References

Chopin T., Buschmann A.H., Halling C., Troell M., Kautsky N., Neori A., Kraemer G.P., Zertuche-Gonzales J.A., Yarish C. and Neefus C. 2001. Integrating seaweeds into marine aquaculture systems: a key toward sustainability. Journal of Phycology 37: 975–986.

Fei X. 2004. Solving the coastal eutrophication problem by large scale seaweed cultivation. Hydrobiologia 512: 145–151.

Friedlander M., Levy I. 1995. Cultivation of *Gracilaria* in outdoors tanks and ponds. Journal of Applied Phycology 7: 315–324.

Hafting J.T. 1999a. Effect of tissue nitrogen and phosphorus quota on growth of *Porphyra yezoensis* blades in suspension culture. Hydrobiologia 398: 305–314.

Hafting J.T. 1999b. A novel technique for propagation of *Porphyra yezoensis* Ueda blades in suspension cultures via monospores. Journal of Applied Phycology 11: 361–367.

Israel A., Katz S., Dubinsky Z., Merrill J.E., Friedlander M. 1999. Photosynthetic inorganic carbon utilization and growth of *Porphyra linearis* (Rhodophyta). Journal of Applied Phycology 11: 447–453.

Katz S., Kizner Z., Dubinsky Z. and Friedlander M. 2000. Responses of *Porphyra linearis* (Rhodophyta) to environmental factors under controlled culture conditions. Journal of Applied Phycology 12: 535–542.

McHugh D.J. 2003. A guide of the seaweed market. FAO Fisheries Technical Paper No. 441, pp. 105.

Mencher F.M., Spencer R.B., Woessner J.W., Katase S.J. and Barclay D.K. 1983. Growth of nori (*Porphyra tenera*) in an experimental OTEC-aquaculture system in Hawaii. Journal of World Mariculture Society 14: 458–470.

Merrill J.E. 1993. Development of nori markets in the western world. Journal of Applied Phycology 5: 194–154.

Miura A., Aruga Y. 1987. Distribution of *Porphyra* in Japan as affected by cultivation. Journal of Tokyo University of Fisheries 74: 41–50.

Neori A., Shpigel M. and Ben-Ezra D. 2000. A sustainable integrated system for culture of fish, seaweed and abalone. Aquaculture 186: 279–291.

Notoya M. 1999. 'Seed' production of *Porphyra* spp. by tissue culture. Journal of Applied Phycology 11: 105–110.

Provasoli L. 1968. Media and prospects for cultivation of marine algae. In: Watanabe A, Hattori, A (eds), Cultures and Collections of Algae. Proceedings U.S.- Japan Conf. Hakonte. Japanese Society of Plant Physiology, pp. 63–75.

Sidirelli-Wolff M. 1992. The influence of temperature, irradiance and photoperiod on the reproductive life history of *Porphyra leucosticta* (Bangiales, Rhodophyta) in laboratory culture. Botanica Marina 35: 251–257.

Tanaka T., Kakino J., Miyata M. 1997. Existing conditions and problems of nori (*Porphyra*) cultivation at the coast of Chiba prefecture in Tokyo bay. Natural History Research 3: 97–109.

Yamamoto M., Watanabe Y., Kinoshita H. 1991. Effects of water temperature on the growth of red alga *Porphyra yezoensis* form *narawaensis* (nori) cultivated in an outdoor raceway tank. Nippon Suisan Gakkaishi 57: 2211–2217.

Yarish C., Wilkes R., Chopin T., Fei X.G., Mathieson A.C., Klein A.S., Friel D., Neefus C.D., Mitman G.G. and Levine I. 1998. Domesticating indigenous *Porphyra* (nori) species for commercial cultivation in northeast America. World Aquaculture 29: 26.

Journal of Applied Phycology (2006) 18: 241–249
DOI: 10.1007/s10811-006-9023-0

Rapid survey technique using socio-economic indicators to assess the suitability of Pacific Island rural communities for *Kappaphycus* seaweed farming development

M.T. Namudu & T.D. Pickering*
Marine Studies Programme, The University of the South Pacific, Private Bag, Suva, Fiji
*Author for correspondence: e-mail: pickering_t@usp.ac.fj

Key words: *Kappaphycus alvarezii*, socio-economic survey technique, seaweed farming, Pacific islands

Abstract

The literature on economic feasibility of farming seaweeds like *Kappaphycus alvarezii* in tropical locations is mainly based upon Asian case studies, and often does not take into account social factors in seaweed farming success. Pacific island countries are culturally and economically distinct from Asia, and efforts are now being made to establish seaweed industries here. Past experiences have showed that social factors often outweigh technical factors in determining the success of rural development projects. In addition, Pacific island communities are very diverse in their socio-economic make-up. The particular community chosen for location of a development project is therefore critical to success. Project managers need to recognize in advance the best type of community for seaweed farming development. The objective of this study was to identify socio economic factors that can be used as predictors of project success or failure. Using results of social survey techniques carried out in eight communities within the Fiji Group, a rapid survey technique has been developed which can enable decisions about whether a community is suitable for farming seaweed or not. Though developed from Fiji case studies, the technique can be applied in other rural Asia/Pacific situations.

Introduction

There is a small but growing literature on the economics of farming the seaweed *Kappaphycus alvarezii* (Doty) Doty ex P.C. Silva, mainly in Asia; for example, see Padilla and Lampe (1989), Alih (1990), Firdausy (1991), and Tseng and Fei (1997). Industry development in Pacific island countries including Fiji has been documented by Ask (2003), Ask et al. (2003a), Ask et al. (2003b), Luxton et al. (1987), Luxton and Luxton (1999), Luxton (2003), South and Pickering (2006), and Pickering (2006).

Selection of places to farm *K. alvarezii* in the Fiji Islands (SW Pacific Ocean) had historically been made on technical grounds related to suitability of sites for good seaweed growth. Socio-economic factors were limited to whether or not there was close proximity to infrastructure like ports or towns (Sam Mario pers. comm.). In recent years however, experiences across the Pacific and elsewhere have pointed to a need for study to identify the critical socio-economic factors (for example, alternative livelihoods, population demographics, geography, local traditions, etc.) for which information must be gained in order to recognize in advance the best types of communities for *Kappaphycus* seaweed projects. These critical socio-economic factors could then be used as predictors of success or failure in *Kappaphycus* seaweed farming development, if incorporated into the design of a rapid survey technique for selection of communities for seaweed farming.

When seaweed farming re-commenced in Fiji in 1997 after five years of no production, seaweed buyers FMC Corporation placed importance upon identifying such predictors of success and provided support for this study to be carried out. "Success" within this context shall mean that, once external assistance to start up a seaweed project is withdrawn, farming activity will continue and sustain itself into the future.

In this study the socio-economic characteristics of several village communities in Fiji are reported, and assessed as indicators for selection of sites to farm seaweed. A rapid survey technique is proposed that could be used to predict the areas best suited for seaweed farming projects.

Materials and methods

Historical records from seaweed production (Sam Mario, unpubl. data) from different parts of Fiji were obtained to find out whether there has been any shift in farming from some parts of Fiji to other parts (Figure 1).Characteristics of the relevant areas were noted, to find out which are associated with the most-recently farmed areas.

Fieldwork was carried out in three broad categories of communities in Fiji (Figure 3). The categories were: Community Type I–communities in which seaweed is currently being farmed; Namuka-i-Lau (visited in May 2002) and Nakobo (visited in July 2002), Community Type II–communities in which seaweed had previously been farmed but now abandoned; Dama (visited in January 2002) and Malake (visited in September 2002), and Community Type III–communities which are soon planning to take up seaweed farming–Yaqeta (visited in August 2002) and Serua (visited in October 2002).

A total of 78 farmers were interviewed, representing about 12% of the total number of farmers currently in Fiji (658 farmers). Interviews took place with all of the farmers available in each community at the time of visit, using the Bauan dialect of the Fijian language. During the interview, each farmer in all communities completed a written questionnaire (also in Bauan) seeking basic household economic information. A second questionnaire with special questions for each of the three community types was then completed, which for Type I communities sought perceptions about *why* seaweed farming was successful, for Type II communities sought perceptions about *what* were the factors behind lack of success, and for Type III communities sought perceptions about *what* they find attractive about the notion of seaweed farming. Lastly, there was a set of questions for the village headman about village infrastructure, traditions, and population demographics. Typically each interview and questionaire took 30–60 min to complete. Figures for currency are reported in Fiji dollars (FJD 1.00 = approx. USD 0.50 at time of writing) unless otherwise stated.

Results

Production figures broken down by district within Fiji from 1985 to the present show that *Kappaphycus* seaweed farming has moved from the Western part of Fiji (Ra Province) to the Central part (Tailevu and Cakaudrove) during 1985–1992, and since 1997 most production has come from remote parts of South Eastern Fiji (Southern Lau Group) (Figure 1 and 2). Based upon interview results, the following socio-economic indicators emerged as likely predictors of success in establishment of seaweed farming projects in Fiji.

Demographics

In Type I communities, active seaweed farmers were between 20 and 70 years old, with ages 30–40 most predominant but other age groups in this range well represented except for ages 60–70. The population recorded in the 1996 census provisional figure obtained from the Fiji Bureau of Statistics (2003), for Namuka-I-Lau and Nakobo was 306 individuals (51 households) and 104 individuals (28 households) respectively. According to the village headmen, the population was similar in 2002 when the study took place. Out of this population, 21 individuals were actively participating in *Kappaphycus* seaweed farming in both villages, assisted by some nuclear-family members. In Nakobo, 14 were male (household heads), and 7 were female (aged between 26–52 years). There were no female farmers at Namuka-I-Lau, though women in both places did assist household heads with land-based aspects such as tying or drying plants. Of the Type II and III communities, Dama, Malake, Yaqeta, and Serua had 35, 66, 65 and 28 households respectively.

Alternative livelihoods

Prior to seaweed farming, Type-I Community Namuka-I-Lau depended for its livelihood on subsistence fishing and reef gleaning, with very limited subsistence agriculture owing to rocky ground, and with steady cash income only from copra and beche-de-mer. Copra fetches $0.22 per kilogram which translates into $20–$ 100 per

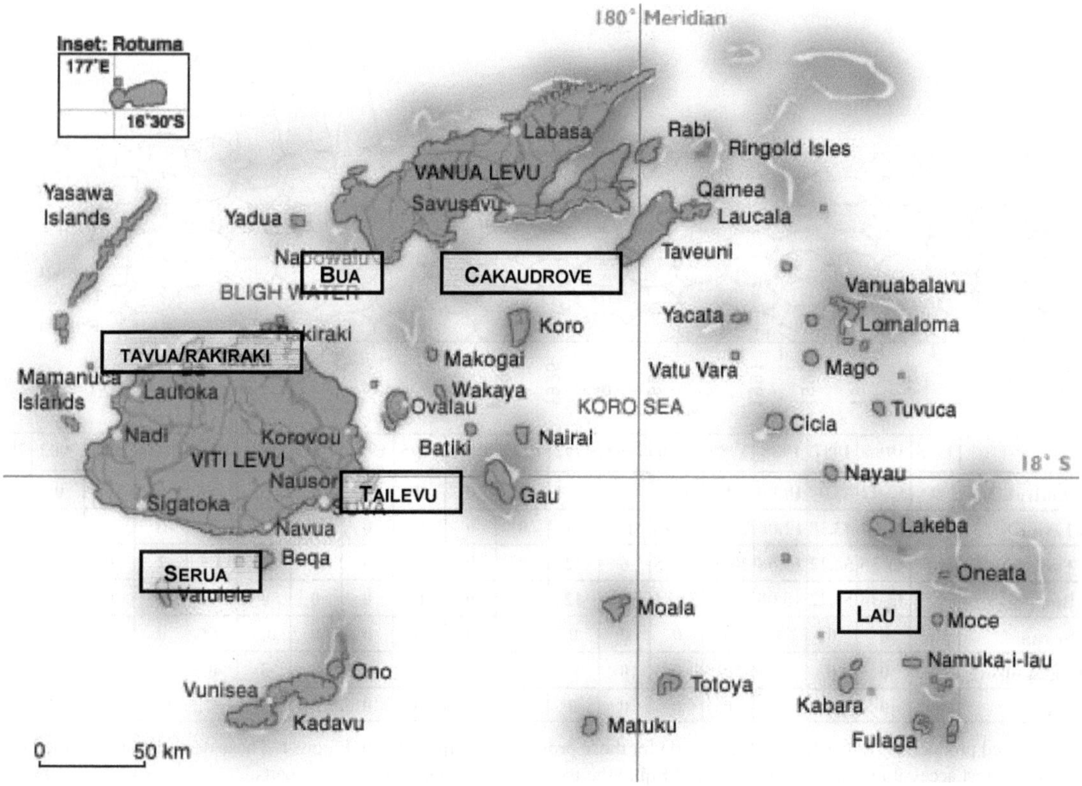

Figure 1. The Fiji Group, showing the location of provinces where seaweed has been farmed in Fiji.

month representing about a week of work per month. Copra production engages the whole family and is a strenuous activity. Beche-de-mer is caught opportunistically while fishing for finfish and is sold fresh to a middleman for drying, typically bringing \$20–\$40 per household over a monthly period. Giant clams are occasionally collected upon request from relatives living in Suva, whereupon a group of young men would be paid \$10–\$20 each to collect sufficient for the order. Manufacture of tapa cloth (*masi*) also occasionally takes place upon request of relatives when the need arises (weddings, birthdays, funerals) and fetches between \$50–\$300 per piece. When the Inter-Island vessel visits the island (twice a month, depending upon weather) any catch of fresh fish available at the time will be sent to Suva for sale, but the amount sold in this way is small.

The village of Nakobo relies on copra, kava, sea cucumber, root crops, freshwater prawns, and fishing for their sources of livelihood. From copra a household can earn between \$20–\$80 for a week's work but this is only done when a need arises (school fees, Christmas, etc.). Kava is harvested when ready and brings \$80–\$ 250 per household for each 2–3 month harvesting cycle. Beche-de-mer, prawns, finfish and root crops bring \$15–\$30, \$10–\$30, \$40–\$80 and \$15–\$50 per month respectively. Sales are not on a weekly or regular basis but occur opportunistically or when there is market demand. A daily truck services the area around Nakobo village thus allowing villagers to visit the town of Savusavu any time they are ready to sell produce, or to go shopping.

In both of these Type-I communities most seaweed farmers were earning on average \$50–\$100 per week from seaweed farming. During 1999–2000 however, when marketing arrangements were going smoothly, technical support was regular and seedstock was readily available, some full-time farmers were earning around \$300 per week from seaweed. For both villages, in times of bad weather, foodstuffs like canned meat or fish, instant noodles etc. are bought to supplement their subsistence diet. Nakobo villagers have an advantage since food on land is in abundance, especially a variety of green leafy vegetables and root crops.

Type-II Community Malake has long been actively engaged in commercial fishing over long distances (as

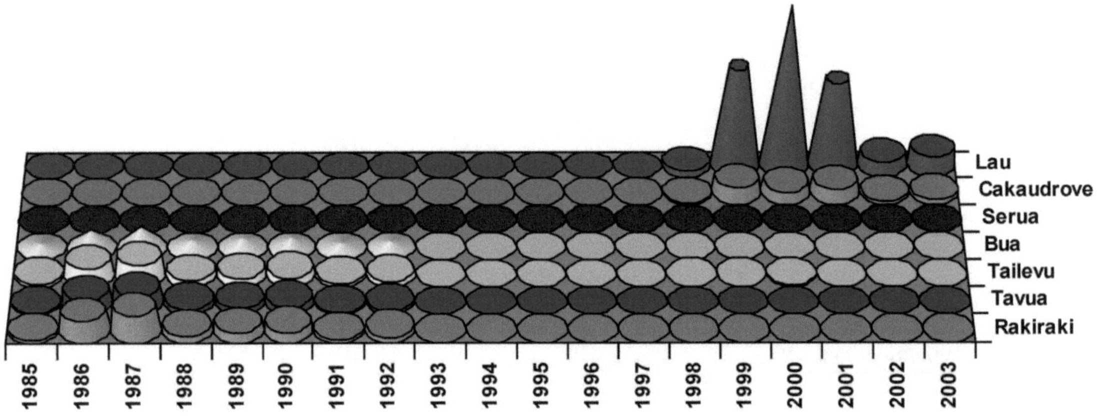

	1985	1986	1987	1988	1989	1990	1991	1992	1993	1994	1995	1996	1997	1998	1999	2000	2001	2002	2003
■ Rakiraki	9	52.02	65.07	18.09	24.1	26.22	7.2	14.4	0	0	0	0	0	0	0	0	0	0	0
■ Tavua	6	34.68	43.38	12.06	16.07	17.48	4.8	9.6	0	0	0	0	0	0	0	0	0	0	0
□ Tailevu	7.5	43.35	54.22	15.08	20.09	21.85	6	12	0	0	0	0	0	1	1.5	2	0	0	0
□ Bua	7.5	43.35	54.22	15.08	20.09	21.85	6	12	0	0	0	0	0	0	0	0	0	0	0
■ Serua	0	0	0	0	0	0	0	0	0	0	0	0	0	0.3	0.5	0	0	0	0
■ Cakaudrove	0	0	0	0	0	0	0	0	0	0	0	0	0	3	40	30	35	8	10
■ Lau	0	0	0	0	0	0	0	0	0	0	0	0	0	15.5	257.5	408	227	37	50

Figure 2. Seaweed production figures from 1985 to 2003 for those provinces of Fiji shown in Figure 1. These demonstrate that the industry has shifted from the most accessible and developed parts of Fiji to the most remote and least-developed parts.

far as Lomaiviti and Yasawa Groups) on trips of a week or more at a time, which brings on average $125 per week to a household. They became interested in seaweed farming at a time when incentives to farmers included presentation of fiberglass punts and 40 hp outboard motors. These were needed to make long journeys of over an hour to reach the closest suitable seaweed farming sites. People at Malake do not engage in subsistence agriculture, but prefer to buy groceries from nearby Rakiraki town. Malake is also close to tourist resorts, which provides cash employment for some village women.

People at Dama Village in Bua rely on copra, reef and mangrove gleaning (e.g. mangrove crabs) and artisanal fishing. Women also make fine mats from a material called *kuta* and sell them ($100–300 per piece) every so often. These activities bring in about FJD50–100 per household per week.

When seaweed farming activity was at its height in Malake and Dama village in the 1980's, farmers were earning between $10–$25 per week. Seaweed was being bought at that time by Coast Biologicals Ltd. at FJD0.35 c per dry kg.

Type-III community Yaqeta is on an isolated small island, but is near a large tourist hotel and the community has also set up its own backpacker lodges. In addition to paid employment in tourism (3/4 of all adults), they also sell fruit and vegetables to the tourist operations. A typical household income is $150–200 for those in employment, while those fishing (e.g. for lobster) or gardening receive $50–120.

The community at Serua Island is involved in copra, artisanal fishing and reef gleaning (fish sold to Navua town). Average weekly household income from these activities is $30–100.

Preferences for particular livelihoods

In the Type-I and Type-III communities the advent of seaweed farming is seen as a relief from the back-breaking work of other livelihoods, and was also welcomed because provision of a boat and outboard also made other livelihoods like fishing easier. In Type-II communities the same feelings applied, however seaweed farming in Malake was abandoned after Coast Biologicals Ltd withdrew in 1987 because the level of

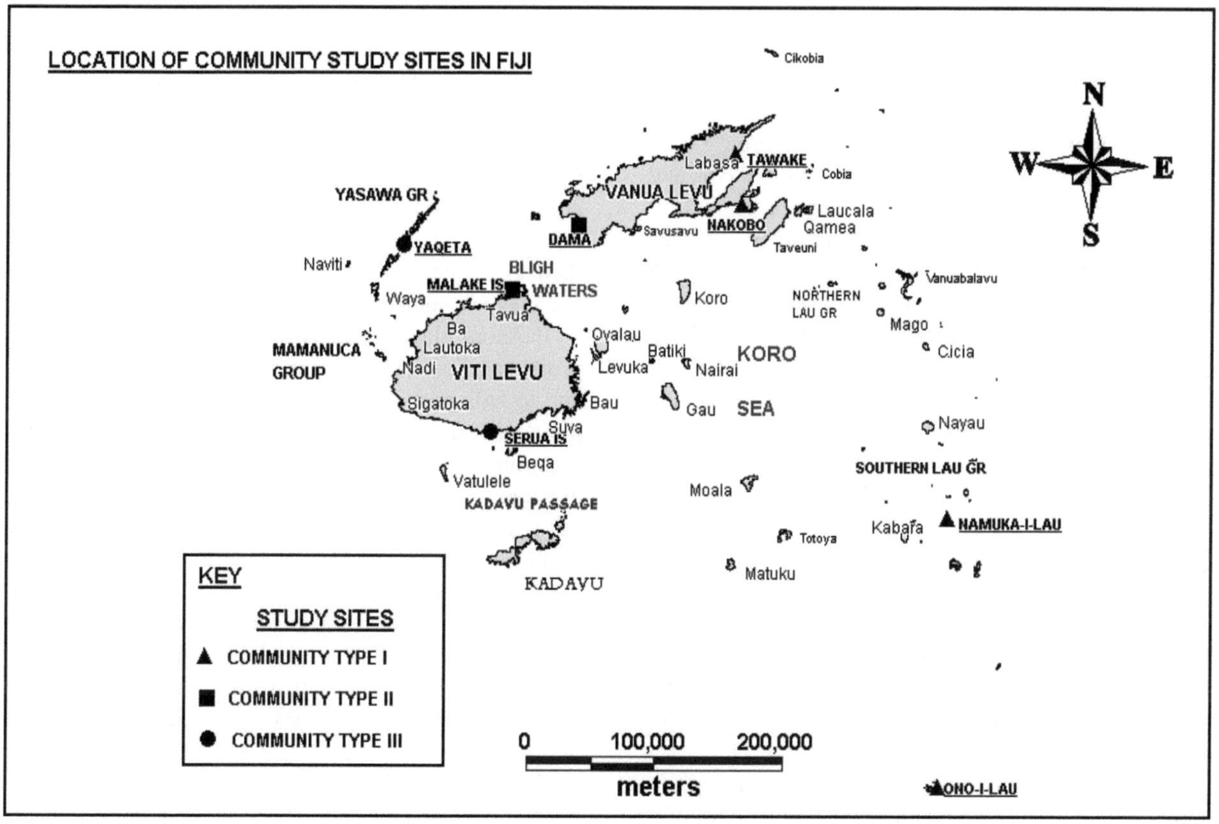

Figure 3. The Fiji Group, showing the location of the six communities compared in this study.

farmer incentives, quick response of technical support from agents/fieldmen, and timeliness of payment, all suffered when government took over the seaweed projects. People at Dama village expressed similar views to those of Malake. In all communities interest in seaweed farming was expressed, but people saw it as an activity to be supplemented by their existing livelihoods, in case things go wrong, rather than as a full-time activity.

Barriers to entry into seaweed farming

Questions about what people perceive as barriers to entry into seaweed farming elicited a range of responses. For all community types, issues which prevented people from becoming farmers were:

1. Lack of *Kappaphycus* seedstock, due to inadvisable over-harvesting and selling by the farmer, plant mortalities caused by high sea-surface temperatures during ENSO events, or plant losses after stormy weather.
2. Suitable farm sites located far from the village and only accessible by boat (takes 20–60 min), but no boats available and none provided by government under the seaweed project. This barrier applied to all communities except Namuka-i-Lau where suitable farm sites are found only 50 m from the village.
3. Lack of knowledge about best-practice farming techniques (this applied to all communities).
4. Lack of technical advice from qualified field personnel to overcome (3) above.
5. Gender, in that women are not taken seriously if they express interest in seaweed farming, and are not supported (e.g. allocated necessary materials) or are even ridiculed in those few instances where women have become active farmers.

One issue which was not a barrier to entry (for male gender) was availability of farm materials; all communities had been supplied with ample stock of ropes and raffia for farm construction. Gender, age and infirmity were not barriers in the sense that all could participate in on-shore activities and all derived benefits from them.

Best business model for income-generating activities

In Fiji, over the years, three main types of seaweed farming business model have been tried; (1) community (e.g. church) or tribe/clan groups farming cooperatively, (2) company farms owned by the buyer and operated by labour hired on a daily wage basis, and (3) individual or household (nuclear family) operated farms. At the time of this study, 100% of farms in all of the communities visited were household-operated farms. People perceived there would be problems (e.g. with farm maintenance, and equitable distribution of benefits) with community-operated farms, and spoke against this type of model. Administration of a household-based business was seen as much easier to achieve.

Support of social organizations within the community for farming activities

All communities appreciate any development projects set up in the rural areas. These developments are viewed as intended to help the villagers economically and socially, and improve their standard of living. Thus for a project like *Kappaphycus* seaweed farming, social organizations within these communities all tended to be in support. All of the seaweed farmers interviewed, along with their families, were part of organized church, youth, men's, or women's groups in the villages studied. They felt their ability to contribute to group activities was enhanced by income derived from seaweed.

Delivery of external support for seaweed farming

It has been government policy to help farmers until their farms are self-supporting, through a "start-up" pack of assistance to interested seaweed farmers (Apisai Sesewa, pers. comm.). From 1997 to 1999 the standard start-up pack included the following equipment required for farming:

1. Seedlings (usually brought from a government nursery at Kiuva village).
2. Nylon ropes/heavy fishing lines as seaweed suspension lines between poles/posts.
3. Drying racks constructed from mangrove posts, timber, nylon mesh screen and plastic covers.
4. Crow bars (to assist in securing wooden posts in sandy or coral-rubble substrates.
5. 20-foot marine ply punt, 15 hp outboard motor, and 10 L fuel.
6. Raffia for tie-ties.

Government placed conditions on provision of this start-up assistance, e.g. the farmer must set up and sell a minimum of 400 lines of seaweed, or else return their boat and engine. There were complaints about the allocation and administration of this assistance, however. One farmer mentioned many cases where boats were not taken back from people who never reached the required number of lines, while others who did reach their target never received any of the promised assistance. No women ever received this start-up assistance.

Farmers place value on regular visits by seaweed project staff as a sign that the project is of importance, and to help sort out any technical problems being experienced. The frequency of such visits was seen as very inadequate, being at roughly 6-month intervals on average, or in the case of Nakobo there was no visit for two whole years. People felt that monthly visits by project staff were most appropriate. At times when there have been visits by district-level project staff, advice given has been incorrect, e.g. about plant size at harvest, farm location and farm orientation. People also reported a lack of warning about impending high sea-surface temperature events, and what to do about it.

Problems in marketing

All communities expressed a wish that the price paid per kilo be raised, but at the same time indicated they would be willing to farm seaweed at the prevailing price in Fiji (FJD 0.50, or USD 0.27) if it were not for one very important point; payment for seaweed was often very late. In some cases, seaweed was paid for 6 months after delivery. Farmers also complained that they have been made responsible for the cost of freight to Suva (FJD 1.50 for each 20 kg sack) when their understanding was that FJD0.50 was to be a beach price. They lose interest in seaweed farming when they see that the buyers of copra, fish, and other marine products are paying people promptly (within 24 h).

Traditions, village politics

A few interviewees complained that they do not have any time for other chores because of seaweed farming. No disputes based upon traditional ownership of land, reef, or resources were reported to us during this study. The support of traditional leaders is regarded by people as essential because, in their thinking and belief, chiefs are revered within the Fijian social stratum. The position chiefs hold is not merely political, but spiritual as well, which accords them a *mana* or spiritual blessing and power.

At the time of this study, in five of the six communities people enjoyed the support and blessing of their chiefs to take part in seaweed projects. In fact, government could not have promoted such a project in these communities if the blessing of chiefs had not been obtained. In Dama Village, however, the chief had determined that seaweed could be farmed only by people of the village in which he himself resided, but not by other villages over which he had dominion. People in these other villages wanted to farm seaweed, but could not.

Discussion

Since seaweed farming began in Fiji the geographical centre of farming effort, initially in the west, has moved eastwards. This initial westward distribution reflected the importance attached to being near transport infrastructure (ports, towns) for baling and export of seaweed. Such communities have a wide range of livelihoods available to them, however. Peoples' main motivation for entering seaweed farming appeared to have been to get a free boat from the seaweed project which they could then use in these other livelihoods. The new approach taken from 1997 onwards has been to target communities with few other sources of income due to their isolation and lack of ready access to markets for perishable produce (Sam Mario pers. comm.). The highest-performing seaweed farming community is now the island of Ono-i-Lau, one of the farthest-flung and remote islands of the entire Fiji Group.

Some trends have emerged from the information and peoples' perceptions collected in the six communities studied. Factors which can mark out a community as suitable (or unsuitable) for seaweed farming are discussed below.

Firstly there should be suitable seaweed farming environments within easy reach of the village. Ideally a boat should not be required, except perhaps for harvesting. Every coastal village in Fiji will have some kind of boat that can be hired for this, as boats are essential for fishing and as a basic mode of transportation. Farming can still take place if a boat is required on a daily basis for farming, and individual households lacking a boat can follow other avenues to obtain one rather than expect a free boat and engine. Acquisition of a boat and engine by those households that lack one is not an easy thing, but government does have some micro-finance schemes available. Interview data showed that, when boats were given free, many households, e.g. in Malake, entered seaweed farming not with the intention to farm seaweed but to get a boat and engine.

Although a few people complained that seaweed farming interferes with other chores, seaweed fieldmen and successful farmers commented that such people are simply not well motivated to do anything. Engaging themselves in the new project will only affect their household plans if they do not have a set timetable to work from. A person's character will also determine whether he is suitable to take up projects for the community like seaweed farming or not, and is a factor that needs to be assessed through brief interviews by any visiting project officer. Some people will work only enough to meet basic needs, while others are willing to work to make money for money's sake.

For any project to be set up in rural areas, people need to be available to take it up. This means that communities need to be of a reasonable size. From the demographic results, one can see that all the communities visited were of similar size, so demography alone does not explain their differences in suitability for seaweed projects. In terms of demography, the majority of interested people were between the ages of 20–50 years. This is the standard age group that the Fisheries Department in Fiji target when they go out to introduce this project. Female participation is presently not significant in any of these communities, so there is wide scope for seaweed production to be increased if women can also be supported and encouraged to farm seaweed.

Options for alternative sources of income differed widely between communities, and even differed between the two Type-I communities. The people who became active farmers in Nakobo did so firstly for the incentives provided (free boats, etc.) and secondly because they could see that others in the village so engaged were now making a useful amount of extra cash income from selling seaweed which supplemented their other livelihoods. It appears that seaweed farming in Fiji can compete against copra, against fishing in those areas with poor market access for fresh fish, against beche-de-mer in those places where it has been over-fished, but can only compete provided payment for seaweed is prompt. Seaweed farming cannot compete against artisanal fishing in places with ready access to main-centre markets, against large-volume sales of kava, or against tourism. Tourism livelihoods are more open to younger people, however, so there is some scope for seaweed farming in such places if carried out by people aged 50–70 years. Seaweed's competitiveness against other livelihoods is greatly enhanced if payments for seaweed are prompt, and suffers badly if it is not.

The chiefs of an area are respected by their people and by government projects, irrespective of their decision. It is seen as a taboo to speak against a chief or an elder. The positive attribute of this chiefly system was found to be that chiefs are open to suggestions about ways to improve the standard of living for their people. In this way, there is a healthy working relationship with various government departments who have responsibility for sustainable use of their resources.

Conclusion – a rapid survey technique for seaweed farming

From the various socio-economic factors discussed above which can indicate the likelihood of success in seaweed farming, a set of questions can be constructed, to which any person visiting a community for the first time should seek answers.

1. Are areas suitable for seaweed farming located nearby?
   This can be judged by technical knowledge about seaweed requirements, and by conducting a range of simple tests and growth trials following methods such as those of Ask (1999). Accessibility of sites must be considered, in particular whether or not a boat and engine is necessary, and whether or not people already have or can possibly obtain their own boats.
2. Is there sufficient population of those age groups and gender groups most likely to be interested and involved in seaweed farming? It is no use spending time developing a place where there are insufficient people able to take up farming and generate the volume of seaweed required to justify investment in that area. Can seaweed farming be adopted by all interested parties in the village, or only by certain clans (e.g. fisher clans)?
3. What alternative livelihoods exist in the village?
   For example, seaweed farming in Fiji can compete against copra, fishing (in areas remote from markets), beche-de-mer (if over-fished) and subsistence agriculture. Seaweed farming cannot readily compete against artisanal fishing with ready market access, large-scale cultivation of kava, or tourism, unless significant demographic groups exist who do not participate in these livelihoods, or they have complementary seasons of activity.
4. What is the strength of feelings for adherence to particular income activities?
   Some communities may have strong preferences toward a particular way of life and have low interest in changing, even if other indicators show up as positive for the project. Are there people present who already display entrepreneurial tendencies in other activities?
5. What are the barriers to entry into seaweed farming? Is a boat and engine really necessary? Is the area sufficiently accessible that project staff will be able to make at least monthly visits to solve problems, and provide advice and encouragement?
6. What are peoples' preference for a suitable unit of production for seaweed farming? Options are: Household; Community co-operatives; Company farms. Experience shows that Household farms are least problematic.
7. Are traditional leaders and social organizations within the community supportive of seaweed farming? Will it cause problems if they are not? Are the traditional leaders liked and respected by the villagers?
8. Will traditional tenure systems be compatible with seaweed farming?
   Traditional disputes over resource ownership or income must either not exist, or have a means available for being readily solved. Are there any current disputes?
9. Is the village sufficiently well-served by road or shipping links for seaweed to be transported to main centres? The village should not be so well-served that they have ready urban-market access for produce like fresh fish on a weekly basis, but sufficiently well-served that seaweed can be taken out in bulk every 2–6 months.

By focusing upon these key indicators, a good idea of the chances of success for seaweed farming within a community can be obtained. National priorities can then be set for the particular areas earmarked to receive investment of time and resources for seaweed projects.

Acknowledgements

The authors thank Erick Ask of FMC Corporation for suggesting this line of research and for providing funding support to carry it out. We are also indebted to the following people for their help in providing valuable information for this paper: Jeff Kinch, Sompert Gereva,

Rory Stewart, Sam Mario, Ratu Manoa (Nakobo), Aliki Ratakele, Mr Bale (Ono-I-Lau), Apisai Sesewa and Tavenisa Vereivalu. Lastly, thanks go to the vanua of Nakobo, Namuka-i-Lau, Yaqeta, Malake, Serua and Dama for hosting the visits to each location.

References

Alih EM (1990) Economics of seaweed *(Eucheuma)* Farming in Tawi-Tawi Islands in the Phillipines. In Hirano R, Hanyu I (eds), The Second Fisheries Forum. Asian Fisheries Society, Manila, Phillipines.

Ask EI (1999) Cottonii and spinosum cultivation handbook. Unpublished Report, FMC Corporation, Philadelphia PA, USA, 52 pp.

Ask EI (2003) Creating a sustainable commercial *Eucheuma* cultivation industry: The importance and necessity of the human factor. In Chapman ARO, Anderson RJ, Vreeland VJ, Davison IR (eds), Proceedings of XVIIth International Seaweed Symposium, Oxford University Press, pp. 13–18.

Ask EI, Batibasaga A, Zertuche-Gonzalez JA, de San M (2003a) Three decades of *Kappaphycus alvarezii* (Rhodophyta) introduction to non-endemic locations. In Chapman ARO, Anderson RJ, Vreeland VJ, Davison IR (eds), Proceedings of XVIIth International Seaweed Symposium, Oxford University Press, pp. 13–18.

Ask EI, Ledua E, Batibasaga A, Mario S (2003b) Developing the cottonii (*Kappaphycus alvarezii*) cultivation industry in the Fiji Islands. Proceedings of XVIIth International Seaweed Symposium, Oxford University Press, Cape Town 81–85.

Firdausy CM (1991) Economic returns from different seaweed (*Eucheuma*) farm operations in Nusa Penida, Bali. IARD Journal 13: 24–31.

Luxton DM (2003) *Kappaphycus* agronomy in the Pacific Islands. In Chapman ARO, Anderson RJ, Vreeland VJ, Davison IR (eds), Proceedings of XVIIth International Seaweed Symposium, Oxford University Press, 41–47.

Luxton DM, Robertson M, Kindley MJ (1987) Farming of *Eucheuma* in the South Pacific islands of Fiji. Hydrobiologia 151/152: 359–362.

Luxton DM, Luxton PM (1999) Development of commercial *Kappaphycus* production in the Line Islands, Central Pacific. Hydrobiologia 398/399: 477–486.

Padilla JE, Lampe HC (1989) The economics of seaweed farming in the Phillipines. Naga, 3–5.

Pickering TD (2006). Advances in seaweed aquaculture among Pacific island countries. J. Appl. Phycol. (in press).

South GR, Pickering TD (2006) The seaweed resources of the Pacific Islands. In Critchley AT, Ohno M, Largo D (eds), Seaweed Resources. Expert Centre for Taxonomic Identification (ETI), Univ. Amsterdam (CD-ROM series).

Tseng CK, Fei XG (1997) Economic aspects of seaweed cultivation: Macroalgal commercialization in the Orient. Proceedings of XIIth International Seaweed Symposium. Hydrobiologia 151/152: 167–172.

Journal of Applied Phycology (2006) 18: 251–257
DOI: 10.1007/s10811-006-9021-2

Artificial seed production and cultivation of the edible brown alga, *Sargassum fulvellum* (Turner) C. Agardh: Developing a new species for seaweed cultivation in Korea

Eun Kyoung Hwang[1,*], Chan Sun Park[2] & Jae Min Baek[1]
[1]*Seaweed Research Center, National Fisheries Research and Development Institute, Mokpo 530–831, Korea;*
[2]*Department of Marine Resources, Mokpo National University, Jeonnam 534–729, Korea*

*Author for correspondence: e-mail: ekhwang@momaf.go.kr/ekhwang@hotmail.com; fax: +82-61-285-1949

Received 20 June 2004; revised and accepted 14 January 2005

Key words: artificial seed production, growth, maturation, *Sargassum fulvellum*

Abstract

Sargassum fulvellum is a brown alga recently introduced to the seaweed cultivation industry in Korea. There is current interest in the commercial scale of aquaculture of this species. For the artificial seeding and cultivation of this alga, growth and maturation were investigated from September 2002 to August 2003. Indoor culture experiments for maturation induction were also conducted at temperatures of 5, 10, 15, 20 and 25 °C and irradiances of 20, 50, 80 and 100 μmol photons m^{-2} s^{-1} under 16:8 h (L:D) photoperiod. Within a given culture test range, higher temperature and irradiance levels favoured the maturation of receptacles in *S. fulvellum.* Using temperature and irradiance control for thalli, artificial seed production of this species could be done one month earlier than thalli matured in nature. Under natural condition, receptacle formation of the plants began in February, and the eggs were released from March to April. For mature thalli of 200 g wet wt., artificial seeding was complete enough for attachment on seed strings of 100 m. Mean production obtained from the artificial seeding technique *in situ* was 3.0 kg wet wt m^{-1} of culture rope during the cultivation period.

Introduction

The perennial brown alga *Sargassum fulvellum* (Turner) C. Agardh has a wide distribution from the south to the eastern coasts of Korea. This species usually grows at depths of 3–5 m or more. There are 28 species of *Sargassum* species reported in Korea (Lee & Kang, 2002). Among them, *S. fulvellum* is the common edible *Sargassum* species used as a seaweed salad. In the local Korean market, the retail price of naturally collected *S. fulvellum* is US$ 2.3–3.8 kg^{-1} fresh wt.; that of *Porphyra* and *Undaria* is 0.5–1.5 and 0.07–0.1 kg^{-1} fresh wt., respectively. The seaweed cultivation industry in Korea depends on a few species such as *Porphyra, Undaria, Laminaria* and *Hizikia* (Sohn, 1993, 1998). *S. fulvellum* can be a potential seaweed species for economic seaweed cultivation by fishermen in Korea. Since the demand for *S. fulvellum* is likely to remain high in the future, it is necessary to develop mariculture. Furthermore, *Sargassum* beds play important ecological roles in the coastal ecosystem due to their large biomass and high productivity. These beds provide nursery areas to commercially important fish species and help to preserve environmental conditions. Therefore, considerable information has been accumulated on their growth, maturation period and cultivation techniques from ecological and industrial viewpoints.

This is the first report on artificial seeding and cultivation of *S. fulvellum.* We report here the growth and maturation period of *S. fulvellum*, and artificial seeding and culture conditions for its commercial cultivation.

Materials and methods

Plants were collected monthly at Wando (34°18′N, 126°45′E) on the southwestern coast of Korea from September 2002 to August 2003. Three 50 × 50 cm quadrats were randomly placed, and all thalli inside the quadrats were collected and carried to the laboratory. Once in the laboratory, thalli were cleaned of epiphytes and rinsed with filtered seawater. All the thalli were measured in length and weighed. The water temperature was measured at the sampling site. For indoor culture experiments, reproductive plants were transported to the laboratory immediately after collection in February 2003. The plants were rinsed in sterile, filtered seawater and the receptacles excised. These were treated by immersion in 1% Betadine (povidone iodine) solution for a few seconds. They were then incubated at 15 ± 0.5 °C in an antibiotic mixture solution (Guillard, 1968) for a day. After being cleaned, the receptacles were cultured in petri dishes (30 explants in each) with 20 ml of PESI culture medium. Water temperatures of 5, 10, 15, 20 and 25 °C were used for the maturation of receptacles. Irradiances were measured on a LI-1000 Data Logger (Li-Cor, U.S.A.) as 20, 50, 80 and 100 μmol photons m^{-2} s^{-1} at the surface on the sterilized petri dishes. In all the cultures, multi-room chamber incubators (HB-302M-4, Hanbaek Co., Korea) were used for the photoperiodic control of 16:8 h (L:D). Total length was expressed as the mean values of the plants in each treatment. Maturation was determined as the percentage of explants in each treatment showing germling release, under microscopic observation ($n = 30$ in each treatment).

The collected thalli were kept and attached on 150 cm lengths of polyvinyl chloride (PVC) pipes for 2 months in a square concrete tank containing 2 tons of seawater, which was continuously aerated. The PVC pipes help to sink the thalli to bottom of the tank so that the thalli cultured under water are not exposed to the air. It was also easy to handle the thalli and select those mature enough to have embryos.

Mature thalli were thinned out from the PVC pipes and then moved to plastic dishes (50 cm in diameter, 20 cm in depth), to gather dense embryo solutions by rubbing the mature thalli which had embryos in their receptacles. The liberated embryos that sunk onto the bottom of the plastic dishes were collected by net (ca. 300–500 μm in mesh size), and washed several times with fresh filtered seawater. As a seeding material for the embryos, PVC frames (ca. 35 cm height × 45 cm width, holding a total length of 100 m of string made of mixed nylon and polypropylene fibers) were used for *Undaria* cultivation, and the collected embryos in suspension were attached onto the seed frames with a paintbrush. At the time of the attachment, all seed fiber had to be dried before attaching the embryos because the dried fiber absorbed the water containing the embryos. Embryos may have been held by the absorption force of the string fiber until they grew their holdfast on the substratum. Seedlings of *S. fulvellum* were reared in an indoor tank for 60 days until they were up to 10–15 mm in length. The tank used for the seedling culture was 80 cm wide, 7 m long and 70 cm deep. Fresh seawater and air were continuously supplied, through a pipe, to the tank. Water temperature was not controlled due to the influx of natural seawater. Illumination was regulated with a shading sheet, to about 60–80 μmol photons m^{-2} s^{-1} (on the water surface, at noon on fine days). During the seeding and culture, the length and number of fronds on the seed string and number of laterals were measured once a week. After one month of tank culture, seedlings were transferred to the nursery culture ground in Wando for one month. After that, culture at sea was carried out using a long-line system, described by Sohn and Kain (1989). A 100 m line (10 mm in diameter) was used with coiled seed strings (3 mm diameter, 50 mm length) every 10 cm. The main culture line was held at 1 m depth, using plastic buoys. Culture ropes were periodically cleaned of fouling. Biological variables such as length of thalli and biomass per culture rope were measured monthly during the culture season.

Results

Water temperature varied from 7.1 to 23.6 °C during the experiment (Figure 1). Maximum water temperature was recorded in September 2002, and the minimum in February 2003. *Sargassum* started to grow when the temperature decreased below 23 °C in September. Growth of *Sargassum* thalli was observed from September in the natural habitat. From February, growth in length of the stipe increased rapidly. It reached a mean maximum of 104.6 ± 20.7 cm in the middle of March, and then started to decrease. In May, main axes were bleached and only holdfast parts were left on the substratum. In nature, receptacle formation was observed from February to April when water temperature was 7.1–12.8°C. The peak period for egg release from female receptacles was from March to April.

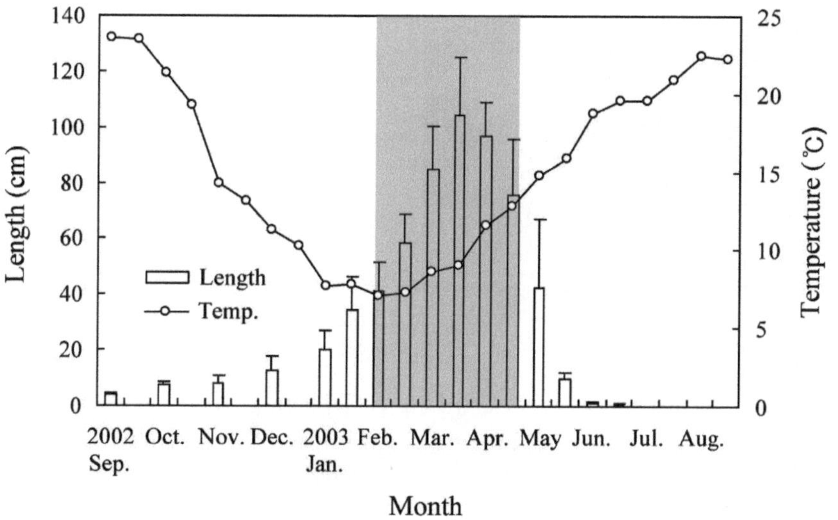

Figure 1. Fluctuations in mean length of *Sargassum fulvellum* and water temperature of the habitat in Wando, Jeonnam from September 2002 to August 2003. Grey area indicates maturation period in nature and vertical bar SD.

Temperature effects on egg release were significantly different between 5 °C and 25 °C. After 9 days culture under 20 °C and 80 μmol m^{-2} s^{-1}, the egg release rate reached a maximum value of 97% (Table 1, one-way ANOVA, $p < 0.01$). However, irradiance effects on egg release were not significantly different between 20 and 100 μmol photons m^{-2} s^{-1} (Table 2, one-way ANOVA, $p > 0.05$).

For artificial seeding, immature thalli were attached to, and maintained on, PVC pipes (Figure 2A) to avoid their exposure to the air. Mature thalli showed egg release on their receptacles (Figure 2B). Embryo suspensions of *S. fulvellum* (Figure 2C and 2D) could be produced at as high densities by rubbing the mature thalli. Diameter of the embryos was ca. 200 μm and they developed rhizoids. These embryos were quickly attached to the surface of seed strings (Figure 2E) by paintbrush. For the seeding of 10,000 m of seed strings, at least 20 kg-fresh wt of mature thalli were needed. Just after seeding, the seed frames were cultured in tanks for one month (Figure 2F). During one month of culture the growth of germlings in tanks produced a mean thallus elongation of 5.2 ± 2.5 mm. Range in thallus length varied between 3 and 6 mm in one month (Table 3). Density of the germlings varied, with the culture period, between 13 and 20 per one centimeter of seed string (Table 3).

The seed frames were transferred to the nursery culture ground in May (Figure 3A). During the nursery culture, germlings grew up to 1–3 cm in length. There were also many epiphytic algae, copepods and hydrozoans on the seed strings during the nursery culture

Table 1. Temperature effect on the percentage of egg release from excised receptacles of *Sargassum fulvellum* during 11 days culture under 80 μmol photons m^{-2} s^{-1} and 16:8 h (L:D).

	Temperature ( )				
Day	5	10	15	20	25
0	0	0	0	0	0
2	0	0	10 ± 1.4	3 ± 1.5	0
4	0	11.0 ± 1.5	50 ± 30.4	20 ± 5.8	3.3 ± 1.0
6	0	24.0 ± 13.4	63.3 ± 30.6	56.7 ± 20.8	11.0 ± 3.5
9	2.1 ± 1.6	32.5 ± 11.5	72.5 ± 20.8	89.4 ± 5.8	26.7 ± 11.5
11	3.3 ± 2.8	36.7 ± 20.8	76.7 ± 30.6	96.7 ± 1.1	36.0 ± 1.5

Values represent mean and standard deviation.

Table 2. Irradiance effect on the percentage of egg release from excised receptacles *of Sargassum fulvellum* during 11 days culture under 20 °C and 16:8 h (L:D).

	Irradiance (μmol m^{-2} s^{-1})			
Day	20	50	80	100
0	0	0	0	0
2	2.1 ± 0.5	2.3 ± 1.2	3.1 ± 1.0	3.1 ± 1.1
4	15.0 ± 1.4	10.4 ± 2.3	18.3 ± 5.4	22.3 ± 4.0
6	35.3 ± 3.2	54.2 ± 10.4	64.0 ± 24.1	68.4 ± 26.1
9	83.4 ± 15.2	82.9 ± 24.2	92.6 ± 10.4	85.2 ± 15.4
11	89.5 ± 10.4	90.4 ± 17.8	95.7 ± 5.4	90.5 ± 9.3

Values represent mean and standard deviation.

Figure 2. Artificial seeding process for *Sargassum fulvellum.* A: Immature thalli attached to PVC pipe which is sunk to the bottom (scale bar 500 μm) of culture tank. B: After egg release, embryos are ready to detach from their receptacles. C: Embryos obtained by rubbing the mature thalli of B (scale bar 500 μm). D: Dense embryo suspension from 20 kg fresh wt of mature thalli. E: Seeding of embryos on a seed frame by paintbrush (frame is 50 cm × 60 cm, holding a total length of 100 m of string made of mixed nylon and polypropylene fibers). F: Tank culture of seed frames just after seeding.

(Figure 3B). A steel roller (Figure 3C) was used to set the seed strings onto the main culture rope. By unwinding the roller as the boat moved along, seed strings were wound onto the main rope (Figure 3D and 2E).

As the temperature decreased after September, thalli grew faster and faster and reached 1.5–2 m in length in December (Figure 3F). The main problem detected in cultivation at sea was the high degree of biofouling observed on the lines within one month of their installation. Smaller thalli were weakened by the settlement of Gammaridea and Caprellidea, which precipitated their grazing death. Thalli that survived were those that were longest when they were transferred to the sea, although even these plants became heavily encrusted with epibionts, which had to be removed whenever the plants were measured. Table 4 shows the rates of growth and biomass. Thalli grew gradually during fall and winter, with mean lengths ranging from 1.2 cm in June to 128.6 cm in January (Table 4). Harvest of the culture ropes in January 2003 gave a mean value for total fresh drained weight of 3.4 ± 0.5 kg m^{-1} of main culture rope (Table 4).

Table 3. Development and growth of germlings of *Sargassum fulvellum* during indoor tank culture from April to May 2003.

Day	Length[a]	Number of laterals	Density[b]
0	0.2 ± 0.1	1	20.7 ± 3.2
10	1.5 ± 1.1	1	16.3 ± 1.2
20	2.8 ± 1.4	1	14.3 ± 1.5
30	5.2 ± 2.5	1.9 ± 0.3	13.3 ± 3.2

[a]Length of germlings (mm).
[b]Number of germlings per one centimeter of seed string.

Discussion

As pointed out by Kapraun (1999), seaweed mariculture generally results in less environmental impact and degradation than the harvesting of wild populations.

Figure 3. Nursery culture and main cultivation of *Sargassum fulvellum*. A: Seed frames were moved to nursery culture ground after one month of tank culture on land. B: Young thalli attached on seed frames with many epiphytic algae and hydrozoans for one-month of nursery culture. C: Steel roller device for setting and moving the seed strings to main culture. D: Attaching the seed strings on the main culture rope by pulling back and rolling the seeding device. E: Young thalli attached to seed string which is wound around main culture rope. F: Harvest of thalli after six months culture.

The present work shows that, in *S. fulvellum*, a method of mass production of embryos at a commercial scale and successful seeding on string (either natural or artificial), is possible. This technology will also permit the cultivation of a seedstock of germlings. In the indoor culture, a high germling survival and a density ranging between 13 and 20 individuals per cm on the seed string were observed, after one month of culture (Table 3). The cultivation of *S. fulvellum* from spores offers a great advantage in seaweed cultivation farms of Korea.

Table 4. Growth and biomass of *Sargassum fulvellum* during the main cultivation.

Month	Length[a]	Biomass[b]
Jul. 2003	1.2 ± 0.4	5.5 ± 1.1
Aug.	3.4 ± 0.7	12.4 ± 2.1
Sep.	4.0 ± 1.5	14.1 ± 1.5
Oct.	15.8 ± 9.4	75.4 ± 12.4
Nov.	43.5 ± 15.6	954.2 ± 70.4
Dec.	85.7 ± 43.2	2,043.5 ± 276.1
Jan. 2004	128.6 ± 57.4	3,404.1 ± 471.3

[a]Length of thalli (cm).
[b]Fresh weight (g) of thalli per one meter of main culture rope.

As in many other seaweeds, onset of reproduction in *S. fulvellum* occurs at the end of the growth period (Figure 1). Because maturation is not simultaneous, at the beginning of the reproductive period the first mature thalli have competed their growth, whereas some smaller non-reproductive thalli could still grow for some time (Norton, 1977; Hales & Fletcher, 1989; Arenas & Fernández, 1998). In our

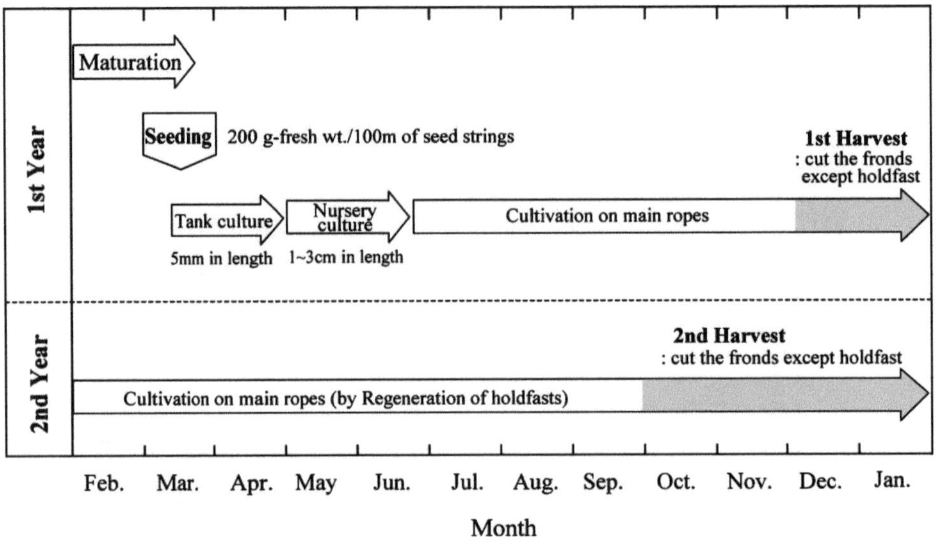

Figure 4. Diagrammatic schedule of artificial seed production and cultivation of *Sargassum fulvellum*. During the first year of the cultivation, labor cost for seed production can be compensated for 2nd harvest by regenerated biomass from holdfast.

results, the maturation peak of *S. fulvellum* was in April (Figure 1) when water temperature was about 12°C. Egg release of *S. fulvellum* was facilitated at 15 °C and 20 °C (Table 1). The lack of significant differences between the percentages of eggs released under different irradiance conditions (Table 2) indicates that irradiance has no effect once temperature requirements have been satisfied.

Hales and Fletcher (1990) also found that the relationship between the growth of germlings and temperature increased positively from 10 to 20 °C. Deysher (1984) noted seawater temperature ranges from 5 to 28 °C in *S. muticum* habitats in Japan. He stated that the reproductive phenology of the species appears to be regulated not by specific temperature levels which trigger the onset of gamete production, but rather by daily accumulation of temperature analogous to the classical 'degree-day' models for reproduction in various land plant systems. The degree-day model best fits *S. fulvellum* as well.

Early growth in frond length of *S. fulvellum* was very similar to that of *Hizikia fusiformis* in the experiment by Hwang et al. (1994). Both early growth and attachment strength of the young plants in the indoor tank of *Sargassum* depend on the ability of the fertilized embryos to settle and germinate under certain environmental conditions. Among the many interacting environmental factors in the indoor tank, light and temperature could play an important role in further development of the plant.

An important role of *S. fulvellum*, beyond that of a seafood, is as a major component of the sea forest. The genus *Sargassum* is the most common brown alga in temperate regions (Yoshida, 1983). *Sargassum* belongs to the Fucales, a group of brown algae without alternating phases in their life history. The plant is diploid and produces gametes that fuse to form a new plant (Guiry and Blunden, 1991). Part of the life history of this species includes a yearly winter "dieback" of the upper part of the plant after reproduction, while the holdfast remains perennially on the substrate and regenerates a few months later. The perennial holdfast is composed of finger-like outgrowths from the basal part of the stem (Yoshida, 1983). *Sargassum* beds play important ecological roles in the coastal ecosystem due to their large biomass and high productivity. Furthermore, these beds are important as nursery areas for commercially important fish species and in their role in the preservation of environmental conditions.

These perennial holdfasts of *S. fulvellum* can also be used for repeated harvests over 2 years in a culture ground. Figure 4 summarizes the artificial seed production and cultivation of *S. fulvellum*. Multi-harvests over a 3–4 year culture period are even possible if holdfast re-use methods such as those in *Hizikia* cultivation are employed (Hwang et al., 1998). The maximum quantity of *S. fulvellum* harvested from cultivation was calculated at approximately 3.4 kg fresh weight per one meter of culture rope. In the first year of the cultivation, harvesting could be possible from December to

January. However, the harvest period could be extended from October to January in the second year, because of the fast growth of the young frond from the regeneration of the holdfast.

Conclusion

Egg release from the receptacles of *Sargassum fulvellum* peaked from March to April. Therefore, artificial seed production can start in April when the seawater temperature is over 10 °C in nature. Growth of germlings in tank culture produced a mean thallus elongation of 5.2 mm during one month from April to May. After then seed frames bearing young thalli were transferred to the *in situ* culture ground for one month of nursery culture. Young thalli grew in length during the fall and winter months, reaching a mean of 128.6 cm and a maximum biomass of 3.4 kg m^{-1} of main culture rope in January. This species offers good potential to diversify seaweed cultivation in Korea.

Acknowledgements

This work is funded by a grant from the National Fisheries Research and Development Institute (RP-05-AQ-2).

References

Arenas F, Fernández C (1998) Ecology of *Sargassum muticum* (Phaeophyta) on the North Coast of Spain. III. Reproductive ecology. Bot. Mar. 41: 209–216.

Deysher LE (1984) Reproductive phenology of newly introduced populations of the brown alga, *Sargassum muticum* (Yendo) Fensholt. Hydrobiologia 116/117: 403–407.

Guillard RRL (1968) A simplified antibiotic treatment for obtaining axenic cultures of marine phytoplankton. Mimeographed document. Woods Hole Oceanography Institute of Marine Biology Laboratory 9 pp.

Guiry MD, Blunden G (1991) Seaweed resources in Europe: Uses and potential. Wiley, Chichester: 203 pp.

Hales JM, Fletcher RL (1989) Aspects of the ecology of *Sargassum muticum* (Yendo) Fensholt in the Solent region of the British Isles. 2. Reproductive phenology and senescence. In JS, Ryland, PA, Tyler (eds.), Reproduction, Genetics, and Distributions of Marine Organisms. Proceedings of the 23rd European Marine Biology Symposium, Olsen and Olsen Publication, Denmark, pp 115–125.

Hales JM, Fletcher RL (1990) Studies on the recently introduced brown alga *Sargassum muticum* (Yendo) Fensholt. V. Receptacle initiation and growth, and gamete release in laboratory culture. Botanica Marina 33: 241–249.

Hwang EK, Park CS, Sohn CH (1994) Effects of light intensity and temperature on regeneration, differentiation and receptacle formation of *Hizikia fusiformis* (Harvey) Okamura. Korean J. Phycol. 9: 85–93.

Hwang EK, Cho YC, Sohn CH (1998) Reuse of holdfasts in *Hizikia* cultivation. J. Korean Fish Soc. 32: 112–116.

Kapraun DF (1999) Red algae polysaccharide industry: Economics and research status at the turn of the century. Hydrobiologia. 398/399: 7–14.

Lee YP, Kang SY (2002) A catalogue of the seaweeds in Korea. Cheju National University Press, Korea, 662 pp.

Norton TA (1977) The growth and development of *Sargassum muticum* (Yendo) Fensholt. J. Exp. Marine Bio. Ecol. 26: 41–53.

Sohn CH (1993) *Porphyra*, *Undaria* and *Hizikia* cultivation in Korea. Korean J. Phycol. 8: 207–216.

Sohn CH (1998) The seaweed resources of Korea. In AT, Critchley, M. Ohno (eds), Seaweed resources of the world. Jap. Int. Coop. Agency 15–33.

Yoshida T (1983) Japanese species of *Sargassum* subgenus *Bactrophycus* (Phaeophyta, Fucales). Journal of Faculty of Science Hokkaido Univ. Series V (Bot.) 13: 99–246.

Journal of Applied Phycology (2006) 18: 259–267
DOI: 10.1007/s10811-006-9025-y

Farming of the giant kelp *Macrocystis pyrifera* in southern Chile for development of novel food products

Alfonso Gutierrez[1], Tomás Correa[1], Verónica Muñoz[1], Alejandro Santibañez[2], Roberto Marcos[3], Carlos Cáceres[4] & Alejandro H. Buschmann[1,*]
[1]*i~mar, Universidad de Los Lagos, Camino Chinquihue km 6, Puerto Montt, Chile;* [2]*Departamento de Gobierno y Empresas, Universidad de Los Lagos, Casilla 557, Puerto Montt, Chile;* [3]*Productos del Pacífico, S.A. de C.V., Ensenada, México;* [4]*Consultora Los Lagos, Población Varmontt Nr. 4, Puerto Varas, Chile*

*Author for correspondence: e-mail: abuschma@ulagos.cl; fax: +56 65 322418

Key words: *Macrocystis pyrifera*, farming, food products, Chile

Abstract

This study explores the potential cultivation of the giant kelp *Macrocystis pyrifera* (L.) C.A. Agardh in southern Chile, for the development of novel food products. The study demonstrates the importance of considering the collection site of the parent sporophytes for successful cultivation. This study also shows that the ropes must be seeded with 10,000 to 40,000 spores ml^{-1}, depending on the culture method used. We also demonstrated that under environmental conditions in southern Chile, the seeded ropes must be put at sea at the latest during autumn (April) in order to reach the harvesting season in December. However, several other management aspects must be considered to improve the quality of the product. Our final estimation indicates that over 14.4 $kg\,m^{-1}$ of rope (fresh weight) can be produced and from this total production, over 70% can reach the quality to produce different food products that are already being introduced in oriental countries. The remaining 30% can be used for abalone feeding and is also available for the organic fertilizer industry located in Chile.

Introduction

California kelp beds started to be harvested as a source of potash during the first decade of the 20th century and commercial interest in the giant kelp *Macrocystis pyrifera* expanded significantly between the 1970s and 1980s (Neushul, 1987; Druehl et al., 1988). This interest was primarily for the production of alginates, but also to produce biomass as a feedstock for methane production as a consequence of the energy crisis at that time (North et al., 1982; Gerard, 1987; Neushul & Harger, 1987). Nevertheless, *M. pyrifera* commercial cultivation for methane production was never a reality. At present, the supply of *M. pyrifera* biomass for the alginate industry relies exclusively on restoration practices and management of natural beds to obtain a sustainable production (North, 1979; McPeak & Barilotti, 1993; Vásquez & McPeak, 1999). After the energy crisis and because of the low price of alginates, farming research on *Macrocystis* declined sharply.

On the other hand, other brown algal species began to be commercially cultivated in Japan, China and Korea, mainly for human consumption (Tseng, 1987; Kaneko, 1999; Hanisak, 1998) while kelp-farming attempts for this purpose have also proved technically feasible in other regions (e.g. Druehl et al., 1988; Kain, 1991; Merril & Gillingham, 1991). Interestingly, the demand for brown algae is also increasing due to the introduction of new uses such as fertilizers, cultivation for bioremediation purposes, and abalone as well as sea urchin feeding among others (Petrell et al., 1993; Vásquez & Vega, 1999; Buschmann et al., 2001c; Ugarte & Sharp, 2001; Chopin et al., 2001). In Chile, despite the commercial importance of various algal species, aquaculture is still limited to the red alga *Gracilaria chilensis* (Buschmann et al., 2001b).

In Chile, Abalone and sea urchin cultures, organic fertilizer production and novel seafoods have created a new niche market for the giant kelp *Macrocystis pyrifera*. Increased harvesting is already causing some deterioration of different kelp populations (Vásquez & Vega, 1999). Considerable information on *Macrocystis* cultivation has been published in the past (North, 1979). However, some basic knowledge necessary to run a successful commercial activity is still lacking, especially with regard to the different environmental conditions and complex morphological and reproduction variability between populations, that can have important commercial consequences.

Considering this new market scenario, the potential impact on natural populations and the lack of biological knowledge necessary to produce a high quality product, this paper deals with the cultivation of *M. pyrifera* in southern Chile. Specifically, the effect of the origin of the parental plants on the survival and growth of young sporophytes cultivated on ropes was tested, in both hatchery and field conditions. Finally, a pilot cultivation was established to determine the potential yields of *M. pyrifera* in southern Chile and we describe some of the food products developed.

Materials and methods

Study sites

Fertile sporophylls of *Macrocystis pyrifera* were collected at six localities in southern Chile: Metri (41°35′S; 72°42′W), Pargua (41°47′S; 73°25′W), Calbuco (41°46′S; 73°08′W), Pucatrihue (40°33′S; 73°43′W), Bahía Mansa (40°34′S; 73°44′W) and Curaco de Velez (42°26′S; 73°35′W) (Figure 1). Site selection was based on the presence of abundant kelp populations and different water movement conditions. The plants were collected by scuba divers and transported, within 6 h, on ice to the seaweed culture laboratory in Metri. All field cultivation experiments were carried out in Metri and the pilot culture in Calbuco (Figure 1).

Cultivation of different populations

The sporophylls collected in Metri, Pargua, Calbuco, Pucatrihue and Bahía Mansa, were washed under tap water and UV treated filtered seawater (0.2 μm) containing commercial iodine (0.5% for 10 s), packed in filter paper, covered with aluminum foil and stored at 15 °C (Figure 2A). After 12 h, 10 to 15 sporophylls were placed in 20 L sterile plastic containers filled with filtered (0.2 μm) and autoclaved seawater (Figure 2B). Sporulation started in all cases after 25 to 35 min and, after 1 h the sporophylls were removed and the water was filtered with a 100 μm mesh. Eight PVC cylinders covered with a 1.5 mm nylon rope were introduced into each of the 20 L containers to allow for spore settlement (Figure 2B). After 12 h, the eight PVC cylinders were removed and placed in a 30 L glass tank filled with autoclaved, filtered, and Provasoli enriched seawater (McLachlan, 1973; Figure 2B). Culture was carried out at a photon flux density of 30–40 μmol $m^{-2}\,s^{-1}$; a temperature of 9–10°C; a salinity of 30% and a pH of 7.8–7.9. Photoperiod was 16:8 (L:D) during the first week; 14:10 (L:D) during the second week; 12:12 (L:D) during the third week and 10:14 (L:D) thereafter (following a previously determined protocol; Buschmann unpublished results). After 44 days, 3 cm pieces of rope were randomly cut off and plant density (number of sporophytes fronds per cm of rope fragment) estimated under a stereomicroscope. Furthermore, the maximum lengths of the juvenile sporophytic fronds were determined using an ocular micrometer. The data were statistically analyzed by a one-way ANOVA after logarithmic transformation to ensure the normality and homocedasticity of the data. If significant differences were detected between treatments, a Tukey *a posteriori*-test (according to Steel and Torrie, 1985) was performed. As the data were obtained from independent tanks, seeded from independent plants and comparisons between times were not considered, no pseudoreplication exists (*sensu* Hurlbert, 1984).

After 60 days in the hatchery (September), seeded ropes were attached to a 3 m long horizontal supporting rope (18 mm diameter) in groups of three placed at 2 m depth (Figure 3) in Metri (Figure 1). Three seeded ropes were used for each one of the five original populations. Plant density and length of the different *Macrocystis pyrifera* populations were estimated, after one and two months in the field, by random sampling of 5 cm seeded rope sections under a stereomicroscope. All data were analyzed as above.

Pilot study

Parent sporophytes were collected in Curaco de Velez (Figure 1) and seeded on ropes following the same methods and culture conditions described above. The ropes were seeded in mid January and were brought to the Calbuco culture site (Figure 1) in March. The initial culture conditions of the sporophytes in the field were

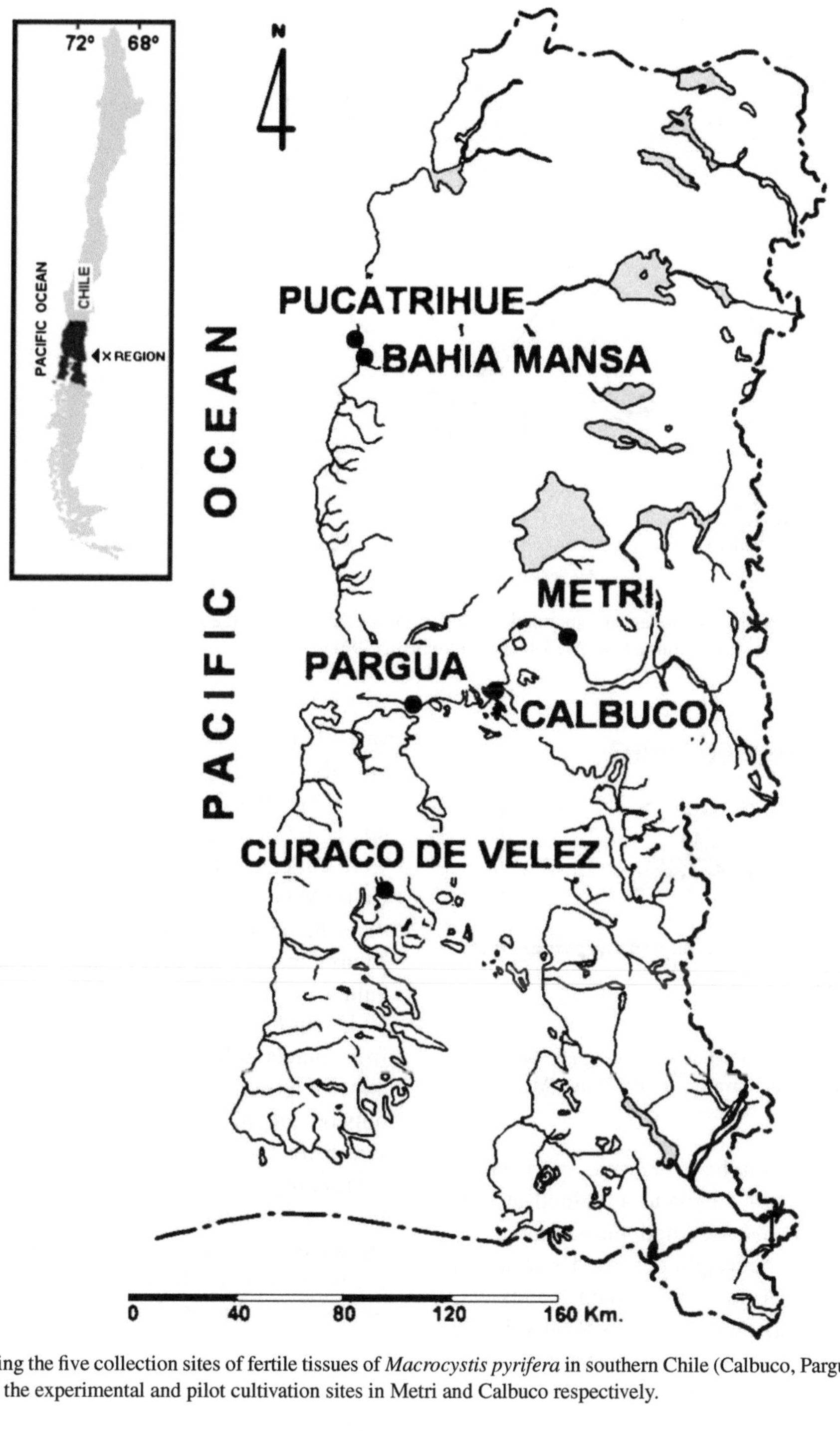

Figure 1. Map indicating the five collection sites of fertile tissues of *Macrocystis pyrifera* in southern Chile (Calbuco, Pargua, Metri, Pucatrihue and Bahía Mansa) and the experimental and pilot cultivation sites in Metri and Calbuco respectively.

1 mm plant length, and a mean density of 51 plants mm^{-1}. The experiment was initiated in March using the horizontal culture system at 1 m depth (Figure 3), based on previous results (Buschmann, unpublished data). The *Macrocystis pyrifera* fresh weight produced per m of long-line was evaluated by taking 10 random 1.5 m rope samples. Total fresh weight was determined and then the blades, stipes and pneumatocyst

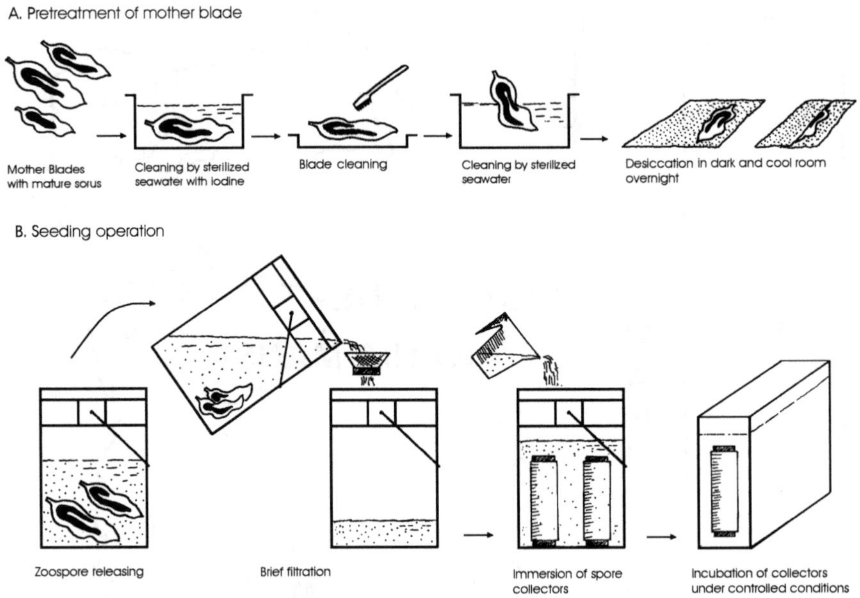

Figure 2. Scheme representing the pretreatment and sporulation induction procedure (A), and the seeding operation (B) for *Macrocystis pyrifera* mass production. The procedure was modified from Merrill and Gillingham (1991).

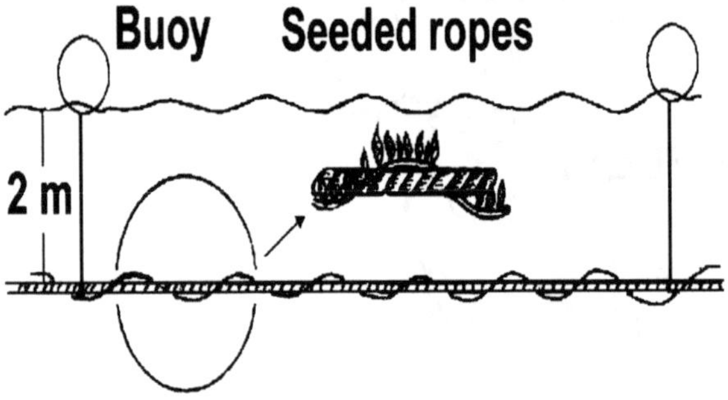

Figure 3. Suspended culture method used in this study: horizontal culture method adapted from Kawashima (1993) and Merrill and Gillingham (1991).

and disposable parts (holdfast, necrotic, perforated, and epiphyted tissues) of 22 plants obtained at random, were separated and weighed individually. Finally, "substantiality" (the weight per frond area; *sensu* Kawashima, 1993) of 30 randomly taken fronds was determined by weighing 1 cm^2 frond discs on a digital balance ($\pm$ 1 mg). Statistical analysis was performed following the above-described protocol.

Results

Cultivation of different populations

In the hatchery, the mean number of sporophytes produced at 9–10°C, 40 μmol photons m^{-2} s^{-1} with a variable photoperiod, varied between 1.0 and 12 individuals mm^{-1} of seeded rope (Figure 4A). The number of sporophytes produced was significantly higher ($p < 0.05$) from fertile sporophytic tissues, which were collected from Metri and Pucatrihue. Pargua and Bahía Mansa showed the lowest seeding success under the same culture conditions, whereas Metri and Bahía Mansa showed the highest size increment, with mean values of 2.85 mm in 44 days (4B).

Under field conditions, all the individuals produced from plants collected in Bahía Mansa and Pucatrihue (wave exposed coast) showed a mortality of 100% after only one month of cultivation (Figure 5). No significant size difference ($p > 0.1$) was found between plants originating from Calbuco, Pargua and Metri during the first month of culture. After two months in culture, no significant differences ($p > 0.08$) were found between the studied populations (Figure 5), although Calbuco plants showed a trend towards increased growth.

Pilot study

The pilot study carried out in Calbuco produced a biomass of 14.4 kg m^{-1} ($\pm$4.8 kg m^{-1}; S.D.) after 7 months of culture, and showed a substantiality value of the harvested fronds of 68.5 mg cm^{-2} ($\pm$1.6 mg cm^{-2}; S.D.). The different tissue types of *Macrocystis pyrifera* produced yields that varied from 1.3 kg m^{-2} (pneumatocyst) up to 4.1 kg m^{-2} (stipe) and 4.8 kg m^{-2}

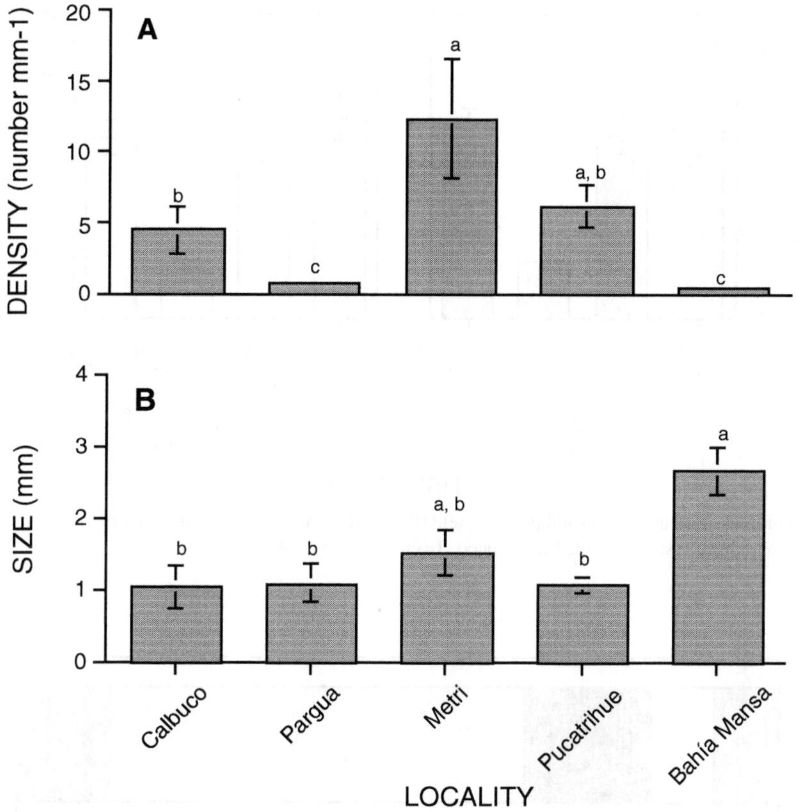

Figure 4. Cultivation of *Macrocystis pyrifera* under controlled conditions, initiated from fertile sporophytes collected in 5 sites of southern Chile: Calbuco, Pargua, Metri, Pucatrihue and Bahía Mansa. (A) Mean density (number mm^{-2}) of sporophytes on nylon ropes; and (C) mean length (mm) of sporophytes attached to nylon ropes after 44 days in the hatchery (9 replicates per location).

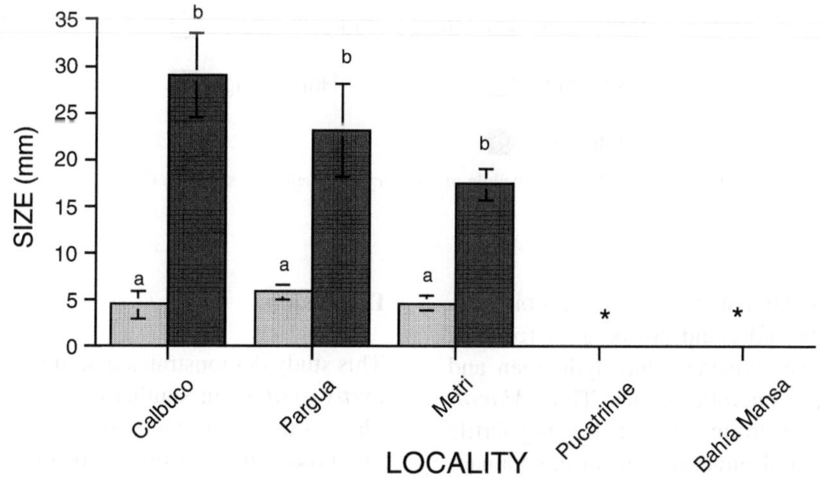

Figure 5. Cultivation of *Macrocystis pyrifera* at a depth of 2 m under field conditions, initiated from fertile sporophytes collected in 5 sites of southern Chile: Calbuco, Pargua, Metri, Pucatrihue and Bahía Mansa. Mean size (mm ± 1 SE) of sporophytes attached to nylon ropes after one month (dotted bars) and two months (grey bars). Letters above the bars indicate no significant differences after a Tukey *a posteriori*-test ($p < 0.05$). Asterisks indicate no survival of the germlings. Three samples were taken at each location.

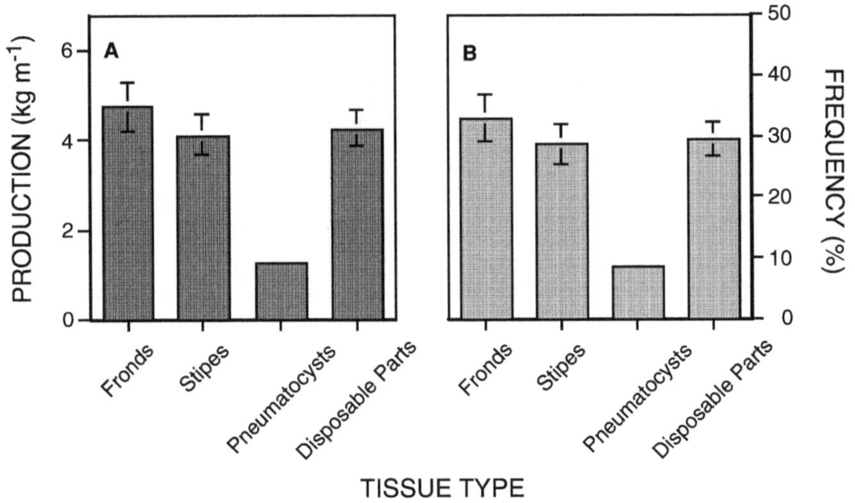

Figure 6. Mean (± 1 SE) biomass production (A) and percent yield (B) of different tissues and low quality tissues of *Macrocystis pyrifera* after seven months in a horizontal culture system installed in Calbuco. Twenty 1-m^2 samples were taken.

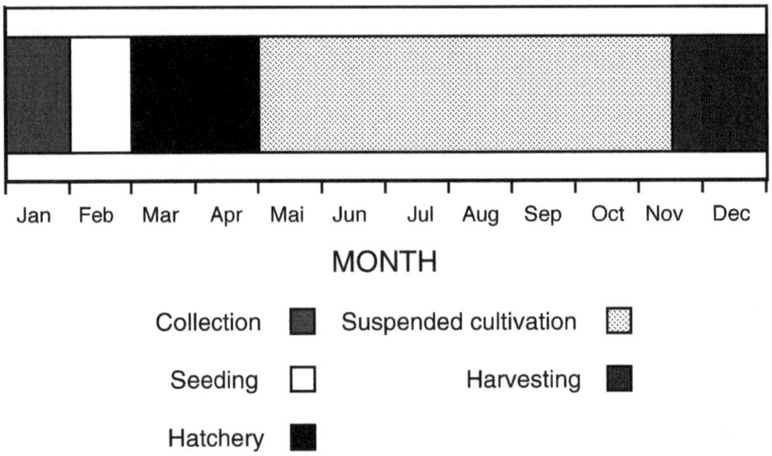

Figure 7. Production schedule of *Macrocystis pyrifera* in southern Chile.

(frond) (Figure 6A). Almost 30% of the total biomass was of lower quality and could not be used for food (Figure 6B), as it was covered with hydrozoan and bryozoan colonies, or necrotic tissues. Thus, *Macrocystis pyrifera* can be produced by collecting fertile tissues from the natural environment during January (summer), whereas rope seeding should be carried out at the latest during February (Figure 7). The hatchery phase takes at least 1.5 to 2 months, which means that the suspended culture at sea can be initiated in April or May (fall) and harvesting is possible in late November or December (Figure 7).

Discussion

This study demonstrates that the cultivation of *Macrocystis pyrifera* in southern Chile is technically feasible and that a productivity of over 14 kg m^{-1} during one production season can be obtained, which seems very promising (Figures 8A and B). The comparison of this value is not easy as most of the production estimates are based on extrapolations from physiological or growth measurements made on parts of kelp plants (Neushul & Harger, 1987). Estimations of 7 g ash free dry wt. m^{-2} day^{-1} (Neushul & Harger, 1987) and of

Table 1. Description of *Macrocystis pyrifera* food products developed.

Food product	Product specification
Fresh Salted Frond (Figure 8 D)	1 × 3 cm salted blade pieces are washed in filtered seawater and blanched at 100°C for 1 min and then mixed with salt (23%). The product has a green color and is placed in 15 or 10 kg plastic bags
Pneumatocyst Rings and Stipe Pieces (Figure 8C and E)	Slices of pneumatocysts and 2-cm pieces of stipes are washed and heated in vinegar with alcohol content not higher than 1% and an acetic acid value not lower than 5%. The products are packed in glass or packs as required. The product must be maintained in cool conditions (2–6°C)

Figure 8. (A) Pilot culture of *Macrocystis* in Calbuco, (B) harvesting *Macrocystis* and products developed for human consumption in Chile: (C) stipe fragments; (D) blade strips; (E) pneumatocyst rings and (F) a general presentation plate of these *M. pyrifera* products.

4 wet kg per production period per m^2 (Coon, 1982) exist, suggesting that our results are encouraging. It is important to mention that over 70% of the harvested biomass can be used to produce high quality food products (Figures 8C, D, E and F; Table 1), which represents a significantly higher success rate for the use proposed in this study when compared to other uses. In addition, the introduction of massive kelp culture in Chile may have other associated benefits such as providing a means of removing nitrogen and phosphorus produced by salmon farming (Buschmann et al., 2001a; Chopin et al., 2001; Troell et al., 2003). It is important to remember that this area of Chile produces over 300,000 tonnes of salmon, thus creating significant environmental impacts and conflicts (Buschmann, 2001).

To date, *Macrocystis pyrifera* has not been used to produce food products in Japan. Traditionally the species used as food in oriental countries are *Laminaria japonica* (Kombu) and *Undaria pinnatifida* (Wakame) (Kawashima, 1993; Ohno & Matsuoka, 1993). In Chile, the bull kelp *Durvillaea antarctica* has been traditionally used as a food source, but has a very low price. Because *Laminaria* and *Undaria* are not present in Chile, the idea was to develop alternative products that could be exported to oriental countries with a tradition in seaweed consumption (Abbott, 1996). However, as the characteristics of color, texture, substantiality, and mucilage content of *Macrocystis* cannot be compared to those of *Laminaria* or *Undaria* we pursued the commercial strategy of creating novel products (Figure 8C to F), which received a positive market response in Asia after a first dispatch and market tour. Thus, because this is a developing market, it is believed that the commercial cultivation of *Macrocystis* is possible in Chile. In addition, alternative use of this kelp as abalone and sea urchin feed, or as organic fertilizer, strengthens its economic feasibility. Lower grade parts of *Macrocystis* can also be used for the other purposes indicated above.

Recently, it has been demonstrated that Chilean stocks of *M. pyrifera* show small genetic differences from other Southern Hemisphere species, but stronger differences from Californian stocks (Coyer et al., 2001). Despite this low genetic diversity, plants collected at various sites in southern Chile showed different potential for use in aquaculture practices. The plants from the most exposed areas cannot be used for farming in the inner seas, of southern Chile, although they presented higher growth under hatchery conditions. In contrast, marginal differences were observed between *M. pyrifera* populations from the inner seas. In contrast to an earlier report that identifies *M. laevis* in this region (Aguilar-Rosas et al., 2003) and indicates that some morphologically distinct plants exist, our six years of observations and experimentation strongly suggest that this smooth bladed plants correspond to *M. pyrifera*. Nevertheless, these results recognize that to some extent each farmer must consider site-specific characteristics of his own licensed location and the morphological characteristics of the parent sporophytes, before starting commercial activities.

Despite the promising production results obtained here, *Macrocystis* production in Chile still needs more research. Several factors remain to be studied, but two aspects are especially important, as has been demonstrated for other algal resources (Wikfors & Ohno, 2001). Firstly, strain selection cannot be overlooked. The manipulation of the growth capabilities of this resource is important, but given that it is required to produce food products, it is important to study aspects that can be used to produce plants with specific morphological characteristics. Unpublished results indicate that morphological variation and life-history variability of *M. pyrifera* in southern Chile is high (Buschmann et al., 2004), but we still do not know how to manipulate morphological variation or how to maintain certain morphological characteristics under culture conditions. Secondly, we wish to highlight management requirements during the field phase that will allow a successful commercial operation. During this phase aspects such as plant density, harvesting strategies, and disease control are extremely relevant, as can be seen for example, during the intensive cultivation of *Laminaria japonica* (Kawashima, 1993). All these aspects remain unknown and will undoubtedly provide excellent material for future studies in southern Chile.

Acknowledgments

This study was financed by a FONDEF grant (D901/1101) and FONDECYT 1010706. The authors acknowledge the help of Mariam Hernández-González, Luis Filún, René Reyes, Rodrigo Martínez and Ricardo Ceña. The collaboration of Marcelo Brintrup is especially acknowledged, as well as the constructive criticism to this manuscript given by D. Varela, R. Stead, D. M. Luxton and an anonymous reviewer.

References

Abbott IA (1996) Ethnobotany of seaweeds: Clues to uses of seaweeds. Hydrobiologia 326/327: 15–20.

Aguilar-Rosas LE, Aguilar-Rosas R, Marcos-Ramírez R, Cáceres-Rubio CF, McPeak RH (2003) New record of *Macrocystis laevis* Hay (Laminariales, Phaeophyta) on the Pacific coast of Chile. In: Chapman ARO, Anderson RJ, Vreeland VJ, Davison IA (eds), Proceedings of the 17th International Seaweed Symposium, Oxford University Press, Oxford, pp. 337–340.

Buschmann AH (2001) Impacto Ambiental de la Acuicultura. El Estado de la Investigación en Chile y el Mundo, Terram Publicaciones, Santiago, pp. 63.

Buschmann AH, Troell M, Kautsky N (2001a) Integrated algal farming: A review. Cah. Biol. Mar. 42: 83–90.

Buschmann AH, Correa JA, Westermeier R, Hernández-González MC, Norambuena R (2001b) Red algal farming: A review. Aquaculture 194: 203–220.

Buschmann AH, Hernández-González MC, Aroca G, Gutiérrez A (2001c) Seaweed farming in Chile: A review. The Global Aquaculture Advocate 4: 68–69.

Buschmann AH, Vásquez JA, Osorio P, Reyes E, Filún L, Hernández-González MC, Vega A (2004) The effect of water movement, temperature and salinity on abundance and reproductive patterns of *Macrocystis* spp (Phaeophyta) at different latitudes in Chile. Mar. Biol. 145: 849–862.

Chopin T, Buschmann AH, Halling C, Troell M, Kautsky N, Amir N, Kraemer GP, Zertuche-González JA, Yarish C, Neefus C (2001) Integrating seaweeds into aquaculture systems: A key towards sustainability. J. Phycol. 37: 975–986.

Coon LM (1982) *Macrocystis* harvest strategy in British Columbia. In: Srivastava LM (ed.), Synthetic and Degradative Processes in Marine Macrophytes, Walter de Gruyter, Berlin, pp. 265–282.

Coyer JA, Smith GJ, Andersen RA (2001) Evolution of *Macrocystis* spp. (Phaeophyceae) as determined by ITS1 and ITS2 sequences. J. Phycol. 37: 574–585.

Druehl LD, Baird R, Lindwall A, Lloyd KE, Pakula S (1988) Longline cultivation of some Laminareaceae in British Columbia. Aquacult. Fish Management 19: 253–263.

Gerard VA (1987) Optimizing biomass production on marine farms. In: Bird KT, Benson PH (eds), Seaweed Cultivation for Renewable Resources, Elsevier Science Publishers, Amsterdam, pp. 95–106.

Hanisak MD (1998) Seaweed cultivation: Global trends. World Aquacult. 29: 18–21.

Hurlbert SH (1984) Pseudoreplication and the design of ecological field experiments. Ecol. Monogr. 54: 187–211.

Kain JM (1991) Cultivation of attached seaweeds. In: Guiry MD, Blunden G (eds), Seaweed Resources in Europe, Uses and Potential. John Wiley and Sons Limited, Chichester, pp. 307–377.

Kaneko T (1999) Kelp cultivation in Japan. World Aquacult. 30: 62–65+68.

Kawashima S (1993) Cultivation of the brown alga, *Laminaria* "Kombu". In: Ohno M, Critchley AT (eds), Seaweed Cultivation and Marine Ranching, Japan International Cooperation Agency, Yokosuka, pp. 25–40.

McLachlan J (1973) Growth media - marine. In: Stein J (ed.), Handbook of Phycological Methods. Culture Methods and Growth Measurements, Cambridge University Press, Cambridge, pp. 25–51.

McPeak R, Barilotti DC (1993) Techniques for managing and restoration *Macrocystis pyrifera* kelp forests in California, USA. Facultad de Ciencias del Mar. Universidad Católica del Norte, Coquimbo. Serie Ocasional 2: 271–284.

Merrill JE, Gillingham DM (1991) Bullkelp Cultivation Handbook, National Oceanic and Atmospheric Administration, Washington, D.C. pp. 70.

Neushul M (1987) Energy from marine biomass: The historical record. In: Bird KT, Benson PH (eds), Seaweed Cultivation for Renewable Resources, Elsevier Science Publishers, Amsterdam, pp. 1–37.

Neushul M, Harger BWW (1987) Nearshore kelp cultivation, yield and genetics. In Bird KT, Benson PH (eds), Seaweed Cultivation for Renewable Resources, Elsevier Science Publishers, Amsterdam, pp. 69–93.

North WJ (1979) Evaluación, manejo y cultivo de praderas de *Macrocystis*. In Santelices B (ed.), Actas Primer Symposium Algas Marinas Chilenas, Ministerio de Economía Fomento y Reconstrucción, Santiago, pp. 75–128.

North WJ, Gerard VA, Kuwabara J (1982) Farming of *Macrocystis* at coastal and oceanic sites. In: Srivastava LM (ed.), Synthetic and Degradative Processes in Marine Macrophytes, Walter de Gruyter, Berlin, pp. 247–262.

Ohno M, Matsuoka M (1993) *Undaria* cultivation "Wakame". In: Ohno M, Critchley AT (eds), Seaweed Cultivation and Marine Ranching, Japan International Cooperation Agency, Yokosuka, pp. 41–49.

Petrell RJ, Tabrizi KM, Harrison PJ, Druehl LD (1993) Mathematical model of *Laminaria* production near a British Columbian salmon sea cage farm. J. Appl. Phycol. 5: 1–14.

Steel RGD, Torrie JH (1985) Bioestadística. Principios y Procedimientos, McGraw-Hill, Bogotá, pp. 622.

Troell M, Halling C, Neori A, Buschmann AH, Chopin T, Yarish C, Kautsky N (2003) Integrated Mariculture: Asking The Right Questions. Aquaculture 226: 69–90.

Tseng CK (1987) Some remarks on the kelp cultivation industry of China. In: Bird KT, Benson PH (eds), Seaweed Cultivation for Renewable Resources, Elsevier Science Publishers, Amsterdam, pp. 147–153.

Ugarte RA, Sharp G (2001) A new approach to seaweed management in Eastern Canada: The case of *Ascophyllum nodosum*. Cah. Biol. Mar. 42: 63–70.

Vásquez JA, Vega A (1999) The effect of harvesting of brown seaweeds: A social, ecological and economical important resource. World Aquacult. 30: 19–22.

Vásquez JA, McPeak RH (1999) A new tool for sea urchin control and kelp restoration. Calif. Fish Game Bull. 84: 149–158.

Wikfors GH, Ohno M (2001) Impact of algal research in aquaculture. J. Phycol. 37: 968–974.

Journal of Applied Phycology (2006) 18: 269–277
DOI: 10.1007/s10811-006-9033-y

Effects of temperature and salinity on the growth of *Gracilaria verrucosa* and *G. chorda*, with the potential for mariculture in Korea

H. G. Choi[1,*], Y. S. Kim[2], J. H. Kim[3], S. J. Lee[4], E. J. Park[4], J. Ryu[4] & K. W. Nam[4,*]
[1]*Faculty of Biological Science and Institute of Basic Natural Sciences, Wonkwang University, Iksan, Jeonbuk 570–749, Korea;* [2]*School of Marine Life Science, Kunsan National University, Kusan 573-701, Korea;* [3]*Jang Heung Fisheries Technology Institute, Jeonnam, 529-801, Korea;* [4]*Department of Marine Biology, Pukyong National University, Busan 608-737, Korea*

*Author for correspondence: e-mail: hgchoi@wonkwang.ac.kr or kwnam@pknu.ac.kr

Key words: *Gracilaria chorda*, *G. verrucosa*, growth, cultivation, salinity, temperature

Abstract

Effects of temperature and salinity on the growth of the two agarophytes, *Gracilaria verrucosa* (Hudson) Papenfuss and *Gracilaria chorda* Holmes were examined in Korea. Both species grew over a wide range of temperatures (10–30 °C) and salinities (5–35‰), and grew well at 17–30 °C and a salinity of 15–30‰. In culture, *G. verrucosa* grew faster than *G. chorda* and their maximum growth rates were 4.95% day^{-1} (30 °C, 25‰) and 4.47% day^{-1} (at 25 °C, 25‰), respectively. In the field population the maximum growth and fertility of *G. chorda* were observed in summer. The growth rate of *G. verrucosa* was slightly higher than that of *G. chorda* for 2 weeks on the cultivation rope and in culture but it was much lower after being contaminated with epiphytes. The biomass of the epiphytes was 0.82 g dry wt. per host plant in *G. verrucosa* and 0.001 g in *G. chorda*. *G. chorda* exhibited resistance to epiphytism and grew 7 times in length and the dry weight increased 15 times after 55 days. In conclusion, *G. chorda* appears to be a good agarophyte with a fast growth rate and resistance to epiphytism, and compared with *G. verrucosa*, has good potential for commercial cultivation.

Introduction

Seaweeds belonging to the genus *Gracilaria* are very important as a food for humans and marine animals, and as a source of industrial agars (Zemke-White & Ohno, 1999). *Gracilaria* is now the most important agarophyte, producing approximately 60% of the agar in the world (Tseng, 2001), and commercial cultivation is performed on a very large scale in several countries such as Chile, China and Taiwan (Dawes, 1995). There is increasing demand for industrial agars for use as materials in electrophoresis and as a culture medium for microbes.

Korea is a major country for algal production but the only species cultivated are *Laminaria japonica* Areschoug, *Undaria pinnatifida* (Harvey) Suringar, and *Porphyra* spp. Their price fluctuates annually with variations in production and weather. Therefore, it is essential to develop cultivation techniques for new species. In Korea, food grade agars are made from *Gelidium* spp., especially *G. amansii* (Lamouroux) Lamouroux. However, industrial agars extracted from *Gracilaria* spp. are imported from Chile and Brazil.

Gracilaria verrucosa, which is one of 8 species found on the Korean coast, occurs in the upper intertidal zone and is distributed widely from estuaries to open sea around the Korean Peninsula. *Gracilaria chorda* occurs in the lower intertidal zone of the southwestern coast of Korea and grows up to 7 m in length on the cultivation ropes of *U. pinnatifida*, which were established 2–3 m below the sea surface. A few studies have been performed on the commercial cultivation of *Gracilaria verrucosa* to develop cultivation techniques, particularly on the environmental conditions to induce release of spores for seedlings in the laboratory and on the growth and reproduction of *G. verrucosa* field populations (Kim et al., 1998, 2001). However, there are no data on the growth and reproduction of *Gracilaria chorda*, even though the alga is of commercial interest as an agarophyte and foodstuff due to its higher production.

For cultivation, the growth responses of *Gracilaria chorda* to temperature and salinity should be studied in order to determine the choice of cultivation area (estuaries or open water) and cultivation season (summer or winter). In addition, data on the growth and reproduction of natural populations of *G. chorda* are needed. Accordingly, the aims of this study were to examine the effects of salinity and temperature on the growth of *Gracilaria chorda* and *G. verrucosa* in laboratory culture and to measure the growth of the two algae on cultivation ropes in order to determine which of the two species would be better for cultivation.

Materials and methods

Study site

Studies were conducted at Ihoijin Jangheung (34°27′N, 126°56′E), on the border between Gogeum Island, Wando, and Gangjin, on the southern coast of Korea. Rock beds lie along the shoreline and a large amount of gravel and vast mud flats are found in the intertidal zone in the study area. The seawater movement is relatively low and the transparency of seawater is approximately 0.8–2.0 m. The average sea surface temperature ranges from 7.11 to 23.73 °C throughout the year, and the salinity ranges from 30.22 ‰ in summer to 34.38 ‰ in winter. The dominant algae were *Enteromorpha prolifera* (Oeder) J. Agardh, *Monostroma nitidum* Wittrock, *Ulva pertusa* Kjellman and *Gloiopeltis furcata* (Postels *et* Ruprecht) J. Agardh, *Porphyra* spp., *Gracilaria verrucosa* and *Sargassum* spp.

Field population structure of Gracilaria chorda

Gracilaria chorda was collected seasonally on cultivation ropes of *Undaria pinnatifida* at Ihoijin, Jangheung, Korea from February 2003 to April 2004. The plants were transported to the marine laboratory of Wonkwang University and length was measured. The plants were dried in an oven at 80 °C for 5 days and then weighed to calculate the growth of the *Gracilaria chorda* field population. The population structures and size distribution were also examined in order to determine the recruitment season of the species.

Laboratory culture

For culture, vegetative plants of *Gracilaria verrucosa* and *G. chorda* were collected from the intertidal zone of Tangjasum and cultivation ropes of *U. pinnatifida* at Ihoijin, Jangheung, Korea in June 2003. The plants were transported to the laboratory and the apical parts of fronds were rinsed several times with filtered seawater to remove diatoms and detritus attached to the fronds. The healthy apical fronds, 10 mm in length, were excised from the vegetative fronds of the two species and kept for 2 days to reduce the negative effects of cutting.

Growth studies were carried out in five incubators at 10, 17, 25, 30 and 35 °C and salinities 5, 15, 25 and 35‰ for 2 weeks. The daylength (12: 12 h light: dark, LD) and irradiance (100 μmol photons m^{-2} s^{-1}) were also kept constant. Five apical fragments of each species were placed in a flask containing 100 mL of PES medium (Provasoli, 1968) and the experiments were replicated three times. The culture medium was changed every 4 days throughout the experimental period. After 2 weeks, the length of the plants was measured and the relative growth rate (RGR %day^{-1}) was calculated with the mean length for each replicate using the following equation (Rueness & Tananger, 1984):

$$\mathrm{RGR}\,(\%\mathrm{day}^{-1}) = 100\,\ln(\mathrm{Lt}/\mathrm{Lo})/\mathrm{t}$$

where Lo is the initial plant length, Lt is the final length after *t* days and *t* is the number of days.

Transplant experiment

Vegetative plants of *Gracilaria verrucosa* and *G. chorda* were collected at the two sites reported above. The fronds were cut into 15 cm lengths and inserted between the braids of polypropylene cultivation rope at 15 cm intervals (100 m for each species). The ropes were suspended at 1 m below the seawater surface at Ihoijin Jangheung from 25 March to 20 May 2004. For each species, 60 plants (20 replicate plants per period) were harvested 13, 27 and 55 days after transplantation. For both species, the plant length, weight and branch number were measured at each sampling date. The relative growth rate (RGR %day^{-1}) was calculated using the mean plant length for each replicate using the equation described above.

When transplanting the thalli of the two species, the mean length, dry weight and branch number of *G. verrucosa* was 15 cm, 0.09 ± 0.01 g (mean ± SE,

$n = 30$) and 6.43 ± 0.97 individuals, respectively. For *G. chorda,* the mean length, dry weight and branch number of 30 plants were 15 cm, 0.10 ± 0.01 g, and 6.07 ± 0.99 individuals, respectively.

After 27 days, epiphytic algae were found on the fronds of *G. verrucosa*. At the end of the experiment, these algae were identified under a microscope, carefully detached from the fronds, and dried in an oven at 80 °C for 5 days and then weighed.

Statistical analyses

Statistical analyses were carried out using STATISTICA version 5.0 software. A two-way ANOVA was used to test the effects of temperature and salinity on the relative growth rate of each species. A one-way ANOVA was used to determine the differences in length, weight, and branch number between the two species for the cultivated plants. A Tukey test was applied when significant differences were detected between the means (Sokal & Rohlf, 1995). Homogeneity of the variance was tested using Cochran's test (Underwood, 1997).

Results

Field population of Gracilaria chorda

Thalli of *Gracilaria chorda* are erect, terete and pink to reddish purple in color. The alga is attached to the substratum by a single, small discoid holdfast. There are 50 to 60 lateral branches (<5 cm) on the main axis of 10 cm but there were few branches longer than 5 cm, unlike *G. verrucosa.*

The mean lengths of *G. chorda* ranged from 85.40 ± 4.14 cm (mean ± SE, $n = 9$) in February to 236.32 ± 14.44 cm ($n = 19$) in June (Figure 1). Maximum growth of *G. chorda* occurred in the summer with an average seawater temperature 18 °C and the alga grew up to 7 m in length on the cultivation ropes of *Undaria pinnatifida*. The mean dry weight of *G. chorda* reached a maximum of 3.18 ± 0.53 g

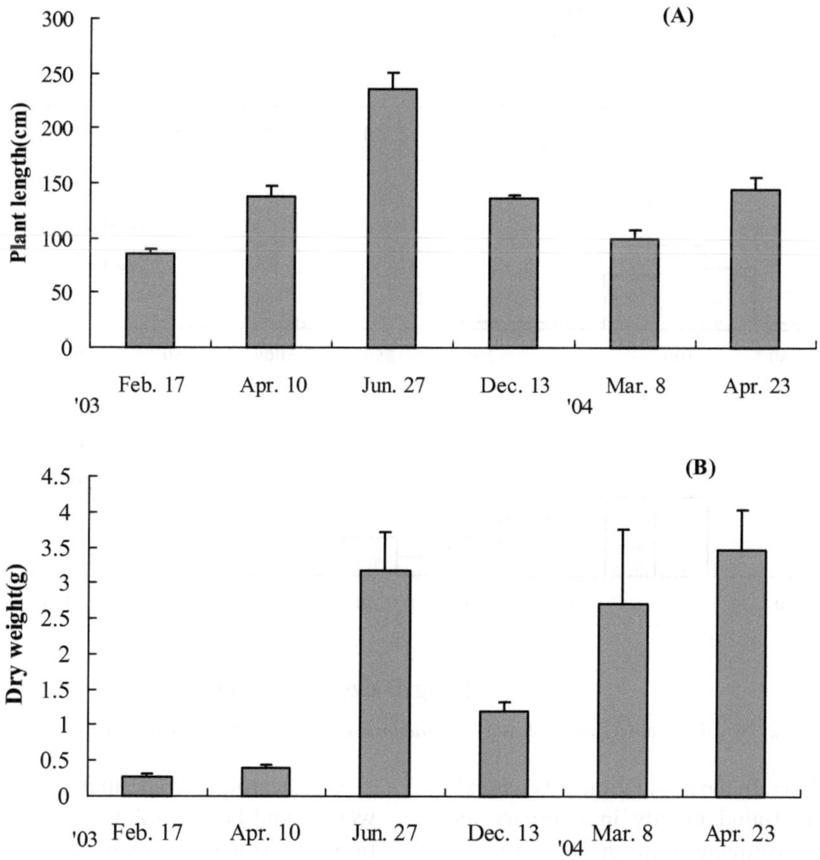

Figure 1. Seasonal variations in mean plant length (A) and dry weight (B) of *Gracilaria chorda.* Bars show standard errors.

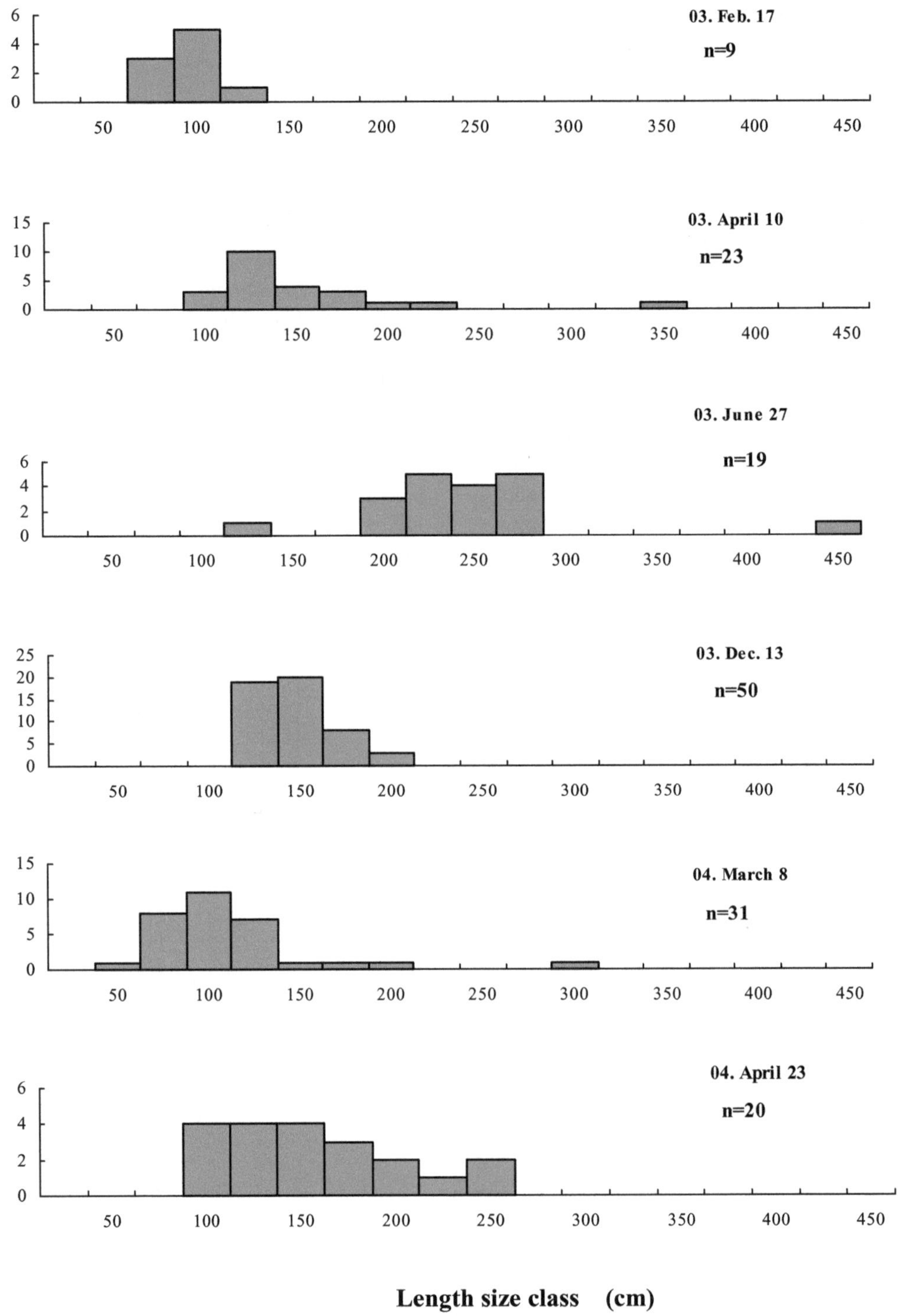

Figure 2. Size distribution (length classes) of *Gracilaria chorda* in plant length over the study period.

$d(n = 19)$ in June 2003. In the population of *G. chorda*, smaller plants were found mainly in February and March, indicating recruitment of the species in winter (Figure 2).

Cystocarpic and spermatangial plants of *G. chorda* were found between April and June, and the proportions of fertile plants increased with increasing seawater temperature. The cystocarp diameter of *G. chorda* ranged from 0.5 to 1.2 mm with a maximal value of 3 mm in June 2003, and the number of cystocarps was 21.70 ± 3.13 individuals (mean ± SD, $n = 10$) on the fronds of 10 cm. Female plants have fewer branches than vegetative plants which have 14–18 individuals on the fronds of 10 cm.

Effects of salinity and temperature on growth

The growth of the two species was significantly affected by salinity and temperature in culture. *Gracilaria chorda* and *G. verrucosa* grew in a wide range of temperatures (10–30 °C) and salinities between 5–35‰ (Figure 3). However, within 4 days at 35 °C, *G. chorda* thalli became completely discolored and those of *G. verrucosa* partially discolored, at all the salinity treatments tested. Therefore, the upper intertidal species, *G. verrucosa* tolerated high temperatures better than *G. chorda*.

Both species grew well at higher temperatures (17, 25, and 30 °C) than at 10 °C (Figure 3). However, the fronds of *G. chorda* grown at 30 °C and 5‰ in culture died within 7 days. The relative growth rates ranged from 0.91 to 4.47% for *G. chorda* and 1.59 to 4.95% day^{-1} for *G. verrucosa*. In general, *G. verrucosa* had a significantly higher growth rate than *G. chorda* in the experimental treatments ($p < 0.01$). There were significant differences in the relative growth rates of each species among the temperature treatments but no differences were found at the higher temperatures ($p > 0.05$). The optimal temperature for growth of the two species was between 17–30 °C and the maximal growth was observed at 25 °C.

Gracilaria verrucosa grew well in a wide range of salinities (5 to 35‰) compared with *G. chorda*, particularly at 5‰ (Figure 3). The growth of the two species was more temperature dependant than salinity dependant. The optimal salinity for growth was 25‰ for *G. chorda* and 15–25‰ for *G. verrucosa*.

Growth of the two species on cultivation ropes

The field populations of the two species differed in length, dry weight and number of branches. The mean length, dry weight and number of branches of *Gracilaria verrucosa* was 43.95 ± 3.70 cm (mean ± SE, $n = 20$), 0.91 ± 0.18 g and 48.45 ± 4.85 ($n = 20$), respectively. The mean plant length, dry weight and number of branches of *G. chorda* was 144.45 ± 10.63 cm ($n = 20$), 3.47 ± 0.57 g, and 71.55 ± 6.05 ($n = 20$), respectively.

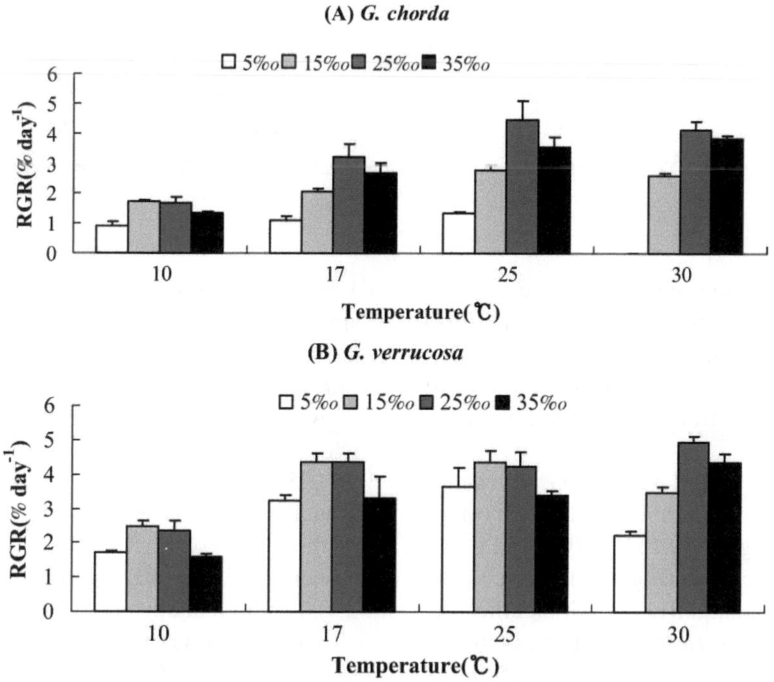

Figure 3. Mean relative growth rates (% d^{-1}) of *Gracilaria chorda* (A) and *Gracilaria verrucosa* (B). Bars show standard errors ($n = 3$).

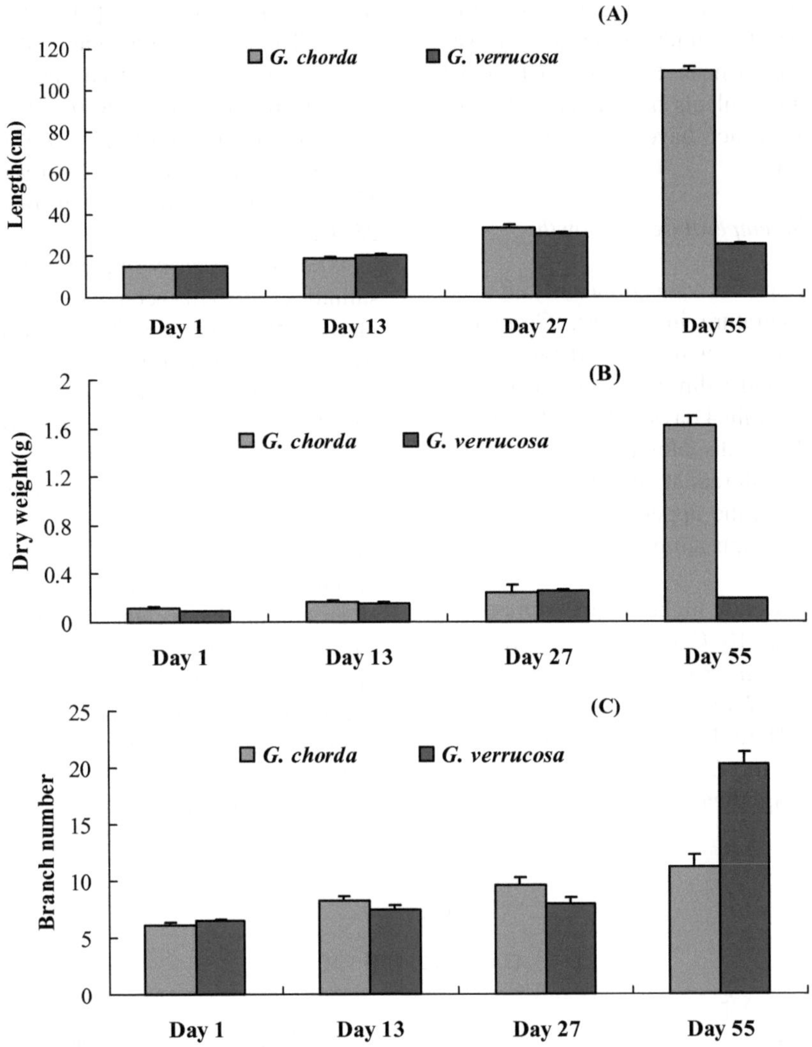

Figure 4. Mean length (A), weight (B) and branch number (C) of *Gracilaria chorda* and *Gracilaria verrucosa* over the cultivation period. Bars show standard errors ($n = 3$).

At the beginning of the experiment, the average length of both species was 15 cm. Fronds of *Gracilaria chorda* and *G. verrucosa* inserted into the cultivation ropes grew well without falling off and a new holdfast of each plant had formed from the part that had been inserted into the rope.

Thirteen days after transplantation, *G. verrucosa* grew to a mean length of up to 19.60 ± 1.25 cm (mean ± SE, $n = 20$) and weight of 0.16 ± 0.01 g and it was slightly longer than *G. chorda* (Figure 4). However, by day 27, *G. chorda* was slightly, but not significantly ($p > 0.05$) longer than *G. verrucosa*. The number of branches increased with time from 6.07 ± 0.34 to 11.17 ± 1.07 in *G. chorda* and from 6.43 ± 0.26 to 20.2 ± 1.13 in *G. verrucosa*. However, *G. chorda* had more smaller branches (<5 cm) than *G. verruca*. Fifty-five days after transplantation, *G. chorda* had grown to 7 times its initial length and 15 times its dry weight.

At the end of the experimental period, the mean width of the main branches in *G. verrucosa* was 1.29 ± 0.05 mm (mean ± SE, $n = 30$) and that of the lateral branches was 1.04 ± 0.04 mm. In addition, the width of the main axes in *G. chorda* was 1.53 ± 0.05 mm (mean ± SE, $n = 30$) and the lateral branch was 1.11 ± 0.05 mm in diameter.

The relative growth rates ranged from 1.54 ± 0.34 to 4.26 ± 0.19% d^{-1} in *G. chorda* and −0.71 ± 0.07 to 3.13 ± 0.63% d^{-1} in *G. verrucosa* (Figure 5). For

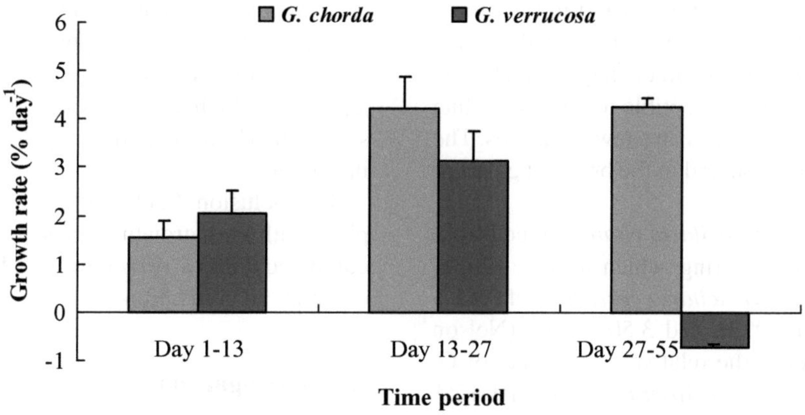

Figure 5. Mean relative growth rates in plant length of *Gracilaria chorda* and *G. verrucosa.*

2 weeks, *G. verrucosa* grew faster than *G. chorda* but then this trend reversed. Between 27 and 55 days, the relative growth rate of *G. chorda* reached a maximum, and was significantly higher than that of *G. verrucosa.* In contrast, the growth of *G. verrucosa* was retarded after 27 days and negative growth was observed after 55 days as a result of heavy epiphytes.

Eight epiphytic algae, including *Enteromorpha compressa, Polysiphonia* sp. and *Ceramium* sp. were observed on the *Gracilaria verrucosa* fronds and the three species on *G. chorda* (Table 1). The biomass of the epiphytes was 0.82 ± 0.20 g dry wt. (mean $\pm$ SE, $n = 10$) in *G. verrucosa* and 0.001 ± 0.0 g in *G. chorda* per host plant (g dry wt.). The epiphytic algae were mainly found where the branches were divided in both species. The number of branches was greater in *G. verrucosa* than in *G. chorda* even though the mean length of the plants was shorter. In addition, *G. chorda* has many lateral branches which are located on the surface of the fronds, unlike *G. verrucosa.*

Table 1. List of algal epiphytes of *Gracilaria verrucosa* and *Gracilaria chorda.*

Gracilaria verrucosa	*Gracilaria chorda*
Goniotrichum alsidii (Zanardini) Howe	*Polysiphonia* sp.
Ceramium sp.	*Enteromorpha compressa* (L.) Greville
Polysiphonia sp.	*Punctaria latifolia* Greville
Enteromorpha compressa (L.) Greville	
Enteromorpha sp.	
Porphyra sp.	
Punctaria latifolia Greville	
Leatheia difformis (L.) Areschoug	

Discussion

The growth patterns of *Gracilaria verrucosa* and *G. chorda* on ropes changed over the study period. Within the first 2 weeks on the rope, *G. verrucosa* grew faster than *G. chorda*, as observed in laboratory culture. However, the growth rate of *G. verrucosa* was lower than *G. chorda* after 4 weeks. Such a growth pattern of *G. verrucosa* may have been caused by its habitat in the upper intertidal zone. The rapid initial growth of *G. verrucosa* may be the result of a sufficient nutrient supply, because the fronds were continually submerging in seawater, resulting in reduced environmental stresses (such as desiccation and high temperatures). However, *G. verrucosa* was soon contaminated by epiphytic algae, which resulted in a decrease in growth, as reported in *Gracilaria chilensis* Bird, McLachlan *et* Oliveira (Buschmann & Gomez, 1993). By contrast, *G. chorda* was comparatively resistant to epiphytes: their biomass was 0.82 g dry wt. in *G. verrucosa* and 0.001 g in *G. chorda* per g dry wt. of host plant. It is unclear if the epiphytic algae attached to the fronds of the *G. verrucosa* before or after the transplantation, and why they prefer *G. verrucosa.* However, it appears that this is also related to their habitat. *G. verrucosa* grows in the upper intertidal zone where it is intermittently exposed to air, which may help to prevent epiphyte infection. Dawes et al. (2000) reported differences in cell-wall structure may determine the relative epiphyte-resistance of two species, *Gracilaria tikvahiae* McLachlan and *G. cornea* J. Agardh. However, the cell wall structures of the two agarophytes were not compared in this study. In addition, the branching pattern of the host algae may play a role. *G. verrucosa*, with many large branches on the main axis, was more susceptible to epiphytes than *G. chorda,* with short branches on a few main axes. The epiphytes are usually observed in the branching part of the host alga.

The growth rates of *Gracilaria chorda* ranged from 4.21 to 4.26% d^{-1} in the spring, which are very similar to the reported rates in *Gracilaria verrucosa* of 4.45% (Chirapart & Ohno, 1993), and 3.50–4.80% (Nelson et al., 1980). However, the relative growth rate of *G. chorda* is low compared to *Gracilaria* spp. (10–12% and 10.5%: (Ren et al., 1984; Hurtado-Ponce, 1990) and it is higher than *G. fisheri* (Xia *et* Abbott) Abbott, Zhang *et* Xia (2.56%), *G. firma* Chang *et* Xia (0.91%) and *G. salicornia* (C. Agardh) Dawson (0.86%) (Chirapart & Ohno, 1993). In culture, the growth rate of *G. chorda* reached a maximum of 4.47% at 25 °C and 25‰. At end of June, when the seawater temperature was approximately 20 °C, the largest plants of 7 m in length were found. These results suggest that *G. chorda* is an agarophyte with a potential for commercial cultivation due to its high growth rate.

Gracilaria verrucosa tolerated the relatively high temperature of 35 °C and a lower salinity of 5‰ better than *G. chorda*. *Gracilaria chorda* grew well at the higher temperatures of 25–30 °C in culture, and the growth of the species reached a peak in summer in the field population. In addition, the growth rate of *G. chorda* was higher at higher salinities (25 and 35‰) than at the low salinities (5 and 15‰) indicating that the alga is adapted to oceanic conditions and can be cultivated in open seawater. In contrast, the upper intertidal alga, *G. verrucosa* has a broad tolerance to salinity and would grow well in estuaries.

According to resource allocation theory, reproduction imposes a cost (Mathieson & Guo, 1992). Thus, it is known that the presence of reproductive structures in *Gracilaria* spp. affects the growth of the vegetative plants. This is supported by the results of Santelices & Varela (1995), who reported that vegetative plants grew faster than cystocarpic and tetrasporic plants of *G. chilensis*. Thus, we believe that mass culture of vegetative plants is likely to be preferable, even though the growth rates of *G. chorda* and *G. verrucosa* in the various reproductive states were not compared in the present study. Fertile plants with cystocarps and spermantangia were mainly found in the summer in both species. On the basis of optimal temperature and salinity for growth, we suggest that the cultivation of *Gracilaria chorda* could begin in the spring, when the seawater temperature is approximately 13 °C and after harvesting of *Undaria pinnatifida* and *Laminaria japonica*, which are cultivated in winter. *G. chorda* should then be harvested during the reproductive period in summer.

In conclusion, *G. chorda* appears to be a good agarophyte with a fast growth rate and resistance to epiphytes compared with *G. verrucosa*, and has the potential for commercial cultivation.

Acknowledgments

We would like to thank two anonymous reviewers for helpful comments which improved the manuscript. Thanks are also due to In Ju Lee and Sung Bae Lee for helping with fieldwork. This paper was supported by a grant from Wonkwang University in 2005.

References

Buschmann AH, Gomez P (1993) Interaction mechanisms between *Gracilaria chilensis* (Rhodophyta) and epiphytes. Hydrobiologia 260/261: 345–351.

Chirapart A, Ohno M (1993) Growth in tank culture of species of *Gracilaria* from the southeast Asian waters. Bot. Mar. 36: 9–13.

Dawes CP (1995) Suspended cultivation of *Gracilaria* in the sea. J. Appl. Phycol. 7: 303–313.

Dawes CJ, Teasdale BW, Friedlander M (2000) Cell wall structure of the agarophyte *Gracilaria tikvahiae* and *G. cornea* and penetration by the epiphyte *Ulva lactuca*. J. Appl. Phycol. 12: 567–575.

Hurtado-Ponce AQ (1990) Vertical rope cultivation of *Gracilaria* (Rhodophyta) using vegetative fragments. Bot. Mar. 33: 477–481.

Kim YS, Choi HG, Kim HG, Nam KW, Sohn CH (1998) Reproductive phenology of *Gracilaria verrucosa* (Rhodophyta) in Cheongsapo near Pusan, Korea. J. Fish. Sci. Technol. 1: 147–151.

Kim YS, Choi HG, Nam KW (2001) Effects of light, desiccation and salinity for the spore discharge of *Gracilaria verrucosa* (Rhodophyta) in Korea. J. Fish. Sci. Technol. 4: 257–260.

Mathieson AC, Guo Z (1992) Patterns of fucoid reproductive biomass allocation. Br. Phycol. J. 27: 271–292.

Nelson SG, Tsutsui RN, Best RR (1980) Evaluation of seaweed mariculture potential in Guam: I. Ammonia uptake by growth of two species of *Gracilaria* (Rhodophyta). University of Guam Marine Laboratory Technology Reports 61: 1–20.

Provasoli L (1968) Media and prospects for the cultivation of marine algae. In Watanabe A, Hattori A(eds), Cultures and Collections of Algae. Proceeding of the Us-Japan Conference, Japanese Society for Plant Physiology, Hakonep: 63–67.

Ren GZ, Wang JG, Chen MG (1984) Cultivation of *Gracilaria* by means of low rafts. Hydrobiologia 116/117: 72–76.

Rueness J, Tananger T (1984) Growth in culture of four red algae from Norway with potential for mariculture. Hydrobiologia 116/117: 303–307.

Santelices B, Varela D (1995) Regenerative capacity of *Gracilaria* fragments: effects of size, reproductive state and position along the axis. J. Appl. Phycol. 7: 501–506.

Sokal RR, Rohlf FJ (1995) Biometry, 3rd edition. W.H. Freeman, New York.

Tseng CK (2001) Algal biotechnology industries and research activities in China. J. Appl. Phycol. 13: 375–380.

Underwood AJ (1997) Experiments in Ecology: Their Logical Design and Interpretation Using Analysis of Variance. Cambridge University Press, Cambridge.

Zemke-White WL, Ohno M (1999) World seaweed utilization: an end-of century summary. J. Appl. Phycol. 11: 369–376.

Journal of Applied Phycology (2006) 18: 279–286
DOI: 10.1007/s10811-006-9027-9

Developmental studies in *Porphyra vietnamensis*: A high-temperature resistant species from the Indian Coast

Dinabandhu Sahoo*, Pooja Baweja & Neetu Kushwah
Marine Biotechnology Laboratory, Department Of Botany, University of Delhi, Delhi – 110007, India

*Author for correspondence: e-mail: dbsahoo@hotmail.com

Key words: *Porphyra*, development, high temperature, India

Abstract

Porphyra vietnamensis Tanaka & Pham-Hoang Ho (Bangiales, Rhodophyta) is a tropical seaweed collected from the west coast of India. Thalli of the blade phase are found growing only during the rainy season between July and September. They grow on rocky intertidal or subtidal substrata or as epiphytes on other seaweeds such as *Enteromorpha flexuosa* and *Chaetomorpha media*. The gametophytic thallus is monostromatic and covered with spines at the base. The species is monoecious. Male gametangia are found in patches that are distributed in the upper part of the thallus. Archeospores are found at the thallus margins and give rise to the blade phase after one week of germination even at 30 °C. Zygotospores germinated at 25 °C into conchocelis within three days from the date of their inoculation. Conchospores were released at 30 °C. The young blades grew at 32 °C in the laboratory.

Introduction

Porphyra (Bangiales, Rhodophyta) is one of the world's most valued maricultured seaweeds, and is primarily used as food in many oriental countries. It is highly prized for its flavour and as a health food as it is rich in proteins and vitamins. Nearly 17 types of free amino acids, including taurine, which controls blood cholesterol levels (Tsujii et al., 1983) can be found within the genus, which has an annual value of over US$ 1.8 billion (Yarish et al., 1999). The biology and ecology of *Porphyra* has been studied more thoroughly than that of any other red algal genus (Tseng & Sun, 1989; Cole, 1990; Hawkes, 1990). Recently, it has been reported that *Porphyra* has much more potential and can be used as an experimental system for modern biological research, like *Arabidopsis thaliana* (Sahoo et al., 2002). *Porphyra* is represented by more than 133 species, which are particularly abundant in cold-temperate and boreal shores of the Northern and Southern Hemispheres (Yoshida et al., 1997). In some areas, individual species grow throughout the year whereas other species are very seasonal, with big crops present on the rocks for only one to two months of the year. So far seven species of *Porphyra*, namely *P. chauhanii* Anil Kumar & Panikkar, *P. indica* V. Krishnamurthy & Baluswami, *P. kanyakumariensis* V. Krishnamurthy & Baluswami, *P. crispata* Kjellman, *P. okhaensis* H. Joshi, Oza & Tewari, *P. suborbiculata* Kjellman, and *P. vietnamensis* T. Tanaka & Pham-Hoang Ho have been reported from different parts of the Indian coast, of which *P. vietnamensis* is the most abundant (Sahoo et al., 2001). Børgesen (1937) reported *P. vietnamensis* as *P. tenera* Kjellman from Madras Harbour. Later, Sreeramulu (1952) originally described the plants as *P. naidum* Anderson. Subsequently, Umamaheswara Rao and Sreeramulu (1963) confirmed the species as *P. vietnamensis*. Lewmanomont and Ogawa (1978) studied the life cycle of *P. vietnamensis* from Songkhla, Thailand but were not successful in obtaining the release of conchospores from the conchocelis phase. Imada and Abe (1980) were able to get conchospores from *P. vietnamensis* by using phytohormones. Lewmanomont and Chittpoolkusol (1993) studied the life history of *P. vietnamensis* from Thailand and were able to complete the life history in the laboratory at 25 °C. Although *P. vietnamensis* has been reported from different parts of the Indian coast, no detailed studies have been undertaken on its development in culture. In the present study, it has been found that a particular strain of *P. vietnamensis* can complete its life history at 32 °C in the laboratory. Since most of the *Porphyra* species grow at lower temperatures, this Indian strain of *P. vietnamensis* could be a potential crop for mass cultivation in tropical seas.

Materials and methods

Porphyra vietnamensis was collected during July 2003 from the rocky coast of Goa in western India. Fertile blades were selected and cleaned with seawater in the field to remove visible epiphytes and other contaminants. The thalli were wrapped in absorbent cotton moistened with seawater, packed in plastic bags and transported to the laboratory in an air-conditioned railway compartment. In the laboratory, the thalli were washed thoroughly in autoclaved seawater and each thallus was observed under a stereo binocular microscope. Thallus surfaces were cleaned with the help of sterilized cheese cloth and epiphytes were removed. Subsequently, the thalli were washed 3–4 times in autoclaved seawater. Then each thallus was blotted with tissue paper. Individual thalli were then wrapped in cheese cloth, put in a polythene bag and kept in a refrigerator. After two hours thalli were taken out and individual thalli were put into petri-dishes (size 90 mm, Polylab India) separately in three different culture media: f/2 (Guillard & Rhyther, 1962), PES medium (Provasoli, 1968) and autoclaved seawater. The petri-dishes were kept at 20 °C, 25 °C, 30 °C and room temperature (32 °C) under cool white 40 W fluorescent lights at an approximate photon fluence rate of 40 μmol photons m^{-2} s^{-1}. The thalli in petri-dishes were observed after every two hours for the release of spores which were collected by means of Pasteur pipette and transferred to new petri-dishes separately in the above mentioned media. A drop of GeO_2 was added to each petri-dish to prevent the growth of diatoms (1 g GeO_2 dissolved in 100 mL of distilled water). All petri-dishes were kept in the above mentioned temperature and light conditions with a photoperiod of 12:12 h light: dark cycle. The salinity of media was maintained at 25 ppt. Culture media were changed after every seven days and observations on spore development were recorded. Photographs were taken through a Nikon E600 photomicroscope (Nikon, Tokoyo, Japan) using ILFORD Black and White film. Spore terminology follows Nelson et al., 1999.

Observations

Porphyra vietnamensis was found growing predominantly on the west coast of India especially in the Provinces of Maharashtra, Goa and Karnataka. The species was strictly seasonal. The leafy thalli appeared on the rocky substratum, oyster shells (*Crassostrea gryphoides*) or as an epiphyte on seaweeds such as *Enteromorpha flexuosa* (Wulfen) J. Agardh and *Chaetomorpha antennina* (Bory de Saint-Vincent) Kützing in the mid-tidal to spray zone from the beginning of July. The blades started degenerating from the end of August and completely disappeared by the end of September. Over the last several years the senior author has observed that the growth of *Porphyra* species in India is associated with the onset of the monsoon season. During this season the seawater is turbid and rich in nutrients due to river and land-surface discharge. The seawater temperature decreases substantially from 30–32 °C to 19–23 °C and salinity from 31–33 ppt to 19–22 ppt during these months. The sky is mostly cloudy during this season, which appears to favour the growth of the species.

Thallus morphology

Considerable variation was observed in the gross morphology of *P. vietnamensis* but in general the blades were monostromatic, membranous, lanceolate to linear-lanceolate and sometimes ribbon-like, purple to pink-purple in colour. The thalli were attached to the substratum with discoid holdfasts. Sometimes thalli were branched from a common base having several bladelets (Figure 1). Usually the thalli were 3–15 cm in height but sometimes grew up to 40 cm. They were 1.5–3 cm broad and 25–32 μm thick. The cells of the basal regions were large, with pigmented pear-shaped heads, and elongated (Figure 2). Margins were undulate with 2–3 celled spines, which were found towards the basal region (Figure 3). The vegetative cells in the central region were vacuolated with a single stellate plastid (Figure 4). The thalli were monoecious with distinct male gametangial and zygotosporangial patches found towards the margin and these were present on the same thallus. The male gametangia could be distinguished as pale-yellow patches at the outer margins. There were 64 spermatia per male gametangium arranged in 4 tiers

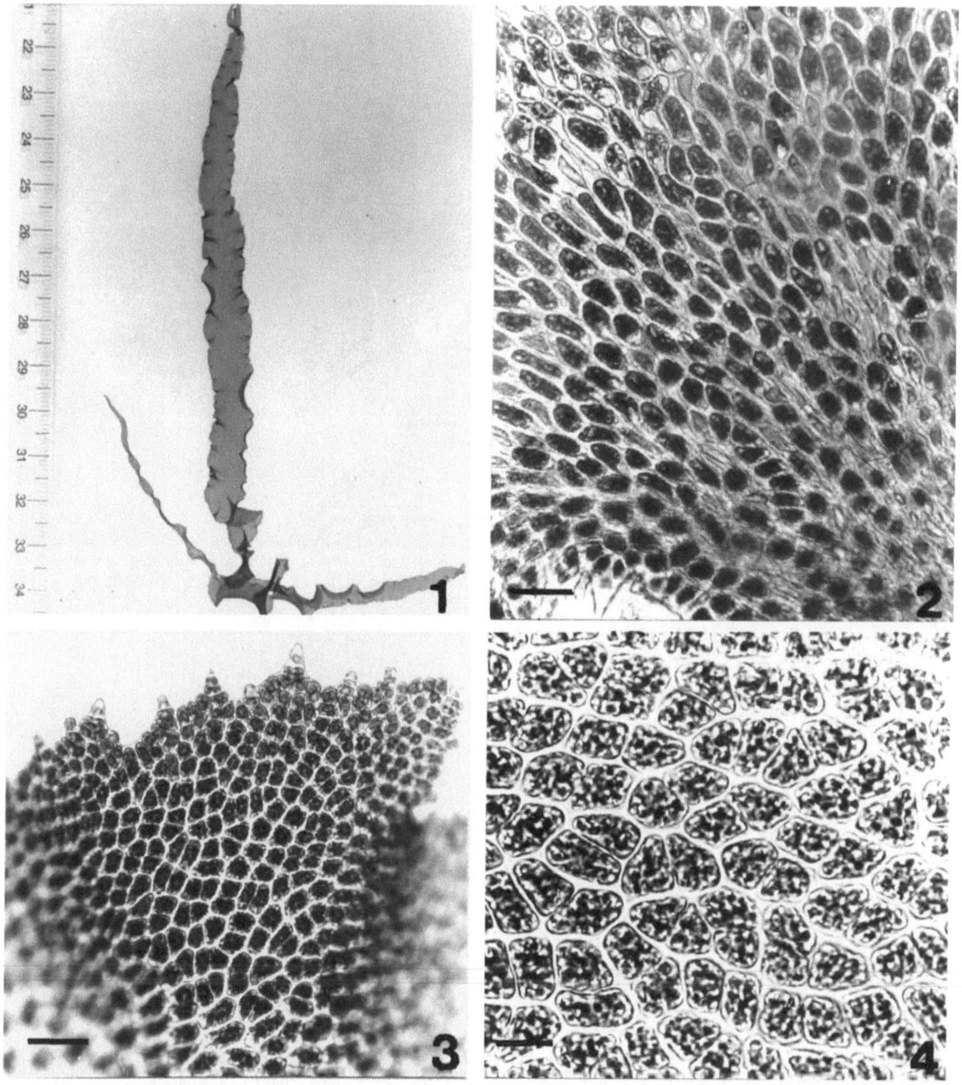

Figure 1–4. Porphyra vietnamensis. Thallus morphology. (1). Mature thallus of *Porphyra vietnamensis* showing branching at the base. (2). Basal portion showing pear shaped cells. Scale bar = 100 μm. (3). Margin showing spines. Scale bar = 100 μm. (4). Surface view of central region showing vegetative cells. Scale bar = 20 μm.

of 16 each, having the spore formula a/4, b/4, c/4. There were 8 zygotospores arranged in two tiers of 4 each, having the spore formula a/2, b/2, c/2.

Developmental studies

In culture the gametophytic thalli produced two types of spores: archeospores which germinated directly to give rise to the young blades, and zygotospores which gave rise to concohocelis. The conchocelis produced conshocporangia which released conchospores that developed into blades.

Archeospore development

Archeospores were released from the margins of the mature blades in all the temperature and light conditions tested. The spores were round, larger in size than zygotospores, vacuolated, thick walled and were between 18–20 μm in diameter (Figure 5). The spores germinated unipolarly after 24 h of release. After 48 h, the spores divided to form a 2-celled structure where one of the cells showed distinct polarity (Figure 6). The spores divided and formed young germlings (minute blades) in between 4–6 days (Figures 7 and

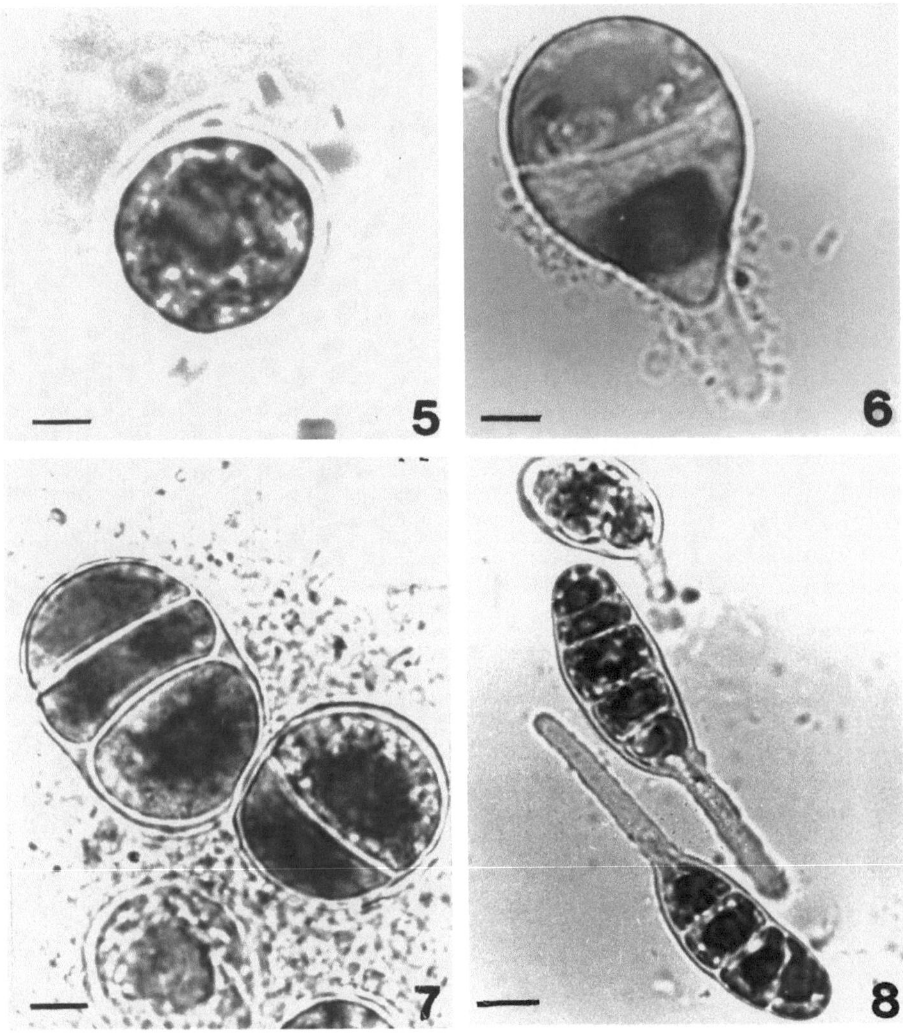

Figure 5–8. Porphyra vietnamensis: archeospore development. (5). A thick walled archeospore. Scale bar = 10 μm. (6). Two celled stage showing polarity. Scale bar = 10 μm. (7). 1–3 celled stages. Scale bar = 20 μm. (8). Young germlings. Scale bar = 20 μm.

8), completing the asexual mode of life history in less than one week. When cultured for a longer period, i.e. 45–60 days, the thalli developed normal morphology (i.e. similar to field material) at 20, 25 and 30 °C (Figure 13) but at 32 °C showed abnormal thallus morphology (Figure 14).

Conchocelis development

Zygotospores were released from the margins of the fertile blade in 24–48 h in all the above temperature and light conditions mentioned earlier. The spores were thick-walled, and between 12–15 μm in diameter, pinkish in colour and each with a distinct, stellate plastid. The zygotospores germinated unipolarly between 48–72 h after release (Figure 9). Subsequently, the spores developed into conchocelis filaments of indeterminate growth with extensive branching (Figure 10). The growth of conchocelis was slower at 20 °C whereas it grew rapidly at 25 and 30 °C but could still survive and remain healthy at 32 °C. When the conchocelis was allowed to grow in the petri-dish, it formed an extensive network of filaments which were attached to the bottom of the dish, whereas it formed spherical colonies or balls when grown in a conical flask. The conchocelis filaments were light pink in colour.

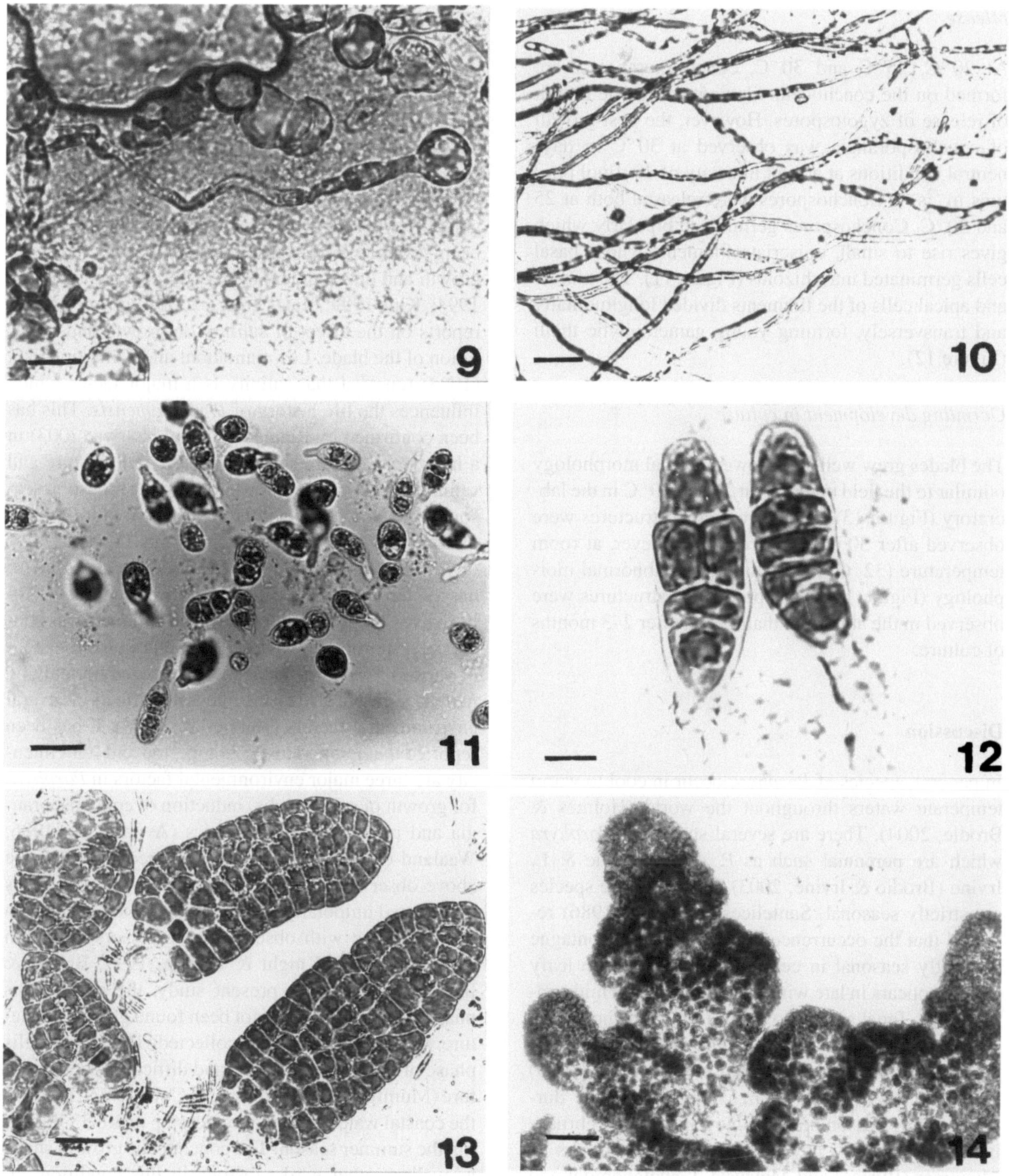

Figure 9–14. Porphyra vietnamensis: development of zygotospores. (9). Zygotospores showing unipolar germination into conchocelis. Scale bar = 50 μm. (10). Free living conchocelis growing at room temperature (32 °C). Scale bar = 50 μm. (11). Conchospores and young germlings in different stages of development at 20 °C–25 °C. Scale bar = 20 μm. (12). Young sporelings showing different planes of cell divisions. Scale bar = 20 μm. (13). Young thalli grown at 25 °C. Scale bar = 50 μm. (14). Formation of abnormal thalli at room temperature (32 °C). Scale bar = 10 μm.

Formation of conchosporangia and conchospore release

At 20 °C, 25 °C and 30 °C, conchosporangia were formed on the conchocelis filaments after one month of release of zygotospores. However, the best growth of conchosporangia was observed at 30 °C in day-neutral conditions at a light intensity of 40 μmol photons $m^{-2}s^{-1}$. Conchospores were released both at 25 and 30 °C. Conchospores germinated bipolarly which gives rise to small, uniseriate filaments whose basal cells germinated into rhizoids (Figure 11). The middle and apical cells of the filaments divided longitudinally and transversely, forming young gametophytic thalli (Figure 12).

Germling development in culture

The blades grew well and showed normal morphology (similar to the field material) at 25 and 30 °C in the laboratory (Figure 13). The reproductive structures were observed after 30 days of growth. However, at room temperature (32 °C) the thalli showed abnormal morphology (Figure 14). No reproductive structures were observed in the abnormal thalli, even after 2–3 months of culture.

Discussion

In general, species of *Porphyra* occur in cool to warm temperate waters throughout the world (Holmes & Brodie, 2004). There are several species of *Porphyra* which are perennial such as *P. dioica* Brodie & L. Irvine (Brodie & Irvine, 2003) whereas some species are strictly seasonal. Santelices and Avila (1986) reported that the occurrence of *P. columbina* Montagne is strictly seasonal in central Chile, where the leafy thallus appears in late winter and grows up to midsummer only. Tanaka and Ho (1962) reported the occurrence of *P. vietnamensis* from the Northwest Pacific region of Vietnam. Ogawa and Lewmanomont (1979) reported growth of *Porphyra vietnamensis* only during the rainy season between November and February in Thailand. They have also found this species in Myanmar (Burma) and Malaysia. Lewmanomont and Chittpoolkusol (1993) also reported the occurrence of this species from Songkhla provinces, Thailand during the rainy season when the seawater temperature is 24–27 °C and salinity between 8–26 ppt. Tseng (1983) and Wu (1988) reported the occurrence of the same species from China. In the present study, it has been observed that *Porphyra vietnamensis* is strictly seasonal in India, where gametophytic thalli are found only during the rainy season when the water temperature is between 19 and 23 °C, the salinity goes down to between 19–22 ppt and the nutrient levels are higher in the sea than in the summer and winter seasons. Further, it has been observed by the senior author over the years that if the onset of the monsoon is delayed then the appearance of the gametophytic thalli is accordingly delayed. There have been several studies on the effect of temperature on growth and development of *Porphyra* (Gargiulo et al., 1994; Kim 1999; Katz, et al., 2000) but only a few reports on the effect of salinity on growth and maturation of the blade. Lewmanomont and Chittpoolkusol (1993) reported that salinity is a major factor which influences the life history of *P. vietnamensis*. This has been confirmed by Ruangchuay and Notoya (2003) in a laboratory culture study. Since the temperature and salinity of sea water are lower during the monsoon season, these two factors might be responsible for growth of *Porphyra* during monsoon season in India.

P. vietnamensis, like most other species of *Porphyra,* has a heteromorphic life history with alternation of a foliose leafy gametophytic thallus and a filamentous conchocelis sporophytic phase. The pattern of life history of various *Porphyra* species is determined not only by genetic traits but also by the integrated effects of several environmental factors (Katz et al., 2000). It has been reported that temperature, photoperiod and light intensity are three major environmental factors in *Porphyra* for growth of conchocelis, induction of conchosporangia and release of conchospores (Avila et al., 1986; Waaland et al., 1990). We are in agreement with the above observations. In *P. vietnamensis*, zygotospores germinated unipolarly and developed into conchocelis in accordance with observations of others (Kapraun & Luster, 1980; Knight & Nelson, 1999; Brodie & Irvine, 2003). In the present study, the conchocelis phase of this species has not been found growing in nature where the plants were collected. The conchocelis phase of *Porphyra* is cryptic and difficult to find in nature (Mumford, 1980; Conway & Cole, 1977). In India, the coastal water temperature goes up to 30–33 °C during the summer season. The conchocelis grow in shells as well as other hard substrata in this condition, although they are not visible to the naked eye in the field. This is substantiated by the present study where the conchocelis grows at 30 °C and even survives at 32 °C. In some *Porphyra* species, conchosporangial production requires a specific environmental stimulus, such as short-day conditions, as in *P. tenera* (Dring, 1967). In the present study, conchosporangia were formed in *P. vietnamensis* without the application of any environmental stimuli, as reported in *P. miniata* (Lyngbe) C. Agardh, *P. angusta* Okamura et Ueda and *P. torta* Krishnamurthy (Chen et al., 1970; Chiang & Wang, 1980). Photoperiod and temperature are the most important factors regulating the formation and release of conchospores in the conchocelis phase of *Porphyra* spp. (Edwards, 1969; Freshwater & Kapraun, 1986; Waaland et al., 1987, 1990; Garguilo et al., 1994). However, in *P. vietnamensis*, the conchospores were released without any change in temperature and photoperiod. Generally, in temperate species the optimum growth temperatures of blade and conchocelis are different. For example, *P. lacerata* Miura (Notoya & Nagaura, 1998), *P. moriensis* Ohmi (Notoya & Miyashita, 1999), *P. pseudolinearis* Ueda and *P. dentata* Kjellman (Kim, 1999) have lower temperature (5–15 °C) requirements for the blades and higher temperature (20 °C) for the conchocelis phase. The present study confirms that *P. vietnamensis* does not require such a wide range of temperature induction. Both the blade and conchocelis phases can be grown at 25–30 °C.

The archeospores germinated directly to give rise to young blades, thus completing the asexual life history within a week. It has been observed that the blades of *P. vietnamensis* can grow well even at a high temperature, i.e. at 32 °C, which is quite uncommon in *Porphyra*. From these results we conclude that *P. vietnamensis* from India is a potential species for large-scale mariculture in tropical water.

References

Avila M, Santelices B, McLachlan J (1986) Photoperiodic and temperature regulation of the life history of *Porphyra columbina* (Rhodophyta, Bangiales) from central Chile. Can. J. Bot. 50: 1867–1872.

Børgesen F (1937) Contribution to a South Indian marine algal flora – II. J. Ind. Bot. Soc. 16: 1–56.

Brodie JA, Irvine LM (2003) Seaweeds of the British Isles. Vol. 1. Rhodophyta. Part 3B. Bangiophycidae. Intercept, Hampshire, U.K. 167 pp.

Chen LC-M, Edelstein T, Ogata E, McLachlan J (1970) The life history of *Porphyra minata*. Can. J. Bot. 48: 385–389.

Chiang Y-M, Wang J-C (1980). A study on the production of conchosporangia in the conchocelis phase of *Porphyra angusta* Okamura et Ueda. Phycologia 19: 20–24.

Cole KM (1990) Chromosomes. In Cole, KM, Sheath, RG (eds.), Biology of the red algae. Cambridge University Press, New York: 73–101.

Conway E, Cole K (1977) Studies in the Bangiaceae: structure and reproduction of the conchocelis of *Porphyra* (Bangiales, Rhodophyceae). Phycologia 16: 205–216.

Dring MJ (1967) Effects of daylength on growth and reproduction of the conchocelis-phase of *Porphyra tenera*. J. Mar. Biol. Ass. U.K. 47: 501–510.

Edwards P (1969) Field and culture studies on the seasonal periodicity of growth and reproduction of selected Texas benthic marine algae. Contrib. Mar. Sci. 14: 59–114.

Freshwater WD, Kapraun F (1986) Field, culture and cytological studies on *Porphyra carolinensis* Coll et Cox (Bangiales, Rhodophyta) from North Carolina. Jpn. J. Phycol. 34: 251–262.

Gargiulo GM, Masi FD, Genovese G, Tripodi G (1994) Karyological and effects of temperature and photoperiod on the life cycle of *Porphyra leucosticta* Thuret in Le Jolis (Bangiales, Rhodophyta) from the Mediterranean Sea. Jpn. J. Phycol. 42: 271–280.

Guillard RRL, Rhyther JH (1962) Studies of marine planktonic diatoms. I. *Cyclotylla nana* Hustedt and *Detumula confervacea*. Can. J. Microbiol. 8: 229–239.

Hawkes MW (1990) Reproductive strategies. In Cole, KM, Sheath, RG (eds.), Biology of the Red Algae. Cambridge University Press, New York: 455–476.

Holmes MJ, Brodie J (2004). Morphology, seasonal phenology and observations on some aspects of the life history in culture of *Porphyra dioica* (Bangiales, Rhodophyta) from Devon, U.K. Phycologia 43: 176–188.

Imada O, Abe T (1980) Formation and germination of conchospores in *Porphyra vietnamensis*. Bull. Jpn. Soc. Sci. Fish. 46(8): 1051.

Kapraun DF, Luster DG (1980). Field and culture studies of *Porphyra rosengurtii* Coll et Cox (Rhodophyta, Bangiales) from North Carolina. Bot. Mar. 33: 449–457.

Katz S, Kizner Z, Dubinsky Z, Friedlander M (2000) Responses of *Porphyra linearis* (Rhodophyta) to environmental factors under controlled culture conditions. J. Appl. Phycol. 12: 535–542.

Kim N-G (1999) Culture studies of *Porphyra dentata* and *Porphyra pseudolinearis* (Bangiales, Rhodophyta), two dioecious species from Korea. Hydrobiologia 398/399: 127–135.

Knight G, Nelson WA (1999) An evaluation of characters obtained from life history studies for distinguishing New Zealand *Porphyra* species. J. Appl. Phycol. 11: 411–419.

Lewmanomont K, Ogawa H (1978) Study of the life cycle of *Porphyra* of Thailand. A report of the Faculty of Fisheries, Kasetsart University, Bangkok.

Lewmanomont K, Chittpoolkusol O (1993) Life cycle of *Porphyra vietnamensis* Tanaka and Pham-Hoang Ho from Thailand. Hydrobiologia 260/261: 397–400.

Mumford, TF Jr (1980) The reproductive biology *in vitro* of nine species of *Porphyra* (Bangiales, Rhodophyta) from Washington. J. Phycol. 16 (suppl.): 29.

Nelson, WA, Brodie, J, Guiry, MD (1999) Terminology used to describe reproduction and life history stages in the genus *Porphyra* (Bangiales, Rhodophyta). J. Appl. Phycol. 11: 407–410.

Notoya, M, Miyashita, A (1999) Life history, in culture, of the obligate epiphyte *Porphyra moriensis* (Bangiales, Rhodophyta). Hydrobiologia 398/399: 121–125.

Notoya, M, Nagaura, K (1998) Life history and growth of the epiphytic thallus of *Porphyra lacerate* (Bangiales, Rhodophyta) in culture. Algae 13: 207–211.

Ogawa H, Lewmanomont, K (1979) The *Porphyra* of Thailand – II. Distribution and seasonal occurrence of *Porphyra vietnamensis* Tanaka et P-H. Ho. Jpn. J. Phycol. 27: 95–98.

Provasoli, L (1968) Media and products for the cultivation of marine algae. In Watanabe, A & A Hattori (eds.), Cultures and Collections of Algae. Japanese Society of Plant Physiology, Tokyo.

Ruangchuay, R, Notoya, M (2003) Physiological responses of blade and conchocelis of *Porphyra vietnamensis* Tanaka et Pham-Hoang Ho (Bangiales, Rhodophyta) from Thailand in culture. Algae 18: 21–28.

Sahoo, DB, Nivedita, Debasish (2001) Seaweeds of Indian Coast, APH Publishing Corporation, New Delhi, India, pp. 283.

Sahoo, DB, Tang X, Yarish, C (2002) *Porphyra* – the economic seaweed as a new experimental system. Current Science 83: 1313–1316.

Santelices, B, Avila, M (1986) Bases biologicas para maximizar cosecha de "Luche" (*Porphyra columbina* Montagne) en Chile Central. Actas II congress. Algas Marina Chilenas : 201–211.

Sreeremulu, T (1952) On a *Porphyra* from Waltair Coast. Science and Culture 18: 285–286.

Tanaka, T, Ho, P-H (1962) Notes on some marine algae from Viet-Nam I. Memoirs of the Faculty of Fisheries, Kagoshima University 11: 24–40.

Tseng, CK (1983) Common seaweeds of China, Science Press, Beijing, pp. 316.

Tseng, CK, Sun, A (1989) Studies on the alternation of the nuclear phases of chromosome numbers in the life history of some species of *Porphyra* from China. Bot. Mar. 32: 1–8.

Tsujii, K, Kchikawa, T, Matusuura, Y, Kawamura, M (1983) Hypercholestrolemic effect of taurocyamine or taurine on the cholesterol metabolism in white rats. Sulphur Amino acids 6: 239–248.

Umamaheswara, Rao, M, Sreeramulu, T (1963) Vertical zonation and seasonal variation in the growth of *Porphyra* on Visakhapatnam coast. Current Science 32: 173–174.

Waaland, JR, Dickson, LG, Carrier, JE (1987) Conchocelis growth and photoperiodic control of conchospore release in *Porphyra torta* (Rhodophyta). J. Phycol. 23 : 399–406.

Waaland, JR, Dickson, LG, Duffield, ECS (1990) Conchospore production and seasonal occurrence of some *Porphyra* species (Bangiales, Rhodophyta) in Washington State. Hydrobiologia 204/205: 453–459.

Wu, C (1988) The seaweed resources of China. In Critchley AT, Ohno M (eds.), Seaweed resources of the world. Kanagawa International Fisheries Training Center. Japan International Cooperation Agency : 34–46.

Yarish, C, Chopin, T, Wilkes, R, Mathieson, AC, Fei, XG, Lu, S (1999) Domestication of nori for Northern America: The Asian experience. Bull. Aquacult. Ass. Canada 1: 11–17.

Yoshida, T, Notoya, M, Kikuchi, N, Miyata, M (1997) Catalogue of species of *Porphyra* in the world, with special reference to the type of locality and bibliography. Nat. Hist. Res. Special Issue 3: 5–18.

Journal of Applied Phycology (2006) 18: 287–294
DOI: 10.1007/s10811-006-9028-8

Spore adhesion and cell wall formation in *Gelidium floridanum* (Rhodophyta, Gelidiales)

Z.L. Bouzon[1,*], L.C. Ouriques[1] & E.C. Oliveira[2]
[1]*CCB, Universidade Federal de Santa Catarina, C. postal 476. CEP 88040-900 Florianópolis, SC, Brasil;*
[2]*Instituto de Biociências, Universidade de São Paulo*

*Author for correspondence: e-mail: zenilda@ccb.ufsc.br

Key words: cytochemical, *Gelidium*, germination, lectins, tetraspores

Abstract

The attachment of spores to a substratum is essential for their germination and, therefore, to the completion of the life cycle of the red algae. In most red algae, spores are liberated without a cell wall, within a sheath of mucilage which is responsible for their primary attachment. Utilizing fluorescent-labeled lectins, we identified carbohydrate residues and their locations in the mucilage and cell walls of spores of *Gelidium floridanum*. Cell wall formation and mucilage composition were studied with calcofluor, toluidine blue – O (AT-O), alcian blue (AB) and periodic acid-Schiff (PAS). In the mucilage we identified α-D mannose, α-D glucose, β-D-galactose, N-acetyl-glucosamine and N-acetyl-galactosamine. The first two sugar residues were not found in the cell wall of the germ tube but they were present on the rhizoid's cell wall indicating their importance to substrate adhesion. A cell wall is produced soon after the spore's attachment, beginning with a polar deposition of cellulose and its gradual spread around the spore as indicated by calcofluor. The cell wall matrix was positive to AB and metachromatic to AT-O, indicating acidic polysaccharides, while cellulose microfibrills were positive to PAS. A polar disorganization of the cell wall triggers the process of germination. As spores are the natural form of propagation of *Gelidium*, the understanding of the mechanisms of spore attachment may contribute to the cultivation of this valuable seaweed.

Introduction

The Gelidiales is a comparatively small order of red algae with about 140 species, some of which are important sources of bacteriological agar and agarose (Bailey, J.C. & D.W. Freshwater, unpublished). The order is characterized by a set of attributes, including a triphasic isomorphic life-history, and an intercalary carpogonium, which upon fertilization produces a gonimoblast connected to nutritive cells. The members of the order have agar in their cell walls and the spores germinate following a typical pattern known as "*Gelidium*-type" (Hommersand & Fredericq, 1988).

Spores are the natural form of dispersal in most red algae and their fixation to a substratum is a fundamental process in the development of the adult thallus (Chamberlain & Evans, 1981). Spores are the obvious link connecting the life-history phases of macroalgae, and their attachment is the first signal to triggering the metabolic changes that lead to germination.

Spores in red algae are released without a cell wall and they are surrounded by an optically transparent mucilage which is responsible for the first attachment to the substrate (Avanzini, 1989). This mucilage is composed of glycoproteins (Chamberlain & Evans, 1973; Pueschel, 1979) or sulphated polysaccharides (Ramus, 1974). Red algal polysaccharides have been characterized under light microscopy using different histochemical techniques (e.g. Gordon & McCandless, 1978; Cole et al., 1985; Rascio et al., 1991). However, most studies have been carried out with vegetative cells and not with spores. The mucilage that surrounds tetraspores of *Champia parvula* (C. Agardh)

Harvey reacts positively to sulphated and carboxylated polysaccharides (Apple et al., 1996).

Cytochemical methods based on the property of lectins to interact specifically with mono- and disaccharides have also been utilized to characterize cell wall compounds (Costas et al., 1993; Costas & López-Rodas 1994; Hori et al., 1996). Lectins conjugated to fluorescent dyes have been used to detect carbohydrate residues in mucilage and cell walls of microalgae (von Sengbusch et al., 1982; von Sengbusch & Müller, 1983; Callow, 1985).

In our study, we utilized cytochemical techniques to characterize the polysaccharides that participate in the attachment and cell-wall formation in the initial phases of tetraspore germination of *Gelidium floridanum* W. R. Taylor, an agarophytic alga of commercial importance in Brazil.

Material and methods

Tetrasporophytic specimens of *Gelidium floridanum* were collected at Ponta do Sambaqui, Ilha de Santa Catarina, in November 2003. Branch tips with tetrasporangia were placed on microscope slides in a petri dish with sterile seawater in the dark at 23 °C. After spore release, the branches were removed and the slides were exposed to fluorescent light (40 μmol photons m^{-2} s^{-1}; 14:10 h light: dark photoperiod), and kept at 23 °C. Periodically, slides, with attached spores, were fixed in 2.5% paraformaldehyde in phosphate buffer 0.2 M (pH 7.2), by dropping fixative on the slides that were covered with parafilm, during 5 h at 4 °C. After that, the slides were washed twice in phosphate buffer for 10 min.

Histochemical staining

Periodic acid-Schiff (PAS) was used to identify neutral polysaccharides, and the control consisted of staining the material without pre-treatment with the periodic acid (Gahan, 1984). Toluidine Blue (AT-O) was used to identify acid polysaccharides through a metachromatic reaction (Gordon & McCandless, 1978). Alcian Blue (AB) was used to identify acid polysaccharides according to Ravetto (1964). Coomasie Brilliant Blue was used to identify proteins according to Gahan (1984). To study the cell wall deposition, the spores were incubated in sterilized seawater containing 10 μg ml^{-1} of Calcofluor White M2R for 15 min (Kim & Fritz, 1993).

Probing with FITC-lectins

Spores were incubated with 100 μg ml^{-1} of FITC-lectins in 0.6 M of sorbitol with 10 mM of $CaCl_2$ diluted in distilled water for 1 h and washed with deionized water. The control was made with FITC-lectin, 0.2 M of glucose, 0.4 M of sorbitol and 10 mM of $CaCl_2$ for 15 min. (Apple et al., 1996). The tested lectins were Con- A, RCA, UEA, WGA and SBA. Chemicals were supplied by Sigma (Saint Louis, USA). All preparations were observed with an epifluorescence microscope with the adequate filters for FITC and calcofluor and photographed with Fujichrome ISO 400.

Results

Tetraspores of *Gelidium floridanum* were liberated without a cell wall within a mucilage layer which is responsible for their primary attachment to the substratum. On release, tetraspores were spherical and 26–30 μm in diameter. The mucilage layer was composed of a mixture of acid polysaccharides as indicated by its positive reaction to AB and its violaceous metachromasia with AT-O (Figures 1 and 2). The spore's surface did not react with CBB and PAS, indicating the absence of neutral proteins and polysaccharides in the mucilage layer (Figures 3 and 4).

Soon after attachment, a thin cell wall was produced around the spore. The germination process was characterized by a polar evagination which gradually elongated, being pushed by the migrating cytoplasm, and giving rise to the germ tube. At this stage, a treatment with AB showed the presence of acid polysaccharides in the cell wall of the germ tube (Figure 5). When stained with AT-O, the metachromasia is restricted to a thin layer around the germ tube differing from what was seen in the mucilage surrounding the spore (Figure 6).

Neither the mucilage nor the cell wall reacted with CBB, indicating the absence of proteins in this region (Figure 7). The presence of neutral polysaccharides in the cell wall was shown by a positive reaction to PAS (Figure 8). As the germination proceeded, the tube became divided into several cells, yielding the same reactions as at the beginning of the germination process (Figures 9–12). The distal cell of the tube elongated and gave rise to the first rhizoid (Figures 13–16) which reacted like the other cells, giving, however, a stronger reaction with AB and showing a higher concentration of sulphated polysaccharides (Figure 13). Its weaker reaction with the PAS showed a reduction in the

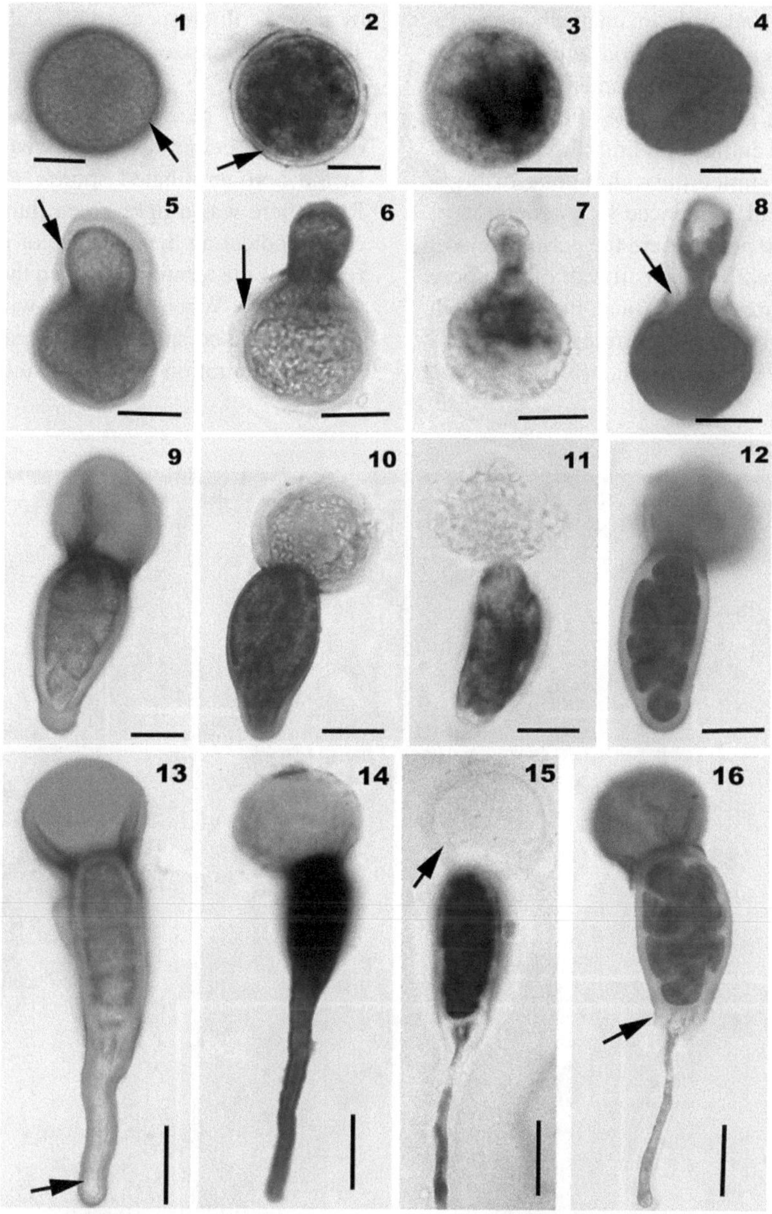

Figures 1–16. Initial germination phases of tetraspores of *Gelidium floridanum* under light microscopy. *Figures 1–4.* Histochemical reactions showing the mucilage and the cell wall stained with AB, AT-O CBB and PAS, demonstrating the presence of neutral and acid polysaccharides. Note the positive reaction to AB (arrow) and its metachromasia with AT-O (arrow). *Figure 5.* First stage in the formation of the germ tube; the layer of mucilage stained with AB is pointed (arrow). *Figure 6.* Elongation of the germ tube showing the migration of the protoplasm stained with AT-O and a metachromatic halo over the body of the spore (arrow). *Figure 7.* Developing germ tube stained with CBB. *Figure 8.* Cell surface stained with PAS showing a rupture of the wall (arrow). *Figures 9–12.* Tetrasporelings showing the cellularization of the germ tube stained with AB, AT-O, CBB and PAS. Scale bar = 10 μm. Figures 13–16 Tetrasporelings of *Gelidium floridanum* bearing rhizoids. *Figure 13.* Tetrasporelings stained with AB showing strong positive reaction around the entire body, including the rhizoid (arrow). *Figure 14.* Tetrasporeling stained with AT-O. *Figure 15.* Tetrasporeling stained with CBB showing the negative reaction on spore body (arrow). *Figure 16.* Tetrasporeling stained with PAS: observe that mucilage is weakly positive (arrow). Scale bar = 10 μm.

amount of neutral polysaccharides in the rhizoidal cell (Figure 16).

The absence of a cell wall in unattached spores was evidenced by a negative reaction with calcofluor (Figure 17). After the primary attachment there was a deposition of cellulose starting at one of the poles and gradually covering the entire sporeling (Figures 18 and 19). The germination started through a polar disorganization of the cell wall, evidenced by a reduction of the fluorescence, at the point where the germ tube will later be formed (Figure 20). The cell wall of the spore became thinner (Figures 21–23) and the tube went through a series of cell divisions forming smaller cells with high fluorescence using calcofluor (Figures 24 and 25). The distal cell of the tube elongated and gave rise to the rhizoidal region, which had little cellulose as shown by a weak fluorescence with calcofluor (Figure 25). Table 1 summarizes the results of the histochemical tests.

The lectin reactions are shown in Table 2. Three of the five lectins tested showed a positive reaction. In the non-germinated spores, after incubation with RCA there was a light-green fluorescence of the mucilage indicating the presence of β-D-galactose as the residue of the terminal sugar in the adhesive substance (Figure 26). When incubated with Con-A FITC, the non-germinated spores fluoresced strongly showing a high concentration of α-D-mannose and α-D-glucose

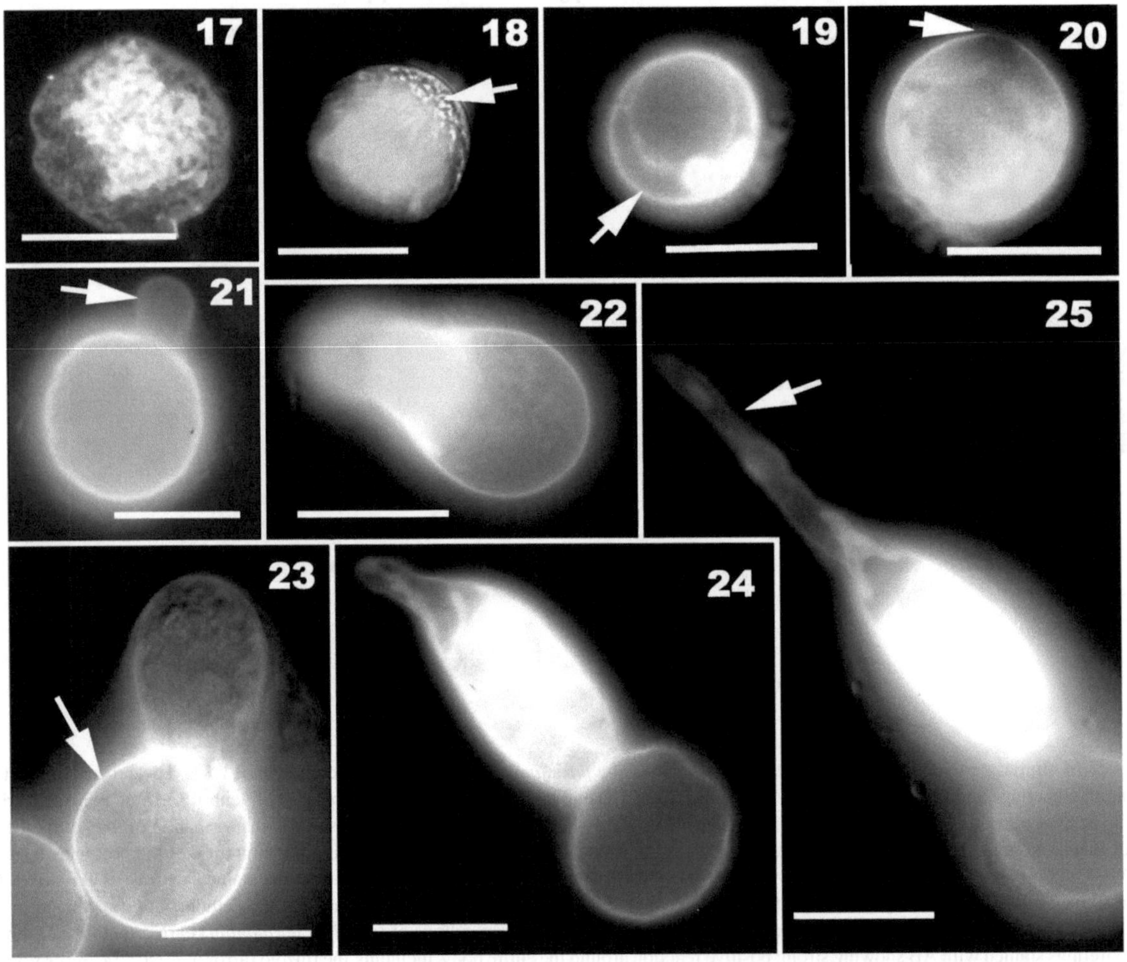

Figures 17–25. Tetraspores and tetrasporelings of *Gelidium floridanum* in different stages of development stained with calcofluor. *Figure 17.* Non-germinated tetraspore without cell wall. *Figure 18.* Beginning of cell deposition on a pole (arrow). *Figure 19.* Tetraspore surrounded by the cell wall. *Figure 20.* Polar disorganization of the cell wall signalling the start of the germination process (arrow). *Figures 21–23.* Germ tube in varying stages of development covered by a thin cell wall. *Figures 24–25.* Tetrasporelings showing a very thin cell wall on the rhizoids (arrow). Scale bar = 20 μm.

Table 1. Reactions of *G. floridanum* tetraspores to different treatments: + + + strongly positive, ++ positive, + weakly positive; – negative.

			Germinated spore		
Reactives	Substances labeled	Spore	Body of the spore	Germ tube	Rhizoid
Alcian blue	Acid polysaccharides	+++	++	+	+++
Toluidine blue	Sulphated polysaccharides	++	+	+	+
Coomassie brilliant blue	Proteins	–	–	–	–
PAS	Neutral polysaccharides	–	–	+	–
Calcofluor	Cellulose	+	++	++	+

Table 2. Fluorescence of the mucilage that surrounds the tetraspores of *Gelidium floridanum* with FITC conjugated with some lectins: + + + strongly positive, ++ positive, + weakly positive; – negative.

Lectin	Specificity	Fluorescence
Con-A	α-D-mannose, $\alpha\alpha\alpha$-D-glucose	+++
RCA	β-D-galactose	++
UEA	α-L-fucose	–
WGA	N-acetyl-α (1, 4)-D-glucosamine	+
SBA	α or β N-acetylgalactosamine, α or β-D-galactose	–

in the mucilage (Figure 27). The mucilage also fluoresced, but only slightly, when incubated with WGA, showing a low concentration of N-acetyl-alpha-(1,4)-D-glucosamine (Figure 28). The other tested lectins (UEA and SBA) did not fluoresce.

In the cell wall of the germ tube, sugar residues were not detected with the lectins employed. However, this result may have been masked by the intense orange auto-fluorescence of the cells of the germ tube (Figures 29–31). At this stage fluorescence was only observed at the distal pole of the germ tube when incubating with Con-A (Figure 30). Carbohydrate residues labelled by Con-A, absent in the germ tube, were detected in the rhizoidal region (Figures 32 and 33). The presence of α-D mannose and α-D glucose in the mucilage and in the rhizoid indicate a relation of these carbohydrates with the adhesion mechanisms.

Discussion

Red algal spores usually do not germinate unless they attach to a substratum. As the spores are liberated without a cell wall, the primary attachment is mediated by the mucilage that surrounds the spores, which is produced during the sporogenesis (Chamberlain & Evans, 1973; Pueschel, 1979). The production of mucilage seems to be associated with the large quantity of dictyosomes that are seen during the ontogeny of the spores. This mucilage has a role in the release of the spores and might also have a role in the protection of the spores until the formation of the cell wall.

The strong positive reaction of the mucilage of *G. floridanum* tetraspores with AB pH 1.0 and AT-O indicated an abundance of sulphated polysaccharides. The adhesive nature of sugars conjugated with proteins has been demonstrated in seaweeds (Chamberlain & Evans, 1973; Pueschel, 1979). However, our tests with CBB did not show the presence of proteins in the *G. floridanum* mucilage. Similar results were obtained by Chamberlain and Evans (1973, 1981) with non-liberated spores of *Ceramium* sp. and by Apple et al. (1996) with spores of *Champia parvula* showing that protein, if indeed it is present at all, is not detected by histochemical methods.

Our results with calcofluor white show that the cellulose deposition begins soon after the spores' adhesion to the substratum. Northcote and Pickett-Heaps (1965) have shown that the synthesis of cellulose is mediated by the Golgi bodies, and Villemez et al. (1968) have shown that it occurs on the plasmatic membrane. The presence of sulphated polysaccharides in the spores' mucilage and in the cell wall of the sporelings was shown by the metachromasia with

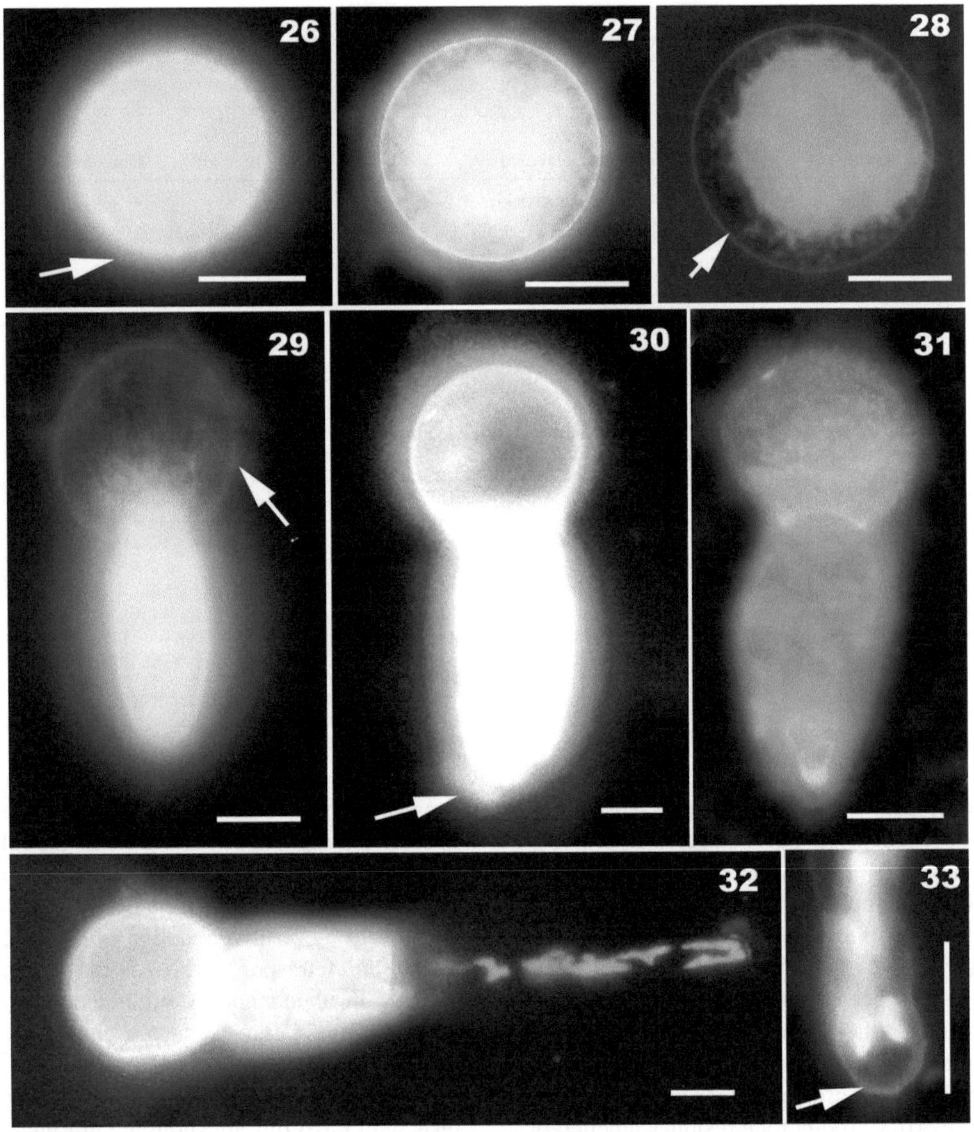

Figures 26–33. Tetraspores and tetrasporelings of *Gelidium floridanum* incubated with lectins conjugated with FITC. *Figure 26.* Non-germinated spore incubated with RCA-FITC showing a light fluorescence in the mucilage. *Figure 27.* Non-germinated spore with Con-A-FITC showing strong fluorescence in the mucilage. *Figure 28.* Another non-germinated spore incubated with WGA-FITC weak fluorescence is seen as a halo over the mucilage (arrow). *Figures 29-31.* Germinated spores incubated with RCA, Con-A and WGA, respectively. Note the fluorescence at the extremity of the germ tube in *Figure 30* with Con-A (arrow). *Figure 32.* Tetrasporeling showing a rhizoid fluorescing with Con-A-FITC. *Figure 33.* Detail of the extremity of the rhizoid fluorescing with Con-A-FITC (arrow). Scale bar = 10 μm.

AT-0. This is expected in agarophytes, where the presence of agar, a complex polymer with a neutral and a sulphated fraction, is well known. Our results show that sulphated polysaccharides are present in the adhesive mucilage, particularly on the rhizoid as well as in the amorphous matrix of the cell wall. Apple et al. (1996) noted that in *C. parvula,* the composition of the adhesive mucilage on the rhizoids appeared to be different from the mucilage that surrounds the spores.

The labelling of the mucilage by lectins conjugated with fluorescent dyes indicates the presence of specific sugars in the adhesive mucilage of algal and fungal spores (Apple et al., 1996). Apparently lectins have a greater affinity for sugars linked to proteins than for sugars alone (von Sengbusch et al., 1982). This suggests that the sugars present in the adhesive mucilage of the spores of *G. floridanum* are components of glycoproteins. Although we did not detect any proteins with CBB, they may be present in small amounts or blocked by polysaccharides. Three of the lectins conjugated with FITC complexed with different sugars on the spores' mucilage (Table 1). The presence of β-D-galactose was detected by FITC-RCA, but only slightly, suggesting that this sugar is not the main component of the mucilage. On the other hand, non-germinated spores fluoresced strongly when incubated with FITC-ConA, showing a high concentration of α-D-mannose and α-D-glucose in the mucilage, although Con-A is specific for α-D-mannose and less so for α-D-glucose (von Sengbusch et al., 1982; Walko et al., 1987). The mucilage showed a fluorescence limited to a thin layer when it was incubated with WGA, indicating a low concentration of N-acetyl-α-(1, 4)-D-glucosamine. No fluorescence was shown when the spores were incubated with UEA or SBA.

No sugar residues were detected on the germ tube by the tested lectins, although the fluorescence may have been masked by the intense auto-fluorescence of the germ tube cells. However, the presence of α-D-mannose and/or α-D-glucose was detected by Con-A at the distal region of the rhizoids. The presence of α-D mannose and/or α-D glucose in the mucilage and in the rhizoids suggests the participation of these sugars in the mechanism of attachment of *G. floridanum* spores.

Our results suggest that: (1) the attachment and germination of the tetraspores of *G. floridanum* depends on the compounds of the extra-cellular matrix formed during the spores' ontogeny; (2) the mucilage that surrounds the spores facilitates its primary interaction with the substratum; (3) the mucilage may have a role in keeping the spores apart from each other, and in protecting them before the formation of the cell wall, functioning as a buffering layer between the naked spore and the substratum, and (4) the same sugars are involved in the adhesion of the naked spores as in adhesion of the rhizoids to the substratum.

Acknowledgments

We are very grateful to the Conselho Nacional de Desenvolvimento Científico e Tecnológico and to the Fundação de Amparo à Pesquisa do Estado de São Paulo.

References

Apple ME, Harlin MM, Norris JH (1996) Characterization of *Champia parvula* (Rhodophyta) tetraspore mucilage end rhizoids with histochemical stains and FITC-labeled lectins. Phycologia 35:245–252

Avanzini A (1989) La ultraestructura de las esporas de Rhodophyta. Insula 19:7–10

Callow JA (1985) Sexual recognition and fertilization in brown algae. J Cell Sci 2:219–232

Chamberlain AHL, Evans LV (1973) Aspects of spore production in the red alga *Ceramium*. Protoplasma 76:139–159

Chamberlain AHL, Evans LV (1981) Chemical and histochemical studies on the spore adhesive of *Ceramium*. In: Fogg, GE, Jones, WE (eds) Proceedings of the 8th International Seaweed Symposium, Menai Bridge, pp 539–542

Cole KM, Park CM, Reid PE, Sheath RG (1985) Comparative studies in the cell walls of sexual and asexual *Bangia atropurpurea* (Rhodophyta): I. Histochemistry of polysaccharides. J Phycol 21:585–592

Costas E, López-Rodas V (1994) Identification of marine dinoflagellates using fluorescent lectins. J Phycol 30:987–990

Costas E, Gonzalez-Chavarri E, Aguilera E, Gonzalez A, Gil S, Lopez Rodas V (1993) Use of lectins to recognize and differentiate unicellular algae. Bot Marina 36:1–4

Gahan PB (1984) Plant histochemistry and cytochemistry: an introduction. Academic Press, London

Gordon-Mills EM, McCandless EC (1978) Carrageenans in the cell walls of *Chondrus crispus* Stack. (Rhodophyceae, Gigartinales): III. Metachromasia and the topootical reaction. Phycologia 17:95–104

Hommersand MH, Fredericq S (1988) An investigation of cystocarp development in *Gelidium pteridifolium* with a revised description of the Gelidiales (Rhodophyta). Phycologia 27:254–272

Hori K, Ogata T, Kamiya H, Mimuro M (1996) Lectin-like compounds and lectin receptors in marine microalgae: hemagglutination and reactivity with purified lectins. J Phycol 32:783–790

Kim GH, Fritz L (1993) Ultrastucture and cytochemistry of early spermatangial development in *Antithamnion nipponicum* (Ceramiaceae, Rhodophyta). J Phycol 29:797–805

Northcote DH, Pickett-Heaps JD (1965) A function of the Golgi apparatus in polysaccharide synthesis and transport in the root cap cells of wheat. Biochem J 98:159–167

Pueschel CM (1979) Ultrastructure of tetrasporogenesis in *Palmaria palmata* (Rhodophyta). J Phycol 15:409–424

Ramus J (1974) In vitro molybdate inhibition of sulfate transfer to *Porphyridium* capsular polysaccharide. Plant Physiol 54:945–949

Rascio N, Mariani P, Vecchia FD, Trevisan R (1991) The vegetative thallus of *Pterocladia capillacea* (Gelidiales, Rhodophyta): II. Pit connections. Bot Marina 34:187–194

Ravetto C (1964) Alcian blue-alcian yellow: a new method for the identification of different acidic groups. J Histochem Cytochem 12:44–45

Villemez CL, MacNab JC, Albersheim P (1968) Formation of plant cell wall polysaccharides. Nature 218:878–880

von Sengbusch P, Müller U (1983) Distribution of glycoconjugates at algal cell surfaces as monitored by FITC–conjugates lectins. Studies on selected species from Cyanophyta, Pyrrhophyta,

Raphidophyta, Euglenophyta, Chromophyta and Chlorophyta. Protoplasma 114:103–113

von Sengbusch P, Mix M, Wachholz I, Manshard E (1982) FITC-labeled lectins and calcofluor white ST as probes for the investigation of the molecular architecture of cell surfaces. Studies on Conjugatophycean species. Protoplasma 111:38–52

Walko RM, Furtula V, Nothnagel EA (1987) Analysis of labeling of plant protoplast surface by fluorophore-conjugated lectins. Protoplasma 141:33–46

Journal of Applied Phycology (2006) 18: 295–299
DOI: 10.1007/s10811-006-9031-0

Forecasting infections of the red rot disease on *Porphyra yezoensis* Ueda (Rhodophyta) cultivation farms

Chan Sun Park[1,*], Makoto Kakinuma[2] & Hideomi Amano[2]
[1]*Department of Marine Resources, Mokpo National University, 61 Torim-ri, Chounggye-myon, Muan-gun, Jeonnam 534–729, Korea;* [2]*Laboratory of Marine Biochemistry, Faculty of Bioresources, Mie University, 1515 Kamihama, Tsu, Mie 514–8507, Japan*

*Author for correspondence: e-mail: cspark85@mokpo.ac.kr; fax: 82-61-452-8875

Key words: *Porphyra yezoensis*, cultivation, red rot disease, *Pythium porphyrae*, forecast, zoospores

Abstract

Pythium porphyrae is a fungal pathogen responsible for red rot disease of the seaweed *Porphyra* (Rhodophyta). Infection forecasts of *Porphyra* by *P. porphyrae* were estimated from the epidemiological observations of *Porphyra* thalli and numbers of zoospore of *P. porphyrae* in laboratory and cultivation areas. Four features of forecasting infections were determined by relating zoospore concentrations to the incidence of thallus infection; infection (in more than 1000 zoospores L^{-1}), microscopic infection [less than 2 mm in diameter of lesion (in from 2000 to 3000 zoospores L^{-1})], macroscopic infection [more than 2 mm in diameter of lesion (in from 3000 to 4000 zoospores L^{-1}), and thallus disintegration (in more than 4000 zoospores L^{-1}). High zoospore concentrations led to more infection. The tendency that zoospore concentration of *P. porphyrae* increased with the rate of infection of *Porphyra* thalli was generally observed in forecasting infections in both the laboratory and in cultivation areas. Based on the *Porphyra* cultivation areas, the accuracy and consistency of forecasting infections suggest that this method could be employed to manage and control red rot disease.

Introduction

Red rot disease (Akagusare), caused by *Pythium porphyrae* (Oomycetes), is one of the most destructive fungal diseases of *Porphyra* and can seriously reduce both yield and quality in *Porphyra* farms every year (Amano et al., 1995). The causative organism of this disease is spread by zoospores released into seawater. After the zoospores attach to *Porphyra* thalli, they form hyphae to penetrate the cell of *Porphyra* thalli and to kill the alga within a few days (Sasaki & Sato, 1969; Fujita, 1990). There is then a massive release of new zoospores. Thus, the pathogen can survive as an endophyte, an epiphyte, or latent infections. The movement of infected asymptomatic *Porphyra* or *Porphyra* parts could serve as a means of introducing this serious disease into other geographic regions (Fujita & Zenitani, 1977; Kerwin et al., 1992). Therefore, it is important to be able to quickly assess the amount of zoospores in seawater prior to an outbreak of the disease at *Porphyra* farms.

To provide efficient protection of *Porphyra* farms from red rot disease, the development of an epidemiologically-based forecasting system for timely preventive application has been proposed by Park et al. (2001a). For example, a forecasting system can detect the disease more quickly than a routine such as visual observations (Sakaguchi et al., 2001; Uppalapati et al., 2001), and would allow farmers to use less fungicides (an organic acid-seawater mixture; pH about 2). This would give the opportunity to reduce the frequency of their treatments, thus lowering risks for the environment, yet still providing adequate or improved protection to *Porphyra* farms.

In the previous study, we designed the species-specific polymerase chain reaction (PCR) primers PP-1 (5′-TGTGTTCTGTGCTCCTCTCG-3′) and PP-2 (5′-CCCAAATTGGTGTTGCCTCC-3′) based on internal transcribed spacers (ITS) rDNA sequences of *P. porphyrae* (Park et al., 2001b). We showed that it was possible to detect a single zoospore of *P. porphyrae* using these primers and to analyze quantitatively zoospores of *P. porphyrae* in the *Porphyra* cultivation farms by competitive PCR (Park et al., 2001a). However, we have yet to determine how many zoospores must be present in the seawater column to initiate an outbreak of the red rot disease.

The objective of this study was to describe the forecasting of infections of red rot disease from zoospore concentrations in the seawater of *Porphyra* cultivation farms by applying the results of susceptibility of *Porphyra* thalli infested artificially with *P. porphyrae* zoospores in the laboratory, and the relationships between the epidemiology of *Porphyra* thalli and the amount of *P. porphyrae* zoospores in seawater during the growing seasons at *Porphyra* farms.

Materials and methods

The incidence and expansion rate of red rot disease to *P. yezoensis* thalli were determined by d artificially infection with zoospores of *P. porphyrae*. Blade discs (4 mm in diameter) of *P. yezoensis* were cut with a cork-borer from uninfected healthy blades cultured in the laboratory. Twenty discs were placed in each beaker containing 1 L of sterile seawater (15 °C) with appropriate salinity (32‰) and pH (7.5). To obtain zoospores, five corn meal agar discs (6 mm in diameter) containing the edge of the *P. porphyrae* growth circle were transferred to Arasaki B liquid medium of 100 mL (Arasaki et al., 1968) and maintained for a further 4 days at 20 °C. Hyphae were gently shaken at 100 rpm on an orbital shaker (Iuchi Co. Ltd., Osaka, Japan) to release zoospores. After about 15 h, zoospores were discharged. The mycelia left were filtered out with a 20 μm nylon mesh. After zoospore concentration was determined by haemocytometer, about 10, 100, 500, 1000, 2000, and 4000 zoospores L^{-1} were inoculated.

The incidences of blade discs infected by *P. porphyrae* were determined by naked eye and using an inverted light microscope every day during the seven days after inoculation of the zoospores into the beaker containing *P. yezoensis* thalli discs. The disease incidence was categorized using an index as follows. No infection, -; infection, + and microscopic infection, ++ (less than 2 mm in diameter of lesion). The expansion of area per lesion on infected discs was determined after staining with 1% Erythrosin B solution using the microscope's digital camera computer system (DP50-A Olympus, Tokyo, Japan). Rates of disease expansion in terms of lesion area in infected discs were calculated from the mean area of lesions per disc measured in a sampling unit of three beakers, at the time of assessment. Three replicate sets of the treatments were completed.

The estimation of zoospores of *P. porphyrae* in the *Porphyra* cultivation areas was conducted at farms which use the floating net cultivation system in Wando, Korea (34°122′11″N; 127°21′35″E) during successive *Porphyra* cultivation seasons (field trial 1 in December 2002, field trial 2 in December 2003) using the methods described in the previous study (Park et al., 2001a). At the same time, to examine the relationship between the number of zoospores and incidence levels of red rot disease, the *Porphyra* thalli from the *Porphyra* nets were sampled and observations were carried out as described above. The experiment was conducted using a completely randomized design and each treatment was replicated in the three stations.

Results and discussion

The outbreak and the rates of disease incidence of red rot disease from zoospores infested artificially in the laboratory experiments were investigated (Figure 1). The time taken before disease outbreak differed with zoospore concentrations. In the case of 2000 and 4000 zoospores L^{-1}, infection occurred approximately 12 h

Number of zoospores L^{-1} \ Days from incubation	0	1	2	3	4	5	6	7
4000	+	+	++	++	++	++	++	++
2000	+	+	+	++	++	++	++	++
1000	-	+	+	+	++	++	++	++
500	-	-	+	+	+	++	++	++
100	-	-	-	+	+	+	++	++
10	-	-	-	-	+	+	+	++

Figure 1. Features of infection of *Porphyra yezoensis* thalli artificially infested by zoospores of *Pythium porphyrae* [- : no infection; + : infection; ++ : microscopic infection (< 2 mm in diameter of lesion).

after zoospore inoculation. In the other hand, in the case of 10 and 100 zoospores, it took three to four days. The time taken to microscopic infection (less than 2 mm in diameter of lesion) after disease infection did not differ with zoospore concentrations. In 10 to 4000 zoospores L^{-1}, microscopic infection was observed in two and three days after disease infection. When the concentrations of fungal zoospores were high, the prevalence of disease was greater at higher concentrations of fungal zoospores than at lower zoospore concentrations. This result indicates that the outbreak of red rot disease was affected by the concentration of fungal zoospores. In the present study, on *in vivo* infection assays, the *Porphyra* thalli infested with high concentration of zoospores showed faster mycelial proliferation than either the *Porphyra* thalli infested with low concentration of zoospores and non-zoospore controls. Similar results were reported by Sakaguchi et al. (2001). These results suggest a possible critical threshold level that poses no disease threat to *Porphyra* farms during the growing season, if the number of *P. porphyrae* zoospores have been exactly estimated in the seawater at these farms.

Figure 2 shows the rates of expansion of lesion area per day in the laboratory experiments. The regression equation between rate of expansion of lesion area per lesion and time is given as: $y = 2689.4X^2 + 1006.8X$ with an r^2 value of 0.96. The expansion of lesion area per lesion was more than 40 000 μm^2 at 4 days after infection. The rate of expansion of lesion area per day was approximately 10 000 μm^2. Lesion area per lesion expanded slowly in the early stage of infection, but rapidly in the late stage of infection. As shown in Figure 1, the time taken to microscopic infection (less than 2 mm in diameter of lesion) was between three and four days after disease infection. Differences in

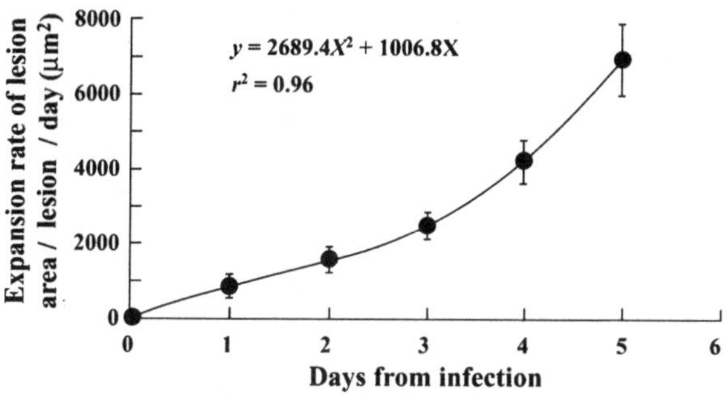

Figure 2. Expansion rates of lesion area per lesion on infected *Porphyra yezoensis* thalli with respect to time after artificial infection by zoospores of *Pythium porphyrae*.

the number of infection sites and the expansion of lesions on different thalli accounted for the difference of susceptibility of the initial steps of host-pathogen reactions of *Porphyra* thalli. Involvement of nutritional and/ or metabolic status of the host plant in the growth of *P. porphyrae* or degree on colonization of *Porphyra* thalli have been described (Arasaki, 1947; Kato et al., 1973; Uppalapati & Fujita, 2001).

In the Wando area, *Porphyra* cultivation starts between early October and late September every year when the temperature of seawater drops to below approximately 20 °C. Zoospores of *P. porphyrae* were detected on 5 December for the first time in both field trials 1 and 2, and the concentration of zoospore was approximately 50 zoospores L^{-1} (Figure 3). Thereafter, the number of zoospores in the seawater gradually increased depending on growing periods of *Porphyra*, and the concentration of zoospores increased rapidly until late December, when it peaked at approximately 4500 zoospores L^{-1}. Even though seawater sampling in the present study was done at one-day intervals, data analyses were based on a convenient, discrete daily period, which was necessary to represent continuous epidemic processes, such as zoospore production and dispersal, and subsequent infection. The results of this study clearly showed that the degree of infection of red rot disease in the *Porphyra* cultivation areas was related to the number of *P. porphyrae* zoospores.

As in the results of epidemiological observation, the infection of *Porphyra* thalli was found on 11 to 13 December for the first time in field trial 1 and 2. The number of zoospores of *P. porphyrae* at that time was approximately 1000 zoospores L^{-1}, and the prevalence of infected thalli (microscopic infection) was visible by microscopy on 14 to 17 December when the number of zoospores was more than 2000 zoospores L^{-1}. Macroscopic infection was clearly visible to the naked eye between 17 and 20 December when the number of zoospores was more than 3000 zoospores L^{-1}. Thus, the time taken from microscopic infection to macroscopic infection (more than 2 mm in diameter of lesion) was three to four days, and the disintegration of *Porphyra* thalli by disease infection occurred six to seven days after macroscopic infection. These results are similar to those of Amano et al. (1996), observed during *Porphyra* cultivation periods, using a monoclonal antibody.

Some methods, including acid treatment, freezing, and exposure of nets, are used to control the spread of red rot disease on *Porphyra* farms. These treatments,

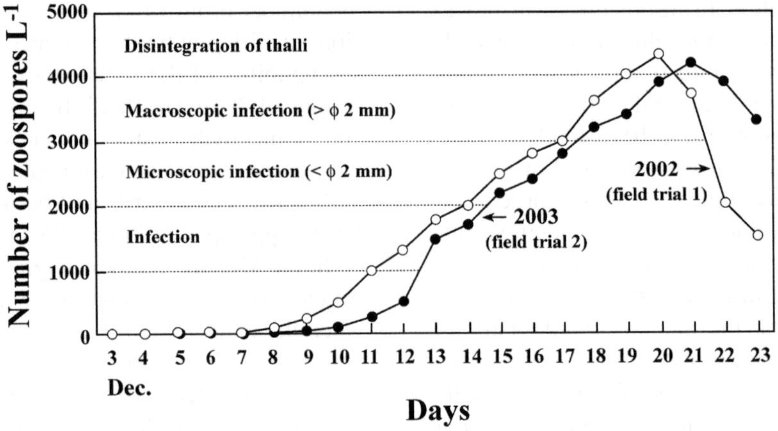

Figure 3. The number of zoospores of *Pythium porphyrae* per litre of seawater and the incidences of observed infection of *Porphyra yezoensis* thalli in *Porphyra* cultivation areas in Wando, Korea during successive *Porphyra* cultivation seasons (field trial 1; December 2002, field trial 2; December 2003) using the methods described in the previous study by Park et al. (2001a).

however, are only effective if they are implemented at the early stages of infection (Arashima et al., 1994; Uppalapati & Fujita, 2000). The results described in this paper are probably more easily used in the field, in *Porphyra* cultivations, because they are based both on data obtained from the laboratory and from cultivation areas. The present study shows that, based on the occurrence of zoospores of *P. porphyrae,* infections of red rot disease on *Porphyra* farms can be detected and forecast earlier than was possible with previous detection methods. It could predict the disease nine to ten days before the microscopic infections could be observed by conventional light microscopy and twelve to thirteen days before the macroscopic infections (more than 2 mm in diameter of lesion) could be observed by the naked eye. Thus, on the basis of the number of zoospores in the seawater, the forecasting system of recognizable thresholds can be applied on a cultivation-by-cultivation basis so that farmers can take appropriate measures where necessary. This means that infections of red rot disease on *Porphyra* farms can be more effectively forecast than using previous methods such as observation by the naked eye.

This study shows that by using competitive polymerase chain reaction techniques it is possible to predict the potential occurrence of fungal infection and the critical threshold levels that pose disease threats to *Porphyra* cultivation. This technique is now being extended as a more simple way to prevent or control red rot disease in seawater at *Porphyra* cultivation farms.

Acknowledgments

This work was supported by grant No. R08-2003-000-10074-0 from the Basic Research Program of the Korea Science & Engineering Foundation.

References

Amano H, Sakaguchi K, Maegawa M, Noda H (1996) The use of a monoclonal antibody for the detection of fungal parasite, *Pythium* sp., the causative organism of red rot disease, in seawater from *Porphyra* cultivation farms. Fish. Sci. 62: 556–560.

Amano H, Suginaga R, Arashima K, Noda H (1995) Immunological detection of the fungal parasite, *Pythium* sp.; the causative organism of red rot disease in *Porphyra yezoensis*. J. Appl. Phycol. 7: 53–58.

Arasaki S (1947) Studies on red rot of *Porphyra tenera*. Nippon Suisan Gakkaishi 13: 74–90.

Arasaki S, Akino K, Tomiyama T (1968) A comparison of some physiological aspects in a marine *Pythium* on the host and on the artificial medium. Bulletin of Misaki Marine Biology Institute of Kyoto University 12: 203–206.

Arashima K, Amano H, Suginaga R, Noda H (1994) Preparation of monoclonal antibodies against the fungal parasite, *Pythium* sp., the causative organism of laver red rot. Fish. Sci. 60: 481–482.

Fujita Y (1990) Introduction to Applied Phycology. In Akatsuka I, Sheath R.G (eds.), Diseases of cultivation *Porphyra* in Japan. SPB Academic Publishing, The Netherlands: 177–190.

Fujita Y, Zenitani B (1977) Studies on pathogenic *Pythium* of laver red rot in Ariake Sea farm-II. Experimental conditions and nutritional requirements for growth. Nippon Suisan Gakkaishi 43: 89–95.

Kato S, Watanabe T, Sato Y (1973) Studies on the diseases of cultural *Porphyra*-VII. A comparison of physiological properties among the different isolates of the causal fungus of the red wasting disease. Nippon Suisan Gakkaishi 39: 859–865.

Kerwin JL, Johnson LM, Whisler HC, Thininga AR (1992) Infection and morphologensis of *Pythium marinum* in species of *Porphyra* and other red algae. Can. J. Bot. 70: 1017–1024.

Park CS, Kakinuma M, Amano H (2001a) Detection and quantitative analysis of zoospores of *Pythium porphyrae*, causative organism of red rot disease in *Porphyra*, by competitive PCR. J. Appl. Phycol. 13: 433–441.

Park CS, Kakinuma M, Amano H (2001b) Detection of red rot disease fungi *Pythium* spp. by polymerase chain reaction. Fish. Sci. 67: 197–199.

Sakaguchi K, Park CS, Kakinuma M, Amano H (2001) Effects of varying temperature, salinity, and acidity in the treatment of *Porphyra* infected by red rot disease. Suisanzoshoku 49: 77–83.

Sasaki M, Sato S (1969) Composition of medium and cultural temperature of *Pythium* sp., a pathogenic fungus, of the 'Akagusare' disease of cultivated *Porphyra*. Bulletin of Tohoku Region National Fisheries Research Institute 29: 125–132.

Uppalapati SR, Fujita Y (2000) Carbohydrate regulation of attachment, encystment, and appressorium formation by *Pythium porphyrae* (Oomycota) zoospores on *Porphyra yezoensis* (Rhodophyta). J. Phycol. 36: 359–366.

Uppalapati SR, Fujita Y (2001) The relative resistance of *Porphyra* species (Banggiales, Rhodophyta) to infection by *Pythium porphyrae* (Peronosporales, Oomycota). Bot. Mar. 44: 1–7.

Uppalapati SR, Kerwin JL, Fujita Y (2001) Epifluorescence and scanning electron microscopy of host-pathogen interactions between *Pythium porphyrae* (Peronosporales, Oomycota) and *Porphyra yezoensis* (Bangiales, Rhodophyta). Bot. Mar. 44: 139–145.

Journal of Applied Phycology (2006) 18: 301–306
DOI: 10.1007/s10811-006-9032-z

Occurrence of *Polysiphonia* epiphytes in *Kappaphycus* farms at Calaguas Is., Camarines Norte, Phillippines

A.Q. Hurtado[1,2,*], A.T. Critchley[3], A.Trespoey[3] & G. Bleicher Lhonneur[3]
[1]*Aquaculture Department, Southeast Asian Fisheries Development Center (SEAFDEC), Tigbauan, 5021 Iloilo, Philippines;* [2]*Integrated Services for the Development of Aquaculture and Fisheries, McArthur Highway, Tabuc Suba, Jaro, Iloilo City 5000 Philippines;* [3]*Degussa Food Ingredients/BL Texturant Systems, Degussa Texturant Systems France SAS, 50500 Baupte, France*

*Author for correspondence: e-mail: annehur@yahoo.com

Key words: 'goose bump', *Polysiphonia*, epiphytes, *Kappaphycus*, water movement

Abstract

This paper describes the occurrence of an epiphyte infestation of *Kappaphycus* farms in Calaguas Is. Camarines Norte, Philippines. In particular, percentage cover of 'goose bump'-*Polysiphonia* and 'ice-ice' disease, and some environmental parameters that influence the thallus condition of *Kappaphycus alvarezii* in Calaguas Is. were assessed during 3 separate visits and are discussed.

Commercial cultivation of *Kappaphycus* at Calaguas Is. began in the early 1990s. After five years of farming, the stock was destroyed by a strong typhoon. The area was re-planted the following year and production increased annually and reached its peak in 1998–1999. However, the following year, the first occurrence of a *Polysiphonia* epiphyte infestation occurred concurrently with an 'ice-ice' disease. Consequently, annual production and the number of seaweed planters declined rapidly, and this situation persists to the present time. This paper highlights the etiological factors and their consequences.

Results show that farm-site selection is critical for the success of *Kappaphycus* production. Characteristics of water movement and light intensity in farming areas contributed to the occurrence and detrimental effect of the phenomenon described as 'goose bumps': a morphological distortion of the host seaweed due to the presence of a *Polysiphonia* sp. epiphyte. A strong inverse correlation was observed between the occurrence of *Polysiphonia* and water movement: areas with low water motion registered a higher % cover (65%) of *Polysiphonia* than those in more exposed areas (17%). Although 'goose bump'-*Polysiphonia* infestation and 'ice-ice' disease pose a tremendous problem to the seaweed farmers, the results of this limited assessment provide a useful baseline for future work.

Introduction

Kappaphycus alvarezii (also colloquially called "cottonii"), a kappa carrageenan-producing seaweed is commercially cultivated in the tropics, notably in the Philippines, Indonesia, Malaysia and Fiji Islands. The increasing demand for carrageenan on world markets due to its diverse product applications, makes *Kappaphycus* an important marine commodity. Despite this seaweed being the Philippines' major carrageenan-containing marine plant, there are still raw material production problems that consequently affect the end product. These problems are mainly 'ice-ice' disease and epiphyte infestation. Local warm-water events are also detrimental to productivity.

'Ice-ice' disease of *Kappaphycus* was reported as early as 1974 in the Philippines by Trono (1974). Uyenco (1977) and Uyenco et al., (1981) described the occurrence of pathogenic micro-organisms and stressed the interplay of ecological and physiological conditions of the seaweed. Their findings were confirmed by Largo et al. (1995a,b) in the laboratory, who reported that *Vibrio-Aeromonas* complex and *Cytophaga-Flavobacterium* complex induce 'ice-ice' disease when the plant is stressed by either low salinity or low light intensity. The lytic activity of the bacterium which digests epidermal cells and destroys chloroplast, resulted in initial bleaching of the infected part. Furthermore, Largo et al. (1995b) reported the gradual hydrolysis of the thallus starting from the cortical layer and ending with the medullary cells, leading to full necrosis (tissue death). Much has been reported on 'ice-ice' disease, however, information on epiphyte infestation is very limited: for references to epiphytic filamentous algae see Ask (1999), Ask and Azanza (2002), and Hurtado et al. (2001). The impact of *Polysiphonia* epiphytes was briefly reported in *Kappaphycus* farms in Calaguas Is., Camarines Norte, Philippines and other parts of the Bicol region by Largo (2002) and Critchley et al. (2004). The filamentous, red *Polysiphonia* epiphyte creates small, elevated pores or 'goose bumps' on the surface that are actually sites of penetration from the cortical to the medullary layers of the host plant. Although superficially morphologically similar to reproductive sporangia, these structures are not reproductive and are perhaps somewhat similar to 'galls' in higher plants.

The occurrence of *Polysiphonia* epiphytes in Calaguas Is., observed since 2000, has resulted in tremendously reduced biomass production of *Kappaphycus* in the formerly productive cultivation area. Even now, only a few people continue to farm *Kappaphycus* since the *Polysiphonia* outbreak. The infestation is persistent rather than periodic. Other than the reports of Ask (1999) and Ask and Azanza (2002) no other documentation of *Polysiphonia* epiphytism is known for other parts of the Philippines. There is also little quantification of the impact on seaweed biomass and crop value which epiphytes may have, although there are anecdotal reports on the occurrence of *Polysiphonia* in western Visayas and Luzon.

The present study results from a call for assistance from the NGO community (US Peace Corps, J. Schubert pers. comm.) to assess the occurrence of this epiphyte further and to determine the environmental conditions that trigger its occurrence. Results of this study will provide benchmark information for future work.

Materials and methods

The study was conducted at Calaguas Is., Camarines Norte, Philippines (14°24′–14°30′N and 122°54′–123°1′E). Calaguas Islands is a group of small islands off Vinzons, Camarines Norte, facing the Pacific Ocean. It is made up of 3 barangays namely, Banocboc, Pinagtigasan and Mancawayan. Two pronounced seasons are experienced in this area: the wet season brought by the south-west monsoon trade wind and the dry season brought by the north-east trade wind. The former season experiences frequent to moderate rainfall except when there are typhoons and calm seas, while the latter experiences moderate to strong wave action brought by north-east trade winds.

Three visits (February, May and November 2003) were made in four farming areas: Banocboc, Sugod (protected and exposed areas) and Pinagtigasan. In each farming area, 10–25 samples of seaweed material were taken randomly, placed in labelled bags and the fresh weight determined using a digital balance. The presence of 'goose bumps', meso-epiphytes (<1 mm long) and 'ice-ice' disease was estimated as % cover using a scale of 1–10 (1 = 10%, 2 = 20% ... 10 = 100%).

Environmental parameters were determined on site, for each sampling time. Light intensity (μmol photons m^{-2} s^{-1}) and water movement (m s^{-1}) were measured with a LI -250 light meter and FP7 water flow meter, respectively. Water temperature, salinity, turbidity, and total dissolved solids were measured using a YSI 650 MDS.

Correlation analysis (R^2) between the percentage cover of meso-epiphytes and some environmental parameters was determined at the 5% level of significance. The colloquial name 'goose bump' *Polysiphonia* is retained since it is descriptive and easily understood by the fisherfolk (Figure 1a and b).

Results

The selection of the farming site was a critical factor in *Kappaphycus* production. Water movement and light intensity in the farming area contributed to the occurrence of 'goose bumps' and *Polysiphonia*. Protected areas (Banocboc, Sugod (protected) and Pinagtigasan) registered a higher % cover (15–65%) of *Polysiphonia* than the exposed areas (Sugod, exposed) (17%) (Figure 2 a–c). A strong correlation between the % cover of 'goose bumps'-*Polysiphonia* and light intensity, R^2 = 0.63 to 1.0) and water movement (R^2 = 0.61 to 1.0) was observed at each of the sampling sites. Other water parameters showed no correlation with the occurrence of 'goose bumps'-*Polysiphonia.* Results of the correlation analysis between light intensity – water movement and 'goose bumps'-*Polysiphonia* infestation showed

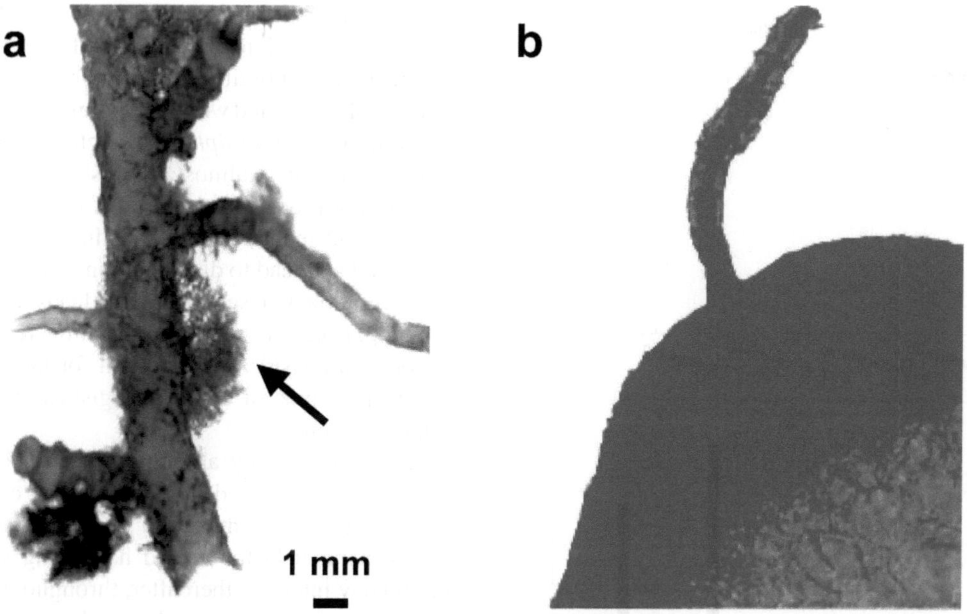

Figure 1. (a). A number of *Polysiphonia* filaments arising from the surface of epiphytised *Kappaphycus.* (b). Emergent *Polysiphonia* filament from a 'goose bump'–like structure made up of the outer tissues of the *Kappaphycus* host.

that the variations in 'goose bumps'-*Polysiphonia* % cover was strongly accounted for in the mean values recorded in light intensity – water movement. If these factors were limiting, then problems with epiphyte infestations arise.

Discussion

Among the four collecting sites, Sugod (exposed) 'goose bumps'-*Polysiphonia* only occurred during the month of November, although at a relatively low %. Despite strong water movement at this site, 'goose bumps'-*Polysiphonia* occurred due to the use of new 'seedlings' for out-planting and on-growing which had been taken from crops where the 'goose bumps' *Polysiphonia* had developed. This use of "infected" material was a desperate action among fishermen in this region and elsewhere, just to have 'seedlings' available for the next culture period (at this stage, they were unaware of the relationship between the 'goose bumps', the *Polysiphonia* and their loss of stock). In fact, the use of seedlings, vegetatively propagated from material already infected with *Polysiphonia,* hastened the proliferation of the epiphytes and caused damage more readily and earlier in the subsequent crop. Relocation of infected seedlings to new cultivation areas was also a vector for rapid dispersal since, under favourable conditions, re-growth of settled spores would produce an increased epiphytic load (see also Buschmann et al., 1997 for an example with *Gracilaria* as the epiphytised host plant).

A high % occurrence of 'ice-ice' was observed in May in Sugod (protected region) (87%) and Pinagtigasan (70%) collecting sites. This appears to be associated with high water temperatures and low water movement (9–32 m s^{-1}). The adverse effect of 'ice-ice' infected thalli has been reported in *Kappaphycus striatum* – *'sacol'* strain by Mendoza et al., (2002). They found that there was a decrease in carrageenan yield, gel strength and viscosity and an increase in syneresis index, which resulted in the material producing poor quality carrageenan of low molecular weight. In Calaguas Is., although 'ice-ice' occurred, it was not as persistent as the 'goose bump'-*Polysiphonia* during the three periods of sampling and therefore, 'goose bump'-*Polysiphonia* infestation is more of a problem than 'ice-ice' disease in the Calaguas Is. region. The effect of 'goose bump'-*Polysiphonia* infestation on the quality of carrageenan is yet to be determined.

Reports on the epiphyte problem of *Kappaphycus* are limited to the identification of the non-epiphytic 'fouling' seaweeds such as *Enteromorpha, Ulva, Hypnea, Dictyota* and *Hydroclathrus* (Ask, 1999;

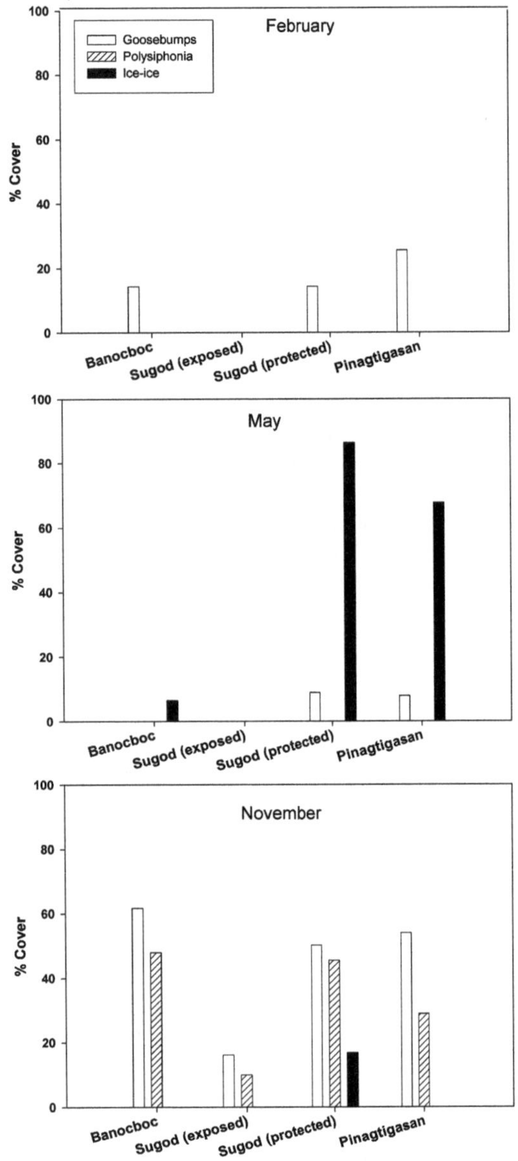

Figure 2. (a–c) Percentage cover of 'goose bump'-*Polysiphonia*-'ice ice' at the four collecting sites over time.

Hurtado et al., 2001; Ask & Azanza, 2002). There are no quantitative reports on the meso-, endo- and macro-epiphytes on cultured *Kappaphycus*. Likewise, reports on 'ice-ice' disease were not quantified (LEAP 2003) except for the earlier work of Uyenco et al. (1981).

Epiphyte control seems to be more easily managed in tank cultivation (Fletcher, 1995) than in open, extensive systems such as those typically used for large seaweed cultivation (Brawley & Fei, 1987) and bottom planting for *Gracilaria* in Chile (Buschmann et al., 1994). Manual removal of macro-epiphytes in the field is possible, but requires a considerable amount of labour and would be extremely tedious work. The removal of *Polysiphonia* which causes the 'goose bump' reaction is almost impossible because of the penetrating foot-like structures of this epiphyte in the cortical and medullary cells of the host plant, which weaken it and lead to discoloration, fragmentation and disintegration. Consequently, this leads to total loss of the crop. Likewise, sourcing of uninfected 'seedlings' becomes a considerable problem for the next season, due to problems of cost and logistics of transport in remote regions.

This brief survey among the seaweed planters in Calaguas Is., revealed that commercial cultivation of *Kappaphycus* farming began in the early 1990's during September-October, and harvesting was done at 45–60 day intervals thereafter, throughout the season. After five years farming, the stock was destroyed by a strong typhoon. Re-planting was done in the area, leading to an influx of people from the mainland to Calaguas Is., As a result, production increased annually and reached its peak in 1998–1999 with an average production of 2,000 t dwt month^{-1}. In 2000 however, the seaweed farmers experienced the first occurrence of *Polysiphonia* infestation and also 'ice-ice' disease, and these problems continue through to the present (2004). Consequently, there has been a remarkable reduction in annual production and inevitably the number of seaweed planters. This decline was ca. 41% up to 2001 (Borja, 2003), with an estimated loss in value of US$ 750 000. As a result, some farmers have transferred to Polilio, Is. Quezon, for seaweed farming. Other former seaweed farmers may have returned to fishing for a livelihood.

There was no zoning of seaweed cultivation observed in this area and as a result, the cultivated areas appeared to be somewhat overcrowded, resulting in impeded water movement. In addition, only suspended (hanging) cultivation is used in the Calaguas Is. However, the styrofoam commonly used for flotation was almost at the same level as the suspended seaweed culture, thereby exposing the crop to the air after it had gained some biomass. Consequently, the seaweeds were exposed to intense sunlight, especially during the summer months (March to May), thus losing pigments which also resulted in thallus discoloration. In 2004 the seaweed cultivators really had only one good harvest, i.e. their first. At the start of the second crop, there was an early occurrence of 'goose bumps' during the second week of cultivation and the seaweed farmers had no choice but to harvest in the 3rd–4th week (i.e. 21–28 days after planting, rather than the normal >45 days), to forestall rotting of the thallus and the total loss of their crop. Sun-drying of the seaweed was either by hanging the whole cultivation line with the seaweeds and floats from a wooden bar or by placing the seaweed directly on the ground (the latter, however, is not acceptable to the purchasers of dried seaweed).

Though the assessment study was limited, this preliminary work shows that 'goose bump'-*Polysiphonia* infestation and 'ice-ice' disease are a significant, detrimental problem to the seaweed farmers of the Calaguas Is. Results of the present study will serve as baseline information to future work. Calaguas Is. has potential to be developed as an area for *Kappaphycus* farming, after considering its topography, climatic conditions and proximity to the main island (1.5 h), and to Manila (7 h) for trade and commerce.

Conclusions and recommendations

Meso-endo epiphytism is a real problem in Calaguas Is. as evidenced from the following: (1) decreased and displaced number of planters (from ~300 to <15 farmers in the Calaguas Is. over 4 years), (2) consistent low production since 2000 (<10 t mo^{-1}), (3) persistent and high % occurrence of 'goose bumps' -*Polysiphonia* throughout the period and to the present day. Water movement is a critical factor in *Kappaphycus* production. It is recommended that rigid site assessment/selection must be done before seaweed farming is started in order to minimise problems that affect production.

Problems with epiphyte infestation can be minimised however, if total crop management is implemented properly in the following manner: (1) careful selection of the farming site with emphasis on water movement and siltation, (2) selection of quality 'seedlings' (free from epiphytes) to initiate cultivation, (3) use of the correct stocking density (g m^{-1}), (4) proper culture technique, with frequent inspections of the material, and (5) good post-harvest management (drying and storage). Education of the farmers and dissemination of information through periodic seminars is essential and must be encouraged in order to discuss progress, problems and solutions for farming issues.

Acknowledgements

The authors are thankful to Amadeo Biter, Ephraim Doroteo, Keneth Tibubos and the seaweed farmers of Calaguas Is. for their co-operation, innumerable instances of assistance and support during this study. We also thank Erick Ask, FMC and Danilo Largo for discussions, Winston Wee for logistical support and the members of Marinalg for their interest in the project.

References

Ask EI (1999) Cottonii and Spinosum Cultivation Handbook. FMC Food Ingredients Division., Philadelphia, (52).

Ask EI, Azanza R (2002) Advances in cultivation technology of commercial eucheumatoid species: A review with suggestions for future reserach. Aquaculture 206: 257–277.

Borja P (2003) Seaweed industry status in Bicol Region – Cluster VI. In Hurtado A.Q, Luhan Ma R.J (eds), Proceedings of the National Seaweed Symposium. Seaweed Industry Asssociation of the Philippines, 17–18.

Brawley SH, Fei XG (1987) Studies of mesoherbivory in aquaria and in an unbarricaded mariculture farm on the Chinese coast. J. Phycol. 23: 614–623.

Buschmann AH, Schultz JA, Vergara PA (1994) Epiphytism and herbivory in an inter-tidal *Gracilaria* (Rhodophyta, Gigartinales) farm in southern Chile. In Koop K. (ed), Ecology of Marine Aquaculture, International Foundation for Science, Stockholm, 48–58.

Critchley AT, Largo D, Wee W, Bleicher Lhonneur G, Hurtado AQ, Schubert J (2004) A preliminary summary on *Kappaphycus* farming and the impacts of epiphytes (extended abstract). Jpn. J. Phycol. 52: 231–232.

Fletcher RL (1995) Epiphytism and fouling in *Gracilaria* cultivation; an overview. J. Appl. Phycol. 7: 325–333.

Hurtado AQ, Agbayani RF, Sanares R, Castro-Mallare MTR de (2001) The seasonality and economic feasibility of cultivating *Kappaphycus alvarezii* in Panagatan Cays, Caluya Antique, Philippines. Aquaculture 199: 295–310.

Largo DB (2002) Recent developments in seaweed diseases. In Hurtado A.Q., Guanzon Jr. NG, de Castro-Mallare TR, Luhan Ma RJ (eds), Proceedings of the National Seaweed Planning Workshop, Southeast Asian Fisheries Development Centre Aquaculture Department, Tigbauan, Iloilo, Phillippines, 35–42.

Largo DB, Fukami K, Nishijima T, Ohno M (1995a) Laboratory-induced development of 'ice-ice' disease of the farmed red algae *Kappaphycus alvarezii* and *Eucheuma denticulatum* (Solieriaceae, Gigartinales, Rhodophyta). J. Appl. Phycol. 7: 539–543.

Largo DB, Fukami K, Nishijima T (1995b) Occasional pathogenic bacteria promoting 'ice-ice' disease in the carrageenan-producing red algae *Kappaphycus alvarezii* and *Eucheuma denticulatum* (Solieriaceae, Gigartinales, Rhodophyta). J. Appl. Phycol. 7: 545–554.

Livelihood Enhancement and Peace Program (LEAP), 2003. Assessment of impact of LEAP program-established seaweed production activities in coralline and similar marine ecosystems. Report on SAF 3030, 45 pp.

Mendoza WG, Montaño NE, Ganzon-Fortes ET, Villanueva RD (2002) Chemical and gelling profile of ice-ice infected carrageenan from *Kappaphycus striatum* (Schmitz) Doty '*sacol*' strain (Solieriaceae, Gigartinales, Rhodophyta). J. Appl. Phycol. 14: 409–418.

Trono GC Jr, (1974) *Eucheuma* farming in the Philippines. U.P. Natural Science Research Centre, Quezon City.

Uyenco FR (1977) Microbiological studies of diseased *Eucheuma* sp. and other seaweeds. National Seaweeds Symposium, Metro Manila, Phillippines.

Uyenco FR, Saniel LS, Jacinto GS (1981) The 'ice-ice' problem in seaweed farming. Proc. Int. Seaweed Symp. 10: 625–630.

Journal of Applied Phycology (2006) 18: 307–314
DOI: 10.1007/s10811-006-9026-x

Culture of *Gigartina skottsbergii* (Rhodophyta) in southern Chile. A pilot scale approach

Héctor Romo[1,*], Marcela Avila[2], Mario Núñez[2], Rodrigo Pérez[1], A. Candia[2] & Gesica Aroca[2]
[1]*Departamento de Oceanografía, Universidad de Concepción, Casilla 160-C, Concepción, Chile;* [2]*División de Investigación Acuícola, Instituto de Fomento Pesquero, Balmaceda 252, Puerto Montt, Chile*

*Author for correspondence: e-mail: hromo@udec.cl

Key words: carrageenophyte, cultivation, *Gigartina*, growth, Rhodophyta

Abstract

In the last 10 years studies on the management and exploitation of Chilean carrageenophytes have proliferated in response to the increasing development of the local processing industry. One of the most important sources of raw material for Chilean carrageenan, *Gigartina skottsbergii* Setchell et Gardner, was the subject of an intensive study to design a commercial cultivation technique which could be an alternative to wild harvest. In this context this pilot study reports the first successful attempt to culture *G. skottsbergii* from spores to harvestable plants. A three-step farming approach was developed: (i) seeding of spores onto scallop shells followed by a two-month nursery period in a greenhouse (until the development of initial upright thalli from the discoid crust occurred), (ii) outplanting juvenile plants on shells in the sea on a long-line system (until thalli attained 3–4 cm diameter) and (iii) detachment of fronds from the shells, fixing of individuals to vertical ropes and growth until commercial size was reached. Additional experiments to compare bottom and suspended growth, cultivation by fragmentation and whole fronds and meristematic activity of different zones of the fronds were performed. This study shows the technical feasibility of culturing *G. skottsbergii* from spores, complemented with growth of vegetative fragments, in order to optimize the management of the culture. In the future, therefore, it may be possible to replace the heavy exploitation of wild beds in southern Chile with farming activities.

Introduction

Recently, several studies on wild populations of *Gigartina skottsbergii* Setchell et Gardner have been performed in order to acquire a basic knowledge on yearly abundance patterns, reproductive phenology and recruitment of young thalli (Zamorano & Westermeier, 1996; Avila et al., 1997, 1999a; Westermeier et al., 1999; Marín et al., 2002). Buschmann et al. (1999, 2001) suggested that *G. skottsbergii* was being overexploited and consequently that the development of culture technologies was urgently required to support the local carrageenan industry. In fact, Avila et al. (2003) reported that the heavy exploitation of natural beds was reaching as far south as Seno Año Nuevo (55°25′S; 69°00′ W). Such harvest displacement to southerly sites (approximately 1500 km south from traditionally harvested sites in Chiloe during the 1990s) was the logical consequence of an increasing *G. skottsbergii* shortage in previously productive beds of Chiloe. Since 2002 to late 2003, exploited beds have shown serious symptoms of depletion. Consequently, a continued decline in standing stocks could be a serious risk for both local and overseas carrageenan industries.

Preliminary experiments done by Buschmann et al. (1999) recorded that the growth of germlings seeded in the laboratory on Petri dishes and transplanted to outdoor tanks reached 1–2 mm in 3 months. Buschmann et al. (2001) reported that the growth of germlings seeded in the laboratory on ceramic plates and later transplanted to the field reached about 5 mm^2 after 5 months in the sea. Growth during 6 months, of 16 fragments of fronds excised from immature wild fronds was reported by Buschmann et al. (1999) and growth of 20 whole young fronds during 12 months was reported by Buschmann et al. (2001). Both experiments used the same methods: fastening the fragments and fronds to ropes settled at 20 cm above the bottom. In addition, Correa et al. (1999) suggested from *in vitro* experiments that mass cultures by vegetative propagation techniques could be a promising tool for establishing future farms.

A greenhouse nursery method for *G. skottsbergii* mass culture consisting of: (a) settlement of spores on natural and artificial substrata; (b) survival of germlings under indoor conditions and then outplanting into the sea, was developed by Avila et al. (2003). Germlings of about 60 to 110 mm^2 were obtained in 15 months. The method consisted of seeding of spores on different types of substrata followed by the development of germlings, for two months in a greenhouse. Plants were then transplanted to suspended systems at different depths in the Bay of Hueihue (41°54′S; 73 °31′W in Chiloé Island). The best depth for early growth of *G. skottsbergii* in the environmental conditions at the Bay of Hueihue was shown to be 3–6 m (see Zamorano & Westermeier, 1996, for general abiotic conditions of Ancud near the study area).

This paper presents for the first time the results of a pilot study of 31 months on commercial culture techniques of *G. skottsbergii,* to produce kappa-carrageenan gametophytes, at the Bay of Hueihue. The culture was based on a tetraspore-seeding method using scallop shells as substrata (Avila et al., 2003). Cultivation was followed by the growth of thalli on suspended cultures in the sea until plants reached of commercial size, complementing the results of Avila et al. (2003). In addition, growth of germlings on bottom and suspended systems, and both whole thalli and fragments were compared. Finally, meristematic activity on different parts of young plants was assessed.

Material and methods

Mass culture starting from spores

All mature sporophytes used as reproducers were collected at the Bay of Ancud (41°52′S; 73°31′W) and at Calbuco Channel (41°45′S; 73°05′W). The spore-seeding method on scallop shells was used (Avila et al., 2003). Scallop shells were one of the best substrata reported and were easily available from farms of scallops near the site of study. Spore seeding, using only tetraspores to produce kappa-carrageenan gametophytes, was done at the greenhouse facilities of the Maricultural Station at the licensed site of the Instituto de Fomento Pesquero in the Bay of Hueihue (41°54′S; 73°31′W). After two months, once the basal crusts developed their uprights of gametophyte fronds, the germlings were transplanted to the sea.

Variations of the floating structures described in Romo et al. (2001b) were used in this study. Cylindrical floats of PVC which separated the lines of the 100 m double long-line and the small 2 L buoys were replaced by 9 polystyrene 200 L buoys for better flotation. The shells, perforated in the centre, with germlings growing in the upper side of the shells, were arranged in sets of ten shells fixed to 1 m polypropylene ropes of 3 mm diameter (Figure 1A). Knots adjusted on and under each shell secured them to the ropes; shells were separated by 10 cm. A total of 1370 shells disposed in 137 ropes, were placed at 6 m depth, one of the most suitable depths for growth reported by Avila et al. (2003). Three metres depth was also good for germling growth but in the present study this depth was not used due to excessive fouling that developed on shells in previous

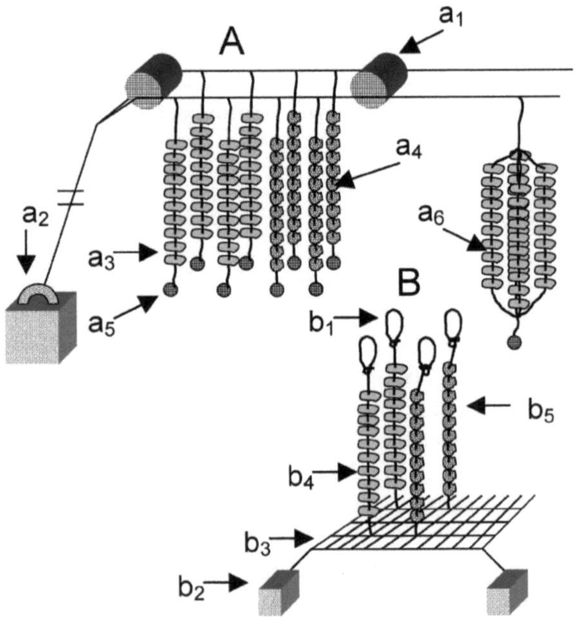

Figure 1. (A) Floating system for *G. skottsbergii* cultivation: a_1 = buoys; a_2 = concrete anchor; a_3 = fronds; a_4 = shells with young thalli; a_5 = sinker; a_6 = four ropes with fronds in a vertical rope. (B) Bottom system, b_1 = buoys; b_2 = concrete anchor; b_3 = net; b_4 = fronds; b_5 = shells with young thalli.

trials. Growth was estimated monthly by measuring frond width, which corresponds to the major diameter of the ellipsoidal thallus. Monthly cleaning to remove fouling and possible grazers was done by using high pressure jets of seawater on the shells. Once the fronds attained 3 to 4 cm diameter they were detached from shells and individually attached to 2 mm diameter polypropylene ropes, at intervals of 10 cm. The attachment was done by means of a 2 mm hole perforated with a cork borer and 10 fronds were threaded in the rope. As in the shells, each young frond was secured to the rope by knots on and under the blade. Once the fronds attained more than 10 cm diameter the 2 mm ropes were replaced by 3 mm ropes and attached with cable-ties on and under each frond until commercial size was reached. In order to optimize the capacity of the floating long-line a minimum of two and maximum of five 1 m ropes with adult fronds were attached to the main vertical ropes (Figure 1A).

Bottom and suspended growth

An experiment to assess the growth of early gametophytes of *G. skottsbergii* growing on shells attached to ropes anchored to the bottom, and also in suspended conditions at 6 m constant depth, was conducted from April to December (autumn to late spring) 2002. The experiment was established in order: (i) to evaluate the performance of both cultivation systems in a site exposed to strong tidal currents like in Chiloe Island and neighbouring areas occur and (ii) to evaluate differences in growth of *G. skottsbergii* in both conditions. A 30 cm stretched-mesh fishing net of 9 m^2 was attached to the bottom of the Bay of Hueihue in a site where there are fluctuating depths from 3 to 9 m (7 m is the maximal tidal amplitude during spring tides). A total of 3610 shells (361 ropes with 10 shells per rope) with thalli of about 0.1 mm initial diameter was placed in ropes maintained in vertical rows by means of 2 L plastic buoys (Figure 1B). Two sets of 8 ropes were identified in suspended and bottom systems, respectively, for monthly measurements of the width of 160 gametophyte germlings growing on shells. While the bottom system was subject to fluctuating depth according to the daily tidal rhythm, the suspended system maintained the plants on the shells at a constant 6 m depth. The Mann Whitney test (Zar, 1999) was used to test whether differences in frond growth in both systems were significantly different.

Whole thalli and fragment growth

An experiment was established to compare growth of whole thalli with growth of fragments. Growth of 10 triangular fragments of young blades per rope on 10 ropes (cutting 2–3 triangular pieces from the basal-hapterial region to the margin of each blade) was compared with ten young fronds per rope on 10 ropes. Ropes were suspended at 6 m depth. As wet weight was different for fragments and whole fronds, monthly growth measurements were converted to daily growth rates (DGR) assuming exponential growth according to Hansen (1980):

$$\text{Daily growth rate } (\%/\text{day}) = 100(\text{Ln}(W_f/W_i))/t,$$

where Ln: natural logarithm; W_f: final wet weight; W_i: initial wet weight and t the time interval (in days) between final and initial measurements. Differences in growth were analyzed at the end of the experiment by the Mann-Whitney test. Some comparisons in the Discussion section were made as monthly growth rate, so the variable t in the formulae in that case corresponded to a monthly interval.

Meristematic activity in different parts of the fronds

In order to investigate zones with faster growth in the fronds and to harvest only old unproductive areas of the thallus with slow growth, the response of different zones of a frond was assessed using a modification of a perforation technique (Norris & Kim, 1972) using a cork borer. Norris & Kim made many perforations at 2–3 mm apart on each frond while we made only 11 perforations (Figure 2). Growth causes the separation of holes in proportion to the amount of growth that has occurred (Figure 2). Growth was assessed during 70 days in two groups, for each of 10 young gametophyte thalli selected among the seeded plants growing in the suspended culture. The size of plants was about 10–15 cm diameter and neither cystocarps nor spermatangia were present. One group had perforations and the other group had none. They were hung from the long-line system at 6 m depth in the same way as described above. Differences in growth rate among marginal, basal zone and whole fronds were analyzed by the Kruskal-Wallis test (Zar, 1999).

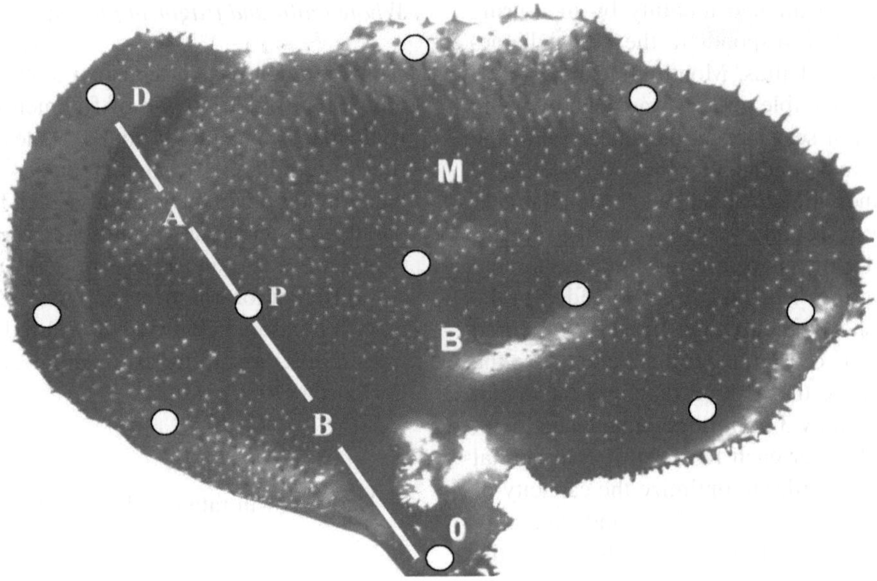

Figure 2. Perforated *G, skottsbergii* frond for measurement of growth in different zones of the thalli. (D) distal point; (P) central point; (0) basal point; (M) marginal zone and (B) basal zone.

Results

Mass culture starting from spores

After the second month in the greenhouse, the germlings were 0.08 to 0.15 mm tall on the shells and were outplanted into the sea at 6 m depth in vertical rows of ten shells per rope. In general, the germlings displayed a very slow growth, especially during the initial 14 months growing in the floating system, when the monthly growth could be measured in millimetres of growth of frond diameter (Figure 3A). Figure 3B shows the continuation of growth during the next months and year reaching about 60 cm width in April 2004. The minimum time to reach a harvestable size of 20–30 cm diameter can be considered to be 26–27 months. Fouling by *Polysiphonia* sp., *Antithamnionella* sp. and undetermined hydrozoans was the major problem when the plants were growing on the shells, but periodical cleaning, using high pressure jets of seawater on the shells, eliminated the fouling. When female fronds developed cystocarps from October and November 2004 until the end of the study, heavy settlement of young commercial mussels, *Aulacomya ater* Molina and *Choromytilus chorus* Molina, occurred on cystocarps as well as on the hapteral area of all fronds, especially during summer.

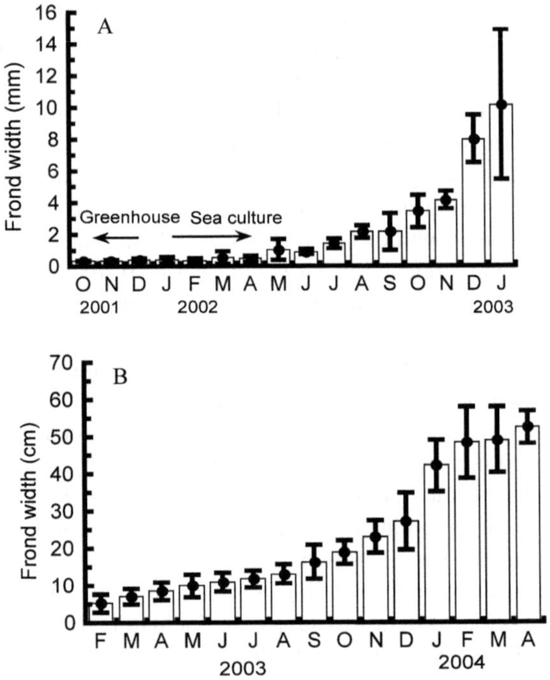

Figure 3. Culture of *G. skottsbergii* suspended at 6 m depth. (A) Growth of very young thalli (frond size is measured in mm width ± S.D.). (B) Growth of young to adult thalli (frond size is measured in cm ± S.D.).

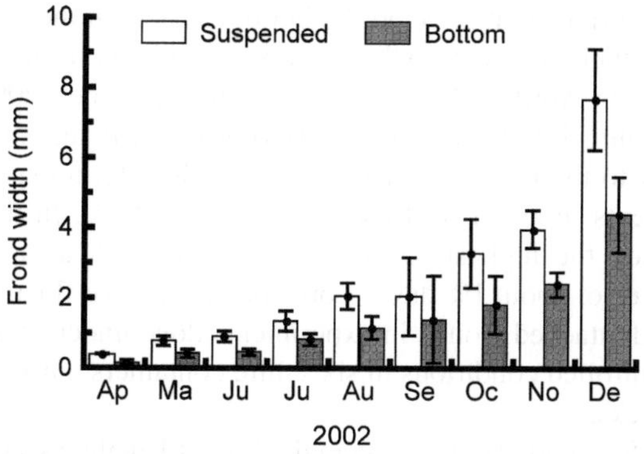

Figure 4. Growth (width ± S.D.) in *G. skottsbergii* suspended at 6 m depth and on the bottom.

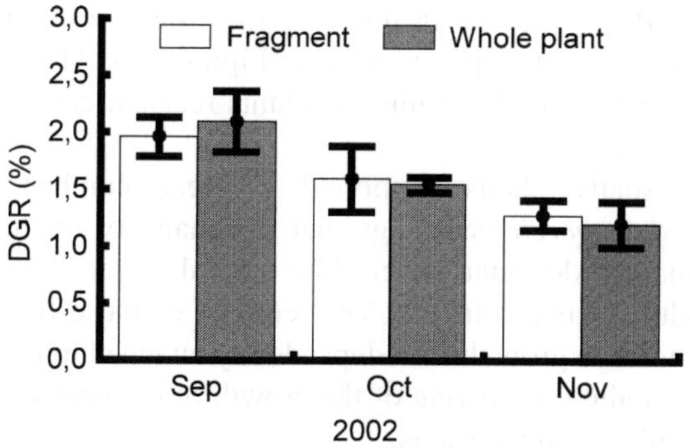

Figure 5. Daily growth rate (± S.D.) of *G. skottsbergii* fragments and whole fronds in the floating system.

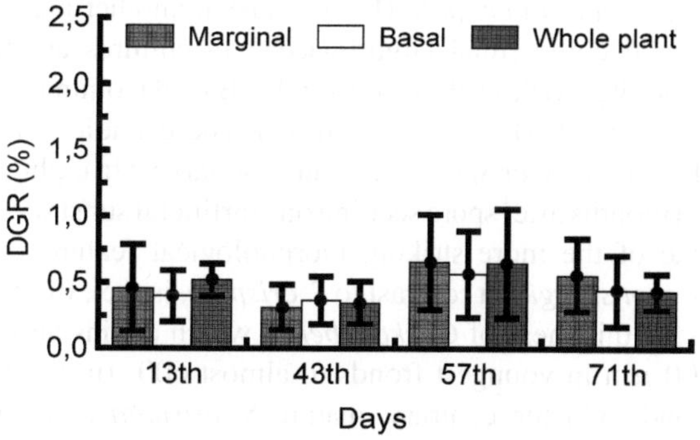

Figure 6. Daily growth rate (± S.D.) of marginal and basal zones of the *G. skottsbergii* frond compared with whole fronds. Thalli were growing at 6 m depth in suspended conditions.

Bottom and suspended growth

There was a significant difference between cultivation methods: the suspended culture of germling gametophytes had significantly greater growth than the bottom culture ($p < 0.05$; Mann-Whitney test, Figure 4). Heavy invasion of drifting *Ulva* sp. on and among the ropes and shells of the bottom system continually occurred. Despite the monthly cleaning by divers, *Ulva* sp. repeatedly invaded the experimental site in few days after each cleaning. The bottom experiment was terminated in November 2002 due to the strong interference by *Ulva* sp. and the severe entangling of the vertical ropes by drifting algae.

Whole thalli and fragmens growth

This experiment, designed to test differences between growth of whole fronds and fragments of young fronds, did not show significant differences between treatments ($p > 0.05$; Mann-Whitney test, Figure 5). Fragments showed total healing along the cut edges before or at the end of the second week of experimentation.

Meristematic activity in different zones of the frond

Growth of marginal and basal zones of the fronds and whole fronds did not differ significantly ($p > 0.05$; Kruskal-Wallis test; Figure 6). The wounds produced by the cork borer healed rapidly and neither promoted nor inhibited the growth around the wounded hole.

Discussion

Mass culture starting from spores

Gigartina skottsbergii belongs to the group of species that must be cultivated in a multi-step sequence (Santelices, 1999). The cultivation process should consider a spore-seeding stage in indoor facilities followed by outplanting to the sea where growth may continue until harvest. This procedure contrasts with the cultivation of *Eucheuma* and *Gracilaria* which can easily be propagated by thallus fragmentation. Our study and Avila et al. (2003) showed that a relatively long time must be spent to obtain visible plants growing on shells during the first year in the sea (Figure 3A). Results reported by Avila et al. (2003) from experiments started on July 1999 (winter) showed 80 mm^2 (about 10 mm width) as the final size at 3 m depth after 13 months in the sea. In the present study, started in October 2001 (mid spring) the mean width recorded on young fronds was also about 100 mm during the same period of time, but in this case the fronds were suspended at 6 m depth in the sea. (Figure 3A). After the following 13–15 months the plants reached commercial size.

Usually shells supported a great number of *G. skottsbergii* germlings, but the majority of them remained dormant, covered by several larger 3–4 cm width dominant fronds. The removal of these fronds in order to grow them independently attached to ropes could allow triggering of the growth of the uncovered germlings on the shells.

Growth rates of *G. skottsbergii,* of between 0.5 to 2.5% monthly, computed in several experiments in this study, are low compared, for instance, to the growth of *Sarcothalia crispata* (Bory) Leister, another blade-shaped commercial Gigartinaceae in cultures at the same site (Avila et al., 1999b) and in Bay of Coliumo, in central Chile (Romo et al., 2001a). These studies concluded that *S. crispata* was ready for harvest in eight to ten months after spore seeding onto artificial substrates. One of the more striking morphological features of *G. skottsbergii,* in contrast to *S. crispata,* is the extreme blade thickness of *G. skottsbergii* which attains about 400 μm in youngest fronds to almost 600 μm in old fronds. On the contrary, mature *S. crispata* is about 80 μm thick. Extremely low photosynthetic rates and high metabolic expenses required to grow and maintain a thick non photosynthetic medullar tissue could be the cause for low growth rates of *G. skottsbergii.* On the other hand, thickness of the fronds permitted the fronds to be re-attached to vertical ropes for suspended culture and subsequently to grow independently from shells.

Bottom and suspended growth

It could be assumed that disturbances produced by heavy and repeated invasions of drifting *Ulva* sp. (forming a thick carpet above the bottom system and entangling the net and vertical ropes) could be one of the causal factors for low growth of bottom fronds. For example, the canopy provided by large fronds of *Ulva* sp. reduced the availability of light for photosynthesis. Although biomass of *Ulva.* sp. was not monitored, the persistent invasion of *Ulva* sp. on the bottom stand of *G. skottsbergii,* despite the monthly clearance of drifted *Ulva* sp. by divers, reinforced our idea that *Ulva* sp. could retard the normal growth of *G. skottsbergii.* After 9 months and several repairs to entangled ropes, the bottom method was discarded for use in commercial farming. Notwithstanding, and considering the worst conditions that affected the bottom system at the Bay of Hueihue, the growth of bottom germlings from 0.5 mm in April to 4.5 mm diameter in December, 2002 (Figure 4) represented about 0.8 daily growth rate similar to the results of several indoor, short-term experiments reported by Buschmann et al. (2004). In that study the highest specific growth rate with a mean value of about 1% day^{-1} corresponded to free-floating and attached fronds in experiments done under semi-controlled conditions in six 2-litre containers during 4 weeks.

Suspended cultivation in the Bay of Hueihue seems to be clearly better for growth than the bottom conditions at Ancud tested by Buschmann et al. (1999). That experiment consisted of “16 fragments excised from immature fronds fastened to 8 m nylon cord” which in 6 months grew from 30 cm in July to 40 cm length in December (Buschmann et al., 1999a, p. 433). Interpolating the data in Figure 8A and assuming exponential growth (see Materials and Methods in the present contribution) calculations revealed a 4.8% monthly growth rate. This was compared with our data on suspended culture in the same period of the year in which the cultivated plants grew from 14 cm width in July to 30 cm width in December 2003 (Figure 3B). The computed growth of suspended plants at Hueihue was 12.7% monthly growth rate, about twice as high as the growth at Ancud. It was not possible to repeat the comparison on data of young fronds, growing on the bottom at Calbuco, reported by Bushmann et al., 2001 (see Figure 7, p 262), because while in the text it is reported that juveniles increased from 100 to 390 cm^2 in 8 months, the graph in their Figure 7 indicates that growth really occurred during 12 months of experimentation (September to August).

Whole thalli and fragment growth

The non-significant difference in growth between whole fronds and fragments confirmed for the first time that fragments could grow in the sea with the same growth rate as intact thalli. This had been suggested by Correa et al. (1999, p. 326). They stated that “vegetative propagation is possible not only because the early stages in the process can be successfully manipulated *in vitro*, but because those plantlets can develop normally in both tank systems and in the sea, growing at the same rates as wild plants (unpublished)”. Surprisingly, the same research group commented later that Westermeier et al. (unpublished data) said that, "vegetative fragments and subsequent regeneration resulted in plants with an elongated habit and a growth potential higher than non fragmented fronds" (see Buschmann et al., 2001, p. 262). At present our findings are solving this discrepancy and, at least in field-suspended conditions, both fragments and whole fronds grew at similar daily growth rates.

Meristematic activity in different parts of the frond

The fact that growth of marginal and basal zones of the fronds did not differ significantly from whole fronds confirms the statement of Norris and Kim (1972) that in the genus *Gigartina* most growth of the young and adult blade is by a diffuse intercalary meristem that is formed as the upright thallus expands to form a blade.

A practical application of these results is: (i) similar rate of growth of fragments and whole fronds and (ii) confirmation of the uniform activity of meristems in different parts of the fronds. These results suggest that opportune fragmentation of adult fronds, for further growth, should prevent their excessive weight which could cause premature detachment from the ropes. This procedure could allow fronds to grow to their maximum size before the harvest. The diffuse intercalary pattern of growth differs from the apical-to-basal gradient in growth rate exhibited by *Eucheuma* or *Kappaphycus*. For about 30 years it was known that farmers usually tie young pieces of thalli with good growth potential onto lines and that they harvest the unproductive old pieces (Doty, 1980). Young fragments function as "seeds" but in the case of *G. skottsbergii,* our findings suggest that the young fragments function as little plants, with uniform meristematic proprieties, that should be harvested once mature. Another farming strategy has been proposed by McNeill et al. (2003) for *Gigartina atropurpurea* from New Zealand. This species shows regenerative activity whereby prunings taken near the base of the frond develop a new frond with higher yield than wild un-pruned plants used as controls. On the contrary, *G. skottsbergii* did not show this pattern of growth.

Finally, growth management of young fronds of *G.skottsbergii*, selected from wild harvests, can be done by attaching them to ropes to grow as described above. This procedure will allow extra growth, after three or more months, in suspended culture, by increasing the yield of young, wild harvested fronds. Young thalli can be considered to be thin plants, less than 20 cm wide, and without signs of gametophytic maturation (for instance plants from October-November 2002 in this study): in three additional months they can reach a commercial size of 40 cm width and either be mature or non-reproductive. This management can complement the farming production of spore-propagated plants. This research proved that it is technically possible to farm *G. skottsbergii* but it is necessary to search for new approaches to improve its growth during the first year and to assess its economic feasibility.

Acknowledgements

The authors acknowledge the financial support of Grants D97 I 1064 and D00 I 1109 of FONDEF, as well as to the firms FMC Corporation, Gelymar S.A., Danisco Chile S.A. Neogel S.A. and Algina S.A. The support of the Dirección de Investigación of the Universidad de Concepción and the field and hatchery support of H. Cortez, H. Pavez, I. Landeros, A. Millaquén, R. Sepúlveda and R. Ruiz. We also thank the helpful criticism of two anonymous reviewers.

References

Avila M, Núñez M, Candia A, Norambuena R (1997) Patrones fenológicos y reproductivos de una población de *Gigartina skottsbergii* (Gigartinaceae, Rhodophyta) en Ancud, Chile. Gayana Oceanologia. 5: 21–32.

Avila M, Candia A, Nuñez M, Romo H (1999a) Reproductive biology of *Gigartina skottsbergii* (Gigartinaceae, Rhodophyta) from Chile. Hydrobiologia 398/399: 149–157.

Avila M, Ask E, Rudolph B, Nuñez M, Norambuena R (1999b) Economic feasibility of *Sarcothalia* (Gigartinales, Rhodophyta) cultivation. Hydrobiologia 398/399: 435–442.

Avila M, Candia A, Romo H, Pavez H, Torrijos C (2003) Exploitation and cultivation of *Gigartina skottsbergii* in southern Chile. In Chapman, ARO, Anderson RJ, Vreeland VJ, Davison IR (eds), Proceedings of the 17th international seaweed symposium: pp. 137–143.

Buschmann AJ Correa A, Westermeier R (1999) Recent advances in the understanding of the biological basis for *Gigartina skottsbergii* (Rhodophyta) cultivation in Chile. Hydrobiologia 398/399: 427–434.

Bushmann AH, Correa JA, Westermeier R, Paredes MA, Aedo D, Potin P, Aroca J, Beltrán J, Hernández-González MC (2001) Cultivation of *Gigartina skottsbergiii* (Gigartinales, Rhodophyta): Recent advances and challenges for the future. J. Appl. Phycol. 13: 255–266.

Buschmann AH, Varela D, Cifuentes M, Hernández González MC, Henríquez L, Westermeier R, Correa JA (2004) Indoor cultivation of the carragenophytic red alga *Gigartina skottsbergii*. Aquaculture 241: 357–370.

Correa JA, Beltrán J, Buschmann J, Westermeier R (1999) Healing and regeneration responses in *Gigartina skottsbergi*

(Rhodophyta, Gigartinales): Optimization of vegetative propagation for cultivation. J. Appl. Phycol. 11: 315–327.

Doty MS (1980) Outplanting *Eucheuma* species and *Gracilaria* species in the tropics. In Abbott IA, Foster MS, Eklund LF (eds), Pacific Seaweed Aquaculture. Pub. by the California Sea Grant College Program. Institute of Marine Resources. University of California, pp. 19–22.

Hansen JE (1980) Physiological considerations in the mariculture of the red algae. In Abbott IA, Foster MS, Eklund LF (eds), Pacific Seaweed Aquaculture. California Sea Grant College Program. Institute of Marine Resources, University of California, pp. 80–91.

Marín SL, Westermeier R, Melipillán J (2002) Simulation of alternative management strategies for red algae, luga roja, (*Gigartina skottsbergii* Setchell and Gardner) in southern Chile. Ecol. Mod. 154: 121–133.

McNeill SE, Page M, Falshaw R (2003) Field trials to optimise the timing and frecuency of pruning for cultivation of a New Zealand carragenophyte, *Gigartina atropurpurea*. J. Appl. Phycol. 15: 391–405.

Norris RE, Kim DH (1972) Development of thalli in some Gigartinaceae. In Abbott IA, Kurogi M (eds), Contributions to the Systematic of Benthic Marine Algae of the North Pacific. Japanese Society of Phycology, Kobe, pp. 265–269.

Romo H, Alveal K, Werlinger C (2001a) Growth of the commercial carragenophyte *Sarcothalia crispata* (Rhodophyta, Gigartinales) on suspended culture in central Chile. J. Appl. Phycol. 13: 229–234.

Romo H, Avila M, Candia A (2001b) Manual de Técnicas de Cultivo y Repoblación de "luga roja" (*Gigartina skottsbergii*). Publicación Técnica, Proyecto FONDEF D97 I1064. IFOP-Universidad de Concepción, Chile: 1–30.

Santelices B (1999) A conceptual framework for marine agronomy. Hydrobiologia 398/399: 15–23.

Westermeier R, Aguilar A, Sigel J, Quintanilla J, Morales J (1999) Biological basis for the management of *Gigartina skottsbergii* (Gigartinaceae, Rhodophyta) in southern Chile. Hydrobiologia 398/399: 137–147.

Zamorano J, Westermeier R (1996) Phenology of *Gigartina skottsbergii* (Gigartinales, Rhodophyta) in southern Chile. Hydrobiologia 326/327: 253–259.

Zar JH (1999). Biostatistical Analysis. Prentice Hall, New Jersey, 663. pp.

Journal of Applied Phycology (2006) 18: 315–321
DOI: 10.1007/s10811-006-9030-1

Can kelp extract (KELPAK®) be useful in seaweed mariculture?

D.V. Robertson-Andersson[1,*], D. Leitao[1], J.J. Bolton[1], R.J. Anderson[2], A. Njobeni[1] & K. Ruck[3]
[1]*Botany Department, University of Cape Town, Rondebosch 7701, South Africa;* [2]*Seaweed Unit, Marine and Coastal Management, Private Bag X2, Roggebaai, 8012, South Africa;* [3]*Jacobsbaai Sea Products, Private Bag X2, Rhine Road, Jacobsbaai 8050, South Africa*

*Author for correspondence: e-mail: droberts@botzoo.uct.ac.za

Key words: integrated aquaculture, kelp extract, *Ulva, Gracilaria, Ecklonia,* Kelpak®

Abstract

The addition of low concentrations of commercial kelp extract (*Ecklonia maxima*: Kelpak®) in addition to fertiliser has proven to be beneficial in agriculture. It triggers rooting in field crops, increases yields and has other useful effects, such as parasite reduction. Its efficacy has been attributed to the fact that Kelpak® is produced by a cold process, and is a high auxin/low cytokinin product. The aim of this study was to investigate if seaweeds (which do not have a root system) grown in culture systems, would benefit from the addition of Kelpak® or a combination of Kelpak® and fertilizer. A preliminary laboratory experiment was carried out by growing excised 15 mm tips of the red alga *Gracilaria gracilis* in culture dishes containing Provasoli Enriched Seawater medium to which various concentrations of Kelpak® were added. *Gracilaria* tips in some of the Kelpak® treatments (1:2500; 1:1000; 1:500) grew significantly better than the control. Further experiments were carried out on a pilot commercial scale at Jacobsbaai Sea Products Ltd. on the South African west coast. *Ulva lactuca* was grown in effluent from fish (turbot) culture, with additions of 1:5000, 1:2500 and 1:500 concentrations of Kelpak® once a week. The intermediate Kelpak® concentration (1:2500) produced the highest growth of *Ulva* in the turbot water, while the highest Kelpak® concentration (1:500) inhibited *Ulva* growth. In another *Ulva* experiment, various combinations of aquaculture effluent water, commercial fertiliser and Kelpak® at 1:2500 were used. Best growth of *Ulva* was obtained in turbot water containing both fertiliser and Kelpak®. The results suggest that Kelpak® could be useful in commercial seaweed mariculture operations.

Introduction

The use of seaweed extracts as soil drenches and foliar sprays on agricultural plants is increasing, even though the literature on seaweed extracts is contradictory. Some studies suggest that seaweed extracts have no effect on plant growth (Verkleij, 1992). In contrast, documented studies on a commercial extract of the brown kelp *Ecklonia maxima* (Kelpak®: Featonby-Smith & van Staden, 1983, 1987; Crouch, 1990) have reported that these seaweed extracts improve the growth rates and yields of crops, as well as preventing pests and improving the overall quality of the product. Many of the physiological responses shown by crop plants treated with seaweed concentrates are thought to be due to cytokinins and auxins, a number of which have been demonstrated to occur in Kelpak® (Stirk & van Staden, 1996, 2004; Crouch et al., 1992). The beneficial effects of this product have been attributed to the plant hormone content of the extract. Since seaweed concentrates are applied in small doses, the active compounds in these concentrates need to be effective at low concentrations. Many studies have looked at the effect of applying plant growth regulators (PGR), such as auxins and gibberellins, on seaweed growth (review in Lobban & Harrison, 1997; Yokoya et al., 1999, 2003).

Kelpak® is a commercially available seaweed extract and is marketed as a plant growth stimulator due to its hormonal content and not its nutrient content

(Featonby-Smith & van Staden, 1983, 1987). It is manufactured by Kelp Products (Pty) Ltd. in Simons Town, South Africa, from epiphyte-free fronds and stipes of the brown alga *Ecklonia maxima* (Osbeck) Papenfuss, using a cold cell-burst process (Verkleij, 1992; Stirk & van Staden, 1996, 2004; Stirk et al., 2004). This process excludes the use of heat, chemicals or dehydration that could affect some organic components of the concentrate (Verkleij, 1992).

The aim of this paper was to test the effects of Kelpak on growth of seaweeds in culture. This was initially done in controlled conditions in the laboratory, using excised tips of *Gracilaria*, which are easy to grow and measure. Subsequently, pilot commercial-scale experiments were carried out to test the effects of the kelp extract on the growth of *Ulva* in tank culture on a commercial abalone/fish farm. *Ulva* was chosen for this, as we have considerable experience in growing *Ulva* in these systems as potential feed for abalone. Nutrient content of the cultured *Ulva* was also measured, with regard to its use as abalone feed. Experiments on *Gracilaria* growth in these systems were not as successful as *Ulva*, particularly due to low temperatures on the farm, and these are not presented.

Björnsäter and Wheeler (1990) and De Busk et al. (1986) showed that additions of fertilizer to nutrient-depleted water significantly increased Specific Growth Rate (SGR) of *Ulva*. Species of *Ulva* have been successfully grown in effluent water (abalone, fish and human) and SGR's are significantly higher compared to *Ulva* sp. grown in seawater (Ryther et al., 1975; Vandermeulen & Gordin, 1990; Cohen & Neori, 1991; Neori et al., 1991; Neori, 1996; Jimenez del Rio et al., 1996; Shpigel et al., 1997; Goldberg et al., 1998). As some land plants grown with fertilizer and Kelpak®, showed significant increases in SGR (Featonby-Smith & van Staden, 1983, 1987; Crouch, 1990), the authors wished to test this observation for seaweeds using an effluent aquaculture medium as the source of nutrients for the seaweeds.

Materials and methods

Four different experiments were run in order to investigate the effects of Kelpak® on cultivated algae. Kelpak® adds only a very small amount of nutrients as a proportion of the total nutrients applied to the seaweeds at these low concentrations: it has an N: P ratio in mg N/P per g DW of Kelpak® of 55.98: 49.15 ($\pm$ 0.01; $n = 6$) (Robertson-Andersson, 2004). This was tested in the first instance in the laboratory, using excised tips of *Gracilaria*, selected for ease of growth and measurement, followed by measurements of growth rates of *Ulva* in an existing experimental system on a commercial abalone farm. Growth of *Ulva* was tested in combinations of seawater, abalone and turbot effluent, with various concentrations of Kelpak®, with and without additional fertilization. Nitrogen content of the seaweed thalli was measured as a physiological parameter of seaweed health.

Laboratory experiments

The material was collected one day prior to the start of the experiment. *Gracilaria gracilis* (Stackhouse) Steentoft, Irvine *et* Farnham was collected from Saldanha Bay on the South African west coast, and was washed with running fresh water and sterile seawater and brushed with a paint brush to eliminate contaminants. The darkest thallus fragments were selected and 15 mm unbranched apical segments cut.

One-third strength standard Provasoli Enriched Seawater medium (PES) was prepared according to a standard recipe (Starr & Zeikos, 1987). The Kelpak® treatments were; 1:100, 1:250, 1:500, 1:1000, 1:2500, and 1:5000 added to one third strength PES. The control consisted of one-third strength PES medium with no Kelpak® added. The culture medium was changed every 2 days. The experiments were carried out at 15 °C temperature with an irradiance 50–80 μmol photons m^{-2} s^{-1} provided by cool white fluorescent tubes and a photoperiod of 16 h. (light) : 8 h. (dark). Culture vessels were 200 cm^3 crystallizing dishes to which 200 mL of PES was added as well as five 15 mm apical segments of *Gracilaria gracilis.* There were four replicates for each treatment. The flasks were moved within the experimental setup on a daily basis to ensure a uniform environment for all flasks. The initial, and on completion of the experiment, final biomass (in fresh weight) was measured, from which the SGR could be calculated using the following formula (Evans, 1972) and calculated as: $SGR = [\ln (W_t/W_0)]/(t_t - t_0)$

Where W_0 and W_t are initial and final wet weights (wwt) in grams and t_0 and t_t are initial and final times in days respectively.

Number of branches per tip was measured for each treatment at the end of the experiment.

Pilot commercial scale experiments

The commercial scale experiments were run on the Jacobsbaai Sea Products (JSP) Aquaculture Farm (west coast, South Africa), a land-based intensive mariculture operation of *ca.* 11 h. The farm predominantly cultivates abalone (*Haliotis midae*) and turbot (*Scopthalmus maximus*). The *Ulva* material used was *U. lactuca* Linnaeus, originally from Simon's Town Harbour near Cape Town, which had been grown in tanks on the farm for a year prior to the experiments. The experimental seaweeds were cultivated in 96 L white PVC (0.60 m × 0.40 m × 0.40 m) tanks, using an air curtain produced by a U-shaped pipe system. Abalone-effluent and turbot-effluent generated on the farm were the two culture media used in these experiments with unfiltered seawater as a control. The experiments ran over winter from May 2002 to August 2002. The average seawater temperature on the farm was 14.6 °C (min 6 °C, max 20 °C) (Robertson-Andersson, 2004).

The first experiment consisted of determining the correct Kelpak® concentration to use in conjunction with fertilizer to obtain optimal SGR. The second and third experiments were run to test the effect of fertilizer and Kelpak® concentration (as determined from the first experiment) in stand alone additions or combined in both effluent culture media. The water volume exchange rate was 20 times per day. The fertilizer used was a combination of Maxiphos® (a commercially available fertilizer) and ammonium sulphate, with a ratio of 10:1 (Duke et al., 1986). Fertilizer and/or Kelpak® was added once a week. After the addition of the fertilizer no water exchange occurred for 20 h., after which a normal water exchange resumed. The seaweeds were acclimatized to the new culture conditions for two weeks after which all material was harvested and then tanks restocked to original stocking density (2 kg m^{-2}) using the harvested material. The experiment was then run for two weeks and the seaweeds collected for data analysis.

To determine the optimum concentration of Kelpak®, a tank system was set up at JSP to run 12 tanks with seaweed on turbot effluent. A control system consisted of a seawater control (four tanks) and a turbot effluent control (three tanks). The remaining 9 tanks were divided into 3 series of three turbot effluent tanks, with the following Kelpak® dilutions (1:500; 1:2500; and 1:5000).

After completion of the first experiment the seaweeds were left for two weeks to acclimatize to normal farm conditions (background nutrient concentrations) after which the algal biomass was harvested and used to restock the tanks. SGR of *Ulva* determined which Kelpak® concentration would be used in the following two effluent experiments. The Kelpak® concentration of 1:2500 showed best results and was used in both the turbot and abalone effluent media. The experiment was then rerun for two periods of two weeks in each effluent medium with the following treatments:

- 3 effluent tanks with fertilizer and a 1:2500 concentration of Kelpak®
- 3 effluent tanks with fertilizer only
- 3 effluent tanks with Kelpak® at 1:2500 concentration only
- 4 seawater and 3 effluent control tanks

A two week break between experiments was left to allow the algae to acclimatize to the new effluent media.

Nutrient analysis

During harvest, samples were taken for wet to dry weight ratio analysis, and tissue total nitrogen was determined using the micro-Kjeldahl technique (Solorzano, 1969).

Statistical analysis

The effect of different concentrations of Kelpak® treatments on the SGR were statistically analyzed using ANOVA, single factorial analysis of variance ($p = 0.05$) (STATISTICA V6.1), to test the null hypothesis that the means of the SGR of all tested Kelpak® concentrations were not significantly different. The least significance difference (LSD) test or planned comparison test was conducted at the 95% confidence level, to distinguish significantly different results. The data were not transformed. The pilot commercial scale ex periments showed that there was a greater effect on the seaweed growth during the second run in all the experiments implying that the algae required two weeks to acclimatize to the conditions. Thus the statistics quoted in this paper are for the second run only.

Results

Laboratory experiment

All concentrations of Kelpak® tested, with the exception of the most concentrated (1:100), significantly increased *Gracilaria* tip growth (Figure 1). G. *gracilis* apical segments had significantly higher SGR after 15 days in PES medium with the 1:1000 Kelpak® dilution, followed by 1:500 and 1:2500 Kelpak® dilution

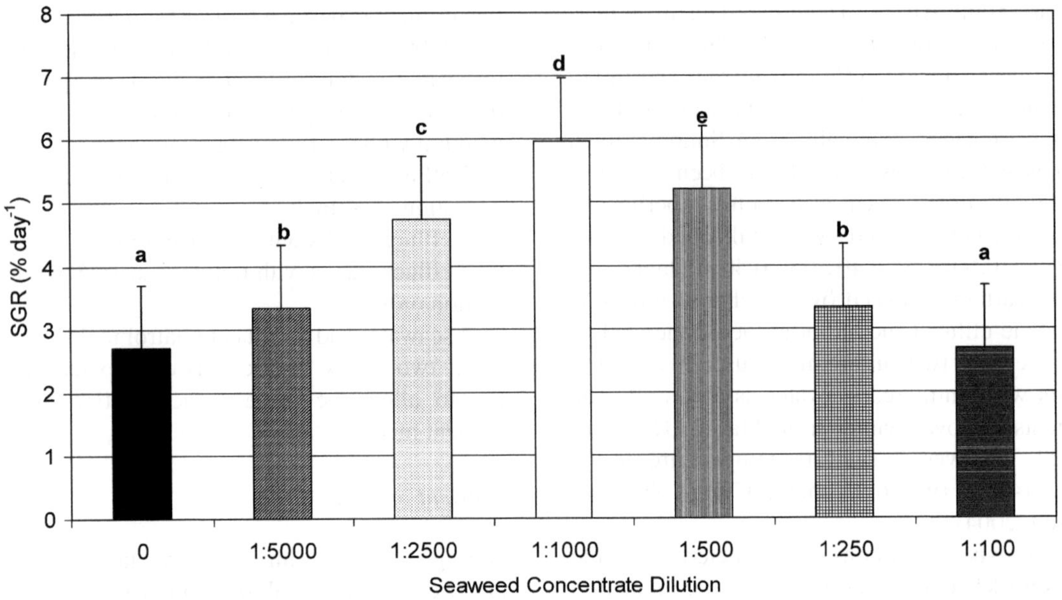

Figure 1. The effect of various Kelpak® dilutions on the SGR (%.d^{-1}) of apical segments of *Gracilaria gracilis* after 15 days growth under laboratory conditions. Different letters indicate significant differences ($p < 0.05$: one-way ANOVA and LSD post-hoc test). Vertical lines represent ± one Standard Error, $p = 0.05$.

compared to the control and other treatments. Apical segments growing at 1:250 and 1:5000 Kelpak® dilution had significantly higher SGR compared to the control and 1:100 Kelpak® dilution. There was no significant difference in the SGR of apical segments growing in 1:100 kelp concentrate dilution and control. Although the 1:5000 treatment showed slightly increased branching, there was no significant change in the number of branches of *G. gracilis* in the different Kelpak® concentrations compared to the control. Branching was, however, significantly reduced in the 1:100 treatment, compared to the 1:5000 treatment (Table 1).

Effect of various Kelpak® concentrations added to turbot effluent in pilot commercial scale experiment

The SGR of the *Ulva* grown using the 1:5000 and 1:2500 Kelpak® concentrations were not significantly different to the turbot control (Table 2). The SGR of the *Ulva* grown in both the seawater and the 1:500 Kelpak® concentration were significantly lower than all other treatments (LSD post-hoc test, $p < 0.01$ in all cases). Material from the seawater treatment had significantly lower tissue nitrogen levels (Table 2), compared to all other treatments (ANOVA, $df = 20$, $p < 0.01$; LSD post-hoc test, $p < 0.01$). The 1:2500 treatment had slightly higher growth and nitrogen content than the other effluent treatments, but not significantly so.

Effect of Turbot effluent - Kelpak® - Fertilizer combination in a pilot commercial scale experiment

In all cases the addition of either fertilizer, or Kelpak® (1:2500) or a combination of both, significantly

Table 1. The effect of various Kelpak® dilutions on the total number of branches and the average branches of apical segments of *Gracilaria gracilis* per dish after 15 days in growth under laboratory conditions.

Treatment	Average no of branches
Control	1.4 ± 1.6 ab
1:5000	1.9 ± 1.6 a
1:1000	1.3 ± 1.8 ab
1:500	0.9 ± 1.7 ab
1:100	0.7 ± 0.9 b

Different letters indicate significant differences ($p < 0.05$: one-way ANOVA and LSD post-hoc test).

Table 2. The effect of Kelpak® dilutions on SGR (% d^{-1}) and tissue nitrogen (mg N g^{-1} DW) of *Ulva lactuca* cultivated in turbot effluent on a pilot commercial scale.

Treatment	SGR (% d^{-1})	Tissue N (mg N g^{-1} DW)
Turbot Effluent		
Turbot control	4.2 ± 1.2 b	58.9 ± 2.0 a
Turbot + Fertilizer	6.1 ± 0.8 a	59.8 ± 5.5 a
Turbot + Fertilizer + Kelpak®	6.6 ± 0.4 a	62.0 ± 6.5 a
Turbot + Kelpak®	5.1 ± 0.3 a	62.6 ± 4.8 a
Seawater control	3.5 ± 0.1 c	48.4 ± 6.1 b

All values are ± one Standard Error, $p = 0.05$ Different letters indicate significant differences ($p < 0.05$: one-way ANOVA and LSD post-hoc test).

Table 3. The effect of combinations of fertilizer and or Kelpak® concentrate (1:2500) on SGR (% d^{-1}) and tissue nitrogen (mg N g^{-1} DW) of *Ulva lactuca* cultivated in turbot and abalone effluent on a pilot commercial scale.

Treatment	SGR (% d^{-1})	Tissue N (mg N g^{-1} DW)
Abalone effluent		
Abalone control	6.1 ± 1.2 a	60.6 ± 3.0 a
Abalone + Fertilizer	6.7 ± 1.0 a	63.7 ± 2.5 a
Abalone + Fertilizer + Kelpak®	7.2 ± 0.0 a	58.0 ± 4.5 a
Abalone + Kelpak®	6.3 ± 0.0 a	59.0 ± 2.3 a
Seawater control	4.9 ± 0.2 b	53.0 ± 6.5 b

All values are ± one Standard Error, $p = 0.05$. Different letters indicate significant differences ($p < 0.05$) (one-way ANOVA and LSD post-hoc test).

increased the SGR of the *Ulva* above the turbot control and the seawater control (Table 3). The combined Kelpak® – fertilizer – turbot effluent media was significantly higher than all other treatments except the turbot – fertilizer treatment (LSD post-hoc test, $p < 0.01$).

The tissue N of *Ulva* in the seawater control (Table 3) was significantly lower than all other treatments (ANOVA, $df = 20$, $p < 0.01$; LSD post-hoc test, $p < 0.01$).

Effect of Abalone effluent – Kelpak® – fertilizer combination in a commercial scale experiment

The results of the abalone effluent experiments were similar to those of the turbot effluent experiment (Table 3). Addition of Kelpak® (1:2500) and/or fertilizer increased the SGR of *Ulva* above that of the abalone effluent and seawater controls. The abalone effluent combined with Kelpak® (1:2500) and fertilizer had a significantly higher SGR than all other treatments (ANOVA, $df = 20$; $p < 0.01$; LSD post-hoc test, $p < 0.01$) except the abalone – fertilizer treatment. In contrast to the turbot experiments, there was no significant difference in SGR between the abalone control and effluent plus Kelpak® treatments in the abalone effluent experiments.

There was a significant decrease in tissue nitrogen of *Ulva* cultivated in seawater compared to all other treatments (ANOVA, $df = 20$; $p < 0.01$; LSD post-hoc test, $p < 0.01$).

Discussion

As far as we are aware, this is the first study to examine the effect of commercial kelp concentrate on seaweed growth. This is surprising, as the beneficial effect of seaweed concentrate on the growth, yield and disease reduction in crop plants has been well documented over the past 30 years (Finnie & van Staden, 1985; Featonby-Smith & van Staden, 1983, 1987; Stirk & van Staden 1996, 2004; Stirk et al., 2004). Finnie and van Staden (1985) demonstrated that the concentration ratio of *Ecklonia maxima* kelp extract is an important factor in controlling its efficiency. In tomato plants, strong concentrations (1:100 seaweed extract: water) were found to have an inhibitory effect upon root growth, whereas weak concentrations (1:600) had a stimulatory effect. The Kelpak® concentrations used in this study are within the range commonly used in land plant studies.

Treatment with 1:5000 Kelpak® concentration did not significantly increase SGR of the seaweed, which is in agreement with Beckett and van Staden (1990), who showed that the growth of wheat was not stimulated by low concentrations (1:1000 retail product–5 times more dilute than commercial product) of Kelpak® compared to the controls. The 1:2500 Kelpak® concentration caused the highest SGR increase in the tank experiments on the aquaculture farm. This is slightly more concentrated than the Kelpak® treatment which produced maximum growth in *Gracilaria* tips in our laboratory experiment. This could be due to the fact that on the commercial farm a once off pulse addition was used, while in the laboratory the segments were constantly exposed to a more dilute concentration. The field experiment results are in agreement with studies in which Kelpak® used at a concentration of 1:2 500 and applied regularly, improved the total biomass of *Beta vulgaris* and *Phaseolus vulgaris* (Crouch, 1990) and the root growth of cucumber plants (Nelson & van Staden, 1984).

A pronounced inhibitory effect was observed using a 1:500 concentration in both the laboratory and the aquaculture farm, for both *Gracilaria* and *Ulva*. This effect was also found in the study of Finnie and van Staden (1985) who reported inhibition of tomato roots at this concentration.

The experiments using abalone and turbot effluent water indicate that additions of both fertilizer and Kelpak® (1:2500 concentration) significantly increase SGR of *Ulva* above that of effluent or seawater controls alone. Moreover, a combination of Kelpak® (1:2500 concentration) with fertilizer significantly increased SGR of *Ulva* over all other treatments. This increase in SGR is similar to that of the kelp concentrate in agriculture, where it is used as a very diluted 'root drench' for a short period, in addition to traditional fertilisers. Therefore the positive effects on agricultural crops are not due to addition of major nutrients from the Kelpak®. This might be the case in the seaweed experiments in this study. The effects on agricultural crops have been reported to be caused by the concentrations of plant growth hormones (especially auxins and cytokinins) in Kelpak®. It is known from the literature that seaweeds contain plant growth substances (Bradley, 1991; Stirk et al., 2004), and that additions of plant growth hormones to media increases growth (callus production) of various red algae, including *Gracilaria* (e.g. Yokoya et al., 1999, 2003).

It was not in the scope of this study to prove that plant growth hormones are the active ingredients in Kelpak®, causing a positive effect on growth in *Gracilaria* and *Ulva.* The reasons for the increase in SGR are not understood, but it is thought that the hormonal content, particularly cytokinin, plays an important role (Feantonby-Smith & van Staden, 1983, 1987; Bradley, 1991). Other studies have looked at the effects of additions of cytokinin (Blunden & Wildgoose, 1977) and synthetic cytokinin (Finnie & van Staden, 1985) on SGR and have come to the same conclusions.

Tissue N values decreased in all treatments between the first and second runs where SGR increased. This relationship follows that described by Rosenberg and Ramus, 1982; Duke et al., 1986, 1987 and 1989a,b, with an increase in SGR causing a decrease in tissue nitrogen. This is due to the faster SGR requiring more nutrients. The increased growth with Kelpak®, compared to the other treatments did not, however, decrease N, meaning that there is potential for using the *Ulva* in bioremediation, and as high quality abalone feed.

All experiments showed that the SGR was slightly higher during the second run compared to the first run. This may indicate that the alga has an acclimatization period to the experimental conditions before an effect can be monitored. Observations (unpublished) showed a similar acclimatization period for *Gracilaria gracilis*, a species cultivated on a pilot commercial scale at JSP. Results for *Gracilaria* are not shown as the SGR was low due to low nightly water temperatures (5.5 °C) present at the time of the study. Branching in *Gracilaria* however, appeared to correlate well with the Kelpak® concentrations similar to that noted in the laboratory culture (unpublished data).

Conclusions

The results of the seaweed cultivation experiments using Kelpak® dilutions are similar to previous findings on the effect of seaweed extract applications on certain field crops (Featonby-Smith & van Staden, 1983). The highest SGR in the pilot commercial experiments was obtained using a concentration of 1:2500 of Kelpak®. The weakest concentration of 1:5000 did not change the SGR. The highest concentration (1:500) used in this study showed inhibitory effects. Thus, Kelpak®, may have commercial potential in the seaweed mariculture industry.

Acknowledgments

The authors would like to thank the staff at Jacobsbaai Sea Products Ltd., as well as Kelp Products (Pty) Ltd. Funding was supplied through the Swedish South African Collaborative Programme (NRF/SIDA), the Department of Environment and Tourism, and the Marine and Coastal Management Branch of DEAT.

References

Beckett RP, van Staden J (1990) The effect of seaweed concentrate on the growth and yield of potassium stressed wheat. Plant and Soil 116: 20–36.

Björnsäter BO, Wheeler PA (1990) Effect of nitrogen and phosphorus supply on growth and tissue composition of *Ulva fenestrata* and *Enteromorpha intestinalis* (Ulvales, Chlorophyta). J. Phycol. 26: 603–611.

Blunden G, Wildgoose PB (1977) The effects of aqueous seaweed extract on sugar beet. J. Sci. Food Ag. 28: 121–125.

Bradley PM (1991) Plant hormones do have a role in controlling growth and development of algae. J. Phycol. 27: 317–321.

Cohen I, Neori A (1991) *Ulva lactuca* biofilters for marine fishpond effluents: I: Ammonium uptake kinetics and nitrogen content. Bot. Mar. 34: 475–482.

Crouch IJ (1990) The effect of seaweed concentrate on plant growth. PhD Thesis. Department of Botany, University of Natal, South Africa.

Crouch IJ, Smith MT, van Staden J, Lewis MJ, Hoad GV (1992) Identification of auxins in commercial seaweed concentrate. Pl. Physiol. 139: 590–594.

De Busk TA, Blakeslee M, Ryther JH (1986) Studies on the outdoor cultivation of *Ulva lactuca* L. Bot. Mar. 29(5): 381–386.

Duke CS, Lapointe BE, Ramus J (1986) Effects of irradiance on growth, RuBPCase activity and chemical composition of *Ulva* species (Chlorophyta). J. Phycol. 22(3): 362–370.

Duke CS, Litaker W, Ramus J (1987) Seasonal variation in RuBP-Case activity and N allocation in the Chlorophyte seaweeds *Ulva curvata* (Kutz.) and *Codium decorticatum* (Woodw.) Howe. J. Exp. Mar. Biol. Ecol. 112: 145–164.

Duke CS, Litaker W, Ramus J (1989a) Effects of the temperature, nitrogen supply and tissue nitrogen on ammonium uptake rates of the Chlorophyte seaweeds *Ulva curvata* and *Codium decorticatum*. J. Phycol. 25: 113–120.

Duke CS, Litaker W, Ramus J (1989b) Effect of temperature on nitrogen limited growth rate and chemical composition of *Ulva curvata* (Ulvales, Chlorophyta). Mar. Biol. 100: 143–150.

Evans GC (1972) The quantitative analysis of plant growth. Studies in Ecology. Blackwell Scientific Publications, Oxford: 247–254.

Featonby-Smith BC, van Staden J (1983) The effect of seaweed concentrate on the growth of tomatoes in nematode infested soil. Scient. Hort. 20: 137–146.

Featonby-Smith BC, van Staden J (1987) Effects of seaweed concentrate on grain yield in barely. S. Afr. J. Bot. 53: 125–128.

Finnie JF, van Staden J (1985) Effect of seaweed concentrate and applied hormones on *in vitro* cultured tomato roots. J. Pl. Physiol. 120: 215–222.

Goldberg R, Clark P, Wikfors GH, Shpigel M (1998) Performance of *Ulva ridgida* as a biofilter in a flow-through mariculture system. J. Shellf. Res. 17: 345–355.

Jimenez del Rio M, Ramazanov Z, Garcia-Reina G (1996) *Ulva rigida* (Ulvales, Chlorophyta) tank culture as biofilters for dissolved inorganic nitrogen from fishpond effluents. Hydrobiologia 326/327: 61–67.

Lobban CS, Harrrison PJ (1997) Seaweed Ecology and Physiology. Cambridge University Press, Cambridge. pp. 366.

Nelson WR, van Staden J (1984) The effect of seaweed concentrate on growth of nutrient stressed cucumbers. Scient. Hort. 19: 81–82.

Neori A, Cohen I, Gordin H (1991) *Ulva lactuca* biofilters for marine fish pond effluents. II. Growth rate, yield and C:N ratio. Bot. Mar. 34: 483–489.

Neori A (1996) The type of N supply (ammonia or nitrate) determines the performance of seaweed biofilters integrated with intensive fish culture. Isr. J. Aquacult. 48: 19–27.

Robertson-Andersson DV (2004) The cultivation of *Ulva lactuca* (Chlorophyta) in an integrated aquaculture system, for the purposes of abalone feed and the bioremediation of aquaculture effluent. M. Sc. Thesis, University of Cape Town, pp. 266.

Rosenberg G, Ramus J (1982) Ecological strategies in the seaweeds *Gracilaria foliifera* (Rhodophyceae) and *Ulva* sp. (Chlorophyceae): Soluble nitrogen and reserve carbohydrates. Mar. Biol. 66: 251–259.

Ryther JH, Goldman JC, Gifford CE, Huguenin JE, Wing AS, Clarner JP, Williams LD, Lapointe BE (1975) Physical models of integrated waste recycling in marine polyculture systems. Aquaculture 5: 163–177.

Shpigel M, Gasith A, Kimmel E (1997) A biomechanical filter for treating fish-pond effluents. Aquaculture 152: 103–117.

Solorzano L (1969) Determination of ammonium in natural waters by the phenol-hypochlorite method. Limnol. Oceanogr. 14: 799–801.

Starr RC, Zeikos JA (1987) UTEX – the culture collection of algae at the University of Texas at Austin. J. Phycol. 23: 38–99.

Stirk WA, Arthur GD, Lourens AF, Novak O, Strnad M, van Staden J (2004) Changes in cytokinin and auxin concentrations in seaweed concentrates when stored at an elevated temperature. J. Appl. Phycol. 16: 31–39.

Stirk WA, van Staden J (1996) Comparison of cytokinin- and auxin-like activity in some commercially used seaweed extracts. J. Appl. Phycol. 8: 503–508.

Stirk WA, van Staden J (2004) Potential new applications for the southern African kelps. S. Afr. J. Bot. 70: 145–151.

Vandermeulen H, Gordin H (1990) Ammonium uptake using *Ulva* (Chlorophyta) in intensive fishpond systems: Mass culture and treatment of effluent. J. Appl. Phycol. 2(4): 363–374.

Verkleij FN (1992) Seaweed extracts in agriculture and horticulture: A Review. Biol. Ag. Hort. 8: 309–324.

Yokoya NS, Kakita H, Obika H, Kitamura T (1999) Effects of environmental factors and plant growth regulators on growth of the red alga *Gracilaria vermiculophylla* from Shikoku Island, Japan. Hydrobiologia 398/399: 339–347.

Yokoya NS, Plastino EM, Artel R (2003) Physiological responses and pigment characterization of two colour strains of the carrageenophyte *Hypnea musciformis* (Rhodophyta). In Chapman A.R.O, Anderson R.J., Vreeland V.J., Davison I.R (Eds), Proceedings of the 17th International Seaweed Symposium. Oxford University Press, Oxford: 425–434.

Journal of Applied Phycology (2006) 18: 323–334
DOI: 10.1007/s10811-006-9029-7

A remote sensing approach to estimating harvestable kelp biomass

M. S. Stekoll[1,*], L. E. Deysher[2] & M. Hess[2]
[1]*Department of Natural Sciences, University of Alaska SE, 11120 Glacier Highway, Juneau, AK 99801 USA;*
[2]*Ocean Imaging, 201 Lomas Santa Fe Drive, Suite 370, Solana Beach, CA 92075*

*Author for correspondence: e-mail: ffmss@uaf.edu; phone: 1-907-465-6279

Key words: kelp mapping, remote sensing, biomass, kelp management, kelp canopy, *Nereocystis*

Abstract

Regulations of the Alaska Department of Fish and Game require that all fisheries in the state have a harvest management plan. In southeast Alaska two species of floating kelps, *Nereocystis luetkeana* and *Alaria fistulosa*, have been commercially harvested since 1992 for use as agrochemicals by the Alaska Kelp Company. However, there is currently no harvest management plan for this fishery. The lack of a formalized management plan is one factor that has kept the kelp industry from expanding in the state. We have employed an aerial digital multispectral imaging system (DMSC) calibrated with ground truthing for performing such an assessment. The system can be flown at varying altitudes to achieve spatial resolutions ranging from 0.5 to 2 m. Rapid ground truthing techniques were developed using morphometric measurements to predict biomass. Analysis of the DMSC imagery showed that good correlations could be developed between the multispectral imagery and kelp biomass estimates collected at the ground-truth sites. Repeatable estimates of kelp bed area derived from the multispectral imagery could be made at varying tidal levels. However, broad scale maps of kelp biomass suitable for estimating harvest rates could not be made at different tide levels. Multispectral imagery suitable for this purpose must be collected at a standard tidal level.

Introduction

Kelp beds with floating canopies cover much of the near shore ocean surface along the West Coast of North America (Foster & Schiel, 1985). There are major beds of these kelps in the Alexander Archipelago of southeast Alaska. A survey of floating kelps in southeast Alaska in the early 1900's found over 100 separate beds containing an estimated seven million metric tons of plant biomass. The majority of the plants were *Nereocystis luetkeana* (Mertens) Postels *et* Ruprecht with *Alaria fistulosa* Postels *et* Ruprecht and *Macrocystis* sp. making up the remainder (Frye, 1915). Subsequent surveys have confirmed the locations and area extent of these beds, but have noted that the dominant species making up the beds has changed since 1915 (van Tamelen & Woodby, 2001). Current accurate estimates of kelp biomass are lacking at this time. Although selected areas of *Macrocystis* have been surveyed in southeast Alaska (van Tamelen & Woodby, 2001) there has been no survey done for *Nereocystis* since 1915.

There are currently two commercial uses for floating kelps in Alaska. The giant kelp, *Macrocystis*, is harvested in southeast, south-central and western Alaska, mainly for the commercial herring roe-on-kelp harvest (Stekoll, 1998). In addition, in southeast Alaska the two other species of floating kelps, *N. luetkeana* and *A. fistulosa*, have been commercially harvested since 1992 for use as agrochemicals. The major harvester of this resource is the Alaska Kelp Company (formerly Pacific Mariculture Company, Inc.) which processes the seaweeds for use as a plant fertilizer supplement. The product (formerly sold as OptiCrop and now as Alaska Kelp or Garden Grog) is used in various agricultural and horticultural applications and has enjoyed moderate commercial success.

In order to ensure the sustainability of *Nereocystis* harvesting, it is necessary to develop a stock assessment method for management of the kelps. Because resource managers normally use estimated biomass to regulate harvests, the key to the harvest management plan is the ability to determine the biomass of the kelp beds. However, estimation of the biomass of floating kelps is problematic with respect to time and expense needed for accurate estimates. The Alaska Department of Fish & Game is currently considering management of these beds using the area covered by the kelp canopy (van Tamelen & Woodby, 2001).

Mapping of kelp beds by remote sensing has primarily used near-infrared aerial photography for delineating surface canopy area. Aerial photography has been used in Canada (Foreman, 1975) and for over 30 years to map the *Macrocystis pyrifera* canopies in California (North et al., 1993). This method is used for determining kelp bed lease data by the California Department of Fish and Game. SPOT satellite imagery was found to be useful for estimating the biomass of floating canopies of *M. pyrifera* (Belsher & Mouchot, 1992). However, work on *Macrocystis* beds in California has shown that only the largest beds in this region can be effectively mapped due to the 20-meter spatial resolution currently available with the SPOT multispectral sensor (Deysher, 1993). The kelp beds in Alaska are comparable to the small and medium sized beds in California. A spatial resolution finer than 20 m is required to quantify the biomass of these beds. In addition, kelp beds in Alaska can be composed of three different kelp species. Kelp monitoring surveys in Washington State using a multispectral imaging system found that a 4 m spatial resolution was not sufficient to distinguish *Macrocystis* from *Nereocystis*.

The research described here addresses the biomass and area assessment aspect of a kelp harvest management plan. The objective of our research was to develop a reliable and cost effective method for estimating the area and biomass of the two species of kelps (*Nereocystis luetkeana*, and *Alaria fistulosa*) in southeast Alaska that have potential for a viable fishery

Methods

Site selection

The study area was a harvest area used by the Alaska Kelp Company. The site lies north of Pt Baker at the south end of Keku Strait along the southwest coast of Kupreanof Island (Figure 1) at approximately 133°24′W and 56°29′N. The study area consisted of approximately 20 distinct kelp bed areas that ranged in size from 0.2 to 12 hectares. Many of the beds fringe the small islands and shallower reefs in the area and comprise mainly *A. fistulosa. Nereocystis* was found on the outside of the *Alaria* beds and on reefs with more wave exposure and current flow.

Ground truthing

In 2002 seven separate kelp beds in the permitted harvest area north of Pt Baker were selected and classified by their estimated kelp densities. A surface buoy was anchored in the middle of the each of the selected sites. Scuba divers swam transects of 20 to 40 m parallel with the outer edge of the kelp canopy, starting at the buoy anchor. Density counts were taken with 1 m square quadrats placed every 4 m along the transect. All *Nereocystis* and *Alaria* were counted. Several *Nereocystis* plants (thalli) to represent all sizes of plants were haphazardly selected near each transect and taken back to Pt Baker for morphometric measurements. In order to determine the consistency of the morphometric correlations from more distant sites, we also collected about 90 *Nereocystis* plants near Juneau, Alaska. Fresh weight was determined for each plant along with the same suite of measurements made on the plants near Pt Baker.

The following measurements were taken on individual plants collected from sites near Pt Baker and in Juneau in the summer of 2002.

Stipe length: measured from the holdfast to the top of the bulb (pneumatocyst).
Blade length: measured from the top of bulb to the end of longest blade.
Bulb diameter: the outside diameter of the bulb.
Sub-bulb diameter: the diameter of the stipe 15 cm below the bottom of the bulb.
Blade weight: the fresh weight of all of the blades.
Total weight: the fresh weight of the blades, stipe and bulb.

Based on correlation results from 2002, in 2003 only the bulb and sub-bulb diameter and the total fresh weight were measured.

In 2003 the density counts were performed by preselecting eight sites of varying density from beds in the study site. The center of each of the selected density sites was marked with a buoy. At the surface four sets of two floating quadrats (0.5 × 2.0 m) were placed orthogonal to each other ("+" shape), centered on the

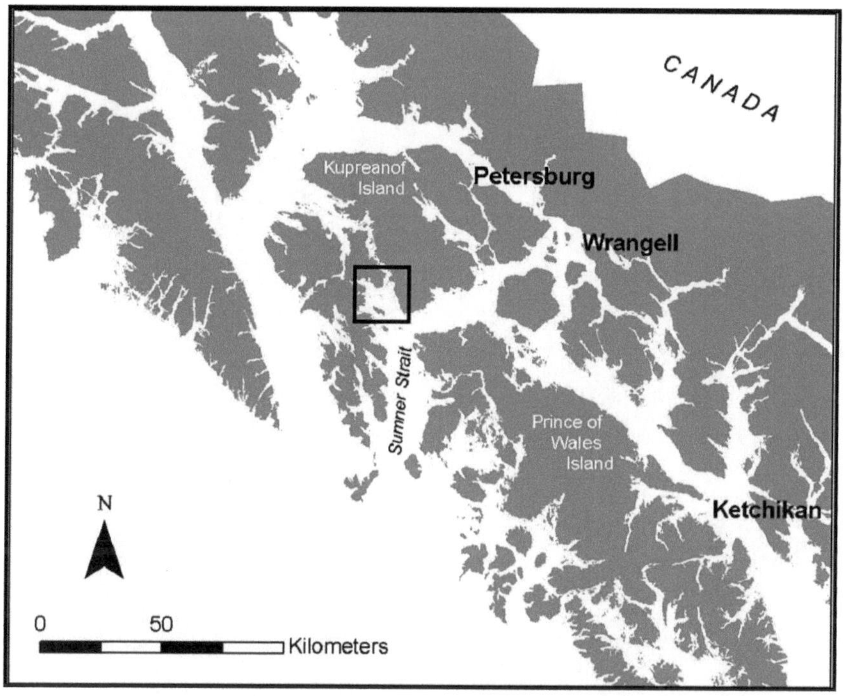

Figure 1. Study area in Keku Strait at the south end of Kupreanof Island.

buoy. In essence we sampled a 4+ m diameter circle along two diameters that were at right angles to each other. The number of *Nereocystis* bulbs in each quadrat was counted and the diameters of the bulb and sub-bulb sections of the thallus were measured using calibrated plastic forceps. Kelp densities on the bottom were estimated by scuba divers counting plants within quadrats (2 × 0.5 m) radiating away from the buoy anchor weight. In addition about 100 *Nereocystis* plants were harvested by scuba and used for additional data for morphometric measurements and biomass.

At a separate bed that was to be commercially harvested, we arbitrarily selected ten locations inside the southeast facing edge of the bed. At each sampling location we measured the diameters of the bulb and the sub-bulb of all *Nereocystis* plants lying within two 2.0 × 0.5 m quadrats. An estimated plant density and biomass per area in the area to be harvested were calculated from these data.

In 2002 the estimated plant biomass in the density sites was determined by multiplying the densities as determined from the scuba transects by the mean plant biomass as determined from the plants harvested near the transects. These estimated biomasses and plant densities were used in the calibration of the digital multispectral camera (DMSC) imagery. In 2003 the biomasses in the density plots were estimated from the morphometric data. Each individual plant weight was estimated using the regression equation for the relationship between the weight of a plant and its sub-bulb diameter. Final biomass estimates for the sampled quadrats were calculated from the sum of the individual plant weights within the quadrats.

Alaria biomass was estimated in 2003 by determining the mean fresh weight per meter length of the blade region and multiplying by the total length of *Alaria* in each quadrat. This number was adjusted by a factor (1.33) to account for the subsurface *Alaria* biomass determined from scuba surveys.

DMSC flights

The initial DMSC image set was collected on July 14, 2002, 9:11–10:27 ADT (17:11–18:27 GMT). The area covered by the DMSC imagery was a 4.8 km^2 rectangle on the east side of the AKC harvest area extending approximately 4 km north from Trouble Island (Figure 2). The flight lines were set up in a north-south direction over seven kelp bed ground-truth stations. The tide levels during the flight were between −0.21 m and the day's low of −0.66 m. Skies were partly cloudy with a ceiling of approximately 823 m. The aircraft altitude

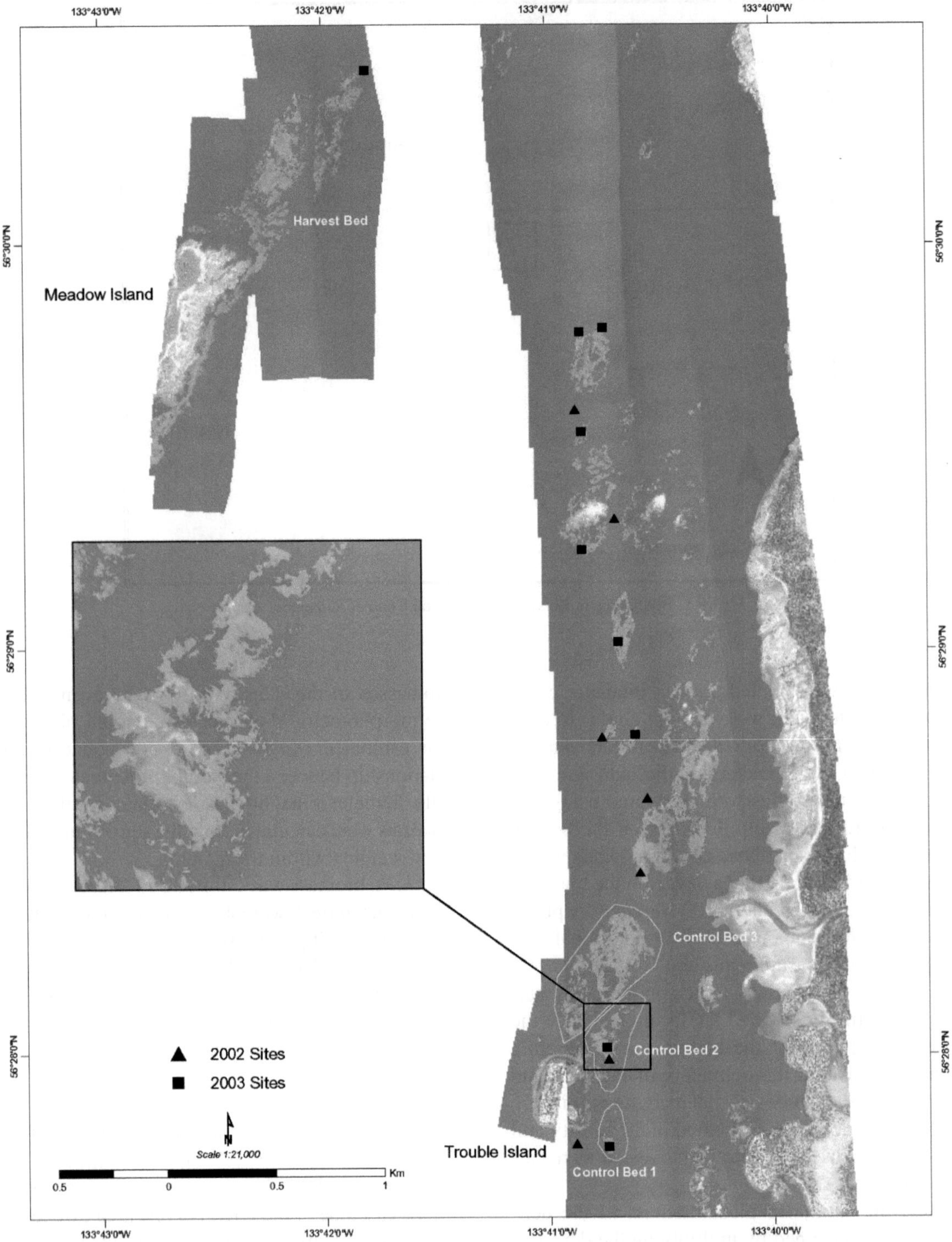

Figure 2. Mosaic of DMSC imagery acquired on the pre-harvest flight on 26 July 2003 showing locations of the ground control sites used for determining kelp biomass in 2002 and 2003 and the control beds used to compare area and biomass estimates between the pre- and post-harvest flights in 2003.

Table 1. Spectral characteristics of the DMSC color filters used for the kelp monitoring flights

	2002		2003	
	Peak (nm)	Bandwidth (nm)	Peak (nm)	Bandwidth (nm)
Band 1	451	20	451	20
Band 2	601	10	551	20
Band 3	643	10	710	10
Band 4	782	20	782	20

was 731 m, which produced a resolution of approximately 0.3 m. The DMSC was fitted with filters to acquire data in four wavelengths (Table 1) that had been used to differentiate three maturity stages of *Macrocystis* in a previous study in Monterey Bay (R. Zimmerman, personal communication).

During the 2003 field season, DMSC imagery was collected before and after a harvest of approximately 6800 Kg of *Nereocystis* by AKC. The kelp was harvested from a single bed located just north of Meadow Island and approximately 1 km northwest of the beds used for collecting the ground truth data in both (Figure 2). The DMSC was set up with a spectral filter combination (Table 1) optimized to differentiate *Nereocystis* from *A. fistulosa*. The bands were selected based on field spectrometry of *Nereocystis* and *A. fistulosa* populations in the Juneau area. The pre-harvest DMSC data were collected on July 26, 2003 between 9:03 and 10:06 AM ADT (17:03–18:03 GMT). Weather conditions were calm with light winds of approximately 2.5–5 m s^{-1}. Tide levels during the time of collection were between +2.11 and +2.64 m. The skies were partly cloudy with a ceiling of approximately 3915 m. Ground sampling distance was 0.39 m.

The post-harvest DMSC data were collected on July 31, 2003 between 2:41 and 3:05 PM ADT (23:41–00:05 GMT). Weather conditions were calm with light winds of approximately 2.5–5 m s^{-1}. Tide levels during the time of image acquisition were between +3.65 and +3.75 m. The skies contained scattered clouds with a ceiling of approximately 850 m. Light levels were sufficient and comparable to those of July 26. Collected data scenes had a ground sampling distance of 0.36 m.

Image processing

Five multi-scene image mosaics were created from the DMSC scenes acquired in 2002 and 2003. Each 4-banded mosaic was created by selecting the best scenes from each of the flight lines and georeferencing them to US Forest Service DOQQs (Digital Ortho Quarter Quads)with a pixel resolution of 1.8 m.

In order to improve water depth penetration and adjust for the different environmental conditions that existed at the time of each acquisition, processing steps were taken to help normalize the data. Natural log transformations of the five 4-banded mosaics were created in order to increase water penetration and decrease the reflectance variance created by the inconsistent flight height and depth of the target plants. Taking the natural log of each band also helped to minimize background noise in the data. Band ratios and "normalized difference" (ND) images were tested to identify band combinations that best separated the kelp from water and other vegetation.

The "normalized difference" raster images were created using an algorithm very similar to the Normalized Difference Vegetation Index (NDVI) commonly used to create indices of vegetation health or vigor for agricultural and terrestrial applications. For example, the normalized difference between DMSC Band 1 (451 nm) and Band 4 (782 nm) was calculated using Equation (1) below:

$$(B4 - B1)/(B4 + B1) \quad (1)$$

For both the 2002 and 2003 data, an unsupervised classification method, ISODATA (Iterative Self-Organizing Data Analysis Technique) (Tou & Gonzalez, 1974), was used to generate thematic maps that segregated kelp canopy from the surrounding water pixels. The unsupervised classification was also used to identify the mats of *Fucus* that collected in the kelp canopy and to try to distinguish between the *Nereocystis* and *A. fistulosa* canopy areas. The bands that showed the best separation when used as input to the ISODATA algorithm were a combination of the natural log transformations of Bands 1 through 4; the normalized difference of Bands 2 and 4, and Bands 1 and 4; and the ratio of the natural log of Band 4 to the natural log of Band 1.

The classification rasters for each area mosaic were used for two purposes. A 20-class raster was merged down to a three-class raster showing *Nereocystis/Alaria, Fucus* and all other classes. This image was then used to compute the area covered by *Nereocystis* and *Alaria* for select kelp beds. The classification was also merged down to generate two-class raster that isolated *Fucus*. The *Fucus*-only raster image was later applied as a mask to the biomass index image in order to exclude *Fucus* dominated pixels in the final products.

The relationships between the DMSC imagery and kelp biomass were determined by calculating the average pixel values of various band combinations for a 4 × 4 m area in the imagery centered on the buoy marking each ground-truth station. This area covered the area sampled by the surface quadrats. The ratio of the log (ln) of band 4 and log (ln) of band 1 yielded the highest correlation coefficient of a number of different band combinations that were regressed with the biomass values. The biomass images were generated by applying the biomass regression equation developed for each year to each pixel of the raster images generated from the band ratios.

Results

Morphometrics and biomass correlations

The best predictor for whole plant wet weights of *Nereocystis* was the weight of the blades from the plant (Table 2). The next best predictor was the diameter of the stipe just below the bulb or the sub-bulb diameter (Table 2). The exponential regression line calculated for the 2002 data had an R^2 of 0.88. Other measurements had poorer correlations with plant weight (Table 2).

The relationship of the sub-bulb diameter to the total plant weight varied among the three separate data sets (Figure 3). In 2003 the best correlation for the regression line was found using a power curve rather than an exponential curve. This regression was strongly influenced by one very large plant measured in 2003.

Nereocystis maximum density in the beds was 9–10 plants per square meter with the biomass per square meter of bottom area as high as 16 kg (Table 3).

Table 2. Regression equations for the different morphometric measurements for *Nereocystis* wet weight estimation. Data for these regressions are from 2002 sampling.

Meristic	Best fit regression equation	R^2
Weight of blade (Kg)	Plant weight = 1.3212 × + 0.1052	0.98
Diameter of 'sub-bulb' (cm)	Plant weight = $0.038e^{0.125\times}$	0.88
Length of blade (m)	Plant weight = $0.0455e^{0.124\times}$	0.85
Diameter of bulb (cm)	Plant weight = $0.0006e^{1.435\times}$	0.78
Length of stipe (m)	Plant weight = $0.0346e^{0.501\times}$	0.65
Total length (m)	Plant weight = $0.0664e^{0.295\times}$	0.33

Table 3. Summary of density and biomass data for *Nereocystis luetkeana* (*Nl*) and *Alaria fistulosa* (*Af*) from the study site north of Pt Baker, Alaska

Site	Species	plants m^{-2}	SE	kg m^{-2}	SE	N*
2002						
1	*Nl*	7.2	1.6	5.62		
2	*Nl*	7.2	2.5	9.92		
3	*Nl*	8.2	1.9	13.64		
4	*Nl*	7.2	2.3	14.10		
5	*Nl*	1.0	0.4	0.62		
6	*Nl*	7.2	1.1	3.65		
7	*Nl*	10.3	1.6	14.18		
2003						
Harvest	*Nl*	3.2	0.70	8.51	2.13	20
1	*Nl*	8.4	1.16	10.95	1.57	8
1'	*Nl*	9.0	2.00	15.95	5.07	2
2	*Nl*	3.5	0.87	5.05	2.02	4
3	*Nl*	0.0	0.00	0.00	0.00	8
4	*Nl*	1.0	0.38	1.75	0.56	8
5	*Nl*	0.6	0.38	2.66	1.98	8
6	*Nl*	0.0	0.00	0.00	0.00	8
7	*Nl*	5.3	0.92	12.56	1.35	8
8	*Nl*	0.9	0.44	0.46	0.32	8
1	*Af*	0.3	0.25	0.03	0.03	8
1'	*Af*	0.0	0.00	0.00	0.00	2
2	*Af*	15.0	1.47	1.81	0.18	4
3	*Af*	0.0	0.00	0.00	0.00	8
4	*Af*	4.3	1.85	0.51	0.22	8
5	*Af*	2.9	1.25	0.35	0.15	8
6	*Af*	6.9	2.17	0.83	0.26	8
7	*Af*	4.4	1.15	0.53	0.14	8
8	*Af*	6.6	2.43	0.80	0.29	8

Data were collected as described in Methods. Site 1' is a revisit to site 1. N* = number of quadrat sampled.
SE = standard error of the mean.

Alaria fistulosa data were obtained in 2003 from 19 samples. The mean weight of a one meter long section of *Alaria* was 182 ± 16 (SE) g. The mean was adjusted by multiplying by a factor of 1.33 which is a rough estimate to correct for the proportion of the frond that was submerged. Mean densities varied from 0 to 15 plants per square meter. Estimated biomass in the kelp canopy was up to 1.8 kg m^{-2} (Table 3).

Kelp bed area

The mosaic of the DMSC imagery for the pre-harvest flight on 26 July 2003 (Figure 2) shows the harvest bed and the kelp beds used for generating the biomass

Table 4. Kelp bed area comparisons for the harvest and control beds including percent changes in the estimates between surveys

	Harvest bed		Control bed 1		Control Bed 2		Control bed 3	
	Area	% Chg	Area	% Chg	Area	% Chg	Area	% Chg
7/14/02	No Data	–	4,465	–	22,537	–	88,210	–
7/26/03	47,742	–	4,588	+2.8	22,913	+1.7	91,902	+4.2
7/31/03	51,150	+7.1	4,991	+8.8	22,850	−0.3	89,667	−2.4

All areas are in m^2.

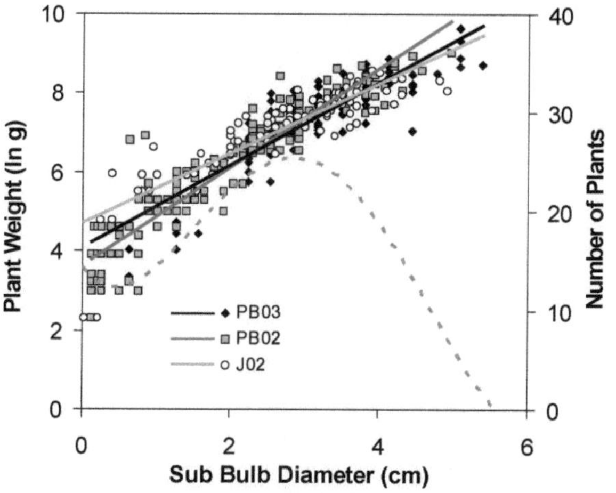

Figure 3. Nereocystis luetkeana wet weight per plant as a function of the diameter of the stipe below the pneumatocyst (sub bulb). Points represent individual plants from beds north of Pt Baker (PB), Alaska in 2002 and 2003 plus plants from Juneau (J), Alaska in 2002. Straight lines are best fit regression lines from the corresponding data sets. The dotted curve is the size-frequency distribution of the *Nereocystis* plants measured for the three data sets.

relationships. The three kelp beds imaged on both the pre-harvest and post-harvest flights, which serve as controls for quantifying errors in areal cover and biomass estimates, are delineated at the south end of the study area. The total kelp bed areas derived from the classification images for the harvest and three control beds are summarized in Table 4. The classification images showing the extent of the harvest bed for the pre- and post-harvest flights are shown in Figure 4. The kelp bed area included both surface canopy and subsurface *Nereocystis* and *Alaria* plants, but excluded areas with *Fucus* cover. The error ranges in the coverage estimates at the control sites between the two flights in 2003 were all within 10%.

The kelp canopy areas were also estimated from a LandSat 7 Enhanced Thematic Mapper (ETM+) satellite image acquired on 4 August 2002. The ETM+

Table 5. Kelp bed area estimates derived from ETM+ (Landsat 7) satellite imagery acquired on 4 August 2002

	Pixel Count	Area (m^2)
Harvest bed	396	80,412.75
Control bed 1	42	8,528.63
Control bed 2	167	33,911.44
Control bed 3	546	110,872.13

panchromatic band has 15 m spatial resolution and the spectral bandwidth extends into the near-IR. The estimates of kelp canopy area obtained from this image are presented in Table 5. The canopy estimates from the ETM+ imagery are approximately 50% greater than the areas derived from the high resolution DMSC imagery.

Biomass relationships with DMSC data

The ratio of the natural log of band 4 to the natural log of band 4 (ln band 4/ln band 1) showed the best correlation of the various band combinations that were compared (Figure 5). Because *Nereocystis* was not distinguishable from *Alaria fistulosa* under all of the imaging conditions, the image data were regressed against the combined estimated biomasses of *Nereocystis* and *Alaria*.

A biomass map was generated for the pre-harvest flight on 26 July 2003 (Figure 6). The beds were largely composed of low and medium biomass classes with patches of high biomass just north of Trouble Island, small patches in the harvest bed, and some larger patches just north of Meadow Island. The biomass estimates for the 2003 flights ranged from 38% to 66% lower than the estimates from the 2002 flight, and estimates for the post-harvest flight in 2003 averaged 8% lower than the pre-harvest flight (Table 6).

Various unsupervised classification algorithms were used to separate *Nereocystis* and *Alaria fistulosa*, the

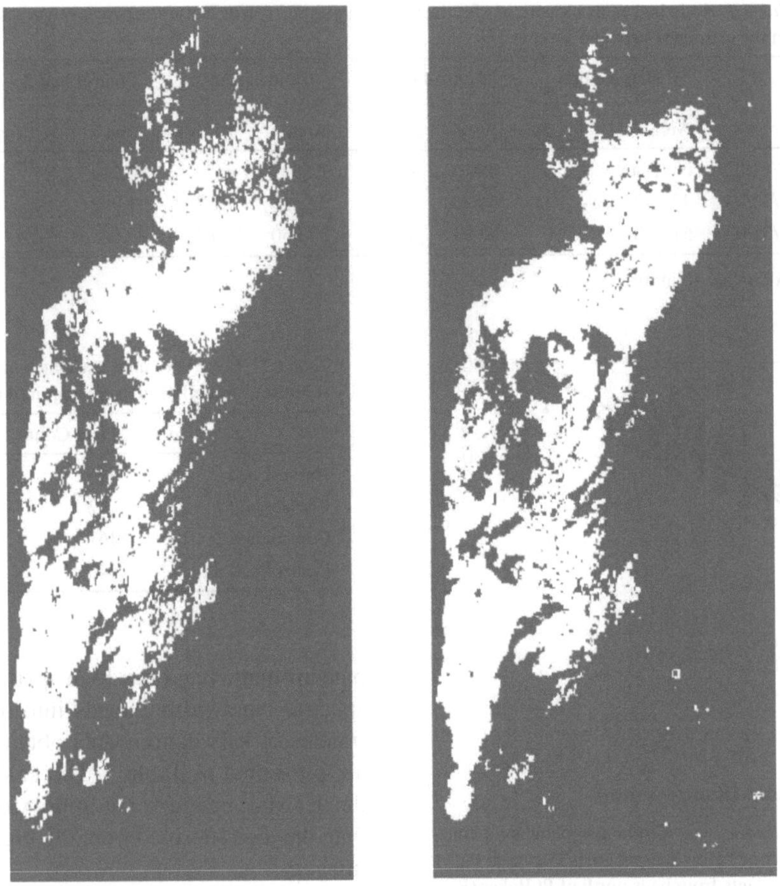

Figure 4. Classified kelp bed areas of the harvest bed on the pre- and post-harvest flights.

two main kelp species that form a surface canopy in this area from masses of drift *Fucus.* The masses of drift *Fucus* were easily separable from the kelp canopy. We produced a *Fucus* mask to exclude these plants from the biomass calculations. The masking procedure most likely led to an underestimation of kelp biomass because in many instances there were kelp plants directly below the *Fucus*.

The unsupervised classification algorithms could not consistently differentiate areas of *Nereocystis* and *A. fistulosa* canopy. Regions of very uniform *Alaria* canopy, such as the canopy adjacent to the shoreline on the north side of Trouble Island were distinguished as a distinct class. However, other areas that we knew were predominantly *Alaria* were classified as *Nereocystis*. Regions of mixed *Nereocystis* and *Alaria* were uniformly classified the same as areas that were predominately *Nereocystis*.

Discussion

Ground truthing biomass for floating kelps is a very time consuming and expensive enterprise. Scuba divers are usually necessary in order to retrieve the whole thallus. In addition the size and morphology of floating kelps make it very difficult to keep plants from becoming entangled and tearing during retrieval for weighing. The size of the plants requires a fairly large vessel for obtaining fresh weights. For these and other reasons we attempted to find a morphometric measurement that would be relatively easy to measure in the field and yet be a reasonable predictor of biomass.

Of the measurements we made, the best predictor of whole plant biomass was the weight of the blade (Table 2). Although it is possible to cut and weigh the blades while in the field without the use of scuba, this process requires too much time and effort. In contrast

Table 6. Kelp bed biomass (kg m^{-2}) comparisons for the harvest and control beds including percent changes in the estimates between surveys

	Harvest Bed		Control bed 1		Control bed 2		Control bed 3	
Date	Biomass	% Chg	Biomass	% Chg	Biomass	% Chg	Biomass	% Chg
7/14/02	No Data	–	658,103	–	2,567,142	–	9,919,719	–
7/26/03	2,843,044	–	222,911	−66	1,266,062	−51	6,141,369	−38
7/31/03	2,708,261	−5	229,675	+3	1,311,964	+4	3,981,751	−35

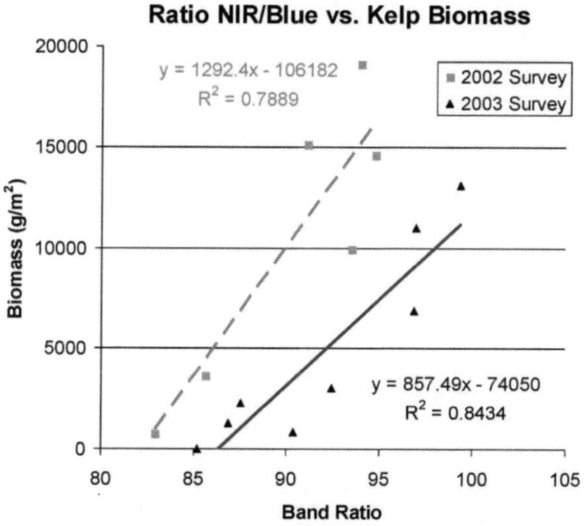

Figure 5. Relationship between kelp wet biomass and the ratio of Bands 1 and 4 from the DMSC imagery. The DN values from the two bands were log (ln) transformed before forming the ratio. The regression lines were significantly different between years (ANCOVA F=21.224, p=0.001).

the diameter of the sub-bulb is relatively easy and quick to measure and has a reasonable correlation to the total fresh weight of the plant.

A drawback of the sub-bulb method is that there are significant variations among different locations and/or different times of the growing season. An analysis of covariance of the $\log_e$ of the plant weight versus the sub-bulb diameter revealed that regression lines are not similar. For the best accuracy in estimating *Nereocystis* biomass, separate ground truthing should be performed in each area to be assessed. On the other hand, the regression equations are in closest agreement in the middle of the size range which includes the majority of the plants (Figure 3). For management purposes a regression equation combining several data sets may be the best compromise.

Other researchers have used various methods to estimate biomass of floating kelps. During investigations on the subtidal effects of the *Exxon Valdez* oil spill the biomass of *Nereocystis* was estimated by measuring the diameter of the stipes one meter above the bottom (Dean et al., 1996). In southeast Alaska van Tamelen and Woodby (2001) found a significant linear correlation between biomass of *Macrocystis* and individual frond lengths. Stekoll and Else (1992) used a similar approach to estimate the biomass of artificially cultured *Macrocystis* near Sitka, Alaska. In Chile *Macrocystis* biomass has been estimated from the diameter of the holdfast (Vasquez & Vega, 2004). In Canada from 1975 to 1996 biomass of the floating kelp beds was assessed using an inventory method (KIM-1) developed by Foreman (1975). The KIM-1 method used near-IR aerial photographs to determine the density and the area of beds containing *Macrocystis* and/or *Nereocystis*. These data were combined with field sampling for plant densities and mean plant weight to obtain species specific biomass estimates for large areas of kelp beds.

The biomass of *Alaria* was estimated assuming that most of the plant can be considered to be a two dimensional plane with a constant weight per meter of plant. This process is similar to that used in the KIM-1 method on other floating kelps (Foreman, 1975). In contrast, the *Nereocystis* estimates we performed were based on individual plant measurements. Although it would not be too difficult to cut and weigh the *Alaria* fronds within each quadrat, the precision gained from on site weighing of all of the fronds of *Alaria* does not justify the extra cost in time and effort.

Comparisons of the areas of the harvest bed and three control beds (Table 4) show that the DMSC multispectral imagery provides very reliable estimates of kelp bed size even when the imagery is acquired at very different tidal levels. The DMSC provides a distinct advantage over traditional near-IR aerial photography that can detect only the portions of the kelp plant directly on the surface. The area of the surface canopy is strongly influenced by tide and current conditions and this will be reflected in kelp bed area estimates made with near-IR aerial photography. The DMSC provides additional

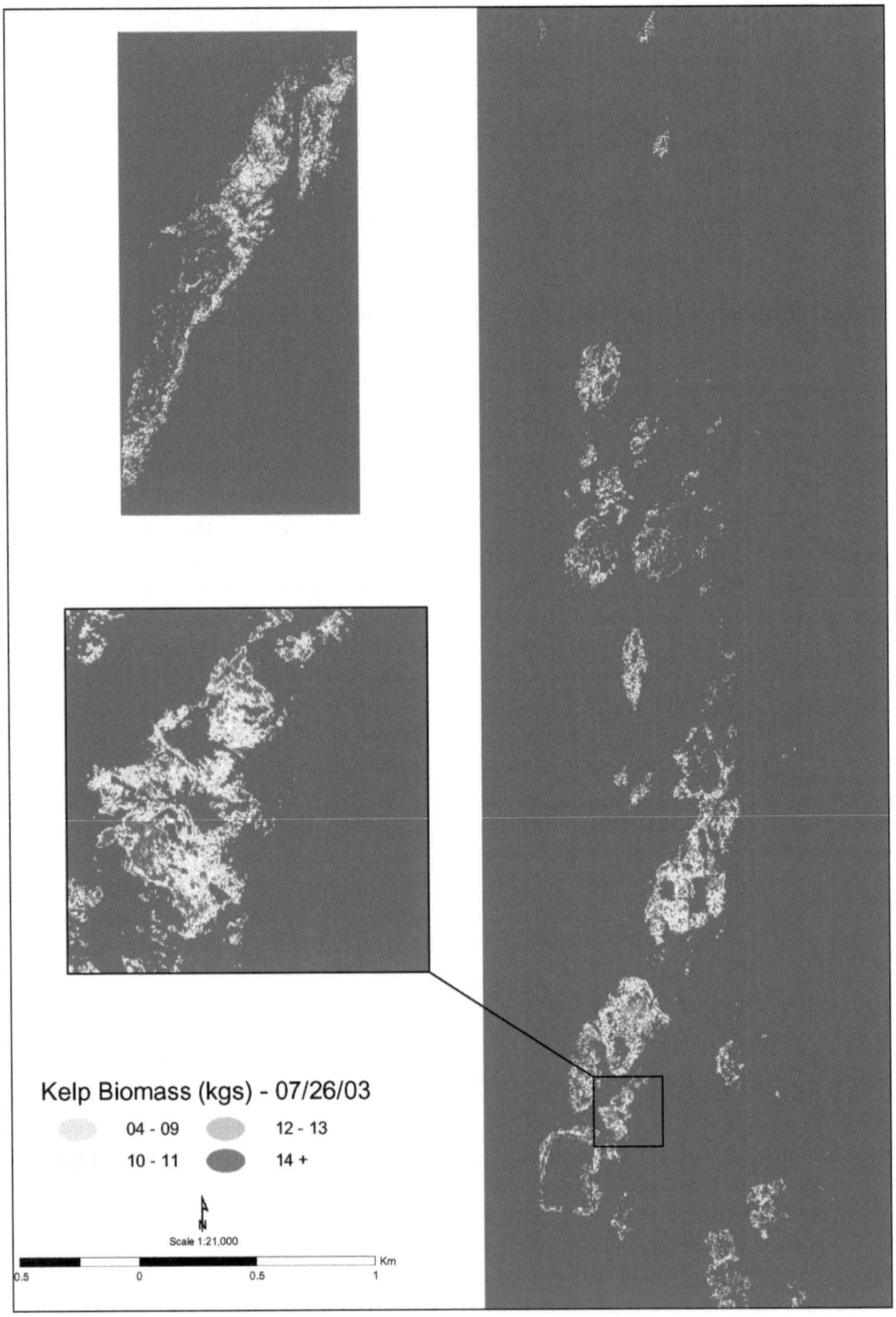

Figure 6. Map of kelp biomass (Kg m^{-2}) for the harvest and control beds calculated from the pre-harvest DMSC imagery collected on 26 July 2003.

spectral bands (451 and 551 nm) that can detect subsurface kelp plants down to 3 m. Secchi disk measurements in this region made during the sampling in 2003 ranged from 4 to 5 m. Classification algorithms using the 4 spectral bands of the DMSC imagery can distinguish both canopy and subsurface kelp plants and thereby provide a very accurate estimate of kelp bed size that is fairly independent of tidal or current conditions.

The sizes of the three control beds used for the 2003 studies were nearly identical to their sizes in 2002. The beds in this area grow on distinct reef structures and it appears that reef size defines the upper limits for the size of the kelp beds. Other beds in the region, however, decreased in size from 2002 to 2003. For reasons not clear at this time, a *Nereocystis* bed on the south side of Trouble Island that we used for a ground truth site in 2002 had only a few plants in 2003.

The radiometric characteristics of the DMSC imagery provided good correlations with the kelp biomass estimates made at the ground truth sites. However, the comparisons of the biomass images between 2002 and 2003 showed much higher biomass values for 2002. The beds did not change in size between the two years and it appears that the higher biomass estimates for 2002 are related to the lower tidal levels at the time the imagery was collected in 2002. Lower tidal levels allow more frond material to float on the surface thereby increasing the overall brightness in the near-IR band, which is the dominant spectral band that we found useful for estimating biomass. Currents can also affect the amount of kelp canopy on the surface and our experience at the AKC harvest site showed that the currents are spatially variable even at the same time in the tidal cycle. The optimum method for estimating biomass over a large harvest site with the multispectral imagery would be to collect the data at a standard tidal level with low currents.

Field spectroscopy studies in Juneau showed small differences between the reflectance spectra of *Nereocystis* and *Alaria fistulosa*. We tried to capitalize on these differences by selecting filters for the DMSC in the spectral areas where the two species showed the most difference. However, we could not differentiate the two species based on the 4 spectral bands of the DMSC imagery. We noted that there was significant variability in the canopy color of the two species in different areas of the study site. *Nereocystis* plants in more protected areas appeared to have a lighter color than plants in more exposed areas. A study in Washington State using a CASI (Compact Airborne Spectrographic Imager) sensor with 11 bands of spectral data could not distinguish *Macrocystis* from *Nereocystis* at 4-m resolution (Pers. Comm. T. Mumford).

We were also not able to differentiate the two species based on any morphological differences of the surface fronds. Even though the general morphology of the two species is quite different, the fronds in the surface canopy for both species are long linear objects with smooth edges. The 0.3 m resolution of the DMSC imagery was not sufficient to show differences in the two canopies. High resolution aerial photography has been shown to differentiate *Macrocystis* and *Nereocystis* canopy in California (Berry et al., 2001), but the surface fronds of these two species are very different.

Weather was the main problem encountered in both years of the study. It was difficult to get even minimum flying conditions (915 m ceiling) during low tide periods and during the optimal windows for sun angle. The lack of a dedicated aircraft for the project also contributed to the logistical constraints because aircraft time could not be scheduled more than 24 h in advance. In addition, the charter pilots who flew these surveys were not accustomed to flying the straight and closely spaced flight lines that are required for this type of survey.

There have been numerous studies using different multispectral sensors to map kelp canopies. The primary problems with the visual satellite sensors are spatial resolution, cloud cover, and cost. Studies in California (Augenstein et al., 1991; Deysher, 1993) have shown that SPOT satellite imagery with a spatial resolution of 20 m is sufficient to map the larger kelp beds in southern California, but is marginal for the smaller beds in this region. The panchromatic band of ETM+ imagery from Landsat 7 has 15 m spatial resolution, which is sufficient to resolve most of the kelp beds in southeast Alaska. However, the overestimation of canopy area (Table 5) as compared to higher resolution imagery must be resolved before this imagery can become a regular tool in harvest management. Multispectral images with spatial resolutions as high as 4 m can be obtained from the Ikonos (Space Imaging), QuickBird (DigitalGlobe), and OrbView-3 (Orbimage) satellites. However, the cost of acquiring imagery over large areas of a region such as southeast Alaska would be prohibitive for only kelp mapping activities. The costs could be reduced if the imagery can be acquired for multiple uses within the same region and a multiagency license can be developed.

There is an inverse relationship between the spatial resolution of satellite images and the time that elapses between repeat passes over an area. Satellites with image resolution sufficient to map kelp canopy generally have a revisit time of between 16 and 25 days. In areas such as southeast Alaska, long periods of cloud cover may preclude the collection of imagery for months at a time. The revisit frequency becomes a problem even in areas of moderate cloud cover if image acquisition during a specific tide period is required.

Since radar frequencies are not blocked by clouds, radar satellite imagery has the advantage of being able to be acquired even during periods of complete cloud cover. However, to be able to discriminate kelp canopy there must be a significant difference in surface roughness between the water and the kelp. There are differences in the surface roughness requirements in the different types of radar imagery (X-band, L-band, and C-band), but it appears that there are many false positives due to areas of increased sea roughness due to wind squalls (X-band) or slicks due to oil or other naturally occurring surfactants (L-band and C-band) (Jensen et al., 1980).

The KIM kelp inventory method developed in Canada (Foreman, 1975) is very similar to the methods used for the present study. The main difference, however, is that the KIM method is dependent on diver surveys for every inventory. The methodology developed by this project is planned to be independent of ongoing field biomass surveys.

Kelp beds in California are inventoried primarily by area determined by aerial near-IR photography and, more recently, digital multispectral imagery. Visual estimates of biomass made by Dale Glantz of ISP Alginates, the largest kelp harvester in California, are also being integrated into the state's kelp bed inventory data. The visual estimates have been calibrated by years of experience comparing visual observations with harvest records.

At this time, we believe that the DMSC imagery combined with the relatively quick method for ground truthing is the best tool for mapping kelp beds for both area and biomass estimates in SE Alaska.

Acknowledgements

The authors would like to thank the Alaska Kelp Company and its employees for providing logistical support for this research. Mike Mortell and Chris Dahl were especially helpful in this respect. Divers for this project were Elizabeth Calvert, Kristen Cecil, and Erin Meyer. The Alaska Department of Fish & Game contributed critical overview of this research.

References

Augenstein EW, Stow DA, Hope AS (1991) Evaluation of SPOT HRV-XS data for kelp resource inventories. Photogramm. Eng. Rem. S. 57: 501–509.

Belsher T, Mouchot, MC (1992) (Use of satellite imagery in management of giant kelp resources, Morbihan Gulf, Kerguelen Archipelago). Evaluation par teledetection satellitaire des stocks de *Macrocystis pyrifera* dans le Golfe du Morbihan (Archipel de Kerguelen). Oceanol. Acta 15: 297–307.

Berry H, Sewell A, Van Wagonen B (2001) Temporal trends in the areal extent of canopy-forming kelp beds along the Strait of Juan de Fuca and Washington's outer coast. Proceeding of the Puget Sound Research Conference 2001. Puget Sound Action Team. Olympia, Washington. http://www.psat.wa.gov/Publications/01_proceedings/PSRC_2001.htm

Dean TA, Stekoll MS, Smith RO (1996) Kelps and oil: the effects of the *Exxon Valdez* oil spill on subtidal algae. In: Rice SD, Spies RB, Wolfe DA, Wright BA (eds), Proceedings of the *Exxon Valdez* oil spill symposium. American Fisheries Society Symposium 18, pp. 412–423.

Deysher LE (1993) Evaluation of remote-sensing techniques for monitoring giant kelp populations. Hydrobiologia 260/261: 307–312.

Foreman RE (1975) KIM-1. A method for inventory of floating kelps and its application to selected areas of Kelp Licence Area 12. Benthic Ecological Research Program Report 75-1. Report to Federal Fisheries and marine Service and Provincial Marine Resources Branch. 81 pp.

Foster MS, Schiel DR (1985) Ecology of giant kelp forests in California: A community profile. National Coastal Ecosystems Team, Slidell, LA., NTIS: 172pp.

Frye TC (1915) Part IV. The kelp beds of Southeast Alaska. In Cameron FK (ed.), Potash from Kelp, U.S.D.A., Washington, DC: 60–104.

Jensen JR, Estes JE, Tinney L (1980) Remote sensing techniques for kelp surveys. Photogramm. Eng. Rem. S. 46: 743–755.

North WJ, James DE, Jones LG (1993) History of kelp beds (*Macrocystis*) in Orange and San-Diego counties, California. Hydrobiologia 2609/261: 277–283.

Stekoll MS (1998) Seaweed resources of Alaska. In Ohno M, Critchley AT (eds), Seaweed Resources of the World, Japan International Cooperation Agency, Tokyo: pp. 258–265.

Stekoll MS, Else PV (1992) The Feasibility of *Macrocystis* Mariculture in Southeast Alaska, The State of Alaska and the Japan Overseas Fisheries Cooperation Foundation, Tokyo, Japan, 171 pp.

Tou JT, Gonzalez RC (1974) Pattern Recognition Principles. Addison-Wesley Publishing Co., Reading, Massachusetts.

van Tamelen PG, Woodby D (2001) *Macrocystis* biomass, quality, and harvesting effects in relation to the herring spawn-on-kelp fishery in Alaska. Alaska Fish. Res. Bull. 8: 118–131.

Vasquez J, Alonso Vega JM (2004) Kelps: yield, harvesting and culture in northern Chile. Abstract Book Aquaculture 2004. World Aquaculture Society, Baton Rouge, Louisiana, 609.

Journal of Applied Phycology (2006) 18: 335–341
DOI: 10.1007/s10811-006-9036-8

The effects of harvesting of the South African kelp (*Ecklonia maxima*) on kelp population structure, growth rate and recruitment

M.D. Rothman[1,2,*], R.J. Anderson[1] & A.J. Smit[2]
[1]*Marine and Coastal Management, Private Bag X2, Roggebay, South Africa, 8012;* [2]*Botany Department, University of Cape Town, Private Bag, Rondebosch, South Africa, 7700 (Current address: School of Biological and Conservation Sciences, University of KwaZulu-Natal, Westville, Pvt Bag X54001, Durban 4000, South Africa)*
*Author for correspondence: e-mail: mrothman@botzoo.uct.ac.za

Key words: commercial harvesting, *Ecklonia maxima*, growth rate, population structure, recruitment, South Africa

Abstract

Ecklonia maxima is an economically important kelp in South Africa. The harvested kelp is used mainly as feed for cultured *Haliotis midae* (abalone) on farms all along the South African South and West Coast. The effects that different harvesting methods have on the growth of sub-canopy kelps, kelp population structure and kelp recruitment were tested in a kelp bed at Bordjies Rif near Cape Town. Two 30 × 10 m sites were set up, about 100 m apart, in near monoculture stands of *E. maxima*. Each 30 × 10 m area was subdivided into three treatments. In treatment 1 (T1) the whole 'head' of each kelp sporophyte that reached the surface was cut off between the bulb and the primary blade ('lethal' method). In treatment 2 (T2) ('non-lethal' method), the secondary fronds of all sporophytes that reached the surface were cut 20–30 cm from their junction with the primary blade, and removed. In the control plot, the kelp plants were not treated. Harvesting treatments were done approximately every four months, at low spring tide, from 3 March 2003 to 3 November 2003 (three treatments). The effects of harvesting on the kelps depend largely on the size of plant and the time the fronds were removed; however, no seasonal pattern could be observed. The different treatments had no effect on the growth rate, population structure or recruitment of the kelp. This means that factors other than light play an important role in the growth, structure and recruitment of the kelp beds in False Bay. Results are discussed in relation to current commercial harvesting practices.

Introduction

The commercial use of seaweeds in South Africa began during the Second World War when agar from Japan became unavailable in Britain (Anderson et al., 1989). This may have been the impetus needed to start the South African seaweed industry in the early 1950's (Isaac & Molteno, 1953). Seaweed was mainly collected as beach-cast material, dried and then exported (Anderson et al., 1989). A number of seaweed species are harvested commercially in South Africa but the kelp *Ecklonia maxima* is harvested in the largest quantities (Anderson et al., 2003). The amount of fresh kelp fronds harvested for abalone feed in South Africa increased exponentially from less than one ton (wet) in 1992 to more than 6000 tons (wet) in 2003 mainly due to the increase in the number of abalone farms along the South African West Coast. Abalone require about 7% of their body mass of kelp per day: to produce 100 t of abalone, to a size between 50–70 mm in diameter, requires 5 t of freshly harvested kelp daily (Levitt et al., 2002). It is likely that the harvesting pressure on kelp beds will increase as more abalone farms are constructed along the South African west coast and existing farms expands.

The South African coastline is divided into 23 concession areas in which successful applicants have the right to harvest one specified seaweed resource. Fourteen of these concession areas have kelp in them, the main species in the south being *Ecklonia maxima*. Levitt et al. (2002) studied the regrowth of *Ecklonia maxima* and the understorey biota after harvesting, but knowledge of the effects of harvesting is still very limited. In South Africa kelp is harvested by various methods. When stipe and fronds are required divers cut the stipe at the bottom just above the holdfast, thus killing the plant. This method is not used for abalone feed and is not considered further here. Only fronds (blades) are used for abalone feed, and there are two methods of frond-harvesting. In the first method the whole 'head' of the kelp sporophyte is cut off between the bulb at the top of the stipe and the primary blade (Figure 2). This is an easy way of harvesting, but it kills the plant. In the second method, the secondary fronds are cut 20–30 cm from the junction with the primary blade (Figure 2). Levitt et al. (2002) showed that the latter type of harvesting does not kill the plant, because the meristematic zone at the base of the secondary fronds is unharmed. The fronds continue to grow, and Levitt et al. (2002) calculated that this non-lethal harvesting method ultimately gives yields that are 4–5 times higher than the 'lethal' method. Both of these frond-harvesting methods alter the state of the canopy in a kelp bed.

Reports from other countries indicate that dense kelp canopy can decrease the amount of light that penetrates to the bottom by more than 90% (Norton et al., 1982; Kimura & Foster, 1984; Schiel & Nelson, 1990). Removing kelp canopies can increase the abundance of understory plants (Kimura & Foster, 1984; Sharp & Pringle, 1990) because light is one of the factors affecting their growth (Schroeter et al., 1995). We therefore hypothesized that removal or thinning (by cutting the distal fronds) of the surface canopy of *E. maxima*, would, through its effect on light penetration, result in an increase in the relative growth rate of the subcanopy kelp plants. Furthermore, we expected an increase in kelp recruits (juvenile sporophytes) and ultimately in the number of subcanopy kelp plants, which would alter the population structure of harvested beds. This study investigated the effects of these two methods of frond-harvesting (of the canopy) on the growth (stipe elongation) rate, recruitment, and population structure of the sub-canopy of the kelp *E. maxima*.

Materials and methods

Study site

A site was selected on the Cape Peninsula, about 60 km south of Cape Town, at Bordjies Rif (18°27′48″E, 34°18′54″S) (Figure 1), where no previous harvesting had taken place. The substratum is of medium relief Table Mountain sandstone with 1–1.5 m gullies and mixed vertical/horizontal aspect. The homogeneous stand of kelp (*Ecklonia maxima*) is fairly dense (~8 plants m^{-2}) and the bed is between 70–100 m wide and 300–400 m long. The study site is in a partially sheltered bay that allows for favorable working conditions most of the time.

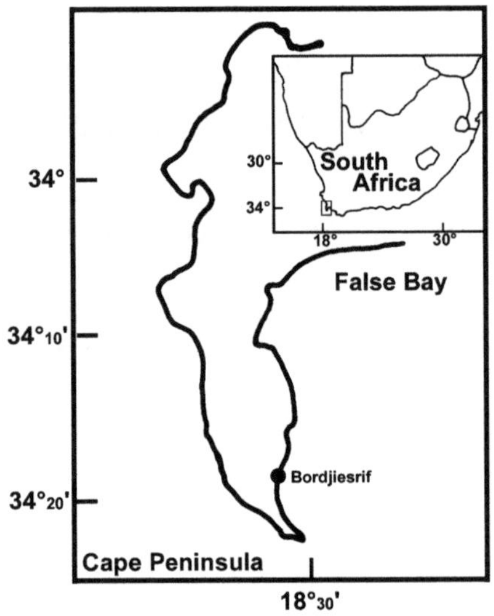

Figure 1. Location of Bordjies Rif in South Africa.

Experimental design

Two areas, each 30 × 10 m, were marked out parallel to the shore (Figure 2; ABCD) with weights and stainless steel eyebolts fixed to the rock with epoxy putty. The corners were marked with small sub-surface buoys. Each 30 × 10 m area was then subdivided

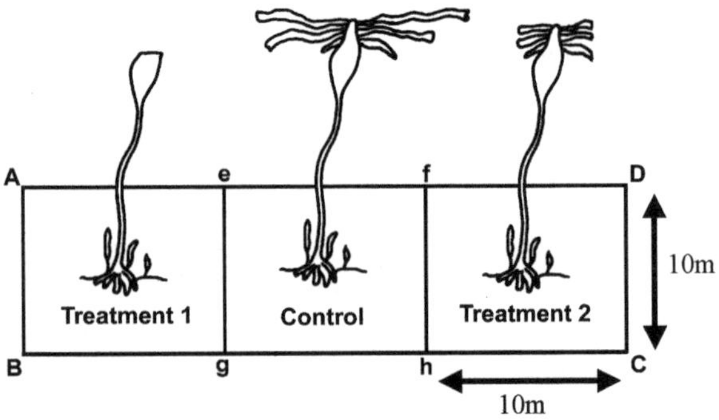

Figure 2. Treatment 1 (T1), where the whole head was harvested just above the bulb (lethal). Treatment 2 (T2), where only the fronds were harvested, 20–30 cm from the primary blade (non-lethal). The centre plot was the control (C, no treatment).

into three 10×10 m plots and marked with eyebolts (Figure 2; efgh). The two replicate, 30×10 m areas were approximately 100 m apart, and both were between 2–5 m deep.

Each 30×10 m area was subdivided into three treatments. In treatment 1 (T1) the whole 'head' of each kelp sporophyte that reached the surface was cut off between the bulb and the primary blade, and the primary blade and the secondary fronds were removed ('lethal' method). In treatment 2 (T2) ('non-lethal' method), the secondary fronds of all sporophytes that reached the surface were cut 20–30 cm from their junction with the primary blade, and removed. In the control plot (C), the kelp plants were not treated. Harvesting treatments were done approximately every four months, at low spring tide, from 3 March 2003 to 3 November 2003 (three treatments). These treatments mimicked the two frond-harvesting methods used commercially. All subsequent sampling was done on sub-canopy kelps (see below).

Growth rates

Divers using SCUBA made measurements once every two months. In each 10×10 m plot (in each treatment, in three different size classes, 10 in each) 30 plants were marked, each with an unique number. The first group of plants, with stipe lengths ranging from 0–10 cm (small plants), were numbered 1–10. The second group, with stipe lengths ranging from 11–50 cm (medium plants), were numbered 11–20. The third group, with stipe lengths of 51–100 cm (long plants), were numbered 21–30. In total, 180 plants (in six plots at two sites) were marked.

Divers measured the stipe length of each marked plant and recorded it on an underwater slate. Two months later divers again measured all marked plants. The tags were then removed and new sets of plants in each size class were marked and measured to keep the samples independent. It is important to note that growth rate was measured as stipe elongation. The relative growth rates (RGR, % day^{-1}) were determined using the formula

$$\mathrm{RGR} = \frac{\ln (W_2/W_1)}{n} \times 100$$

where W_2 and W_1 are the final and initial mass, respectively, and n is the number of days. An ANOVA and a Tukey post-hoc comparison were done to determine the effect that the different treatments had on growth. By using a 3-way analysis of variance, the growth rates were compared over time, treatment and between the different size classes.

Plant density, recruitment and population structure

Divers haphazardly placed ten 1 m^2 quadrats within each of the six blocks. The number of plants within six size classes (0–1 cm, 2–10 cm, 11–25 cm, 25–50 cm, 50–100 cm and >100 cm) was recorded in each 1 m^2 quadrat. For this study kelp plants with a length of $\leq$10 cm were regarded as recruits. The effects of treatments (T1, T2 and C) on population structure and recruitment were tested statistically using a 2-way ANOVA and Tukey's *post-hoc* test.

Water temperature

Water temperature was measured using a Starmon-mini temperature recorder with a Hart 1504 thermometer at 8 m depth placed on a concrete block approximately 300 m from the site at Bordjies Rif. Temperatures were recorded every 10 min and daily means calculated.

Light

Light was measured using a LI-COR LI-1000 Data Logger and SA: LI-193SA Underwater Type Spherical Quantum Sensor. Eight to twelve readings were made at each of the following depths: above the surface, 10 cm under the surface, 1 m, 2 m, 3 m, and at the bottom and between 10h00 and 12h00, with good visibility (about 15 m) on a cloudless day in November. Mean values were calculated for each depth and expressed relative to the surface value (1). The light was measured before and after experimental harvesting in a treatment 1 plot, a treatment 2 plot, and once in a control.

Results

Growth rate

Treatment had no significant effect on growth rate in all three size classes of sub-canopy kelp (Figure 3). Long plants (50–100 cm) grew faster than short plants (0–10 cm) in all treatments (Figures 3 and 4). Growth rates of medium plants (11–50 cm) were intermediate between those of long and short plants (Figures 3 and 4) and often overlapped with one or both (Figure 3). A

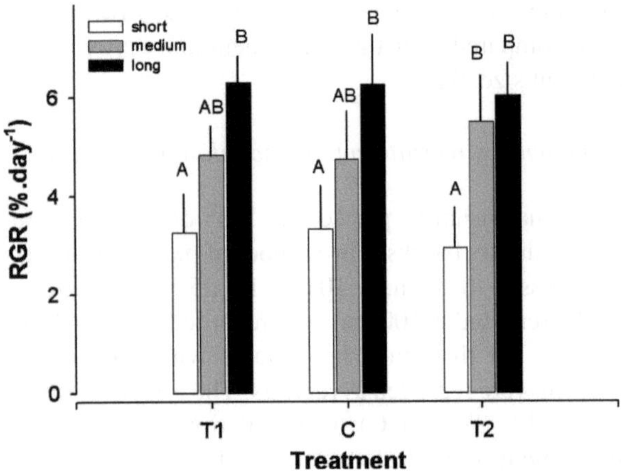

Figure 3. Relative Growth Rate (stipe elongation) of the different size classes of *Ecklonia maxima* under different harvesting conditions. Bars denote 95% confidence intervals. Letters denote statistical grouping (ANOVA and Tukey *post-hoc* test).

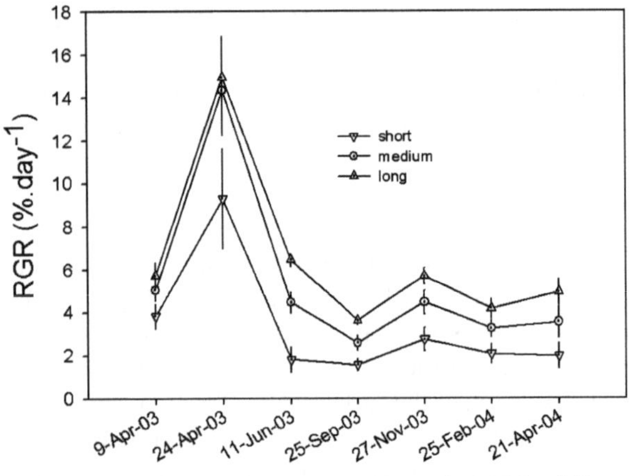

Figure 4. Relative Growth Rate (stipe elongation) of the different size classes of *Ecklonia maxima* over time. Bars denote 95% confidence intervals.

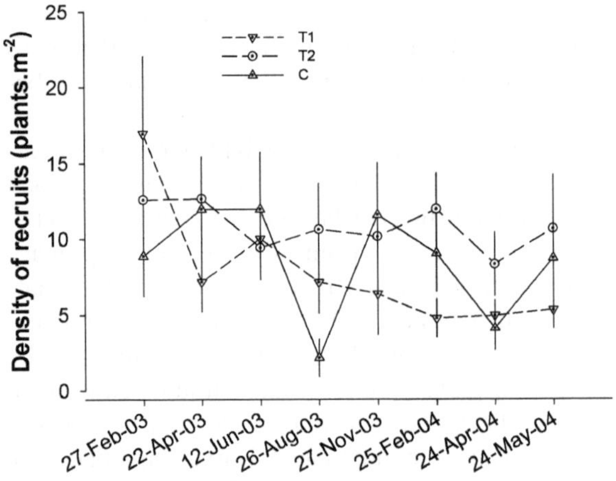

Figure 5. Mean densities of recruits (sporophytes with stipe length under 10 cm) over time under different harvesting conditions. Bars denote 95% confidence intervals.

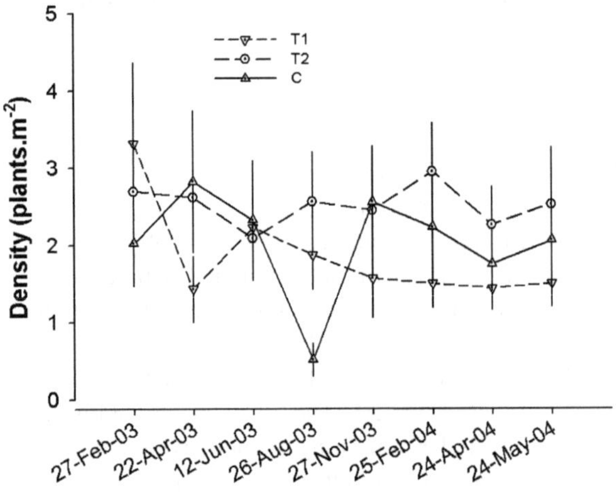

Figure 6. Mean densities of sporophytes (sporophytes with stipe length >10cm) over time under different harvesting conditions. Bars denote 95% confidence intervals.

peak in growth rate of all three size classes of sporophytes was observed in April 2003 (Figure 4).

Plant density and recruitment

Treatment had no effect on density of recruits ($n = 960$, $p = 0.8796$) although the density of the recruits varied significantly between sampling periods ($n = 960$, $p < 0.0001$: Figure 5). The density of sporophytes with stipe length between 11 and >100 cm (Figure 6) different between treatments and size classes ($n = 5280$, $p < 0.0001$) but densities at each sampling were variable and showed no consistent pattern. Note that densities of recruits (plants with stipe <10 cm: Figure 5) were consistently higher than those of the longer sporophytes (Figure 6), but showed similar patterns of temporal fluctuation.

Population structure

Recruits (plants with stipe length <10 cm) constituted the largest proportion of the kelp bed population (Figure 7). Once sporophytes reached the 10–25 cm size class, decline in density (mortality) was relatively gradual (Figure 7).

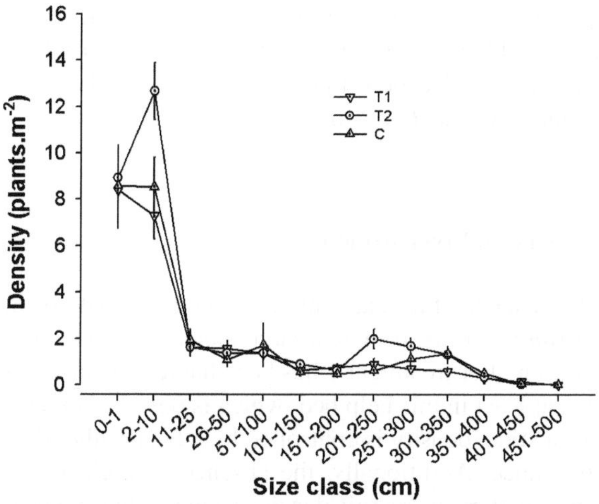

Figure 7. Mean densities of sporophytes of different size classes under different harvesting conditions. Bars denote 95% confidence intervals.

Water temperature

There was a seasonal pattern in the mean daily seawater temperatures, with consistently cooler winter months (mid-May to end August), while the summer water temperatures (September to May) were 2–5 °C higher but a lot more variable (Figure 8). In February 2003 to March 2003 and November 2003 to March 2003 water temperatures were above 17 °C for long periods but also occasionally fell to below 14 °C but only for short periods (Figure 8). In April 2003 temperatures fluctuated regularly (every 5–7 days) between about 14 and 16 °C, but were on average about 2 °C lower than during the previous 2 months.

Light

In the control and before harvesting in both treatments, 60–80 % of light incident on the surface was lost in the first 10 cm (in the kelp canopy) (Figure 9). At 1 m depth, over 90% of the light was lost and light remained low down to the bottom at 3–3.4 m (between 27.5 and 54 μmol photons m^{-2} s^{-1}). Harvesting treatment 1 (removal of entire kelp heads), increased light penetration at all depths: the mean amount of irradiance at the bottom increased 12 times from 27.5 to 330.8 μmol photons m^{-2} s^{-1} (Figure 9).

Discussion

Growth rates

Contrary to our hypothesis, removal of either the whole surface canopy (T1) or only distal portions of the secondary fronds (T2) had no effect on rate of growth (stipe elongation) of sub-canopy kelps despite a 12-times increase in bottom irradiance after harvesting in T1. We considered this to show that growth is limited by a factor other than light in this kelp bed. Treatment 2 left a considerable biomass of fronds intact and shading at depths below 1 m remained similar to the control.

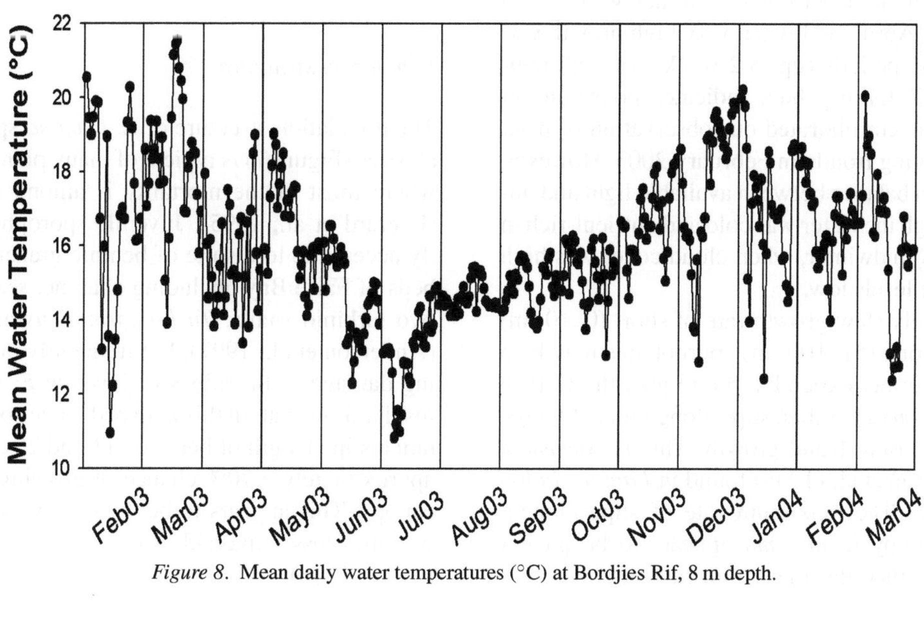

Figure 8. Mean daily water temperatures (°C) at Bordjies Rif, 8 m depth.

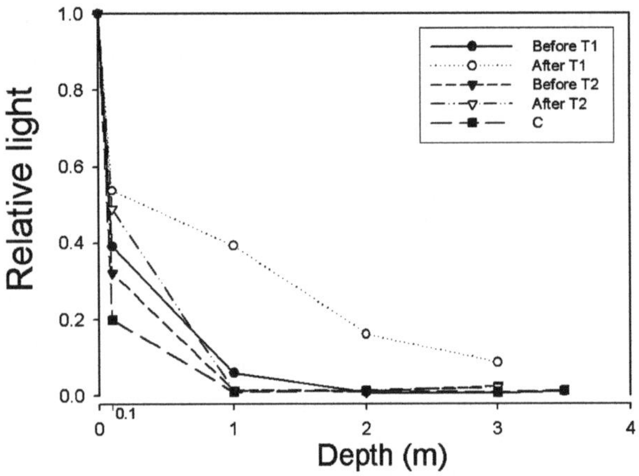

Figure 9. Relative light, before and after harvesting, with increase in depth in the kelp bed at Bordjies Rif. Mean values were calculated for each depth and expressed relative to the surface value (1).

In treatment 1 (after harvesting) the bulbs protruded above the surface of the water and light penetration on the bottom increased visibly.

In the South African west coast upwelling system, water temperature is inversely related to nutrient levels, with temperatures below 14 °C generally associated with nutrient-rich conditions (Andrews, 1974). Bolton and Anderson (1987) showed that, under nutrient-sufficient culture conditions, small sporophytes grew well over a range of temperatures between 8 and 18 °C, with the best growth achieved at 12 °C. It is thus likely that kelp growth may be faster when there is frequent import of cold, nutrient-rich water interspersed with short periods of warmer water, such as occurred in April 2003 when very high growth was recorded. Long periods (up to 2 weeks) of high temperatures (>17 °C) may have indicated poor nutrient conditions: this corroborated our observation of pale, unhealthy looking fronds in February 2003. However, there must be a balance between available light and nutrients: although the water was cold and nutrient-rich in June, this was midwinter, when cloud cover was high and irradiance levels low.

The relatively slow growth rate of short (0–10 cm) compared to long (51–100 cm) sporophytes may be a result of the former expending more growth effort in frond production rather than stipe elongation, although we did not measure frond growth. This is consistent with what Sjøtun et al. (1998) found in *Laminaria hyperborea* beds. The maximum rate of stipe elongation of subcanopy *E. maxima* appears to be greater (57 mm/week) than the maximum rate of stipe elongation in *Laminaria hyperborea* (13 mm/week: Kain, 1979). This is perhaps not surprising for a sporophyte with a gas-filled bulb that suspends its fronds at the surface, such as *E. maxima.*

Density and recruitment

The absence of a treatment effect on the density of *E. maxima* recruits, despite a large increase in light penetration after treatment 1 has been harvested, suggests that at least in this kelp bed, which is only 3–5 m deep at low tide, density is controlled by a factor other than irradiance. Additionally, the absence of a clear seasonal pattern in recruit density suggests that recruitment here may be stochastic, as has been shown in the study of *E. maxima* (Levitt et al., 2002) and in California *Macrocystis* beds (Deysher & Dean, 1986). In the latter study, recruitment was shown to be episodic rather than seasonal, and resulted from a combination of several factors creating an 'environmental window' (Deysher & Dean, 1986). Factors which may control the density of recruits in South African kelp beds include grazers (Fricke, 1978), and the availability of substratum (Anderson et al., 1997). The fecundity of the kelp population also plays a role as well as suitable environmental conditions and the time it takes for the propagules to settle and to become established (Reed et al., 2004). It appears likely, therefore, that the availability of primary space may be the most important factor for recruitment in *E. maxima* beds where light is not limiting.

Population structure

The population structure of *E. maxima* sporophytes at all sites (Figure 7) is typical of many plant populations where most of the mortality is among the juveniles (Everard et al., 1995). Juvenile sporophytes are easily accessible to a suite of benthic grazers in the kelp beds of False Bay, including abalone, sea urchins and two turbinid snails: *Turbo sarmaticus* and *T. cidaris* (Anderson et al., 1997). We frequently observed grazing damage to the stipes of juvenile *E. maxima.* Our results show that, in this kelp bed, once sporophytes attain a stipe length of between 11 and 25 cm, they have approximately a 70% chance of growing to reach the canopy. This appears to be the size where *E. maxima* becomes less vulnerable to grazing, as suggested by Fricke (1978).

Conclusions

Current frond-harvesting methods (lethal and non-lethal) do not affect the growth (stipe elongation) rate of sub-canopy *E. maxima* plants, their density or recruitment of juveniles in a shallow-water, dense kelp bed in False Bay. Management should thus continue to allow both methods of harvesting to be used when ease of harvesting is more important than obtaining maximum yields. Future research should investigate the effects of canopy clearing on deeper and/or denser kelp beds, as well as trying to determine what factors may affect kelp recruitment.

Acknowledgments

We thank Marine and Coastal Management for funding this project. We furthermore thank the University of Cape Town. We are grateful to G.M. Branch for helpful discussions on this project. Thank you to Messrs D. Kemp, C. Boothroyd, M. Noffke, G. Fridjhon and E. Tronchin who assisted with the diving work.

References

Anderson RJ, Simons RH, Jarman NG (1989) Commercial seaweeds in South Africa: A review of utilization and research. S. Afr. J. Mar. Sci. 8: 277–299.

Anderson RJ, Carrick P, Levitt GJ, Share A (1997) Holdfasts of adult *Ecklonia maxima* provide refuges from grazing for recruitment of juvenile kelps. Mar. Ecol. Prog. Ser. 159: 265–273.

Anderson RJ, Bolton JJ, Molloy FJ, Rotmann KWG (2003) Commercial seaweeds in southern Africa. In: Chapman, ARO, Anderson RJ, Vreeland VJ,, Davison IR (eds.), Proceedings of the 17th International Seaweed Symposium, Oxford University Press, Oxford, pp. 1–12.

Andrews WRH (1974) Selected aspects of upwelling research in the southern Benguela current. Tethys 6: 327–340.

Bolton JJ, Anderson RJ (1987). Temperature tolerance of two southern African *Ecklonia* species (Alariaceae: Laminariales) and of hybrids between them. Mar. Biol. 96: 293–297.

Deysher LE, Dean TA (1986) In situ recruitment of sporophytes of the giant kelp, *Macrocystis pyrifera* (L.) C.A. Agardh: Effects of physical factors. J. Exp. Mar. Biol. Ecol. 103: 41–63.

Everard DA, Midgley JJ, van Wyk GF (1995) Dynamics of some forests in Kwa Zulu-Natal, South Africa, based on ordinations and size-class distributions. S. Afr. J. Bot. 61(6): 283–292.

Fricke AH (1978) Kelp grazing by the common sea urchin *Parechinus angulosus* Leske in False Bay, Cape. South African Journal of Zoology 14(3): 143–148.

Isaac EW, Molteno CJ (1953) Seaweed resources of South Africa. J. S. Afr. Bot. 19: 85–92.

Kain JM (1979). A view of the genus *Laminaria*. Oceanogr. Mar. Biol. Ann. Rev. 17: 101–161.

Kimura RS, Foster MS (1984). The effects of harvesting *Macrocystis pyrifera* on the algal assemblage in a giant kelp forest. Hydrobiologia 116/117: 425–428.

Levitt GJ, Anderson RJ, Boothroyd CJT, Kemp FA (2002) The effects of kelp harvesting on its regrowth and the understorey benthic community at Danger Point, South Africa, and a new method of harvesting kelp fronds. S. Afr. J. Mar. Sci. 24: 71–85.

Norton TA, Mathieson AC, Neushul M (1982) A review of some aspects of form and function in seaweeds. Bot. Mar. 25: 501–510.

Reed DC, Schroeter SC, Raimondi PT (2004) Spore supply and habitat availability as sources of recruitment limitation in the giant kelp *Macrocystis pyrifera* (Phaeophyceae). J. Phycol. 40: 275–284.

Schiel DR, Nelson WA (1990) The harvesting of macroalgae in New Zealand. Hydrobiologia 204/205: 25–33.

Schroeter SC, Dean TA, Thies K, Dixon JD (1995) Effects of shading by adults on the growth of blade-stage *Macrocystis pyrifera* (Phaeophyta) during and after the 1982–1984 El Niño. J. Phycol. 31: 697–702.

Sharp GJ, Pringle JD (1990) Ecological impact of marine plant harvesting in the northwest Atlantic: A review. Hydrobiologia 204/205: 17–24.

Sjøtun K, Fredriksen S, Rueness J (1998) Effects of canopy and wave exposure on growth in *Laminaria hyperborea* (Laminariaceae: Phaeophyta). Europ. J. Phycol. 33: 337–343.

Journal of Applied Phycology (2006) 18: 343–349
DOI: 10.1007/s10811-006-9037-7

Harvesting of the kelp *Ecklonia maxima* in South Africa affects its three obligate, red algal epiphytes

R.J. Anderson[1,*], M.D. Rothman[1], A. Share[1] & H. Drummond[2]
[1]*Seaweed Unit, Marine and Coastal Management, Pvt Bag X2, Roggebaai 8012, South Africa;* [2]*Botany Department, University of Cape Town, Rondebosch 7701, South Africa*

*Author for correspondence: e-mail: Anderson@botzoo.uct.ac.za

Key words: *Ecklonia maxima, Carpoblepharis flaccida, Gelidium vittatum, Suhria vittata, Polysiphonia virgata,* kelp, epiphytes, resource management

Abstract

In South Africa, more than 7000 t (f wt) of kelp (*Ecklonia maxima*) fronds are harvested annually to feed cultured abalone. *Carpoblepharis flaccida*, *Gelidium vittatum* and *Polysiphonia virgata* are conspicuous red algal epiphytes on older kelps and provide habitat and food for numerous animals. Over 4.5 y, we examined the effects of one destructive harvest of *E. maxima* on these 3 epiphytes. Two 20 × 20 m plots of kelp with similar epiphyte loads were demarcated. In one, all *E. maxima* sporophytes with stipes longer than 50 cm were harvested. The other plot served as a control. After 2.5 y the biomass of *E. maxima* in the harvested plot had recovered to control levels, but the epiphyte load (g epiphytes. kg kelp^{-1}) was statistically lower in the harvested plot after 2.5 and 3.5 y, and only recovered after 4.5 y. While most commercial harvesters cut through the "heads" (primary blades) of the kelp, effectively killing them, a new, non-lethal method removes secondary blades 20–30 cm from their bases, leaving the meristems and primary blades intact. At 5 sites studied, *G. vittatum* and *P. virgata* were found almost entirely on stipes and primary blades, and harvesting only distal parts of secondary blades limited losses to about 50% of *C. flaccida* biomass. To protect epiphytes, non-lethal harvesting is recommended and permanent non-harvest zones have been established in addition to limiting kelp yields and disallowing harvesting in Marine Protected Areas.

Introduction

The kelp *Ecklonia maxima* (Osbeck) Papenfuss occurs along the cool-temperate west coast of South Africa, where it dominates the surface canopy of kelp beds between Cape Agulhas and at least Cape Columbine (Figure 1). It has been collected as beach-cast since the 1950's (Anderson et al., 1989) and harvested since the 1970's for the production of a plant-growth stimulant. Since the early 1990's, increasing amounts of *E. maxima* have been harvested as feed for abalone cultured in land-based farms (Anderson et al., 2003). In 2003 more than 7000 t of fresh fronds were harvested from *E. maxima* beds, and demand is increasing as abalone farms expand.

The effects of harvesting on the *Ecklonia* plants and understorey communities have been studied in the past (Levitt et al., 2002) and are being studied now (M. Rothman pers. comm.). However, these studies do not consider effects on the 3 macroalgae that are obligate epiphytes on the stipes and fronds of *Ecklonia*. These 3 rhodophytes, *Gelidium vittatum* (Linnaeus) Kuetzing, *Polysiphonia virgata* (C. Agardh) Sprengel and *Carpoblepharis flaccida*, (C. Agardh) Kuetzing attain significant biomasses (see later) and were shown by Allen and Griffiths (1981) to bear at least 27 species of invertebrates. Furthermore, *C. flaccida* forms a significant part of the diet of the commercially important line-fish *Pachymetopon blochii* (Val.) (Pulfrich & Griffiths, 1988). *G. vittatum* was formerly called *Suhria vittata* (Linnaeus) Endlicher (see Tronchin et al., 2002) and has been considered as a potential commercial agarophyte, but because of its epiphytic nature, would be difficult to obtain in sufficient quantities unless it could be

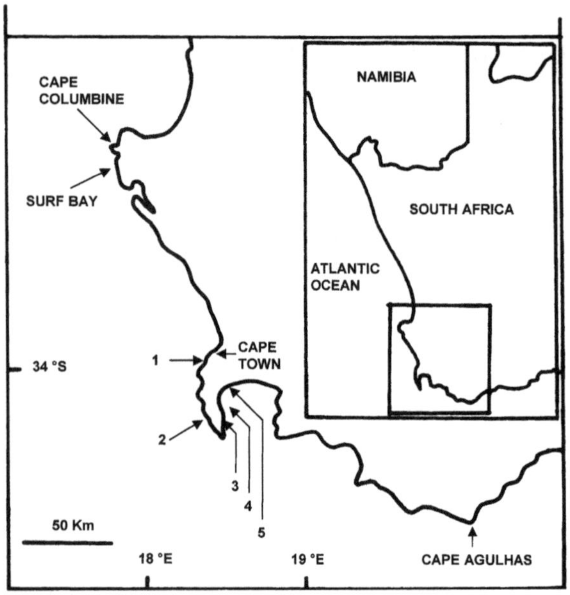

Figure 1. Map showing location of study sites on the South African west coast. Epiphyte survey sites on the Cape Peninsula, are: 1 Oudekraal, 2 Soetwater, 3 Buffelsbaai, 4 Glencairn, 5 Dalebrook.

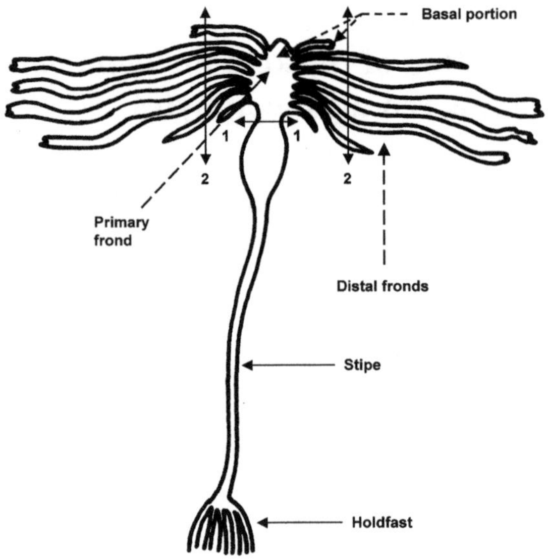

Figure 2. Diagram of *Ecklonia maxima* sporophyte to show parts referred to in text.

cultivated (Anderson & Bolton, 1985; Anderson et al., 1989; Anderson, 1994).

How harvesting affects epiphytes depends on the harvesting methods and the position of the epiphytes on the kelp. If whole kelps are removed, all attached organisms will be lost. In Norway, after harvesting of *Laminaria hyperborea* (Gunn.) Foslie by trawling, young kelps grew up rapidly to replace the mature sporophytes, but epiphytes and holdfast fauna populations took significantly longer to recover (Christie et al., 1998). Because epiphyte populations take time to become established, they are more abundant on older kelp plants (Whittick, 1983; Christie at al., 1998). Epiphytic macroalgae are often an important habitat for small invertebrates that may be ecologically important in the kelp–bed system (Christie et al., 1998; Allen & Griffiths, 1981). Furthermore, it is reasonable to assume that the abundance of such invertebrates will increase with the biomass of the epiphytes, as shown in Norway (Christie, 1995).

The 3 epiphytes in this study were known to have somewhat different distributions on the sporophytes, but these have never been quantitatively established. *Gelidium vittatum* is found on stipes or on the limpet *Cymbula compressa* that in turn grows only on these stipes. *Polysiphonia virgata* grows on the stipes. While *Carpoblepharis flaccida* was known to grow on fronds, it was not clear where on the primary or secondary fronds it is concentrated.

It is important to know where the epiphytes occur on the kelp sporophytes, because there are basically 3 harvesting methods used in South Africa. Kelp harvested for the extraction of a plant-growth stimulant is cut at the base of the stipe, and stipes and fronds used. The holdfast subsequently dies and rots off the substratum. Kelp for abalone feed is harvested in one of two ways. In the first method, the whole "head" (primary frond and attached secondary fronds – see Figure 2) is cut off, and either the stipe and holdfast die and rot off, or the stipes are cut off by divers a few days later and collected and sold for alginate extraction. In the second method, only the distal parts of the secondary fronds are cut off. This method does not kill the sporophyte: the remaining basal parts of the secondary fronds continue to grow, and all other parts are unharmed. The main advantage of this "non-lethal" method is that a substantially higher yield of kelp fronds can be obtained from a given area of kelp bed, because the replacement of biomass does not involve going through the whole life-history of the kelp: the secondary fronds continue to grow from their basal meristems (Levitt et al., 2002).

Most of the commercial harvesters supplying abalone feed prefer to cut the whole head off the *Ecklonia* sporophyte because it is easier and yields a high return per effort during each boat trip. However, on some areas of the coast, the demand for kelp fronds is now threatening to exceed the limits set by management, and on this basis alone, it may become necessary to ban lethal harvesting for abalone feed (generally only the fronds are used) in order to increase overall annual yields.

This study had two main aims. The first was to determine how long it takes for epiphyte populations to recover after *Ecklonia* sporophytes are harvested. The second was to determine the distribution of the 3 epiphytes on *Ecklonia* in order to assess the relative effects of lethal versus non-lethal (distal frond) harvesting methods.

Methods and materials

Harvesting experiment

The harvesting experiment was done at Surf Bay (32°58′70″S, 17°53′00″E), about 120 Km north of Cape Town, between May 1995 and November 1999. This area was chosen because it had never been harvested. A large and apparently uniform kelp bed was selected by visual inspection from the shore at LWS and by SCUBA inspection, and two 400 m^2 (20 × 20 m) areas (one harvest area, one control) marked out with subsurface buoys. In order to measure epiphyte abundance but limit destruction in the control at the start of the experiment, we did not collect all kelps from quadrats (see later) but randomly collected 20 sporophytes with stipes longer than 2 m, by cutting the base of the stipe. SCUBA was used for all sampling. Holdfasts were not removed from the rock, and plant and animal epiphytes on holdfasts were ignored. The sporophytes and their epiphytes (if present) were weighed individually. We then statistically compared epiphyte loads (as g epiphytes per g kelp) in the two areas using a *t*-test for independent sample means, after establishing homogeneity of variances using Levene's test (all statistics were done on Statistica 6, Statsoft). The harvest area was then cleared of all kelps with stipes longer than 50 cm (a normal commercial method).

The site was inspected periodically, and 2.5 y later, in November 1997, the kelp in the harvest plot was judged to have recovered to a visually similar biomass to the control. We then placed fifteen 1 m^{-2} quadrats haphazardly in each plot, and collected from them all kelp sporophytes with stipes longer than 25 cm. The sporophytes were taken ashore, weighed (fresh weight) and the fresh weight of all epiphytes on the stipes and fronds of each kelp recorded. All comparisons (data shown in Figures 3–4) were done using *t*-tests, after

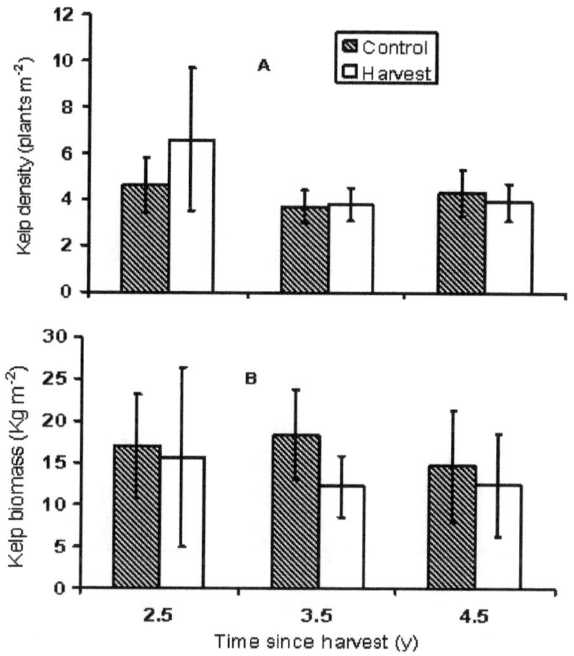

Figure 3. Mean density (A) and mean fresh biomass (B) of *Ecklonia* sporophytes at Surf Bay, in harvested and control areas, at various intervals after harvesting, with 95% confidence limits of means.

testing for homogeneity of variances (Levene's Test). We compared the harvest and control plots with respect to mean kelp biomass, mean kelp density, mean biomass of epiphytes and mean weight of epiphytes per kelp plant. This sampling method was repeated in November 1998 (3.5 y) and November 1999 (4.5 y).

Epiphyte survey

Five sites were sampled, at spring low tides, on the Cape Peninsula (Figure 1), between April and September in 2001. At each site ten 1 m^2 quadrats were placed at approximately equal intervals along a line from 1 m depth to the edge of the kelp that reached the surface. All *Ecklonia* sporophytes with stipes longer than 50 cm were cut above the holdfast and taken ashore. Each plant was cut into 3 parts (see Figure 2): stipe, primary blade with the first 30 cm of the secondary blades attached (referred to as "basal fronds") and the remaining portion of the secondary blades ("distal fronds"). All epiphytic macroalgae were removed from each of the above portions of the sporophytes, identified and weighed wet. Each portion of sporophyte was weighed wet.

All tests were done with Stastistica 6, and a critical significance of $p = 0.05$ was assumed. Chi-square

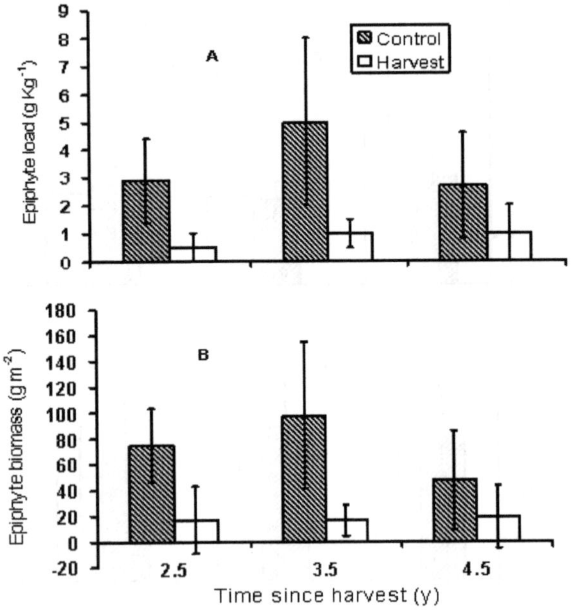

Figure 4. Mean load of total epiphytes (A) and mean total epiphyte biomass (B) for all 3 species (*C. flaccida*, *G. vittatum*, and *P. virgata*) at Surf Bay, in harvested and control areas, at various intervals after harvesting. A: units are g fresh mass of epiphytes per kg fresh mass of kelp. B: units are g m^{-2} of substratum. Vertical lines show 95% confidence limits of means.

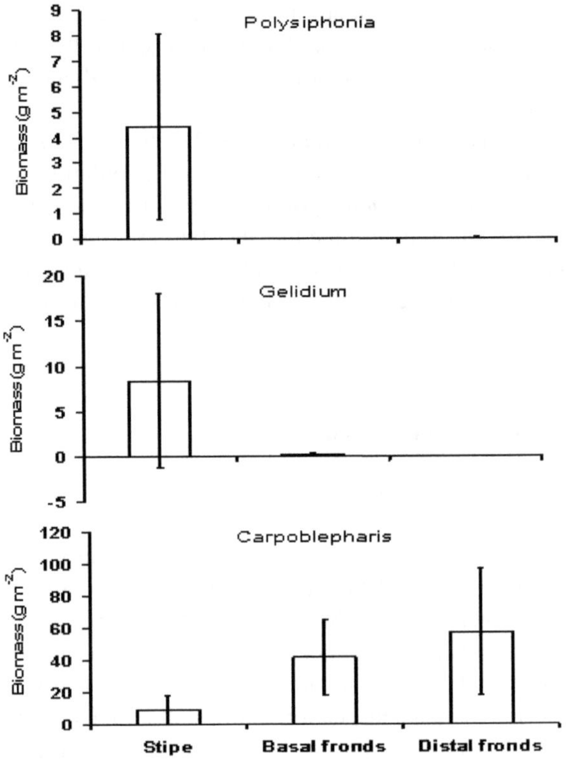

Figure 5. Mean fresh biomass of *Polysiphonia virgata*, *Gelidium vittatum* and *Carpoblepharis flaccida* on different parts of *Ecklonia*, for all 5 survey sites. Vertical lines show 95% confidence limits of means.

tests were used to compare the presence/absence of epiphytes on the different parts of the kelp and mean biomass values are shown in Figure 5. A *t*-test was used to compare the biomass of *Carpoblepharis* on the distal fronds with the biomass remaining on the basal fronds, using log-transformed data to satisfy conditions for normal distribution.

Results

Harvesting experiment

At the start of the experiment, the 20 *Ecklonia* sporophytes from the harvest plot had a mean epiphyte load of 11.64 (±7.26: 95% confidence limits of mean) g epiphyte per kg kelp, while mean load for the 20 sporophytes from the control was 10.29 ± 7.13 g kg^{-1}. These were statistically similar (*t*-test; $p = 0.942$). There was no significant difference between the densities of the kelp sporophytes in the harvest and control plots (Figure 3A) after 2.5 y ($p = 0.207$), 3.5 y ($p = 0.895$) and 4.5 y ($p = 0.761$). Similarly, there were no significant differences in mean kelp biomass (Figure 3B) after 2.5 y ($p = 0.827$), 3.5 y ($p = 0.385$) and 4.5 y ($p = 0.579$).

The mean epiphyte: kelp biomass ratios in the treatment and control areas (Figure 4A) were different after 2.5 y ($p = 0.040$) and 3.5 y ($p = 0.018$), but similar after 4.5 y ($p = 0.104$). The mean biomass of epiphytes (per m^2 of substratum; Figure 4B) was very variable but followed a similar pattern to the epiphyte/kelp ratio, with statistically different means after 2.5 y ($p = 0.039$) and 3.5 y ($p = 0.012$) but similar means after 4.5 y ($p = 0.169$).

Epiphyte survey

Seventeen species of seaweeds, besides the 3 epiphytes under study, were found on the stipes and 13 on the basal fronds of *Ecklonia*, but the distal fronds bore only *Carpoblepharis flaccida*. Because all species except these 3 epiphytes were never abundant on *Ecklonia* but are common in the understorey of these kelp beds, we do not consider them to be threatened by kelp-harvesting, and they were omitted from further analyses.

Table 1. Two-way summary of the occurrence (presence/absence) of each epiphyte species on the different parts of *Ecklonia,* based on combined samples for all 5 sites. For each species df = 2, *p* indicates observed vs expected probabilities according to the null hypothesis that epiphytes are equally distributed on all parts of kelps

Epiphyte	Part of kelp	Present	Absent	Total	Pearson Chi-square	Probability
Polysiphonia virgata	Stipe	28	193	221	47.57	$p = 0.00001$
	Basal	2	219	221		
	Distal	1	220	221		
	Total	31	632	663		
Gelidium vittatum	Stipe	35	186	221	61.49	$p = 0.00001$
	Basal	3	218	221		
	Distal	0	221	221		
	Total	38	625	663		
Carpoblepharis flaccida	Stipe	16	205	221	111.69	$p = 0.00001$
	Basal	92	129	221		
	Distal	117	104	221		
	Total	225	438	663		

Most of the biomass of *P. virgata* and *G. vittatum* occurs on the stipes of *Ecklonia* (Table 1, Figure 5A and B). However, almost all *C. flaccida* is found on the distal and basal fronds of *Ecklonia,* with very little on the stipes (Table 1 and Figure 5C).

If only distal fronds are harvested, almost all of the *P. virgata* and *G. vittatum* will be left behind. However, the biomass of *C. flaccida* on the distal fronds (mean and 95% confidence limits = 58 ± 31 g m^{-2}) was similar to the total remaining on the basal fronds and stipes (55 ± 24 g m^{-2})($t = 0.0898$; $n = 50$; $p = 0.929$).

Discussion

This study shows that while *Ecklonia* has recovered from harvesting after 2.5 y, the 3 obligate epiphytes take more than 3.5 y (up to 4.5 y) to recover, both in terms of total biomass (g m^{-2} of substratum) and biomass per kelp biomass. These results are consistent with the findings of Christie et al. (1998) on *Laminaria* in Norway. There the kelps recovered biomass and many of their stipes some epiphyte cover, within 2–3 y after a trawl harvest, but the relative abundance of epiphytes had not recovered before the next trawl 5 y later.

From the available evidence it appears that the cover of algal (and faunal) epiphytes on kelps is mainly related to the age of the host plants, as reported by Whittick (1983) in a study of *Laminaria hyperborea* in Scotland. Jennings and Steinberg (1997), in Australia, found that when *Ecklonia radiata* (C. Agardh) J. Agardh tissue was suspended in the water, epiphyte abundance correlated positively with time of exposure. They also found that most of the variation in epiphyte distribution on *E. radiata* was explained by the increase in epiphyte loading on older tissue. They ascribed this to either simple accumulation, or the fact that older tissue is higher in the water column. However, in *E. maxima*, the biomass of *C. flaccida* was similar on distal and basal portions of fronds. Basal and distal fronds *E. maxima* are at a similar level, because the fronds are long and stream out in the water. Thus epiphyte load is unlikely to depend on a higher position (and more light) in this case. Also, it does not appear to depend on the relative age of the frond portion, because although the distal fronds are older, the basal portion as defined here consists young parts of secondary fronds (the bases) and the primary frond (which is older).

In *E. maxima*, the meristem of each secondary frond lies near its base and junction with the primary frond, and is very high in polyphenols (Tugwell & Branch, 1989). Although Russell (1983) considered phenols in the meristem of *Laminaria* to prevent the settlement of *Ectocarpus*, Jennings and Steinberg (1997) discount these substances as deterrents of algal epiphytes, because of their solubility in water. Our results suggest that *Carpoblepharis* is not inhibited by phenols in the younger parts of *Ecklonia* fronds. While it is not possible to age individual *Ecklonia maxima* sporophytes, we have repeatedly observed that the highest epiphyte loads are borne on apparently old plants: those with stiff, dark stipes and broad, often tattered fronds. Our results agree with those of Whittick (1983), indicating that age is probably the main determinant of epiphyte load on these kelps.

We also provide quantitative proof that the 3 obligate epiphytes are distributed differently on the kelp sporophytes, with *Polysiphonia virgata* and *Gelidium vittatum* essentially limited to stipes, and *Carpoblepharis flaccida* distributed equally on "basal" and "distal" fronds, but with very little on stipes. The reasons for these distribution patterns are unknown, but may be related to how suitable the different substrata are for attachment of epiphytes. The stipes are relatively rigid, and subject to far less mutual abrasion than the fronds. We found a total of 21 macroalgal species (including these 3 obligate epiphytes) to be epiphytic on *Ecklonia* stipes in this study, and Stegenga et al. (1997) recorded 50 species on stipes of *E. maxima*, attesting to their suitability as a substratum. Christie et al. (2003) reported a higher diversity and abundance of fauna on the stipes than on the fronds of *Laminaria hyperborea*. Why is only *Carpoblepharis flaccida* found on distal fronds? In a physiological study, Stacey (1985) showed that the closely related *Carpoblepharis minima* is partly parasitic on its host, *Laminaria pallida,* in that photosynthetic assimilates are transferred from the host to *C. minima*. However, there is no transfer from *E. maxima* to *G. vittatum* (as *S. vittata*), which is thus fully autotrophic (Stacey, 1985). Although the *C. flaccida*/ *E. maxima* relationship has not been studied, it is possible that it is similar, implying a partly parasitic and perhaps closer link (in evolutionary terms) between *C. flaccida* and *E. maxima*.

The results of this study are important for management of the harvesting of *Ecklonia*. The non-lethal harvesting method has the least ecological effect on epiphytes because only distal fronds are cut and only about 50% of the biomass of *C. flaccida* is removed, while the other two species, and any of the other numerous macroalgae growing epiphytically on the stipes in particular, are essentially unaffected. Clearly this method will have the least effect on fauna that inhabit the epiphytes (Allen & Griffiths, 1981) and on *Pachymetopon* (Pulfrich & Griffiths, 1988) and other fish that feed on and among the epiphytes. Furthermore, this method is also predicted to yield up to 5 times more frond biomass per area of substratum over time (Levitt et al., 2002) and so when only frond material is required, it is clearly preferable on ecological grounds. However, many commercial operators are reluctant to harvest this way, claiming that it is difficult and expensive. Harvesters operate at low tides, and must lean out of a boat and gather the distal fronds of each plant, then cut them about 30 cm from the base and pull the slippery mass of loose fronds aboard. It is easier to take hold of secondary fronds, lift the "head" or primary blade out of the water, and cut it off, with all the secondary fronds attached. Also, each cut then yields more material, because no basal frond portions are left behind.

South African kelp harvesting is managed on an area basis, with a single company having the right to harvest in each Concession Area (Anderson et al., 2003), with a maximum annual yield permitted. This maximum is based on the biomass in the Concession Area. Up to now the kelp-frond harvesters in 13 of the 14 kelp areas have been free to choose their harvesting method. In the remaining area, where demand is most intensive, the o perator has only been allowed to harvest distal fronds, but was granted twice the relative yield because kelp recovery was presumed to be much faster (Levitt et al., 2002). After 3–4 years the success of this operation will be evaluated. Meanwhile, Marine and Coastal Management (the controlling authority) has set aside about 10% of each Concession Area as a "kelp reserve zone", where no harvesting may be done. This is intended mainly to protect some populations of "old" kelps and their epiphytes. However, parts of all of the harvested zones are in fact not accessible, because of shallow reefs or pinnacles, and many sub-surface kelps bear epiphytes but cannot be reached from a boat. General diving observations suggest that there are still healthy epiphyte populations even in harvested kelp beds, but a quantitative survey would be difficult because of the scale of the harvesting operations and the spatial and geographical variations in epiphyte biomass.

In areas where there are many abalone farms, *Ecklonia* beds are harvested whenever the surface canopy appears to have recovered: in areas where kelp heads are cut off, this means effectively about every 2 y. Because this interval is too short to allow full recovery of epiphyte populations, it is likely that they are being reduced, with ecological consequences that are difficult to predict. However, the results of this study clearly support the introduction of a "non-lethal", fronds-only method of harvesting *Ecklonia* in South Africa.

Acknowledgments

This study was supported by Marine and Coastal Management and H Drummond received support from the National Research Foundation. We thank Chris Boothroyd and Derek Kemp for technical assistance and AJ Smit, Geoff Fridjhon and Dylan Anderson for diving assistance.

References

Allen JC, Griffiths CL (1981) The fauna and flora of a kelp bed canopy. S. Afr. J. Zool. 16: 80–84.

Anderson RJ (1994) *Suhria* (Gelidiaceae, Rhodophyta). In Akatsuka, I. (ed.), Biology of Economic Algae, SPB Academic Publishing, The Hague: pp. 345–352.

Anderson RJ, Bolton JJ (1985) Suitability of the agarophyte *Suhria vittata* (L.) J. Agardh (Rhodophyta: Gelidiaceae) for mariculture: geographical distribution, reproductive phenology and growth of sporelings in culture in relation to light and temperature. S. Afr. J. Mar. Sci. 3: 169–178.

Anderson RJ, Simons RH, Jarman NG (1989) Commercial seaweeds in southern Africa: A review of utilization and research. S. Afr. J. Mar. Sci. 8: 277–299.

Anderson RJ, Bolton JJ, Molloy FJ, Rotmann K.W.G (2003) Commercial seaweeds in southern Africa. In Chapman ARO, Anderson RJ, Vreeland VJ, Davison I (eds), Proceedings of the 17th International Seaweed Symposium, Oxford University Press, Oxford: pp. 11–10.

Christie H (1995) Description of the kelp forest fauna at Froan, mid Norway; variation in an exposure gradient. NINA Oppdragsmelding 368: 1–22. (In Norwegian with English summary).

Christie H, Fredriksen S, Rinde E (1998) Regrowth of kelp and colonization of epiphyte and fauna community after kelp trawling at the coast of Norway. Hydrobiologia 375/376: 49–58.

Christie H, Jørgensen NM, Norderhaug KM, Waage-Nielsen E (2003) Species distribution and habitat exploitation of fauna associated with the kelp (*Laminaria hyperborea*) along the Norwegian coast. J. Mar. Biol. Ass. U.K. 83: 687–699.

Jennings JG, Steinberg PD (1997) Phlorotannins versus other factors affecting epiphyte abundance on the kelp *Ecklonia radiata*. Oecologia 109: 461–473.

Levitt GJ, Anderson RJ, Boothroyd CJT, Kemp FA (2002) The effects of kelp harvesting on its regrowth and the understorey benthic community at Danger Point, South Africa, and a new method of harvesting kelp fronds. S. Afr. J. Mar. Sci. 24: 71–85.

Pulfrich A, Griffiths CL (1988) Feeding biology of the hottentot, *Pachymetopon blochii* (Val.), with an estimate of daily ration. S. Afr. J. Zool. 23: 196–207.

Russell G (1983) Formation of an ectocarpal epiflora in blades of *Laminaria*. Mar. Ecol. Prog. Ser. 11: 181–187.

Stacey VJ (1985) The physiology and biochemistry of the *Laminaria pallida/Carpoblepharis minima* and *Ecklonia maxima/Suhria vittata* associations from South-Western Cape waters, South Africa. PhD Thesis, University of Cape Town, 152 pp.

Stegenga H, Bolton JJ, Anderson RJ (1997) Seaweeds of the South African west coast. Contrib. Bolus Herb. 18: 655.

Tronchin EM, Freshwater DW, Bolton JJ, Anderson RJ (2002) A reassessment and reclassification of species in the genera *Onikusa* Akatsuka and *Suhria* J. Agardh *ex* Endlicher (Gelidiales, Rhodophyta) based on molecular and morphological data. Bot. Mar. 45: 548–558.

Tugwell S, Branch GM (1989) Differential phenolic distribution among tissues in the kelps *Ecklonia maxima*, *Laminaria pallida* and *Macrocystis angustifolia* in relation to plant-defence theory. J. Exp. Mar. Biol. Ecol. 129: 219–230.

Whittick A (1983) Spatial and temporal distribution of dominant epiphytes on the stipe of *Laminaria hyperborea* (Gunn.) Fosl. (Phaeophyta: Laminariales) in S. E. Scotland. J. Exp. Mar. Biol. Ecol. 73: 1–10.

Journal of Applied Phycology (2006) 18: 351–359
DOI: 10.1007/s10811-006-9044-8

Changes in the brown seaweed *Ascophyllum nodosum* (L.) Le Jol. plant morphology and biomass produced by cutter rake harvests in southern New Brunswick, Canada

Raul A. Ugarte[1,*], Glyn Sharp[2] & Bruce Moore[1]
[1]*Acadian Seaplants Limited, 30 Brown Avenue, Dartmouth, N.S. Canada, B3B 1X8;* [2]*Department of Fisheries & Oceans. Bedford Institute of Oceanography, 1 Challenger Drive, Dartmouth, N.S. Canada, B2Y 4A2*
*Author for correspondence: e-mail: rugarte@acadian.ca

Key words: *Ascophyllum*, rockweed harvest, habitat, invertebrates, cutter rake, management

Abstract

Shoots and clumps of shoots of the commercial brown seaweed *Ascophyllum nodosum* ("rockweed") add to the benthic complexity of the intertidal environment, providing an important habitat for invertebrates and vertebrates. To protect the structure of this habitat, management plans for the rockweed harvest of southern New Brunswick include restrictions on gear type and exploitation rates limited to 17% of the harvestable biomass. However, owing to physical and environmental factors, the harvest is not homogeneous, creating patches of exploitation ranging from 15 to 50%.

A direct relationship existed between clump vulnerability, weight and length in a controlled harvest at 50% exploitation within 8 m by 8 m plots. At this exploitation rate, the gear rarely impacted clumps below 50 g or 60 cm in length. Clumps larger than 300 g and 130 cm were reduced by up to 55% of their length and 78% of their biomass. The overall impacts of the harvest on intertidal habitat is however of short duration as biomass recovers after a year of the experimental harvest. The rapid recovery is mostly due to a stimulation of growth and branching of the suppressed shoots of the clumps. Some harvested plots showed biomass even higher than initial levels, suggesting an increase in productivity at least during the first year after the harvest.

Introduction

The brown seaweed *Ascophyllum nodosum* (L.) Le Jol. ("rockweed") dominates the rocky intertidal of the Atlantic shores of Nova Scotia and New Brunswick, Canada. The rockweed plant is an assemblage (clump) of dichotomously branching dominant shoots and basal or suppressed shoots arising from a common holdfast and floated by vesicles (Cousens, 1984; Sharp, 1986). The buoyant biomass creates a dense canopy as the tide rises. The high density of branching and suppressed shoots in a clump and the distribution and biomass of clumps in the intertidal create also a complex habitat for invertebrates and fishes during the tidal cycle (Rangeley & Kramer, 1998).

Ascophyllum nodosum's economic value as a raw material for fertilizer, animal fodder and alginate led to its commercial harvest in the Maritimes in 1959 (Sharp, 1986). Harvesting techniques have ranged from simple knives to sophisticated and expensive vessels with hydraulically driven suction cutters (Sharp et al., 1995). Although over the past 15 years the rockweed harvest of the Maritimes has expanded in quantity and extent, the harvesting technique has evolved from harvesting machines to manual cutter rakes (Ugarte & Sharp, 2001).

The cutter rake is attached to a 3 m pole and is equipped with sharp tooth-shaped blades held in a rake-head protected by three guides (Figure 1A). The shoots are cut by the blades and the tines of the rake gather the cut shoots while the guides prevent the blade from contacting the substratum. Harvesters work during the falling and rising tides, with vessel having a 4 to 6 t capacity (Figure 1B). The rake is drawn through the

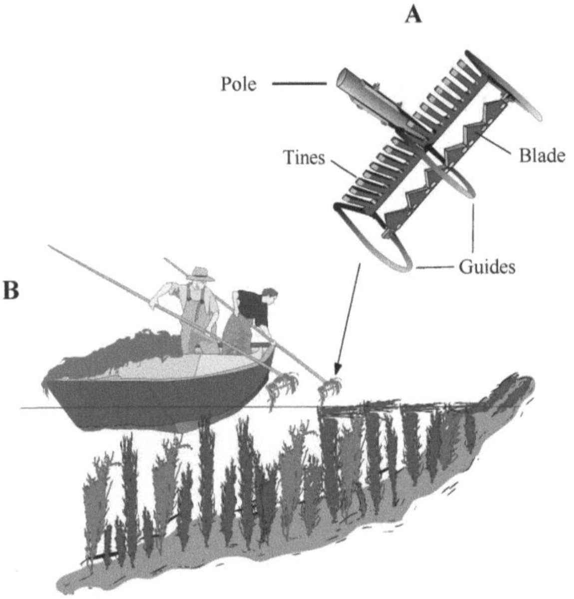

Figure 1. Harvest method of *Ascophyllum nodosum* in southern New Brunswick, Canada. (A) The manual cutter rake (B) Harvesting *Ascophyllum nodosum* from a 6 to 7 m vessel with 4 to 6 ton capacity on the falling and rising tide.

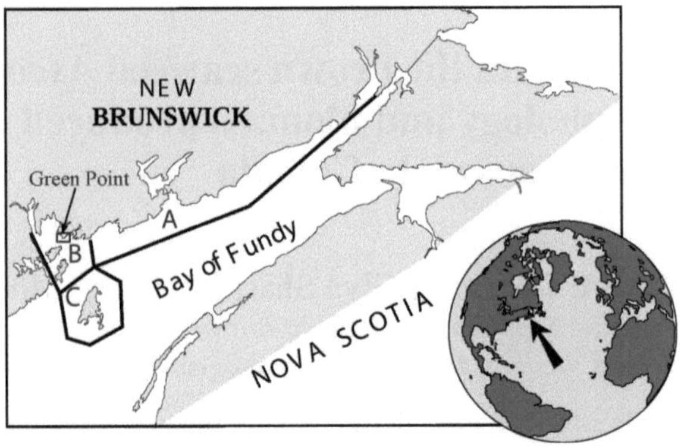

Figure 2. Ascophyllum nodosum harvesting areas and study site location in southern New Brunswick, Canada.

floating canopy at a 45 to 60 degree angle. Once the harvester reaches the rockweed bed, the vessel moves up and down the intertidal zone with the tide and along the bed with the wind and current. The harvester can choose areas of the bed to harvest but cannot directly control the cutting height of the clumps.

In southern New Brunswick regulations restrict gear type and the exploitation rate is limited to 17% of the harvestable biomass in order to protect the structure of this habitat (Ugarte and Sharp, 2001). However, owing to physical and environmental factors, the harvest is not homogeneous, resulting in patches of exploitation ranging from 15 to 50% (DFO, 1999).

The present paper examines the length and biomass structure of *Ascophyllum. nodosum* clumps immediately before and after a harvest and recovery in experimentally harvested plots. To date experimental harvests using various gear types have only examined shoots, and clump length and density (Ang et al., 1993; Lazo & Chapman, 1996).

Methods

The study area was located in Harvesting Area B, which produces the highest landing of the three harvesting areas in the northern shore of the Bay of Fundy in southern New Brunswick, Canada (Figure 2). A closed site (previously non-harvested) located at Green Point, inside Area B, was the location of the experimental harvest (Figure 2). This site was semi-exposed, with a boulder substratum, 15°–30° slope and 100% rockweed cover.

During the summer of 2001 the population structure of rockweed in Harvesting Area B was determined by sampling 16 sites within this area. All clumps were removed from fifteen 0.25 m^2 randomly-placed quadrats without evidence of harvest, along a 30 m transect set in the middle of the rockweed zone at each site. Clumps were bagged and immediately refrigerated (5 °C) and analyzed within two days. The total length, wet weight and number of shoots were measured in 1,196 clumps. From these, 482 clumps were randomly selected, dissected into 10 cm segments from base to tip and the wet weight of each segment measured to 0.1 g.

To measure the impact of the harvest on the population structure, five plots (8 m × 8 m) were permanently marked with rock anchors in the middle of the rockweed zone located at the Green Point study site (Figure 2). Thirty quadrats (each 0.125 m^2) were evenly spaced along five parallel transects in each plot. All clumps in a quadrat were tagged using permanent tags (Sharp and Tremblay, 1985). A total of 1,256 clumps was tagged at the start of the experiment but only 1,137 (90.5%) provided reliable data throughout the experiment. Three plots were randomly selected as treatments and two as controls. Each tagged clump was measured for length to 0.1 cm using a flexible tape. The wet weight (accurate to 0.1 g) for each clump was measured using a low-profile base electronic scale (Acculab VI-600). The clump was carefully piled inside a tared

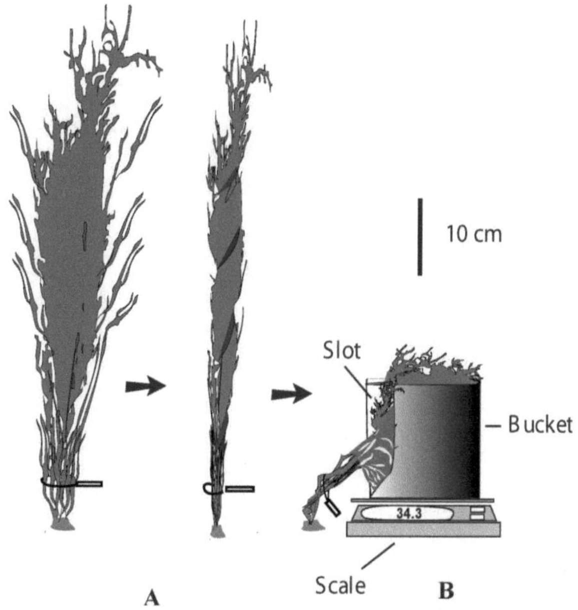

Figure 3. Measurement of *Ascophyllum nodosum* clump weight. (A) Twisting the clump, (B) Rolling the clump inside a tared bucket over a low-profile base scale.

plastic bucket with a vertical slot to weigh it as close to the holdfast level as possible (Figure 3). As the aim of this measurement was to quantify changes of clump weight after the harvest, the small portion of the clump biomass close to the holdfast and outside the bucket (Figure 3) was considered constant before and after the harvest. To avoid variations, all weight measurements during the study were done by one trained person. Variation due to desiccation was avoided by measuring the clumps within an hour of being exposed. In addition, 30 suppressed shoots in three size classes (10–20, 21–40 and 41–60 cm) were individually tagged in each plot to evaluate their response to harvesting. The use of the scale was difficult in the lower size class so their initial weight (pre-harvest) was obtained through a regression with their length. The relation was highly significant ($p < 0.001$, $n = 350$) with $R = 0.93$ in this small unbranched class (Ugarte, unpublished data).

To determine the rockweed biomass in the study site, fifteen 0.25 m^2 quadrats were randomly sampled along a 10 m transect set in the middle of the bed. These data were used to determine the biomass removal needed to reach the target 50% exploitation rate in the experimental harvest. A commercial harvester with a conventional cutter rake then harvested the required biomass in each treatment plot. The tagged clumps were re-measured for length and weighed immediately after the harvest. The clumps were re-measured three times during the first year of the experiment: October 2001, April 2002 and August 2002. Another measurement was carried out in August 2003, two years after the harvest. Mean differences were compared using *t*-tests. Biomass values were log transformed.

Results

Length and biomass structure of rockweed clumps

The distribution of biomass within a clump changes with the total length of the clump. At smaller size classes the biomass is proportionately closer to the bottom (Figure 4). Fifty percent of the biomass is within the lower half of the plant up to the 90 cm length class. In the 130 cm length class 50% of the biomass is distributed in the upper one third of the clump (Figure 4).

Clump length is normally-distributed and ranges from 12 cm to 143 cm, with a mean of 74.5 cm (SD $\pm$ 27.6 cm) (Figure 5). Clump weight varied from 1.0 g to 765.3 g, with a mean of 68.2 g (SD $\pm$ 112.2). 83% of the clumps in the stand are under 100 cm; however, they only contribute 50.8 % of the total weight. The remaining 17.1% of the clumps over 100 cm contribute 49.2% of the stand biomass (Figure 5).

Effect of harvest on clump length and biomass structure

The impact of the harvest increased with clump size. Length was significantly reduced ($p < 0.01$) in those clumps over 70 cm. Clumps over 90 cm and 130 cm lost 35% and 45% of their original length, respectively (Figure 6A). The rake did not reduce the length significantly in those clumps under 60 cm ($p > 0.01$) (Figure 6A).

A more dramatic effect of the harvest was evident when measuring the clump biomass. The biomass of clumps in the 90.1–99.0 cm category was reduced by 56.5%. Clumps over 130 cm lost 78% of their biomass. However, as with length, those clumps below 70 cm were not reduced in biomass ($p > 0.05$) (Figure 6B).

Biomass recovery

The mean clump biomass of unharvested rockweed increased in the fall due to vegetative and reproductive growth. Wet weight reached a peak by late April when receptacles had a high water content. Clump weight

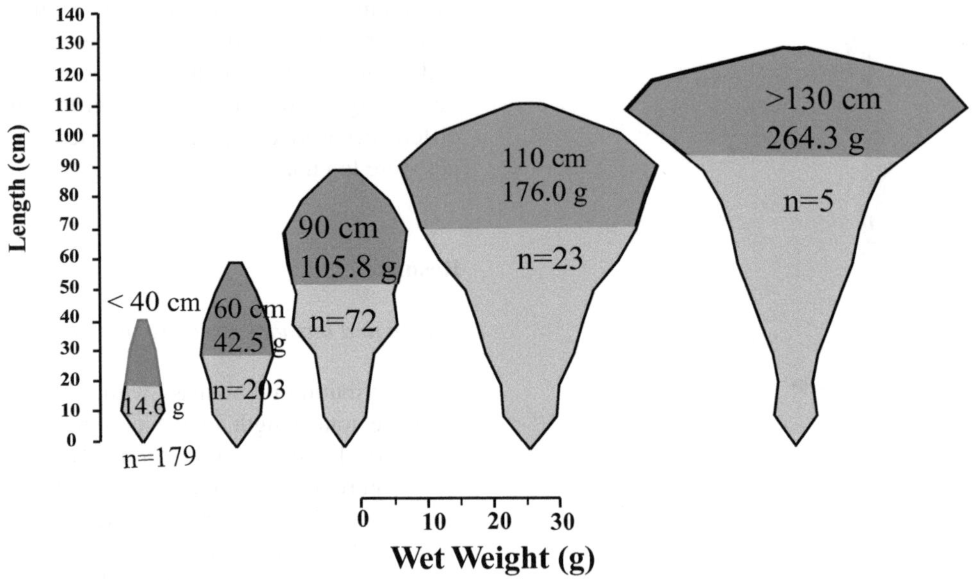

Figure 4. Mean distribution of biomass in 10 cm increments within individual *Ascophyllum nodosum* clumps of 5 size classes. The shaded areas in each size class represent 50% of the biomass.

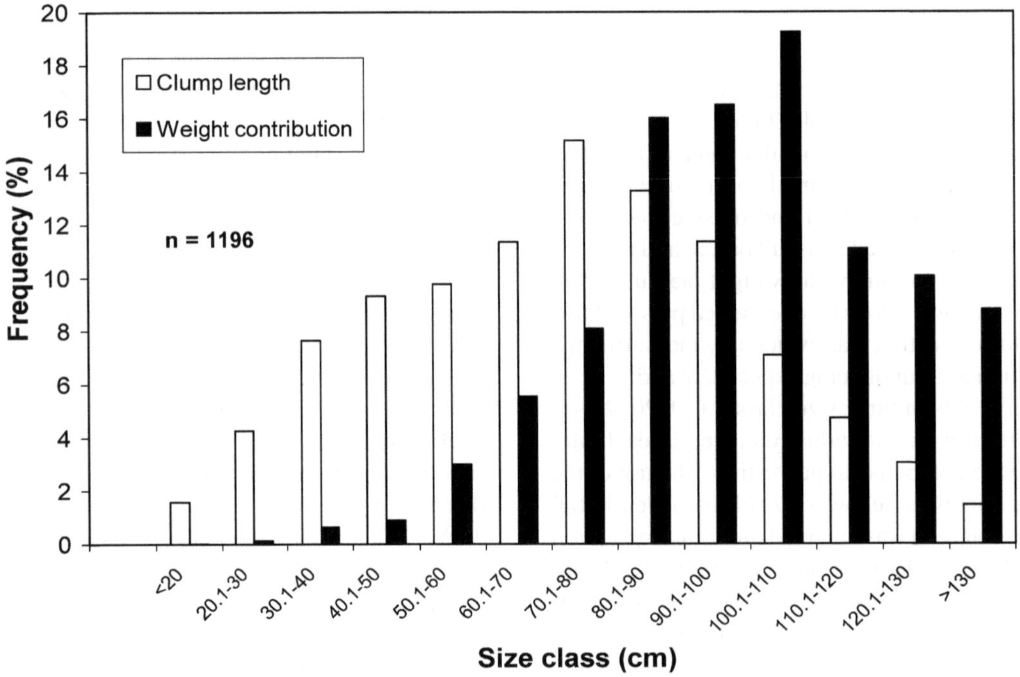

Figure 5. Length frequency and weight contribution of *Ascophyllum nodosum* clumps in study area.

dropped suddenly in mid May, after the breakdown of the receptacles and reached its previous year's level in July (Figure 7).

After the reduction in their mean biomass in early August 2001, clumps in all harvested plots showed an increase in their mean biomass (Figure 7). Here, the growth rate was higher than the control after October 2001 and until April 2002 (Figure 7). A year after the harvest in July 2002 harvested clumps had a 85% biomass recovery in plot 6, a total recovery in plot 3,

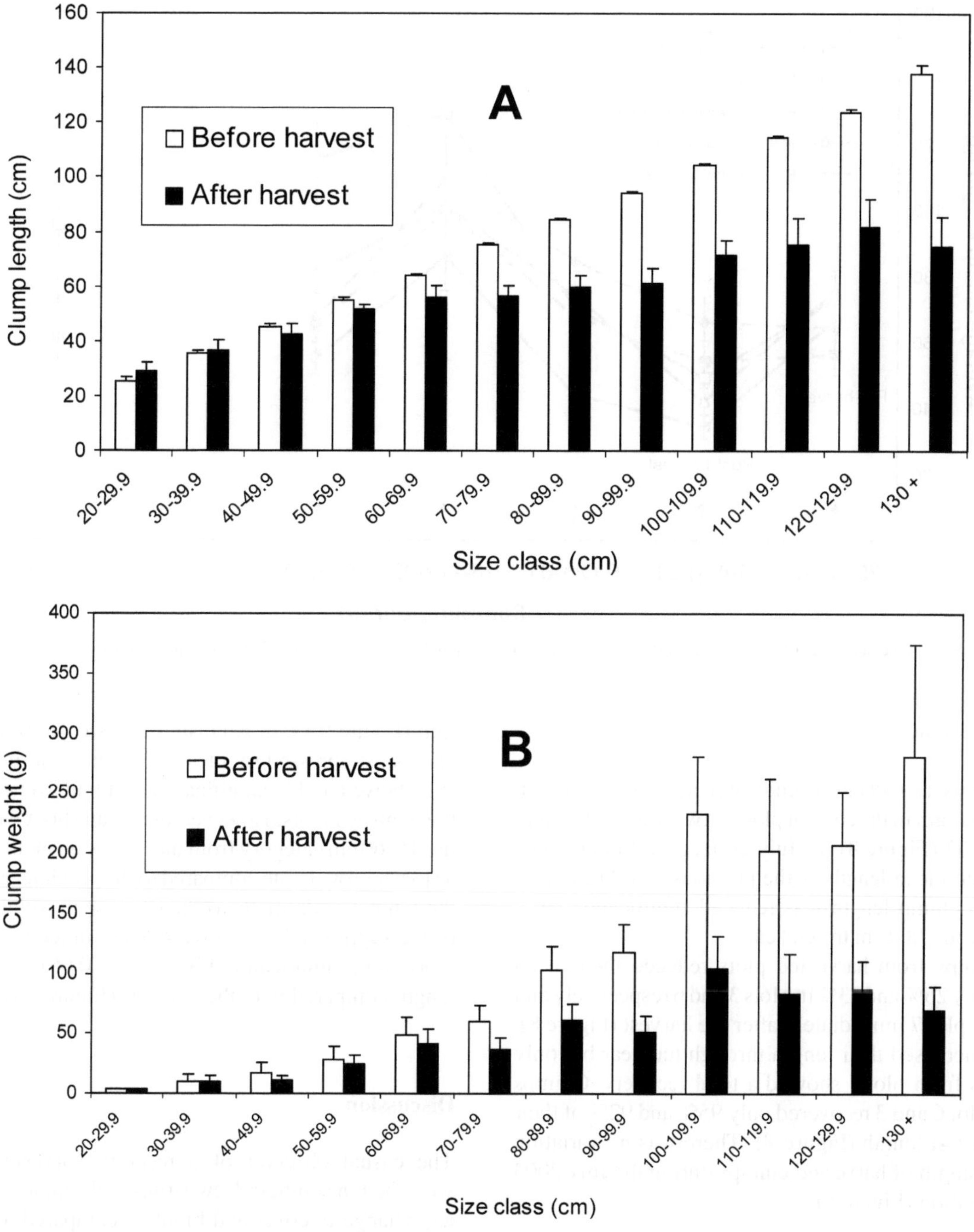

Figure 6. Changes in clump structure of *Ascophyllum nodosum* in 12 different size classes after a 50% exploitation rate harvest in experimental site at Green Point, southern New Brunswick ($N = 672$; Vertical bars are $\pm$ 2 standard errors). (A) Changes in mean length of clumps (B) Changes in mean weight of clumps.

and a 52% increase in plot 7 in comparison to the original biomass (Figure 7). By July 2003, two years after the harvest, clumps of plot 6 had totally recovered their original biomass, clumps in plot 3 had increased their original mean biomass by 22%, while those in plot 7 were down from the July 2002 biomass but still maintained a 23% increase from their original biomass. Control clumps were not significantly higher in their original mean biomass ($p > 0.05$) when re-weighed in July 2003 (Figure 7).

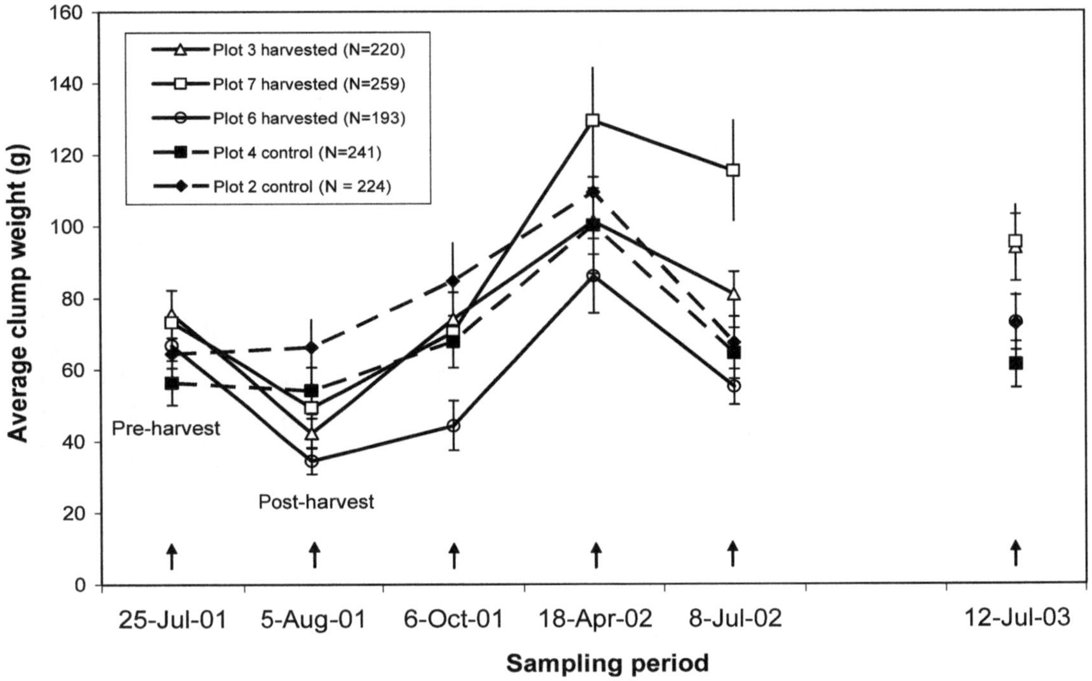

Figure 7. Seasonal changes in average clump weight of *Ascophyllum nodosum* harvested and control plots (vertical bars are ± 1 standard error).

Length recovery

There was a small but significant increase ($p < 0.01$) in clump length in the control plots between July 2001 and July 2002 (Figure 8). In July 2003, plot 2 maintained the same mean length as the previous year but in plot 4 mean clump length was reduced significantly ($p < 0.01$) from 78.2 cm to 65.4 cm.

Clumps from harvested plots reduced their mean length by 25% and 23% in plots 3 and 6 respectively and 12% in plot 7 immediately after the harvest (Figure 8). They increased their length through the year but only clumps from plot 7 showed a total recovery. Clumps from plot 6 and 3 recovered only 95% and 92% of their pre-harvest length (Figure 8). There was no variation in the length of harvested clumps during the July 2003 examination (Figure 8).

Suppressed shoots

Shoots between 21–40 cm and 41–60 cm in the harvested plots increased their biomass by 131% and 249%, respectively over the control plot shoots after the first year of the harvest (Figure 9A). There was no significant difference ($p > 0.05$) between treatment and control for shoots of the 1–20 cm class during this period (Figure 9A). In August 2003, two years after the harvest, suppressed shoots over 20 cm of initial length still showed highly significant weight differences from the control shoots. However, the mean shoot weight in the 41–60 cm category from the harvested plots showed a slower growth rate compared to the previous year. Although not as dramatic as the increase in weight, most of the suppressed shoots over 20 cm in the treatments plots had a significantly higher ($P < 0.05$) increase in length compared with the controls (Figure 9 B).

Discussion

The casual observer of a recently harvested rockweed bed in southern New Brunswick cannot perceive any change in cover and biomass compared to undisturbed beds. It appears counter-intuitive that 12,000 t of biomass can be removed from the accessible resource without obvious signs. Our examination of biomass distribution in the bed and in the clumps provides an explanation. Harvesting of rockweed with a cutter rake at or below the target 17% exploitation rate will impact patches of rockweed habitat within beds. In these patches, harvesting reduces the biomass and total length of selected clumps by cutting a portion of

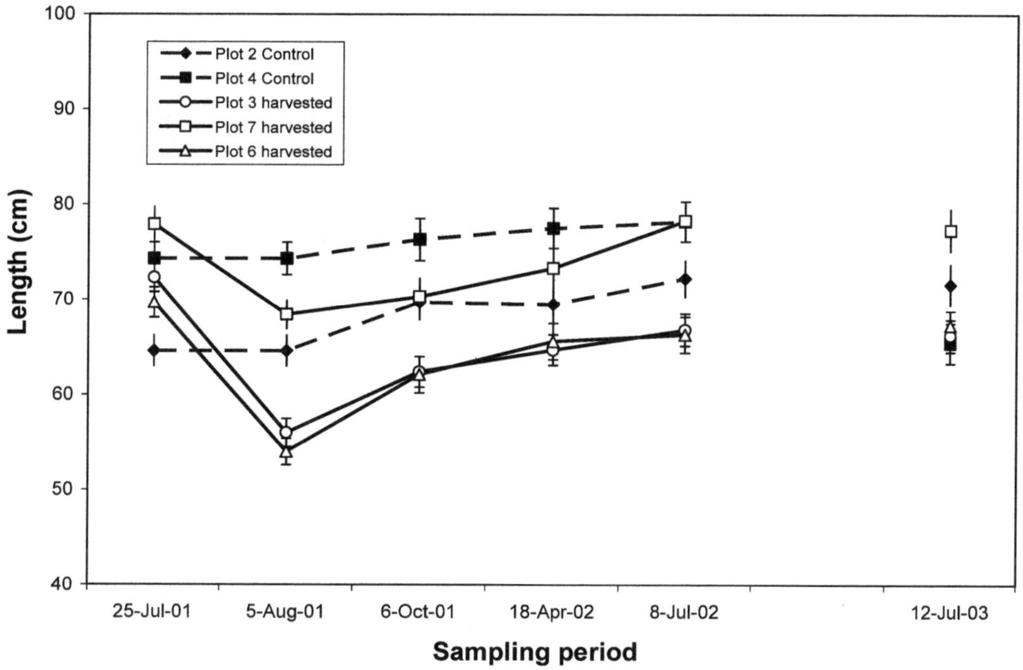

Figure 8. Seasonal changes in average clump length of *Ascophyllum nodosum* from harvested and control plots (vertical bars are ± 1 standard error).

their shoots. Due to the skewed distribution of biomass in the clump and the stand, small changes in clump length can result in localized exploitation rates of 50%. The exponential relationship between shoot length and weight shows that most of the biomass is in the distal portion of the clump. The harvester is able to direct his rake to the larger clumps that form the canopy of the stand. The diagonal, basal to distal cutting action of the rake removes the upper, heavier part of these clumps, changing significantly their complexity and spatial structure. However, the largest net change in the harvested patches is in clump biomass not length or the number of shoots. Though tedious, measuring clump biomass before and after the harvest is the best way to determine any structural change. Measuring frond length and circumference and obtaining their correlation with biomass (Cousen, 1984; Aberg, 1990) cannot be used in this case as the volume and biomass relationship is lost after the harvest.

The reduction in the complexity and spatial structure in the harvested patches could potentially affect both the abundance of associated invertebrates and the abundance and behaviour of vertebrates. The body size and abundance of metazoans in small tufted algae are affected by the size and structural variety of the algal species (Gee and Warwick, 1994; Pavia et al., 1999). Moderate changes in shape and branching within the structural units of the red alga *Gracilaria* did not affect predation on amphipods (Masterson, 1998). However, reduction in biomass within rockweed clumps can potentially affect those species most closely related to the plant surface, such as *Littorina obtusata* (Johnson and Scheibling, 1987). Micro-spatial complexity is directly affected as the amount of epiphyte biomass is reduced. Invertebrate species abundance and diversity in *A. nodosum* epiphytes is linearly related to this level of complexity (Hicks, 1980). Algal cover also affects schooling behaviour of juvenile pollock, which use it to avoid predation (Rangeley and Kramer, 1998). The behaviour of eider ducklings can be also affected by the structure of the *A. nodosum* stand (Hamilton, 2001).

Although the current scale of harvesting in New Brunswick does not alter shoot or clump density or bed cover, the overall structural complexity is altered. However, net changes in canopy height or complexity of clumps quickly become diluted to small differences between harvested and un-harvested stands when placed in the context of the entire bed and the intertidal landscape. The question remains whether this change in the canopy or height structure in harvested patches causes

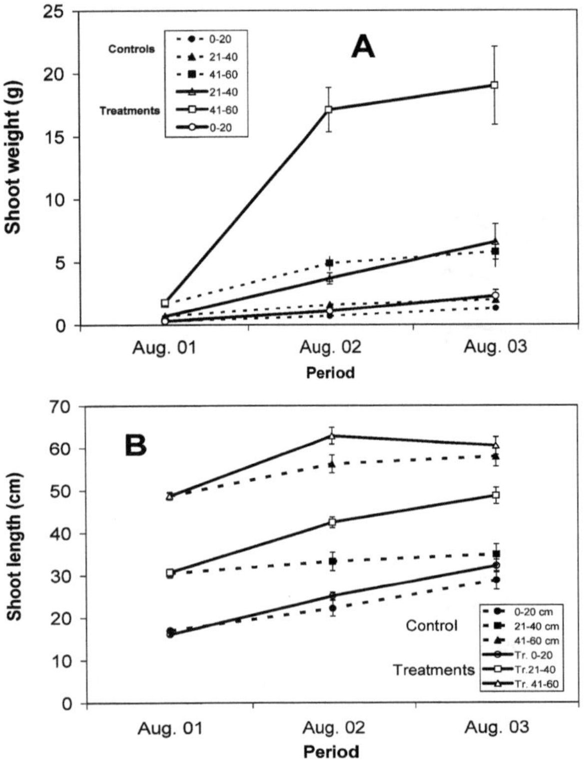

Figure 9. Changes in suppressed shoots of *Ascophyllum nodosum* after a 50% exploitation-rate harvest in experimental site at Green Point, southern New Brunswick (Vertical bars are ± 1 standard error). (A) Changes in average weight of shoot. (B) Changes in average length of shoot.

any significant reduction in the value of a stand as a habitat and whether this alters critical environmental factors for fauna. In our southern New Brunswick experiments, the structural changes produced by rockweed harvesting in the habitat are short lived as the reduction in standing crop at this scale of harvest is compensated for by the overall production during the summer and fall months. The removal of the canopy enhances growth and production by the initiation of new laterals from cut or basal shoots (Lazo & Chapman, 1996), thus redeveloping the complexity of the clump within a year.

Changes in biomass also become less significant as we move from the stand to the sector, to the harvest area, and then the coast or the bay. The change is a function of the degree of harvest, the amount of accessible rockweed compared to the total rockweed in the system and the importance of macrophyte production to the total primary production in the system.

The current quota of 12,000 t is spread over the southern New Brunswick rockweed resource and is 7.5% of the total rockweed biomass of 159,000 t (DF0, 1999). Annual production to biomass ratios of rockweed are 0.4 to 0.5 depending on the method of calculation (Cousens, 1984). According to this information, the annual productivity of rockweed in New Brunswick would range from 64,000 t to 79,500 t. Consequently, this harvest does not diminish the standing stock of rockweed as it takes 15.1% to 18.7% of annual production in southern New Brunswick.

The question of cumulative effect is very relevant. Harvesting has been most intensive in harvesting area B (Ugarte & Sharp, 2001), because of resource abundance as well as easier access to the resource both for harvesters and materials-handling issues for the company. The goal is to spread harvesting evenly between sectors as well as within sectors. The harvester expects a minimum catch-per-unit of effort in a bed and if this is not reached, he will move to another bed. The very large tidal range also prevents harvesters from remaining in one place for more than a few minutes as the tide rises or falls. Harvesters do not normally return to the same patch in the same year as their catch per unit effort could not be sustained in an area that is still recovering from harvest. The harvest within sectors is not controlled to the level of a bed and re-harvest of a patch is possible. However, in theory, at a 17% exploitation rate of the harvestable biomass, it could take 6 years before it is 100% probable that all harvested patches in a bed will be re-harvested.

Acknowledgements

The authors extend their appreciation to J. Sharp and J. Bettle for their extensive field assistance. This study was supported and entirely financed by the Department of Research and Development of Acadian Seaplants Limited.

References

Aberg P (1990) Measuring size and choosing category size for a transition matrix study of the seaweed *Ascophyllum nodosum*. Mar. Ecol. Prog. Ser. 63: 281–287.

Ang PO, Sharp GJ, Semple RE (1993) Changes in the population structure of *Ascophyllum nodosum* (L.) Le Jolis due to mechanical harvesting. Hydrobiologia 260: 321–326.

Cousens R (1984) Estimation of annual production by the intertidal brown alga *Ascophyllum nodosum* (L.) Le Jolis. Bot. Mar. 27: 217–227.

DFO (1999) The impact of rockweed harvest on the habitat of southwest New Brunswick. DFO Maritimes Regional Habitat Status Report 99/2E: 1–9.

Gee JM, Warwick RM (1994) Metazoan community structure in relation to the fractal dimensions of marine macroalgae. Mar. Ecol. Prog. Ser. 103: 141–150.

Hamilton DJ (2001) Feeding behavior of common eider ducklings in relation to availability of rockweed habitat and duckling age. Waterbirds 24: 233–241.

Hicks GRF (1980) Structure of phytal harpacticoid copepod assemblages and the influence of habitat complexity and turbidity. J. Exp. Mar. Biol. Ecol. 44: 157–192.

Johnson SC, Scheibling RE (1987) Structure and dynamics of epifaunal assemblages on intertidal macroalgae *Ascophyllum nodosum* and *Fucus vesiculosus* in Nova Scotia, Canada. Mar. Ecol. Prog. Ser. 37: 209–227.

Lazo L, Chapman ARO (1996) Effects of harvesting on *Ascophyllum nodosum* (L.) Le Jol. (Fucales, Phaeophyta): A demographic approach. J. Appl. Phycol. 8: 87–103.

Masterson J (1998) Investigation of the effects of macrophyte structure, food resources and health on habitat selection and refuge value in vegetated aquatic systems. Dissertation Abstracts International Part B: Science and Engineering 58: 4019.

Pavia H, Carr H, Aberg P (1999) Habitat and feeding preferences of crustacean mesoherbivores inhabiting the brown seaweed *Ascophyllum nodosum* (L.) Le Jol. and its epiphytic macroalgae. J. Exp. Mar. Biol. Ecol. 236: 15–32.

Rangeley RW, Kramer DL (1998) Density-dependent antipredator tactics and habitat selection in juvenile pollock. Ecology 79: 943–952.

Sharp GJ (1986) *Ascophyllum nodosum* and its harvesting in Eastern Canada. In: Case studies of seven commercial seaweed resources. FAO Technical Report 281:3–46.

Sharp GJ, Ang PO, McKinnon D (1995) Rockweed harvesting in Nova Scotia, Canada: Its socio-economic and biological implications for coastal zone management. In Wells PG, Ricketts P (eds), Proceedings of the Coastal Zone Conference: 1632–1644.

Sharp GJ, Tremblay DM (1985) A tagging technique for small macrophytes. Bot. Mar. 28: 549–551.

Ugarte R, Sharp GJ (2001) A new approach to seaweed management in eastern Canada: The case of *Ascophyllum nodosum*. Cah. Biol. Mar. 42: 63–70.

Journal of Applied Phycology (2006) 18: 361–368
DOI: 10.1007/s10811-006-9039-5

Carrageenans from cystocarpic and sterile plants of *Chondrus pinnulatus* (Gigartinaceae, Rhodophyta) collected from the Russian Pacific coast

I.M. Yermak[1,*], A.O. Barabanova[1], V.P. Glazunov[1], V.V. Isakov[1], Kim Yong Hwan[2], Shin Kwang Soon[2], T.V. Titlynova[3] & T.F. Solov'eva[1]
[1]*Pacific Institute of Bioorganic Chemistry, Far East Branch of the Russian Academy of Sciences, Vladivostok, 690022, Russia;* [2]*Department of Food Science & Biotechnology, Kyonggi University, Suwon, 442-760, Korea;* [3]*Institute of Marine Biology, Far East Branch of the Russian Academy of Sciences, Vladivostok, Russian*

*Author for correspondence: e-mail: iren@piboc.dvo.ru

Key words: life history of algae, carrageenan, structure, biological activities

Abstract

The chemical structure, gel properties and biological activity of the carrageenans isolated from cystocarpic and sterile plants of *Chondrus pinnulatus* were investigated. The total carrageenan content of the sterile plant was observed to be twice that of the cystocarpic plants. According to data obtained by ^{13}C-NMR and FT IR, the gelling polysaccharides from cystocarpic and sterile plants of *C. pinulatus* have similar structures and were identified as κ/ι-carrageenans. The difference between these polysaccharides was in the ratio of the κ- and ι-segments, with a predominant content of κ-segments in cystocarpic plants (80%). Moreover, KCl-insoluble fractions possibly contain hetero-disperse μ/ν precursor: amounts of this in the polysaccharide from sterile plants were more than that extracted from the cystocarpic plants. The KCl-soluble fractions (non gelling) were λ-carrageenans with another carrageenan type that had a low amount of 3,6-anhydrogalactose. Carrageenans from cystocarpic stages showed good gelling properties, whereas those from sterile plants formed a very weak gel. Structural differences and molecular weight of carrageenans obviously determine the biological activity of the polysaccharides. Non gelling-carrageenans from both types of *C. pinnulatus* plants showed high macrophage-phosphatase activity and κ/ι-carrageenan from cystocarpic plant possessed a potent anti-coagulant activity, which was extremely strong in a low concentration of $100\,\mu g\ mL^{-1}$.

Introduction

Carrageenans are a complex family of water-soluble galactans extracted from marine red algae that have applications as gelling, thickening and suspending agents in food processing. Carrageenans have pronounced biological activities or other properties useful in the biomedical field. These polysaccharides are composed of alternating α-(1–3) and β-(1–4) linked D-galactosyl residues and several types of carrageenan are identified on the basis of the modification of the disaccharide repeating unit by sulphate esters and by the presence of 3,6 -anhydrogalactose as 4-linked residue. At least 16 types of carrageenans have been defined, some of which have no commercial importance so far (Knutsen et al., 1994). Native carrageenans are often hybrids of more than one type of repeating unit (Craigie, 1990; Knutsen et al., 1994). Variations in carrageenan structures are known to occur not only between different species of the Gigartinaceae, but also between different life stages of the same species (McCandless et al., 1973; Craigie, 1990; Stortz & Cerezo, 1993; Falshaw & Furneaux, 1994). It has been shown that members of the family Gigartinaceae yield different carrageenans from karyologically different generations (Craigie, 1990). Close studies of the carrageenans from cystocarpic and tetrasporic plants of *Iridaea undulosa* (Stortz & Cerezo, 1993) and *Gigartina skottsbergii* (Matulewicz et al., 1989) showed that the cystocarpic plants exhibit two major products separable by potassium chloride precipitation: one of the products is a soluble, partially cyclisized μ/ν- carrageenan and the other is a gelling κ/ι-carrageenan. The tetrasporic plants produce sets of λ-carrageenans gelling at high concentrations of potassium chloride. The polysaccharides that have been extracted from tetrasporic plants of *Gigartina clavifera* and *G. alveata* are predominantly ξ-carrageenans (Falshaw & Furneaux, 1995). This suggests the taxonomic position of algae does not provide full information on the type of polysaccharide they contain. Therefore, the establishment of the polysaccharide composition of an alga in relation to the phase of its life cycle remains an important research objective.

Carrageenans have a wide spectrum of biological action (Lahaye & Kaeffer, 1997; Yermak & Khotimchenko, 2003), depending on the polysaccharide structure (Güven et al., 1990). *Chondrus pinnulatus* is abundant in the Russian Far-Eastern seas. This species of marine alga grow mostly on lower eulittoral and upper sublittoral rocks and stones, on open coasts with low wave action. The system of carrageenans from *C. pinnulatus* has been studied with material extracted from unsorted (mixed phase) samples (Yermak et al., 1999). The aim of the present study was to investigate the structure, gel properties and biological activity of the systems of carrageenans isolated from cystocarpic and sterile plants of *C. pinnulatus* (Gigartinaceae) collected from the Great Peter Bay (the Sea of Japan).

Materials and methods

Algae

Chondrus pinnulatus (Harv.) Okam. is widespread in the Sea of Japan. Material was harvested in The Great Peter Bay at the end of August and separated into sterile plants (SP) and plants with cystocarps (CP).

Extraction

Dried and milled algae (50 g) were suspended in hot water (1.5 L) and the polysaccharides were extracted at 90 °C for 2 h in a boiling water bath. The suspensions were centrifuged (2 500 g, 20 min, 20 °C) and the algal residues were re-extracted twice with water for 2 h in a boiling water bath. The supernatants were pooled. The polysaccharides were separated into the gelling-KCl-insoluble (a) and non-gelling the KCl-soluble fractions (b) as described previously (Yermak et al., 1999).

Analytical methods

The total amount of carbohydrates was estimated using the phenol-sulphuric acid method, using D-galactose as standard (Dubois et al., 1956). Monosaccharides as alditols acetate derivatives (Englyst & Cummings, 1984) were identified by GLC using an Agilent 6850 gas chromatograph equipped with a HP-5MS capillary column (30 m × 0.25 mm) with 5% Phenyl Methyl Siloxane and flame-ionization detector. The analyses were carried out at temperature programming from 175 to 220 °C with 3 °C min^{-1}. The content of 3,6-anhydrogalactose was determined according to the method of Usov and Elashvili (1991). The protein content of samples was determined according to the method of Lowry et al. (1951) using crystalline bovine serum albumin as the standard protein. The content of ash in the polysaccharides was determined gravimetrically after incineration of samples at 550 °C for 16 h followed by 2 h at 900 °C. The sulphate ester content of polysaccharide was determined according to the method Lahaye and Axelos (1993) by HPLC equipped (conductivity detector Waters 431) with a IC-Park A Anion column (50 × 4.6 mm 10 μm, Waters), eluted by 2 mM borate/gluconate eluent (flow rate: 1.0 mL min^{-1}).

Infrared spectroscopy and nuclear magnetic resonance spectroscopy

Films of polysaccharides for infrared analysis were obtained by drying in polyethylene molds (about 0.5 cm deep, 2.5 cm diameter) at 35–40 °C, 2 mL of an aqueous solution containing 5–7 mg of polysaccharide. The polysaccharide film was clamped between NaCl windows and the IR spectrum was recorded in the 4000–600 cm^{-1} region using a Bruker Vector 22 instrument, taking 240 scans with a resolution of 2 cm^{-1}. ^{13}C-NMR spectra of polysaccharide solutions in D_2O were recorded with a DXR-500 spectrometer operating at 60 °C and 62.9 MHz. Chemical shifts were determined from CD_3OD assigned at 50.15 ppm and used as an internal standard.

Sedimentation

The molecular weights of carrageenan solutions (0.1% w/v) in 0.1 mol L^{-1} NaCl were determined using an analytical ultracentrifuge 3130 MOM (Hungary) at 48 × 1,000 g. The apparent molecular weights were calculated by the method of Archibald (Elias, 1961).

Rheological measurement

Gel strength measurement was performed using a Sun-Rheometer Compact-100 (Japan) at 20 °C. KCl-insoluble solutions (0.5 to 2.5% w/v) were obtained by heating the polysaccharides in water at 70 °C for 30 min. Solutions were placed in cylindrical glass bottles (diameter 25 mm, sample height 19 mm) and KCl was added to prepare aliquots with a concentration of 1% (w/v) and kept at 5 °C for 24 h. The gels were warmed to room temperature for approximately 2 h, before the gel strength measurements were made using plunger diameter-12 mm, gel thickness – 0.8 mm, moving distance-4 mm and table speed-60 mm min^{-1}. The gel breaking strength of the samples was the force required to break the gel surface.

Macrophage-phosphatase enhancing activity

(a) Animal: male ICR mice, weighing 30–35 g were purchased from Samyuk Co. (Korea), and were housed and maintained at 24 h constant humidity (55%). They had free access to food (SAM 31, Samyuk Co., Korea) and water from the beginning of the experiment.
(b) Preparation of the macrophage monolayer: thioglycollate-elicited peritoneal macrophages were obtained from the mice, and the macrophage monolayer was prepared by the method of Matsumoto et al. (1990). More than 95% of the adherent cells showed typical macrophage morphology with characteristic staining.
(c) Determination of phosphatase activity in macrophages was performed according to the method of Suzuki et al. (1990). Briefly, macrophages were cultured in the absence or presence of test samples for 15 h at 37 °C. Thereafter, the macrophage monolayer was solubilized by addition of 25 of 0.1% Triton X-100, then 150 of 10 nM *p*-nitrophenol phosphate and 50 μL of 0.1 M citrate buffer (pH 5.0) were added to each well and incubated for 1 h at 37 °C and then 50 μL of 0.2 M borate buffer (pH 9.8) was added. After 10 min extinction due to phosphatase activity of macrophages was measured at 405 nm.

Anticoagulant activity

The anticoagulant effects of carrageenan samples were assessed using APTT (activated partial thromboplastin time) assay with citrated plasma sample (1:10 v/v, 3.8% sodium citrate) obtained from Fox et al. (1993). Coagulation time assays were performed semi-automatically with a blood coagulation analyzer (BC2210, Kyoto-Daiichi Science, Japan). APTT assays were performed with activated Cephaloplastin Reagent (Dade®, Actin®, Dade Co. Ltd., USA).

Anti-complementary activity

Anti-complementary activity was determined by the method of Kabat and Mayer (1964) with slight modification. Various dilutions of carrageenans in water (50 μL) were mixed with 50 μL of normal human serum (NHS) and 50 μL of gelatin veronal buffered saline (pH 7.4, GVB^{++}). The mixtures were preincubated at 37 °C for 30 min and 350 μL of GVB^{++} was added. 250 μL of IgM-sensitized sheep erythrocytes (Nippon Biotest Laboratory Inc. Japan) at 1×10^8 cells/mL^{-1} were added to the mixtures and diluted serially (10–160 folds) and then incubated at 37 °C for 60 min. After the addition of phosphate-buffered saline (PBS, pH 7.2) and centrifugation, the absorbance of the supernatants was measured at 412 nm. NMS was incubated with water and GVB^{++} as a control. The anti-complementary activity was expressed as the percent inhibition of the total complement hemolysis (TCH_{50}) of the control ($ITCH_{50}$).

Results

The polysaccharides were extracted from the reproductive (cystocarpic) and sterile forms of seaweed and separated by 4% KCl into KCl-insoluble (a) and KCl-soluble (b) fractions. The crude polysaccharides samples contained proteins and they could not be fractionated by KCl. The yields and the chemical composition of these fractions are listed in Table 1. The yield of polysaccharide from SP of *C. pinnulatus* was higher, than from CP.

The FT IR-spectra of all examined samples showed a strong and broad absorption at 1250 cm^{-1}, characteristic for total sulphate esters (Figure 1). Infrared spectra of the KCl-insoluble fractions from cystocarpic and sterile plants of samples showed peaks at 930 cm^{-1} that are characteristic of 3,6-anhydro-D-galactose residues

Table 1. Characterization of carrageenan fractions of *Chondrus pinnulatus*

			Content (%) dry weight							
Sample	KCl Solubility	Yield% algal dry	Sugars	3,6-anhydro Gal	Sulphate	Protein	Ash	Gel strength (Pa)	Apparent mol.weight kDa
Sterile plants	a	40.0	32.9	22.0	23.9	2.8	23.0	70.42	290
	b	9.4	36.5	–	24.9	3.1	26.1	–	220
Cystocarpic plants	a	20.5	40.8	38.0	22.9	4.7	21.0	133.2	420
	b	18.2	35.5	–	27.0	5.7	25.0	–	389

a – KCl insoluble; b – KCl soluble.

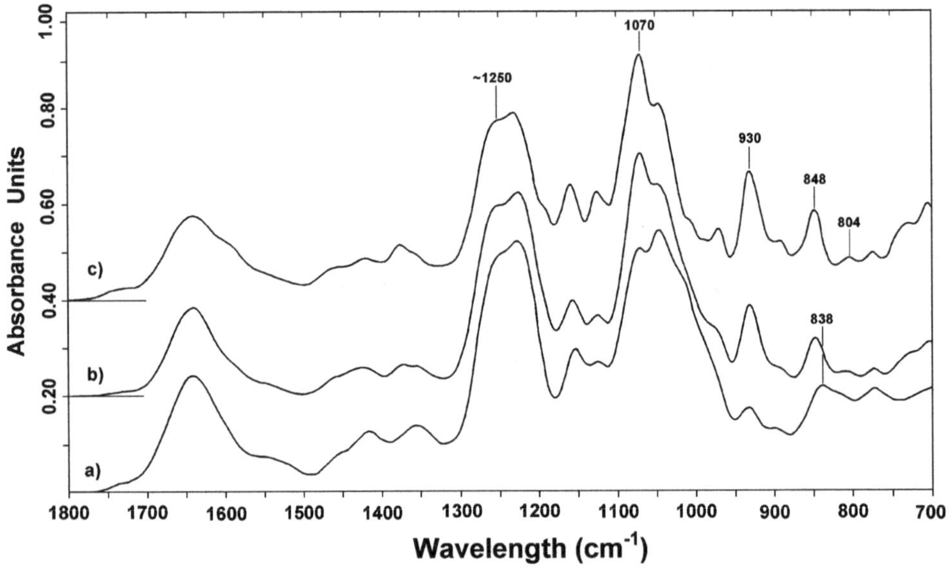

Figure 1. Infrared spectra of polysaccharides from cystocarpic and sterile forms of *Chondrus pinnulatus*. (a) KCl-soluble fraction polysaccharides from sterile form of *Chondrus pinnulatus.* (b) KCl-insoluble fraction polysaccharides from sterile form of *Chondrus pinnulatus*. (c) KCl-insoluble fraction polysaccharides from *Chondrus pinnulatus* with cystosarpic.

(Figure 1) as found in k- and i-carrageenan (Stancioff & Stanley, 1969). Strong absorption at 848 cm^{-1} (the secondary axial sulphate on C-4 of galactose) that is characteristic of k- and i-type carrageenans and absorption at 804 cm^{-1} (2-sulphate on 3,6-anhydro-D-galactose), that indicated the presence of ι-carrageenan, was observed in both KCl-insoluble samples.

^{13}C-NMR spectra of KCl-insoluble polysaccharide samples from sterile and cystocarpic plants are identical and contain more than 12 signals. In both case 4 anomeric signals at 103.2, 95.9, 92.7 and 95.3 ppm (Figures 2 and 3) were observed. Double signals at 103.2 ppm are characteristic of the C1 of galactose residues of κ- and ι-carrageenans, signals at 95.9 and 92.7 ppm are assigned to the C1 of 3, 6-anhydrogalactose residues of κ- and ι-carrageenans (Usov et al., 1983; Falshaw et al., 1996). However some additional signals were detected in these spectra at 105.4 ppm and 71–73 ppm region. We assign this to precursor elements of gelling carrageenan μ, ν-type (Ciancio et al., 1993; Van de Vede et al., 2002). The resonances in downfield were typical for κ- and ι-carrageenans. Thus, according to ^{13}C-NMR and FT IR spectroscopy data of the gelling (KCl-insoluble) fractions from both plants suggested k/ι hybrid carrageenan. The difference between polysaccharides from CP and SP was related of k- and ι-segments. For the sterile form this ratio was 60:40, while for carrageenan from cystocarpic plants it was 80:20.

The FT IR spectra of KCl-soluble polysaccharide fractions from the cystocarpic and sterile plants showed a broad, asymmetric band in the 800–838 cm^{-1} region due to sulphate ester groups on galactose residues (Chopin & Whalen, 1993; Greer & Yaphe, 1984). The

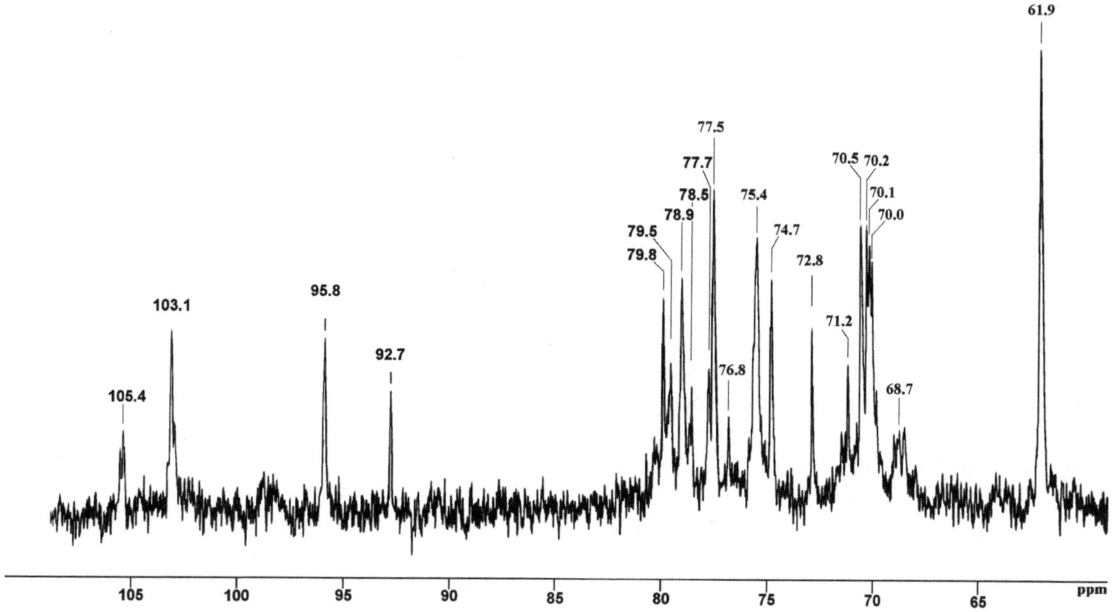

Figure 2. ^{13}C-NMR spectrum of KCl-insoluble fraction polysaccharide from sterile form of *C. pinnulatus.*

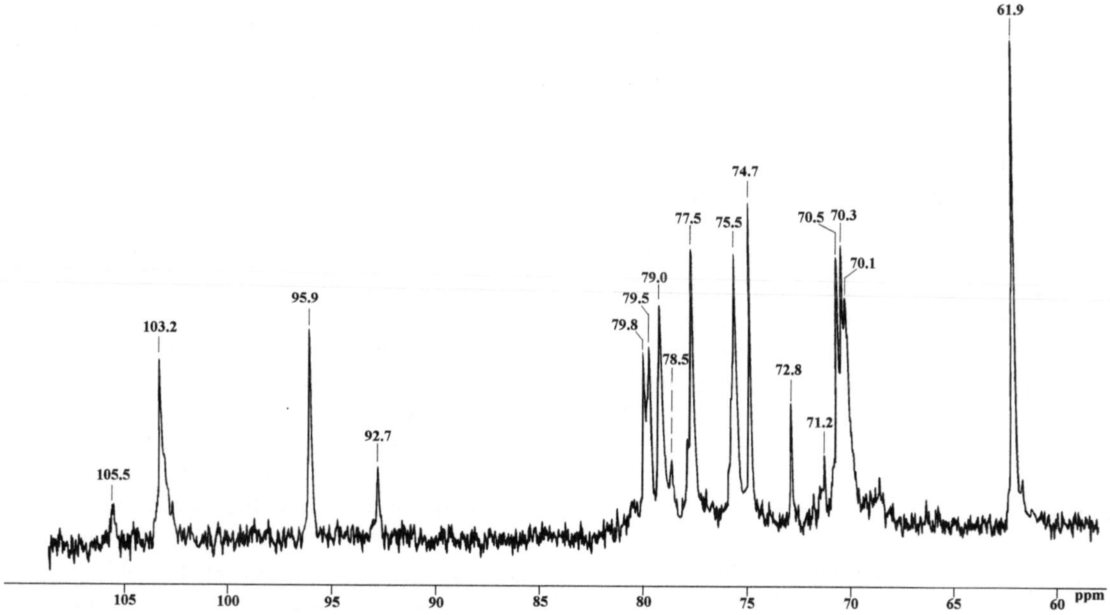

Figure 3. ^{13}C-NMR spectrum of KCl-insoluble fraction polysaccharide from cystocarpic *C. pinnulatus.*

spectrum of CP had a band at 838 cm^{-1} (equatorial secondary sulphate on C-6 of galactose) and a broad shoulder, which may be characteristic for λ- or ν-type carrageenans (Craigie & Leigh, 1978). In the case of carrageenan from sterile plants the absorption bands were weaker and broader. The spectra of SP and CP also showed weak absorption at 930 cm^{-1}. This is characteristic of 3,6-anhydrogalactosyl units that are not normally associated with λ-carrageenan. However, such absorbance has been observed in tetrasporic stages of seaweeds of the Gigartinaceae from New Zealand (Falshaw & Furneaux, 1995).

The high viscosity of KCl-soluble fractions, even when hot, makes it difficult to obtain well resolved

^{13}C-NMR spectra. We used ultrasonication to reduce the viscosity of the samples from sterile and cystocarpic plants prior to NMR spectroscopy. The chemical shifts obtained for both samples are 103.5 and 92.6 and 64.8 ppm and may be characteristic of λ-carrageenan (Stortz et al., 1994; Falshaw & Furneaux, 1995). According to these results the KCl-soluble polysaccharide fractions were mainly non gelling λ - carrageenans mixed with other types of carrageenan that contain low amounts of 3,6-anhydrogalactose.

Rheological properties of κ/ι-carrageenan gels from both plants and the effect of the carrageenans concentration on the rheological properties of their gel in the presence of 1% (w/v) KCl were investigated. The gel strength obtained with the carrageenan from CP was 133.2 Pa (at a 2.5% of concentration of polymer), whereas the carrageenan from SP formed a weaker gel (70.4 Pa) at the same concentration.

Using sedimentation analysis apparent molecular weights of the carrageenans were determined. The values of molecular weights appeared to be 420 to 220 kDa. The highest values of apparent molecular weight were obtained for fractions of carrageenans from CP.

Three different assays that included the anti-complementary, the macrophagephosphatase and anti-coagulant activity have been used for the determination of biological activity of carrageenan. As shown in Table 2, all samples have diverse biological activities. Non-gelling type carrageenans from sterile and cystocarpic plants showed high macrophage-phoshatase and anti-coagulant activity, respectively. The anti-coagulant activity of the KCl-insoluble fraction from cystocarpic was extremely strong at a low concentration of 100 μg mL^{-1}.

Discussion

Carrageenophytes of the Gigartinaceae biosynthesise different structures according to the phase of the life-cycle. For *G. skottsbergii* and *I. undulosa* (Stortz et al., 1993, 1994), it has been proved that carrageenans of the κ-family are produced by cystocarpic plants, whereas tetrasporic plants yield λ-carrageenans. It was shown that cystocarpic *G. skottsbergii* biosynthesise not only the major κ/ι - and μ/ν -carrageenans previously reported (Matulewicz et al., 1989; Ciancia, 1993), but also small amounts of a polymer that would be a hybrid containing κ- (41%), ι- (18%) and λ- (29%) structures and single stubs of galactose (4.5%) (Ciancia et al., 1997). This clear-cut scheme has been altered recently by reports of the biosynthesis of minor quantities of L-galactose containing galactans by cystocarpic carrageenophytes (Ciancia et al., 1993). In addition, these molecules could be hybrids built up by structures of the κ- and λ-types.

The system of carrageenans extracted from unsorted samples of *C. pinnulatus* collected from the Russian Pacific coast was studied by Yermak et al. (1999). According to our current data, gelling polysaccharides from cystocarpic and sterile plants of *C. pinnulatus* have similar structures and were identified as κ/ι-carrageenans. The difference between these polysaccharides was in the ratio of the κ- and ι-segments, with a predominant content of κ-segments in cystocarpic plants (80%). Moreover, KCl-insoluble fractions possibly contain a heterodisperse μ/ν precursor, with higher amounts of this in the polysaccharide from sterile plants. Although the yield of KCl-insoluble carrageenan from sterile plants of *C. pinnulatus* was twice as high as from cystocarpic plants, it only formed a very weak gel. Hybrid κ/ι-carrageenan from the cystocarpic plants showed better gelling properties and the different rheological behavior may be connected with the chemical structure of carrageenans. The lower gel strength of the κ/ι carrageenan from sterile plants is most likely due to its low average molecular weight. A relationship between the mechanical properties and molecular weight of κ-carrageenan has been demonstrated (Rochas et al., 1990). The KCl-soluble fractions (non gelling) were λ-carrageenans with other carrageenan types that had low amounts of 3,6-anhydrogalactose. The difference in the chemical structure and physical-chemical properties among the carrageenans may explain their different biological activities. The λ-type of carrageenan from both plants possesses high macrophage-phosphatase activity, while the λ-type and κ/ι-carrageenans from cystocarpic plants show high anti-coagulant activity. It was shown that λ-carrageenan had greater antitrombic activity than κ-carrageenan, probably, to its higher sulphate content (Anderson & Duncan, 1965). In our case, all fractions of carrageenans contain substantial amounts of proteins, which could also account for the different biological activities. Preliminarily, we have not found a correlation between anticoagulant activity of carrageenans and contents of sulphate and proteins (Tables 1 and 2). Anticoagulant activities of carrageenans decreased with the molecular weight of polysaccharides. The extremely strong anticoagulant activity of κ/ι carrageenan from cystocarpic plants may be due to the high molecular weight of

Table 2. Biological activity of carrageenan fractions of *Chondrus pinnulatus*

Sample	KCl-solubility	Anti-complementary ($ITCH_{50}$)	Macrophage-phosphatase(%)	Anti-coagulant (APTT, sec)
Sterile plants (SP)	a	17	159	187
	b	0	212	59.7
Cystocarpic plants (CP)	a	16	151	≥600
	b	12	184	584
Control	–	0	100	58.7
Sample concentration		at 1000 μg/mL	at 1000 μg/mL	at 100 μg/mL

a – KCl insoluble; b – KCl soluble.

this polysaccharide (Yermak & Khotimchenko, 2003). Further studies on these aspects are currently being conducted.

Acknowledgements

This work was supported financially by the Program "Physical and chemical biology" and "Foundation sciences to medicine" for Basis Research of the Russian Academy of Sciences and by grant from Presidium of Far East Branch of RAS – Russian Science Foundation.

References

Anderson W, Duncan JGC (1965) The anticoagulant activity of carrageenan. J. Pharm. Pharmacol. 17: 647–654.

Ciancia M, Matulewicz M, Cerezo A (1997). A L-Galactose-containing carrageenan from cystocarpic *Gigartina skottsbergii.* Phytochem. 45: 1009–1013.

Ciancia M, Matulewicz MC, Finch P, Cereso AS (1993) Determination of the structure of cystocarpic carrageenans from *Gigartina scottsbergii* by methylation analysis and NMR spectroscopy. Carbohydr. Res. 238: 241–248.

Chopin T, Whalen E (1993) A new and rapid method for carrageenan identification by FT IR diffuse reflectance spectroscopy directly on dried, ground algal material. Carbohydr. Res. 246: 51–59.

Craigie JS,. Leigh C (1978) Carrageenans and agars. In Hellebust JA, Craigie JS (eds), Handbook of phycological methods, physiological and biochemical methods. Cambridge University Press, Cambridge: 109–131.

Craigie JS (1990) Cell walls. In Cole KM, Sheath RG (eds), Biology of the Red Algae. Cambridge University Press, Cambridge: 221–257.

Dubois M, Gilles KG, Hamilton JK, Rebers PG, Smith F (1956) Colorimetric method for determination of sugars and related substances. Anal. Chem. 2: 350–356.

Elias HG (1961) Ultracentrifugen-methoden. Beckman Instruments, Munchen.

Englyst HN, Cummings JH (1984) Simplified method for the measurement of total non-starch polysaccharides by liquid chromatography of constituent sugars as alditol acetates. Analyst 109: 937–942.

Falshaw R, Furneaux RH (1994) Carrageenan from tetrasporic stage of *Gigartina decipiens* (Gigartinaceae, Rhodophyta). Carbohydr. Res. 252: 171–182.

Falshaw R, Furneaux RH (1995) Carrageenan from tetrasporic stages of *Gigartina clavifera and Gigartina alveata* (Gigartinaceae, Rhodophyta). Carbohydr. Res. 276: 115–165.

Falshaw R, Furneaux RH, Wong H, Liao ML, Bacic A, Chandrkrachang S (1996) Structural analysis of carrageenans from Burmese and Thai samples of *Catenella nipae* Zanardini. Carbohydr. Res. 285: 81–98.

Greer CW, Yaphe W (1984) Characterization of hybrid (beta-kappa-gamma) carrageenan from *Eucheuma gelatinae* J. Agardh (Rhodophyta, Solieriaceae) using carrageenases, infrared and ^{13}C-nuclear magnetic resonance spectroscopy. Bot. Mar. 27: 473–478.

Fox I, Dawson A, Loynds P, Eisner J, Findlen K, Levin E, Hanson D, Mant T, Wagner J, Maraganore J (1993) Anticoagulant activity of hirulog, a direct thrombin inhibitor, in humans. Blood Coagul. Fibrinolysis 69: 157–163.

Güven KC, Güvener B, Guler E (1990) Pharmacological activities of marine algae. In Akatsuka J (ed.), Introduction to Applied Phycology. Academic Publishing, Hague, Netherlands, pp. 67–92.

Kabat EE, Mayer MM (1964) Complement and complement fixation. In Charles CT (ed.), Experimental Immunochemistry, Springfield IL.

Knutsen SH, Myslabodski DE, Larsen B, Usov AI (1994) A modified system of nomenclature for red algal galactans. Bot. Mar. 37: 163–169.

Lahaye M, Kaeffer B (1997) Seaweed dietary fibres: Structure, physico-chemical and biological properties relevant to intestinal physiology. Sciences des Aliments 17: 563–584.

Lahaye M, Axelos MAV (1993) Gelling properties of water-soluble polysaccharides from proliferating marine green seaweeds (*Ulva spp.*). Carbohydr. Polym. 22: 261–265.

Lowry OH, Rosebrough NJ, Farr AL, Randall RJ (1951) Protein measurement with the Folin phenol reagent. J. Biol. Chem. 193: 265–275.

Matsumoto T, Tanaka M, Yamada H, Cyong JC (1990) A new photometric microassay for the quantitation of macrophage Fc receptor function. *In vitro* enzyme-containing immune complexes clearance (EIC) assay. J. Immunol. Meth. 129: 283–290.

Matulewicz MC, Ciancia M, Nozeda MD, Cerezo AS (1989) Carrageenan systems from tetrasporic and cystocarpic stages of *Gigartina skottsbergii*. Phytochem. 28: 2932–2941.

McCandless EL, Craigie JS, Walter JA (1973) Carrageenans in the gametophytic and sporophytic stages of *Chondrus crispus*. Planta 112: 201–212.

Rochas C, Rinaudo M, Landry S (1990) Role of the molecular weight in the mechanical properties of kappa carrageenan gels. Carbohydr. Polym. 12: 255–266.

Stancioff DJ, Stanley NF (1969) Infrared and chemical studies on algal polysaccharides. Proc. Int. Seaweed Symp. 6: 595–609.

Stortz CA, Cerezo AS (1993) The systems of carrageenans from cystocarpic and tetrasporic stages from *Iridaea undulosa*: Fractionation with potassium chloride and methylation analysis of the fractions. Carbohydr. Res. 242: 217–227.

Stortz CA, Bacon BE, Cherniak R, Cerezo AS (1994) High-field NMR spectroscopy of cystocarpic and tetrasporic carrageenans from *Iridaea undolosa*. Carbohydr. Res. 261: 317–326.

Suzuki I, Tanaka H, Kinoshita A, Oikawa S, Osawa M, Yadomae T (1990) Effect of orally administered beta-glucan on macrophage function in mice. Int. J. Immunopharmacol. 12: 675–684.

Usov AI, Ivanova EG, Shashkov AS (1983) Polysaccharides of algae. 33. Isolation and ^{13}C–NMR spectral study of some new gel forming polysaccharides from Japan Sea seaweeds. Bot. Mar. 26: 285–294.

Usov AI, Elashvili MYs (1991) Quantitative determination of 3, 6-anhydrogalactose derivatives and specific fragmentation of the red algal galactans under reductive hydrolysis conditions. Bioorganicheskaya Khimiya 6: 839–848.

Van de Velde F, Knutsen SH, Usov AI, Rollema HS, Cerezo AS (2002) Trends in Food Sci. Technol. 13: 73–92.

Yermak IM, Kim YH, Titlynov EA, Isakov VV, Solov'eva TF (1999) Chemical structure and gel properties of carrageenan from algae belonging to the Gigartinaceae and Tichocarpaceae, collected from the Russian Pacific coast. J. Appl. Phycol. 11: 41–48.

Yermak IM, Khotimchenko YS (2003) In Fingerman M, Nagabhushanam R (eds.), Recent Advances in Marine Biotechnology. Science Publishers, Enfield NH 9: 207–255.

Journal of Applied Phycology (2006) 18: 369–373
DOI: 10.1007/s10811-006-9042-x

A comparative study of specificity of fucoidanases from marine microorganisms and invertebrates

M.I. Kusaykin[1,*], A.O. Chizhov[2], A.A. Grachev[2], S.A. Alekseeva[1], I. Yu Bakunina[1], O.I. Nedashkovskaya[1], V.V. Sova[1] & T.N. Zvyagintseva[1]
[1]*Pacific Institute of Bioorganic Chemistry, Far-East Branch of RAS, 690022, Vladivostok, 159 Prospect 100-letya Vladivostoku;* [2]*N.D. Zelinsky Institute of Organic Chemistry, RAS, 119991, Moscow, 47 Leninsky Prospect*

*Author for correspondence: e-mail: mik@piboc.dvo.ru; fax: (4232) 314-050

Key words: fucoidan, fucoidanase, marine invertebrates, marine bacteria, seaweed

Abstract

Specificities of actions of fucoidanases from the marine microorganism *Pseudoalteromonas citrea* KMM 3296 and the marine mollusk *Littorina kurila* were studied. The enzymes possess similar specificities and catalyze the cleavage of accessible α-(1→3)-fucoside bonds in fucoidans with highly sulfated α-(1→4; 1→3)-L-fucooligosaccharides. A high degree of sulfation of the fucose residues in fucoidans makes α-(1→3)-L-fucoside bonds inaccessible for the action of the studied enzymes. The maximum degree of cleavage of fucoidan was achieved by the fucoidanase from the marine bacterium *Pseudoalteromonas citrea* KMM 3296.

Introduction

Fucoidans, highly sulfated polysaccharides of brown algae, posses diverse biological activities. The most interesting are antitumor, anticoagulant, and antiviral activities, e.g., against HIV, hepatitis virus, and herpes virus (McClure et al., 1992; Nishino et al., 1991). For the last decade, the structure of these polysaccharides has been extensively studied. A close correlation between structural characteristics of fucoidans and the taxonomy of the corresponding brown algae was hypothesized: it is known that α-(1→3)-L-fucans are found in *Laminaria*, whereas species of *Fucus* genus mainly contain α-(1→3, 1→4)-L-fucans (Bilan et al., 2002). Structure/activity correlations for these polysaccharides are poorly studied. Usually fucoidans have a high d. p., so depolymerisation is needed for medicinal applications.

The enzymes degrading polysaccharides are widely used in structural studies, in studies of biological activities, and in preparation of drugs (Zvyagintseva et al., 1995). Fucoidanases are reported found only from marine organisms, and their activities are usually extremely low (Burtseva et al., 2000; Kusaykin et al., 2003; Bakunina et al., 2002). There are only a few studies on isolation and characterization of fucoidanases (Berteau & Mulloy, 2003). Information on the specificity of fucoidanases is scarce (the type of the glycosidic bond cleaved, and the influence of degree of sulfation of a substrate on the catalytic activity of these enzymes). Nevertheless, a fucoidanase from *Flavobacterium* sp. SA-0082 has been reported, and is already used for depolymerisation of fucoidan in the preparation of fucoidan-containing foods and beverages (Umeda et al., 1998). The most valuable sources of these enzymes from a technology standpoint, are still to be found.

The characterisitcs of enzymatic action of fucoidanases from a marine mollusc *Littorina kurila* and a marine bacterium *Pseudoalteromonas citrea* are presented in this paper.

Materials and methods

Analytical procedures

Neutral carbohydrates were quantified by the phenol-sulfuric acid method (Dubois et al., 1956); reducing carbohydrates were determined according to Nelson (1944). Oligosaccharide composition was analyzed with a Jeol-JLC-6AH liquid chromatograph (Jeol, Japan) and a Bio Gel P-2 column (1 × 100 cm) eluted with 0.02 M acetate buffer, pH 5.4 at 16 mL/h^{-1}, orcinol- sulfuric acid assay. Monosaccharide composition was determined by HPLC with a LC-5001 carbohydrate analyzer (a Durrum DA-X8-11 column (385 × 3.2 mm) (Biotronik), bicinchoninate assay, and a C-R2 AX integrating system (Shimadzu)). The content of protein was determined by the method of Lowry (1951).

Substrates

Fucoidans from the brown algae *Laminaria cichorioides* and *Fucus evanescens* were isolated as described by Zvyagintseva et al. (1995).

Fucoidan from *F. evanescens* was purified as follows. To remove alginic acid, 100 mL of acetic acid was added to 300 mL of a solution of the fucoidan (50 mg mL^{-1}) and the precipitate formed was immediately centrifuged (9000 g, 10 min). The supernatant was neutralized with a solution of NaOH and the salt formed was removed by ultrafiltration at a 1 kDa cutoff (Sigma) using stepwise dilution. The resulting solution of fucoidan was applied to a column with DEAE-cellulose (Sigma) (20 × 30 cm) equilibrated with 0.01 M HCl and then eluted with a stepwise gradient of a NaCl solution (0.35, 0.5, 0.75, 1, 1.5, 2, and 3 M). The concentration of fucoidan was monitored by the phenol-sulfuric acid method (Dubois et al., 1956). The corresponding carbohydrate-containing fractions were pooled and dialyzed, then concentrated by ultrafiltration (1 kDa cutoff) and lyophilized.

Enzyme

Acidic (pH optimum at 5.4) and basic (pH optimum at 8.5) fucoidanases from a hepatopancreas of *L. kurila*, were isolated as described previously (Kusaykin et al., 2003). Fucoidanase from the bacterium *Pseudoalteromonas citrea* KMM 3296 was prepared as described by Bakunina et al. (2002).

Activities of enzymes

The activities of fucoidanases were determined by an increase of the amount of reducing sugars (Nelson et al., 1944). The incubated mixture contained 100 μL of the enzyme, 200 μL of a solution of fucoidan (4 μg mL^{-1}), and 200 μL of the corresponding buffer (0.05 M succinate buffer containing 0.2 M of NaCl, pH 5.4, or 0.05 M borate buffer, pH 8.5, or 0.01 M phosphate buffer, pH 7.2). The time of incubation did not exceed that needed to cleave 10% of the substrate in the incubated mixture. The amount of the enzyme which catalyzed the formation of 1 nmol of fucose for 1 h under conditions of determination was accepted as a unit of activity.

Preparation of products of enzymatic degradation of fucoidan

Fucoidanases from hepatopancreas of *L. kurila* were prepared as follows. Dry fucoidan (200 mg) was added to a solution of fucoidanases (20 mL, 10^{-2} units) in 0.05 M succinate buffer, pH 5.4 with 0.2 M of NaCl or in 0.02 M borate buffer, pH 8.5. After dissolution of the substrate, the mixture was incubated for 72 h at 37 °C. The reaction was stopped by boiling. High molecular weight products of the reaction were precipitated with ethanol (1:4, v/v). The fraction containing low molecular products of the reaction was evaporated to dryness *in vacuo* and then analyzed with an automatic liquid analyzer Jeol-JLC-6 AH. The product obtained using fucoidanase at pH 8.5 was separated by gel filtration on Bio Gel P-2, giving two fractions, P-1-L and P-2-L. Fraction P-1-L was subjected to ultrafiltration, 1 kDa cutoff. Non-dialyzable fraction (P-1-L) was analyzed.

Fucoidanase from bacterium KMM 3296 were prepared as follows. To a solution of fucoidanase (20 mL, 10^{-2} units) in 0.05 M phosphate buffer, pH 7.2, 200 mg of dry fucoidan (*F. evanescens*) was added. After dissolution of the substrate, the mixture was incubated for 7 days at 37 °C under sterile conditions. The reaction was stopped by boiling. The resulting products were separated on DEAE-cellulose (1 × 15 cm), the carbohydrate-containing fractions were desalted on Sephadex G-10 (1 × 50 cm), evaporated to dryness *in vacuo* and analyzed an automatic liquid analyzer Jeol-JLC-6 AH on a Bio Gel P-2 column (1 × 100 cm).

Desulfation of fucoidans and the product of their enzymatic cleavage

Fucoidan (50–100 mg) was transformed to a pyridinium form (Zvyagintseva et al, 2003) and dissolved in 18 mL of DMSO and 2 mL of pyridine by stirring then heating for 10 h at 100 °C. The solution was poured into water and DMSO was removed by ultrafiltration on a Millipore membrane with 1000 Da cutoff. The aqueous solution was concentrated and lyophilyzed.

Methylation of fucoidans and preparation of partially methylated polyol acetates was carried out as reported previously (Chizhov et al., 1999 and references therein). GLC-MS analysis of partially methylated polyol acetates was done with a Finnigan MAT ITD-700 (ion trap detector) mass spectrometer coupled with a Carlo Erba series 4200 gas chromatograph (capillary column column Ultra-1, Hewlett Packard, crosslinked polymethylsiloxane, 25 m length, 0.25 mm internal diameter, 0.33 μ liquid film thickness). Temperature program: isotherm 150 °C (1 min), then ramp 5 °C/min to 280 °C. Helium was used as a carrier gas. The component ratios were approximated by total ion current (TIC).

Results and discussion

It was shown previously that the marine bacterium *Pseudoalteromonas citrea* KMM* 3296 and the marine mollusk *Littorina kurila* have significant activities of fucoidanases (Burtseva et al., 2000; Bakunina et al., 2002). Fucoidanases from these resources have been partially purified and their properties studied (Kusaykin et al., 2003). Here we present the results of a comparative study of the specificity of three fucoidanases; i.e., basic (pH optimum at 8.5) and acidic (pH optimum at 5.4) from hepatopancreas of *L. kurila* and fucoidanases from *P. citrea* KMM 3296 (pH optimum at 7.2). Notably, the starting level of fucoidanase activity in the microbial source was one order of magnitude higher than in hepatopancreas of *L. kurila.*

Fucoidans from the brown algae *Fucus evanescens* and *Laminaria cichorioides* were used as substrates. Fucoidan from *F. evanescens* contained Fucose (95% on neutral carbohydrates content), Xylose (2.8%), Mannose (0.2%), Glucose (2%), and the molar ratio fucose:SO_4^{2-} was equal to 1:0.43. For the desulfated sample of this fucoidan, methylation analysis gave the ratio of acetates as 2,3,4-tri-O-methyl-:2,3-di-O-methyl-:2,4-di-O-methyl:2-O-methyl:3+4-O-methylfucitols as follows: 23:11:39:9:18. So, fucoidan from *F. evanescens* used as a substrate is a partially sulfated α-(1→3; 1→4)-L-fucan (linkage ratio 1→3:1→4 is 3.5:1). This fucoidan fraction substantially differs from that isolated previously (Bilan et al., 2002) from *F. evanescens*, which is a linear polymer with alternating (1→3)- and (1→4)-linked fucose residues sulfated mainly by C-2 and partially acetylated by other hydroxy groups. Fucoidan from *L. cichorioides* is almost totally sulfated α-(1→3)-L-fucan (Zvyagintseva et al., 2003).

The characteristics of the products of exhaustive enzymolysis of substrates (fucoidans from *L. cichorioides* and *F. evanescens*) are given in Table 1. The maximum degree of cleavage was obtained for fucoidan from *F. evanescens* by fucoidanase from *P. citrea* KMM 3296. When using acidic fucoidanase from *L. kurila* to cleave fucoidan from *L. cichorioides,* formation of low-molecular products is three times lower than for fucoidan from *F. evanescens* (Table 1). This fact may be explained by the greater accessibility of O-glycosidic bonds in low sulfated fucoidan from *F. evanescens* in comparison to highly sulfated fucoidan from *L. cichorioides*.

The action of the basic form of fucoidanase from *L. kurila* on fucoidan from *F. evanescens*, yielded three times more low-molecular products than the action

Table 1. The products of enzymatic cleavage of fucoidans by fucoidanases

		Characteristics			
Enzyme (pH-optimum)	Substrate (m wt kDa)	HMP, yields[a], %	n[b]	LMP, yields[c], %	n
Acidic fucoidanase (5.4)	Fucoidan from *F. evanescens*, 60	85	$n > 7$	15	$7 > n > 2$
	Fucoidan from *L. cichorioides*, 20	95	$n > 7$	5	$7 > n > 2$
Basic fucoidanase (8.5)	Fucoidan from *F. evanescens*, 60	55	$n > 7$	45	$7 > n > 2$
Fucoidanase from *P. citrea* KMM 3296 (7.2)	Fucoidan from *F. evanescens*, 60	30	$n > 7$	70	$5 > n > 2$

[a]HMP: highly molecular products obtained by precipitation with 80% aqueous ethanol (in % of total amount of products).
[b]n: degree of polymerization.
[c]LMP: low molecular products.

Table 2. The characteristics of low-molecular products of enzymatic cleavage of fucoidan from *F. evanescens* by action of fucoidanase from hepatopancreas of *L. kurila* and *P. citrea* KMM 3296

A source of enzyme	Products	% from Yield, the starting substrate	M. wt., kDa or n^a	Carbohydrate composition, %						Molar ratio $Fuc:SO_4^{2-}$
				Fuc	Gal	Xyl	Rha	Glc	Man	
Pseudoalteromonas	P-1-Ps	26	$5 \geq n \geq 2$	96	4	0	0	0	0	1:0.31
citrea KMM 3296	P-2-Ps	8	2–3	97.2	0.4	2.1	0.3	0	0	1:0.53
Hepatopancreas	P-1-L	30	3–10	92	1	1.8	0	2.5	3.7	1:0.59
Littorina kurila,	P-2-L	8	$7 \geq n \geq 2$	50	0	0	0	50	0	0
	P-1-1-L	17	3–10	92	1	1.8	0	2.5	3.7	1:0.59

[a]n: degree of polymerization of products.

of the acidic form of the fucoidanase on the same fucoidan.

To study enzymatic transformation in detail, fucoidan from *F. evanescens* as a substrate and two enzymes (basic fucoidanase from *L. kurila* and fucoidanase from *P. citrea* KMM 3296) were chosen. The products obtained from the action of fucoidanases from *P. citrea* KMM 3296 on this fucoidan, as separated by ion exchange chromatography on DEAE-cellulose, yielded the two fractions (P-1-Ps and P-2-Ps) shown in Table 2. Acid hydrolysis of the products showed that they consist mainly of fucose. All fractions obtained by transformation of fucoidan with the microbial enzyme had sulfate groups and the total content of sulfate remained practically constant in comparison to starting fucoidan.

Gel chromatography on Bio Gel P-2 of the fraction P-1-Ps showed that it is a mixture of di-, tri-, tetra-, and pentafucooligosaccharides. This fraction (yield 26%, Table 2) has 30% by wt. of sulfate, which corresponds to sulfation one of two hydroxyls in the fucose residues. Solvolytic desulfation followed by methylation analysis of the fraction P-1-Ps gave the ratio of acetates of 2,3,4-tri-O-methyl-:2,3-di-O-methyl-:2,4-di-O-methyl:2-O-methyl-:3- and 4-O-methylfucitols equal to 11:50:26:13: none. Thus, the ratio of (1→4)- and (1→3)-linked fucosyl residues changed from 3.5 in the starting fucoidan to 0.5 in P-1-Ps, which demonstrates the predominant cleavage of α-(1→3)-glycosidic bonds by the fucoidanase from *P. citrea* KMM 3296.

The products formed by the action of basic fucoidanase from *L. kurila* (Table 2) were separated as follows. High molecular weight products of enzymatic cleavage of fucoidan were precipitated with 80% aqueous ethanol. Low molecular weight products remaining in the solution were separated on Bio Gel P-2, giving two fractions, P-1-L and P-2-L. Fraction P-1-L was subjected to ultrafiltration on a membrane with 1 kDa cutoff. Results of the analysis of the non-dialyzable fraction (P-1-1-L) are given in Table 2. Total acid hydrolysis gave Fucose (92%), Xylose (1,8%), Mannose (3,71%), and Glucose (2,5%); sulfate content was 40% by wt. in P-1-1-L. Solvolytic desulfation followed by methylation analysis gave the ratio of acetates 2,3,4-tri-O-methyl-:2,3-di-O-methyl-:2,4-di-O-methyl:2-O-methyl:3- and 4-O-methylfucitols equal to 10:41:31:11:7. The data show that the ratio of (1→4)- and (1→3)-linked fucosyl residues changed from 3.5 to 0.75 in P-1-1-L, which also demonstrates the predominant cleavage of α-(1→3)-glycosidic bonds by the fucoidanase from *L. kurila*. In the ^{13}C NMR spectrum of P-1-1-L, the most intense signals at 96.5 (C1), 69.1 (C2), 70.2 (C3), 81.1 (C4), and 68.8 ppm (C5) are preliminarily assigned to the fragment →4)-α-L-FucP-(1→ and weaker signals at 97.3 (C1), 67.5 (C2), 76.6 (C3), 69,8 (C4), and 67.7 ppm (C5) may be assigned to the →3)-α-L-FucP-(1→ link; in addition, the following signals were putatively assigned to the →3, 4)-α-L-FucP-(1→ fragment ((C-3, C-4)-branching points): 101.5 (C1), 68.4 (C2), 77.2 (C3), 70.2 (C4), and 67.7 ppm (C5).

Conclusions

The fucoidanases from the marine mollusk *L. kurila* and the marine bacterium *P. citrea* KMM 3296 have a similar specificity: they catalyze the predominant cleavage of α-(1→3)-glycosidic bonds between fucose residues in the polysaccharide. In contrast to fucoidanase from *L. kurila*, the bacterial fucoidanase cleaves fucoidan forming mainly di-, tri-, tetra-, and pentafucooligosaccharides, whereas the action of the basic form of fucoidanase from *L. kurila* yields higher molecular weight products of 3–10 kDa (Table 2).

Probably, these differences are related to structural peculiarities of active centers of enzymes and the mechanism of action of the enzymes on the polymer substrate.

Acknowledgments

The work was supported by RFBR (Projects No. 03-04-49534, 05-04-48211), FEB RAS grants and the FCB RAS program.

References

Bakunina IYu, Nedashkovskaya OI, Alekseeva SA, Ivanova EP, Romanenko LA, Gorshkova NM, Iskov VV, Mikhailov VV (2002) Degradation of fucoidan by the marine proteobacterium *Pseudoalteromonas citrea*. Microbiology (Moskow) 71: 49–55.

Berteau O, Mulloy B (2003) Sulfated fucans, fresh perspectives: Structures, functions, and biological properties of sulfated fucans and an overview of enzymes active toward this class of polysaccharide. Glycobiology 13: 29R–40R.

Bilan MI, Grachev AA, Ustuzhanina NE, Shashkov AS, Nifantiev EN, Usov AI (2002) Structure of a fucoidan from the brown seaweed *Fucus evanescens*. Carbohydr. Res. 337: 719–730.

Burtseva YuV, Kusaykin MI, Sova VV, Shevchenko NM, Skobun AS, Zvyagintseva TN (2000) Distribution of fucoidan-hydrolase and some glucosidase among marine invertebrates. Biologiya Morya 26: 429–432.

Chizhov AO, Dell A, Morris HR, Haslam SM, McDowell RA, Shashkov AS, Nifant'ev NE, Khatuntseva EA, Usov AI (1999) A study of fucoidan from the brown seaweed *Chorda filum*. Carbohydrate Research 320: 108–119.

Dubois M, Gilles KA, Hamilton J, Robers PA, Smith F (1956). Colorimetric method for determination of sugars and related substances. Anal. Chem. 28: 350–356.

Kusaykin MI, Burtseva YuV, Svetasheva TG, Sova VV, Zvyagintseva TN (2003) Distribution of O-glycosylhydrolases in marine invertebrates. Enzymes of the marine mollusk *Littorina kurila* that catalyze fucoidan transformation. Biochemistry (Moskow) 68: 384–392.

Lowry OH, Rosebrough NJ, Farr AL, Randall RJ (1951) Protein measurement with Folin phenol reagent. J. Biol. Chem. 193: 265–275.

McClure MO, Moore JP, Blanc DF (1992) Investigations into the mechanism by which sulfated polysaccharides inhibit HIV-infection *in vitro*. AIDS Research Human Retroviruses 8: 18–26.

Nelson TE (1944) A photometric adaptation of the Somogy method for the determination of glucose. J. Biol. Chem. 153: 375–381.

Nishino T, Nagumo T, Kiyohara H, Yamada H (1991) Structural characterization of a new anticoagulant fucan sulfate from the brown seaweed *Ecklonia kurome*. Carbohydr. Res. 211: 77–90.

Umeda Y, Kihara H, Ikai K, Kato I (1998) Fucoidan-containing foods or beverages. Chemical Abstracts 128: 47606d.

Zvyagintseva TN, Elyakova LA, Isakov VV (1995) Enzyme transformation of laminarans into $1\rightarrow3;1\rightarrow6$-$\beta$-D-glucans, having immunostimilating action. Bioorganicheskaya Khimya. 21: 218–225.

Zvyagintseva TN, Shevchenko NM, Chizhov AO, Krupnova TN, Sundukova EV, Isakov VV (2003) Water-soluble polysaccharides of some far-eastern brown seaweeds. Distribution, structure, and their dependence on the developmental conditions. J. Exp. Mar. Biol. Ecol. 294: 1–13.

Journal of Applied Phycology (2006) 18: 375–380
DOI: 10.1007/s10811-006-9043-9

Comparative characterization of laminarinases from the filamentous marine fungi *Chaetomium indicum* Corda and *Trichoderma aureviride* Rifai

Yulia Burtseva*, Natalia Verigina, Victoria Sova, Mikhail Pivkin & Tatiana Zvyagintseva
Pacific Institute of Bioorganic Chemistry, 690022, Vladivostok, Russia

*Author for correspondence: e-mail: Zvyag@piboc.dvo.ru

Key words: marine mycelial fungi, *Chaetomium indicum, Trichoderma aureviride,* laminarinase, O-glycosylhydrolases

Abstract

Marine filamentous fungi (103 strains) isolated from various marine habitats were studied for their ability to produce extracellular O-glycosylhydrolases. Cultural filtrates of these strains were shown to contain a series of glycanases (laminarinases, amylases, cellulases, pustulanases) and glycosidases (β-glucosidases, N-acetyl-β-glucosaminidases, β-galactosidases, α-mannosidases). Two species of marine fungi from different habitats were chosen for isolation of laminarinases and detailed study on enzyme properties. The fungus *Chaetomium indicum* associated with the alga *Fucus evanescens* C. Agardh was collected near the Kuril Islands, and *Trichoderma aureviride* was sampled from bottom deposits of South China Sea. Properties of extracellular laminarinases were similar: temperature optimums (40–45 °C), molecular masses (54–56 kDa), K_m (0.1–0.3 mg mL^{-1}). Temperature stability of laminarinase of *C. indicum* was significantly higher than those from *Trichoderma aureveride*. It is shown that these enzymes are specific to β-1,3-bonds in glucans, release predominantly glucose from laminaran and do not catalyze reaction of transglycosylation. Accoding to these data enzymes are exo-1,3-β-D-glucan-glucanohydrolases (EC 3.2.1.58). Inhibitor analysis demonstrated the significant role of tryptophan and tyrosine residues in the catalytic activity of enzymes. Molecules of *T. aureviride* laminarinase contained the functionally important thiol group.

Introduction

Laminaran, 1,3-β-linked glucan, is an important storage polysaccharide of many brown seaweeds (Phaeophyta) and is produced by these algae in large quantities. Seaweeds, including the Phaeophyta, are well known to be colonized by marine fungi. These microorganisms are capable of degrading laminaran and other carbohydrates, and may therefore have a role in the breakdown of seaweeds.

Systematic analysis of the composition and level of O-glycosylhydrolase activity of marine filamentous fungi have not previously been conducted. However, effective producers of carbohydrate metabolism enzymes are found among these microorganisms (Grant & Rhodes, 1992; Pointing et al., 1999; Lee, 2000; Burtseva et al., 2003).

The aim of our work was to study the distribution of some O-glycosylhydrolases (glycosidases and glycanases) in filamentous fungi inhabiting the marine environment, and to isolate and to characterize the properties, specificity and type of action of laminarinases from the marine facultative fungi *Chaetomium indicum* and *T. aureviride.*

Materials and methods

Fungal strains and the growth of fungi

The studied fungal strains were obtained from the Collection of Marine Microorganisms (KMM) of the Pacific Institute of Bioorganic Chemistry (PIBOC), Far

East Branch of the Russian Academy of Sciences. All these fungi were collected during marine expeditions aboard the research vessel "Akademik Oparin" near the Kuril Islands and Pociet Bay of the Sea of Japan, as well as the South China Sea.

Fungi were grown on modified Tubaki's medium, which contained (g L^{-1} sea water): non-purified sheet agar (3.0), peptone (1.0), KH_2PO_4 (1.0), yeast extract (0.5), $MgSO_4{\cdot}7H_2O$ (0.5), $FeSO_4$ (0.02), pH 7.0. The fungus *Chaetomium indicum,* inhabiting the brown alga *Fucus evanescens*, and the fungus *T. aureviride* from the bottom sediments, were used as producers of laminarinases.

Determination of activity

A standard reaction mixture contained the enzyme solution (20 μL) in succinate buffer (pH 5.2) and the substrate solution (500 μL, 1 mg mL^{-1}). Glycanases were assayed by accumulation of the reducing sugars after incubation with the corresponding polysaccharides at 37 °C for 20 min. Glycosidases were determined under the same conditions with *p*-nitrophenyl derivatives of the corresponding sugars as the substrates. The amount of the enzyme catalyzing the formation of 1 μmole of a reaction product (glucose or *p*-nitrophenol) per one minute under these conditions was taken as one unit of activity. Specific activity was defined as one unit of enzyme per one mg of protein. Activity of the enzymes in a cultural liquid was determined in a stationary phase of the growth of these fungi.

Principle analytical method

Reducing sugars were assayed by the method of Nelson (1944). Protein concentration in solution was determined by the method of Lowry et al. (1951). Liquid chromatography of sugars was performed using a JEOL-JLC-6AH automatic liquid analyser (Japan) on a Biogel P-2 column (0.9 × 90 cm) in 0.05 M sodium acetate buffer (pH 5.2) containing 0.2 M NaCl at a flow rate of 7–9 mL h^{-1}. Carbohydrates were determined with the orcinol-sulfuric acid reagent. Products of transglycosylation were determined by HPLC method using a Du Pont 8800 chromatograph with an Ultrasil-NH_2 column (10 × 25 mm). The column was eluted with acetonitrile:H_2O, 80:20 (v/v). Oligosaccharides were detected at 300 nm.

Isolation and purification of laminarinases

The following steps were used for purification of laminarinases: ultrafiltration on a membrane PM-10, PM-30, gel filtration on Biogel P-200, Sepharose CL-6B, Superdex 75 HR 10/30 columns, cation exchange chromatography on CM-cellulose, 15 *Q* PE, 15 *S* PE columns, hydrophobic chromatography on Phenyl-Sepharose 6.

Estimation of molecular mass

Molecular mass of laminarinases was estimated by gel filtration on Biogel P-200 (*C. indium*) and Sepharose CL-6B columns (*T. aureviride*).

Michaelis constants

Michaelis constants were calculated according to the Lineweaver-Burk method (Dixon & Webb, 1958).

Results and discussion

One hundred and three strains of marine fungi isolated from various marine habitats were studied for their ability to produce extracellular enzymes. Distribution of O-glycosylhydrolases in fungi of various genera is presented in Figure 1. It has been established that glycosidases are widely distributed in cultural filtrates of these fungal strains: β-glucosidases (in 47 samples), N-acetyl-β-glucosaminidases (in 36 samples), β-galactosidase (in 9 samples), α-mannosidases (in 5 samples). Among enzymes degrading polysaccharides, amylases (in 38 samples) and laminarinases (in 33 samples) are most widespread, whereas the enzymes splitting pustulan (in 6 samples) and CM-cellulose (in 4 samples) are rare. The enzymes hydrolysing agar and fucoidan were not found under the conditions described.

Two species of marine fungi from different habitats were chosen for isolation of laminarinases and detailed study of enzyme properties. *C. indicum* associated with the alga *Fucus evanescens* was collected near the Kuril Islands and *T. aureviride* was sampled from bottom deposits of the South China Sea. Composition of O-glycosylhydrolases produced by these two species was nearly identical: laminarinase, amylase, N-acetyl-β-D-glucosaminidase, β-D-gluco- and galactosidase. Cellulase in addition to these enzymes was found in cultural liquid of *C. indicum*, but pustulanase was found in cultural liquid of *T. aureviride*.

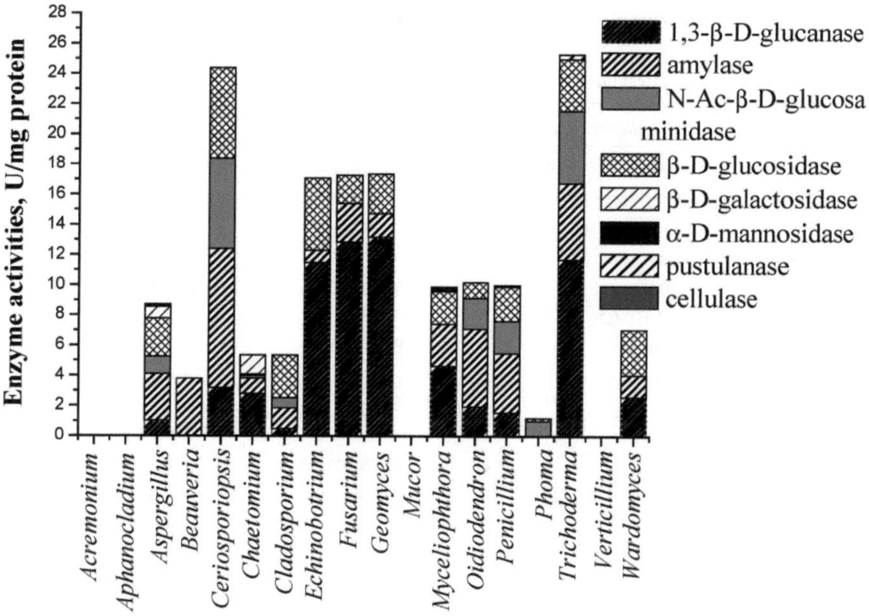

Figure 1. Distribution of O-glycosylhydrolases in marine fungi by genera: 1) β-1,3-glucanase; 2) amylase; 3) N-acetyl-β-glucosaminidase; 4) β-glucosidase; 5) β-galactosidase; 6) mannosidase; 7) pustulanase; 8) cellulase. Average values were taken.

Combining methods of ultrafiltration, hydrophobic interaction chromatography, gel filtration and ion exchanging chromatography, preparations of laminarinases with 1–2% yields were isolated without impurities of other enzymes. Some principal properties of the enzymes were studied.

Extracellular laminarinases of *C. indicum* and *T. aureviride* showed best activity in a low acidic range, and possessed high pH stability in range pH from 4.5 to 7.5. Laminarinase of *T. aureviride* was more stable at pH 3.5, than laminarinase of *C. indicum.* Temperature optima of laminarinases of *C. indicum* and *T. aureviride* were 45 °C and 40 °C, respectively. Temperature stability of laminarinase of *C. indicum* was significantly higher than laminarinase of *T. aureveride*. Molecular masses of enzymes determined by gel chromatography were close (56 kDa for laminarinase of *T. aureviride* and 54 kDa for laminarinase of *C. indicum*), as well as values K_m (0.3 mg mL^{-1} for *T. aureviride* laminarinase and 0.1 mg mL^{-1} for *C. indicum* laminarinase).

A method of inhibitory analysis was applied to elucidate the role of some functional groups in the catalytic activity of laminarinases. *p*-Chloromercuribenzoate, a reagent for the thiol group, had no effect on the activity of *C. indicum* laminarinase, but inhibited the laminarinase of *T. aureviride* by 66%. Chesters and Bull (1963) suggested that thiol groups were involved in the formation of the enzyme–substrate complex in fungal laminarinases: the enzyme activity was inhibited by phenylmercury nitrate, and the exo activity was inhibited to a greater extent than the endo activity. Three forms of exo-β-1,3-glucanase from *A. persicinum* were inhibited by *p*-hydroxymercury benzoate by 25–30% (Pitson et al., 1995). Our data suggest that the molecule of *T. aureviride* laminarinase contains a functionally significant free thiol group. *N*-Bromosuccinimide, capable of specific oxidation of tryptophan residues, and completely inactivates *T. aureviride* and *C. indicum* laminarinases. This fact indicates that tryptophan residues were essential for the catalytic activity of the enzymes. Loss of activity after chemical modification of these residues was characteristic for most known laminarinases from various organisms (Svetasheva et al., 1984). Acetylimidazole, which acylates the phenol group of tyrosine, completely inhibited laminarinase of *T. aureviride,* and laminarinase of *C. indicum* by 50%. Modification of histidine residues could be a side reaction. However, diethyl pyrocarbonate, a specific reagent to histidine, did not reduce enzyme activity. It is likely that modification with acetylimidazole was directed to the tyrosine residue involved in the activity of laminarinases. Thus, inhibitor analysis demonstrated the role of tryptophan and tyrosine residues in the catalytic activity of

Table 1. Effect of chemical reagents on laminarinases from *C. indicum* and *T. aureviride*

Reagent	Reagent concentration, M	Residual activity, %	
		C. indicum	*T. aureviride*
p-Chloromercuribenzoate	5×10^{-3}	100	34 ± 0.02
N-Bromosuccinimide	10^{-2}	0	0
N-Ethylmaleimide	10^{-2}	–	100
EDTA	5×10^{-3}	100	–
Sodium azide	10^{-2}	100	–
Acetylimidazol	10^{-2}	50 ± 0.02	0
CME-carbodiimide	10^{-2}	100	–
CME-carbodiimide and glycine methyl ester	10^{-2}	100	–
Diethylpyrocarbonate	10^{-2}	94 ± 0.07	100
Halistanol sulfate	6×10^{-5}	100	113 ± 0.03
Inhibitor of *L. cichorioides*	3×10^{-7}	100	100

Table 2. Substrate specificity of laminarinases from *C. indicum* and *T. aureviride*

Substrate	Type of bond, ratio	Relative hydrolysis rate, %	
		C. indicum	*T. aureviride*
Laminaran	β-1,3; β-1,6 90:10	100	100
Translam	β-1,3; β-1,6 75:25	80 ± 0.02	69 ± 0.02
Yeast glucan	β-1,3; β-1,6 90:10	3	3
Pachyman	β-1,3; β-1,6 98:2	0.02	–
CM-pachyman	β-1,3; β-1,6 98:2	0.13	1.2
Zymosan	β-1,3; β-1,6	0.4	–
Aubasidan	β-1,6; β-1,3 75:25	0.08	–
Lichenan	β-1,3; β-1,4 70:30	1	0
Pustulan	β-1,6	0	0
Amilopectin	α-1,4	0	0
Xylan	β-1,4	0	–
CM-cellulose	β-1,4	0	0
p-Nitrophenyl-acetyl-β-D-glucosaminide	β	0.05	0
p-Nitrophenyl-β-D-glucopyranoside	β	0.05	0
p-Nitrophenyl-β-D-galactopyranoside	β	0	0
p-Nitrophenyl-α-D-mannopyranoside	α	0	0

laminarinases from the marine fungi *C. indicum* and *T. aureviride*.

The proteinaceous inhibitor of endo-laminarinases of marine mollusks isolated from the brown alga *Laminaria cichorioides* Miyabe did not affect the activity of either of the two laminarinases (Yermakova et al., 2002). Halistanol sulfate did not decrease activity of *C. indicum* laminarinase and showed a weak activating effect on the activity of *T. aureviride* laminarinase (Table 1). It has been demonstrated that sulfated polyoxysteroids, to which halistanol sulfate belongs, efficiently inhibit endo-laminarinases of marine mollusks but not the exo-laminarinase of a land mollusks, or even activate it (Zvyagintseva et al., 1986).

Using a series of glucans and glycosides with various types of bonds, these laminarinases were established to be specific to β-1,3-bonds in glucans. Enzymes hydrolysed relatively low molecular mass glucans (laminaran and translam) at a rather high rate; a serieses of slightly soluble glucans with mixed type of bonds (yeast glucan, lichenan, zymosan and some others) were hydrolysed significantly slower (Table 2). Pustulan, lichenan and glycosides were practically not hydrolysed.

Some differences in the rates of hydrolysis of laminaran and translam depend on peculiarities of their structures. These glucans differ in the number and location of β-1,6-bonds, as well as in molecular mass. β-1,6-Bonded residues of glucose (10% of β-1,6-bonds) are present in laminaran as branches of the main chain and distributed evenly along the chain of the glucan. Whereas about one third of β-1,6-bonded glucose residues in translam (containing 25% of β-1,6-bonds), is included in the main chain. The β-1,6-bonded glucose residues are mainly located in a non-reducing end of the molecule. Exo-type enzymes, which cleave mono- or oligosaccharides from the non-reducing end of the polymer, are sensitive to the substituting groups located at this site (Sova et al., 1997).

Kinetics of hydrolysis of laminaran with laminarinases of *C. indicum* and *T. aureviride* was studied by the accumulation of the reaction products. The results obtained were characteristic for exo-enzymes. Liquid chromatography showed glucose to be a main hydrolysis product in both cases (Figure 2).

To reliably confirm the type of action, we attempted to determine the transglycosylation activity of laminarinase. Endo-glycanases, cleaving internal bonds in polymer molecule, are known to catalyse

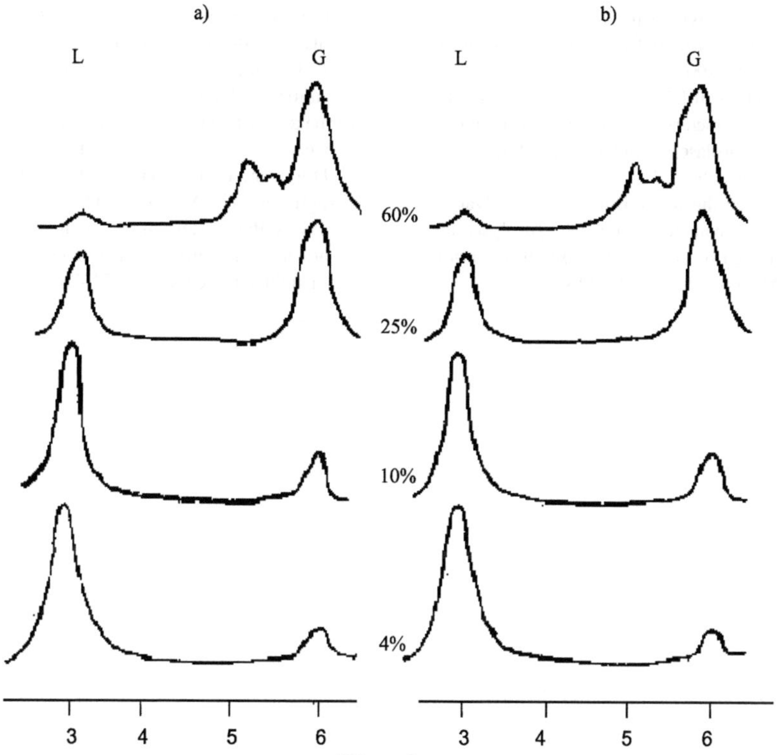

Figure 2. Gel filtration of the products of laminaran hydrolysis with β-1,3-glucanases from *C. indicum* (a) and *T. aureviride* (b) on Bio-Gel P-2 (Jeol-JLC-6AH, liquid chromatograph): L-laminaran; G-glucose.

both hydrolysis and transglycosylation, whereas exoglycanases catalyse hydrolysis only. To determine the transglycosylation activity of laminarinases from *C. indicum* and *T. aureviride*, a mixture containing a donor (laminaran) and an acceptor (*p*-nitrophenylglucoside) was used. The HPLC method revealed no products of transglycosylation in the reaction mixture. This result suggests both laminarinases to be exo-enzymes.

Thus, fungal laminarinases are specific to β-1,3-bonds in glucans, release predominantly glucose from laminaran and do not catalyze the reaction of transglycosylation. According to these data, the enzymes are exo-1,3-β-D-glucan-glucanohydrolases (EC 3.2.1.58).

Acknowledgements

This work was supported by the Grants of the Russian Foundation of Fundamental Research No. 05-04-48291, No. 03-04-49534, Grants from Presidium of the FEB RAS 06-III-B-05-127 Presidium of the Russian Academy of Sciences "Molecular and Cell Biology", "Bioresource", Program "Physical and chemical biology" for Basic Research of the Russian Academy of Science.

References

Burtseva YuV, Verigina NS, Sova VV, Pivkin MV, Zvyagintseva TN (2003) Filamentous marine fungi as producers of O-glycosylhydrolases. β-1,3-Glucanase from *Chaetomium indicum*. Mar. Biotech. 5: 349–359.

Chesters CGC, Bull AT (1963) The enzymic degradation of laminarin. Biochem. J. 86: 28–45.

Dixon M, Webb EC (1958) Enzymes. Longmans, Green and co., London, New York, Toronto.

Lee YS (2000) Qualitative evaluation of ligninolytic enzymes in xylariaceous fungi. J. Microbiol. Biotech. 10: 462–469.

Lowry OH, Rosebrough NJ, Farr AL, Randall RJ (1951) Protein measurement with Folin phenol reagent. J. Biol. Chem. 193: 265–275.

Nelson N (1944) A photometric adaptation of the Somogyi method of determination of glucose. J. Chem. 153: 375–381.

Grant WD, Rhodes LL (1992) Cell-bound and extracellular laminarinase activity in *Dendryphiella Salina* and 5 other marine fungi. Bot. Mar. 35: 503–511.

Pointing SB, Buswell JA, Jones EBG, Vrijmoed LLP (1999) Extracellular cellulolytic enzyme profiles of five lignicolous mangrove fungi. Mycol. Res. 103: 696–700.

Pitson SM, Seviour RJ, McDougall BM, Woodward JR, Stone BA (1995) Purification and characterization of 3 extracellular (1 → 3)-β-D-glucan glucohydrolases from the filamentous fungus *Acremonium persicinum.* Biochem. J. 308: 733–741.

Sova VV, Zvagintseva TN, Svetasheva TG, Burtseva YuV, Elyakova LA (1997) Comparative characterization of hydrolysis and transglycosylation catalyzed by β-1,3-glucanases from various sources. Biokhimiya (Moscow) 62: 1113–1118.

Svetasheva TG, Sova VV, Shevchenko NM, Elyakova LA (1984) Study of functional groups essential for the catalytic activity of β–1,3-glucanase from *Chlamys albidus*, using chemical modification. Biokhimiya (Moscow) 49: 1762–1768.

Yermakova SP, Sova VV, Zvyagintseva TN (2002) Brown seaweeds protein as inhibitor of marine mollusk endo-(1→ 3)-β-D-glucanases. Carboh. Res. 337: 229–237.

Zvyagintseva TN, Makar eva TN, Stonik VA, Elyakova LA (1986) The sulfated steroids of sponges of family Halichondriidae are natural inhibitors of endo-1→3–β-D-glucanases. Khimya Prirodnych Soedineny 1: 71–78.

Journal of Applied Phycology (2006) 18: 381–387
DOI: 10.1007/s10811-006-9034-x

Seasonal variation in the chemical composition of tropical Australian marine macroalgae

Susan M. Renaud & Jim T. Luong-Van*
Faculty of Science and Primary Industries, Charles Darwin University, Darwin, Northern Territory 0909, Australia

*Author for correspondence: e-mail: jim.luong-van@cdu.edu.au

Key words: tropical macroalgae, seasonal variation, carbohydrate, lipid, protein

Abstract

The proximate chemical composition (ash, soluble carbohydrate, lipid and protein) was determined in 30 common species of tropical Australian marine macroalgae from Darwin Harbour (12°26′S, 130°51′E), in summer (hot and wet) and winter (cool and dry). There was a wide diversity of species in both seasons (19 species in summer and 20 species in winter). In most species, the major component was soluble carbohydrate (chlorophytes range 2.5–25.8% dry weight (dw), phaeophytes range 8.4–22.2% dw, rhodophytes range 18.7–39.2% dw) with significantly higher ($p < 0.05$) percentages only in winter season rhodophytes. Highest percentages of protein were found in rhodophytes collected in the summer (range 4.8–12.8% dw), with significantly lower percentages ($p < 0.05$) during winter. All species had lipid contents within the range 1.3–7.8% dw, with highest percentages in summer phaeophytes, but no significant differences between species or season. Most species had moderate to high ash contents (24.2–89.7% dw), with the highest percentages during summer. Compared with summer samples, macroalgae collected in winter had higher energy value and slightly lower percentages of inorganic matter. The variation of algal groups and chemical composition may influence the availability of the food source for the majority of herbivores, which in turn is likely to effect their ecology and community structure.

Introduction

Knowledge of the chemical composition of marine macroalgae is both important for the assessment of nutritional value to marine invertebrate or vertebrate herbivores (Hawkins & Hartnoll, 1983), and for the evaluation of potential sources of protein, carbohydrate and lipid for commercial use (Chapman & Chapman, 1980) or for possible human consumption (Abbott, 1988). Seasonal variations in the chemical composition and nutritive value have been reported in common marine macroalgae from Hong Kong (Kaehler & Kennish, 1996), coastal India (Kumar, 1993) and Ireland (Mercer et al., 1993), but little is known of temporal variations in chemical composition of tropical Australian macroalgae.

Wynne and Luong-Van Thinh (1997) identified 76 species of chlorophytes, phaeophytes and rhodophytes collected from Darwin Harbour on the north coast of Australia, and 10 species were subsequently analysed for the proximate chemical composition (carbohydrate, lipid and protein (Renaud et al., 1997).

The aim of the present study was to compare the chemical composition, including ash, soluble carbohydrate, lipid and protein, of common inter-tidal tropical Australian marine macroalgae, in the dry winter season and wet summer season.

Materials and methods

Macroalgal collection

Triplicate samples of each of thirty species of marine macroalgae, including 7 chlorophytes, 8 phaeophytes and 15 rhodophytes were collected from the surface

Table 1. Chemical composition (ash content, soluble carbohydrate, total lipid, total protein) and calculated energy value, of tropical chlorophytes, collected from Darwin Harbour, Northern Territory, Australia (mean, $n = 3$). Coefficients of variation: ash ±2%; carbohydrate ±4%; lipid ±5%; protein ±4%.

	Code	(% dry weight) Ash	Soluble CHO[1]	Lipid	Protein	Energy (kJ g^{-1})
Summer						
Anadyomene brownii (J.E. Gray) J. Agardh	CC9	24.4	25.8	6.2	9.0	8.8
Caulerpa racemosa (Forsskal) J. Agardh	CC8	42.2	16.6^S	3.8	6.8	5.9
Halimeda macroloba Decaisne[2]	CC6	74.4^S	4.7	2.3	6.6	3.2
H. opuntia (Linn.) Lamouroux[2]	CC5	86.0	2.7	2.3	3.2	2.1
Neomeris van-bosseae Howe[3]	CC18	55.4	15.2^S	2.7	1.4	3.9^S
Mean		*56.5*	*13.0*	*3.5*	*5.4*	*4.8*
Without calcified species:		33.3	21.2	5.0	7.9	7.4
Winter						
Caulerpa lentillifera J. Agardh	CC36	48.9	12.8	2.7	6.6	4.8
C. racemosa (Forsskal) J. Agardh	CC49	47.7^W	14.7	4.4	6.9	5.8
Enteromorpha intestinales (Linn.) Link	CC42	49.5	18.7	1.8	3.2	4.6
*Halimeda macroloba*Decaisne[2]	CC51	64.4	2.7	2.5	4.6	2.5
H. opunta (Linn.) Lamouroux[2]	CC50	89.7^W	2.5	2.9^W	3.2	2.2^W
Neomeris van -bosseae Howe[3]	CC28	57.8^W	8.3	2.6	1.5	2.7
Mean		*55.7*	*10.0*	*2.8*	*4.3*	*3.8*
Without calcified species:		48.7	15.4	3.0	5.6	5.1
Overall mean:		58.2	11.3	3.1	4.8	4.2
Without calcified species:		42.5	17.7	3.8	6.5	6.0

[1]CHO = carbohydrate.
[2]Highly calcified species.
[3]Moderately calcified species.
W signifies significantly higher in winter season (ANOVA, $p < 0.05$).
S signifies significantly higher in summer season (ANOVA, $p < 0.05$).

rocks and coral reefs of the intertidal zone off Channel Island, East Point, Nightcliff Beach and Rapid Creek, in Darwin Harbour (12°26′S,130°51′E), Northern Territory, Australia. The study area has a monsoonal climate, with 97% of the annual average 1670 mm rain falling in the October–April summer season, when the winds are frequently from northerly directions. Very little rain falls during the May–September winter season, with predominantly south-easterly winds. Temperatures are high year round, with monthly means for Darwin ranging from 29.2 °C in November to 24.9 °C in July (Commonwealth Bureau of Meteorology, 1998).

There was a wide diversity of species in both seasons (summer 19 species, winter 20 species), but with a predominance of rhodophytes during the summer and phaeophytes during the winter tab (Tables 1, 2 and 3). Nine species were collected in both seasons, including the chlorophytes *Caulerpa racemosa*, *Halimeda macroloba*, *H. opunta* and *Neomeris van-bosseae*, phaeophytes *Dictyota ciliolata*, *Padina boryana* and *Rosenvingea nhatrangensis*, and rhodophytes *Acanthophora muscoides* and *Hypnea* sp.

Seaweed samples were collected into plastic bags, stored on ice and transported to the laboratory, where they were washed with distilled water to remove sand and surface debris, and holdfasts and epiphytes removed. Samples were then rinsed with 0.5 M ammonium formate, freeze-dried, ground and stored at −75 °C prior to chemical analysis.

Analytical methods

For each species, duplicate analyses were averaged for each of the triplicate samples for soluble carbohydrate, total protein, total lipid and total ash (inorganic matter). Soluble carbohydrates were determined by the colorimetric method of Dubois et al. (1956), after extraction with 0.5 M H_2SO_4. Total lipid was analysed gravimetrically after extraction with chloroform-methanol (2:1) by the method of Bligh and Dyer (1959). Total ash

Table 2. Chemical composition (ash content, soluble carbohydrate, total lipid, total protein) and calculated energy value, of tropical phaeophytes from Darwin Harbour, Northern Territory (mean, $n = 3$ except where indicted). Coefficients of variation as in Table 1.

	Code	(% dry weight)				
		Ash	Soluble CHO[1]	Lipid	Protein	Energy (kJ g^{-1})
Summer						
Dictyota ciliolata Kutz	CC55	47.2^{S}	15.2	7.8^{S}	4.1	6.4
Padina boryana Thivy[2]	CC10/15	36.5^{S}	19.3	4.4	6.4	6.4
Rosenvingea nhatrangensis Dawson	CC56	45.2	12.6^{S}	2.6	3.4	4.0
Mean		43.0	15.7	4.9	5.0	5.6
Winter						
Dictyota ciliolata Kutz	CC40	33.0	20.3^{W}	7.1	10.7^{W}	8.6^{W}
Feldmannia indica (Sond.) Womerley&A. Bailey[2]	CC30/34	45.1	18.7	3.6	7.4	6.3
Hydroclathrus clathratus (Bory) Howe	CC26	49.4	18.3	2.9	4.2	5.2
Padina boryana Thivy	CC20/32	33.5	18.4	5.2^{W}	10.6^{W}	7.6^{W}
Rosenvingea nhatrangensis Dawson	CC31	56.6^{W}	8.4	3.1	6.6^{W}	4.2
Sargassum decurrens (Turner) C. Agardh[2]	CC22/23	30.4	22.2	3.3	7.1	6.7
Sargassum filifolium C.Agardh	CC41	28.2	21.4	4.0	10.2	7.6
Turbinaria conoides (J. Agardh) Kützing	CC38	34.4	19.7	2.3	5.9	5.6
Mean		*38.8*	*18.4*	*3.9*	*7.8*	*6.5*
Overall mean:		38.7	17.9	4.0	6.9	6.2

[1]CHO = carbohydrate.
[2]$n = 6$.
W signifies significantly higher in winter season (ANOVA, $p < 0.05$).
S signifies significantly higher in summer season (ANOVA, $p < 0.05$).

was determined gravimetrically after heating at 550 °C for 18 h in a muffle furnace (Heraeus Thermicon). (For more details of these, methods see Renaud et al., 1994). Total nitrogen was determined by Flow Injection Analysis (Lachat 8000). For all samples, total protein was calculated from total Kjeldahl nitrogen (%) × 6.25 (Renaud et al., 1999). The energy content of macroalgal biomass was determined by multiplying the values obtained for protein, carbohydrate and lipid by 23.86, 17.16 and 36.42 kJ g^{-1}, respectively (Brett and Groves, 1979).

Statistical analysis

Ash, carbohydrate, lipid and protein data were treated statistically by one-way analysis of variance (ANOVA) with species as the source of variance. For species that occurred both in summer and winter, the chemical data were analysed by ANOVA with season as the source of variance. Equality of variance and normality were checked by Bartlett's test. Pairwise comparisons after ANOVA were made using Tukey's test. Hierarchical cluster analysis (Euclidean distance) was used to identify natural groupings in the data.

Results

Chemical composition

Considering all species collected over both seasons, the major biochemical component was soluble carbohydrate in all except the calcareous *Halimeda* species (Tables 1, 2 and 3). The percentage of soluble carbohydrate in the rhodophytes (overall mean 26.7% dry weight; range 18.7–39.2% dw) (Table 3) was significantly higher (ANOVA, $p < 0.05$) than the phaeophytes (mean 17.9%, range 8.4–22.2%) (Table 2) and the chlorophytes (mean 11.3%, range 2.7–25.8%) (Table 1) (ANOVA, $p < 0.05$ in both cases). However, if the chlorophytes with high calcification (*Halimeda* spp. and *N. van-bosseae*) were not included, then the overall mean for that class was 17.7% dw, which is similar to 17.9% dw for phaeophytes.

The highest percentages of protein were found in rhodophytes (mean 8.0% dw, range 4.8–12.8% dw), with higher than 10% dw in *A. muscoides*, *B. tenella*, *L. majuscula* and *W. plumosa* (Table 3). There were significantly lower percentages of protein in the other two macroalgal classes (phaeophytes: mean 6.9% dw;

Table 3. Chemical composition (ash content, soluble carbohydrate, total lipid, total protein) and calculated energy, of tropical rhodophytes from Darwin Harbour, Northern Territory (mean, $n = 3$, except where indicated). Coefficients of variation as in Table 1.

	Code	(% dry weight) Ash	Soluble CHO [1]	Lipid	Protein	Energy (kJ g^{-1})
Summer						
Acanthophora muscoides (Linn.) Bory	CC19	45.0^S	29.5	2.7	10.0	8.4
Botrycladia leptopoda (J. Agardh) Kylin	CC2	59.1	23.1	1.4	7.1	6.2
Eucheuma denticulatum (N.L. Burman) Collins et Hervey	CC1	43.8	30.6	1.6	5.0	6.7
Gracilaria salicornia(C. Agardh) Dawson	CC14	49.3	24.4	1.3	6.0	6.1
Gracilaria sp. [2]	CC11/13	53.1	21.6	1.9	7.0	5.7
Hypnea sp.	CC16	37.5^S	33.0	2.4	6.3	8.0
Laurencia majuscula (Harv.) Lucas	CC12	42.2	18.8	5.1	12.5	8.1
Portieria hornemannii (Lyngbye) P.C. Silva	CC3	37.4	21.8	5.3	9.8	8.0
Soliera robusta (Grev.) Kylin	CC4	58.1	22.5	3.4	4.8	5.2
Wrangelia plumose Harvey	CC7	35.1	22.3	5.6	12.8	8.9
Mean		41.5	24.8	3.3	8.4	7.2
Winter						
Acanthophora muscoides (Linn.) Bory [2]	CC25/33	42.4	32.6^W	2.2	9.0	7.9
Bostrychia tenella (J.V. Lamouroux) J. Agardh	CC29	24.2	31.2	4.5	10.8	10.3
Champia sp.	CC39	58.1	23.4	2.1	6.1	6.2
Gracilaria crassa Harvey ex J. Agardh	CC37	52.3	18.7	1.9	6.4	5.4
Hypnea sp.	CC21	34.7	31.7	3.4^W	6.9	13.8^W
Spiridia sp.	CC43	28.9	39.2	1.9	4.9	8.6
Tolypiocladia calodictyon (Harvey ex Kützing) P.C. Silva	CC24	44.9	26.7	3.3	8.8	7.9
Mean		*40.8*	*29.1*	*2.8*	*7.6*	*8.6*
Overall mean:		43.5	26.7	3.0	8.0	7.8

[1]CHO = carbohydrate.
[2]$n = 6$.
W signifies significantly higher in winter season (ANOVA, $p < 0.05$).
S signifies significantly higher in summer season (ANOVA, $p < 0.05$).

chlorophytes: mean 4.8%) (ANOVA, $p < 0.05$ in all cases) (Tables 1 and 2, respectively). Again, if the highly calcified chlorophytes are excluded, then the overall percentage of protein was similar to phaeophytes (6.5 and 6.9% dw, respectively).

All species had lipid contents within the range 1.4–7.8% dw. The highest percentages were in *D. ciliolata* (7.8% dw) and in the phaeophytes (mean 4.0%; range 2.0–7.8% dw) (Table 2), followed by the chlorophytes (mean 3.1; range 1.8–6.2% dw) (Table 1) and then the rhodophytes (mean 3.0; range 1.4–5.6% dw) (Table 3). However, ANOVA showed that these differences were not significant.

There was a wide range of ash contents (24.2–89.7% dw), with the highest percentages (64.4–89.7% dw) in the heavily calcified chlorophytes, *Halimeda* spp. (Table 1). ANOVA demonstrated that the ash contents of the chlorophytes (range 24.4–89.7% dw) (Table 1) were significantly higher than those of the phaeophytes (range 28.2–56.6%) (Table 2) and the rhodophytes (range 24.2–59.1%) (Table 3) ($p < 0.05$ in each case).

Cluster analysis of the chemical composition (ash, carbohydrate, lipid and protein) of all species over both seasons indicated four main groups of macroalgae (Figure 1). Group I was made up of a single species, the calcareous chlorophyte *Halimeda*, which was low in all three biochemical components, carbohydrate, lipid and protein. Group II included species with moderate levels of carbohydrate and protein, and accounted for all phaeophyte species, together with most chlorophyte species. Group III included those rhodophyte species which had high percentages of carbohydrate, together with moderate to high percentages of protein and low percentages of lipid. Group IV included the rest of the rhodophytes, with lower percentages of carbohydrate, but which were similar to Group III in terms of protein and lipid. One chlorophyte species,

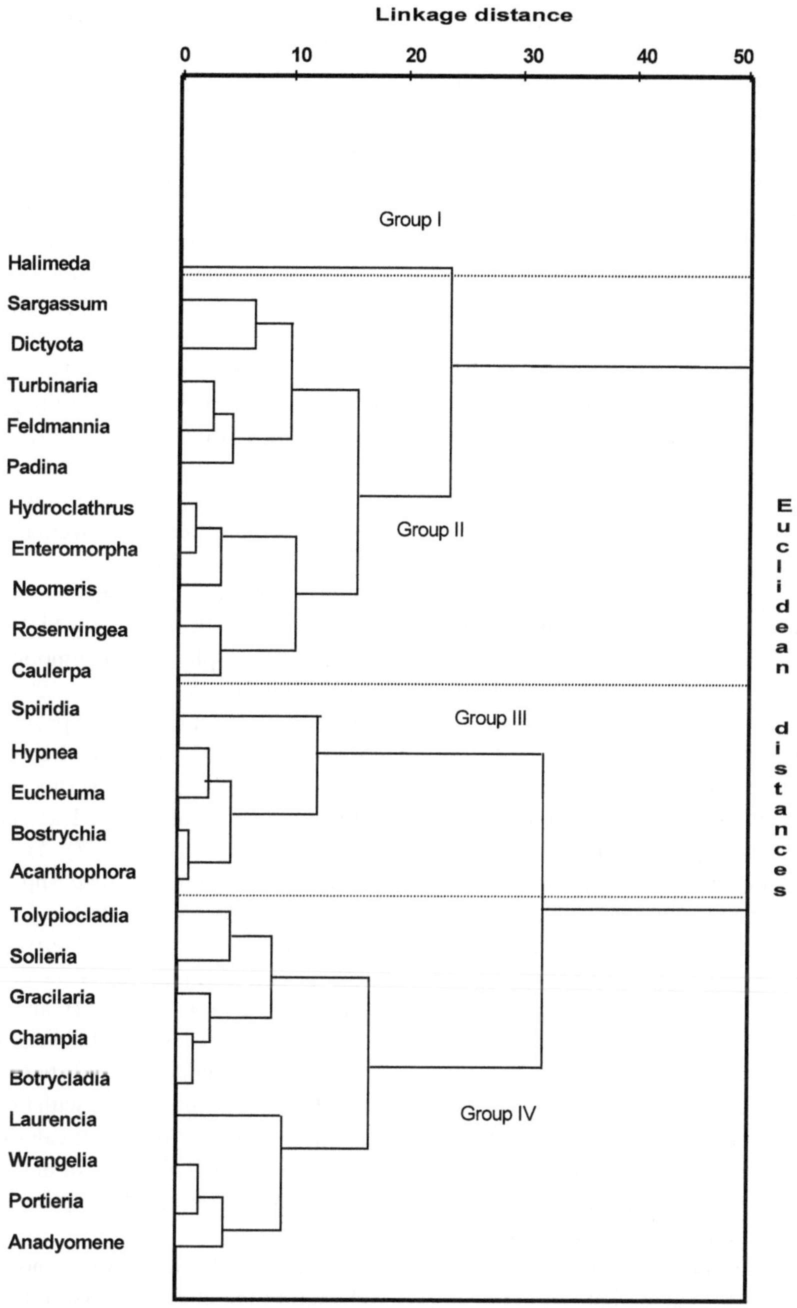

Figure 1. Hierarchical cluster analysis (complete linkage, Euclidean distances) of the chemical composition (carbohydrate, lipid, protein), of 30 species of tropical marine macroalgae from Darwin Harbour, North Australia.

Anadyoneme, which had low percentages of ash, lipid and protein, was also included in Group IV.

Seasonal variation in chemical composition

Nine species were collected in both summer and winter seasons (Tables 1, 2 and 3). There was a trend of significantly higher percentages of carbohydrate in 3 species collected in the summer (Table 1: *C. racemosa*, *N. vanbosseae* and Table 2: *R. nhatrangensis*) (ANOVA, $p < 0.05$ in all cases). However *A. muscoides* (Table 3) and *D. ciliolata* (Table 2) had significantly higher percentages of carbohydrate in the winter (ANOVA, $p > 0.05$), while all other species had no temporal variation in carbohydrate content. There was no significant difference in total lipid content with collection season for the all species except *Hypnea* sp. (Table 3) which had a significantly higher lipid content in the winter (ANOVA, $p < 0.01$). There was no significant variation in protein content of chlorophytes and rhodophytes with season, but the phaeophytes *D. ciliolata*, *P. boryana* and *R. nhatrangensis* (Table 2) had significantly higher percentages of protein in the winter (ANOVA, $p < 0.001$ in all cases). There was no overall trend in ash content, as five species, including *H. macroloba* (Table 1), *D. ciliolata* and *P. boryana* (Table 2), *A. muscoides* and *Hypnea* sp. (Table 3), had significantly higher percentages of ash during the summer and the other 4 species had significantly higher percentages in the winter (Tables 1 and 2) (ANOVA, $p < 0.05$ for all species).

Calculated energy

Highest calculated energy values were found in the rhodophyte species *Hypnea* sp. and *B. tenella* (Table 3 :13.8 and 10.3 kJ g^{-1}, respectively), followed by *W. plumose* (Table 3), *A. brownii* (Table 1), *D. ciliolata* (Table 2) and *A. muscoides* (Table 3) (8.8–8.6 kJ g^{-1}). Amongst the three of these species that were collected in both seasons, *D. ciliolata* (Table 2) and *Hypnea* sp. (Table 3) had significantly higher nutritive value ($p < 0.05$), in terms of higher calculated energy values when collected in the winter, but *A. muscoides* (Table 3) did not have significant energy difference with season.

Discussion

The percentage of soluble carbohydrate in *Hypnea* sp. was 1.8–7.5 times higher than those of several seaweed species of the same genus collected in Darwin Harbour (Renaud et al., 1997), the Indian Tuticorin Coast (Kumar, 1993), and coastal Hong Kong (Wong & Cheung, 2000). The soluble carbohydrate content of *C. racemosa* was consistent with an earlier report for this species (Kumar, 1993) but was five times higher than the percentage reported for the same species collected in the Mexican Yucatan peninsula (Robledo & Pelegrin, 1997). In the present study, the lowest percentages of soluble carbohydrate were found in *Halimeda opunta* and *H. macroloba* (2.5 and 2.7% dw, respectively) (Table 1). These results were five to eight times lower than previous reports for non-calcified members of *H. borneensis* (Renaud et al., 1997) and *H. tuna* (Kumar, 1993).

The highest amounts of protein were in members of the rhodophytes, while lowest percentages were found in the chlorophytes, which are of the same order as results for 11 chlorophyte, phaeophyte and rhodophyte species (range 6.4 to 8.0%) from Hong Kong (Kaehler & Kennish, 1996) and 5 species from Mexico (Robledo & Pelegrin, 1997), but slightly higher than results for 21 species of Indian macroalgae (mean 3.6%) (Kumar, 1993). The protein content of *Gracilaria* spp. was of the same order as reported for the Indian *G. cortica* (Kumar, 1993), and the Mexican *G. cornea* (Robledo & Pelegrin, 1997). On the other hand, the percentages of protein in *Hypnea* spp. in the present study and an earlier study (Renaud et al., 1997) were 30–80% lower than those reported for *H. japonica* and *H. charpoides* (Wong & Cheung, 2000) but were about twice those reported for *H. musciformis* (Kumar, 1993) and *H. valentiae* (Banaimoon, 1992).

The percentages of lipid found in this study were comparable with or slightly higher than previous reports for phaeophytes (Banaimoon, 1992; Mercer et al., 1993; Kaehler & Kennish, 1996; Robledo & Pelegrin, 1997), for chlorophyte species (Banaimoon, 1992; Mercer et al., 1993; Kaehler & Kennish, 1996; Robledo & Pelegrin, 1997), and for rhodophytes (Banaimoon, 1992; Mercer et al., 1993; Kaehler & Kennish, 1996; Robledo & Pelegrin, 1997). For example, the total lipid content of *Gracilaria* spp. (1.9% dw) was of the same order as reports for *G. cortica* (2.1% dw), *G. canaliculta* (1.4%), *G. foliifera* (0.7%), *G. textroii* (0.9%) and *G verrucosa* (1.6%) collected in the Arabian sea (Banaimoon, 1992), and the Gulf of Mexico (*G. cornea* 0.3% dw) (Robledo & Pelegrin, 1997).

The results in the present study for the highly calcified *Halimeda* spp., concur with the findings of Kaehler and Kennish (1996) that all calcified seaweed species were high in ash and low in nutrients, and they were dissimilar to non-calcified species, regardless of taxonomic group.

Overall, the study found no single trend in the change of chemical composition with season. Carbohydrate content was significantly higher in summer in 3 of the 9 species collected in both seasons, which concurs with the report of higher percentages of soluble carbohydrate in *Caulerpa racemosa*, *Enteromorpha tuberosa*, *Padina pavonica*, *Gracilaria corticata* and *Hypnea musciformis* collected in the summer (Kumar, 1993). However, Kaehler and Kennish (1996) found no difference in the soluble carbohydrate contents of 2 species of rhodophytes and 2 species of phaeophytes collected in both the summer and winter.

The present study found that there were significant seasonal differences in percentages of protein in all phaeophyte species. Mercer et al. (1993) reported significantly higher protein in the temperate phaeophyte *Alaria esculenta* collected in Ireland in May (summer) but the opposite trend in *Laminaria digitata* collected in February (winter). However, Kumar (1993) reported little difference in the percentage of protein in each of 21 species of tropical macroalgae collected monthly throughout a one-year period.

While the majority of species had no significant difference in the percentage of lipid with season, *Hypnea* sp. had a significantly higher lipid content in the winter. Mercer et al. (1993) reported significantly higher percentages of lipid in 3 phaeophyte and 1 chlorophyte species collected in the Irish winter, but Kaehler and Kennish (1996) reported no significant difference in the lipid contents of 6 species collected in both summer and winter in Hong Kong.

The study found that members of the rhodophytes were the most nutritionally rich species, in terms of carbohydrate, protein and calculated energy value. However, it is important to note that the nutritional values here are based on chemical analysis only. Biological analysis using animal feeding trials would be required to establish the nutritional value of these seaweeds.

Acknowledgments

This research was supported by Northern Territory University.

References

Abbott IA (1988) Food and food products from algae. In: Lembi CA, Waaland JR (eds), Algae and Human Affairs. Cambridge Univ. Press, Cambridge, pp. 135–147.

Banaimoon SA (1992) Fatty acids in marine macroalgae from Southern Yemen (Hadramout) including occurrence of eicosatetraenoic (20:4) and eicosapentaenoic (20:5) acids. Bot. Mar. 33: 165–168.

Bligh EG, Dyer WJ (1959) A rapid method of total lipid extraction and purification. Can. J. Biochem. Physiol. 37: 911–917.

Brett JR, Groves TD (1979) Physiological energetics. In: Hoar WS, Randall DJ (eds), Fish Physiology Vol. VIII, Academic Press, London, pp. 279–351.

Chapman VJ, Chapman DJ (1980) Seaweeds and Their Uses. Chapman and Hall, London.

Commonwealth Bureau of Meteorology, Australia. 1998. http://www.bom.gov.au/climate/averages/tablesa/cw_014015.shtml

Dubois M, Giles KA, Hamilton KS, Rebers PA, Smith F (1956) Colorimetric method for the determination of sugar and related substances. Anal. Chem. 18: 350–356.

Hawkins SJ, Hartnoll RG (1983) Grazing of intertidal algae by marine herbivores. Oceanogr. Mar. Biol. Ann. Rev. 21: 195–282.

Kaehler S, Kennish R (1996) Summer and winter comparisons in the nutritional value of marine macroalgae from Hong Kong. Bot. Mar. 39: 11–17.

Kumar V (1993) Biochemical constituents of marine algae from Tuticorin coast. Indian J. Mar. Sci. 22: 138–140.

Mercer JP, Mai KS, Donlon J (1993) Comparative studies on the nutrition of two species of abalone, *Haliotis tuberculata* Linnaeus and *Haliotis discus hannai* Ino. I. Effects of algal diets on growth and biochemical composition. Invertebrate Reprod. Develop. 23: 2–3.

Renaud SM, Parry DL, Luong-Van T (1994) Microalgae for use in tropical aquaculture I: Gross chemical and fatty acid composition of twelve species of microalgae from the Northern Territory, Australia. J. Appl. Phycol. 6: 337–345.

Renaud SM, Lambridinis G, Luong-Van T, Parry DL, Lee C (1997) Chemical composition of algae for use in *Trochus niloticus*. studies. In: Lee CL, Lynch PW (eds), Trochus: Status, Hatchery Practice and Nutrition, Australian Centre for International Agricultural Research, Canberra, pp. 88–96.

Renaud SM, Luong-Van T, Parry DL (1999) The gross chemical composition and fatty acid composition of 18 species of tropical Australian microalgae for possible use in mariculture. Aquaculture 170: 147–159.

Robeldo D, Pelegrin YF (1997) Chemical and mineral composition of six potentially edible seaweed species of Yucatan. Bot. Mar. 44: 301–306.

Wong KH, Cheung PCK (2000) Nutritional evaluation of some subtropical red and green seaweeds Part I – proximate composition, amino acid profiles and some physico-chemical properties. Food Chem. 71: 475–482.

Wynne MJ, Luong-Van Thinh J (1997) A report on collections of benthic marine algae from Darwin, Northern Australia. In: Lee CL, Lynch PW (eds), Trochus: Status, Hatchery Practice and Nutrition, Australian Centre for International Agricultural Research, Canberra, pp. 81–87.

Journal of Applied Phycology (2006) 18: 389–398
DOI: 10.1007/s10811-006-9035-9

Tissue nitrogen and phosphorus in seaweeds in a tropical eutrophic environment: What a long-term study tells us

Sergio O. Lourenço,* Elisabete Barbarino, Andyara Nascimento, Joana N.P. Freitas & Graciela S. Diniz
Departamento de Biologia Marinha, Universidade Federal Fluminense, Caixa Postal 100644, CEP 24001-970, Niterói, RJ, Brazil

*Author for correspondence: e-mail solourenco@yahoo.com; fax: +55 21 2629 2292

Key words: dissolved nutrients, seaweeds, trophic status, tissue nitrogen, tissue *N:P* ratio, tissue phosphorus

Abstract

Percentages of nitrogen and phosphorus in 10 species of seaweeds (6 green and 4 red algae) were monitored from 1997 to 2004 by seasonal sampling in Guanabara Bay, South-eastern Brazil. The species did not show consistent variations in tissue *N*, *P* and *N*:*P* that related to annual cycles. Throughout this study, higher percentages of tissue *N* and *P* were found in *Bostrychia radicans* and *Grateloupia doryphora* (red algae) and lower in *Cladophora rupestris* and *Codium decorticatum* (green algae). In November 1999, the Icaraí Submarine Sewage Outfall became operational, resulting in a reduction of visual pollution in the area and an improvement in the local quality of seawater for recreational use. Measurements of dissolved nutrients at the sampling site did not indicate significant changes in concentrations after the commissioning of the submarine sewage outfall; however, tissue *P* and *N*:*P* ratio of most of species were significantly lower than in the first two years of this survey. Variations in tissue nitrogen throughout this study were not significant, except for *G. doryphora* in some comparisons. Results show that seaweeds function very well as monitors of environmental changes in Guanabara Bay. Experimental data are needed to identify possible environmental processes which are promoting changes in chemical composition of the local seaweed populations.

Introduction

Anthropogenic inputs of nutrients have had remarkable impacts on marine organisms in coastal areas (Clark, 2001). Increased abundance of opportunistic seaweeds is among the general consequences of nutrient loading in coastal areas (Rivers & Peckol, 1995). Macroalgae respond to nutrient enrichment by taking up nutrients, growing, and storing "excess" nutrients for future growth (Fujita, 1985; Björnsäter & Wheeler, 1990). The proliferation of opportunistic seaweeds affects local biodiversity and may promote a decrease in concentrations of dissolved nutrients in the water column (Rivers & Peckol, 1995; Valiela et al., 1997).

Concentrations of tissue nutrients reflect the environmental conditions of the site, providing a useful indicator of local nutrient status (Fong et al., 1994). In addition, total nutrient concentration in the algal tissue provides an integrated measurement of nutrient regime over time (Wheeler & Björnsäter, 1992; Villares & Carballeira, 2003). Monitoring of tissue nutrients to detect enrichment can be undertaken at less frequent intervals than monitoring of the water-column nutrients, and allows a more accurate evaluation of the nutrient status of the macroalgae. Studies on tissue *N* and *P* content of macroalgae predominantly in temperate coastal environments (Wheeler & Björnsäter, 1992; Peckol et al., 1994) reveal wide fluctuations in the tissue content of *N* and *P* related to seasonal changes and nutrient availability. By comparison, information on tissue *N* and *P* of algae from tropical environments is relatively scarce (Schaffelke, 1999; Fong et al., 2001), and more data are needed from those regions.

Guanabara Bay, Brazil, is a eutrophic coastal environment connected to the sea by a narrow mouth, which partially restricts water exchange. The Bay receives substantial river runoff relative to the total water volume, making it similar to a large estuary in its inner parts (Kjerfve et al., 1997). The Bay is located in a very populated urban area, and long-term cultural eutrophication has generated an environment with permanent high concentrations of dissolved nutrients due to output of both domestic and industrial wastewater (Mayr et al., 1989). Considering these characteristics, we hypothesised that the seaweeds of Guanabara Bay would present permanently high concentrations of tissue *N* and *P* and show no significant variations in their tissue nutrients throughout the year and no inter-annual changes in tissue nutrients.

In this study we report on the seasonal variations of tissue *N*, *P* and *N*:*P* atomic ratio of ten abundant macroalgal species of Guanabara Bay. Comparisons were made between algal *N* and *P* contents and the concentrations of dissolved nutrients in the system in this 7-year assessment. In addition, during this study a submarine sewage outfall was built in the study area and its possible effects on the tissue composition of the macroalgal flora was evaluated.

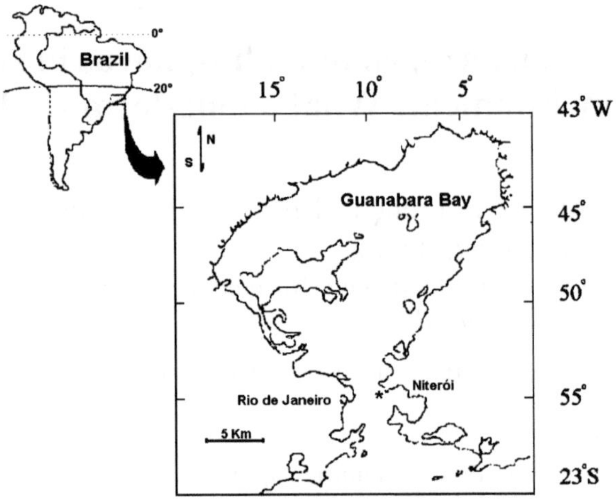

Figure 1. Sampling site in Guanabara Bay. * = Sampling site.

Materials and methods

Study area

The sampling site is located at Boa Viagem Beach (23°04′S, 43°08′W), in Guanabara Bay. The site is in the urban area of Niterói City, and it is located near the entrance of the Bay (Figure 1), which promotes a local dilution in the typical high levels of pollution of the Bay. The Bay shows a low water exchange rate (Mayr et al., 1989) due to geomorphological features and human occupation of coastal areas. Guanabara Bay comprises an area of 381 km^2 and an estimated 2 billion m^3 of water. The catchment area (4000 km^2) includes 35 rivers that contribute substantially to the freshwater input. The mean depth is 7.7 m, varying from 50 m (main channel) to less than 1 m in the inners parts close to the internal margins. The area of Guanabara Bay comprises 15 municipalities, with a population of ca. 7.6-milion inhabitants (FEEMA, 1999). According to Paranhos et al. (2001), ca. 470 t of BOD and 150 t of industrial sewage are disposed of daily into the Bay. CIBG (2004) indicates that in 1994 ca. 8 t of oil derivatives and 55 kg of heavy metals were disposed of daily into the Bay by the 55 main industrial plants in the area. An ongoing program has been reducing many sources of pollution into the Bay, but the total amount of pollutants that enters the system daily is still high (CIBG, 2004).

Taouil and Yoneshigue-Valentin (2002) classified the sampling site as moderately affected by wave action, with an unusual abundance of pebbles and large-grain sand in the intertidal area.

Algae studied

In this study ten macroalgal species were analysed. The identification of the macroalgae was carried out following the checklist of Wynne (1998). Experts were consulted to confirm our identifications.

Chlorophyta: *Chaetomorpha antennina* (Bory) Kützing, *Cladophora rupestris* (L.) Kützing, *Codium decorticatum* (Woodw.) M. Howe, *Enteromorpha flexuosa* (Wulfen) J. Agardh, *Ulva fasciata* Delile, and *Ulva lactuca L.*

Rhodophyta: *Bostrychia radicans* (Mont.) in Orbigny, *Chondracanthus teedii* (Mertens ex Roth) Fredericq, *Grateloupia doryphora* (Montagne) M. Howe, and *Gymnogongrus griffthsiae* (Turner) Mart.

The species are found attached to the rocks and pebbles of the sampling site (water column between 0.4 to 0.8 m). Samples of two species (*B. radicans* and *C. anteninna*) were collected at Itapuca Stone, a site located 400 m from Boa Viagem Beach, where they were more abundant (attached to vertical rock surfaces). We assume that both sampling sites have virtually the same environmental characteristics (temperature, salinity, pollution, dissolved nutrients, etc. – data not shown), except for more water movement at Itapuca Stone.

Sampling

Sampling began in June 1997 (austral autumn) and continued through June 2004. Samples were collected seasonally for a total of 29 field trips, and each sampling occurred in the last 3 weeks of each season.

Whole thalli of adult plants were collected in the early morning and washed in the field with seawater to remove epiphytes, sediment and detritus. At least 15 whole plants of each species were collected, independent of the size of each seaweed. All species were typically found at the same specific points in the site throughout the study (e.g. *C. antennina* was sampled always at the same rocks near the Itapuca Stone; *C. decorticatum* was found always attached to the pebbles near the Boa Viagem Island). The plants were placed in plastic bags, and kept on ice until return to the laboratory (less than one hour). In the laboratory, samples were gently brushed under running seawater, rinsed with distilled water, and dried at 60 °C for at least three days and until constant weight. The dried material was ground into a powder and kept in desiccators containing silica-gel at room temperature until chemical analysis. At the time of each collection of macroalgae, four 250 ml-water samples ($n = 4$) for dissolved nutrient analysis were taken from 15–20 cm below the water surface, as well as measurements of local temperature at the same depth. The samples of water were filtered through cellulose membrane filters (Millipore® HAWP 0.45 μm pore) and kept at −20 °C for spectrophotometric determinations of ammonium, nitrate, and nitrite (Parsons et al., 1984), phosphate and urea (Grasshoff et al., 1983). Each sample was measured at least three times to obtain accurate results, and the results showed in this study represent mean values for four independent samples collected in the field for each sampling.

Tissue analysis

Total N and P were determined in algal tissue after peroxymonosulphuric acid digestion, using a Hach digestor (Digesdhal®, Hach Co.) (Hach et al., 1987). Samples containing 50 to 200 mg (dry matter) were digested with 4 ml concentrated sulphuric acid (Merck Co.) at 440 °C and treated with 17 ml of 30% hydrogen peroxide (Merck Co.). Total nitrogen and phosphorus contents in the samples were determined spectrophotometrically after specific chemical reactions. See Lourenço et al. (2005) for analytical details.

For each species and sampling at least four to six independent (from different plants) measurements of tissue N and P were performed ($4 \leq n \leq 6$).

Statistical analysis

The results for each species separately and for total measurements of all species combined were analysed by single-factor analysis of variance (ANOVA) with significance level $\alpha = 0.05$ (Zar, 1996), followed with a Tukey's multiple comparison test. Suitable transformations of data (e.g. log of the actual data) were made when necessary. Time was the only factor considered in ANOVA.

Results

Nitrite and urea showed the lowest concentrations among dissolved N sources, typically lower than 3.0 μM. Ammonium and nitrate showed higher concentrations, varying in most of the observations between 5.0 and 15.0 μM. Variations of phosphate concentrations fluctuated between 0.4 and 2.6 μM. Wide variations in total dissolved nitrogen and N:P ratio were recorded, but no seasonal trend was detected throughout the study. Salinity fluctuated around 31 psu throughout the survey (Table 1).

Small variations in tissue nitrogen were recorded in the species, with high values (>5.0%) throughout the study. This general description is exemplified by *C. decorticatum*, *G. doryphora* and *U. fasciata* (Figure 2A), visually the most abundant species in the sampling site. Among all species, *Bostrychia radicans* and *Grateloupia doryphora* (red algae) exhibited the highest values for tissue N and *Chaetomorpha antennina*, *Cladophora rupestris* and *Codium decorticatum* the lowest (Table 2). The tissue N content of all species were not significantly different for the vast majority of the paired comparisons tested (ANOVA, $F_{28,907} = 3.289$) (Figure 2B); some comparisons involving data of 1997 (higher values) and 2003–2004 (lower values) exhibited significant differences.

Values for tissue P showed wider variations among the species than the values of tissue N. Most of the comparisons showed significant differences, with a consistent trend of higher values for tissue P in 1997–1999 and lower values in 2000–2004 (Figure 3A, B).

Table 1. Measurements of salinity and dissolved nutrients at the sampling site in Guanabara Bay. Data represent the mean ± standard deviation ($n = 4$) and are expressed as μM, except for salinity (psu).

Sampling	*N*-ammonium	*N*-nitrite	*N*-nitrate	*N*-Urea	*P*-phosphate	Total dissolved *N*	*N*:*P* Ratio	Salinity
Autumn 1997	12.9 ± 4.67	1.46 ± 0.53	0.35 ± 0.07	0.18 ± 0.04	0.96 ± 0.31	14.9 ± 4.80	15.5 ± 4.59	34.8
Winter 1997	7.30 ± 0.78	1.73 ± 0.15	0.85 ± 0.21	0.31 ± 0.05	1.39 ± 0.02	10.2 ± 0.73	7.30 ± 0.50	32.0
Spring 1997	13.2 ± 2.61	1.77 ± 0.09	2.72 ± 0.46	0.41 ± 0.08	1.97 ± 0.09	18.1 ± 3.01	9.19 ± 1.24	32.0
Summer 1998	16.0 ± 2.26	1.06 ± 0.22	1.69 ± 0.30	0.39 ± 0.08	2.04 ± 0.21	19.4 ± 2.44	9.51 ± 2.08	31.3
Autumn 1998	11.6 ± 4.34	2.32 ± 0.57	3.59 ± 1.78	1.10 ± 0.21	1.26 ± 0.47	18.6 ± 2.64	14.8 ± 5.49	29.8
Winter 1998	6.81 ± 1.93	1.87 ± 0.22	5.48 ± 0.97	1.20 ± 0.40	1.38 ± 0.42	15.4 ± 2.15	11.1 ± 4.02	31.0
Spring 1998	5.25 ± 2.29	1.33 ± 0.81	2.55 ± 1.72	0.82 ± 0.12	2.62 ± 1.57	9.95 ± 3.40	3.80 ± 1.72	32.5
Summer 1999	4.10 ± 2.06	1.18 ± 0.40	1.53 ± 0.92	0.38 ± 0.20	1.49 ± 0.33	7.19 ± 1.56	4.83 ± 2.19	28.5
Autumn 1999	10.8 ± 3.99	3.87 ± 1.57	1.24 ± 0.45	1.02 ± 0.21	1.81 ± 0.49	16.9 ± 5.08	9.35 ± 3.39	31.9
Winter 1999	7.84 ± 2.23	1.33 ± 0.15	3.54 ± 0.62	2.81 ± 0.36	1.23 ± 0.20	15.5 ± 2.51	12.6 ± 4.60	31.8
Spring 1999	4.63 ± 1.15	0.78 ± 0.23	1.44 ± 0.56	2.49 ± 0.43	2.64 ± 1.22	9.34 ± 2.30	3.54 ± 1.54	30.1
Summer 2000	3.53 ± 0.96	1.12 ± 0.51	2.63 ± 1.35	3.22 ± 0.24	1.84 ± 0.34	10.5 ± 2.76	5.71 ± 0.43	32.6
Autumn 2000	9.05 ± 1.58	1.10 ± 0.58	11.0 ± 1.27	1.42 ± 0.23	1.70 ± 0.37	22.6 ± 12.8	13.3 ± 2.07	30.8
Winter 2000	16.0 ± 1.10	2.03 ± 0.38	9.80 ± 2.57	1.59 ± 0.35	2.05 ± 0.30	29.4 ± 3.32	14.4 ± 2.35	33.4
Spring 2000	11.9 ± 3.95	0.78 ± 0.11	4.62 ± 0.43	1.22 ± 0.24	2.37 ± 0.42	18.5 ± 3.98	7.81 ± 2.23	32.7
Summer 2001	2.27 ± 0.67	0.36 ± 0.07	2.39 ± 0.89	1.36 ± 0.26	0.87 ± 0.47	6.38 ± 1.23	7.33 ± 4.69	29.9
Autumn 2001	2.40 ± 1.89	1.03 ± 0.18	9.77 ± 3.42	3.45 ± 0.76	0.43 ± 0.05	16.7 ± 3.17	38.7 ± 5.10	31.7
Winter 2001	12.3 ± 1.14	1.15 ± 0.16	7.47 ± 1.79	3.59 ± 0.87	1.26 ± 0.17	24.5 ± 0.60	19.5 ± 3.5	32.0
Spring 2001	7.41 ± 1.11	1.68 ± 0.07	9.28 ± 0.43	3.69 ± 0.32	1.89 ± 0.11	22.1 ± 1.94	11.7 ± 1.6	31.4
Summer 2002	3.90 ± 1.93	1.08 ± 0.41	6.12 ± 1.68	3.03 ± 0.35	1.41 ± 0.14	14.1 ± 2.25	10.0 ± 1.8	32.5
Autumn 2002	6.01 ± 3.68	1.24 ± 0.41	6.33 ± 4.14	5.80 ± 0.89	1.84 ± 0.80	19.4 ± 5.90	10.5 ± 5.9	33.4
Winter 2002	5.36 ± 2.68	1.73 ± 0.37	10.7 ± 3.55	1.79 ± 0.13	1.76 ± 0.33	19.6 ± 6.50	11.1 ± 1.43	33.8
Spring 2002	5.03 ± 1.53	0.73 ± 0.24	6.71 ± 1.52	1.54 ± 0.20	1.38 ± 0.42	14.0 ± 3.04	10.2 ± 2.36	31.6
Summer 2003	8.83 ± 0.77	0.88 ± 0.11	3.15 ± 0.47	3.27 ± 0.92	1.54 ± 0.08	16.1 ± 2.53	10.5 ± 1.44	33.3
Autumn 2003	19.6 ± 1.08	1.93 ± 0.11	5.97 ± 1.21	1.54 ± 0.45	1.50 ± 0.13	29.0 ± 1.37	19.4 ± 2.12	33.9
Winter 2003	16.0 ± 1.67	2.99 ± 0.04	6.60 ± 1.18	1.34 ± 0.15	1.59 ± 0.07	28.5 ± 1.31	17.9 ± 1.4	34.4
Spring 2003	12.1 ± 2.15	2.39 ± 0.39	7.42 ± 1.96	1.68 ± 0.30	1.86 ± 0.47	23.6 ± 1.86	12.7 ± 3.5	31.5
Summer 2004	11.7 ± 1.38	2.37 ± 0.17	11.2 ± 1.65	1.53 ± 0.27	1.37 ± 0.21	26.8 ± 3.03	19.6 ± 3.3	27.1
Autumn 2004	15.0 ± 4.24	3.64 ± 1.01	7.88 ± 3.79	1.35 ± 0.41	1.92 ± 0.67	27.9 ± 9.26	14.5 ± 1.07	30.5

Among all species *Chondracanthus teedii* and *Enteromorpha flexuosa* exhibited the highest values for tissue *P* and *Cladophora rupestris* and *Codium decorticatum* the lowest (Table 2). Overall trends of *P* concentrations in tissues of all species and in the three dominant species are the same, with a significant decrease in the values in the last three years (ANOVA, $5.306 \leq F_{28,907} \leq 14.993$, $0.05 < p \leq 0.001$) (Figure 3A and B, Table 2).

Tissue *N*:*P* ratio was predominantly >20:1 throughout the study. In the first 30 months of the study the average (*N*:*P*) ratio was ca. 23:1, increasing to ca. 28:1 from summer 2001 until the end of the survey (Figure 4A, B). Similarly, the combined (*N*:*P*) analysis indicated significantly different between these two contrasting periods (ANOVA, $5.242 \leq F_{28,907} \leq 10.530$, $0.05 < p \leq 0.001$). *Ulva lactuca* presented the highest (*N*:*P*) ratio among all species, in spring 2001 (56.8:1), and *Chondracanthus teedii* the lowest, in spring 1998 (13.6:1) (Table 2).

In December 1999 the Icaraí Submarine Sewage Outfall became operational. Since then, most of the local sewage receives secondary treatment and it is released ca. 2.5 km from the sampling site, and close (ca. 1.0 km) to the entrance of Guanabara Bay. Despite the fact that there was no difference in the concentrations of dissolved nutrients recorded in the sampling site during the study (Table 1), the initiation of this facility coincides with the main divergence among sets of results for tissue *P* and (*N*:*P*) ratio.

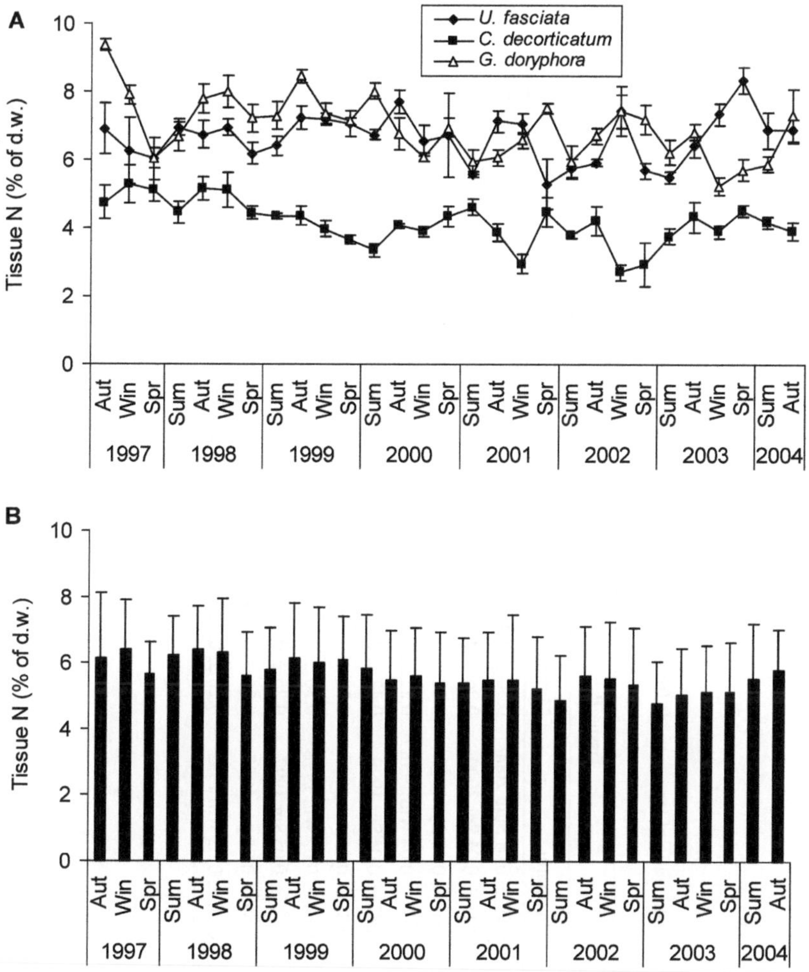

Figure 2. Seasonal fluctuations in the content of nitrogen in *Codium decorticatum*, *Grateloupia doryphora*, and *Ulva fasciata* (A), and mean values of N in the tissues of all macroalgae (B) collected in 29 seasonal samplings in Guanabara Bay. Data are expressed as percentage of the dry weight (d.w.). In (A) each point represents the mean of four to six replicates ± standard deviation ($4 \leq n \leq 6$). In (B) each bar represents the mean of 112 to 144 measurements ± standard deviation ($112 \leq n \leq 1.44$).

Discussion

Dissolved nutrients were detected at medium to high concentrations throughout this study and our data show that no seasonal enrichment of nutrients occurred at the sampling site (e.g. upwelling events, seasonal increase of the volume of sewage, etc.). The high values of ammonium and urea are consistent with the discharge of large volume of domestic sewage in the area (Lavrado et al., 1991; Paranhos et al., 1997). This interpretation is supported by studies by other authors, who confirm the large amount of domestic sewage released into the Bay (Paranhos et al., 1995; FEEMA, 1999). The seaweeds at the sampling site showed high contents of tissue nitrogen and phosphorus by comparison with other studies done in tropical environments (e.g. Fong et al., 2003; Hwang et al., 2004) and similar to results obtained with seaweeds growing in an excess of nutrients (Lapointe et al., 2004).

In the case of partially closed systems, the water turnover rate is comparatively low, which in turn leads to deteriorating water quality in response to even modest pollution loading. Guanabara Bay is connected to the coastal ocean via a 4-km mouth, and has a flushing half-life of 6.5 d, considerably longer than for many other coastal bays (Kjerfve et al., 1997). Inputs of pollution may lead to long-term cumulative effects given the slow water exchange in the Bay. However, the sampling site is close to the entrance of the Bay, where the water turnover is faster than in other parts of the

Table 2. Maximum and minimum mean values for tissue *N*, *P* and *N:P* atomic ratio of ten seaweeds from Guanabara Bay, in 29 seasonal sampling. Data represent percentage of the dry weight, except for (*N:P*) atomic ratio (no units).

Species	Tissue *N*		Tissue *P*		(*N:P*) ratio	
	Maximum	Minimum	Maximum	Minimum	Maximum	Minimum
B. radicans	9.04 ± 0.83 Win 2001	6.65 ± 0.18 Win 2003	0.79 ± 0.01 Sum 2000	0.30 ± 0.04 Spr 2003	49.7 ± 5.48 Spr 2003	21.9 ± 0.53 Sum 2000
C. anteninna	5.26 ± 0.34 Aut 1998	3.04 ± 0.28 Win 2001	0.48 ± 0.05 Spr 2000	0.26 ± 0.01 Sum 2001	27.7 ± 0.67 Win 2002	14.5 ± 1.77 Spr 2001
C. decorticatum	5.27 ± 0.56 Win 1997	2.71 ± 0.23 Win 2002	0.65 ± 0.07 Spr 1997	0.22 ± 0.01 Spr 2002	42.4 ± 7.02 Spr 2001	17.2 ± 1.19 Win 2000
C. rupestris	6.92 ± 0.80 Win 2001	2.42 ± 0.17 Win 2003	0.66 ± 0.04 Win 1998	0.22 ± 0.01 Aut 2003	37.6 ± 5.46 Aut 2001	15.0 ± 1.62 Aut 2002
C. teedii	5.62 ± 0.36 Win 1998	3.25 ± 0.20 Sum 2004	0.82 ± 0.04 Spr 1999	0.42 ± 0.02 Win 2001	25.2 ± 1.31 Spr 2003	13.6 ± 0.77 Spr 1998
E. flexuosa	7.75 ± 0.61 Win 1998	3.16 ± 0.30 Spr 2001	0.83 ± 0.05 Aut 1999	0.20 ± 0.02 Aut 2001	49.1 ± 6.78 Aut 2001	17.1 ± 1.17 Spr 2003
G. griffthisiae	7.08 ± 0.07 Spr 1999	3.65 ± 0.17 Spr 2003	0.64 ± 0.05 Win 1998	0.30 ± 0.02 Sum 2004	38.1 ± 4.38 Sum 2004	17.9 ± 1.36 Aut 2001
G. doryphora	9.35 ± 0.17 Aut 1997	5.23 ± 0.26 Win 2003	0.74 ± 0.05 Spr 1997	0.26 ± 0.04 Sum 2004	49.8 ± 5.05 Sum 2004	18.0 ± 1.75 Spr 1997
U. fasciata	8.33 ± 0.38 Spr 2003	5.29 ± 0.76 Spr 2001	0.64 ± 0.05 Aut 1997	0.35 ± 0.02 Aut 2002	44.7 ± 3.59 Aut 2001	23.4 ± 1.80 Spr 1998
U. lactuca	7.51 ± 0.38 Spr 2002	4.10 ± 0.32 Win 2001	0.76 ± 0.02 Win 1999	0.28 ± 0.01 Spr 2001	56.8 ± 2.49 Spr 2001	19.7 ± 1.02 Win 1999

Data represent the mean of four determinations ± standard deviation ($n = 4$).

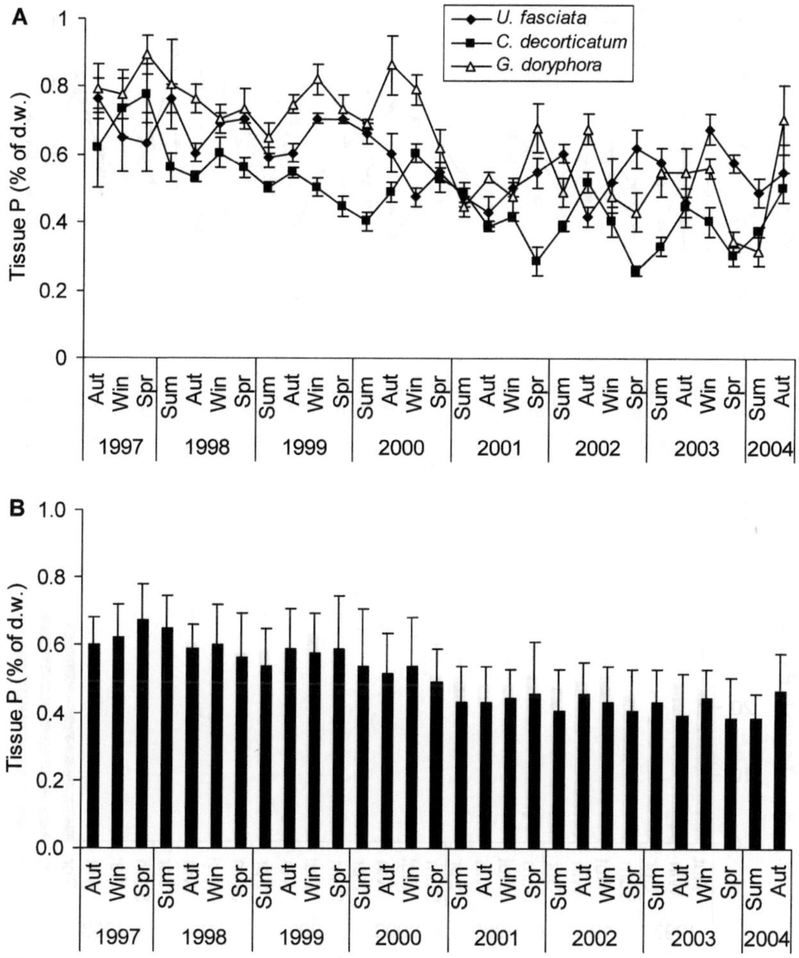

Figure 3. Seasonal fluctuations in the content of phosphorus in *Codium decorticatum*, *Grateloupia doryphora*, and *Ulva fasciata* (A), and mean values of *P* in the tissues of all macroalgae (B) collected in 29 seasonal samplings in Guanabara Bay. Data are expressed as percentage of the dry weight (d.w.). In (A) each point represents the mean of four to six replicates ± standard deviation ($4 \leq n \leq 6$). In (B) each bar represents the mean of 112 to 144 measurements ± standard deviation ($112 \leq n \leq 1.44$).

system. Increased flushing facilitates the replacement of the nutrient-rich polluted waters with nutrient-poor oceanic waters every tidal cycle (Paranhos et al., 2001). This means that the impact of sewage-derived pollutant at the sampling site is substantially lower that in other parts of the Bay. This is probably one of the reasons for the greater species diversity of seaweeds at the study site than in the inner parts of the Bay (Teixeira et al., 1987).

Phytoplankton of Guanabara Bay is dominated by small-sized species (cyanobacteria and nanoplanktonic species), which are strong competitors for nutrients and achieve high biomass (Valentin et al., 1999). Shading by phytoplankton and particulate matter in deeper waters and the lack of rocky substrates in shallow areas restrict the proliferation of seaweeds (Mayr et al., 1989). These factors make the macroalgal biomass in the Bay relatively small compared to other eutrophic systems.

According to the Björnsäter & Wheeler's (1990) classification of macroalgal nutrient status based on *N*:*P* ratio of tissues, a *N*:*P* ratio < 16 indicates *N*-limitation; a *N*:*P* ratio 16–24 indicates *N*-sufficiency and *P*-sufficiency – i.e. no limitation and *N*:*P* > 24 indicates *P*-limitation. Applying this classification to our data we conclude that the macroalgal community in the sampling site is permanently *N*-sufficient and almost permanently *P*-deficient, with few exceptions. However, the *N*:*P* ratio must be evaluated with care, as it may obscure trends for the individual elements. The

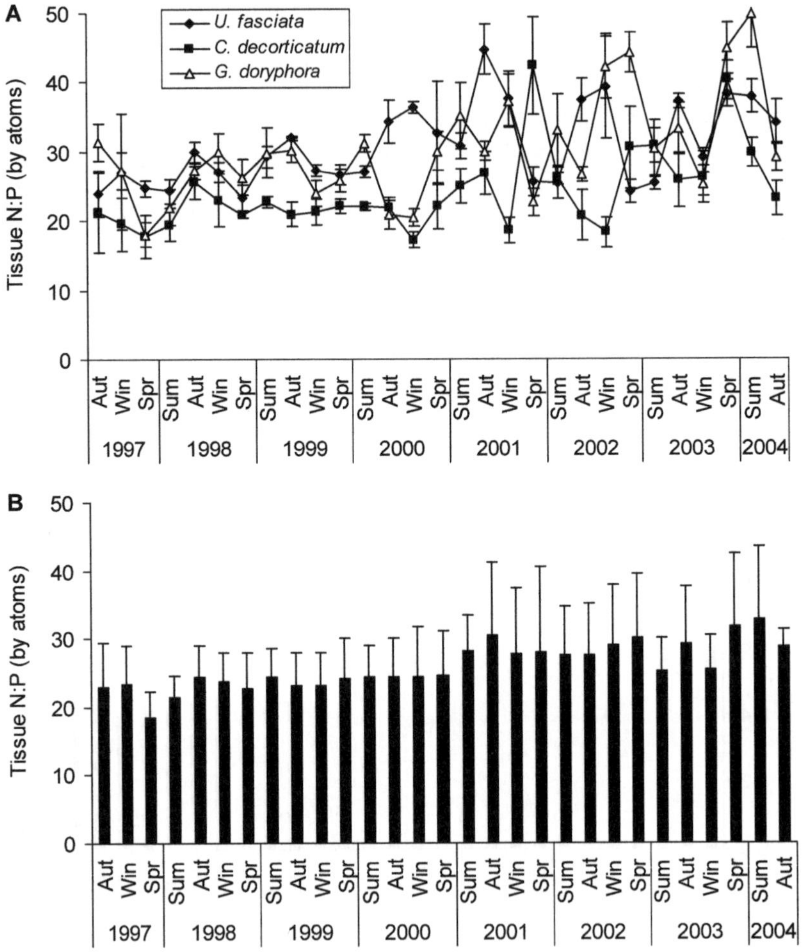

Figure 4. Seasonal fluctuations in tissue (*N:P*) ratio in *Codium decorticatum*, *Grateloupia doryphora*, and *Ulva fasciata* (A), and mean values of *N:P* in the tissues of all macroalgae (B) collected in 29 seasonal sampling in Guanabara Bay. In (A) each point represents the mean of four to six replicates ± standard deviation ($4 \leq n \leq 6$). In (B) each bar represents the mean of 112 to 144 measurements ± standard deviation.

overall mean values for tissue nitrogen and phosphorus in all algae collected in summer 2004 were 5.52 ± 1.69 and 0.38 ± 0.07% of d.w. ($n = 43$), respectively: the lowest mean value for phosphorus throughout this study. Howevere, 0.38% of tissue *P* does not represent a low level, and is actually higher than values found in many other algae from tropical environments (See Fong et al., 2003; Hwang et al., 2004). The high mean overall *N:P* ratio observed in Guanabara Bay in summer 2004 (32.7 ± 10.7, $n = 43$) is strongly affected by the high concentrations of nitrogen and is not necessarily indicative of *P* limitation. Thus, the classification of Björnsäter and Wheeler (1990) must be considered with caution, because the ranges may not be suitable for macroalgae from polluted tropical environments such as Guanabara Bay. In addition, further investigations are needed to test the suitability of that classification for tropical environments, where seaweeds typically grow well with low concentrations of dissolved nutrients and normally have lower tissue *N* and *P* compared to species from temperate environments. In this context the high *N:P* obtained for most of our measurements (typically >24:1) may not represent limitation of macroalgal growth by *P* at the study site.

A comparison of tissue nutrients and dissolved nutrients shows large differences in terms of *N:P* ratio, with higher values for the seaweeds. This apparent contradiction may be explained by the very limited usefulness of our data on dissolved nutrients. Monitoring dissolved nutrients is time-demanding and a reliable assessment needs a large data set, since many variables affect the results. In addition, our field samples were always collected at low tides, when local sources of pollution at the site would be concentrated in the waters. Despite the construction of the Icaraí Submarine Sewage Outfall many sources of domestic sewage still exist and release their contents directly into the Bay. This still happens closer than 100 m to the sampling site. For this reason, our measurements of dissolved nutrients do not represent exactly the actual dynamics of nutrients at the sampling site.

The design of this descriptive study does not allow us to identify clearly the effects of the local sewage outfall since 1999. Our measurements of dissolved nutrients at the sampling site show no obvious pattern over the 7-year assessment (Table 1). However, other studies at different sites in the Bay have shown a significant decrease in the concentrations of dissolved nutrients following the construction of the Icaraí Submarine Sewage Outfall (unpublished data). If this is occurred at our site, it could explain the significant decrease of tissue *P* and increase of tissue *N:P* ratio of the species during this study.

In a related study, Lourenço et al. (2005) studied the seasonal variations of tissue *N* and *P* of eight macroalgal species of Araruama Lagoon, a hypersaline environment of Rio de Janeiro State. Remarkable seasonal variations in tissue nutrients for the seaweeds were found, with higher values in autumn and lower in spring for most of the species. The authors also considered that seaweeds are drastically affected by high temperatures in part of the spring and in the summer. Thermal damage could lead to the loss of tissue and nutrients to the environment (Hanisak, 1993; Menéndez et al., 2001). Typical summer temperatures in Guanabara Bay (>25 °C, data not shown) could potentially result in tissue loss. However, the available data do not support such an interpretation, because high values for tissue *N* and *P* were found during the warmest periods in spring and summer in some sampling. No seasonal variations for tissue nutrients were found, probably a consequence of permanent high concentrations of dissolved nutrients available to the species.

Levels of tissue *N* detected in the seaweeds suggest that the species are permanently saturated with nitrogen, even in periods when lower percentages of tissue *N* were measured. According to Hanisak (1979), maximum growth rate for the green alga *Codium fragile* can be achieved if the species has 2% of tissue nitrogen. Among the ten species studied here, the lowest tissue *N* value was 2.42%, measured in *Cladophora rupestris*, which suggests that the seaweeds did not experience limitation of growth by *N* at the site. The excess of nitrogen available for the seaweeds could stimulate most of them to a luxury consumption of nitrogen, generating high concentrations of tissue N, as demonstrated by Gordon *et al.* (1981) for *Cladophora* in cultures. Those authors determined that the critical tissue *N* and *P* concentrations for growth of *Cladophora* are 2.1% and 0.33%, respectively. Luxury consumption of nitrogen is more pronounced than the consumption of excess phosphorus (Gordon *et al.*, 1981), and this possible trend could account for the high *N:P* ratio seen in this study for most of the measurements done. This argument also points to the absence of phosphorus limitation in the site, especially because the values measured in the tissues were not low: tissue *P* was >0.40% for ca. 85% of all measurements. Our data indicate that seaweeds would not be limited by dissolved *N* and *P* in the site, and possible increments in algal biomass would be controlled by other factors such as herbivory (Lotze & Schramm, 2000) or lack of suitable substrate (Bokn et al., 2003).

In conclusion, we confirm that tissue *N* and *P* of the macroalgal species do not show any seasonal variation. In addition, tissue *N* and *P* of the species tested show high concentrations in most of the observations. However, our hypothesis regarding interannual consistency in tissue *P* and (*N:P*) ratio is rejected, as these have been decreasing over recent years.

We are currently evaluating the tissue *N* and *P* composition of many other seaweed species of Brazil, from coastal oligotrophic environments. These results will hopefully contribute towards a better understanding of the nutrient metabolism of tropical seaweeds.

Acknowledgments

We are indebted to FAPERJ (Foundation for Research Support of Rio de Janeiro State) and National Council for the Development of Science and Technology (CNPq) for the financial support of this study. Thanks are due to Diretoria de Hidrografia e Navegação (Marinha do Brasil) for supplying us with the 7-year data of temperature in Guanabara Bay. A special acknowledgement is due to Dr Yocie Yoneshigue-Valentin (Universidade Federal do Rio de Janeiro) for offering us laboratory facilities, field support for performing this study, and for the identification of part of the samples. The authors are grateful to Dr Lisia M.S. Gestinari and MSc Joel C. De-Paula for their assistance in the identification of the species. S.O.L. acknowledges CNPq and FAPERJ that provided him research fellowships.

References

Björnsäter BR, Wheeler PA (1990) Effect of nitrogen and phosphorus supply on growth and tissue composition of *Ulva fenestrata* and *Enteromorpha intestinalis* (Ulvales, Chlorophyta). J. Phycol. 26: 603–611.

Bokn TL, Duarte CM, Pedersen MF, Marba N, Moy FE, Barron C, Bjerkeng B, Borum J, Christie H, Engelbert S, Fotel FL, Hoell EE, Karez R, Kersting K, Kraufvelin P, Lindblad C, Olsen M, Sanderud KA, Sommer U, Sorensen K (2003) The response of experimental rocky shore communities to nutrient additions. Ecosystems 6: 577–594.

CIBG (Centro de Informações da Baía de Guanabara), (2004) Available at http//:www.cibg.rj.gov.br. Accessed in 28 Oct 2004.

Clark RB (2001) Marine Pollution. 5th ed. Oxford University Press, Oxford, 237 pp.

FEEMA (Fundação Estadual de Engenharia do Meio Ambiente) (1999) Qualidade de água da baía de Guanabara 1990/1997. Special Report. FEEMA, Rio de Janeiro, 29 pp.

Fong P, Kamer K, Boyer KE, Boyle KA (2001) Nutrient content of macroalgae with different morphologies may indicate sources of nutrients for tropical marine systems. Mar. Ecol. Prog. Ser. 220: 137–52.

Fong P, Boyer KE, Kamer K, Boyle KA (2003) Influence of initial tissue nutrient status of tropical marine algae on response to nitrogen and phosphorus additions. Mar. Ecol. Prog. Ser. 262: 111–123.

Fong P, Foin TC, Zedler JB, Donohoe RM (1994) A simulation model of lagoon algal communities based on competition for nitrogen and internal storage. Ecol. Monogr. 64: 225–247.

Fujita RM (1985) The role of nitrogen status in regulation transient ammonium uptake and nitrogen storage by macroalgae. J. Exp. Mar. Biol. Ecol. 92: 283–301.

Gordon DM, Birch PB, McComb AJ (1983) Effects of inorganic phosphorus and nitrogen on the growth of an estuarine *Cladophora* in culture. Bot. Mar. 24: 93–106.

Grasshoff K, Ehrhardt M, Kremling K (1983) Methods of Seawater analysis. 2nd ed. Verlag Chemie, Weinheim, 419 pp.

Hach CC, Bowden BK, Kopelove AB, Brayton ST (1987) More powerful peroxide Kjeldahl digestion method. JAOAC 70: 783–787.

Hanisak MD (1979) Nitrogen limitation of *Codium fragile* ssp. *tomentosoides* as determined by tissue analysis. Mar. Biol. 50: 333–337.

Hanisak MD (1993) Nitrogen release from decomposing seaweeds: Species and temperature effects. J. Appl. Phycol. 5: 175–181.

Kjerfve B, Ribeiro CHA, Dias GTM, Filippo AM, Quaresma VS (1997) Oceanographic characteristics of an impacted coastal bay: Baía de Guanabara, Rio de Janeiro, Brazil. Cont. Shelf Res. 17: 1609–1643.

Lapointe BE, Barile PJ, Matzie WR (2004) Anthropogenic nutrient enrichment of seagrass and coral reef communities in the Lower Florida Keys: Discrimination of local versus regional nitrogen sources. J. Exp. Mar. Biol. Ecol. 308: 23–58.

Lavrado HP, Mayr LM, Carvalho V, Paranhos R (1991) Evolution (1980–1990) of ammonia and dissolved oxygen in Guanabara Bay, RJ, Brazil. Coastal Zone 7: 3234–3245.

Lourenço SO, Barbarino E, Nascimento A, Paranhos R (2005) Seasonal variations in tissue nitrogen and phosphorus of eight macroalgae from a tropical hypersaline coastal environment. Cryptogam. Algol. 26(4): 355–371.

Mayr LM, Tenenbaun DR, Villac MC, Paranhos R, Nogueira CR, Bonecker SLC, Bonecker AC (1989) Hydrological characterization of Guanabara Bay. In Magoo OT, Neves C (eds), Coastlines of Brazil. American Society of Civil Engineering, New York: pp. 129–139.

Menéndez M, Martínez M, Comín FA (2001) A comparative study of the effect of pH and inorganic carbon resources on the photosynthesis of three floating macroalgae species of a Mediterranean coastal lagoon. J. Exp. Mar. Biol. Ecol. 256: 123–136.

Paranhos R (1997) Water pollution trends in Guanabara Bay, Brazil. Proceedings of the Stockolm Water Symposium, Stockholm, pp. 205–212.

Paranhos R, Andrade L, Mendonça-Hagler LC, Pfeiffer WC (2001) Coupling bacterial abundance with production in a polluted tropical coastal bay. Oecologia Brasiliensis 9: 117–132.

Paranhos R, Nascimento SM, Mayr LM (1995) On the faecal pollution in Guanabara Bay, Brazil. Fresen. Env. Bull. 4: 352–357.

Parsons TR, Maita Y, Lalli CM (1984) A manual of chemical and biological methods for seawater analysis. Pergamon Press, Oxford, 173 pp.

Peckol P, Demeo-Andersen B, Rivers J, Valiela I, Maldonado M, Yates J (1994) Growth, nutrient uptake capacities and tissue constituents of the macroalgae *Cladophora vagabunda* and *Gracilaria tikvahiae* related to site-specific nitrogen loading rates. Mar. Biol. 121: 175–185.

Rivers JS, Peckol P (1995) Interactive effects of nitrogen and dissolved inorganic carbon on photosynthesis, growth, and ammonium uptake of the macroalgae *Cladophora vagabunda* and *Gracilaria tikvahiae*. Mar. Biol. 121: 747–753.

Schaffelke B (1999) Short-term nutrient pulses as tools to assess responses of coral reef macroalgae to enhanced nutrient availability. Mar. Ecol. Prog. Ser. 182: 305–310.

Taouil A, Yoneshigue-Valentin Y (2002) Alterações na composição florística das algas da Praia de Boa Viagem (Niterói, RJ). Rev. Bras. Bot. 25: 405–412.

Teixeira VL, Pereira RC, Marques Junior AN, Leitão Filho CM, Silva CAR (1987) Seasonal variations in infralitoral seaweed communities under a pollution gradient in baía de Guanabara, Rio de Janeiro (Brazil). Ciência e Cultura 39: 423–428.

Valentin JL, Tenenbaun DR, Bonecker ACT, Bonecker SLC, Nogueira CR, Villac MC (1999) O sistema planctônico da Baía de Guanabara: Síntese do conhecimento. Oecol. Bras. 7: 35–59.

Valiela I, Mcclelland J, Hauxwell J, Behr PJ, Foreman K (1997) Macroalgal blooms in shallow estuaries: Controls and ecophysiological and ecosystem consequences. Limnol. Oceanogr. 42: 1105–18.

Villares R, Carballeira A (2003) Seasonal variation in the concentrations of nutrients in two green macroalgae and nutrient levels in sediments in the Rias Baixas (NW Spain). Estuar. Coast. Shelf Sci. 58: 887–900.

Wheeler PA, Björnsäter BR (1992) Seasonal fluctuations in tissue nitrogen, phosphorus, and *N:P* for five macroalgal species common to the Pacific northwest coast. J. Phycol. 28: 1–6.

Wynne MJ (1998) A checklist of benthic marine algae of the tropical and subtropical western Atlantic: First revision. Nova Hedwigia 116: 1–155.

Zar JH (1996) Biostatistical analysis. Prentice Hall, NJ, pp. 920.

Journal of Applied Phycology (2006) 18: 399–408
DOI: 10.1007/s10811-006-9040-z

Element concentrations in some species of seaweeds from La Paz Bay and La Paz Lagoon, south-western Baja California, Mexico

Ana P. Rodríguez-Castañeda[1,*], Ignacio Sánchez-Rodríguez[1], Evgueni N. Shumilin[1] & Dmitry Sapozhnikov[2]
[1]*Centro Interdisciplinario de Ciencias Marinas (CICIMAR-IPN), Av. IPN s/n Col. Playa Palo de Santa Rita, Apdo Postal 592, La Paz, Baja California Sur, México. Becarios COFAA-IPN, EDI-IPN;* [2]*Vernadsky Institute of Geochemistry and Analytical Chemistry, Russian Academy of Sciences, Moscow, Russia*

*Author for correspondence: e-mail:aprodri@ipn.mx

Key words: elements, trace elements, seaweeds, instrumental neutron activation analysis, La Paz Bay, Baja California Peninsula

Abstract

La Paz Bay and La Paz Lagoon are water bodies of the Gulf of California that are influenced by waste water discharges from the City of La Paz and from activities of the phosphorite mining company "Rofomex". Because seaweeds concentrate elements from the water and are used as effective indicators of contamination by metals, we investigated their usefulness in this region. Concentrations of certain major elements (Ca, Fe, K and Na) and trace elements (As, Ba, Co, Cr, Cs, Hf, Rb, Sb, Sc, Se, Sr, Ta, Th, U, Zn and Zr) were determined in 12 species of seaweeds from La Paz Bay and La Paz Lagoon using instrumental neutron activation analysis. The contents of trace elements of environmental importance (As, Co, Cr, Fe, Sb, Se and Zn) in all studied samples are within the range of typical levels for a pristine environment not subjected to anthropogenic impact. Somewhat higher concentrations of Cr (81 mg kg^{-1}), Hf (4 mg kg^{-1}), Rb (48 mg kg^{-1}), Sc (6.3 mg kg^{-1}), Ta (0.95 mg kg^{-1}), Th (6.8 mg kg^{-1}), U (33 mg kg^{-1}) and Zn (90 mg kg^{-1}) were found in the green seaweed species *Ulva* (formerly *Enteromorpha) intestinalis*, whereas such elements as As (77 mg kg^{-1}), Sb (1.4 mg kg^{-1}) and Se (1.8 mg kg^{-1}) were mainly concentrated in the species *Sargassum sinicola*, *Codium cuneatum* and *Padina mexicana* respectively. Because of their higher abundance and heterogeneity in elemental composition the seaweeds species *Ulva intestinalis* and *Caulerpa sertularioides* seem to be more suitable for further biomonitoring of heavy metal pollution of the coastal waters in this zone.

Introduction

The development of marine ecosystems is strongly controlled by the biogeochemical cycles of chemical elements, which depend on their interactions with each other and the geological, climatic, physical, chemical and biological processes that occur in the water column and on the interfaces with sediments and the atmosphere (Chester, 2003; De La Lanza & Cáceres, 1994). In coastal marine areas, it is important to understand the biogeochemical cycles of both the major and trace elements because of the possibility of the changes which may occur in them as a result of either natural or man-made alterations to the environment.

Bays, being partly enclosed water bodies, often show strong variations in terms of sediments and chemical composition of the water column. In particular, coastal marine sediments are usually made up of both terrigenous and marine biogenic materials, and their composition can vary depending on hydrodynamic and climatic conditions, the type and strength of material inputs, distance from source and the extent of dilution of natural terrigenous or anthropogenic components, usually enriched in many elements, by silica and carbonates of marine biogenic origin (Chester, 2003).

Some elements are mainly present in the dissolved fraction (ions and molecules) in the sea water, whereas others are incorporated into either colloidal or particulate matter. Particulate and/or dissolved material may interact with each other and alter the biogeochemical fate of the elements in the marine environment. During early diagenesis elements stored in the marine sediments because of the changes in the sediment's pH and redox potential Eh can be released into the interstitial and overlying sea water by dissolution, desorption or autolytic biological processes (Chester, 2003). Any disturbance caused by either physical factors (e.g. currents) or biological factors (e.g. movement of organisms) will stimulate exchanges between the elements in the sediments and those in the seawater.

A variety of organisms, such as seaweeds, can also transfer and accumulate trace elements in the sea (Kennish, 1997). Seaweeds take up metal elements from the aquatic environment, depending on species, exposure time, type of metal and its oxidation state, pH, salinity and presence of organic pollutants (Bernhard & Zattera, 1975; Hassett et al., 1980; Jensen et al., 1976; Myklestad et al., 1978; Phillips, 1977). Contamination of the seaweed surface from simple contact with the elements dissolved in sea water has been observed in both unicellular and pluricellular algae, while metal ions, some of which are essential elements, are also taken up by algae through pores in their cell walls. Consequently, the cell components as well as the composition and structure of the cell walls are important factors in determining the ability of a seaweed species to absorb metals (Kuyucak & Volesky, 1990). For example, in brown seaweeds, the alginates of the cell walls and of the intracellular spaces regulate the exchange of ions, showing an affinity for metals in the following decreasing order: Pb > Cu > Cd > Ba > Sr > Ca > Co > Ni > Zn > Mn > Mg.

Many studies of contaminants and their effects on marine macroalgae have been published since the beginning of the 1960's (see Lobban & Harrison, 1994). Other data have shown that seaweeds can absorb metals such as Pb and Sr (Eide et al., 1980). For example, Ho (1990) found that the seaweed *Ulva lactuca* is an important bioindicator of Cu, Zn and Pb present in sea water. Similar studies were recently done for the coastal zone of Mexico. For example, Robledo & Freile Pelegrín (1997) reported the chemical composition of six species of edible macroalgae from the Yucatan region. Closer to La Paz Bay, Sánchez-Rodríguez et al. (2001) reported the concentrations of elements in various seaweeds from the almost pristine Bay of Loreto, in the central Baja California peninsula.

La Paz Bay and its smaller component, La Paz Lagoon, are particularly interesting for environmental studies because of their proximity to the City of La Paz, the oil reservoirs of Petroleos Mexicanos (PEMEX), the electrical plant owned by the Compañía Federal de Electriciadad, and the activities of the mining company "Roca Fosfórica Mexicana, S.A. de C.V." ("ROFOMEX"), located near the San Juan de la Costa in the western coast of the La Paz Bay. Because of the dry and arid climate of the region, terrigenous material is carried into La Paz Bay and La Paz Lagoon mainly by wind or with episodic discharges of the ephemeral water streams ("arroyos") only after rare but heavy rains. The characteristics of these inputs into the coastal marine environment are largely determined by the different types of rocks (sedimentary rocks, igneous rocks and alluvium, a product of the weathering of the rocks of San Gregorio and San Isidro Formation, and the Comondú geological formation) in the surrounding areas (Figure 1, Hausback, 1984).

The high productivity of coastal waters of La Paz Bay and La Paz Lagoon makes it more interesting to determine the concentrations of elements in the seaweeds, because some of them are edible or could be used as food additives for domesestic animals. It is also necessary to determine which seaweed species are most suitable for future biomonitoring of heavy metal pollution in these areas.

Taking all of this into consideration, as well as the need to increase the use of the region's natural resources in a controlled manner, the present study aimed to determine the concentrations of major and trace elements in some species of seaweeds that occur in La Paz Bay and La Paz Lagoon, and to select species suitable for further biomonitoring of heavy metal pollution of these areas.

Materials and methods

In August 1998, 35 samples of seaweeds were taken from a boat by scuba diving at 19 different locations in La Paz Bay, between Punta Tarabillas and the Espiritu Santo Island (Figure 1, Table 1). The seaweeds were collected by hand directly from the substrate and put into identified plastic bags for later analysis. In the laboratory, they were washed with tap water to get rid of any residues such as sand or shells, and then sorted according to sampling station and species. The seaweeds were identified using taxonomic keys (Abbott & Hollenberg, 1976; Norris, 1975; Silva et al., 1996; Taylor, 1945). They were then left to dry at room temperature, on absorbent paper. Once completely dried, each sample was crushed, sieved and

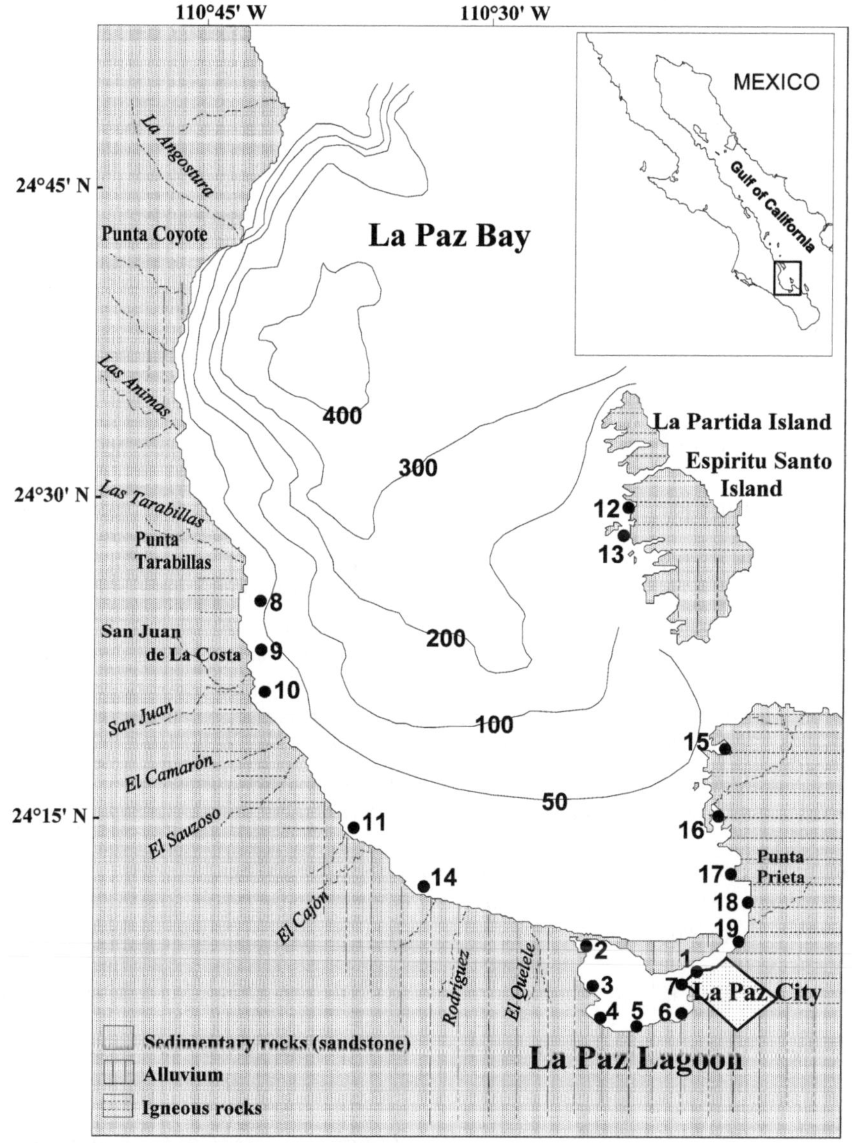

Figure 1. The study area and location of the stations of the seaweed sampling in the La Paz Bay and La Paz Lagoon.

stored for further chemical analysis. Element contents in sub-samples of seaweeds were determined at the V.I. Vernadsky Institute of Geochemistry and Analytical Chemistry of the Russian Academy of Sciences, using instrumental neutron activation analysis (INAA). A 100 mg fraction of each sample was taken, then folded in aluminium paper and irradiated jointly with standard reference materials SRM 1646a "Estuarine sediment", IAEA-356 "Polluted marine sediment" and SD-N-1/2 "Contaminated marine sediment" (IAEA) by thermal neutrons (2.8×10^{13} n $s^{-1}cm^{-2}$). The induced radioactivity of each sample was measured using a semi-conductor gamma ray spectrometer, supplied with a high resolution Ge(Li) detector coupled with a 4096 channel "Nokia" analyzer. Gamma ray spectrometric standard reference sources were used to calibrate the instrument (^{152}Eu) (Sánchez-Rodríguez et al., 2001; Shumilin et al., 2000). In this manner, the content of the following elements in the seaweed dry tissues was determined: As, Ba, Br, Ca, Co, Cr, Cs, Fe, Hf, K, Na, Rb, Sb, Sc, Se, Sr, Ta, Th, U, Zn, and Zr.

Table 1. The study area, location of the stations and species of seaweeds collected in the La Paz Bay and La Paz Lagoon in August 1998.

	Coordinate of station		
Station	Latitude, °N	Longitude, °W	Specie
1	24°08′04″	110°21′00″	*Enteromorpha intestinalis* (Linnaeus) Link; *Caulerpa sertularioides* (S.G. Gmelin) Howe
2	24°10′18″	110°25′81″	*Enteromorpha intestinalis* (Linnaeus) Link; *Caulerpa sertularioides* (S.G. Gmelin) Howe
3	24°08′25″	110°25′36″	*Enteromorpha intestinalis* (Linnaeus) Link; *Caulerpa sertularioides* (S.G. Gmelin) Howe
4	24°06′76″	110°25′08″	*Caulerpa sertularioides* (S.G. Gmelin) Howe
5	24°06′67″	110°23′16″	*Caulerpa sertularioides* (S.G. Gmelin) Howe; *Halimeda discoidea* Decaisne *Spyridia filamentosa* (Wulfen) Harvey
6	24°07′05″	110°21′34″	*Enteromorpha intestinalis* (Linnaeus) Link
7	24°08′49″	110°21′28″	*Enteromorpha intestinalis* (Linnaeus) Link; *Dictyota divaricata* Lamouroux; *Padina mexicana* Dawson1944
8	24°25′08″	110°40′00″	*Enteromorpha intestinalis* (Linnaeus) Link; *Dictyota divaricata* Lamouroux; *Sargassum sinicola* Setchel & Gardner *Amphiroa beauvoisii* Lamourox
9	24°23′64″	110°40′00″	*Enteromorpha intestinalis* (Linnaeus) Link; *Codium cuneatum* Setchell & Gardner; *Sargassum sinicola* Setchel & Gardner
10	24°21′69″	110°40′84″	*Gelidiopsis tenuis* Setchell & Gardner
11	24°15′52″	110°36′76″	*Dictyota divaricata* Lamouroux; *Padina mexicana* Dawson; *Sargassum sinicola* Setchel & Gardner
12	24°30′10″	110°23′08″	*Gelidiopsis tenuis* Setchell & Gardner
13	24°29′23″	110°23′46″	*Gelidiopsis tenuis* Setchell & Gardner
14	24°12′06″	110°33′28″	*Codium cuneatum* Setchell & Gardner; *Gelidiopsis tenuis* Setchell & Gardner
15	24°19′11″	110°19′27″	*Caulerpa sertularioides* (S.G. Gmelin) Howe; *Spyridia filamentosa* (Wulfen) Harvey
16	24°16′23″	110°19′78″	*Caulerpa sertularioides* (S.G. Gmelin) Howe
17	24°13′27″	110°18′73″	*Spyridia filamentosa* (Wulfen) Harvey; *Digenia simplex* (Wulfen) C. Agardh
18	24°12′16″	110°18′04″	*Galaxaura oblongata* (Ellis & Solander) Lamouroux; *Amphiroa beauvoisii* Lamourox
19	24°10′47″	110°18′45″	*Caulerpa sertularioides* (S.G. Gmelin) Howe; *Halimeda discoidea* Decaisne; *Padina mexicana* Dawson

Results

Seaweed species

The location of sampling stations where seaweeds were collected is shown in Table 1. Twelve species were found, belonging to all three divisions (Chlorophyta, Phaeophyta and Rhodophyta). The greatest variety of species was found at Stations 8 and 11, while Stations 3, 6, 13 and 16 were poorest in species abundance, with only one species present. The green macroalgae *Enteromorpha intestinalis* and *Caulerpa sertularioides* were the most widespread during the observation period, being found in 7 stations.

Major elements (Ca, Fe, K and Na)

The concentrations of the major elements in the samples vary depending on algal species and sampling location (Table 2, Figures 1–3). Ca was present at high levels in *Halimeda discoidea* (30.2%) collected at the Station 5 in the southern part of the bay in the La Paz Lagoon area. Two species were found to have a high concentration of Fe: *Padina mexicana*, in samples taken at the south-eastern end of the bay at the Station 11 located in front of the El Cajón Arroyo (1.6%), and *Enteromorpha intestinalis*, in samples taken from a Station 8 located south of San Juan de La Costa (1.4%). The highest accumulation of K was found in samples of *Enteromorpha intestinalis* which were collected in

Table 2. Concentrations of the elements in the samples of macroalgae collected in the La Paz Bay and the Lagoon of La Paz.

						Content of elements (mg kg^{-1})																
Specie	Station	Na (%)	K (%)	Ca (%)	Fe (%)	Rb	Cs	Th	Sr	Ba	Sc	U	Zr	Hf	Ta	Cr	Co	Zn	As	Sb	Se	Br
E. intestinalis	1	0.8	0.1	5.5	1.2	33	3.3	2.2	650	245	4.8	7.5	50	1.2	0.2	15	3.5	50	16	0.5	0.7	4.3
	2	0.7	–	5.0	0.7	18	2.2	1.3	670	150	2.7	1.6	50	0.6	0.1	10	2.9	40	9	0.6	0.4	10.6
	3	0.3	–	2.5	0.9	15	2.3	1.3	395	23	3.7	5.0	50	1.5	0.1	13	4.7	70	3	1.0	1.1	2.7
	6	0.5	–	2.8	1.2	18	2.8	1.3	185	165	3.6	6.2	48	1.2	–	11	4.2	90	3	0.4	0.6	1.7
	7	0.3	–	2.7	0.8	15	2.5	2.2	210	87	3.8	1.0	–	0.8	0.3	11	4.2	50	6	0.4	0.5	1.3
	8	1.0	3.1	8.5	1.4	22	2.4	3.2	645	390	5.1	0.9	230	4.0	0.3	57	3.9	50	13	0.2	0.5	3.1
	9	1.9	–	12.2	1.3	49	3.5	6.9	700	425	6.3	33.3	35	2.7	0.9	81	3.4	50	8	0.3	1.4	2.0
C. cuneatum	9	3.1	0.8	2.2	0.4	6	0.7	0.7	200	140	1.4	0.9	35	0.7	–	15	2.3	30	31	1.4	0.7	5.5
	14	15.2	0.9	4.0	0.1	–	0.1	0.1	630	43	0.2	12.6	4	0.1	–	2	0.4	10	48	0.3	0.4	6.7
C. sertularioides	1	0.4	0.1	2.3	0.4	3	1.1	1.0	215	90	2.0	3.1	60	1.2	0.2	24	1.7	30	21	0.4	1.2	1.2
	2	0.4	0.1	3.0	0.2	2	0.7	0.4	255	50	0.9	6.6	45	0.2	0.2	5	1.1	20	13	0.7	0.7	1.9
	3	0.3	0.1	1.0	0.3	9	1.4	0.8	145	–	1.5	2.8	–	0.4	0.1	4	2.4	40	13	0.4	0.7	1.1
	4	0.3	–	1.0	0.3	4	1.4	0.8	140	55	1.4	3.0	–	0.4	0.1	5	2.1	30	6	0.2	0.5	1.3
	5	0.4	–	2.5	0.6	4	1.4	0.9	360	66	2.6	5.4	–	0.5	–	8	3.5	50	5	0.1	0.2	2.8
	15	0.9	–	3.7	0.3	1	0.3	0.4	420	40	1.0	4.0	34	0.2	–	4	1.2	20	16	0.5	0.8	2.4
	16	3.5	–	6.8	0.4	11	0.6	0.9	910	65	1.5	10.3	38	0.6	0.1	5	1.3	40	11	0.6	0.4	5.7
	19	0.8	–	5.7	0.4	12	1.2	0.8	1160	82	1.5	0.3	76	0.3	–	6	1.4	50	20	1.0	1.0	1.3
H. discoidea	5	1.0	0.3	30.2	0.6	2	1.3	0.9	7245	105	2.5	4.3	10	1.0	–	14	3.0	40	18	0.5	1.2	6.3
	19	2.1	–	26.2	0.2	2	1.2	0.6	6485	95	0.9	1.8	38	0.2	–	5	1.4	20	9	0.7	0.4	7.2
D. divaricata	8	0.7	–	5.1	0.6	15	0.9	2.2	1230	280	2.3	6.0	80	1.1	0.1	23	2.0	40	33	0.2	0.4	2.4
	11	0.7	0.4	6.0	0.8	20	0.8	2.2	1175	330	4.2	6.5	90	1.4	0.5	19	3.2	30	28	0.4	1.4	1.6
P. mexicana	11	1.0	–	14.7	1.6	25	1.6	4.1	3275	515	5.9	4.8	69	3.2	0.5	49	5.5	50	20	0.1	1.8	1.7
	18	0.1	0.1	22.4	1.0	13	2.1	1.7	4755	240	3.8	4.3	43	0.9	–	14	3.5	40	17	0.7	0.7	1.4
	19	1.0	0.2	20.4	0.6	16	2.0	1.2	4345	220	2.4	2.1	29	0.5	–	9	2.0	40	18	0.4	0.3	1.4
S. sinicola	8	0.7	0.4	3.3	0.1	8	0.7	0.2	1735	73	0.5	7.2	15	0.8	–	4	1.4	30	77	0.2	1.6	4.9
	9	0.6	–	4.1	0.2	8	0.4	0.5	1775	74	0.6	1.4	45	0.2	–	8	1.2	30	55	–	1.2	7.1
	11	0.5	–	5.0	0.1	10	0.3	0.3	2355	130	0.5	1.5	20	0.1	–	5	1.1	20	45	0.5	0.3	6.6
G. oblongata	17	0.7	0.7	21.8	0.3	8	0.8	0.5	5750	81	0.9	10.20	28	0.2	–	5	2.1	40	15	0.2	1.0	9.4
A. beauvoisii	8	0.4	–	26.5	0.1	6	0.5	0.3	6620	72	0.3	0.2	10	0.1	–	3	0.2	10	8	0.5	0.2	1.6
G. tenuis	10	0.6	–	2.9	0.1	6	0.4	4	335	120	0.3	4.3	–	0.3	–	4	1.4	20	3	0.3	0.9	5.5
	12	2.4	0.4	3.4	0.2	4	0.5	0.6	380	10	0.8	4.5	39	0.2	0.3	8	1.9	30	13	0.5	0.7	9.5
	13	2.3	0.3	3.5	0.2	4	0.4	0.4	320	25	0.7	4.2	30	0.1	–	3	1.8	20	12	0.7	0.9	10.2
	14	2.3	0.2	5.2	0.1	1	0.3	0.3	715	130	0.2	5.8	32	0.1	–	2	0.7	30	14	1.0	1.6	12.4
S. filamentosa	5	0.5	–	2.4	1.2	17	4.0	1.8	225	80	5.1	6.3	65	1.1	–	12	7.2	60	6	0.2	1.2	3.7
	16	0.6	–	7.7	0.5	3	1.2	0.8	990	63	1.9	2.5	45	0.5	0.1	7	1.6	20	3	0.3	0.5	4.2
	17	0.7	–	8.1	0.6	11	1.7	1.5	1100	205	2.2	3.3	22	1.5	0.2	10	2.3	80	3	0.4	0.7	6.2
D. simplex	17	1.2	1.4	9.2	0.1	33	2.7	1.7	1150	110	2.8	11.5	10	1.1	0.3	14	4.0	90	28	0.5	0.5	7.1

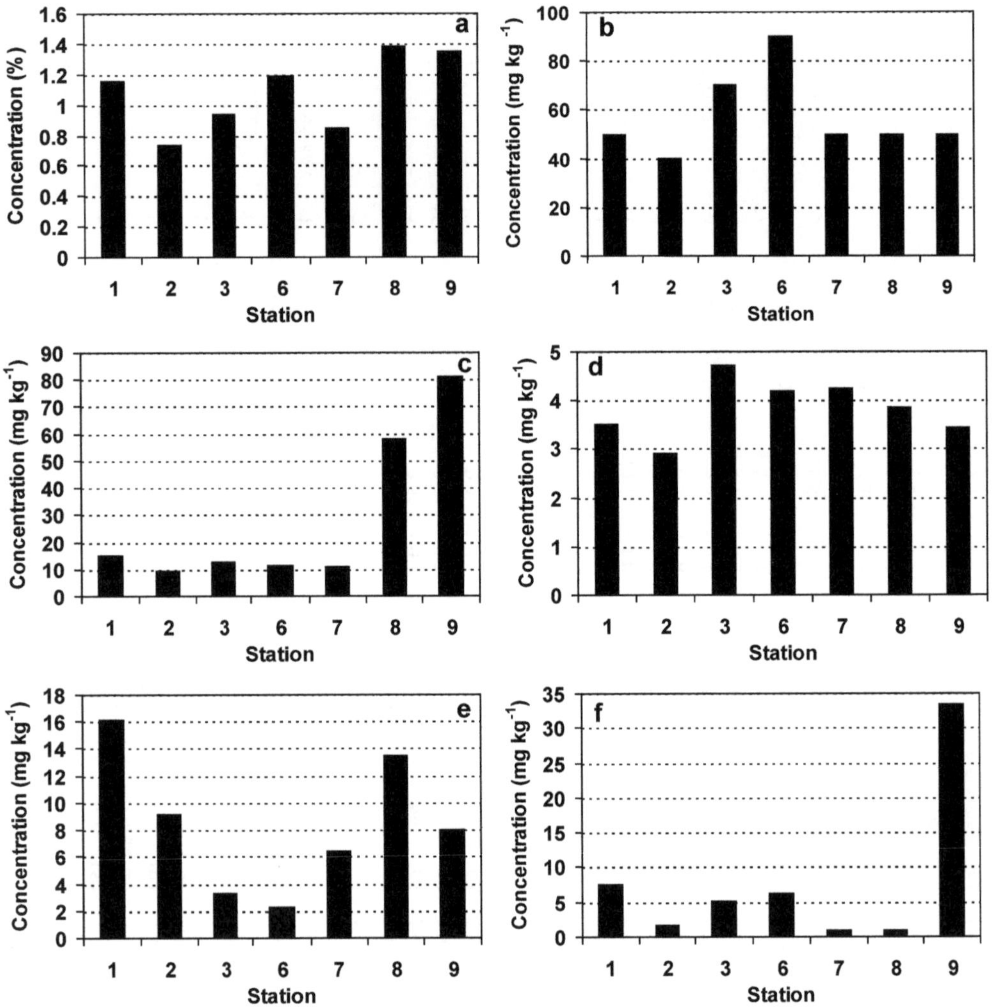

Figure 2. The spatial distribution of the concentration of selected elements in the samples of the seaweed *Enteromorpha intestinalis* from the La Paz Bay and La Paz Lagoon: (a) iron; (b) zinc; (c) chromium; (d) cobalt; (e) arsenic and (f) uranium.

La Paz Bay at Station 8. The highest concentration of Na (15.2%) was detected in *Codium cuneatum* which was collected at Station 14 in the southern part of the La Paz Bay (Figure 1, Table 2).

Trace elements (As, Ba, Rb, Co, Cr, Cs, Hf, Sc, Ta, Sb, Se, Sr, Th, U, Zn and Zr)

Strontium had the same tendency to accumulate as Ca, displaying the highest concentration (7245 mg kg^{-1}) in *Halimeda discoidea* from Station 5 (Table 2). In general, the highest concentrations of most of the trace elements (Cr, Hf, Rb, Sc, Se, Ta, Th, U and Zr) were detected in *Enteromorpha intestinalis* collected at Stations 8 and 9 near San Juan de la Costa on the western side of La Paz Bay. The species *Enteromorpha intestinalis* from Station 9, in La Paz Bay in front of the mining area of San Juan de la Costa, displays the highest concentrations of uranium (33 mg kg^{-1}) and of chromium (80 mg kg^{-1}) (Table 2, Figure 2). Nevertheless, higher concentrations of Co were found in *Spyridia filamentosa* (7.2 mg kg^{-1}) and *Enteromorpha intestinalis* (4.7 mg kg^{-1}) from Stations 3 and 6 respectively, located in the semi-closed La Paz Lagoon (Table 2, Figure 2) and in *Padina mexicana* (5.5 mg kg^{-1}) from Station 11.

Maximum values of 90 mg kg^{-1} of Zn were found in *Enteromorpha intestinalis* and *Digenia simplex*, in areas near the City of La Paz, to the south (Station 6) and south-east (Station 17) of La Paz Bay (Table 2).

Arsenic revealed a different pattern, showing the highest levels in *Sargassum sinicola* collected at Stations 8 (77 mg kg^{-1}) and 9 (55 mg kg^{-1}) in La Paz Bay, in the areas of Punta Tarabillas and San Juan de la Costa respectively. Concentrations of antimony, another environmentally important element, were highest in *Codium cuneatum*, reaching 1.4 mg kg^{-1} in samples from the western side of La Paz Bay (Station 9).

As can be seen from Tables 1–2 and Figures 2–3, the seaweeds *Enteromorpha intestinalis* and *Caulerpa sertularioides* were the most frequent and widespread in the study area, and showed a good range in the contents of accumulated heavy metals.

Discussion

This study shows that concentrations of elements in the seaweeds collected in La Paz Bay and La Paz Lagoon vary depending on species and sampling location, probably because many variables affect the accumulation of elements in algae including the abundance of these elements in the surrounding water (Barnett and Ashcroft, 1985; Sánchez-Rodríguez et al., 2001). Strong correlations have been demonstrated between the levels of dissolved Cu, Pb, Ni and Cr in the water and in algae (Haritonidis & Malea, 1995; Jordanova et al., 1999; Seeliger & Edwards, 1977). Biological,

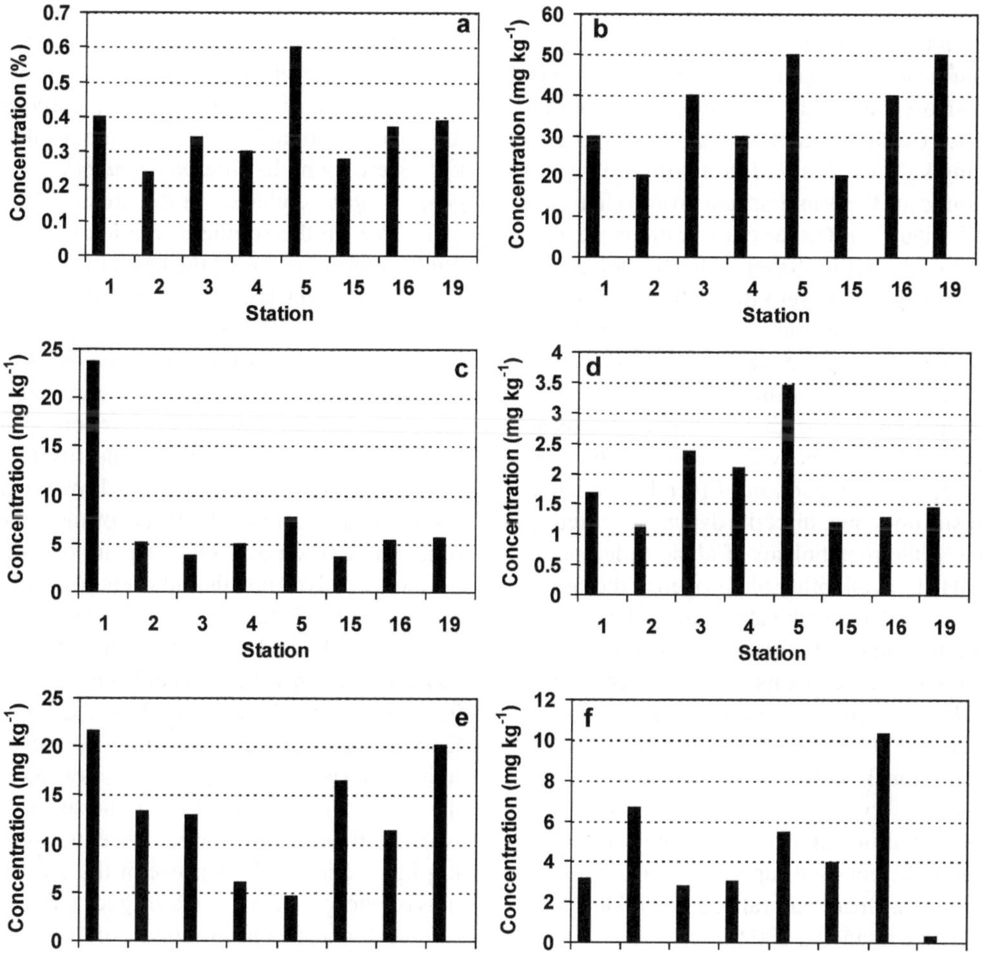

Figure 3. The spatial distribution of the concentration of selected elements in the samples of the seaweed *Caulerpa sertularioides* from the La Paz Bay and La Paz Lagoon: (a) iron; (b) zinc; (c) chromium; (d) cobalt; (e) arsenic and (f) uranium.

physical and chemical conditions affect both the distribution and the role of the elements, compounds and residues in a system (Hassett et al., 1980). Furthermore, seaweeds have a high potential capacity for storing trace metals, depending on the species of alga and the metal (Phillips, 1977; Myklestad et al., 1978). Previous studies (Khristoforova et al., 1983; Ostapczuk et al., 1997; Sánchez-Rodríguez et al., 2001; Sueur et al., 1982) have shown that members of the Phaeophyta have the highest capacity for storing metals. In this study, however, algae belonging to the Chlorophyceae (e.g. *Enteromorpha intestinalis*) from the area near San Juan de la Costa generally accumulated most elements (Cr, Hf, Rb, Sc, Se, Ta, Th, U and Zr). This distinctive feature of this area is probably a result of the weathering of the natural rocks of the drainage basins (mainly sedimentary and volcanic rocks), as well as due to the influence of nearby phosphorite mining operations on the seawater and the marine sediment composition. It was clear that near the mouths of the Las Tarabillas Arroyo and Arroyo San Juan, at Station 9, in front of San Juan de La Costa, concentrations of 33 mg kg^{-1} of U and 80 mg kg^{-1} of Cr in *E. intestinalis* are higher than those found in the same specie from other stations (Table 2, Figure 2). These concentrations are also higher than those found by Sánchez-Rodríguez et al., (2001) for macroalgae in Loreto Bay, where maximum values of ~4 mg kg^{-1} for uranium and of 36 mg kg^{-1} for chromium were found in *Sargassum sinicola*.

A possible explanation for this selective accumulation of metals may be that fact that the initial process of rapid absorption of elements by the seaweeds could be a result of electrostatic attraction of metal ions, since this mechanism does not directly depend on factors which influence the metabolism of algae (a temperature, a light, pH, the availability of nitrogen or the age of organisms), but is related to the abundance of elements in the surrounding water. This could then be followed by the active uptake of metal ions which are transported across the cellular membrane and introduced into the cytoplasm (Crist et al., 1988, 1990; Levine, 1984). To verify this, additional studies need to be carried out, with systematic observations of macroalgae and corresponding concentrations of elements in the water, taking into account the possible supply of metals into the marine environment from natural sources (Hausback, 1984; Rodríguez Castañeda, 2002).

The highest amounts of both Ca and Sr were found in the green alga *Halimeda discoidea* collected in the southern part of the bay, near La Paz Lagoon, and in La Paz Bay, south of Punta Prieta. Macroalgae incorporate Sr through a process involving the exchange of intracellular polysaccharides, whereas Ca ions are used to maintain membranes and the cell wall (Lobban & Harrison, 1994). In fact it has been observed that during ionic exchanges, polysaccharides in brown seaweeds have an affinity for divalent cations such as Ca^{2+} (Karez & Pereira, 1995). This would explain the high concentrations of calcium found in samples of *Padina mexicana* collected in Stations 18 and 19, in the area of Punta Prieta. Both species, *Halimeda*, and to a much lesser extent *Padina*, accumulate calcium carbonate (Lobban & Harrison, 1994).

Selenium, on the other hand, reached its highest concentration (1.8 mg kg^{-1}) in *Padina mexicana* collected at Station 11, in front of the El Cajón Arroyo mouth, which receives terrigenous sediments from a drainage basin of volcanic rocks (Figure 1, Table 2). Since Se is an enzymatic cofactor, it is likely that its accumulation in this seaweed is regulated by metabolic processes (Lobban & Harrison, 1994).

As for Zn, an element that is frequently used to monitor strongly polluted areas, maximum values of 90 mg kg^{-1} were found in *Enteromorpha intestinalis* and *Digenia simplex* collected in the areas close to the City of La Paz, in the southern (Station 6) and southeastern (Station 18) parts of the bay. However, these levels were lower than levels in seaweeds in areas of high human impact (1000 to 2000 mg kg^{-1}). It has been suggested that Zn is taken up both by absorption and by active transport, since it is an important nutrient in algal metabolism (Lobban & Harrison, 1994).

This information is not only useful in determining the pollution status of the La Paz Bay. It is known that seaweeds help reduce the levels of metals in the environment, and removal rates depend on the concentrations of dissolved metals and water pH value (Bernhard & Zattera, 1975; Jensen et al., 1976).

Arsenic showed a preference for the seaweed *Sargassum sinicola* which was collected in Stations 8 and 9, located near Punta Tarabillas and San Juan de la Costa respectively. This feature of As can be attributed to its affinity with this species of seaweed as well as to the spatial distribution of As in the surface sediments, which in these areas showed concentrations of 10 to 20 mg kg^{-1}, apparently because of the weathering products of phosphatic rocks (Rodríguez Castañeda, 2002).

The highest concentrations of the antimony, another important environmental indicator, were found in *Codium cuneatum*, with maxima in samples from the western side of La Paz Bay (Station 9): this can be related to the fact that sediments supplied to this area are influenced by the weathering products of igneous rocks from the surrounding region.

On the basis of our results we conclude that La Paz Bay has not suffered seriously from human impacts, but that the geological characteristics of this region encourage a natural increase in certain elements in the sediments, which is then reflected in the macroalgae. Several macroalgae have been described as excellent bio-indicators because the levels of metals in their tissues are proportional to the concentrations of metals in the surrounding waters (Bryan & Hummerstone, 1973; Försberg et al., 1988; Föster, 1976; Fuge & James, 1973). Results obtained so far from seaweeds collected in the coastal waters of La Paz Bay show that there is localized variation in each of the different areas of the bay, and that the accumulation of some elements is probably determined by their relative concentration in the surrounding water, by a species' particular metabolic processes, and by local environmental conditions.

Conclusions

We found that seaweeds belonging to the Chlorophyceae accumulated the highest contents of the elements studied, with *Enteromorpha intestinalis* as an example for Ba, Cr, Cs, Hf, Rb, Sc, Ta, Th U, and Zr. The variations in the concentrations of these elements in the algae can be related to the influence of local factors such as naturally occuring higher contents of some elements in the water and sediments of certain parts of the bay, or the localized and limited effect of mining operations in the area near San Juan de la Costa.

Of the major and trace elements found in the sediments and seaweeds of the La Paz Bay, the contents of As, Cr, Sb, Se and Zn, usually associated with an intense human activity, did not suggest the existence of such impact on the environment, but did reflect the geological composition of the rocks of the region and the particular characteristics of this water body. Because of their abundance and good range of elemental content the seaweeds *Enteromorpha intestinalis* and *Caulerpa sertularioides* appear most suitable for further biomonitoring of the heavy metal pollution of the coastal waters in this zone.

Acknowledgments

This study was supported by grant # 27728-T (1999–2001) of Consejo Nacional de Ciencia y Tecnología of Mexico, as well by the Coordinación General del Posgrado e Investigación (CGPI) of the Instituto Politécnico Nacional of Mexico (project # 20040093). Rodríguez-Castañeda A. P., I. Sánchez-R. and E. Shumilin were fellows of COFAA-IPN. The authors are greatly indebted to Mrs. Danielle Maither L. for the editng the English text.

References

Abbott I, Hollenberg GJ (1976) Marine algae of California. Stanford University Press, Stanford, CA, pp. 827.

Barnett BE, Ashcroft CR (1985) Heavy metals in *Fucus vesiculosus* in the Humber Estuary. Env. Pollut. 9: 193–213.

Bernhard M, Zattera A (1975) Major pollutants in the marine environments. In Pearson EA, Frangipane ED (eds), Marine Pollution and Marine Waste Disposal. Pergamon Press, New York: 195–300.

Bryan GW, Hummerstone LG (1973) Brown seaweed as indicator of heavy metals in estuaries in southwest England. J. Mar. Biol. Ass. UK 53: 705–720.

Chester R (2003) Marine Geochemistry. Blackwell Publishers, Oxford, pp. 506.

Crist RH, Oberholser K, Schwartz D, Marzoff D, Ryder JD (1988) Interactions of metals and protons with algae. Env. Sci. Technol. 22: 755–760.

Crist RH, Martin R, Guptill PW, Eslinger JM, Crist DR (1990) Interaction of metals and protons with algae. 2. Ion exchange in adsorption and metal displacement by protons. Env. Sci. Technol. 24: 337–342.

De la Lanza EG, Cáceres MC (1994) Lagunas costeras y el litoral mexicano. Universidad Autónoma de Baja California Sur. La Paz, México, pp. 525 (in Spanish).

Eide I, Myklestad S, Melson S (1980) Long-term uptake and release of heavy metals by *Ascophyllum nodosum* (Phaeophyceae) *in situ*. Env. Pollut. 23: 19–28.

Försberg A, Söderlund S, Frank A, Petersson LR, Pedersén M (1988) Studies on metal content in the brown seaweed *Fucus vesiculosus*, from the Archipelago of Stockholm. Env. Pollut. 49: 245–263.

Föster P (1976) Concentration and concentration factors of heavy metals in brown algae. Env. Pollut. 10: 45–53.

Fuge R, James KH (1973) Trace metal concentrations in brown seaweeds in Cardigan Bay, Wales. Mar. Chem. 1: 281–293.

Haritonidis S, Malea P (1995) Seasonal and local variation of Cr, Ni and Co concentrations in *Ulva rigida* C. Agardn and *Enteromorpha linza* (Linnaeus) from Thermaikos Gulf, Greece. Env. Pollut. 89: 319–327.

Hassett JM, Jennett JC, Smith JE (1980) Heavy metal accumulation by algae. In Baker RA (ed.), Contaminants and Sediments. Vol.2. Analysis, Chemistry, Biology. Ann Arbor Science, Michigan: 409–424.

Hausback BP (1984) Cenozoic volcanism and tectonic evolution of Baja California Sur, Mexico. In Frizzel VA (ed.), Geology of the Baja California Peninsula. Society of Economic Paleontologists and Mineralogists, Pacific Section, Tulsa, 219–236.

Ho YB (1990) Metals in *Ulva lactuca* in Hong Kong intertidal waters. Bull. Mar. Sci. 47: 79–85.

Jensen AB, Rystad B, Nelson S (1976) Heavy metal tolerance of marine phytoplankton. II. Copper tolerance of three species in dialysis and batch cultures. J. Exp. Mar. Biol. Ecol. 22: 249–256.

Jordanova A, Strezov A, Ayranov M, Petkov N, Stoilova T (1999) Heavy metal assessment in algae, sediment and water from the Bulgarian Black Sea coast. Wat. Sci. Technol. 39: 207–212.

Karez CS, Pereira RC (1995) Metal contents in polyphenolic fractions extracted from de brown alga *Padina gimnospora*. Bot. Mar. 38: 151–155.

Kennish MJ (ed.), (1997) Practical Handbook of Estuarine and Marine Pollution. CRC Press, Boca Raton, Florida, 524 pp.

Khristoforova NK, Bogdanova NN, Tolstova LM (1983) Metals present in *Sargassum* (Brown) algae of de Pacific Ocean as related to the problem of water pollution monitoring. Oceanology 23: 200–204.

Kuyucak N, Volesky B (1990) Biosorption by algal biomass. In Volesky B (ed.), Biosorption of Heavy Metals. CRC Press, Boca Raton, Florida: 173–198.

Levine HG (1984) The use of seaweeds for monitoring of coastal waters. In Shubert LE (ed.) Algae as Ecological Indicators. Academic Press, London:189–210.

Lobban CS, Harrison PJ (1994) Seaweed Ecology and Physiology. Cambridge University Press, Cambridge.

Myklestad S, Eide I, Melsom S (1978) Exchange of heavy metals in *Ascophyllum nososum* (L.) *in situ* by means of transplanting experiments. Env. Pollut. 16: 277–284.

Norris JN (1975) Marine Algae of the Northern Gulf of California. Ph.D. Dissertation. University of California, Santa Barbara.

Ostapczuk P, Burow M, May K, Mohl C, Froming M, Süßenbach B, Waidmann E, Emons H (1997) Mussels and algae as bioindicators for long-term tendencies of element pollution in marine ecosystems. Chemosphere 34: 2049–2058.

Phillips DJ (1977) The use of biological indicator organisms to monitor trace metal pollution in marine and estuarine environments – a review. Env. Pollut. 13: 281–315.

Robledo D, Freile Pelegrín Y (1997) Chemical and mineral composition of six potentially edible seaweed species of Yucatán. Bot. Mar. 40: 301–306.

Rodríguez Castañeda AP (2002) Elementos Mayores y Traza en Sedimentos y Macroalgas de La Bahía de La Paz, Baja California Sur, México. Tesis de Maestría. CICIMAR-IPN, La Paz, 170 pp. (in Spanish).

Sánchez-Rodríguez I, Huerta-Díaz MA, Choumiline E, Holguín-Quiñones O, Zertuche-González JA (2001) Elemental concentrations in different species of seaweeds from Loreto Bay, Baja California Sur, Mexico: implications for the geochemical control of metals in algal tissue. Env. Pollut. 114: 146–160.

Seeliger U, Edwards P (1977) Correlation coefficients and concentrations factors of copper and lead in seawater and benthic algae. Mar. Pollut. Bull. 8: 19–19.

Shumilin E, Kalmykov St, Sapozhnikov D, Nava-Sánchez E, Gorsline D, Godínez-Orta L, Sapozhnikov Yu, Holguín Quiñones O, Rodríguez –Castañeda A (2000) Major and trace element accumulation in coastal sediments along the southeastern Baja California studied by instrument neutron activation analysis and Pb-210 dating. J. Radioanal. Nucl. Chem. 246: 533–541.

Silva PC, Basson PW, Moe RL (1996) Catalogue of the Benthic Marine Algae of the Indian Ocean. University of California Press, Los Angeles, pp. 1259.

Sueur S, Van den Berg CMG, Ripley JP (1982) Measurement of the metal complexing ability of exudates of marine macroalgae. Limnol. Oceanogr. 27: 536–543.

Taylor WR (1945) Pacific Marine Algae of the Allan Hancock Expeditions to the Galapagos Islands. Allan Hancock Expeditions 12: 1–528.

Journal of Applied Phycology (2006) 18: 409–412
DOI: 10.1007/s10811-006-9038-6

Formation of aldehyde flavor (*n*-hexanal, 3*Z*-nonenal and 2*E*-nonenal) in the brown alga, *Laminaria angustata*

Kangsadan Boonprab[1,*], Kenji Matsui[2], Yoshihiko Akakabe[2], Miyuki Yoshida[2], Norishige Yotsukura[3], Anong Chirapart[4] & Tadahiko Kajiwara[2]
[1]*Department of Fishery Products, Faculty of Fisheries, Kasetsart University, Bangkok, 10900, Thailand;* [2]*Department of Biological Chemistry, Faculty of Agriculture, Yamaguchi University, Yamaguchi, 753-8515, Japan;* [3]*Institute of Algological Research, Faculty of Science, Hokkaido University, Hokkaido, 051-0003, Japan;* [4]*Department of Fishery Biology, Faculty of Fisheries, Kasetsart University, Bangkok, 10900, Thailand*

*Author for correspondence: e-mail: ffisksb@ku.ac.th, bffisksb@yahoo.com; fax: +66(0)29428363

Key words: *Laminaria angustata*, lipoxygenase, fatty acid hydroperoxide lyase, *n*-hexanal, 3Z-nonenal, 2*E*-nonenal

Abstract

2*E*-Nonenal and *n*-hexanal are the major and minor flavor compounds in the edible brown alga, *Laminaria angustata*, respectively. They are believed to characterize the flavor of this alga. However the metabolism of the two compounds is not precisely known. The pathways were clarified by elucidation of the intermediate structure through purification of the intermediate compounds from an enzymatic reaction and identification using HPLC and GC-MS techniques. Formation of *n*-hexanal, 3*Z*-nonenal and 2*E*-nonenal are proposed to be *via* two cascades from unsaturated fatty acids. They are C18:2(*n*-6), linoleic acid cascade and C20:4(*n*-6), arachidonic acid cascade through their hydroperoxides as intermediates by the lipoxygenase/fatty acid hydroperoxide lyase pathway.

Introduction

Biogeneration of aldehydes in higher plants is generally known to be accomplished *via* one of the oxylipin pathways (lipoxygenase/fatty acid hydroperoxide lyase). C9 aldehydes like nonenal and nonadienal are formed from 9-hydroperoxides of unsaturated C18 fatty acids through a cleavage by fatty acid hydroperoxide lyase. On the other hand, in animals such as fish, these aldehydes could be formed *via* C20 or C22 fatty acid (Cadwallader, 2000). Marine algae have C20, C22 and C18 unsaturated fatty acids, and they can produce both plant (C18) and animal type (C20 and C22) fatty acid hydroperoxides. From this it has been postulated that nonenal may be formed from C18 and/or C20 unsaturated fatty acids *via* their hydroperoxides (Gerwick, 1994; Fujimura & Kawai, 2000). The pathway *via* the animal type hydroperoxides was then proposed in marine algae (Kajiwara, 1997). However, convincing evidence to support their enzymatic formation has not been obtained.

Not only nonenal is formed as a major component by the brown alga, *Laminaria angustata* (the well-known edible alga in Japan and other Asian countries) but *n*-hexanal (green, fresh flavor) is also formed and was reported as a minor component (Kajiwara et al., 1996). In higher plants, *n*-hexanal is formed *via* the lipoxygenase/fatty acid hydroperoxide lyase (LOX/HPL) system through 13-hydroperoxy linoleic acid (Blée, 1998). Thus, it was suggested that this pathway might exist also in the algae. Therefore, this study was aimed to clarify the metabolic pathway of the volatile aldehydes [C6 aldehyde (*n*-hexanal) and C9 aldehyde (3*Z*- and 2*E* – nonenal)] in *L. angustata*. We identified precursors for the aldehyde formation in homogenates of *L. angustata* as hydroperoxides of linoleic acid and arachidonic acid. Furthermore, two enzymatic pathways to generate C6 and C9 aldehydes through linoleic acid and arachidonic acid were proposed. The finding is interesting because the brown alga can use different precursors for the production of short chain aldehydes, probably through different types of LOX/HPL systems. The experiments were performed and described in detail by the study of biogeneration of C6 and C9 aldehyde, biosynthesis of C6 aldehyde (*n*-hexanal) from linoleic acid and biosynthesis of C6 aldehyde (*n*-hexanal) and C9 aldehydes (*n*-hexanal, 3*Z*-nonenal and 2*E*-nonenal) from arachidonic acid, and the following results obtained.

Pathways of aldehyde formation

Biogeneration of C6 and C9 aldehydes

In this study, an enzymatic reaction was performed by the incubation of frond homogenate at 4 °C, for 80 min, to form volatile compounds that were explored by using simultaneous distillation extraction (SDE) and solid phase micro extraction (SPME) techniques. An increase in the compounds after incubation was observed, which suggested that they were formed by an enzymatic reaction, especially C6 aldehyde (*n*-hexanal) and C9 aldehyde (2*E*-nonenal). In the reaction with crude enzyme and unsaturated fatty acid as the substrate, C9 aldehydes (3*Z*-nonenal and 2*E*-nonenal) are mainly formed from arachidonic acid, while C6 aldehydes (*n*-hexanal) are formed from either C18 or C20 fatty acids (Boonprab et al, 2003b). This indicates that *Laminaria angustata* has at least two metabolic pathways to form short chain aldehydes.

Biosynthesis of C6 aldehyde (n-hexanal) from linoleic acid

The results from the above biogeneration study of C6 and C9 aldehydes indicate that *L. angustata* could form relatively high amounts of C6 and C9 aldehydes. When linoleic acid was added to a homogenate prepared from the fronds of this alga, formation of *n*-hexanal was observed. When glutathione peroxidase was added to the reaction mixture together with glutathione, the formation of *n*-hexanal from linoleic acid was inhibited, and oxygenated fatty acids accumulated. By chemical analyses, one of the major oxygenated fatty acids was shown to be (*S*)-13 hydroxyoctadecadienoic acid. Therefore, it is assumed that *n*-hexanal is formed from linoleic acid *via* a sequential action of LOX and HPL, by a similar pathway as the counterpart found in higher plants. HPL partially purified from the fronds has a rather strict substrate specificity, and only 13-hydroperoxide of linoleic acid, and 15-hydroperoxide of arachidonic acid are the essentially suitable substrates for the enzyme. (Boonprab et al., 2003a)

Biosynthesis of C6 aldehyde (n-hexanal) and C9 aldehydes (n-hexanal, 3Z-nonenal and 2E-nonenal) from arachidonic acid

In higher plants, C6 and C9 aldehydes are formed from C18 fatty acids, such as linoleic acid or linolenic acid, through the formation of 13- and 9-hydroperoxides, followed by their stereospecific cleavage by fatty acid hydroperoxide lyases. Some marine algae can also form C6 and C9 aldehydes, but the precise biosynthetic pathway has not been fully elucidated. According to the biogeneration of C6 and C9 aldehydes study, *L. angustata* could generate C6 and C9 aldehydes enzymatically. C9 aldehydes were formed exclusively from the C20 fatty acid, arachidonic acid, while C6 aldehydes are derived either from C18 or from C20 fatty acid. Thus experiments to identify the intermediates in the reaction were set up. The intermediates were trapped using a glutathione/glutathione peroxidase system, and subjected to structural analyses by co-injection with the standard hydroperoxide compounds, and by GC-MS for their mass spectrum. Formation of (*S*)-12-, and (*S*)-15-hydroperoxy arachidonic acids [12(*S*) hydroperoxyeicosatetraenoic acid and 15(*S*) hydroperoxyeicosatetraenoic acid] from arachidonic acid could be found and confirmed by chiral HPLC analyses (Boonprab et al., 2003b). This accounts respectively for the formation of C9 and C6 aldehydes. The fatty acid hydroperoxide lyase that catalyzes formation of C9 aldehydes from 12(*S*) hydroperoxyeicosatetraenoic acid seems highly specific for hydroperoxides of C20 fatty acids.

Conclusion

Based on these results it is proposed that there are at least two pathways to form volatile aldehydes in *L. angustata* as shown in Figure 1. The brown algae can use different precursors for the production of short chain aldehydes, probably through different types of LOX/HPL systems.

The marine algae are major components of the earth's biomass, responsible for significant carbon fixation, and occupy an extreme diversity of climatic

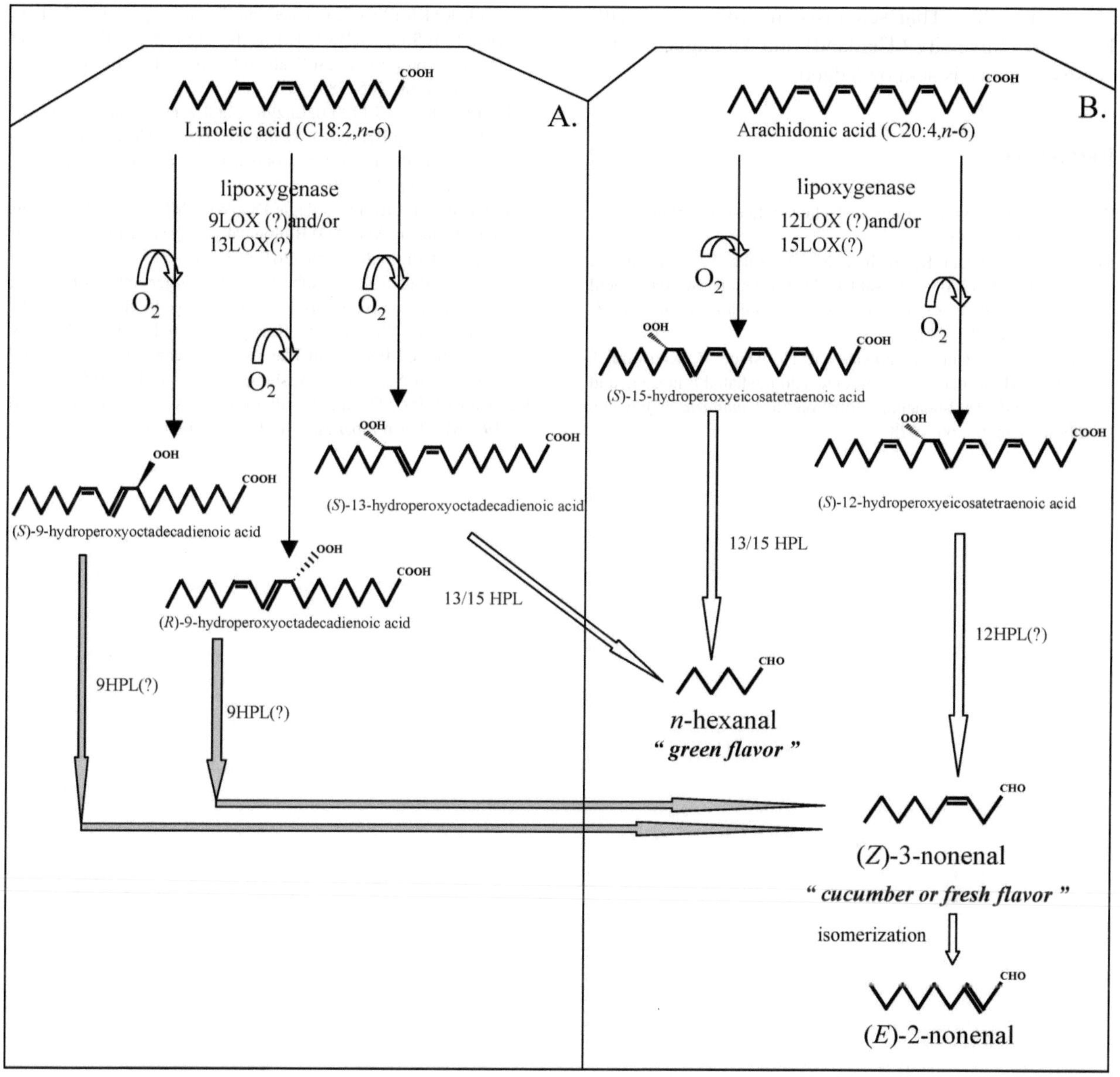

Figure 1. The proposed pathway for the metabolism of linoleic acid and arachidonic acid mediated C6 (*n*-hexanal) and C9 aldehydes [(*Z*)-3-nonenal and (*E*)-2-nonenal] branch of oxylipin pathway in the brown alga, *L. angustata*.
LOX: Lipoxygenase
HPL: Fatty acid hydroperoxide lyase
⇨ Minor pathway
⇨ Major pathway
A. was reported by Boonprab et al. (2003a)
B. was reported by Boonprab et al. (2003b)

niches. Further studies would provide insight into the physiology or regulation of these pathways, which may be involved in growth development, chemical defense, oxidative stress or other mtabolic functions in algae.

Acknowledgements

The support of the JSPS-NRCT Core University Program on "Development of thermotolerant microbial resources and their applications" under the cooperation of Japanese and Thai scientists, in association with Kasetsart University (Thailand) and Yamaguchi University (Japan), is acknowledged.

References

Blée E (1998) Phytooxylipins and plant defense reactions. Prog. Lipid Res. 37: 33–72.

Boonprab K, Matsui K, Yoshida M, Akakabe Y, Chirapart A, Kajiwara T (2003a) C6-aldehyde formation by fatty acid hydroperoxide lyase in the brown alga *Laminaria angustata*. Z. Naturf. 58c: 207–214.

Boonprab K, Matsui K, Akakabe Y, Norishige Y, Kajiwara T (2003b) Hydroperoxy-arachidonic acid mediated *n*-hexanal and (*Z*)-3- and (*E*)-2-nonenal formation in *Laminaria angustata*. Phytochem. 63: 669–678.

Cadwallader KR (2000) Enzymes and Flavor Biogenesis in Fish. In Haard NF, Simpson BK (eds.), Seafood Enzymes. Utilization and Influence on Postharvest Seafood Quality, Marcel Dekker, Inc., New York: pp. 365–383.

Fujimura T, Kawai T (2000). Enzymes and seaweed flavor. In Haard N.F., Simpson BK (eds.), Seafood Enzymes. Utilization and Influence on Postharvest Seafood Quality, Marcel Dekker, Inc., New York, pp. 385–407.

Gerwick WH, Proteau PJ., Nagle DG, Wise ML, Jiang ZD, Bernart MW, Hamberg M (1993) Biologically active oxylipins from seaweeds. Hydrobiologia 260/261: 653–665.

Kajiwara T, Matsui K, Akakabe Y (1996) Biogeneration of volatile compounds via oxylipins in edible seaweeds. In Takeoka GR, Teranishi R, Williams PJ, Kobayashi A (eds.), Biotechnology for Improved Foods and Flavors, American Chemical Society Symposium Series 637, Washington, DC, pp. 146–166.

Kajiwara T (1997) Dynamic studies on bioflavor of seaweed. Koryo 196: 61–70. (in Japanese with English summary)

Journal of Applied Phycology (2006) 18: 413–422
DOI: 10.1007/s10811-006-9046-6

Antimicrobial browning-inhibitory effect of flavor compounds in seaweeds

Tadahiko Kajiwara, Kenji Matsui, Yoshihiko Akakabe, Takushi Murakawa & Chikako Arai
Department of Biological Chemistry, Faculty of Agriculture, Yamaguchi University, Yamaguchi 753-8515, Japan

Key words: seaweeds, essential oils, volatile compounds, tyrosinase inhibitory activity, antimicrobial activity, L-DOPA

Abstract

Since ancient times, the antimicrobial properties of seaweeds have been recognized. However, antimicrobial activities of volatile compounds in seaweeds have not been explored so far. Here, essential oils from seaweeds including green, brown and red algae such as *Laminaria japonica*, *Kjellmaniella crassifolia*, *Gracilaria verrucosa* and *Ulva pertusa* were prepared by using SDE (simultaneous distillation and extraction) apparatus. Volatile compounds in the essential oils were identified as aldehydes, ketones, carboxylic acids, alcohols and hydrocarbons by comparison of GC-retention times and MS data with those of authentic specimens. Flavor compounds such as (3*Z*)-hexenal, (2*E*)-hexenal and (2*E*)-nonenal in some essential oils showed strong antimicrobial activities against *Escherichia coli* TG-1, and *Erwinia carotovora*. Inhibition of browning can be achieved during either of two stages, namely, oxidation reaction by tyrosinase or subsequent non-enzymatic polymerization. Tyrosinase activity was measured by monitoring absorbance at 475 nm originating from dopachrome formed from L-DOPA. Many kinds of aliphatic carboxylic acids, aldehydes and alcohols were used as inhibitors for PPO activity. The results indicated that the α,β-unsaturated carbonyl compounds strongly inhibit tyrosinase activity. When seaweeds are damaged or macerated, the α,β-unsaturated aldehydes such as (2*E*)-hexenal and (2*E*)-nonenal are biosynthesized *via* the corresponding (3*Z*)-unsaturated aldehydes from linolenic acid and arachidonic acid. The flavor compounds that are formed could be valuable as safe antimicrobial browning-inhibitory agents of edible seaweed origin.

Introduction

Since ancient times, antimicrobial properties of herbs and spices have been used for food preservation (Zaika, 1988; Conner, 1993). Naturally occurring antimicrobial agents reported date back to more than a century (Maruzzella & Sicurella, 1960). Antimicrobials in extracts from seaweeds have been explored since the 1950s (Glombitza, 1979). A renewed interest in natural preservation appears to be stimulated by present food safety concerns, growing problems with microbial resistance, and a rise in production of minimally processed food together with "green image" policies of food industries (Suhr & Nielsen, 2003).

In recent years, cut vegetables have proven convenient, and also to reduce the amount of domestic garbage produced. These benefits are welcomed by consumers, and the demand for cut vegetables has increased. However, because the cut surface is exposed to air they are more prone to browning, which decreases their nutritional and market values (Friedman, 1996). The growing need for new and safe antimicrobial agents from edible plants, combined with recent vegetable-poisoning incidents in Japan due to *Escherichia coli* O157: H7, led us to study antimicrobial browning inhibitory effects of flavor compounds in essential oils from edible seaweeds.

Generally, polyphenol oxidase (PPO) occurs in most vegetables and fruits, together with polyphenols (Mayer, 1987). Polyphenols and PPO react in the presence of oxygen, to cause browning, when the tissues of vegetables and fruits are damaged. Tyrosinase, a known PPO, is a copper-containing enzyme that is widely distributed in plants, animals and microorganisms (Whitaker, 1995). Efficient inhibition of the browning process would be useful for improving the processing of cut vegetables. The aim of this study was to find safe and efficient antimicrobial inhibitors of PPO from flavor compounds in edible seaweed essential oils.

Materials and methods

Materials

Antimicrobial activities of algal flavor compounds were tested against *Escherichia coli* TG-1, and *Erwinia carotovora*. Fresh seaweeds were collected along the Aio coast and Hikoshima beach in Yamaguchi, southern Japan, and along Charatsunai coast in Muroran, northern Japan. L-DOPA and DMSO (dimethyl sulfoxide) were purchased from Wako Pure Chemicals (Osaka, Japan). The mushroom tyrosinase (EC: 1.14.18.1) used for the bioassay was purchased from Sigma (St. Louis, MO, USA).

Preparation of essential oils

Cleaned fresh fronds (100 g wet wt) were homogenized in distilled water (100 mL). The homogenates were extracted by a simultaneous distillation extraction (SDE) apparatus for 15 min with pentane-CH_2Cl_2 (7:3, 20 ml) (Schultz et al., 1977). The extracts were dried over Na_2SO_4 and concentrated *in vacuo* to leave essential oils.

Identification of volatile compounds in essential oils

Volatile compounds in the essential oils were identified by comparison of Kovats indices and GC-MS data of synthetic compounds. The GC (a Hewlett Packard 5840 A) was equipped with a FID and a fused silica capillary column (Durabond column DB-1, 0.25 mm i.d. × 60 m). The column temperature was held at 50 °C for 5 min and programmed to increase at 3 °C min^{-1} from 50–240 °C. GC-MS was recorded on a Hitachi H-80A instrument equipped with a fused silica capillary column (Durabond column DB-1, 0.25 mm i.d. × 50 m). The column temperature was held at 75 °C for 5 min and programmed to increase at 3 °C min^{-1} from 75–240 °C. The ionization voltage was 20 eV.

Growth inhibitory effects of essential oils and flavor compounds

Each bacterial strain was incubated in nutrient broth No. 2 at 37 °C overnight (14 h), and test bacterial solutions were prepared with thesame broth to give a concentration of 10^6 cells mL^{-1} by using a hemacytometer. A serial 20-fold dilution of oils and flavor compounds (100, 50, 25, 12.5, 6.25 μg mL^{-1}) was prepared using 50% DMSO, which showed no effects against any bacterial strain tested. Twenty μL of each was added to 160 μl of nutrient broth No. 2 in a 96-well plate with 300 μl volume wells (Millipore, Tokyo, Japan). Finally, 20 μl aliquots of 10^6 cells mL^{-1} bacterial solution were inoculated into the wells and incubated at 37 °C for 24 h. Bacteriostatic activities of oils and flavor compounds were examined by turbidity (OD at 660 nm).

Bactericidal effects of essential oils and flavor compounds

Bactericidal effects of essential oils and flavor compounds were assessed on *E. coli* TG-1 and *Erwinia carotovora*. The test strains were harvested from cultures held overnight in nutrient broth No. 2 by centrifugation and were re-suspended in sterile 50 mM potassium phosphate buffer (pH 7.0) after being washed twice with the same buffer. Washed cell preparations were diluted to 10^6 cells mL^{-1} by using a hemacytometer. The essential oils and flavor compounds were diluted with 50% DMSO to prepare 100, 20, 10, 1, and 0.1 μg mL^{-1} solutions. Twenty μL of the diluted chemicals was mixed well with 160 μL of phosphate buffer (pH 7.0) in a sterile microtube with a volume of 1.5 mL. Next, 20 μL of the 10^6 cells mL^{-1} bacterial cell suspension was added into the tube and subsequently incubated at 37 °C for 1 h (Nakamura et al., 2002). After incubation, decimal dilutions of the sample were carried out up to × 10^4 using physiological saline adjusted to pH 7.2. One hundred μL of diluted cell suspension was spread onto a Mannitol-salt agar plate. All plates were incubated at 37 °C for 24 h, and surviving cells counted according to the colonies appearing on the plate. The percentage of survivors was presented with respect to the control mixture. The experiment was performed in triplicate.

Tyrosinase inhibitory assay by the spectrometer method

The flavor compounds (except those that are water soluble) were first dissolved in DMSO, to concentrations of 500 mM. The enzyme activity was measured by the spectrometric method by reading at A475 nm to detect dopachrome formation (Kubo et al., 1999). First, 33 μL of 1380 units/mL tyrosinase in 0.1 M K-Pi buffer (pH 6.8) solution was mixed with 0.1 M K-Pi buffer (pH6.8) to 1 mL of total volume. Then, 2 or 10 μl of each sample solution (final concentrations are 1 and 5 mM, respectively) and 330 μL of 2.5 mM L-DOPA were added in this order to the mixture. The change at A475 nm was monitored 90 s after addition of L-DOPA.

Tyrosinase inhibitory assay by the oxygen electrode method

The compounds tested were suspended in 5% Gum Arabic to be 50 mM.

The enzyme activity was expressed as the oxygen consumption, which was measured with a Clark type oxygen electrode (YSI 5331; Yellow Springs, Co.) at 25 °C in 0.1 M K-Pi buffer (pH 6.8) with 1.75 ml of the total volume. The reaction was started by the addition of the enzyme to the reaction mixture containing 500 μl of 2.5 mM L-DOPA and 17.5 μl of each sample solution (final concentration, 500 μM). The activity (1 kat) was defined as the quantity of enzyme catalyzing the consumption of 1 mol O_2 s^{-1} at 25 °C (Kermasha et al., 1993).

Results and discussion

Flavor compounds from edible seaweeds

The kelps *Laminaria* and *Undaria* are generally called "kombu" and "wakame" in Japanese, respectively. After kombu and wakame, a red seaweed *Porphyra* sp. (asakusa-nori) is the most popular seaweed in Japanese foods. Many green seaweeds such as *Enteromorpha* sp. (ao-nori), *Ulva* sp. (aosa) and *Monostroma* sp. (hitoegusa) are also used for food. These seaweeds have been eaten since ancient times and are highly favored for their flavors, tastes, and textures (Kajiwara et al., 1993).

Recently, we have explored volatile compounds in fifty or more species of wet and un-decomposed seaweeds in Japan: green seaweeds *Ulva pertusa*, *Monostroma nitidum*, and *Enteromorpha clathrata*; brown seaweeds *Laminaria ungustata* and *Undaria pinnatifida*, and red seaweeds *Porphyra tenera* and *Porphyra yezoensis*, by GC and GC-mass spectrometry (MS) (Kajiwara et al., 1996).

U. pertusa was collected along the Hikoshima coast, Yamaguchi, southern part of the Japan sea, and *L. japonica* and *Kjellmaniella crassifolia* along the Charatsunai coast, Muroran, Hokkaido, northern Pacific Ocean. The SDE-distillates of homogenates of each fresh seaweed gave essential oils in the amounts: *U. pertusa* 4.23 × 10^{-2}%; *L. japonica* 4.55 × 10^{-2}%; and *K. crassifolia* 6.13 × 10^{-2}%. The characteristic odorous oils were analyzed by GC and GC-MS equipped with fused silica capillary columns (SF-96 and DB-1). Among volatile compounds detected in the oils of the green seaweeds, thirty one compounds were identified as volatile components of the Ulvales.

With essential oils of the wet and un-decomposed edible kelps, *L. japonica*, and *K. crassifolia*, from the northern part of Japan, fifty three compounds including alcohols, aldehydes, esters, ketones, hydrocarbons, and carboxylic acids were identified by comparison of Kovats indices and MS data with those of authentic compounds. The nor-carotenoids such as β-cyclocitral, β-homocyclocitral, β-ionone, and dihydroactinidiolide, which have been reported as flavor components of an edible red seaweed, *P. tenera*, seem to be important constituents of some brown algae such as *Costaria costata* and *Alaria crassifolia*. The sesquiterpene alcohol, cubenol, was detected in the volatile oils of all of the submitted kelps, whereas the stereoisomer, epicubenol was detected only in *L. japonica* and *K. crassifolia* in small amounts. (2*E*, 6*Z*)-Nonadienal, (3*Z*, 6*Z*)-nonadienal, (2*E*)-nonenal, and the corresponding alcohols, which are well known as flavor constituents of cucumber, melons, and fish, were found to be the principle odor contributors in some brown seaweeds. The C_9-aldehydes and alcohols particularly were at their highest concentration in *L. japonica* (Table I). Recently, (2*E*, 6*Z*)-nonadienal has been found in *Cymathere triplicata*, a large brown kelp, in Northern Washington.

In the large thalli of the red seaweed *P. yezoensis*, fatty aldehydes such as *n*-pentadecanal, (2*E*, 6*Z*)-nonadienal, and (2*E*)-nonenal, nor-carotenoids and sesquiterpene alcohols were identified. An essential oil of cultivated conchocelis-filaments of the seaweed contained cubenol, phytol, palmitic acid and the long-chain fatty aldehydes such as tetradecanal, pentadecanal, (7*Z*,10*Z*)-hexadecadienal, (8*Z*,11*Z*)-heptadecadienal and (8*Z*)-heptadecenal. Flament and Ohloff (1985) have reported on the identification of more than 100 volatile constituents of dried thalli of *P. tenera*: nor-carotenoids (α-ionone, β-ionone, dihydroactinidiolide etc.) and unsaturated fatty short-chain aldehydes [(2*E*,4*Z*)-decadienal, (2*E*,4*E*)-heptadienal etc.]. However, these compounds were detected only as minor components in wet and un-decomposed conchocelis-filaments. As Table I shows, the aldehydes (C_{10}–C_{17}) were characteristic components in the Ulvales oils. Particularly, the C_{17}-trienal accounted for 35% of the oils of *Ulva* obtained along the Pacific coast of Hokkaido, Muroran and 8–11% along the Yamaguchi coast of the Sea of Japan. A homolog of the C_{16}-trienal, (7Z,10Z,13Z)-hexadecatrienal, was also found in *U. pertusa*. The characteristic aldehyde of *M. nitidum* and *E. clathrata* was (2*E*,4Z,7Z)-decatrienal. The C_{17}-aldehydes and the characteristic C_{10}-trienal were identified in seaweeds. However, the C_{17}-unsaturated aldehydes have been reported as volatiles of an aqueous cucumber homogenate and green leaves of tobacco at flowering time just after topping. Closely related aldehydes such as (3Z,6Z)-3,6,11-dodecatrienal, (2*E*,4Z,7Z)-2,4,7,12-tridecatetraenal, and (3Z,6Z,9Z)-3,6,9,14-pentadecatetraenal have been reported to possess "characteristic seaweed or algae odor".

Antimicrobial effects of essential oils and flavor compounds

It has been reported that some volatile compounds released from spices and herb extracts showed wide antimicrobial activities against fungi (Davis & Smoot, 1972; Vaughn et al., 1993) and bacteria (Morris et al., 1979). Flavor compounds of seaweeds such as (3Z)-hexenal, (2*E*)-hexenal and (2*E*)-nonenal showed potent antimicrobial activities against *E. coli* TG-1 and *Erw. carotovora*. Among the aldehydes, (3Z)-hexenal exhibited the highest bacteriostatic activity with the growth-inhibitory concentration of 50 μg mL^{-1} for *E. coli* TG-1. No activity of *n*-nonanal at 100 μg mL^{-1} was observed. The order of the growth inhibitory effect for *E. coli* was (3Z)-hexenal > (2*E*)-hexenal > (2*E*)-nonenal > *n*-nonanal as shown in Figure 1. There were no large differences in effect against *Erw. carotovora* in these aldehydes. This result is in general agreement with that reported earlier (Nakamura, 2002). Similar activity of (2*E*,4*E*)-decadienal against *E. coli* was reported (Kubo et al., 1995). Essential oils from *Laminaria* at 400 μg mL^{-1} exhibited weak antimicrobial activities against *E. coli* and *Erw. carotovora*. With *Ulva* oils at the same concentration, slight or nonexistent activities were observed.

Antimicrobial browning-inhibitory effect of flavor compounds

In most foods, the browning process consists of two components, enzymatic and non-enzymatic oxidation

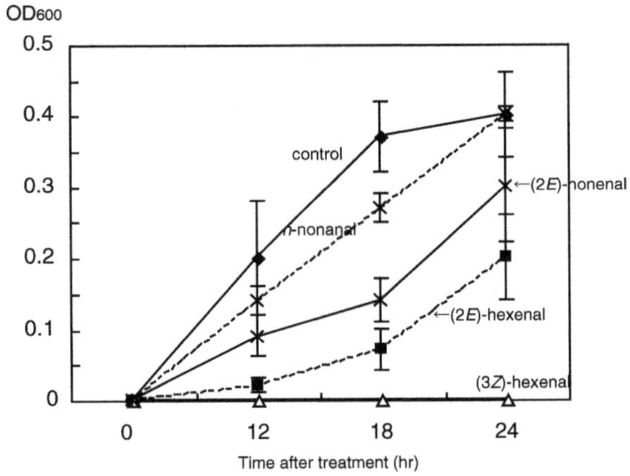

Figure 1. Bacteriostatic effects of 50 μg mL^{-1} of aldehydes against *E. coli* TG-1.

(Figure 2). This unfavorable darkening caused by oxidation generally results in a loss of nutritional and market values (Friedman, 1996). The enzymatic oxidation can be prevented by tyrosinase (EC: 1.14.18.1) inhibitors, and the non-enzymatic oxidation can be protected against by antioxidants. Tyrosinase is known as a polyphenol oxidase (PPO), and is a copper-containing enzyme which catalyzes reactions involving molecular oxygen, and which is widely distributed in microorganisms, animals and plants. Plant PPO contributes negatively to the color quality of plant-derived foods, sea food products (Ogawa et al., 1984) and beverages. This unfavorable darkening from enzymatic oxidation has been of some concern. It is responsible for not only browning in plants but also melanization in animals.

Generally, PPO oxidizes polyphenols under the presence of oxygen when vegetables and fruits are damaged. As a result of cutting, tyrosinase causes the browning of some vegetables and fruits. Such inhibitors can be used to prevent the cut section from browning.

The spectrometric method

In a preliminary screening, N-acetyl-L-cysteine, traumatic acid and tiglic acid were found to inhibit the oxidation of L-DOPA catalyzed by tyrosinase. A more detailed study with N-acetyl-L-cysteine was conducted, and it was found that at lower concentrations of the inhibitor, the inhibitory activity decreased with time (Figure 3). With L-cysteine almost similar results were obtained. These results indicated that these SH reagents did not inhibit the oxidation of phenolics by PPO; rather, they prevented the subsequent polymerization

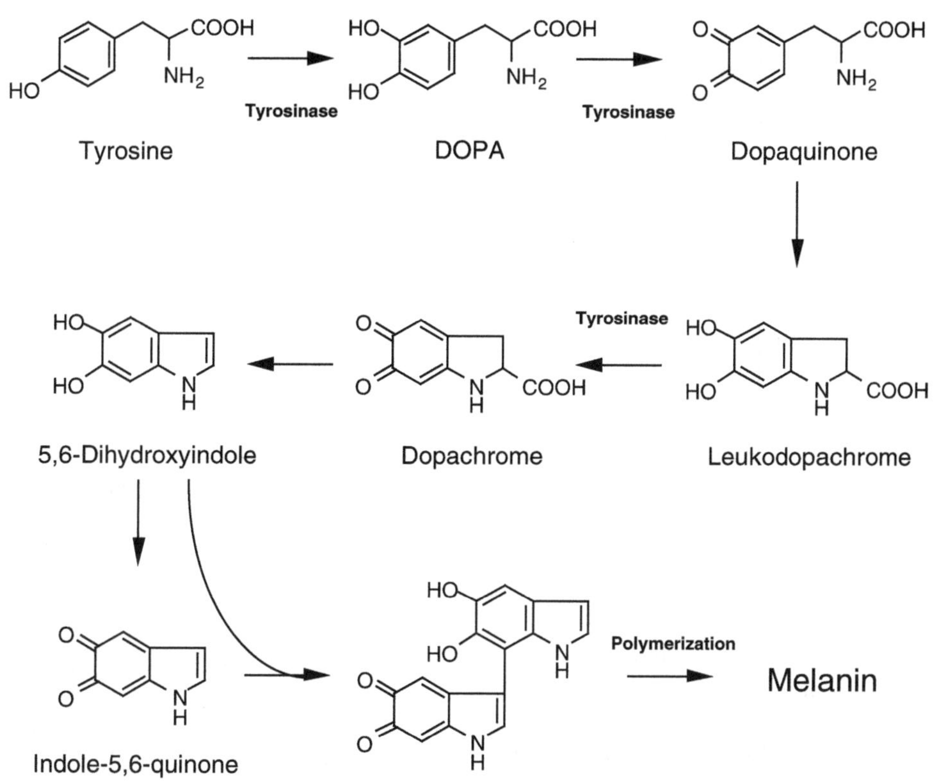

Figure 2. A pathway to form melanin by tyrosinase.

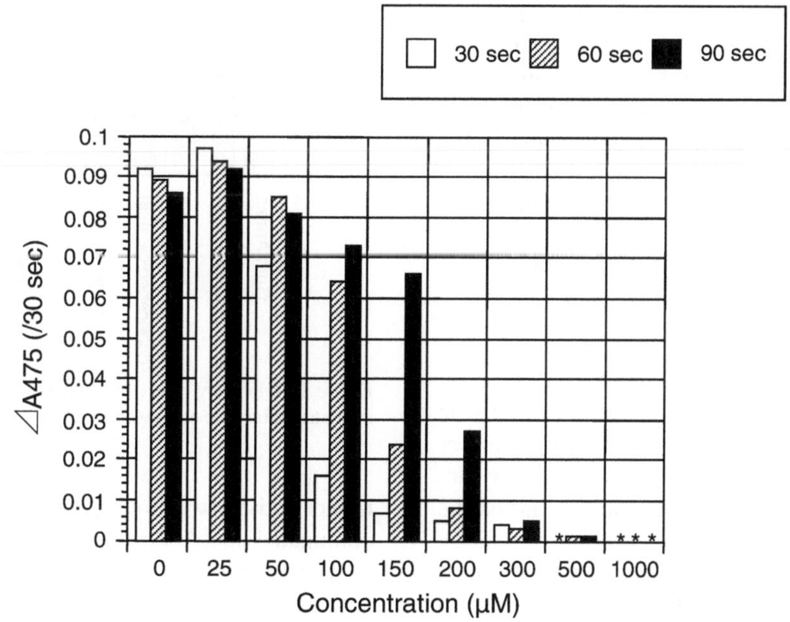

Figure 3. Effect of concentration of N-acetyl-L-cysteine on tyrosinase activity. * ; not detected.

of phenolics, which results in browning (Negishi *et al.*, 2000; Kermasha et al., 1993). On the contrary, traumatic acid was found to exhibit a concentration-dependent inhibitory effect on the oxidation (Figure 4). Thus, it was suggested that it inhibited the oxidation. More than 50% of the initial activity was inhibited with 500 μM of traumatic acid. This inhibition by traumatic acid, C12 dicarboxylic acid having the double-bond at

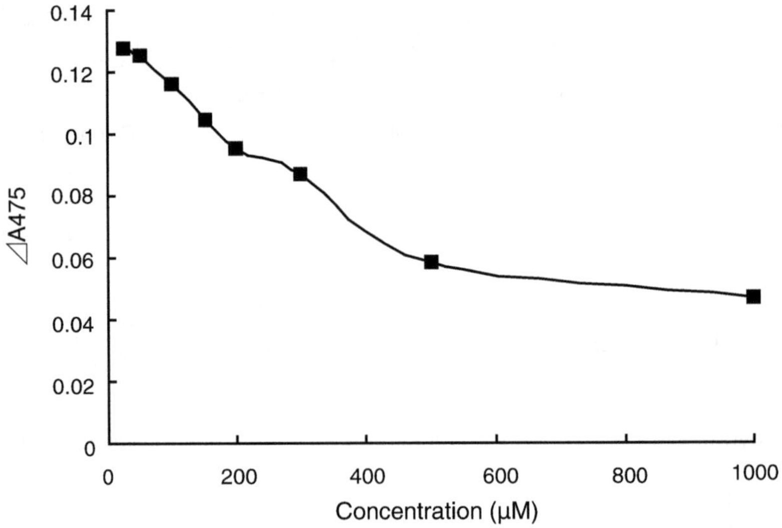

Figure 4. The concentration-inhibition profile of tyrosinase by traumatic acid.

the α, β position, is the first to be reported for carboxylic acids other than aromatic carboxylic acid such as gallic acid (Kubo et al., 2000).

In order to reveal the structure-activity relationship in terms of the inhibitory activity, various aliphatic carboxylic acids of different carbon number and/or different position of a double-bond, were used for the analysis (Kubo et al., 1999; Kubo et al., 2000). In many cases, aliphatic carboxylic acids used as inhibitors could not be used for the spectrometric assay, because of the turbidity caused by the hydrophobic compounds. Therefore, inhibitory effects of the aliphatic carboxylic acids were examined by the oxygen electrode method.

The oxygen electrode method

The aliphatic carboxylic acids could inhibit PPO activity (Figure 5). In comparison with traumatic acid, lauric acid and dodecanedioic acid showed lesser inhibitory activity. Thus, it was expected that the inhibitory activity was not essentially caused by the two carboxylic groups, but that the double-bond at the α, β position was important in exerting the inhibitory activity. Also, the inhibitory activity was not essentially related to the length of carbon chain. Among crotonic acid, tiglic acid, sorbic acid and isobutyric acid, sorbic acid was revealed to be the most potent inhibitor, which suggested that nucleophilicity of a double-bond is an important factor in inhibiting the enzyme. This is also the case in comparison with hexenoic acid and sorbic acid, and

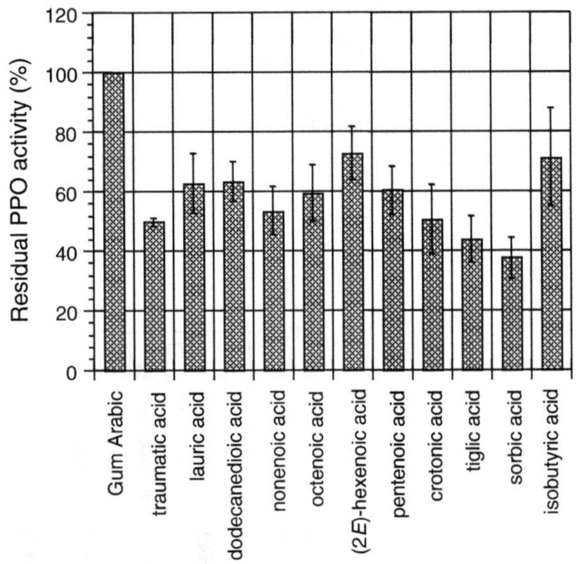

Figure 5. Inhibitory effect of aliphatic carboxylic acids. 500 μM aliphatic carboxylic acids were used.

higher activity was found with a compound having a conjugated double-bond (Figure 6).

The kinetics of inhibition by traumatic acid and sorbic acid were analyzed with a Lineweaver-Burk plot as shown in (Figure 7). The three lines, obtained from the uninhibited enzyme and from the enzyme with traumatic acid and sorbic acid, intersected on the vertical axis. The result demonstrates that traumatic acid and sorbic acid inhibit the oxidation of L-DOPA competitively.

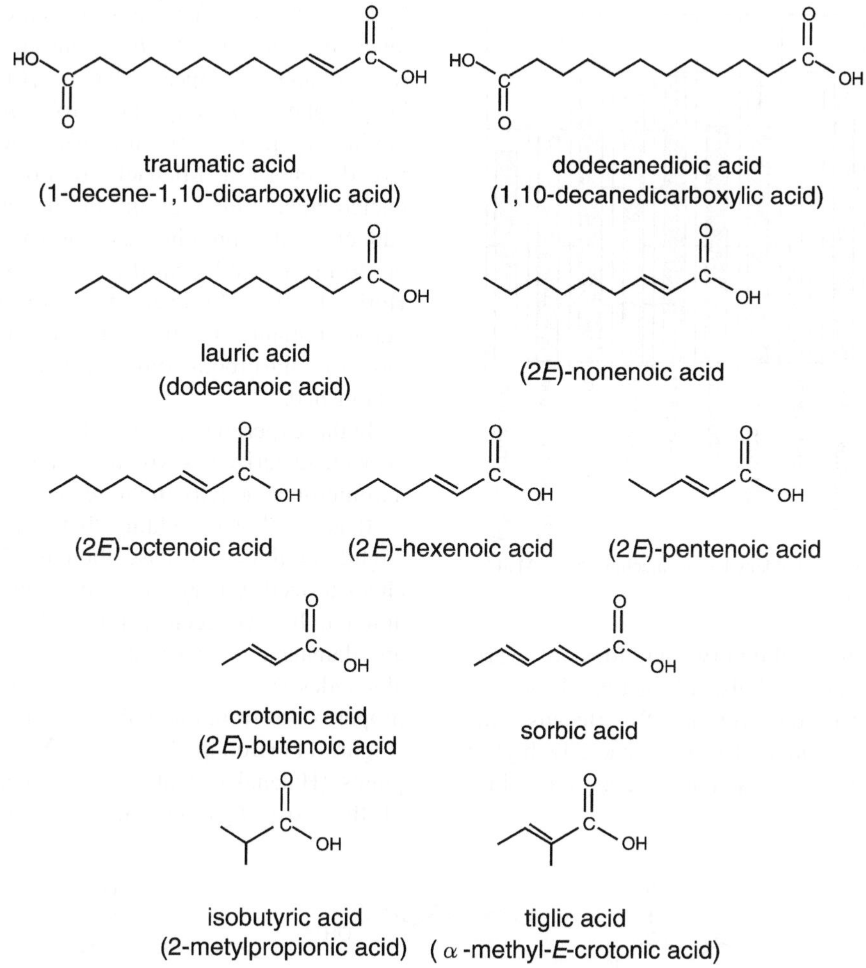

Figure 6. Structures of aliphatic carboxylic acids.

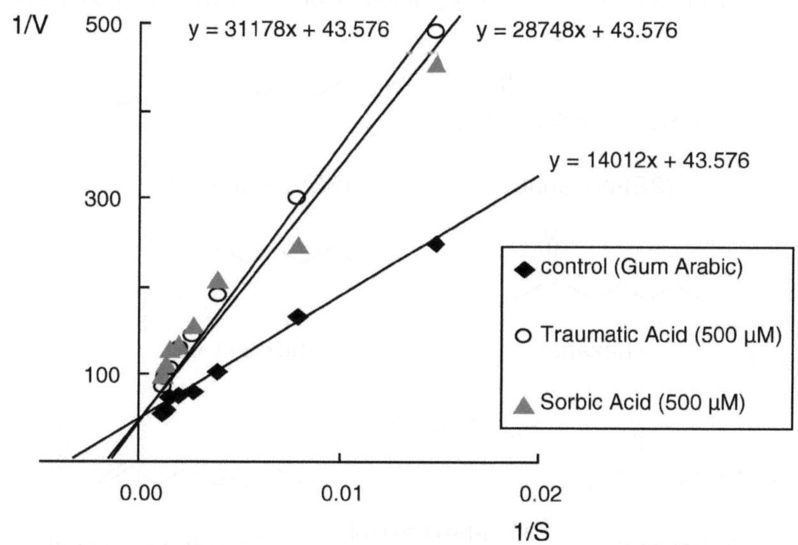

Figure 7. Lineweaver-Burk plots of tyrosinase inhibition with control (5% Gum Arabic: ■) and 500 μM of traumatic acid (○) and 500 μM of sorbic acid (▲).

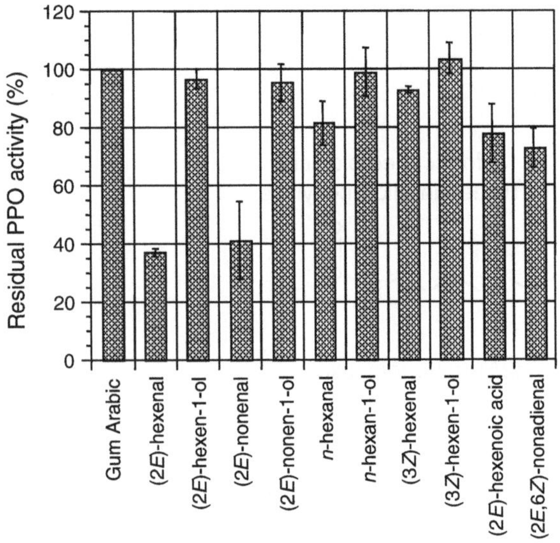

Figure 8. Inhibitory effect of aldehydes and alcohols. 500 μM aldehydes and alcohols were used.

In addition, the inhibitory activities of various aldehydes and alcohols, of saturated or unsaturated, are examined (Figure 8). Because inhibitory activity of aldehydes was generally higher than the corresponding alcohols, and the aldehydes with α, β-unsaturation showed higher potency than those with β, γ-unsaturation and saturated chains, it appears that α, β-unsaturated carbonyl plays an important role in exerting the inhibitory activity (Figure 9). These compounds would act as a Michael addition acceptor of a nucleophile, amino group in the enzyme to form adducts with proteins as shown in (Figure 10), or form a stable Shiff's base (Kubo & Kinst-Hori, 1999). However, the conclusive interpretation remains to be clarified since the structure of mushroom tyrosinase has not yet been established.

In this experiment, seaweed flavor compounds have been characterized as tyrosinase inhibitors having high antimicrobial activity in the seaweed.

It is worthwhile adding that *Laminaria* oil itself slightly inhibits tyrosinase, though (2*E*)-alkenals were characterized as a key flavor compound but in a minute amount. It was recently found that when seaweeds are damaged or macerated, these (2*E*)-unsaturated aldehydes and oxo-enoic acids are produced *via* hydroperoxides from linolenic acid and arachidonic acid (Figure 11) (Boonprab et al., 2003a,b), like in higher plants (Hatanaka et al., 1995). Chemical structures of the oxo-(2*E*)-unsaturated carboxylic acids have

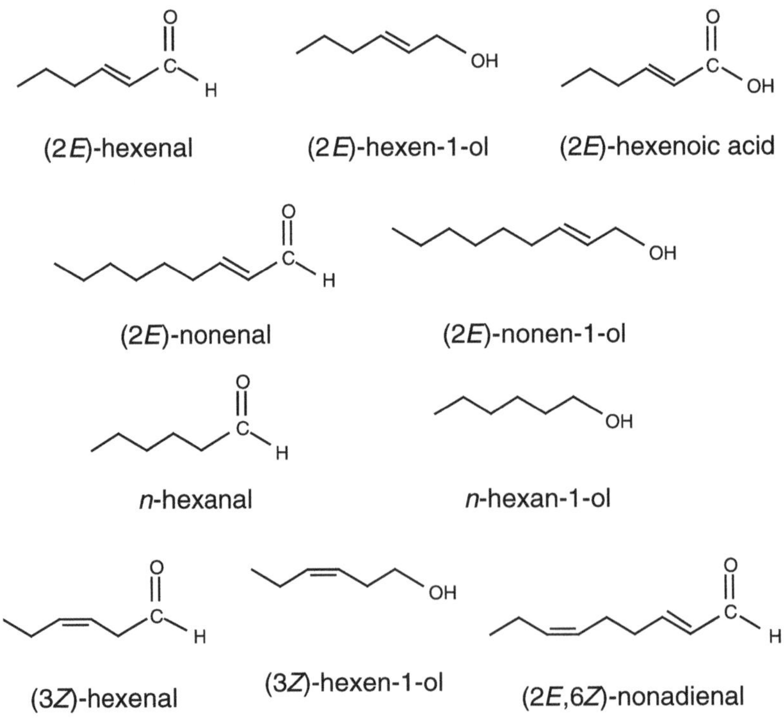

Figure 9. Structures of aldehydes and alcohols.

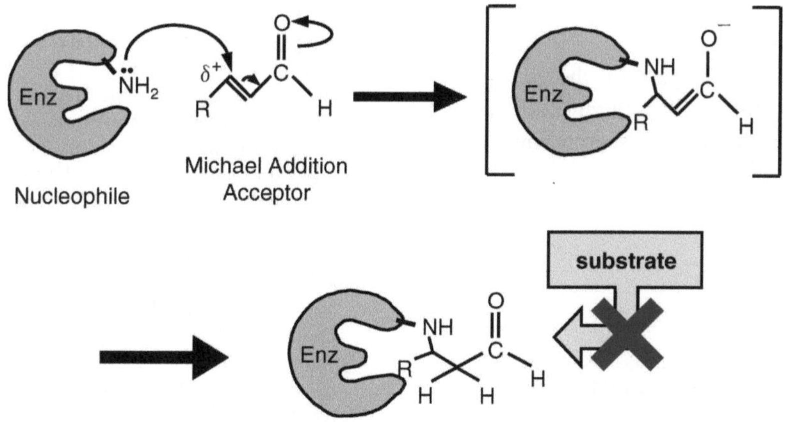

Figure 10. A possible PPO inhibition mechanism of α,β-unsaturated carbonyl compounds.

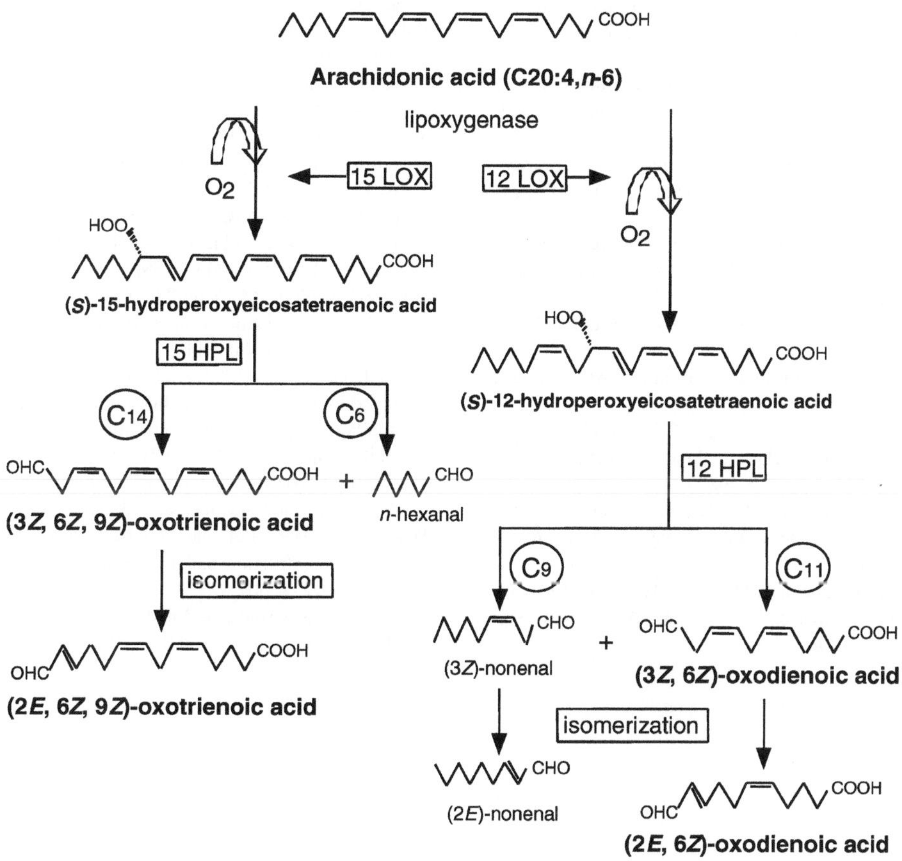

Figure 11. Proposed biosynthetic pathway for C6 and C9 aldehydes and C9–C14 oxoacids in brown algae.

both the combined structures of (2*E*)-unsaturated aldehyde moiety and (2*E*)-traumatic acid like moiety. This therefore clear that the treatment of cut vegetables with edible seaweeds juice incubated with polyunsaturated fatty acids such as linolenic acid, arachidonic acid, and icosapentaenoic acid and the combined treatment with synthetic (2*E*)-alkenals, oxoenoic acids and its analogues are worthy of further studies for prevention of vegetable browning and poisoning.

Acknowledgments

This work was supported in part by the San-Eigen Foundation for Food Chemical Research (2002–2003) and by the Japan Food Chemical Research Foundation (2004).

References

Boonprab K, Matsui K, Akakabe Y, Yotsukura N, Kajiwara T (2003a) Hydroperoxy-arachidonic acid mediated *n*-hexanal and (*Z*)-3- and (*E*)-2-nonenal formation in *Laminaria angustada*. Phytochemistry 63: 669–678.

Boonprab K, Matsui K, Yoshida M, Akakabe Y, Chirapart A, Kajiwara T (2003b) C6-Aldehyde formation by fatty acid hydroperoxide lyase in the brown alga *Laminaria angustata*. Z. Naturf. 58c: 207–214.

Conner DE (1993) Naturally occurring compounds. In Davidson, PM, Branen AL (eds), Antimicrobials in Food, Marcel Dekker, New York, pp. 441–468.

Davis PL, Smoot JJ (1972) Germination of *Penicillium digetatum* spores as affected by solution of volatile compounds of citrus fruits. Phytochemistry 62: 488–489.

Flament I, Ohloff G (1984) Volatile constituents of algae. Odoriferous constituents of seaweeds and structure of nor-terpenoids identified in Asakusa-nori flavor. In Adda J (ed.), Progress in Flavor Research, Elsevier Science Publishers, Amsterdam B. V., pp. 281–300.

Friedman H (1996) Food browning and its prevention: An Overview. J. Ag. Food Chem. 44: 630–653.

Glombitza KW (1979) Antibiotics from algae. In Hoppe HA, Levring T., Tanaka T (eds), Marine Algae in Pharmaceutical Science, Walter de Gruyter, Berlin, pp. 303–342.

Hatanaka A, Kajiwara T, Matsui K (1995) The biogeneration of green odor by green leaves and its physiological functions-past, present and future. Z. Naturf. 50: 467–472.

Kajiwara T, Kodama K, Hatanaka A, Matsui K (1993) Volatile compounds from Japanese marine brown algae. Am. Chem. Soc. Symp. Ser. 525: 103–120.

Kajiwara T, Matsui K, Akakabe Y (1996) Biogeneration of volatile compounds via oxylipins in edible seaweeds. Am. Chem. Soc. Symp. Ser. 637: 146–166.

Kermasha S, Goetghebeur M, Monfette A, Metche M, Rovel B (1993). Inhibitory effects of cysteine and aromatic acids on tyrosinase activity. Phytochemistry 34: 349–353.

Kubo A, Christopher SL, Kubo I (1995) Antimicrobial activity of the olive oil flavor compounds. J. Ag. Food Chem. 43: 1629–1633.

Kubo I, Kinst-Hori I (1999) Tyrosinase inhibitory activity of the olive oil flavor compounds. J. Ag. Food Chem. 47: 4574–4578.

Kubo I, Kinst-Hori I, Kubo Y, Yamagiwa Y, Kamikawa T, Haraguchi T (2000) Molecular design of antibrowning agents. J. Ag. Food Chem. 48: 1393– 1399.

Maruzzella JC, Sicurella NA (1960) Antibacterial activity of essential oil vapors. J. Am. Pharmaceut. Ass. 49: 692–694.

Mayer AM (1987) Polyphenol oxidase in plants-recent progress. Phytochemistry 26: 11–20.

Morris JA, Khettry A, Seitz EW (1979) Antimicrobial activity of aroma chemicals and essential oils. JAOCS 56: 595–603.

Nakamura S, Hatanaka A (2002) Green-leaf derived C6-aroma compounds with potent antibacterial action that act on both gram-negative and gram-positive bacteria. J. Ag. Food Chem. 50: 7639–7644.

Negishi O, Ozawa T (2000) Inhibition of enzymatic browning and protection of sulfhyryl enzymes by thiol compounds. Phytochemistry 54: 481–487.

Ogawa M, Perdigao NB, Santiago ME, Kozima TT (1984) On physiological aspects of black spot appearance in shrimp. Bull. Japn. Soc. Sci. Fish. 50: 1763–1769.

Schultz TH, Flath RA, Mon TK, Eggling SB, Teranishi R (1977) Isolation of volatile compounds from a model system. J. Ag. Food Chem. 25: 446– 449.

Suhr KI, Nielsen PV (2003) Antifungal activity of essential oils evaluated by two different application techniques against rye bread spoilage fungi. J. Appl. Microbiol. 94: 665–674.

Vaughn SF, Spencer GF, Shacha BS (1993) Volatile compounds from raspberry and strawberry fruit inhibit postharvest decay fungi. Journal of Food Science 58: 793–806.

Whitaker JR (1995) Polyphenol oxidase. In Wong DWS. (ed.), Food Enzymes, Structure and Mechanism, Chapman & Hall, New York: pp. 271–307.

Zaika LL (1988) Spices and herbs: their antimicrobial activity and its determination. J. Food Safety 9: 97– 118.

Journal of Applied Phycology (2006) 18: 423–435
DOI: 10.1007/s10811-006-9053-7

Effect of detachment on the palatability of two kelp species

Eva Rothäusler[1,2] & Martin Thiel[2,3,*]
[1]*Institut für Aquatische Ökologie, Albert-Einstein-Str.3, 18057 Rostock, Germany; e-mail: evarot@gmx.de;* [2]*Facultad Ciencias del Mar, Universidad Católica del Norte, Larrondo 1281, Coquimbo, Chile;* [3]*Centro de Estudios Avanzados en Zonas Áridas (CEAZA), Coquimbo, Chile*

*Author for correspondence: e-mail: thiel@ucn.cl; fax: +56 51209812

Key words: defence, dislodgement, detachment, floating, macroalgae, palatability

Abstract

Many species of macroalgae survive after becoming dislodged from their primary substratum, but little is known about their capacity to express anti-herbivore defences after detachment. We examined the effect of detachment on the relative palatability of the two kelp species *Lessonia nigrescens* and *Macrocystis integrifolia* to mesograzers. Laboratory and field experiments were conducted on the northern-central coast of Chile to investigate whether (i) time after detachment and (ii) grazing on detached and attached algae could trigger internal defence mechanisms in the algae, which may have acted as deterrents to grazing. In order to examine palatability, feeding assays were run after each experiment using fresh algal pieces and artificial food. Time after detachment had a significant influence on palatability of *L. nigrescens* but not of *M. integrifolia*. During the first 12 days of detachment, detached *L. nigrescens* held in grazer-free laboratory tanks were not significantly more palatable than attached conspecifics from the field but thereafter detached individuals became more palatable. Floating individuals of *M. integrifolia* showed no effect of detachment, indicating that this alga maintains its defence after detachment. An experiment conducted in the field confirmed these results for *M. integrifolia*. An additional laboratory experiment confirmed that attachment status plays an important role on algal defence reaction for *L. nigrescens* when exposed to grazers. Detached and previously grazed individuals of this species were less palatable than grazer-free control algae, but grazing had no effect on palatability of attached algae. Our results indicate that kelps have varying capacities for development of anti-grazing responses once they become detached, possibly depending on their capacity to float and survive after detachment.

Introduction

Defence mechanisms against herbivorous grazers have been reported for numerous species of marine macroalgae (e.g. Van Alstyne, 1988; Paul & Van Alstyne, 1992; Sotka et al., 2002). These mechanisms include (i) morphological defences such as calcification of structures, increase in tissue toughness and/or modification of growth form (Littler & Littler, 1980; Hay et al., 1988; Hay, 1991) and (ii) chemical defences including the production of noxious or unpalatable chemical compounds, which are termed secondary metabolites, because usually they are not associated with the primary metabolism of the plant (Bazzaz et al., 1987; Duffy & Hay, 2001). Among chemical defences three mechanisms have been recognized: (1) constitutive defence, where secondary metabolites are produced continuously, independent of attack or presence of grazers (e.g. Pavia & Toth, 2000), (2) inducible defence, where algae enhance production of secondary metabolites when under attack by herbivores (e.g. Sotka et al., 2002), and (3) activated defence triggered by injury and acting extremely rapidly (seconds to minutes) by converting a less potent stored secondary metabolite to a more potent one (e.g. Paul & Van Alstyne, 1992; Cetrulo & Hay, 2000). The three mechanisms function as herbivore deterrents, and they have been reported from a wide diversity of different macroalgae, including brown, green, and red algae (e.g. Steinberg, 1984; Paul & Fenical, 1986; Van Alstyne, 1988; Peckol et al., 1996; Cetrulo & Hay, 2000; Pavia & Toth, 2000; Van Alstyne et al., 2001; Sotka et al., 2002; Taylor et al., 2002).

Two classical approaches have been employed to test for the presence of chemical defences in marine macroalgae. The first is based on an empirical method where algae that were exposed to either: (i) naturally occurring herbivory, (ii) experimental grazing levels or (iii) artificial injury *in situ*, are collected from the field and evaluated for the presence of chemical or morphological defences, or (iv) they were injured after collection (e.g Van Alstyne, 1988, 1989; Paul & Van Alstyne, 1992; Peckol et al., 1996; Cronin & Hay, 1996a,b; Pavia et al., 1997; Hammerstrom et al., 1998; Cetrulo & Hay, 2000). In the second type of studies, algae are usually maintained detached in experimental outdoor flow-through systems, where they are grown in tanks of variable sizes and are treated with different grazing regimes and after exposure tested for the presence of defensive mechanisms (e.g. Toth & Pavia, 2001; Sotka et al., 2002; Taylor et al., 2002). Although this allows all algae in a tank to be exposed to the same light and nutrient regimes over time, it does not necessarily reflect the natural condition where algae are growing attached to a firm substratum.

Benthic algae, particularly in shallow areas, are exposed to varying degrees of wave and water currents, which may dislodge them from the substratum. This detachment can cause physical stress for the plants, possibly altering seaweed palatability, as is known in cases of desiccation (Renaud et al., 1990). Stressed algae often show a limited potential for chemical defence (Renaud et al., 1990; Cronin & Hay, 1996b), and it is expected that this is also true for detached algae. However, little is known about the effect of detachment on the presence of defence mechanisms in marine macroalgae even though this appears important in understanding the mechanisms of chemical defence in a natural situation. Understanding the relationship between detachment and defence appears important since assemblages of unattached seaweeds are commonplace in nature (e.g. Benz et al., 1979; Ólafsson et al., 2001; Hirata et al., 2001; Thiel & Gutow, 2004). Storms frequently cause detachment of benthic macroalgae (Norton & Mathieson, 1983), which then contribute to floating or drifting populations. Some macroalgae possess gas bladders (e.g the giant kelp *Macrocystis pyrifera*) or a plant body that temporarily acts as a balloon (e.g. the entire thallus of *Colpomenia perigrina*) (Norton & Mathieson, 1983), allowing them to float at the sea-surface. Other species of algae have no floating potential and after detachment sink to the sea-floor where they might contribute to a species-rich assemblage of drifting macroalgae in shallow waters (Benz et al., 1979; Norton & Mathieson, 1983; Norkko & Bonsdorff, 1996).

After detachment, macroalgae may be exposed to intense herbivory (biological stress). For example, floating macroalgae harbour many animals, including herbivores, that have originally been living on them and thus can quickly consume them (Ingólfsson, 1995, 1998). Also drift-algae in shallow waters are commonly inhabited by a wide diversity of mesograzers, in particular amphipods and isopods (Inglis, 1989; Geertz-Hansen et al., 1993; Ingólfsson, 2000; Brooks & Bell, 2001). Floating algae may survive for extended periods at the sea surface (Hobday, 2000) despite high abundances of grazers, but drifting algae often face high grazing pressure from a wide diversity of benthic grazers and may have little chance to survive for long (see e.g. Rodriguez, 2003). Thus, it can be hypothesized that kelp species that float after detachment may be capable of maintaining their defence after detachment, while kelp species that sink after detachment may lose their defence capacity shortly after detachment.

Here we tested whether two kelp species from the SE-Pacific, that differ in their behaviour (floating or drifting) after detachment, maintain their defences, or whether they loose them due to physiological alterations caused by detachment. Plants of *Lessonia nigrescens* are negatively buoyant and sink to the seafloor after detachment where they may contribute to a large pool of drifting algae. In contrast, *Macrocystis integrifolia* floats after detachment and may travel for a long time with ocean currents (e.g. Helmuth et al., 1994). Field and laboratory experiments were conducted on the northern temperate coast of Chile in order to learn how detachment affects the palatability and thus the defence capacity of these two macroalgae.

Materials and methods

Both laboratory and field experiments were carried out to test the changes in palatability following extended detachment of the two kelp species. Laboratory experimentation was done with *Lessonia nigrescens* and *Macrocystis integrifolia* while field experimentation was carried out with *M. integrifolia* only. An additional laboratory experiment was conducted with *L.*

nigrescens to test whether defence can be induced in attached as well as in detached plants of this species.

Collection and culture conditions of macroalgae

Kelps used in the laboratory were collected by hand during low-tide in the vicinity of Coquimbo, Chile (Figure 1). After collection, algae were kept in a cooler at their ambient temperature and immediately transferred to seawater tanks at the Marine Seawater-Laboratory of Universidad Católica del Norte, Coquimbo. Grazers and epiphytes were carefully removed by hand from test algae prior to experiments. Laboratory experiments were either conducted in a large flow-through seawater tank, containing ≈ 1800 L of seawater, or in plastic aquaria measuring 10 × 19 × 13 cm and containing ≈ 1.5 L of seawater. The tank was supplied with an air pipe and with flowing seawater. The large tank was used to test for the effect of extended detachment in *L. nigrescens* and *M. integrifolia* whereas the small aquaria were used for the induction experiment with *L. nigrescens*.

The small aquaria received filtered (10 μm cotton cartridge) seawater that was continuously pumped from the shallow subtidal waters of Bahía La Herradura into 4 plastic reservoirs (70 L) from where it was then redistributed. Flow regulators were used to supply each aquarium with an individually-controlled flow rate (≈ 0.1 L h^{-1}). Aquaria were additionally maintained with continuous aeration. All laboratory experiments were conducted in outdoor tanks and aquaria with algae were shaded with a black plastic cloth in order to protect them from direct sunlight.

The field experiment with *M. integrifolia* was carried out at Isla Damas (Figure 1) under natural conditions. With this field experiment we also tested the effect of extended detachment.

Mesograzers used in experiments and feeding assays

The consumers used to determine algal palatability after the laboratory and field experiments were either the amphipods *Parhyalella ruffoi* and *Hyale hirtipalma* or the isopod *Isocladus bahamondei*. The amphipod *P. ruffoi* was also used in the grazing treatment in the induction experiment. All mesograzers were previously observed living and feeding on a variety of different macroalgae (e.g. Thiel, 2002), indicating that they are generalist grazers not specialized to a particular algal species. Amphipods and isopods were collected from mixed assemblages of drift algae at Playa Guayacán and from intertidal algae in La Pampilla (Figure 1).

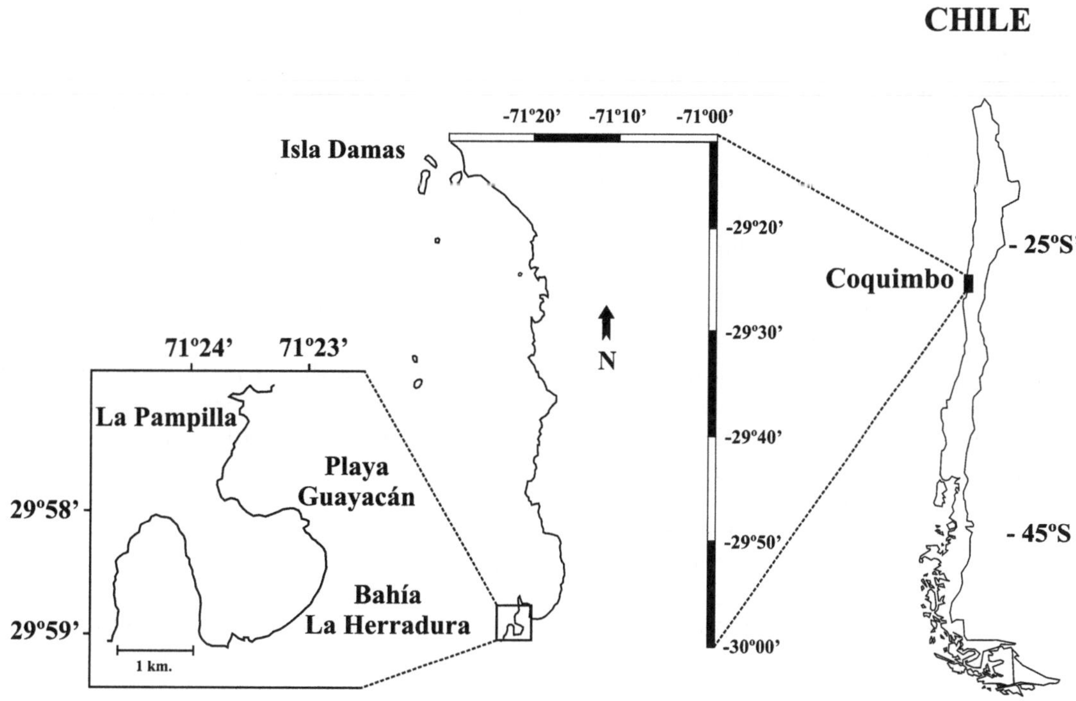

Figure 1. Map of Chile with the locations of the sampling sites in the vicinity of Coquimbo.

Freshly collected individuals of all grazers (*P. ruffoi, H. hirtipalma and I. bahamondei*) were used in feeding assays after each experiment in order to examine the palatability and thereby the defensive mechanisms of the selected macroalgae.

General design of the feeding assays to examine algal palatability

All feeding assays were conducted in a culture room (12 h L: 12 h D; 15 ± 1 °C; light intensity 40 ± 10 μm photons m^{-2} s^{-1}). Choice and no-choice feeding assays were done mainly with fresh algal pieces but also with artificial agar-based food (see below). In choice feeding assays the grazers were offered two algal pieces simultaneously, such that they had the possibility to express a dietary choice. In the no-choice feeding assays grazers could only feed on the offered alga or they would starve. Feeding assays were conducted in petri-dishes (diameter ≈ 8.8 cm; volume ≈ 30 mL). For feeding assays with fresh algal pieces, we determined the amount consumed in mg, and for assays with artificial pieces we counted the number of squares consumed (see below for details). All feeding assays were carried out with five replicates ($n = 5$), unless noted otherwise.

Preparation of artificial food

After completing the respective algal treatments, a lipophilic crude extract was obtained from one piece of each alga. The extraction was done for 48 h with dichloromethane according to a 1:2 ratio (1 g algae wet weight: 2 mL dichloromethane). Dichloromethane extracts only the lipophilic compounds, and thus any non-lipophilic compounds, which may also contain feeding deterrents, are disregarded by this procedure. The lipophilic extract was dropped on freeze-dried and finely powdered *Ulva lactuca.* The dichloromethane then evaporated leaving behind the compounds extracted from the algal tissues. To determine the amount of *Ulva* powder for every sample, algae were weighed before adding the dichloromethane. A ratio 3:1 (g alga wet weight: g dried *Ulva* powder) was used in order to obtain approximately similar proportions between the dried powder and the wet mass of the algae. After evaporation of the lipophilic extract, 8 mL distilled water was added to the *Ulva* powder. A specific amount of agar (0.36 g) was mixed with 10 ml distilled water and heated to boiling point in a microwave. After the agar cooled down to 40 °C the *Ulva* powder with the lipophilic crude extract was added. The agar/*Ulva* mix was poured into a mold lying over a fly mesh (mesh size 1 mm^2) and 200 squares were cut out after hardening. The artificial food went directly into feeding assays.

The experiments

1. *Effect of extended detachment on the palatability of two kelps in an experimental tank*: This laboratory experiment was run in the large outdoor tank between March and May 2003. The aim was to examine whether kelps detached from their primary substratum become more palatable to mesograzers after 3, 6, 12, 24 and 42 days of detachment compared to attached conspecifics taken directly from their natural substratum in the field. Detached algae were held in a grazer-free environment in the laboratory tank, while attached conspecifics in the field were exposed to the natural grazing regime.

Apical parts of *M. integrifolia* were collected from Isla Damas and those of *L. nigrescens* from La Pampilla. Algae were maintained together with two other detached algal species (*Ulva* sp. and *Cryptomenia obovata*) in the large tank since detached algae also accumulate in multi-species assemblages in the field. The algae in the tank (laboratory treatment) were stirred briefly, twice a day, to provide some water movement. Palatability was tested after 3, 6, 12, 24 and 42 d in choice feeding assays with detached individuals from the tank (laboratory treatment) and attached individuals from the field (field treatment) that were collected on the same day. In each petri-dish ($n = 5$) we placed three amphipods (*P. ruffoi*) and allowed them to feed for three days.

2. *Effect of detachment on the palatability of Macrocystis integrifolia in the field*: In order to test the detachment effect on the palatability of apical parts of *Macrocystis integrifolia*, a field experiment was carried out from January to May 2003 at Isla Damas. Both floating and natural attached plants were exposed to natural grazer conditions in the field. Algae were detached from the primary substratum in January 2003, placed in wide-mesh bags (mesh size: 65 mm), and tethered at the sea surface for the duration of the experiment. Palatability was tested after 6, 24, 42, 67 and 97 d in choice feeding assays with apical blades from tethered floating and from natural attached plants. In each petri-dish ($n = 5$) we placed five amphipods (*H. hirtipalma*) and allowed them to feed for three days. Additionally, we conducted identical feeding assays in which we used six isopods (*I. bahamondei*) instead of the amphipods.

3. *Induction experiment with attached versus detached Lessonia nigrescens*: An outdoor laboratory experiment was conducted during austral fall (May) 2003 to examine whether attached or detached *L. nigrescens* plants are more susceptible to amphipod grazing. A total of 35 complete juvenile plants (≈18 g) of the brown alga *Lessonia nigrescens* were sampled randomly during low tide in the intertidal zone of La Pampilla (see Figure 1). During collection, all juvenile plants were carefully removed with their complete holdfast from the substrate using a scalpel. The holdfasts of these juvenile plants had a diameter of approximately 8 cm and a blade length (5 – 8 blades per plant) of 10 – 15 cm. Five plants were frozen at −40°C immediately after sampling in order to represent the natural level of defence (natural attached algae). The remaining 30 plants were distributed individually over 30 aquaria, each with a volume of 1.5 L and continuously flowing seawater. To each of the aquaria we added one complete juvenile plant of *L. nigrescens* but in one half of the aquaria ($n = 15$) the algae were maintained as detached individuals whereas in the other half ($n = 15$) the juvenile plants were carefully sewn with their holdfasts onto the aquaria wall, representing attached plants. All algae continued to grow during the experiment, and the "attached" plants firmly grew onto the plastic wall of the aquarium.

The experiment was separated into an acclimation and a treatment phase, each lasting 10 d. The acclimation phase was included to adjust the defence level after an unknown consumption history in the field. After the acclimation phase, small apical blades were taken from 5 attached and from 5 detached plants and stored in the freezer for 2 d at −40°C. All frozen algae pieces were later compared in artificial food feeding assays (see above).

In the following treatment phase, the factor direct grazing with *P. ruffoi* was tested to examine whether grazing attacks by amphipods might induce defences in detached as well as in attached algae. The algae were exposed to two different grazing levels (direct grazing and grazer-free control): 10 amphipods *P. ruffoi* each were added to one half of the detached *L. nigrescens* ($n = 5$) and to one half of the attached *L. nigrescens* ($n = 5$), while the other half of the detached ($n = 5$) and attached *L. nigrescens* ($n = 5$) were left without grazers as control treatments. After the treatment phase, small apical blades were cut off from all algae in order to carry out feeding assays with artificial food as well as with fresh algae pieces.

We conducted choice-feeding assays with 4 individuals and no-choice feeding assays with 2 individuals of *P. ruffoi*. Choice-feeding assays with artificial food (made from the frozen apical blades) were carried out after the acclimation phase to examine for differences between (A1) the natural attached and the control detached algae pieces, and between (A2) natural attached and control attached pieces. After the treatment phase, choice-feeding assays were carried out (T1) between grazed detached and control detached pieces, (T2) between grazed attached and control attached algae pieces, (T3) between control attached and control detached, and (T4) between grazed attached and grazed detached. Additionally no-choice feeding assays were done (A3; T5). All feeding assays with artificial food lasted for 2d. The fresh algae choice and no-choice feeding assays were conducted with the same treatment combinations as described for the artificial food assays, with the exception that the assays were terminated after 3d.

Statistical analysis

The same statistical procedure was used for the laboratory and the field experiment, testing for the effect of extended detachment and grazing on algal palatability. Choice feeding assays, testing for differences in palatability (consumption) between attached and detached algae were analyzed with a t-test for dependent samples for each sampling date, rather than with a 2-way ANOVA, because some data sets contained negative consumption values caused by algal growth during the assays. Prior to analysis the data were inspected for normality, using Cochran's test. When data were normal a t-test was used, or alternatively a non-parametric Wilcoxon matched pairs test was conducted.

The consumption data from the induction experiment with *L. nigrescens* were examined for normality using the Cochran's test and ln $(x + 1)$ transformed if necessary. Choice feeding assays were analyzed with a t-test for dependent samples. No-choice feeding assays were analyzed with a 2-way ANOVA, with the fixed factors grazing and attachment status, or with the corresponding non-parametric Kruskall-Wallis test. When the ANOVA revealed significant differences, a post-hoc Tukey HSD was applied.

Results

1. *Effect of extended detachment on the palatability of two kelps in an experimental tank*: In the assays designed to measure the effect of detachment

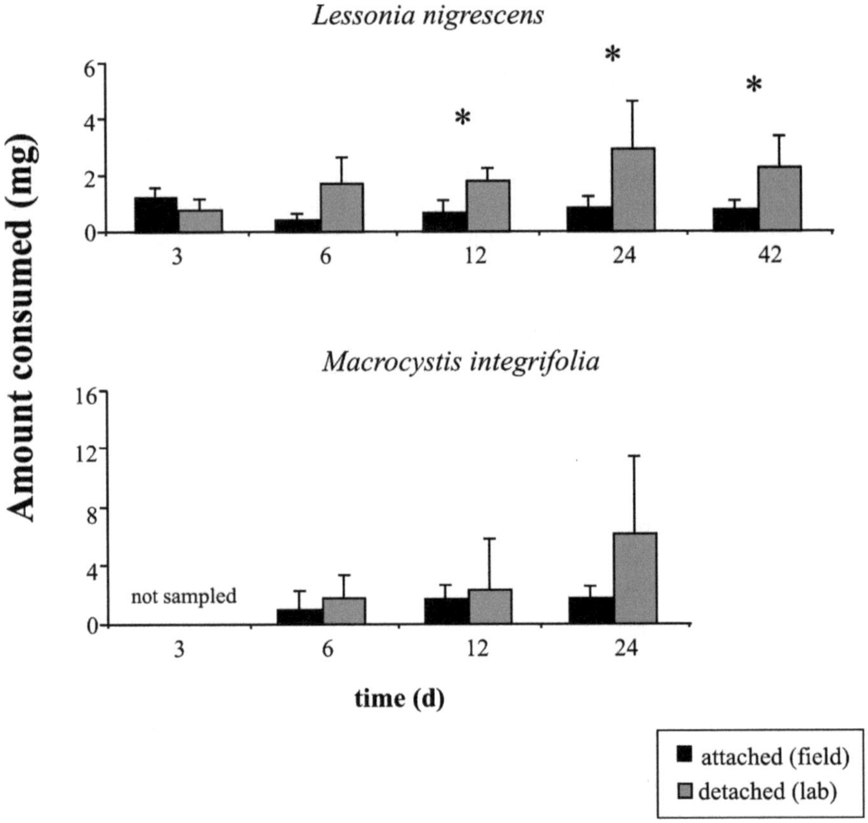

Figure 2. Mean consumption (mg) by the amphipod *Parhyalella ruffoi* of *Lessonia nigrescens* and *Macrocystis integrifolia* after different days of detachment. At each sampling date, grazers were offered a choice between plants from the field (attached) and plants from the experimental tank (detached); error bars represent + 1 SD ($n = 5$). ($^*p < 0.05$).

on palatability of the two kelps *Lessonia nigrescens* and *Macrocystis integrifolia,* a clear tendency was evident for *L. nigrescens* (Figure 2). The palatability of detached *L. nigrescens* held in the grazer-free laboratory tank increased over time. At days 12, 24 and 42 after detachment, the grazer *P. ruffoi* consumed significantly more from the detached pieces (tank) than from the attached pieces (*t*-test for dependent samples, $t = -7.097$, df = 4, $p = 0.002$, Wilcoxon matched pairs test $p = 0.043$ and Wilcoxon matched pairs test $p = 0.043$, respectively, $n = 5$) (Figure 2). Consequently, for *L. nigrescens* the time after detachment appeared to play an important role in the expression of defensive mechanisms. In contrast, the giant kelp *M. integrifolia* did not reveal any significant differences during any of the sampling dates (*t*-test for dependent samples, $p > 0.05$).

2. *Effect of detachment on the palatability of Macrocystis integrifolia in the field*: The time after detachment (6, 12, 24, 42, 67 and 97 days) had no apparent effect on the palatability of *M. integrifolia* blades from detached compared to attached plants (Figure 3). No statistical differences were detected between the algae tips neither for the feeding assays using the amphipod *H. hirtipalma* nor for those with the isopod *I. bahamondei* (Table 1).

Table 1. Result of *t*-test for dependent samples for the mean consumption (mg) of *Macrocystis integrifolia* by the two grazers after different time intervals. wx = Value from Wilcoxon matched pairs test

Time intervals (d)	d.f.	*Hyale hirtipalma*		*Isocladus bahamondei*	
		t	*p*	*t*	*p*
6	4	0.007	0.995	2.347	0.079
12	4	1.173	0.306	1.273	0.272
24	4	−0.074	0.945	0.227	0.831
42	4	0.038	0.971	0.524	0.627
67	4	1.632	1.178	1.608	0.183
97	4	0.865	0.436	wx	0.08

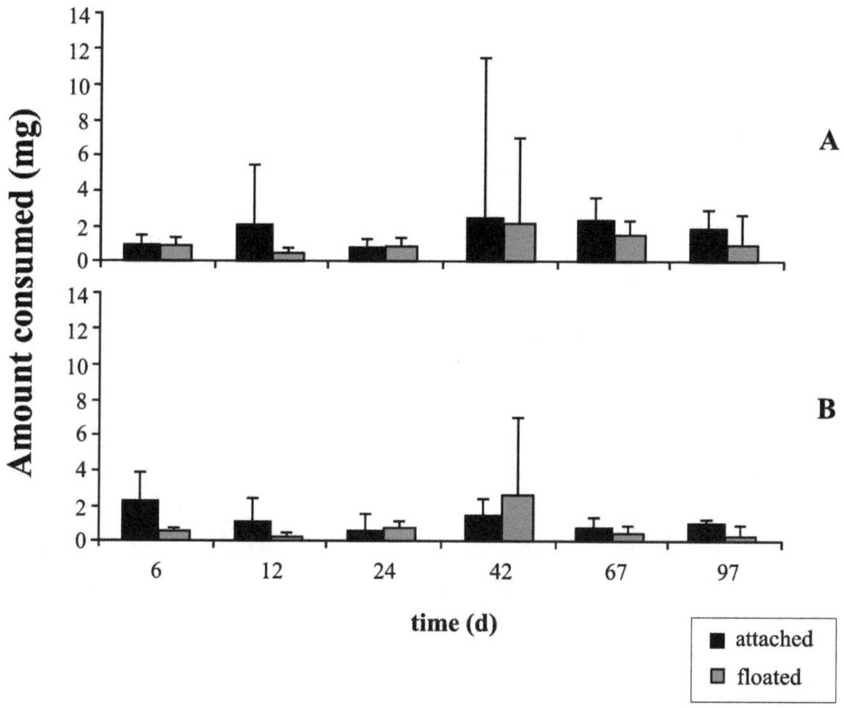

Figure 3. Mean consumption (mg) of *Macrocystis integrifolia* by the grazers: (A) *Hyale hirtipalma* and (B) *Isocladus bahamondei* after different sampling dates (d). Grazers were offered simultaneously an artificially floated alga piece and an alga piece from plants growing on substratum (attached); error bars represent +1 SD ($n = 5$).

3. *Induction experiment with attached versus detached Lessonia nigrescens*: After the acclimation phase, when offering the amphipod *P. ruffoi* a choice between artificial food made from attached and detached algae (A1), and between those made from control or natural algal pieces (A2), no statistical differences were detected (Figure 4A, *t*-test for dependent samples, $t = 0.507$, df $= 4$, $p = 0.639$ and Figure 4B, *t*-test for dependent samples, $t = 0.040$, df $= 4$, $p = 0.97$). The same result was found for the no-choice feeding assays with artificial food (A3) (Figure 4C, Kruskal-Wallis test, $p = 0.685$).

In the assays designed to measure the palatability in response to direct grazing after the treatment phase, we found statistical differences for both choice feeding assays (T1) with artificial and fresh algal pieces. Amphipods *P. ruffoi* consumed in the two assays (Figures 5A and 6A, respectively) significantly more of the detached ungrazed control pieces than from the detached grazed pieces (artificial food: Figure 5A, *t*-test for dependent samples, $t = -3.249$, df $= 3$, $p = 0.047$, $n = 4$; fresh algal pieces: Figure 6A, *t*-test for dependent samples, $t = -3.766$, df $= 3$, $p = 0.032$, $n = 4$). Thus, if differences in palatability were detected, these occurred in the treatments where detachment and grazing were combined. No further significant preferences were detected for the remaining artificial food and live algal choice feeding assays (T2–T4) (Figures 5B–D and 6B–D, respectively). The grazer *P. ruffoi* did not discriminate between attached or detached algal pieces treated by grazing, or controls. The two artificial food and fresh no-choice feeding assays (T5) (5E and 6E, respectively) did not display any significant differences in the palatability concerning the factors attachment status or grazing (Table 2).

Discussion

Our results indicate that extended detachment influenced the palatability of *Lessonia nigrescens* but not of *Macrocystis integrifolia*. Grazing also seemed to have an effect on the palatability of detached *L. nigrescens*, which were less palatable when previously exposed to grazers compared to grazer-free controls. For attached plants of *L. nigrescens* no such differences in palatability were found, which suggests that time after detachment may have an influence on defensive responses

Table 2. Results of 2-way ANOVA after the treatment phase for the mean consumption (mg) of *Lessonia nigrescens* from no-choice feeding assays by the grazer *Parhyalella ruffoi*

No-choice feeding assays treatment phase	Artificial food df	MS	F	P	Fresh alga df	MS	F	P
Attachment status (A)	12	52.874	0.352	0.854	15	0.000053	0.077	0.786
Grazing (G)	12	715.534	0.476	0.503	15	0.000001	0.001	0.972
G ∗ A	12	474.439	0.316	0.584	15	0.000042	0.061	0.809

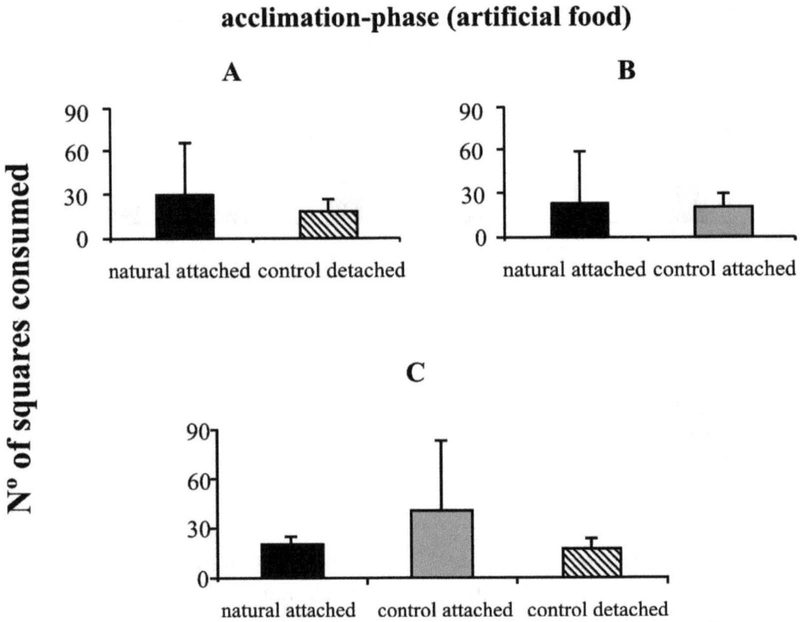

Figure 4. Mean (±SD) number of squares consumed of artificial *Lessonia nigrescens* by *Parhyalella ruffoi* after the acclimation phase. (A & B) choice feeding assays, and (C) no-choice feeding assays; error bars represent +1 SD ($n = 5$).

(and possibly other physiological processes) in marine macroalgae.

Detachment effect

Some species of macroalgae only exist as populations of detached individuals, such as for example the brown algae *Sargassum natans* and *S. fluitans*. Parr (1939) mentioned that both Fucales show no signs of attachment and lack reproductive organs. Their floating thalli represent an effective long-distance dispersal mechanism (Deysher & Norton, 1982) and may support survival of populations via asexual reproduction. Also, kelps that usually grow as attached individuals may persist after detachment. Detachment does not mean death for these plants, which may become entangled in kelp forests or float freely for extended periods (Hobday, 2000). During this floating time some (e.g. the kelp *M. integrifolia*) may even be reproductive (Macaya et al., 2005). Other algae sink to the seafloor after detachment, where they may form dense accumulations of drift algae (Norton & Mathieson, 1983; Ingólfsson, 1995; Norkko & Bonsdorff, 1996). Following detachment, algae may start to deteriorate (e.g. Hobday, 2000), which could be due to loss of antiherbivore defence making detached algae more palatable compared to attached conspecifics.

We found that the detached kelp *L. nigrescens* became more palatable to mesograzers 12 d after detachment if held without grazers (Figure 2). One reason for this could be that algae invested more energy to growth, which resulted in fewer resources available to the production of deterrents (Herms & Mattson, 1992). Rapid growth might also be a mechanism to tolerate future grazer attack. Detached algal species are exposed to a variety of environmental conditions (e.g. light, nu-

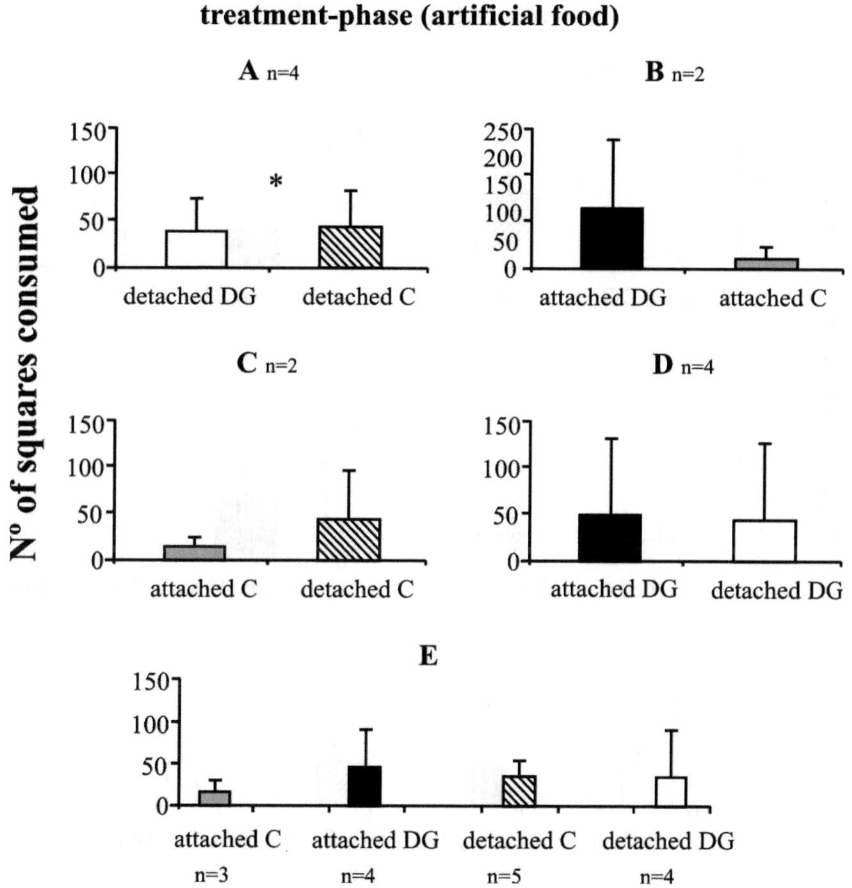

Figure 5. Mean (±SD) number of squares consumed of artificial *Lessonia nigrescens* by *Parhyalella ruffoi* after the treatment phase. Choice feeding assays are represented by the graphs A, B, C and D, while no-choice feeding assays are represented by the graph E. (DG = direct grazing treatment, C = control treatment). Error bars represent +1 SD Significance values are from *t*-test for dependent samples.

trients, temperature, herbivory) differing from those affecting attached plants (Norton & Mathieson, 1983). Possibly the conditions encountered by detached individuals are sub-optimal and thus they may have few extra resources available for defence. *Lessonia nigrescens* does not possess floating structures and consequently will sink to the bottom after detachment. Similarly, as in the natural environment after detachment, in the deep tank used in the first experiment, the detached thalli of *L. nigrescens* faced subtidal conditions while normally they are growing in the intertidal zone (Hoffmann & Santelices, 1997) where they are exposed to the air during each low tide. Thus plants may have experienced physiological stress in the tank after 12 d because they lay at the bottom of the tank at a depth of about 1 m. As a result, the detached plants in the tank received less light than attached control plants in the natural environment. If these changes in palatability after 12 d of detachment indeed were stress-induced changes then this process would be consistent with the result from Renaud et al. (1990) who found that desiccation of the unpalatable brown alga *Padina gymnospora* increased its palatability to sea urchins via loss of chemical defence. Physical features such as, for example, tissue toughness (Watson & Norton, 1985) may also affect herbivore choice. Possibly, over the extended detachment time in the experimental tank the detached individuals lost their toughness (resistance to penetration) and were therefore more susceptible to the grazer than freshly collected attached algae from the field. The results from Littler et al. (1983) indicated that herbivore preference is a function of the degree of seaweed toughness.

In contrast to these results, where an increase in palatability was observed beginning at day 12 after detachment, no detachment effect was observed for *L. nigrescens* after the 10 d of detachment during the acclimation phase (A1) in the small containers

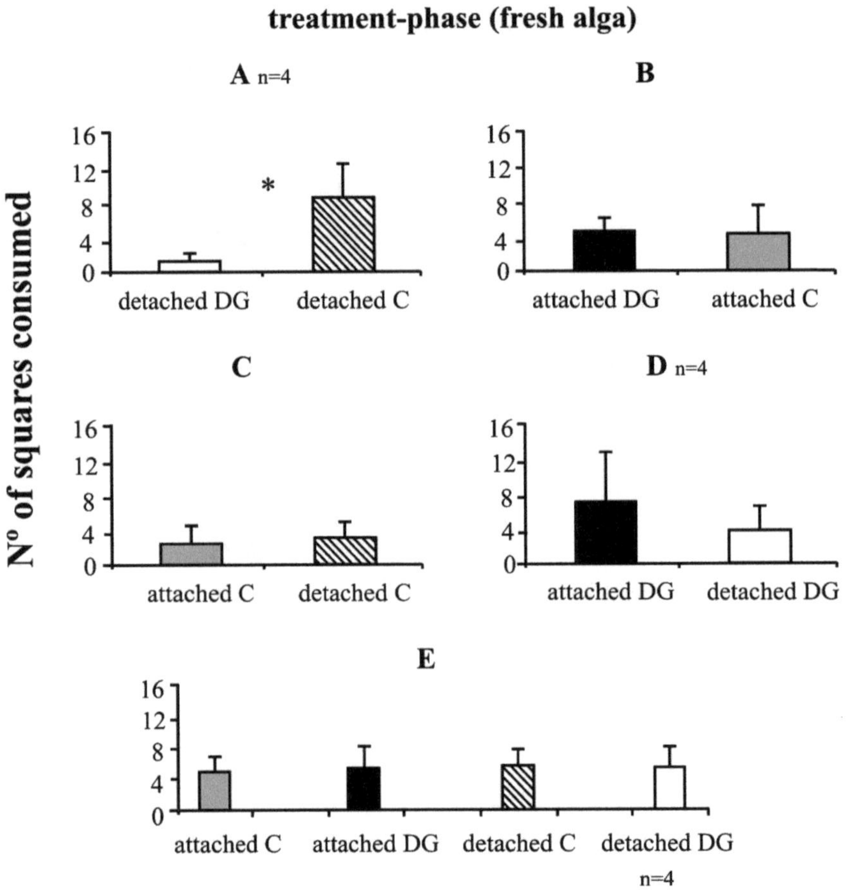

Figure 6. Mean consumption (mg) of fresh *Lessonia nigrescens* after the treatment phase by *Parhyalella ruffoi*. Choice feeding assays are represented by the graphs A, B, C and D , while no-choice feeding assays are represented by the graph E. (DG = direct grazing treatment, C = control treatment). Error bars represent +1 SD ($n = 5$). Significance values are from *t*-test for dependent samples. (*If the number of replicates are not noted above or below bars they are present with $n = 5$)

(Figures 4A). Since later a detachment effect (increase in palatability) occurred in this experiment, detachment effects apparently become important around 10–14 days after detachment.

No effect of extended detachment was observed for the palatability of the kelp *M. integrifolia* after 24 and 97 days of floating, neither in laboratory (Figure 2) nor in field experiments (Figure 3), suggesting that both detached and attached individuals were (i) either undefended or (ii) maintained their defence. We suggest that a permanently expressed defensive capacity exists in the apical tips of *M. integrifolia*. *Macrocystis pyrifera* plants that detach from their substratum float to the surface and may remain buoyant for a maximum of between 65 and 109 days before beaching or sinking (Hobday, 2000). Floating of this and other kelp plants is made possible by gas-filled pneumatocysts (Lobban & Harrison, 1994). Detached individuals continue to function physiologically for some time after detachment (Hobday, 2000). It has also been shown that individuals of *M. integrifolia* can reproduce after becoming detached (Macaya et al., 2005). Consequently, it might be advantageous for this macroalga to maintain its defensive capacity even after becoming detached. Zubia et al. (2003) reported that detached individuals of two floating kelp species, in comparison to attached conspecifics, had almost similar chemical composition (lipids, proteins, minerals, amino acids). This suggests that macroalgae, which have the capacity to float, may possess and maintain anti-herbivore defences after detachment.

Grazing effect

Grazing may have a strong effect on the community structure of macroalgae in coastal marine systems

(Lubchenco & Gaines, 1981). Algae frequently suffer attack by herbivores but many attached macroalgae can suppress palatability in order to deter herbivores. However, how do algae react against grazers after becoming detached from their primary substratum and while drifting on the sea floor or floating on the water surface? Detached algae are often colonized by dense assemblages of mesograzers, in particular amphipods and isopods (Geertz-Hansen et al., 1993; Ingólfsson, 2000; Brooks & Bell, 2001). It could thus be expected that these detached algae may be either (i) very susceptible to these grazers because of the physiological stress caused by detachment, or (ii) use defensive strategies in order to avoid high levels of herbivory.

In our first experiment, grazers were excluded during the entire duration of the extended detachment time, while in the attached control (plants from the field) grazers might have been present. The holdfasts of attached *L. nigrescens* for example, usually have many internal cavities used as a habitat by diverse invertebrate species (Hoffmann & Santelices, 1997; Thiel & Vásquez, 2000). This suggests that during the entire experiment the attached individuals of *L. nigrescens* (Figure 2), which were collected freshly from the field, may have been well defended against the tested grazer irrespective of the grazer pressure in the field (constitutive defence). However, detached plants may not have diverted extra-energy to production of deterrents when grazers were absent and thus lost their defensive capability. A similar pattern was detected for *L. nigrescens* (T1) in the induction experiment (Figures 5A and 6A). Detached plants, previously exposed for 10 days to grazers, were less palatable than grazer-free individuals. This supports our hypothesis that detached plants maintain their defensive capacity when grazers are present but lose it when grazers are absent.

Grazing appeared to have no influence on the palatability of apical parts from attached and detached kelp *Macrocystis integrifolia* since grazers exhibited no preferences in any of the tested choice-combinations (Figures 2 and 3). A comparatively homogeneous palatability was obtained in the laboratory experiment (Figure 2) between the grazer-free detached apical parts from the tank and the attached apical parts from the field where grazers usually are present. A similar pattern was observed in the field experiment (Figure 3) where both detached and attached algae were exposed to possible grazer attacks for 97 days. This suggests that detached and attached plants are equally defended, irrespective of the grazer intensity (constitutive defence). *Macrocystis* kelp forests are inhabited by a variety of invertebrate and vertebrate herbivores, many of which use the kelp plants as food (North, 1994). The tested plant parts, namely the upper parts of the plants, are principally consumed by fishes. A constitutive defence would be advantageous against herbivores that are very mobile, such as fish, that move to more palatable food after a few bites of distasteful food (Paul & Van Alstyne, 1992; Hay, 1996).

Conclusions

We report evidence that detached kelps *L. nigrescens* and *M. integrifolia* are defended against herbivores. Detached plants of *L. nigrescens* only maintained defence when grazers were present. In the absence of grazers, detached plants of *L. nigrescens* might invest energy for survival or attempt to overcome physiological stress, while attached individuals may have sufficient energy to maintain constant defence. In contrast to *L. nigrescens*, where detached plants may lose their defensive capability, apical parts of *M. integrifolia* apparently maintain their defence even after being detached for 97 days. In summary, our results indicated that kelps, once detached from their natural substratum, can respond with defence against herbivorous enemies but they responded differently, possibly depending on their capacity to float (in the case of *M. integrifolia*) or the presence of grazers (in the case of *L. nigrescens*).

Acknowledgments

We are extremely grateful to AS Chapman for many constructive comments that helped to improve the manuscript substantially. Financial support was obtained via fellowships from the GAME-project to ER (project leaders Martin Wahl and Markus Molis at UKiel) and through project FONDECYT 1010356 to MT. We are grateful to the staff of the botany laboratory at UCN for their continuous support during this study. Also special thanks to all the biologists from the BEDIM laboratory at UCN.

References

Bazzaz FA, Chiariello NR, Coley PD, Pitelka LF (1987) Allocating resources to reproduction and defense. BioScience 37: 58–67.

Benz CB, Eiseman NJ, Gallaher EE (1979) Seasonal occurrence and variation in standing crop of a drift algal community in the Indian River, Florida. Bot. Mar. 22: 413–420.

Brooks RA, Bell SS (2001) Mobile corridors in marine landscapes: Enhancement of faunal exchange at seagrass/sand ecotones. J. Exp. Mar. Biol. Ecol. 264: 67–84.

Cetrulo GL, Hay ME (2000) Activated chemical defenses in tropical versus temperate seaweeds. Mar. Ecol. Prog. Ser. 207: 243–253.

Cronin G, Hay ME (1996a) Induction of seaweed chemical defenses by amphipod grazing. Ecology 77: 2287–2301.

Cronin G, Hay ME (1996b) Susceptibility to herbivores depends on recent history of both the plant and animal. Ecology 77: 1531–1543.

Deysher L, Norton TA (1982) Dispersal and colonization in *Sargassum muticum* (Yendo) Fensholt. J. Exp. Mar. Biol. Ecol. 56: 179–195.

Duffy JE, Hay ME (2001) The ecology and evolution of marine consumer-prey interactions. In Bertness MD, Hay ME, Gaines SD (eds), Marine Community Ecology: Sinauer Associates, Sunderland, MA, USA, pp. 131–159.

Geertz-Hansen O, Sand-Jensen K, Hansen DF, Christiansen A (1993) Growth and grazing control of abundance of the marine macroalga, *Ulva lactuca* L in a eutrophic Danish estuary. Aquatic Bot. 46: 101–109.

Hammerstrom K, Dethier MN, Duggins DO (1998) Rapid phlorotannin induction and relaxation in five Washington kelps. Mar. Ecol. Prog. Ser. 165: 293–305.

Hay ME, Paul VJ, Lewis SM, Gustafson K, Tucker J, Trindell R (1988) Can tropical seaweeds reduce herbivory by growing at night? Diel patterns of growth, nitrogen content, herbivory, and chemical versus morphological defenses. Oecologia 75: 233–245.

Hay ME (1991) Fish-seaweed interactions on coral reefs: Effects of herbivorous fishes and adaptations of their prey. In Sale PF (ed.), The ecology of fishes on coral reefs, Academic Press, California, pp. 371–413.

Hay ME (1996) Marine chemical ecology: What's known and what's next?. J. Exp. Mar. Biol. Ecol. 200:103–134.

Helmuth BS, Veit RR, Holberton R (1994) Long-distance dispersal of a subantarctic brooding bivalve (*Gaimardia trapesina*) by kelp rafting. Mar. Biol. 120: 421–426.

Herms DA, Mattson WJ (1992) The dilemma of plants to grow or defend. Q. Rev. Biol. 67: 283–335.

Hirata T, Tanaka J, Iwami T, Ohmi T, Dazai A, Aoki M, Ueda H, Tsuchiya Y, Sato T, Yokohama Y (2001) Ecological studies on the community of drifting seaweeds in the south-eastern coastal waters of Izu Peninsula, central Japan. I. Seasonal changes of plants in species composition, appearance, number of species and size. Phycol. Res. 49: 215–229.

Hobday AJ (2000) Age of drifting *Macrocystis pyrifera* (L.) C. Agardh rafts in the Southern California Bight. J. Exp. Mar. Biol. Ecol. 253: 97–114.

Hoffmann A, Santelices B (1997) Flora marina de Chile central, Ediciones Universidad Católica de Chile, Santiago, Chile, 434 pp.

Inglis G (1989) The colonisation and degradation of stranded *Macrocystis pyrifera* by the macrofauna of a New Zealand sandy beach. J. Exp. Mar. Biol. Ecol. 125: 203–217.

Ingólfsson A (1995) Floating clumps of seaweed around Iceland: natural microcosms and means of dispersal for shore fauna. Mar. Biol. 122: 13–21.

Ingólfsson A (1998) Dynamics of macrofaunal communities of floating seaweed clumps off western Iceland: A study of patches on the surface of the sea. J. Exp. Mar. Biol. Ecol. 231: 119–137.

Ingólfsson A (2000) Colonization of floating seaweed by pelagic and subtidal benthic animals in southwestern Iceland. Hydrobiologia 440: 181–189.

Littler MM, Littler DS (1980) The evolution of thallus form and survival strategies in benthic marine macroalgae: Field and laboratory tests of a functional form model. Am. Nat. 116: 25–44.

Littler MM, Littler DS, Taylor PR (1983) Evolutionary strategies in a tropical barrier reef system: Functional-form groups of marine macroalgae. J. Phycol. 19: 229–237.

Lobban CS, Harrison PJ (1994) Seaweed ecology and physiology, Cambridge University Press, Cambridge, 366 pp.

Lubchenco J, Gaines SD (1981) A unified approach to marine plant-herbivore interactions. I. Populations and communities. Ann. Rev. Ecol. Syst. 12: 405–437.

Macaya E, Boltaña S, Hinojosa I, Macchiavello J, Valdivia N, Vásquez N, Buschmann A, Vásquez J, Vega J, Thiel M (2005) Presence of sporophylls in floating kelp rafts of *Macrocystis* spp. (Phaeophyceae) along the Chilean Pacific Coast. J. Phycol. 41: 913–922.

Norton TA, Mathieson AC (1983) The biology of unattached seaweeds. In Round FE, Chapman DJ (eds.), Progress in Phycological Research: Elsevier SP, Amsterdam, pp. 333–376.

North WJ (1994) Review of *Macrocystis* biology. In Akatsuka I (ed.), Biology of Economic Algae. SPB Academic Publishing, The Hague, Netherlands, pp. 447–527.

Norkko A, Bonsdorff E (1996) Population responses of coastal zoobenthos to stress induced by drifting algal mats. Mar. Ecol. Prog. Ser. 140: 141–151.

Ólafsson E, Ingólfsson A, Steinarsdóttir MB (2001) Harpacticoid copepod communities of floating seaweed: Controlling factors and implications for dispersal. Hydrobiologia 453/454: 189–200.

Parr AE (1939) Quantitative observations on the pelagic *Sargassum* vegetation of the western North Atlantic. Bull. Bingham Oceanogr. Collection 6: 1–94.

Paul VJ, Fenical W (1986) Chemical defense in tropical green algae, order Caulerpales. Mar. Ecol. Prog. Ser. 34: 157–169.

Paul VJ, Van Alstyne KL (1992) Activation of chemical defenses in the tropical green algae *Halimeda* spp. J. Exp. Mar. Biol. Ecol. 160: 191–203.

Pavia H, Cervin G, Lindgren A, Åberg P (1997) Effects of UV-B radiation and simulated herbivory on phlorotannins in the brown alga *Ascophyllum nodosum*. Mar. Ecol. Prog. Ser. 157: 139–146.

Pavia H, Toth GB (2000) Inducible chemical resistance to herbivory in the brown seaweed *Ascophyllum nodosum*. Ecology 81: 3212–3225.

Peckol P, Krane JM, Yates JL (1996) Interactive effects of inducible defense and resource availability on phlorotannins in the North Atlantic brown alga *Fucus vesiculosus*. Mar. Ecol. Prog. Ser. 138: 209–217.

Renaud PE, Hay ME, Schmitt TM (1990) Interactions of plant stress and herbivory: Intraspecific variation in the susceptibility of a palatable versus an unpalatable seaweed to sea urchin grazing. Oecologia 82: 217–226.

Rodríguez SR (2003). Consumption of drift kelp by intertidal populations of the sea urchin *Tetrapygus niger* on the central Chilean coast: Possible consequences at different ecological levels. Mar. Ecol. Prog. Ser. 251: 141–151.

Sotka EE, Taylor RB, Hay ME (2002) Tissue-specific induction of resistance to herbivores in a brown seaweed: The importance of direct grazing versus waterborne signals from grazed neighbors. J. Exp. Mar. Biol. Ecol. 277: 1–12.

Steinberg PD (1984) Algal chemical defense against herbivores: Allocation of phenolic compounds in the kelp *Alaria marginata*. Science 223: 405–406.

Taylor RB, Sotka E, Hay ME (2002) Tissue-specific induction of herbivore resistance: Seaweed response to amphipod grazing. Oecologia 132: 68–76.

Thiel M (2002) The biogeography of littoral algal-associated peracarids along the Chilean coast. J. Biogeogr. 29: 999–1008.

Thiel M, Vásquez JA (2000) Are kelp holdfasts islands on the ocean floor? Indication for temporarily closed aggregations of peracarid crustaceans. Hydrobiologia 440: 45–54.

Thiel M, Gutow L (2004) The ecology of rafting in the marine environment. I. The floating substrata. Oceanogr. Mar. Biol. Ann. Rev. 42: 181–264.

Toth GB, Pavia H (2001) Lack of phlorotannin induction in the kelp *Laminaria hyperborea* in response to grazing by two gastropod herbivores. Mar. Biol. 140: 403–409.

Van Alstyne KL (1988) Herbivore grazing increases polyphenolic defenses in the intertidal brown alga *Fucus distichus*. Ecology 69: 655–663.

Van Alstyne KL (1989) Adventitious branching as a herbivore-induced defense in the intertidal brown alga *Fucus distichus*. Mar. Ecol. Prog. Ser. 56: 169–176.

Van Alstyne KL, Wolfe GV, Freidenburg TL, Neill A, Hicken C (2001) Activated defense systems in marine macroalgae: Evidence for an ecological role for DMSP cleavage. Mar. Ecol. Prog. Ser. 213: 53–65.

Watson D, Norton TA (1985) The physical characteristics of seaweed thalli as deterrents to littorine grazers. Bot. Mar. 28: 383–387.

Zubia M, Payri CE, Deslandes E, Guezennec J (2003) Chemical composition of attached and drift specimens of *Sargassum mangarevense* and *Turbinaria ornata* (Phaeophyta: Fucales) from Tahiti, French Polynesia. Bot. Mar. 46: 562–571.

Journal of Applied Phycology (2006) 18: 437–443
DOI: 10.1007/s10811-006-9045-7

A comparison of various seaweed-based diets and formulated feed on growth rate of abalone in a land-based aquaculture system

Krishni Naidoo[1], Gavin Maneveldt[1,*], Kevin Ruck[2] & John J. Bolton[3]
[1]*Department of Biodiversity and Conservation Biology, University of the Western Cape, Private Bag X17, Bellville 7535, South Africa;* [2]*Jacobsbaai Sea Products Farm, Private Bag X2, Rhine Road, Jacobsbaai 8050, South Africa;* [3]*Department of Botany, University of Cape Town, Rondebosch 7701, South Africa*

*Author for correspondence: e-mail: gmaneveldt@uwc.ac.za

Key words: abalone, diet, growth, seaweed, *Ecklonia*, *Haliotis midae*

Abstract

The effects of different diets on growth in the cultured South African abalone, *Haliotis midae* (Linnaeus), was investigated. Growth of juvenile *Haliotis midae* was monitored on a commercial abalone farm over a period of 9 months in an experiment consisting of 9 treatments with 4 replicates ($n = 250$ individuals per replicate). The treatments were: fresh kelp (*Ecklonia maxima*) blades (seaweed control); Abfeed® (formulated feed control); kelp + Abfeed®; dried kelp pellets; dried kelp blades; dried kelp stipes; fresh kelp with the epiphyte *Carpoblepharis flaccida*; a mixed diet (*Gracilaria gracilis*, *Ulva lactuca*, and kelp) and a rotational diet (abalone were fed 1 of the 9 treatments for the first week and them kelp for the next 3 weeks). Results show that abalone grow well on all fresh seaweed combinations, but grow best on a mixed diet. The likely reason for the success of the mixed diet is that the red and green seaweed was farm grown, with an increased protein content. Dried kelp in any form produced poor growth. Abalone fed on the mixed diet grew at 0.066 mm day^{-1} shell length and 0.074 g day^{-1} body weight; this corresponds to 24.09 mm shell length and 27.01 g body weight increase per annum. Abalone fed on dried kelp grew at only 0.029 mm day^{-1} shell length and of 0.021 g day^{-1} body weight. Abalone grown on Abfeed® grew at 0.049 mm day^{-1} shell length and 0.046 g day^{-1} body weight which corresponds to 17.88 mm and 16.79 g increase per annum; this is better than the dried seaweed feeds, but poorer than the fresh seaweed combinations. This study shows that seaweed diets, particularly if the diets include seaweeds grown in animal aquaculture effluent, are good substitutes for the formulated feed generally used today.

Introduction

The South African abalone, *Haliotis midae* Linn., is a highly sought-after delicacy in the Far East, which is the destination of 90% of the product from the local fishery (Britz et al., 1994). Of 6 abalone species in South Africa, only *H. midae* is presently fished commercially. Although the South African abalone fishery has existed since 1949, the first attempts at cultivating *H. midae* commercially were only made in 1981 when captured specimens were successfully spawned to produce spat and juvenile abalone (Genade et al., 1988). In 2001, twelve abalone farms, with an estimated investment of US$ 12 million, had been established on the South African coast (Sales & Britz, 2001). By 2003, this had increased to 18 farms, with a projected production of 527 and 700 tons per annum for 2003 and 2004 respectively (Gerber, 2004).

The proper nutrition and the resulting growth of cultured abalone are critical factors in the successful culture of this animal. While *H. midae* can reach a maximum size of about 200 mm shell length at an age of over 30 years in the wild, farm production is aimed towards an average size of only 100 mm, which is currently achieved after 5 years (Sales & Britz, 2001). Abalone growth is extremely slow and often varies with size and age. Diet is therefore very important and it has been shown that different diets produce different growth rates (Leighton, 1974; Britz, 1996a; Guzmán & Viana, 1998; Shpigel et al.,

1999; Boarder & Shpigel, 2001; Bautista-Teruel et al., 2003).

Abalone begin to feed immediately after larval settlement, initially consuming benthic diatoms (Tutschulte & Connell, 1988). As they grow, they begin feeding on macroalgae and in the wild may change from one species of macroalga to another as they mature (Stepto & Cook, 1993). Preferences exist, with red algae being favoured by a number of different abalone species (Tutschulte & Connell, 1988; Shepherd & Steinberg, 1992; Stepto & Cook, 1993; Fleming, 1995). Juveniles begin to eat macroalgae at about 10 mm shell length and will eat from 10 to 30% of their body weight in algae each day and have high feeding rates that are due to the high water content and relatively low protein content of macroalgae (Hahn, 1989).

Research thus far has dealt mainly with the natural diet of wild abalone, single-species diets in culture, and more recently, the production of formulated diets (Simpson & Cook, 1998; Sales & Janssens, 2004). Wild abalone generally feed on a broad selection of algae, normally with at least two species being found in the gut at any one time (Barkai & Griffiths, 1986). This implies that abalone typically select more than just a single species and preferentially choose a mixture of algae. In this study we test the effects of various diets on the growth of juvenile abalone in commercial aquaculture systems, including a formulated feed, dry and fresh kelp, and a mixture of kelp, kelp epiphytes, and farm grown seaweed.

Materials and methods

Experimental animals

Abalone of a specific age class often vary in size because of their different feeding rates. For this reason, juvenile abalone of the same age and similar size were chosen as test animals. Hatchery-reared animals (from the Jacobsbaai Sea Products farm), spawned in September 2002, approximately 22 months old, 34.7 ± 5.8 mm in shell length, and 7.8 ± 3.8 g in body weight, were used to test the growth response of juvenile abalone fed on 9 different diets. Flow-through seawater (700 ± 100 L h^{-1}), moderately aerated, was supplied at a temperature of 15.5±2.5 °C in the holding tanks. Abalone were grown in culture baskets, with a stocking density of 5 kg (±500 individuals) per basket. Each basket was subdivided, using mesh, to produce 2 replicates (a stocking density of ±250 individuals per replicate) and two baskets were used for each treatment, i.e. $n = 4$ replicates. Growth was monitored over a 9-month period.

Diets

The 9 diets consisted of: fresh kelp (*Ecklonia maxima* [Osbeck] Papenf.) blades (seaweed control); Abfeed® (formulated feed control); kelp + Abfeed®; dried kelp pellets; dried kelp blades; dried kelp stipes, kelp with a red algal epiphyte (*Carpoblepharis flaccida* [C.Ag.] Kütz.); a mixed diet (*Gracilaria gracilis* [Stackhouse] Steentoft, Irvine et Farnham, *Ulva lactuca* L. and kelp); and a rotation diet (where the abalone were fed 1 of the 9 treatments for the first week, and then kelp for the following 3 weeks). Abfeed® (Sea Plant Products Ltd, South Africa) is a formulated feed containing fishmeal (55%), starch, *Spirulina* spp. (10%), vitamins and minerals (Fleming et al., 1996). The approximate analysis of Abfeed® is 34.6% protein, 43.3% carbohydrates, 5.3% fat, 1.2% crude fibre, 5.7% ash and ~10% moisture (Sea Plant Products Ltd, pers. com.). All kelp was harvested locally. Kelp was chosen as a seaweed control because it is most commonly used as fresh abalone feed in South Africa. Abfeed® was used as an formulated control feed because it is the most common artificial food pellet used on commercial abalone farms in South Africa. *Ulva lactuca* and *G. gracilis* for the mixed diet were obtained on the farm from a cultured stock grown in abalone and fish (turbot) effluent. These seaweeds grown in abalone and turbot effluent have considerably higher nitrogen content than seaweed collected from local seashores (Robertson-Andersson, 2004; Robertson-Andersson et al., 2006). *Ulva lactuca* grown in these systems has an average protein content of 33.4% when grown in abalone waste, and 36.6% when grown in turbot waste as opposed to 3.7–19.9% in wild *U. lactuca* (Robertson-Andersson, 2004). No protein values were available for *G. gracilis* but it is assumed that farm grown *G. gracilis* will also have considerably higher protein content than wild *G. gracilis*.

Representative animals were selected from each treatment ($n = 30$ at 0–2 months, $n = 40$ at 3–8 months and $n = 50$ at 9 months to compensate for differential growth). Abalone shell length and body weight were measured once a month for 9 months. Daily growth rates in terms of body weight (DGBW) and shell length

(DGSL) were calculated as follows:

$$DGBW = (W_1 - W_0)/t$$
$$DGSL = (L_1 - L_0)/t$$

$W0$: mean initial weight, W_1: mean final weight, L_0: mean initial length, L_1: mean final length, and t: time in days.

Body weight/shell length ratio

The body weight-to-shell length ratio (BW/SL) was calculated for all 9 diet treatments. The BW/SL ratio (Mean final weight / Mean final length) gives an indication of the flesh volume per unit shell length growth for each of the 9 diet treatments. BW/SL rations are important in that they indicate the mass of abalone per unit shell length. Thus at marketable size (80–100 mm), the value of an abalone priced by weight will depend on the BW/SL ratio. Certain diets will therefore produce more valuable abalone.

Statistical analysis

All data are expressed as means ± SE. A two-way analysis of variance (ANOVA: Zar, 1984) was used to compare and analyze the effect of the various treatments on shell length and body weight over time. Differences among treatment means were considered significant at $P<0.05$.

Results

Diets

Dried kelp in any form (blades, stipes, and pellets) produced poor growth in both weight ($P < 0.05$) and shell length ($P < 0.05$) when compared against the fresh seaweed treatments and Abfeed® (Figures 1 and 2). In comparing the remaining treatments, shell length produced no meaningful comparison, as there was no significant difference between any of the fresh seaweed treatments and the Abfeed® ($P = 0.37$). Body weight, however, produced more meaningful differences.

There were significant differences in body weight between treatments after only 5 months ($P < 0.05$). The mixed diet (0.074 g day^{-1}) produced the best growth followed by the rotation diet and epiphyte treatments (0.059 g day^{-1}), and then the fresh kelp diet (0.056 g day^{-1}). A surge in growth can be seen after month 7 (Figure 2). This is due to a thinning of the sample size (n=150 per replicate) as growth of the sample had become density dependent by month 7. Generally those diets that contained more than one seaweed and those that included Abfeed®, produced good growth rates (Tables 1 and 2; Figures 1 and 2). The mixed diet, ranked 1, produced the best growth rates for both shell length (0.066 mm day^{-1}) and body weight (0.074 day^{-1}). Abfeed® (0.049 mm day^{-1} shell length & 0.046 g day^{-1} body weight), ranked 6, did not perform as well as the combination feeds, but still performed better than the dried seaweed treatments. Dried kelp (pellets and blades) produced the lowest growth (Tables 1 and 2; Figures 1 and 2).

BW/SL ratios

The mixed diet produced the highest BW/SL ratio (0.528 g mm^{-1}) and the dried blade treatment the lowest (0.306 g mm^{-1}) (Table 2). Again the BW/SL ratios were generally higher for the diet treatments that contained a combination of feeds i.e., the mixed diet (0.528 g mm^{-1}, ranked 1), the rotation diet (0.469 g mm^{-1}, ranked 2), the kelp + epiphyte diet (0.464 g mm^{-1}, ranked 3), and the kelp + Abfeed® diet (0.463 g mm^{-1}, ranked 5). The kelp only diet produced a comparatively good BW/SL ratio (ranked 4 at 0.463 g mm^{-1}).

Discussion

Good growth rates are important in ensuring that cultured animals reach a marketable size and condition within a time that is economically viable. Previous studies on *Haliotis* sp. have shown that "mixed" diets produce better growth rates than single-species diets (Owen et al., 1984; Day & Fleming, 1992; Fleming, 1995, Simpson & Cook, 1998). Wild *H. midae*, in particular, are found to have a variety of algae in their guts, with *E. maxima* forming the larger percentage of the gut contents, followed by red or green algae (Newman, 1968; Barkai & Griffiths, 1986, 1987). This suggests that abalone are naturally selecting a mixture of seaweeds.

Our data are consistent with previously published works in that abalone grown on a combination of different seaweeds perform better than those grown on a single species. Single-species diets, and in particular

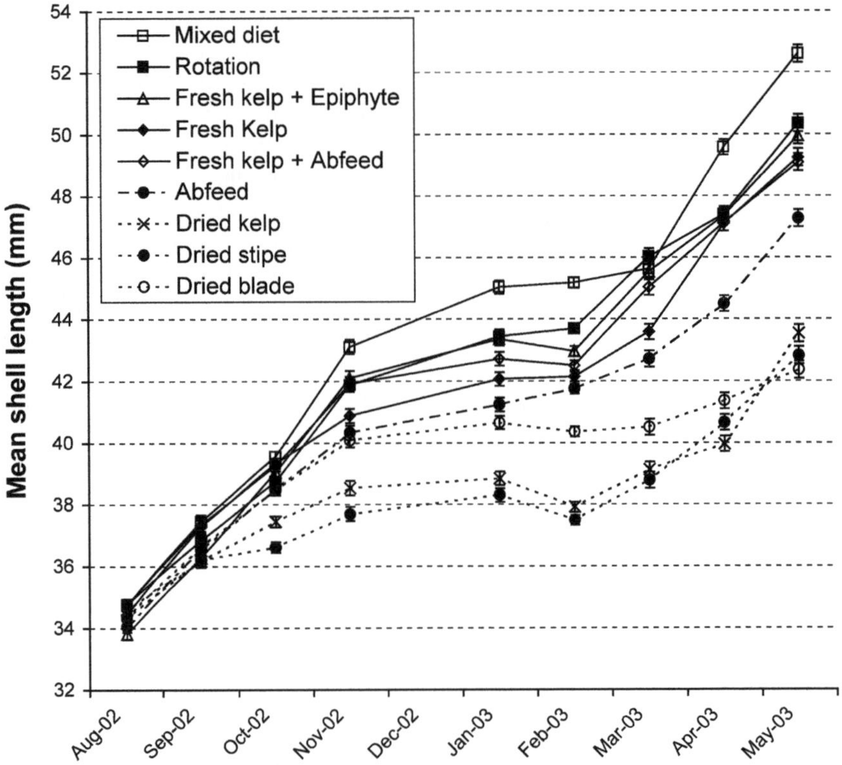

Figure 1. Growth in abalone (*Haliotis midae*) shell length for all diet treatments over the 9-month period.

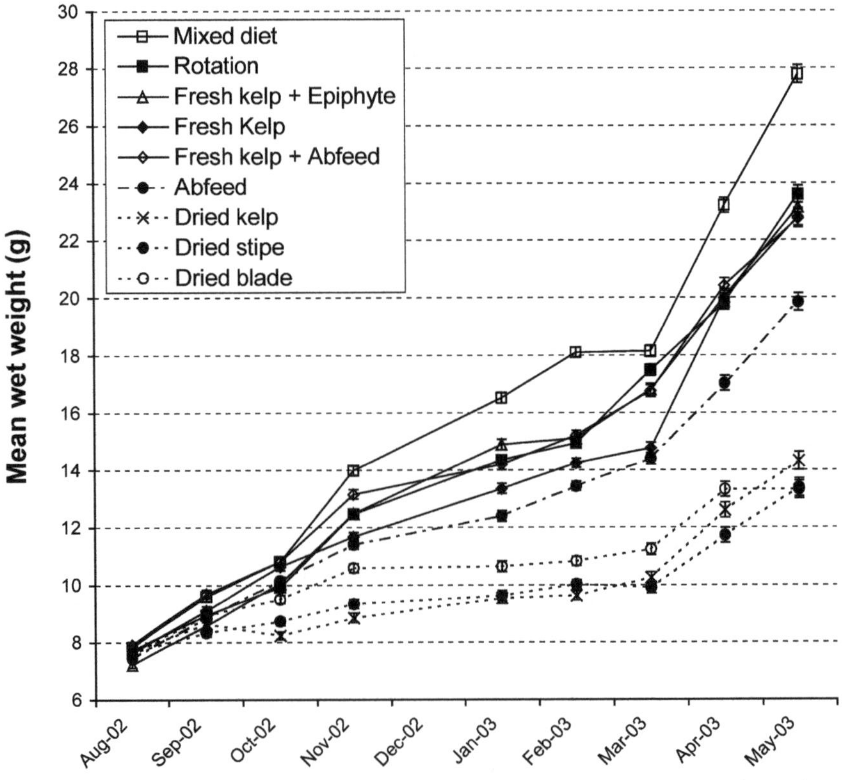

Figure 2. Growth in abalone (*Haliotis midae*) body weight for all diet treatments over the 9-month period.

Table 1. Mean shell length and growth rates of juvenile abalone (*H. midae*) fed various diets over a 9-month growth period (270 days).

Diet treatment	Mean initial length (mm ± se)	Mean final length (mm ± se)	Growth rate (mm·day^{-1})	Growth rate (mm·year^{1})
Mixed diet	34.74 ± 0.20	52.60 ± 0.24	0.066	24.09
Fresh kelp + Epiphyte	33.81 ± 0.18	49.94 ± 0.25	0.060	21.90
Rotation	34.76 ± 0.20	50.35 ± 0.23	0.058	21.17
Fresh kelp	34.45 ± 0.21	49.24 ± 0.24	0.055	20.07
Fresh kelp + Abfeed®	34.75 ± 0.18	49.09 ± 0.25	0.053	19.34
Abfeed®	34.04 ± 0.18	47.27 ± 0.25	0.049	17.88
Dried blade	34.34 ± 0.19	43.53 ± 0.19	0.034	12.41
Dried stipe	34.31 ± 0.18	42.82 ± 0.21	0.032	11.31
Dried kelp pellets	34.58 ± 0.20	42.36 ± 0.18	0.029	10.58

Table 2. Mean wet weight, growth rates and BW/SL ratios of juvenile abalone (*H. midae*) fed various diets over a 9-month growth period (270 days), by rank.

Diet treatment	Mean initial weight (g ± se)	Mean final weight (g ± se)	Growth rate (g·day^{1})	Growth rate (g·year^{1})	BW/SL (g·mm^{-1})	Rank order
Mixed diet	7.83 ± 0.13	27.78 ± 0.36	0.074	27.01	0.528	1
Rotation	7.72 ± 0.12	23.61 ± 0.36	0.059	21.54	0.469	2
Fresh kelp + Epiphyte	7.22 ± 0.13	23.15 ± 0.32	0.059	21.54	0.464	3
Fresh kelp	7.61 ± 0.11	22.81 ± 0.26	0.056	20.44	0.463	4
Fresh kelp + Abfeed®	7.92 ± 0.12	22.75 ± 0.28	0.055	20.08	0.463	5
Abfeed®	7.45 ± 0.11	19.85 ± 0.27	0.046	16.79	0.420	6
Dried stipe	7.66 ± 0.12	14.31 ± 0.18	0.025	9.13	0.334	7
Dried kelp pellets	7.65 ± 0.11	13.39 ± 0.16	0.021	7.67	0.316	8
Dried blade	7.61 ± 0.11	13.32 ± 0.17	0.021	7.67	0.306	9

dried feeds, produce poor growth in abalone. Duncan and Klekowski (1975) stated that essential nutrients might become limiting in experiments where animals are fed single-species diets, which could result in poor growth rates. Various single-species algal diets have been tested in an attempt to identify those diets that maximize the growth rate of abalone in culture. Day and Fleming (1992) demonstrated that on some single-species algal diets *H. rubra* initially grew at a steady rate then failed to grow for the remainder of the trial, suggesting that nutrients were lacking in the diet. Stuart and Brown (1994) suggested that a variety of algae are better able to meet the preferences and nutritional requirements of cultured abalone over extended periods of time.

Kelp (*E. maxima*) is the most abundant algal species along the southwest coast of southern Africa, and is the most likely food source for abalone farms developing along this area of the coast (although in the northern west coast of South Africa and in Namibia it is largely replaced by *Laminaria pallida* (Grev. Ex J. Ag) (Stegenga et al., 1997). However, Stepto and Cook (1993) found that *E. maxima* was the least preferred of three algae fed to *H. midae*. They suggested that this might be due to the high phenolic levels. Simpson and Cook (1998) also stated that in a mixed diet it is likely that *Ecklonia* spp. would be avoided, which would negate the purpose of feeding *Ecklonia* spp. as part of a diet. Additionally, kelp is low in protein content (5%) and high in water content (68–83%) (Hahn, 1989; Robertson-Anderson, 2004). All of this probably accounts for the relatively low growth obtained by the kelp only diet when compared against the other fresh seaweed combinations. In this study fresh kelp, however, is able to produce better growth than all dried feed including the formulated feed, Abfeed®. This is in contrast to some studies (eg. Hahn, 1989; Britz, 1996b; Bautista-Teruel et al., 2003) who showed that growth rate of abalone fed formulated diets containing a combination of plant and animal protein sources (such as Abfeed®) perform better than those fed diets with protein sources of plant origin only.

What then determines the feeding preference? Fleming (1995) suggested that preference for certain algae might be due to the presence of essential nutrients not available in other algae. A number of authors (Day & Fleming, 1992; Fleming, 1995; Simpson & Cook, 1998) showed that algae like *Ulva* and *Gracilaria* species constitute a poor diet when supplied singly, but they may be of great value when supplied as part of a mixed diet, thereby supplying essential nutrients to the diet. These studies however, referred to wild stocks and not protein-enriched *Ulva* and *Gracilaria* species as were used in this study. This probably accounts for the higher abalone growth on our mixed diet.

A number of previous studies have come to similar conclusions. Neori et al. (1998), for example, showed that a number of *Ulva* spp. are able to remove up to 90% of dissolved nitrogen from aquaculture effluent. The culture of *Ulva* spp. in nutrient-rich water increases their protein content roughly 3 to 10 fold (Shpigel et al., 1999; Boarder & Shpigel, 2001; Robertson-Andersson, 2004; Robertson-Andersson et al., 2006). This enriched *Ulva* has subsequently been shown to improve growth in, for example *H. tuberculata* (Neori et al., 1998; Shpigel et al., 1999), *H. discus hannai* (Shpigel et al., 1999), and *H. roei* (Boarder & Shpigel, 2001).

The BW/SL ratio is an important determinant of the economic viability of different diets. The mixed diet has the greatest ability to increase the ratio of body weight to shell length and our results have shown that although shell length growth rates of abalone fed on combination diets remained relatively constant over time, the body weight growth rates showed a significant increase over time. This is important for the commercial farmer, in that growth depends on the feed used for abalone aquaculture. The BW/SL ratio gives an indication of what feeds are more likely to produce better growth.

Natural diets of fresh seaweed, either as kelp alone or combinations of kelp and other seaweeds, produced the best growth in abalone, with the mixed diet performing best. Dried seaweed in any form produced the lowest growth. Although not as good as the fresh feeds, the formulated diet Abfeed® as a single feed performs better than the dried seaweed feeds.

Acknowledgments

We would like to thank the Department of Biodiversity and Conservation Biology at the University of the Western Cape and the Jacobsbaai Sea Products farm for providing funding, research facilities and technical support. We also thank the South African National Research Foundation (NRF/Sweden collaboration), the Department of Environmental Affairs and Tourism, and the International Ocean Institute of Southern Africa for research funding. Deborah Robertson-Andersson provided valuable discussion. We thank Max Troell (University of Stockholm, Sweden) for the collaboration that has made this project a reality.

References

Barkai R, Griffiths CL (1986) Diet of the South African abalone *Haliotis midae*. South African Journal of Marine Science 4: 37–44.

Barkai R, Griffiths CL (1987) Consumption, absorption efficiency, respiration and excretion in the South African abalone *Haliotis midae*. South African Journal of Marine Science 5: 523–529.

Bautista-Teruel MN, Fermin AC, Koshio SS (2003) Diet development and evaluation for juvenile abalone, *Haliotis asinine* and plant protein sources. Aquaculture 219: 645–653.

Boarder SJ, Shpigel M (2001) Comparative growth performance of juvenile *Haliotis roei* fed on enriched *Ulva rigida* and various artificial diets. Journal of Shellfish Research 20: 653–657.

Britz PJ (1996a) Effects of dietary protein level on growth performance of South African abalone, *Haliotis midae*, fed on fishmeal based semi-purified diets. Aquaculture 140: 55–61.

Britz PJ, (1996b) The suitability of selected protein sources for inclusion in formulated diets for the South African abalone *Haliotis midae*. Aquaculture 140: 63–73.

Britz JP, Hecht T, Knauer J, Dixon MG (1994) The development of an artificial feed for abalone farming. South African Journal of Science 90: 7–8.

Day RW, Fleming AE (1992) The determinants and measurement of abalone growth. In Shepherd SA, Tegner MJ, Guzmán Del Próo SA (eds), Abalone of the World. Biology, Fisheries and Culture. Fishing News Books, Oxford, pp. 141–168.

Duncan A, Klekowski RZ (1975) Parameters of an energy budget. In Grodzinski W, Klekowski RZ, Duncan A (eds), Methods for Ecological Bioenergetics. IBP Handbook. Blackwell Scientific Publications, Oxford, pp. 97–148.

Fleming AE (1995) Growth, intake, feed conversion efficiency and chemosensory preference of the Australian abalone, *Haliotis rubra*. Aquaculture 132: 297–331.

Fleming AE, van Barneveld RJ, Hone PW (1996) The development of artificial diets for abalone: A review and future directions. Aquaculture 140: 5–53.

Genade AB, Hirst AL, Smit CJ (1988) Observations on the spawning, development and rearing of the South African abalone *Haliotis midae* Linn. South African Journal of Marine Science 6: 3–12.

Gerber WH (2004) Enhancing the competitive advantage of the South African cultivated abalone industry. M.Sc. thesis, University of Stellenbosch, South Africa.

Guzmán JM, Viana MT (1998) Growth of abalone *Haliotis fulgens* fed diets with and without fish meal, compared to a commercial diet. Aquaculture 165: 321–333.

Hahn KO (1989) Nutrition and growth of abalone. In Hahn KO (ed), Handbook of Culture of Abalone and Other Marine Gastropods. Florida, pp. 135–156.

Leighton DL (1974) The influence of temperature on larval and juvenile growth in three species of Southern California abalones. Fisheries Bulletin 72: 1137–1145.

Newman GG (1968) Growth of the South African abalone *Haliotis midae*. Republic of South Africa Department of Industries, Division of Sea Fisheries, Investigational Report 67: 1–24.

Neori A, Ragg NLC, Shpigel M (1998) The integrated culture of seaweed, abalone, fish and clams in modular intensive land-based systems: II. Performance and nitrogen partitioning within an abalone (*Haliotis tuberculata*) and macroalgae culture system. Aquacultural Engineering 17: 215–239.

Owen B, Disaivo LH, Ebert EE, Fonck E (1984) Culture of the California red abalone *Haliotis rufescens* Swanson (1822) in Chile. Veliger 27: 101–105.

Robertson-Andersson DV (2004) The cultivation of *Ulva lactuca* (Chlorophyta) in an intergrated aquaculture system, for the production of abalone feed and the bioremediation of aquaculture effluent. M.Sc. thesis, University of Cape Town, South Africa.

Robertson-Andersson DV, Leitao D, Bolton JJ, Anderson RJ, Njobeni A, Ruck K (2006) Can kelp extract (KELPAK®) be useful in seaweed mariculture? Journal of Applied Phycology DOI: 10.1007/s10811-006-9030-1.

Sales J, Britz PJ (2001) Research on abalone (*Haliotis midae* L.) cultivation in South Africa. Aquaculture Research 32: 863–875.

Sales J, Janssens GP (2004) Use of feed ingredients in artificial diets for abalone: a brief update. Nutrition Abstracts and Reviews: Series B 74: 13N–21N.

Shepherd SA, Steinberg PD (1992) Food preference of three Australian abalone species with a review of the algal food of abalone. In Shepherd SA, Tegner MJ, Guzmán Del Próo SA (eds), Abalone of the World: Biology, Fisheries and Culture. Blackwell Scientific Publications, Oxford, pp. 169–181.

Simpson JA, Cook PA (1998) Rotation Diets: A method of improving growth of cultured abalone using natural algal diets. Journal of Shellfish Research 17: 635–640.

Shpigel M, Ragg NL, Lapatsch I, Neori A (1999) Protein content determines the nutritional value of the seaweed *Ulva lactuca* L. for the abalone *Haliotis tuberculata* L. and *H. discus hannai* Ino. Journal of Shellfish Research 18: 227–233.

Stegenga H, Bolton JJ, Anderson RJ (1997) Seaweeds of the South African West Coast. Contributions from the Bolus Herbarium, University of Cape Town, 18: 655 pp.

Stepto NK, Cook PA (1993) Feeding preferences of the juvenile South African abalone *Haliotis midae* (Linnaeus, 1758). Journal of Shellfish Research 15: 653–657.

Stuart MD, Brown MT (1994) Growth and diet of cultivated black-footed abalone, *Haliotis iris* (Martyn). Aquaculture 127: 329–337.

Tutshulte TC, Connell JH (1988) Feeding behaviour and algal food of three species of abalones (*Haliotis*) in southern California. Marine Ecology 49: 57–64.

Zar JH (1984) Biostatistical Analysis. 2nd edition. Prentice Hall, New Jersey, 718pp.

Journal of Applied Phycology (2006) 18: 445–450
DOI: 10.1007/s10811-006-9048-4

A simple 96-well microplate method for estimation of total polyphenol content in seaweeds

Qing Zhang, Junzeng Zhang*, Jingkai Shen, Angelica Silva, Dorothy A. Dennis, & Colin J. Barrow
Ocean Nutrition Canada Ltd., 101 Research Drive, Dartmouth, Nova Scotia, B2Y 4T6, Canada
*Author for correspondence: e-mail: jzhang@ocean-nutrition.com

Key words: seaweed, marine algae, total polyphenol, estimation, 96-well microplate

Abstract

Seaweed polyphenols are potent antioxidants and have also been shown to have α-glucosidase inhibiting activity. In our continuous efforts to develop new marine-based nutraceuticals and functional food ingredients, we have investigated many algal species collected on the Atlantic coast of Canada. A simple method for estimating the total polyphenol content in seaweeds and their extracts was developed based on the classic Folin-Ciocalteau colorimetric reaction. By using the 96-well microplate and a microplate reader, this new method saves experimental time, significantly reduces the amount of sample required, handles large number of samples in one experiment, and also improves the repeatability of the results.

A number of algal samples collected on the seashore of Nova Scotia, Canada, were analyzed for their levels of polyphenol content using this microplate-based method. The antioxidant activity of these samples was also assessed by using DPPH (2, 2-diphenyl-1-picrylhydrazyl) radical scavenging assay. The results showed that there is a strong correlation between the total polyphenol content and the potency of antioxidant effect.

Introduction

Polyphenol has been emerging as one major category of natural products that is important to human health (Shahidi & Naczk, 2004; Frei, 1994). Increasing scientific evidence shows that polyphenols are good antioxidants, are effective in preventing cardiovascular and inflammatory diseases, and can also be used as chemopreventing agents for cancer. Some polyphenol products, such as the ones from tea and grape, are becoming popular in the market place. In addition to the polyphenols from terrestrial plants, seaweeds have been shown as another source of polyphenols with unique structural properties. Phlorotannin is one such special type of polyphenolic compound widely distributed in brown algae. It consists of 1,3,5-trihydroxybenzenoid, or phloroglucinol, as the structural unit, with a molecular size ranging from several thousands to tens of thousands for high molecular weight polyphloroglucinols. Its antioxidant activity is one of those most studied, and several species of brown algae such as *Sargassum kjellmanianum* (Yan et al., 1996; Yan et al., 1997; Wei & Xu, 2003), *Eisenia bicyclis* (Nakmura et al., 1996), Cystoseira sp. (Chkhikvishvili & Ramazanov, 2000), *Fucus sp.* (Jimenez-Escrig et al., 2001), and Ecklonia stolonifera (Kang et al., 2003a, 2004) have been reported with respect to free radical scavenging and the inhibition of total reactive oxygen species generation by phlorotannin compounds. More recently, polyphloroglucinols from brown algae were revealed to have bactericidal properties (Nagayama et al., 2002), hyaluronidase inhibition (Shibata et al., 2002), alpha-glucosidase inhibition (Kurihara et al., 2002), chemopreventive properties (Kang et al., 2003b), secretory phospholipase A2s, lipoxygenases, and cyclooxygenase inhibition (Shibata et al., 2003), and HIV-1 reverse transcriptase and protease inhibition effects (Ahn et al., 2004) as well as antioxidative, anti-inflammatory, alpha- and glucosidase inhibitory activities. Brown algal polyphenols have been investigated for their potential functional applications in beverages and edible oils (Yan et al., 1998; Nagai & Yukimoto, 2003).

Among the several assays available to quantify total polyphenols, the Folin-Denis method (Folin & Denis, 1912) is one of the most commonly used. It is based on a color reaction between easily oxidized polyphenols or hydroxylated aromatic compounds and phosphotungsten-polymolybdic acid. It was later improved by Folin and Ciocalteu (1927) with an addition of lithium sulfate to the reagent to prevent precipitation in the reaction in order to increase the sensitivity. Now, Folin-Ciocalteu reagent is commercially available for polyphenol quantification and is found to be a preferred assay (Singleton et al., 1999). The two assays have been widely used in estimating the level of phlorotannin content in algal materials (Van Alstyne, 1995; Jimenez-Escrig et al., 2001).

Although the polyphenol quantification method is well established, there are still opportunities for improvements in terms of time for sample color intensity measurement, the amount of sample, regents and solvents required, and analysis and management of data. We present here a 96-well microplate Folin-Cioalteu assay for estimating polyphenol content in marine algae. This modified Folin-Cioalteu method incorporates the convenience of spectrometric measurement using 96-well microplate, so that it consumes much less reagents and solvents, and it runs more efficiently in a plate that handles more samples in small quantities.

Materials and methods

Chemicals and reagents: Folin-Ciocalteu reagent was purchased from Sigma (Catalogue # F-9252, 500 mL). Sodium carbonate anhydrous: Sigma, S-6139, 500 g, 99.9%. Phloroglucinol dihydrate was from Aldrich (P3, 800-5, 25 g, 97%).

Instrument: 96-Well microplate reader (Molecular Devices Spectra MAX 190, Sunnyvale, CA, U.S.A.).

Algal material: Marine macroalgae collected on the coast of Nova Scotia during 2000–2003 were analyzed. These included the brown algae *Alaria esculenta, Ascophyllum nodosum, Fucus distichus, F. evanescens, F. vesiculosus,* and *Laminaria saccharina*, the red alga *Polysiphonia stricta* and the green alga *Codium fragile.*

Standard solution: 10 mg phloroglucinol (calculated as anhydrous) was dissolved in 100 mL distilled water and used as a stock solution (100 $\mu g\, mL^{-1}$) to make serial dilutions and obtain the standard solution at the concentration of 100, 50, 25, 12.5, and 6.25 $\mu g\, mL^{-1}$.

Sample preparation: Sample solution was prepared from either dry algal powder or algal extract.

Algal powder: weigh out 0.5–1 g in a test tube, add 20 mL MeOH-water (1:1), adjust pH to 2 and shake for 1 h (150 rpm) at room temperature. Centrifuge at 12,000 × *g* for 10 min and recover the supernatant. The residue is then extracted with acetone-water (7:3) under the same conditions and centrifuged. The two liquid extracts are pooled together and mixed well. Take 100 μL of this solution to make a 1:10 dilution with water and use as the sample solution. If the absorbance of the final sample solution is not in the range of the standard curve, further dilution may be required.

Algal extract: Weigh out 1–5 mg of MeOH or water extracts of dry algae, add 1 mL MeOH or water, and agitated by vortex for 30 seconds or until the sample is thoroughly dissolved. The solution is centrifuged for 10 min at 12,000 × *g*. Take 100 μL of the supernatant and dilute with water (1:10, can be changed based on the absorbance of the final sample solution. If the absorbance is not in the range of the standard curve, make 2–10 times further dilution). Use this as the sample solution.

Measurement: Load 20 μL of each sample solution and the serial standard solution on a 96-well microplate (as shown in Figure 1). Add 100 μL Folin-Ciocalteu reagent, mix well and wait 5 min. Add 80 L of 7.5% sodium carbonate solution and mix well. Cover the plate and leave it in the dark at room temperature for 2 h. Measure absorbance at λ 750 nm with a spectrophotometric microplate reader (set auto mix for 60 s before reading).

Distilled water was used as a blank. A reagent blank was carried out using the same procedure and was also measured against the water blank. Each standard solution and sample solution was run in triplicate, and the latter was assayed against sample control (i.e., sample solution without Folin-Ciocalteu reagent and sodium carbonate).

Blank: 200 μL distilled water.

Standard: The concentrations of phloroglucinol (anhydrous) for serial dilutions were: Std01: 100 $\mu g\, mL^{-1}$; Std02: 50 $\mu g\, mL^{-1}$; Std03: 25 $\mu g\, mL^{-1}$; Std04: 12.5 $\mu g\, mL^{-1}$; Std05: 6.25 $\mu g\, mL^{-1}$; Std06: 0 $\mu g\, mL^{-1}$.

Sp1–16: Samples 1–16.

CSp1–16: Control of samples 1–16, i.e., samples without 80 μL 7.5% sodium carbonate and 100 μL Folin-Ciocalteu reagent but using 180 μL distilled water instead. This is used to measure the background absorbance caused by the sample solution.

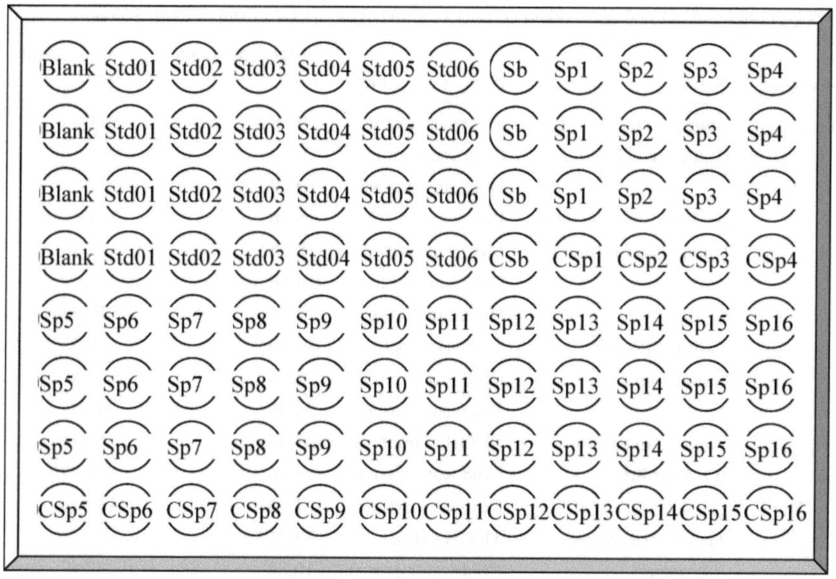

Figure 1. 96-well microplate template for standard phloroglucinol and seaweed sample solutions.

Sb: Sample blank or reagent blank. This is carried through the sample preparation process but without sample.

CSb: Control of sample blank. Sample blank without 80 μL 7.5% sodium carbonate and 100 μL Folin-Ciocalteu reagent while using 180 μL distilled water instead.

Calculation of total polyphenol content:

For algal powder: PGE% = {[(Norm × 40 × 10)÷ 1000] ÷ weight} × 100%

For algal extract: PGE% = {[(Norm × 1 × 10)÷ 1000] ÷ weight} × 100%

PGE: Phloroglucinol equivalents.

Norm: Mean result of sample-mean result of sample blank, where mean result of sample is the average of the triplicate results of each sample, and the result refers to the assay value obtained from the calibration curve (μg mL^{-1}).

Weight: Weight of algal powder or extract used (mg).

Please note that if additional dilution is made, Norm should be further multiplied by the dilution factor.

DPPH antioxidant assay: a 96-well microplate method (Lee et al., 1998; Fukumoto & Mazza, 2000) was used to measure the scavenging activity toward the 2,2-diphenyl-1-picrylhydrazyl (DPPH) free radical.

Results and discussion

UV-Visible spectrum and standard curve: The absorbance in the range of 350–800 nm for Folin-Ciocalteu reagent and phloroglucinol standard solution, *A. nodosum* sample solution from combined MeOH-water (1:1, pH = 2) and acetone-water (7:3) extraction, and *A. nodosum* sample solution from a double MeOH-water (1:1, pH = 2) extraction were recorded and are shown in Figure 2. *A. nodosum* sample has a maximum absorbance at λ 750 nm, which is same as that of the phloroglucinol, so that it is appropriate to use the later as the standard for quantification of total polyphenol level. The UV-Visible spectrum of the combined acidic aqueous MeOH and aqueous acetone extract is basically the same as that of the extract obtained only with acidic aqueous MeOH, but the absorbance is higher. We therefore used the combined aqueous MeOH and acetone method to prepare sample solutions from algal samples.

The calibration curve of standard phloroglucinol solutions is shown in Figure 3. The curve is linear when the concentration of phloroglucinol is in the range of 0–100 μg mL^{-1} ($R \geq 0.998$).

Polyphenol content and antioxidant activity of some algal extracts from Atlantic Canada: The polyphenol content of *A. nodosum* powder and the MeOH, cold and hot water extracts of several other macroalgae from Atlantic Canada are shown in Table 1. Polyphenol content varies between extracts obtained by different solvents,

Table 1. Polyphenol content and antioxidant activity of some algae species from Atlantic Canada

Type of seaweed	Name	Powder or extract	Polyphenol content (PGE%)	Antioxidant potency* (EC_{50}, μg/mL)
Brown algae	*Alaria esculenta*	MeOH extract	9.58	ND
	(Linnaeus) Greville	Cold water extract	0.54	–
		Hot water extract	0.68	–
	Ascophyllum nodosum	Powder	5.26	ND
	(Linnaeus) Le Jolis	MeOH extract	38.95	9.96
		Cold water extract	14.80	33.90
		Hot water extract	12.36	29.97
	Fucus distichus	MeOH extract	30.40	10.32
	(Linnaeus)	Cold water extract	25.96	7.85
		Hot water extract	24.99	9.04
	Fucus evanescens	MeOH extract	23.85	19.43
	(C. Agardh)	Cold water extract	4.86	148.47
		Hot water extract	2.95	188.60
	Fucus vesiculosus	MeOH extract	23.21	ND
	(Linnaeus)	Cold water extract	10.84	37.47
		Hot water extract	12.51	32.97
	Laminaria saccharina	MeOH extract	2.17	–
	(Linnaeus)	Cold water extract	0.34	–
	J.V. Lamoroux	Hot water extract	0.44	–
Red algae	*Polysiphonia stricta*	MeOH extract	12.40	27.8
	(Dillwyn) Greville	Cold water extract	1.03	–
		Hot water extract	1.05	–
Green algae	*Codium fragile*	MeOH extract	1.15	–
	spp *tormentosoides*	Cold water extract	0.53	–
	(van Goor) P.C. Silva	Hot water extract	0.46	–

*BHA was used as a reference, EC_{50}=2.14–4.2 μg mL^{-1}; ND: not determined; "–": no activity.

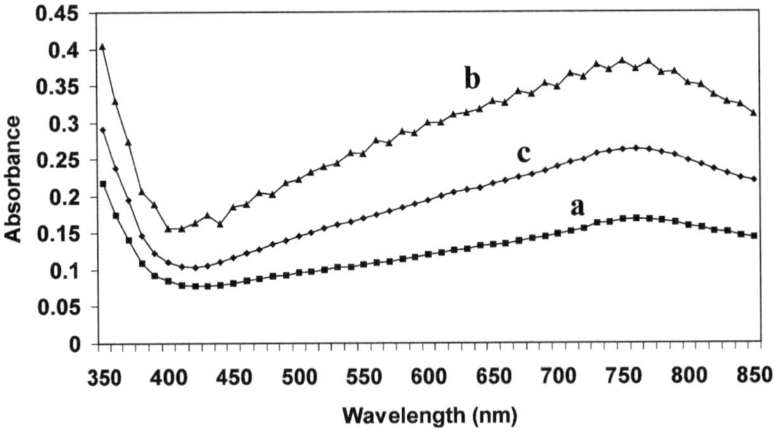

Figure 2. UV-visible spectra for phloroglucinol and *Ascophyllum nodosum* extract solutions. **a**: phloroglucinol; **b**: *A. nodosum* acidic MeOH-water (1:1, pH 2) and acetone-water (7:3) extract; **c**: *A. nodosum* acidic MeOh-water (1:1, pH 2).

such as MeOH, cold water and hot water. For most algal polyphenols, MeOH or aqueous alcohols are the preferred solvents for extraction. Solubility of polyphenols in water is fairly low, and more water-soluble polysaccharides or other components are extracted as well. If seaweed material, not its extract, is the target of analysis we would recommend acidic MeOH-water and acetone-water as the solvents for extraction

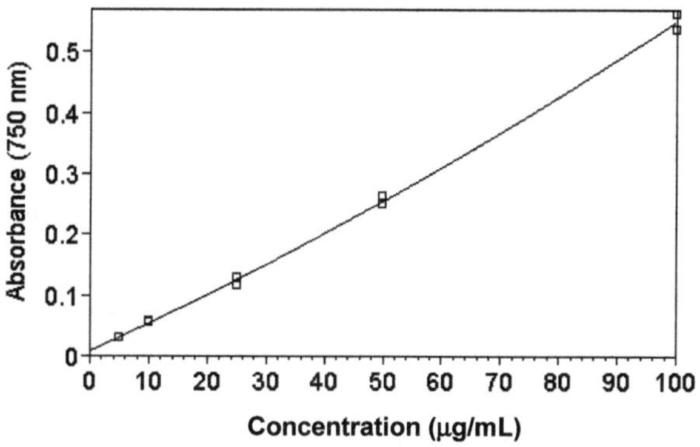

Figure 3. Phloroglucinol standard calibration curve (concentration within 0–100 μg mL^{-1}).

as described above; acidic aqueous MeOH and acetone are more powerful in recovering polyphenolics than the neutral solvents. The DPPH radical scavenging activities of those extracts indicate a good correlation between polyphenol content and antioxidant activity.

Absolute polyphenol content: Due to the complexity of the chemical nature of phenolic components in seaweeds, no method is regarded to be a perfect approach to determine the total level of phenolic compounds. However, some studies on brown algae species, such as *A. nodosum* and *F. vesiculosus*, indicate that by employing a conversion factor (or estimation factor, EF) obtained by using gravimetric methods, one can calculate the absolute polyphenol content by multiplying the EFs with the relative content value from Folin-Denis colorimetric determination (Ragan & Jensen, 1977 and 1978). This approach gives a relatively accurate and absolute value, and makes possible direct comparisons between different algal species. However, EF is sample dependent and varies to some degree according to species, collection time, and location. Determination of EFs is also tedious and time consuming. Using the phloroglucinol equivalents obtained from the Folin-Ciocalteu colorimetric method as a relative determination of polyphenol level appears to be practical for routine QC purpose and is reasonably reliable.

Bioassay directed fractionation and structure characterization of the bioactive polyphenols are in progress.

Acknowledgments

We thank Dr. Carolyn Bird (Institute of Marine Biosciences, National Research Council of Canada, Halifax) for help in sample collection and identification. Angelica Silva is grateful for a NSERC Postdoctoral fellowship.

References

Ahn MJ, Yoon KD, Min S-Y, Lee JS, Kim JH, Kim TG, Kim SH, Kim N-G, Huh H, Kim J (2004) Inhibition of HIV-1 reverse transcriptase and protease by phlorotannins from the brown alga *Ecklonia cava*. Biol. Pharmacol. Bull. 27: 544–547.

Chkhikvishvili ID, Ranazanov ZM (2000) Phenolic substances of brown algae and their antioxidant activity. Prikladnaya Biokhimiya I Mikrobiologiya 36: 336–338.

Folin O, Ciocalteu V (1927) On tyrosine and tryptophane determinations in proteins. J. Biol. Chem. 73: 627–650.

Folin O, Denis W (1912) On phosphotungstic-phosphomolybdic compounds as color regents. J. Biol. Chem. 12: 239–243.

Frei B (ed.) (1994) Natural Antioxidants in Human Health and Disease. Academic Press, San Diego.

Fukumoto LR, Mazza G (2000) Assessing antioxidant and prooxidant activities of phenolic compounds. J. Ag. Food Chem. 48: 3597–3604.

Jimenez-Escrig A, Jimenez-Jimenez I, Pulido R, Saura-Calixto F (2001) Antioxidant activity of fresh and processed edible seaweeds. J. Sci. Food Agric. 81: 530–534.

Kang HS, Chung HY, Jung JH, Son BW, Choi JS (2003a) A new phlorotannin from the brown alga *Ecklonia stolonifera*. Chem. Pharmaceut. Bull. 51: 1012–1014.

Kang HS, Chung HY, Kim JY, Son BW, Jung HA, Choi JS (2004) Inhibitory phlorotannins from the edible brown alga *Ecklonia stolonifera* on total reactive oxygen species (ROS) generation. Arch. Pharmacol. Res. 27: 194–198.

Kang K, Park Y, Hwang HJ, Kim SH, Lee JG, Shin H-C (2003b) Antioxidative properties of brown algae polyphenolics and their perspectives as chemopreventive agents against vascular risk factors. Arch. Pharmacol. Res. 26: 286–293.

Kurihara H, Ayaki T, Takahashi Y, Sasaki S, Ota T, Obori T, Yoshikawa S (2002) α-Glucosidase inhibitors containing brown algae. Japanese Patent Applicn. 2001045778, 6 pp.

Lee SK, Mbwambo ZH, Chung HS, Luyengi L, Gamez EJC, Mehta RG, Kinghorn AD, Pezzuto JM (1998) Evaluation of the antioxidant potential of natural products. Combinat. Chem. High Throughput Scr. 1: 35–46.

Nagayama K, Iwamura Y, Shibata T, Hirayama I, Nakamura T (2002) Bactericidal activity of phlorotannins from the brown alga *Ecklonia kurome*. J. Antimicrob. Chemother. 50: 889–893.

Nagai T, Yukimoto T (2003) Preparation and functional properties of beverages made from sea algae. Food Chem. 81: 327–332.

Nakamura T, Nagayama K, Uchida K, Tanaka R (1996) Antioxidant activity of phlorotannins isolated from the brown alga *Eisenia bicyclis*. Fish. Sci. 62: 923–926.

Ragan MA, Jensen A (1977) Quantitative studies on brown algal phenols. I. Estimation of absolute polyphenol content of *Ascophyllum nodosum* (L.) Le Jol. and *Fucus vesiculosus* (L.). J. Mar. Biol. Ecol. 30: 209–221.

Ragan MA, Jensen A (1978) Quantitative studies on brown algal phenols. II. Seasonal variation in polyphenol content of *Ascophyllum nodosum* (L.) Le Jol. and *Fucus vesiculosus* (L.). J. Mar. Biol. Ecol. 34: 245–258.

Shahidi F, Naczk M (2004) Phenolics in Food and Nutraceuticals. CRC Press, Boca Raton.

Shibata T, Fujimoto K, Nagayama K, Yamaguchi K, Nakamura T (2002) Inhibitory activity of brown algal phlorotannins against hyaluronidase. Int. J. Food Sci. Technol. 37: 703–709.

Shibata T, Nagayama K, Tanaka R, Yamaguchi K, Nakamura T (2003) Inhibitory effects of brown algal phlorotannins on secretory phospholipase A2s, lipoxygenases and cyclooxygenases. J. Appl. Phycol. 15: 61–66.

Singleton VL, Orthofer R, Lamuela-Raventos RM (1999) Analysis of total phenols and other oxidation substrates and antioxidants by means of Folin-Ciocalteu reagent. In Abelson JN, Simon M, Sies H (eds), Methods in Enzymology. Vol. 299, Oxidants and antioxidants, Part A. Academic Press, Orlando, pp. 152–178.

Singleton VL, Rossi JA, Jr (1965) Colorimetry of total phenolics with phosphomolybdic-phosphotungstic acid reagents. Amer. J. Enol. Viticul. 16: 144–158.

Swain T, Hillis WE (1959) The phenolic constituents of *Prunus domestica.* I.-The quantitative analysis of phenolic constituents. J. Sci. Food Agric. 10: 63–68.

Van Alstyne KL (1995) Comparison of three methods for quantifying brown algal polyphenolic compounds. J. Chem. Ecol. 21: 45–59.

Wei Y, Xu Z (2003) Studies on antioxidative activity of high molecular weight polyphenols from two kinds of brown algae. Zhongcaoyao 34: 3170–319.

Yan X, Fan X, Lou Q, Fang G, Chen Y, Song H (1998) Preparation and application of seaweed polyphenol antioxidant. Faming Zhuanli Shenqing Gongkai Shumingshu (Chinese Patent Application), No. 98110252.2, 4 pp.

Yan X, Li X, Fan X, Zhou C (1997) Studies on extraction procedure and antioxidative activity of phlorotannins from *Sargassum kjellmanianum.* Chinese J. Oceanol. Limnol. 15: 42–45.

Yan X, Li X, Zhou C, Fan X (1996) Prevention of fish oil rancidity by phlorotannins from *Sargassum kellmanianum.* J. Appl. Phycol. 8: 201–203.

Journal of Applied Phycology (2006) 18: 451–459
DOI: 10.1007/s10811-006-9051-9

Effects of UVB radiation on the initial stages of growth of *Gigartina skottsbergii, Sarcothalia crispata* and *Mazzaella laminarioides* (Gigartinales, Rhodophyta)

Mansilla, Andrés[1,*], C. Werlinger[2], M. Palacios[1], N.P. Navarro[1] & P. Cuadra[1]
[1]*Facultad de Ciencias, Universidad de Magallanes, Casilla 113-D, Punta Arenas, Chile;* [2]*Dpto. Oceanografía, Universidad de Concepción, Chile*

*Author for correspondence: andres.mansilla@umag.cl

Key words: growth, macroalgae, Magellanic Region, UV

Abstract

The effects of UVB radiation on the growth of macroalgal thalli were evaluated using tetrasporophytic fronds of the Rhodophytes *Gigartina skottsbergii, Sarcothalia crispata* and *Mazzaella laminarioides.* The tetrasporophytic fronds were collected from nature and the tetrasporophyte sporelings grown in a temperature regulated chamber at 8 ± 2 °C with a 12L:12D (Light: Dark) photoperiod, Photosynthetically Active Radiation (PAR) of 55 μmol photons m^{-2} s^{-1} and seawater enriched with 20 mL L^{-1} of Provasoli medium. We exposed the thalli of these macroalgae to PAR (55 μmol photons m^{-2} s^{-1}) and three treatments using a combination of PAR with three different levels of UVB radiation (0.10, 0.15 and 0.23 W m^{-2} for *G. skottsbergii* and *S. crispata* and 0.02, 0.05 and 0.10 W m^{-2} for *M. laminarioides*) during a period of 71 days. Growth of thalli was quantified by measuring their length using digitized photographs of samples.

Important differences were detected in the growth of individuals cultured under the effects of UVB radiation, when compared to the control (i.e. plants exposed to PAR only). In the case of *G. skottsbergii* and *S. crispata* higher levels of UVB radiation resulted in slower growth of thalli. In nearly all measurements for the first two species, UVB radiation levels of 0.1 W m^{-2} induced differences in thallus growth, while for *M. laminarioides* levels of UVB radiation of 0.1 W m^{-2} were effective only after a prolonged period of exposure.

Differential effects of UVB radiation on *G. skottsbergii, S. crispata* and *M. laminarioides* could interfere with the natural populations of these economically important macroalgal species in southern Chile, where they occur under the annual influence of the Antarctic Ozone Hole and the general thinning of the ozone layer.

Introduction

One of the most recognized atmospheric changes during the last decade is the thinning of the stratospheric ozone layer (Kirchhoff et al., 1997). This phenomenon is particularly intense during the austral spring (September to November) in the Magellanic Region (53°S; 70.9°W), the southernmost tip of the American Continent. This results in increased levels of ultraviolet-B radiation (UVB: 280–320 nm) reaching the Earth's surface (Seckmeyer & McKenzie, 1992). UVB is known as the most biologically active type of UV radiation (Komhyr et al., 1989) and is considered harmful to living beings (Diffey, 1991).

Research on UVB radiation and terrestrial plants has demonstrated diverse effects, ranging from variations in leaf morphology to changes in processes such as growth, photosynthesis and flowering (Barnes et al., 1990; Teramura & Sullivan, 1991; Cuadra et al., 1997). Since UV radiation also penetrates the water column (Figueroa, 2002), marine organisms are exposed to its harmful effects, as well. In the intertidal zone, sessile organisms, such as macroalgae, are exposed to high intensities of solar radiation, especially during low tides

(Franklin & Forster, 1997). The negative effects of UVB radiation on algae result in an increase in the concentration of chlorophyll and carotenes (Grobe & Murphy, 1998) and the synthesis of UV screening compounds, such as mycosporines (Franklin et al., 2001; Huovinen et al., 2004). Negative effects are also evident in cellular development and the ultrastructure of the macroalgae (Poppe et al., 2003; Navarro, 2004), as well as in damage to nuclear DNA, principally at the level of nitrogenated bases (Buma et al., 2000; van de Poll et al., 2001).

Physiological processes, such as photosynthesis, are also affected by the action of UV radiation on macroalgae (Dring et al., 1996; Bischof et al., 1998). The photosynthetic rate is seriously affected as the result of variations in the concentration of photosynthetic pigments (Bischof et al., 2000; Yakovleva & Titlyanov, 2001), affecting the productive capacity of energetic molecules like ATP, or by serious damage at the level of electron transport of Photosystem II (Renger et al., 1989; Nedunchezhian & Kulandaivelu, 1991; Kolli et al., 1998). On the other hand, studies conducted in high latitudes demonstrate that UV light can also cause decreased primary productivity, thereby changing the composition of autotrophic communities (Worrest, 1983; Vincent & Roy, 1993; Bischof et al., 1998).

Some researchers have shown that UV radiation affects the absorption of Photosynthetic Active Radiation (PAR) in phytoplankton, due to a decrease in the content of photosynthetic pigments (Montecinos & Pizarro, 1995). UV radiation has also been shown to interfere in the development of chloroplasts and to decrease photosynthetic rate by photoinhibition, in various macroalgal species (Larkum & Wood, 1993; Maegawa et al., 1993; Meindl & Lütz, 1996; Poppe et al., 2003). This in turn affects cellular expansion, and growth, as has been observed in *Ulva expansa* (Grobe & Murphy, 1997, 1998).

In Chile, studies regarding the effects of UVB radiation on algae are scarce, and they have focused on aspects such as phytoplankton acclimation to UV radiation (Montecinos & Pizarro, 1995), the ability of apical segments, cystocarps and thallus fragments of *Gracilaria chilensis* to acclimate to UV radiation (Molina & Montecinos, 1996) and the mycosporine content in red algae of southern Chile (Huovinen et al., 2004). Here, we evaluate the effects of UVB radiation on thallus growth of *Gigartina skottsbergii* Setchell & Gardner, *Sarcothalia crispata* (Bory) Leister and *Mazzaella laminarioides* (Bory) Fredericq under laboratory conditions. All of these carragenophytes are economically important to the Magellanic Region.

Materials and methods

Biological material

Fertile tetrasporophytic fronds of *Gigartina skottsbergii, Mazzaella laminarioides* and *Sarcothalia crispata* were collected from the intertidal and subtidal zone in the Strait of Magellan and transported in glass containers with marine water to the Marine Biology Laboratory of the Department of Natural Sciences and Resources, Faculty of Science, University of Magallanes, Chile.

Visible and UVB radiation source

Three levels of exposure to UVB were used in laboratory experiments (0.1, 0.15 and 0.23 $W\,m^{-2}$ for *G. skottsbergii* and *S. crispata* and 0.02, 0.05 and 0.10 $W\,m^{-2}$ for *M. laminarioides*), and were nominated as UVB1, UVB2 and UVB3 respectively (Table 1), which were supplied by three artificial UVB tubes (TL 20 W/12RS, Philips) with an output at 312 nm. The levels were obtained by adjusting the height of the UVB tubes above the dishes. UVC light was filtered out with cellulose diacetate foil (0.075 mm thick), which allowed 0% transmission below 280 nm. The filters were replaced after 85 h of use to avoid degradation.

For the control and all UVB treatments, PAR tubes (Philips TLT 20 W/54 daylight fluorescent) were used. PAR was kept low and constant during the entire experimental period at 55 μmol photons $m^{-2}\,s^{-1}$.

Experimental treatment and culture condition

Fragments of tetrasporophytic fronds of each of the three species (*G. skottsbergii, M. laminarioides* and *S. crispata*) were carefully washed with tap water and distilled water to eliminate epiphytes and remnants of organic matter. Sporulation, was induced in test tubes containing 50 mL of seawater; test tubes were periodically shaken to avoid spore settlement. For each species, 12 glass slides were inoculated.

Once spores settled on the glass slides, the gametophytes germlings were divided into four groups (3 slides per group) for each species. Slides from the same treatment group were placed in plastic containers with filtered and sterilized seawater and enriched

Table 1. Radiation treatments and intensity of PAR and UVB light utilized in cultures of *G. skottsbergii, S. crispata* and *M. laminarioides.*

	Light Intensity		
Treatments Control (PAR)	*G. skottsbergii* (55 μmol photons m^{-2} s^{-1})	*S. crispata* (55 μmol photons m^{-2} s^{-1})	*M. laminarioides* (55 μmol photons m^{-2} s^{-1})
Treatment 1* (UVB1)	0.10 W m^{-2}	0.10 W m^{-2}	0.02 W m^{-2}
Treatment 2* (UVB2)	0.15 W m^{-2}	0.15 W m^{-2}	0.05 W m^{-2}
Treatment 3* (UVB3)	0.23 W m^{-2}	0.23 W m^{-2}	0.10 W m^{-2}

*In all UVB treatments also used PAR (55 μmol photons m^{-2} s^{-1}).

with 20 mL L^{-1} of Provasoli. Both seawater and Provasoli enrichment were renewed weekly, following the criteria of Romo and Paula (1995). Each of the containers with inoculated glass slides were immediately deposited in a temperature controlled chamber. The chamber was divided into four compartments, each with a different regime of radiation, corresponding to the experimental treatments (Table 1).

Growth measurements

We quantified the increase in thallus length using the image processing software Image-Pro version 4.1. All slides were photographed underwater and immediately returned to the appropriate culture chamber. On each image we measured a total of 30 thalli; if there were fewer than 30 surviving thalli, we measured all individuals present on the slide. All measurements were made between 41–43, 48–50, 54–57, 63–64 and 71 days, with the exception of *G. skottsbergii* for which measurements were not made on day 71.

To estimate the daily growth rate (DGR) of individual thalli, we used the formula of Hansen (1980).

$$\mathrm{DGR}(\%\,\mathrm{perday}) = \frac{100\left[\ln\left(\frac{N_t}{N_o}\right)\right]}{t}$$

where: N_o, Initial size; N_t, Final size; *t*, Time interval in days.

Statistical analysis

A two factor ANOVA analysis was used to evaluate the differences in growth rates between treatment groups and control. For daily growth rates we applied an arcsine transformation prior to a one-way ANOVA. In both analyses, when significant differences were found, we applied a Tukey test (Spjotvoll & Stoline, 1973; Zar, 1999) to locate the differences between the various experimental treatments. Prior to analysis, we verified assumptions of homogeneity of variances using a multivariate test (Anderson, 1958).

Results

Daily growth rate

After 63 days, in *G. skottsbergii* we observed a significantly higher daily growth rate for thalli exposed to control and UVB3 radiation compared with other radiation treatments (Figure 1). In contrast, after 71 days of culture, *S. crispata* only presented a higher DGR for thalli cultured under PAR (control condition), while

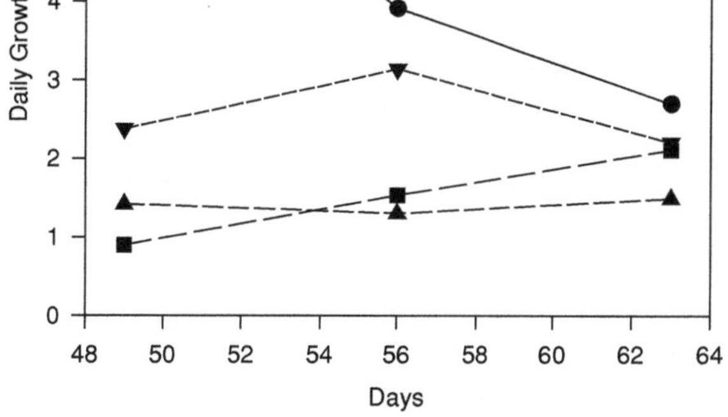

Figure 1. Daily growth rate (% per day) of *G. skottsbergii* thalli under different treatments of UVB radiation (UVB1 = 0.10 W m^{-2}, UVB2 = 0.15 W m^{-2}, UVB3 = 0.23 W m^{-2} and PAR = control). Data are expressed as mean values ($n = 3$).

Table 2. Analysis of variance for average daily growth rates of *G. skottsbergii*, *S. crispata* and *M. laminarioides* under different intensities of UVB radiation

Species	df effect	MS effect	df error	MS error	F	p-level
G. skottsbergii	3	17.5	8	2.7	6.4	0.016
S. crispata	3	15.3	12	4.3	3.6	0.047
M. laminarioides	3	0.3	12	2.9	0.1	0.949

Table 3. Tukey test for average daily growth rates of *G. skottsbergii* and *S. crispata* grown under different intensities of UVB radiation (UVB1 = 0.10 W m^{-2}, UVB2 = 0.15 W m^{-2}, UVB3 = 0.23 W m^{-2} and PAR = control)

Species	PAR (Control)		UVB1		UVB2		UVB3
G. skottsbergii	4.44	=	2.57	≠	1.51	=	1.40
	PAR		UVB1		UVB2		UVB3
S. crispata	3.36	≠	1.45	=	1.44	=	1.30

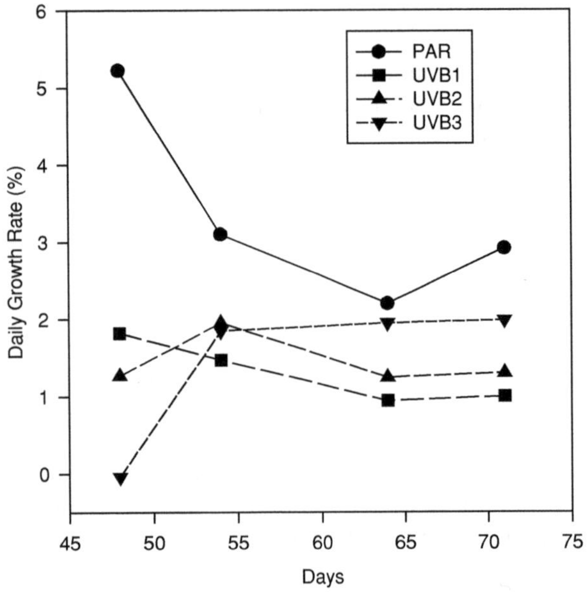

Figure 2. Daily growth rate (% per day) of *S. crispata* thalli under different treatments of UVB irradiation (UVB1 = 0.10 W m^{-2}, UVB2 = 0.15 W m^{-2}, UVB3 = 0.23 W m^{-2} and PAR = control). Data are expressed as mean values (n = 3).

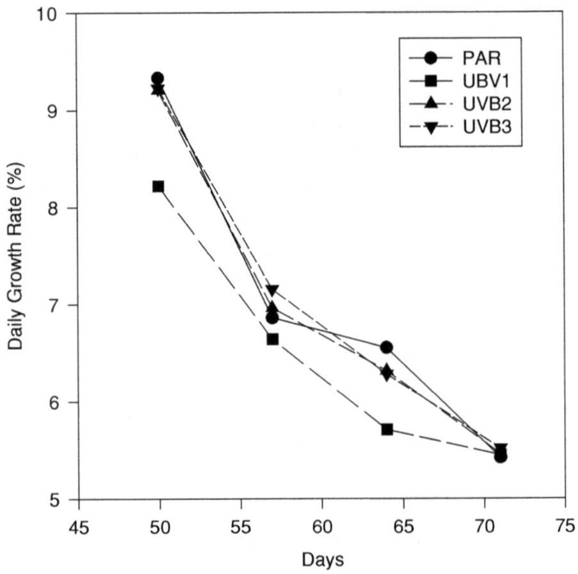

Figure 3. Daily growth rate (% per day) of *M. laminarioides* thalli under different treatments of UVB radiation (UVB1 = 0.02 W m^{-2}, UVB2 = 0.05 W m^{-2}, UVB3 = 0.10 W m^{-2} and PAR = control). Data are expressed as mean values (n = 3).

thalli under a combination of PAR radiation and different levels of UVB radiation did not significantly differ (Figure 2). The growth rates of these plants did not show a clear tendency, rather successive measurements alternating between high and low values (Figures 1 and 2). In the case of *M. laminarioides,* daily growth rates measured over 71 days presented a sustained decrease with no differences between radiation treatments (Figure 3).

Analysis of variance between average daily growth rates (Table 2) indicated significant differences between plants grown under different UVB radiation, in both *G. skottsbergii* (p = 0.016) and *S. crispata* (p = 0.047). In contrast, in *M. laminarioides* DGR did not differ with treatment (p = 0.949).

We observed significantly higher growth rates for *G. skottsbergii* plants exposed to the control and plants exposed to UVB3, compared to plants exposed to UVB1 and UVB2. The UVB1 and UVB2 treatment groups did not differ significantly. In *S. crispata*, only controls were significantly different (higher DGR) than all other treatment groups, among which there were no differences (Table 3).

Thallus growth

The analysis of variance in the three cases indicated that while there were significant differences in the growth of fixation discs – as we can see for *G. skottsbergii* in Table 4, for *S. crispata* in Table 5 and for *M. laminarioides* in Table 6 – these variations were a consequence of the different radiation treatments (with p = 0.000 in the three cases), and did not occur in the same way in all treatments of light intensity, as we can see in the

Table 4. Analysis of variance between thalli of *G. skottsbergii* grown under different treaments of UVB radiation

	df effect	MS effect	df error	MS error	F	p-level
Radiation	3	4.2×10^6	534	6.7×10^5	6.2	0.000
Days	3	42.6×10^6	534	6.7×10^5	62.3	0.000
Interaction	9	1.4×10^6	534	6.7×10^5	2.0	0.034

Table 5. Analysis of variance between thalli of *S. crispata* grown under different treatments of UVB radiation

	df effect	MS effect	df error	MS error	F	p-level
Radiation	4	1.9×10^7	995	3.7×10^5	54.1	0.000
Days	3	11.7×10^7	995	3.7×10^5	318.2	0.000
Interaction	12	0.5×10^7	995	3.7×10^5	12.5	0.000

Table 6. Analysis of variance between thalli of *M. laminarioides* grown under different treatments of UVB radiation

	df effect	MS effect	df error	MS error	F	p-level
Radiation	4	113.1×10^5	1163	1.6×10^4	695.3	0.000
Days	3	47.9×10^4	1163	1.6×10^4	29.4	0.000
Interaction	12	4.6×10^4	1163	1.6×10^4	2.8	0.001

Table 7. Tukey test of *G. skottsbergii* thalli growth on different day and under different treatments of UVB radiation (UVB1 = 0.10 W m^{-2}, UVB2 = 0.15 W m^{-2}, UVB3 = 0.23 W m^{-2} and PAR = control)

Measurement (# of days)	PAR (Control)		UVB1		UVB2		UVB3
42	1265	=	754.25	=	879.38	=	630
49	2024.27	≠	833	=	936.49	=	743.75
56	2187.14	≠	905	=	1089.65	=	977.77
63	2229.88	≠	1031.58	=	1369.23	=	1000

Table 8. Tukey test of *S. crispata* thalli growth on different days and under different treatments of UVB radiation (UVB1 = 0.10 W m^{-2}, UVB2 = 0.15 W m^{-2}, UVB3 = 0.23 W m^{-2} and PAR = control)

Measurement (# of days)	PAR (Control)		UVB1		UVB2		UVB3
41	1573.33	≠	1028.29	=	945.59	=	606.76
48	2268.40	≠	1168.08	=	1033.45	≠	605.04
54	2354.89	≠	1244.99	=	1218.89	≠	771.49
64	2609.19	≠	1276.93	=	1258.95	=	949.27
71	3773.57	≠	1386.67	=	1396.25	=	1099.20

Table 9. Tukey test of *M. laminarioides* for thalli growth on different days and under different treatments of UVB radiation. (UVB1 = 0.02 W m^{-2}, UVB2 = 0.05 W m^{-2}, UVB3 = 0.10 W m^{-2} and PAR = control)

Measurement (# of days)	PAR (Control)		UVB1		UVB2		UVB3
43	170.4	=	159.9	=	140.3	=	141.8
50	327.5	=	284.4	=	267.2	=	270.4
57	445.3	=	405.3	=	371.8	=	386.0
64	674.8	≠	530.4	=	528.3	=	530.2
71	778.8	=	734.9	≠	646.7	=	664.2

results of Tukey test for different treatments of UVB radiation (Tables 7–9).These differences also depended on the number of days to which the individuals were exposed to the radiation treatment.

The initial thalli of all three algal species (*G. skottsbergii, S. crispata* and *M. laminariodes*) presented sustained growth during the entire observation period (Figures 4–6).Plants in the controls grew fastest, while those exposed to different levels of UVB did not show significant differences in growth, although there was a slight tendency for less growth under the greater intensities of UVB.

In *G. skottsbergii* and *S. crispata*, at the end of the experiment the thalli that grew under only PAR radiation lacking UVB were longer (average 2230 and 3774 μm, for each species respectively). Among thalli maintained under different levels of UVB (UVB1, UVB2 and UVB3), the greatest growth for *G. skottsbergii* occurred under UVB1 radiation (mean of 1369 μm), while the greatest growth for *S. crispata* occurred under UVB2 radiation (mean of 1396 μm). For both of these algae, the highest intensities of UVB radiation produced the lowest growth (1000 μm in *G. skottsbergii* and 1099 μm in *S. crispata* (Figures 4 and 5). In *M. laminarioides* there was no significant difference in thallus length, although the mean for control plants was always greater than for plants exposed to UVB radiation. At the end of the experiment, plants without UVB exposure reached an average length of 778 μm, compared with 735 μm for thalli under UVB1 and 647 μm for plants exposed to UVB2, corresponding to the plants with the lowest average size for all species and treatments (Tables 2–4).

The Tukey test indicated (Table 7) that in *G. skottsbergii* there were significant differences in the control group, but only after 49 days. Plants from this group were significantly larger, on average, than plants from the other treatment groups. For plants irradiated

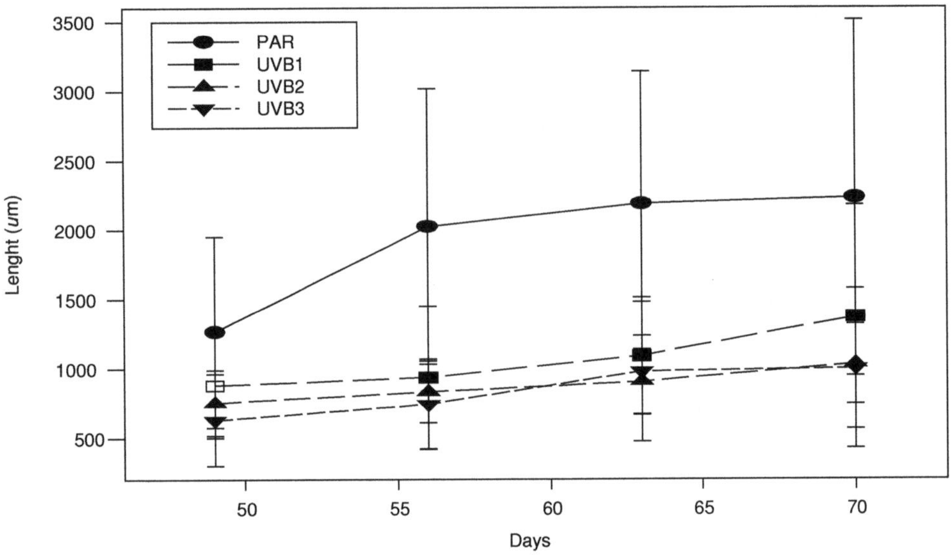

Figure 4. Average thalli growth of *G. skottsbergii* under different treatments of UVB radiation (UVB1 = 0.10 W m^{-2}, UVB2 = 0.15 W m^{-2}, UVB3 = 0.23 W m^{-2} and PAR = control). Data are expressed as mean values ± S.D. (n = 3).

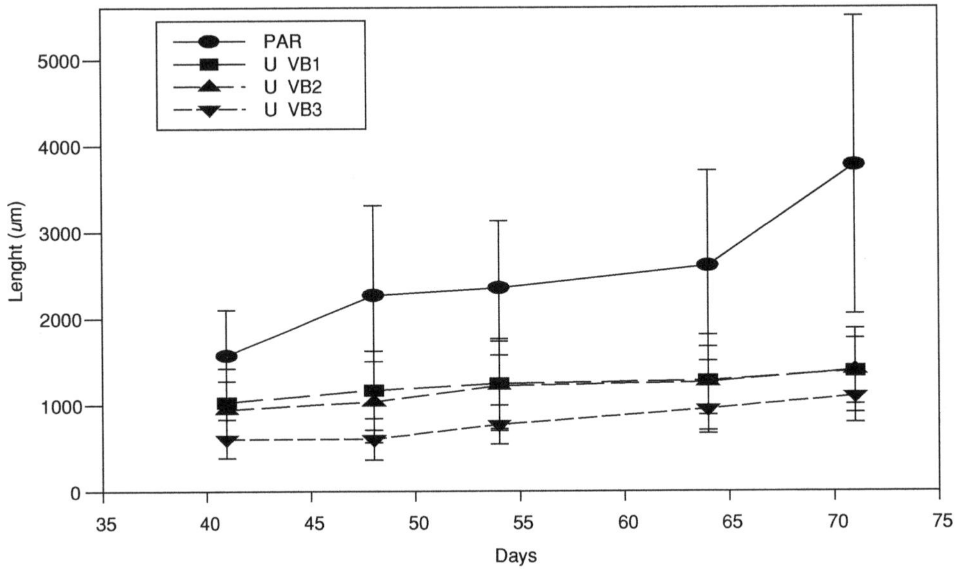

Figure 5. Average thalli growth of *S. crispata* under different treatments of UVB radiation (UVB1 = 0.10 W m^{-2}, UVB2 = 0.15 W m^{-2}, UVB3 = 0.23 W m^{-2} and PAR = control). Data are expressed as mean values ± S.D. (n = 3).

with UVB final thalli were significantly smaller, although there were no significant differences between the UVB treatment groups (Table 7). In the case of *S. crispata*, thalli sizes in the controls were significantly larger. These differences were evident from the first measurements (day 41); furthermore, in this species there was significantly lower growth in plants exposed to the highest UVB intensities, but only on days 48 and 54 (Table 8). For *M. laminarioides*, on the other hand, differences in growth were only manifested after 64 days for the control group, with significant increases in size compared with the plants under the UVB radiation treatments. Furthermore, in the control at 71 days, both the plants under PAR and those at lower UVB radiation levels (i.e. UVB1 and UVB2) had significantly greater average sizes than plants maintained at the highest levels of UVB radiation (UVB3) (Table 9).

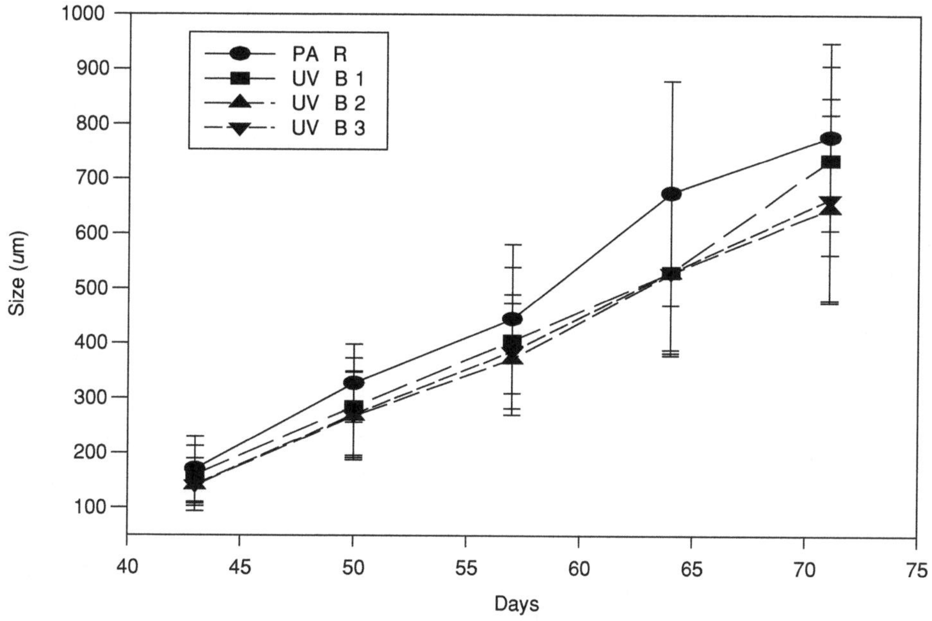

Figure 6. Average thalli growth of *M. laminarioides* under different treatments of UVB radiation (UVB1 = 0.02 W m^{-2}, UVB2 = 0.05 W m^{-2}, UVB3 = 0.10 W m^{-2} and PAR = control). Data are expressed as mean values ± S.D. ($n = 3$).

Discussion

This is the first study to provide quantitative evidence about the negative effects of UVB radiation on the growth of macroalgae in the Magellanic Region of Chile (53°S, 70.9°W), where highs levels of UVB radiation are observed during the austral spring (Casiccia et al., 2003). Significant differences were detected in the growth of individuals cultured under UVB radiation plus PAR (treatment) and those cultured under PAR only (control). Clearly, UVB radiation had a negative effect on sporelings of *G. skottsbergii* and *S. crispata*. In fact, irradiated samples of *G. skottsbergii* and *S. crispata* exihibited 42.1 and 56.8% inhibition respectively, when compared to controls. The growth differences between control thalli of *G. skottsbergii* and *S. crispata* compared with thalli from the treated group has also been found in other macroalgae (van de Poll et al., 2001). These authors measured exponential growth rates in macroalgae, observing that in treatments with UVB radiation the growth rate decreased by 25–40% of the values in macroalgae treated only with PAR radiation.

Similar results are described by Makarov (1999) for nine species of macroalgae from the Barents Sea, where a significant decrease in growth was induced by UVB radiation for all of the analyzed species. In particular, *Ulvaria obscura* was the most affected, showing a decrease in growth rate of 54%. Other studies have shown reductions in growth caused by UVB of up to 31% for the brown alga *Dictyota dichotoma* and 46–70% for the red alga *Chondrus crispus*, following three weeks of treatment, and 9% in the green alga *Ulva expansa,* after three months of exposure (Kuhlenkamp et al., 2001; Franklin et al., 1999; Grobe & Murphy, 1998).

These alterations in growth may be related to inhibition of biosynthetic pathways (Cuadra & Harborne, 1996), cell division processes, chloroplast and thylakoid membrane damage and DNA damage. All these changes may explain the effects found in this study on individuals cultured under UVB radiation (Poppe et al., 2003; Buma et al., 1995, 2000; Navarro, 2004). The low growth rates observed in UVB irradiated thalli of *G. skottsbergii* and *S. crispata* may represent some of the harmful effects that this type of radiation has on diverse cell mechanisms, such as the inhibition of the activity of certain enzymes which participate in algae nitrogen metabolism, as pointed out by Döler (1996) and Döler et al. (1995).

In *G. skottsbergii* and *S. crispata* UVB radiation levels of 0.10 W m^{-2} (i.e. the lowest intensity of UVB radiation utilized in this study) severely inhibited plant growth (Tables 4 and 5). In contrast, for *M. laminarioides* UVB did not cause statically significant differences (Table 6). An explanation of these effects can be found in the habitat requirements of these species. In their natural environments, littoral species are more exposed to solar radiation and therefore, more adapted to UVB exposure. On the other hand, subtidal macroalgae are less exposed and may be not adapted to enhanced levels of this radiation. Similar findings were reported by van de Poll et al. (2001), who observed that littoral species presented a greater tolerance to UVB radiation than subtidal taxa.

In relation to the different UVB treatments (UVB1, UVB2 and UVB3), the fact that higher doses did not increase the inhibitory effects may be attributed to a photoacclimation process. This adaptive mechanism generally involves the biosynthesis of mycosporine-like substances (Huovinen et al., 2004). They may also be responsible for the greater tolerance observed in intertidal species. In this context, Karsten et al. (1999) and Figueroa et al. (2003) found that algae exposed to sunlight presented elevated levels of mycosporines compared to populations of the same species or congeners grown under shade conditions. This may favor growth and survival in irradiated environments. Likewise, the higher DGR values observed in *G. skottsbergii* compared to *S. crispata* under the UVB3 treatment may be explained by the amount of mycosporine-like substances accumulated by the former. As reported by Karsten et al. (1999), macroalgae could present greater or lesser diversity in their mycosporine composition. Besides, species from the same genus may not necessarily be able to produce a similar composition of these compounds (Huovinen et al., 2004).

Another important factor to consider is related morphological aspects, such as thallus thickness. Thicker plants may be better protected against UVB radiation than thinner species. Grobe and Murphy (1994, 1998), demonstrated that in species such as *Ulva expansa*, UVB radiation induces changes in the cell wall by reducing cell expansion and growth rate.

Finally, this study provides experimental evidence about the potential effects that this radiation has on natural ecosystems at high latitudes, like the Magellanic Region, where it can interfere with macroalgal populations of these economically important species. If complementary studies are developed focused on more ultrastructural, physiological or molecular aspects, a better understanding about the effects of this radiation in populations of *G. skottsbergii*, *S. crispata* and *M. laminarioides* may be achieved.

Acknowledgements

We are grateful to the Ministerio de Educación de Chile – PR-344 and the Universidad de Magallanes – 269 PR-F4–01RN-2000 for their financial support. We are also grateful to Christopher B. Anderson and Paula Niell for critically reading the manuscript.

References

Anderson TW (1958) An Introduction to Multivariate Statistical Analysis. John Wiley & Sons, Inc, New York.

Barnes P, Flint S, Cadwell M (1990) Morphological responses of crop and weed species of different growth forms to ultraviolet-B radiation. Am. J. Bot. 77: 1354–1360.

Bischof K, Hanelt D, Wiencke C (1998) UV-radiation can affect depth-zonation of Antarctic macroalgae. Mar. Biol. 131: 597–605.

Bischof K, Hanelt D, Wiencke C (2000) Effects of ultraviolet radiation on photosynthesis and related enzyme reactions of marine macroalgae. Planta 211: 555–562.

Buma AGJ, van Oijen T, van de Poll W, Veldhuis MJW, Gieskes WWC (2000) The sensivity of *Emiliana huxleyi* (Prymnesiophyceae) to ultraviolet-B radiation. J. Phycol. 36: 296–303.

Buma AGJ, Van Hannen EJ, Roza L, Veldhuis MJW, Gieskes WWC (1995) Monitoring ultraviolet-B-induced DNA damage in individual diatom cells by immunofluorescent thymine dimer detection. J. Phycol. 31: 314–321.

Casiccia C, Kirchhoff VWJH, Torres A (2003) Simultaneous measurements of ozone and ultraviolet radiation: Spring 2000, Punta Arenas, Chile. Atmospheric Environment.

Cuadra P, Harborne J, Waterman P (1997) Increases in flavonoid and photosynthetic pigments in *Gnaphalium luteo-album* in response to UV-B radiation. Phytochemistry 45: 1377–1383.

Diffey B (1991) Solar ultraviolet radiation effects on biological system. Physics in Medicine and Biology 36: 299–328.

Döler G (1996) Effect of UV irradiance on utilization of inorganic nitrogen by the Antarctic diatom *Odontella weissflogii* (Janisch) Grunow. Botanica Acta 109: 35–42.

Döler G, Hagmeier E, David C (1995) Effects of solar and artificial UV irradiation on pigments and assimilation of 15N ammonium and 15N nitrate by macroalgae. J. Photochem. Photobiol. 32: 107–113.

Dring M, Makarov V, Schoschina E, Lorenz M, Lüning K (1996) Influence of ultraviolet-radiation on chlorophyll fluorescence and growth in different life-history stages of three species of *Laminaria* (Phaeophyta). Mar. Biol. 126: 183–191.

Figueroa FL (2002) Bio-optical characteristics of Gerlache and Bransfield Strait waters during an Antarctic summer cruise. Deep Sea Res. II 49: 675–691.

Figueroa FL, Escassi E, Pérez-Rodruíguez, Korbee N, Giles AD, Johnsen G (2003) Effects of short-term irradiation on photoinhibition and accumulation of mycosporine-like amino acids in sun and shade species of the red algae genus *Porphyra*. J. Photochem. Photobiol. B: Biol. 69: 21–30.

Franklin LA, Forster RM (1997) The changing irradiance environment: Consequences from marine macrophyte physiology, productivity and ecology. Eur. J. Phycol. 32: 207–232.

Franklin LA, Yakovleva I, Karsten U, Lüning K (1999) Synthesis of mycosporine-like-amino acids in *Chondrus crispus* (Florideophyceae) and the consequences for sensitivity to ultraviolet B radiation. J. Phycol. 35: 682–693.

Franklin LA, Kräbs G, Kuhlenkamp R (2001) Blue light and UV-A radiation control the synthesis of mycosporine-like amino acids in *Chondrus crispus* (Florideophyceae). J. Phycol. 37: 257–270.

Grobe C, Murphy T (1994) Inhibition of growth of *Ulva expansa* (Cholorophyta) by ultraviolet-B radiation. J. Phycol. 30: 783–790.

Grobe C, Murphy T, (1997) Artificial ultraviolet-B radiation and cell expansion in the intertidal alga *Ulva expansa* (Setch.) S. & G. (Chlorophyta). J. Exp. Mar. Biol. Ecol. 217: 209–223.

Grobe C, Murphy T (1998) Solar ultraviolet-B radiation effects on growth and pigment composition of the intertidal alga *Ulva expansa* (Setch.) S. & G. (Chlorophyta). J. Exp. Mar. Biol. Ecol. 225: 39–51.

Hansen JE (1980) Physiological considerations in the mariculture of red algae. In Abbott IA, Foster MS, Ekund LF (eds), *Pacific Seaweed Aquaculture*. Publications of the California Sea Grant Program, La Jolla, pp. 80–91.

Huovinen P, Gómez I, Figueroa FL, Ulloa N, Morales V, Lovengreen C (2004) Ultraviolet-absorbing micosporine-like amino acids in red macroalage from Chile. Botanica Marina 47: 21–29.

Karsten U, Sawall T, Wiencke C (1998). A survey of the distribution of UV-absorbing substances in tropical macroalgae. Physiol. Res. 46(4): 271–279.

Karsten U, Bischof K, Hanelt D, Tüg H, Wiencke C (1999) The effects of ultraviolet radiation on photosynthesis and ultraviolet-absorbing substances in the endemic Arctic macroalga *Devaleraea ramentacea* (Rhodophyta). Physiol. Plant 105: 58–66.

Kirchhoff VWJH, Zamorano F, Casiccia C (1997) UVB enhancements at Punta Arenas, Chile. J. Photochem. Photobiol. B: Biol. 38: 174–177.

Kolli B, Tiwari S, Mohanty P (1998) Ultraviolet-B induced damage to photosystem II in intact filaments of *Spirulina platensis*. Z. Naturf. Sect. C 53(5–6): 369–377.

Komhyr W, Grass R, Leonard R (1989) Total ozone, ozone vertical distributions, and stratospheric temperatures at South Pole, Antarctica, in 1996 and 1997. J. Geophys. Res. 94: 11429–11436.

Kuhlenkamp R, Franklin LA, Lüning K (2001) Effect of solar UV radiation on growth in the marine macroalgae *Dicyota dichotoma* (Phaeophyceae) at Helgoland and its ecological consequences. Helgoland Mar. Res. 55: 77–86.

Larkum A, Wood W (1993). The effect of UVB radiation on photosynthesis and respiration of phytoplankton benthic macroalgae and seagrasses. Photosynth. Res. 36: 17–23.

Maegawa M, Kunieda M, Kida W (1993) The influence of ultraviolet radiation on the photosynthetic activity of several red algae from different depths. Jpn. J. Phycol. 41: 207–214.

Makarov M (1999) Influence of ultraviolet radiation on the growth of the dominant macroalgae of the Barents Sea. Chemosphere: Global Change Science 1: 461–467.

Meindl U, Lütz C (1996) Effects of UV irradiation on cell development and ultrastructure of the green alga *Micrasterias*. J. Photochem. Photobiol. 36: 285–292.

Molina X, Montecino V (1996) Acclimation to UV irradiance in *Gracilaria chilensis* Bird, Mc Lachlan & Oliveira (Gracilariles, Rhodophyta). Hydrobiologia 326/327: 415–420.

Montecino V, Pizarro G (1995) Phytoplankton acclimation and spectral penetration of UV irradiance of the central Chilean coast. Mar. Ecol. Prog. Ser. 121: 1–3.

Navarro NP (2004) Efeitos da radiação UVB em *Iridaea cordata* (Gigartinales, Rhodophyta). Dissertação apresentada ao Instituto de Biociências da Universidade de São Paulo, para a obtenção do título de Mestre em Ciências, na Área da Botânica. 87 pp.

Nedunchezhian N, Kulandaivelu G (1991) Evidence for the ultraviolet-B (280–320) radiation induced structural reorganization and damage photosystem II polypeptides in isolated chloroplasts. Physiol. Plantarum. 81: 558–562.

Poppe F, Schmidt RA, Hanelt D, Wiencke C (2003) Effects of UV radiation on the ultrastructure of several red algae. Phycol. Res. 51:11–19.

Renger G, Völker M, Eckert H, Fromme R, Hohm-Veit S, Gräber P (1989) On the mechanism of photosysthem II deterioration by UVB irradiation. Photochem. Photobiol. 49: 97–105.

Romo H, Paula EJ (1995) Métodos Experimentales para el cultivo de Porphyra. Alveal En K, Ferrario M, Oliveiray E, Sar E (eds.), Manual de Métodos Ficológicos, Universidad de Concepción, Editorial Anibal Pinto S.A., Chile, pp. 55–562.

Seckmeyer G, McKenzie R.L (1992) Elevated ultraviolet radiation in New Zealand (45°S) contrasted with Germany (45°S). Nature 359: 135–137.

Spjotvoll E, Stoline MR (1973) An extension of the *T*-method of multiple comparison to include the cases with unequal sample sizes. J. Am. Stat. Assoc. 68: 976–978.

Teramura A, Sullivan J (1991). Potential impacts of increased solar UVB on global plant productivity. In Riklis E (ed), Photobiology, Plenum Press New York: 625–634.

van de Poll WH, Eggert A, Buma GJ, Breeman AM (2001) Effects of UVB induced DNA damage and photoinhibition on growth of temperate marine red macrophytes: habitat-related differences in UVB tolerance. J. Phycol. 37: 30–37.

Vincent W, Roy S (1993) Solar ultraviolet-B radiation and aquatic primary production: Damage, protection, and recovery. Env. Rev. 1: 1–12.

Worrest RC (1983) Impact of solar ultraviolet-B radiation (290–320 nm) upon marine microalgae. Physiol. Plantarum 58: 428–434.

Yakovleva IM, Titlyanov EA (2001) Effects of high visible and UV irradiance on subtidal Chondrus crispus: Stress, photoinhibition and protective mechanisms. Aquatic Bot. 71: 47–61.

Zar JH (1999). Biostatistical Analysis, 4th Edition, Prentice Hall, Upper Saddle River, NJ, 941 p.

Journal of Applied Phycology (2006) 18: 461–467
DOI: 10.1007/s10811-006-9050-x

Photosynthesis and UV-B tolerance of the marine alga *Fucus vesiculosus* at different sea water salinities

C.A. Nygård* & N.G.A. Ekelund
Mid Sweden University, Department of Natural Sciences, SE-851 70 Sundsvall, Sweden

*Author for correspondence: e-mail: lotta.nygard@miun.se

Key words: *Fucus vesiculosus*, photosynthesis, dark respiration, UV-B, salinity, pigments, oxygen evolution

Abstract

The marine algal species in the Baltic Sea are few due to the low sea water salinity. One of the few species that can be found is *Fucus vesiculosus*. Even this species is affected by the low salinity and becomes smaller in size in the Baltic. In present work the photosynthesis of *F. vesiculosus* in the northern Baltic (Bothnian Sea) was compared to the photosynthesis of *F. vesiculosus* in the Atlantic. Oxygen evolution was measured before and after exposure to 2.3 W of UV-B (280–320 nm) radiation for 5 h, as well as after 48 h recovery in low light. The plants were kept in their own sea water salinity as well as in a changed salinity, this to examine possible correlations between salinity and photosynthesis. The results show a significant higher initial maximal photosynthesis (P_{max}) for Atlantic plants (10.3 nmol O_2 g^{-1} FW s^{-1}) compared to Baltic plants (4.0 nmol O_2 g^{-1} FW s^{-1}). The Baltic plants were found more sensitive to UV-B with a 40–50% decrease of P_{max} as well as a lower degree of recovery (60–70% compared to 75–95% for the Atlantic plants). The higher salinity (35 psu) had a positive effect on the Baltic *F. vesiculosus* with increased P_{max} as well as increased tolerance to UV-B. The lower salinity (5 psu) had a negative effect on the Atlantic plants with a decreased P_{max} as well as a lower tolerance to UV-B. Pigment content was found higher in Atlantic *F. vesiculosus*. The pigment content decreased then the Atlantic plants were transferred to 5 psu. The concentration of Chl a as well as the total content of violaxanthin, diadinoxanthin and zeaxanthin in Baltic plants increased when transferred to 35 psu. The Atlantic *F. vesiculosus* can not survive the low salinity in the northern Baltic (died within 8 weeks). It is likely that a long time acclimation or adaptation to low salinity has taken place for *F. vesiculosus* in northern Baltic. If this is an ecotypic or genotypic development it is too early to say.

Introduction

The marine brown alga *Fucus vesiculosus* L. (Phaeophyceae) is one of few marine species in the brackish water of the Baltic Sea. The low salinity gives the plants a different appearance in the Baltic with a smaller thallus and absence of the characteristic air-bladders in the northern Baltic (Waern, 1952). The northern distribution limit in the Baltic is found at 4 psu (Kalvas & Kautsky, 1998). This intertidal species grows subtidally in the Baltic, due to the absence of tides (Wallentinus, 1979). The plants extend to greater depths in the Baltic than in the Atlantic, mainly due to the lower occurrence of other macroalgae. Accordingly, the difference in depth distribution might give a different tolerance to high light and UV-radiation.

Populations of species can exist in diverse habitats either because they have diverged genotypically and/or because they have the ability to acclimatize phenotypically to the environment. Which is the case with *F. vesiculosus* is so far unknown. One way to try to distinguish these kinds from each other is to perform experiments with measurement of immediate physiological responses to different environmental factors (Dawes et al., 1988).

In the present work *F. vesiculosus* from the northern Baltic (Bothnian Sea) and the Atlantic were compared regarding their photosynthesis and tolerance to

UV-B radiation. The plants were cultivated at different seawater salinities to examine possible correlations between salinity and tolerance to UV-B radiation. The low salinity in the Baltic might exert a stressful effect on *F. vesiculosus*. If that is the case, the Baltic *F. vesiculosus* will show a lower tolerance limit to environmental disturbances such as UV-B radiation than the Atlantic *F. vesiculosus*.

Materials and methods

F. vesiculosus was collected at the small island Åstön (62°24′N, 17°45′E) in the Bothnian Sea, northern Baltic Sea (salinity 5 psu) at a depth of 1 m and at the island Hitra (63°39′N; 9°11′E) in the Atlantic Ocean (salinity 35 psu) in the intertidal zone. The algae were placed in aquariums with aeration at 4 °C and a light/dark cycle of 12/12 h with 35 μmol photons $m^{-2}\,s^{-1}$ of PAR (photosynthetic active radiation, 400–700 nm). The Atlantic plants were considerably bigger than the Baltic (1 m compared to 15 cm), the latter with airbladders absent. Algae from these two sites were either cultivated in their own natural seawater or transferred to sea water of the other site, with accordingly less or higher salinity. One week of acclimation to a stepwise changed salinity was used.

Measurement of oxygen evolution was conducted with a Light Pipette (Brammer, Illuminova, Sweden) consisting of a light source, a cuvette for samples (2 ml), an oxygen electrode, a temperature-controlled waterbath, and a control unit with computer. The light source (PAR) was programmed to deliver an increasing irradiance of 20, 100, 200, 300, 400, 500, 600, 700, 800 and 900 μmol photons $m^{-2}\,s^{-1}$, with 2 min at each irradiance. The computer receives data with 2 s intervals from the electrode in the cuvette. Photosynthetic measurements were conducted before and after 5 h exposure to 2.3 W m^{-2} of UV-B (280–320 nm) radiation (Westinghouse SunLamp, FS 20, 20 W). The ultraviolet radiation was filtered through a cellulose acetate film (0.13 mm thickness) to remove shorter-wavelength components(less than 290 nm) not encountered in nature. The temperature was controlled with a water bath during the exposure and the radiance was measured with an IL 1400A broad band Radiometer (International light Inc., Newburyport, MA; USA). A disc of thalli (distal parts taken below the second frond from the tip, approximately 2 years old) with a diameter of 1.5 cm was placed together with filtered seawater in the chamber and kept in darkness for 10 min before measurement. The experiments were carried out at 4 °C. After measurement the alga was placed between two soft papers and the fresh weight (FW) was noted. The maximal photosynthesis (P_{max}) was obtained from the photosynthesis versus irradiance curves at 500 μmol photons $m^{-2}\,s^{-1}$. The dark respiration was calculated as a mean value during the last 5 min of the initial dark adaptation. Five replicates from different thallus were used, and mean values with 95% confidence limits are presented. Statistical analysis was performed with Minitab ANOVA. Tukey's family error rate was used as a post-hoc test. Only significant results are reported in the text.

Pigment analyses were performed with HPLC technique (Wright et al., 1991) by the Water Quality Institute (VKI) in Denmark. Presented pigments are chlorophyll a (Chl a), chlorophyll c (Chl c), fucoxanthin, beta-carotene and the total content of violaxanthin, diadinoxanthin and zeaxanthin.

Results

Photosynthesis

F. vesiculosus in the Atlantic possesses a significantly higher P_{max} than *F. vesiculosus* in the Baltic; 10.3 nmol $O_2\,g^{-1}$ FW s^{-1} compared to 4.0 nmol $O_2\,g^{-1}$ FW s^{-1} (Figure 1). The transfer of plants from Baltic to Atlantic water significantly increased the P_{max} to 8.3 nmol $O_2\,g^{-1}$ FW s^{-1}, an increase with 100% (Figure 1). This increase occurred within 1 week time. The transfer of Atlantic *F. vesiculosus* to Baltic water instead decreased the photosynthesis, which after 7 weeks was as low as 3.5 nmol $O_2\,g^{-1}$ FW s^{-1} (Figure 1). The Atlantic plants died after 8 weeks in Baltic water.

The exposure to UV-B radiation decreased the P_{max} of Atlantic *F. vesiculosus* to 7.0 nmol $O_2\,g^{-1}$ FW s^{-1} (a decrease with 30%) and to 2.2 nmol $O_2\,g^{-1}$ FW s^{-1} (a decrease with 45%) for the Baltic *F. vesiculosus* (Figure 1). The Atlantic plants showed a full recovery of P_{max} after 48 h in low light, while this could not be found for the Baltic plants (Figure 1).

The initial slopes were found to be the same for Atlantic and Baltic *F. vesiculosus* (Figure 2). The negative effects of UV-B radiation on the initial slope were greater for Baltic plants. The Atlantic plants recovered completely after a 48 h recovery in low light, but this was not the case for the Baltic plants (Figure 2). When *F. vesiculosus* from the northern Baltic was transferred to Atlantic water, not only the P_{max} increased, even the tolerance to UV-B radiation. This higher tolerance

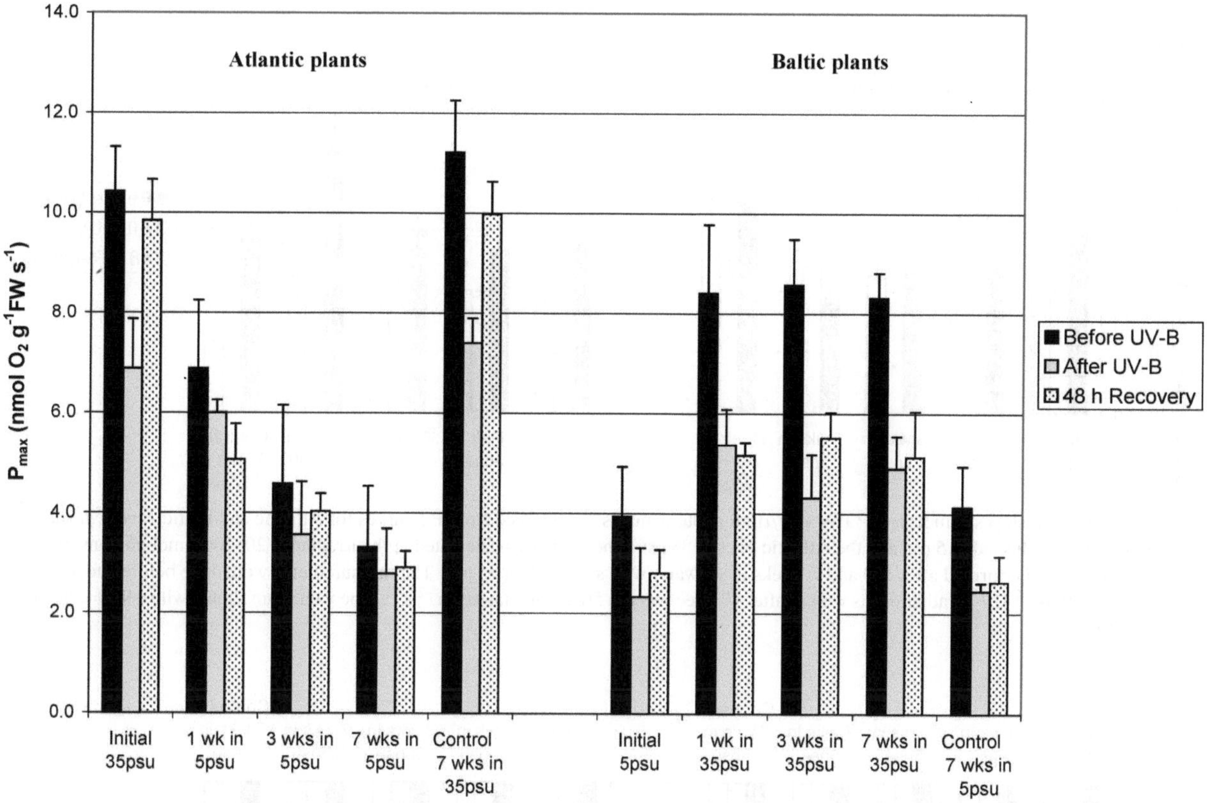

Figure 1. Photosynthetic maximum (P_{max}) of Atlantic and Baltic *F. vesiculosus* after 1, 3 and 7 weeks in seawater with salinity 5 or 35 psu. The Baltic plants were collected at 5 psu and the Atlantic ones at 35 psu. P_{max} is obtained from P:I curves at 500 μmol photons m^{-2} s^{-1}. The photosynthetic measurements were performed before and after a 5 h exposure to 2.3 W m^{-2} UV-B radiation, as well as after 48 h recovery in 50 μmol photons m^{-2} s^{-1}. The values are shown with a 95% confidence limit (n = 5).

to UV-B could be observed in both P_{max} and initial slope (Figures 1 and 2). The Atlantic plants showed a general decrease in both P_{max} and initial slope when transferred to the low salinity of Baltic water. These transferred plants even showed a decreased tolerance to UV-B, but this could only be observed in P_{max} and not in initial slope (Figures 1 and 2).

Baltic *F. vesiculosus* showed a higher dark respiration than Atlantic plants (Figure 3). The dark respiration increased when Atlantic *F. vesiculosus* were transferred to Baltic water. The Baltic plants showed no significant differences while transferred to Atlantic water (Figure 3).

Pigments

All measured pigments were initially found to be significantly higher in Atlantic *F. vesiculosus* (Figure 4a–e). While the Baltic *F. vesiculous* contained 4 mg Chl a g^{-1} DW tissue, the Atlantic contained as much as 18 mg (Figure 4a). The tissue content of the most common accessory pigment in Fucoids, the fucoxanthin, was 1.2 mg g^{-1} DW in Baltic plants and 3.8 mg g^{-1} DW in Atlantic plants (Figure 4b). Also the other measured xantophylls (total content of violaxanthin, diadinoxanthin and zeaxanthin) showed higher concentrations in the Atlantic plants (1.55 mg g^{-1} DW) compared to the Baltic (0.25 mg g^{-1} DW) (Figure 4d). The content of the photoprotective pigment beta-carotene was 0.18 mg g^{-1} DW in Baltic *F. vesiculosus*, while it was as high as 0.79 mg in Atlantic plants (Figure 4c). Even the amounts of Chl c occurred in lower amounts in the Baltic plants (0.22 mg g^{-1} DW) than in the Atlantic plants (0.75 mg g^{-1} DW) (Figure 4e). The relationship between Chl and total carotenoid content showed a higher value for Atlantic plants (3.2) than for Baltic (2.4) (Figure 4f). The content of pigments changed with transfer to another salinity of seawater. This was most obvious for the Atlantic *F. vesiculosus*, which showed decreased contents of all measured pigments when transferred to water from the northern Baltic. The Baltic *F. vesiculosus*

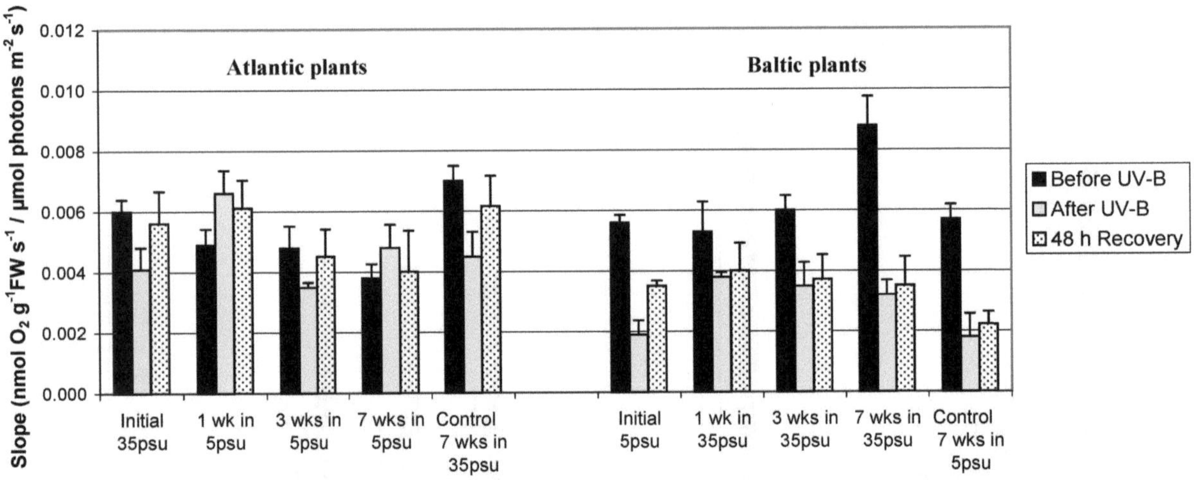

Figure 2. The initial slopes (nmol O_2 g^{-1} FW s^{-1}/μmol photons m^{-2} s^{-1}) obtained from P:I curves for Atlantic and Baltic *F. vesiculosus*. The Baltic plants were collected at 5 psu and the Atlantic ones at 35 psu. The slope was calculated at the irradiance 20–100 μmol photons m^{-2} s^{-1}. Measurements were performed after 1, 3 and 7 weeks in seawater with salinity 5 or 35 psu. The measurements were done before and after a 5 h exposure to 2.3 W m^{-2} UV-B radiation, as well as after 48 h recovery in 50 μmol photons m^{-2} s^{-1}. The values are shown with a 95% confidence limit ($n = 5$).

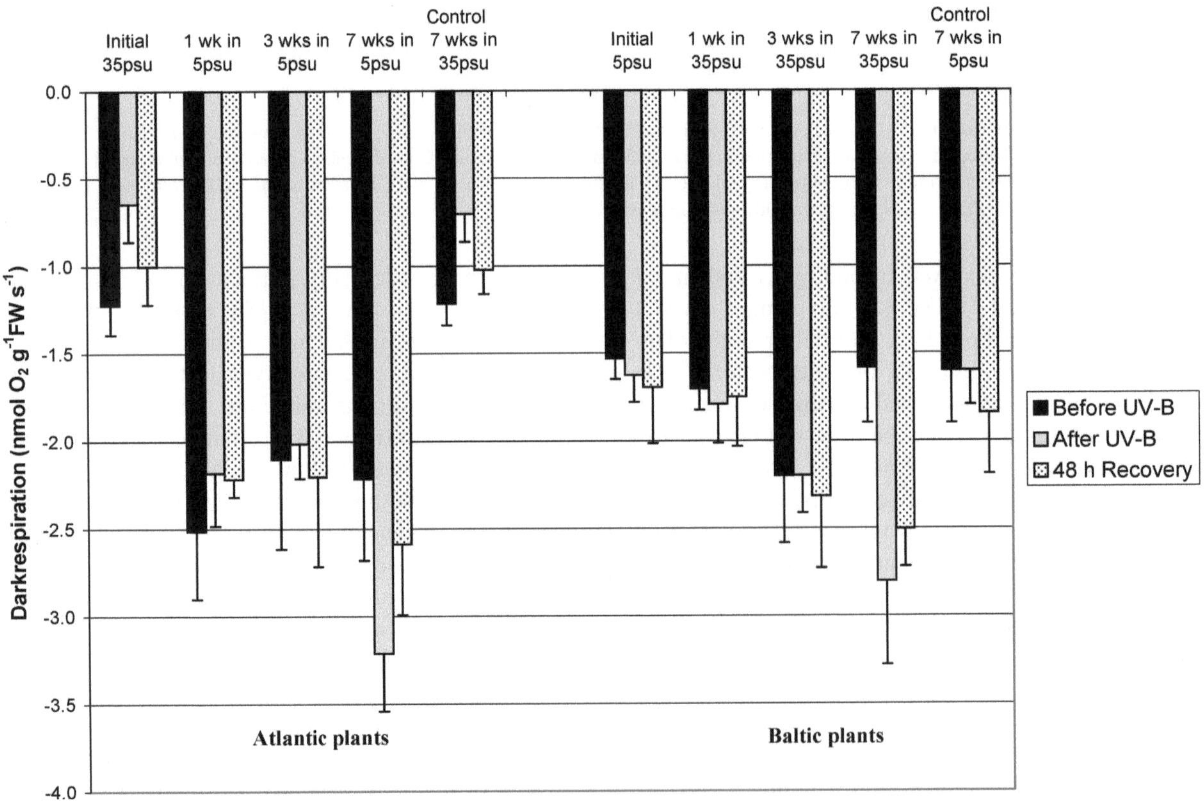

Figure 3. Dark respiration (nmol O_2 g^{-1} FW s^{-1}) of Atlantic and Baltic *F. vesiculosus* after 1, 3 and 7 weeks in seawater with salinity 5 or 35 psu. The Baltic plants were collected at 5 psu and the Atlantic ones at 35 psu. The dark respiration is a mean value during the last 5 min of the initial dark adaptation. The measurements were performed before and after a 5 h exposure to 2.3 W m^{-2} UV-B radiation, as well as after 48 h recovery in 50 μmol photons m^{-2} s^{-1}. The values are shown with a 95% confidence limit ($n = 5$).

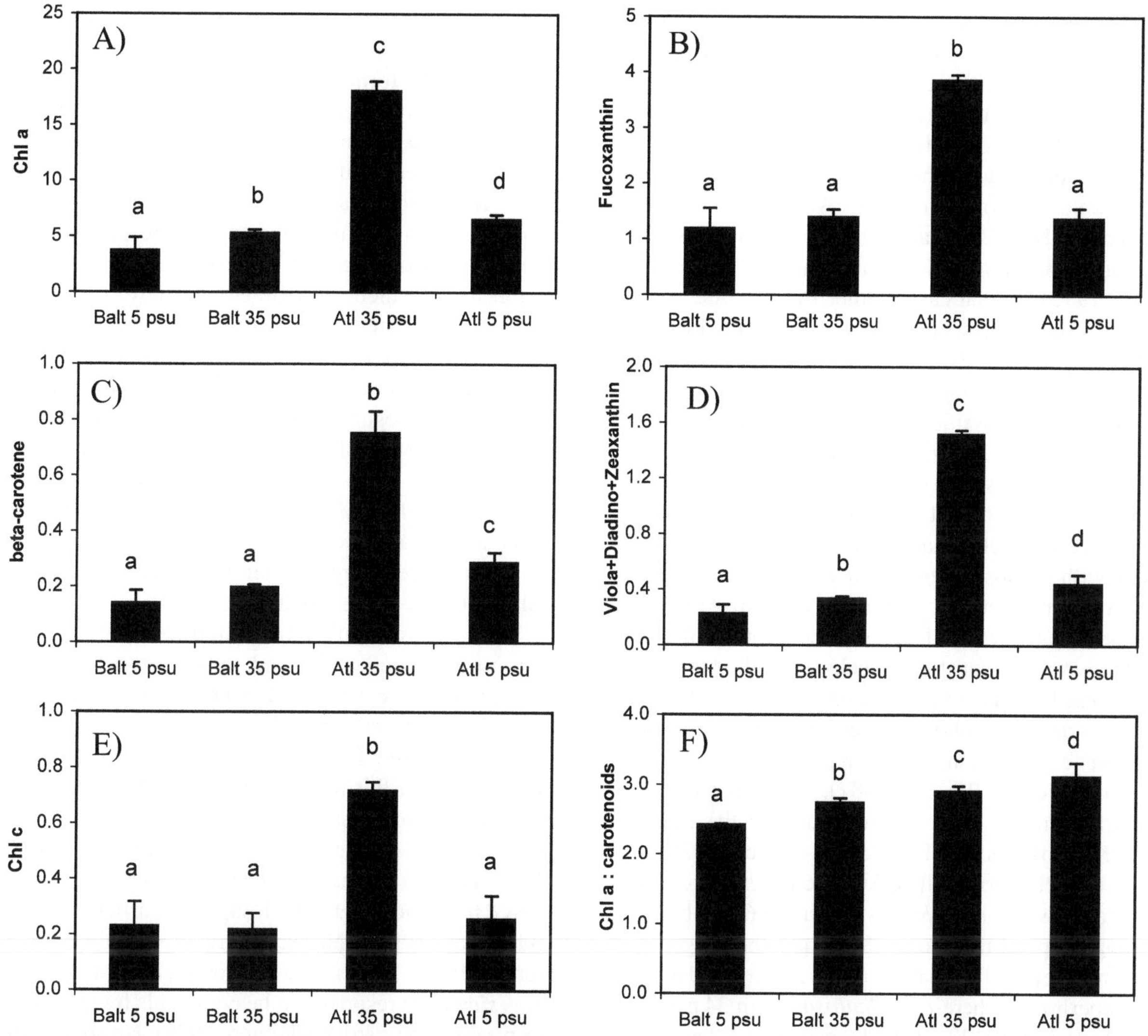

Figure 4. Pigment content (mg g^{-1} DW) of Atlantic (Atl.) and Baltic (Balt.) *F. vesiculosus*. The graphs show the content of: (a) chlorophyll a (Chl a), (b) fucoxanthin, (c) beta-carotene, (d) the sum of violaxanthin, diadinoxanthin and zeaxanthin, (e) chlorophyll c (Chl c), (f) Chl a versus total content of carotenoids. Initial pigment content as well as pigment content after 7 weeks in a changed salinity are shown. The error bars represent a 95% confidence limit ($n = 5$). Similar letters indicate a non-significant difference.

instead increased the Chl a content as well as total amount of violaxanthin, diadinoxanthin and zeaxanthin when moved to Atlantic water. The other pigments did not change in concentration with the transfer (Figure 4).

Discussion

Considering the significantly lower photosynthesis in the Baltic *F. vesiculosus*, it seems likely that the brackish water has a considerable affect on this species. The Atlantic plants died within 2 months after transfer to Baltic water and were thereby not able to acclimatize to the low salinity, which indicates that adaptation has taken place in the Baltic *F. vesiculosus*. Such adaptation might include genetic adaptation. Since no such study took place here, we will concentrate on the physiological differences between the Baltic and Atlantic *F. vesiculosus*.

The thallus of *F. vesiculosus* in a low salinity environment (like the Baltic) is smaller than in marine populations (Kalvas & Kautsky, 1998). This statement was in accordance even in the present study, where the plants were found to be much smaller in size in the Bothnian Sea compared to the Atlantic (data not shown). The Atlantic *F. vesiculosus* also grow faster than the Baltic plants (Bäck et al., 1992b). The higher photosynthesis by the Atlantic *F. vesiculosus* in the present study contributes to a higher growth rate. This higher photosynthesis (P_{max}) indicates a more favourable environment in the Atlantic. It has been shown that both Baltic and Atlantic *F. vesiculosus* experience maximum growth in a salinity of 12 psu (Bäck et al., 1992). In that case neither the Atlantic nor the northern Baltic waters possess an optimal environment for this species. Other works have showed that optimum salinity for photosynthesis in *F. vesiculosus* differs between the Baltic (6 psu) and the Atlantic (12 to 34 psu) (Bäck et al., 1992a; Russell, 1988). Bäck et al. (1992b) showed with a transfer of Atlantic *F. vesiculosus* to Baltic water that the growth rate decreased, but still remained higher than the Baltic plants. The present work with photosynthesis as measuring parameter did not support this trend. The transfer from Atlantic to northern Baltic water decreased the photosynthesis to such low values that after 3 weeks it equalled the Baltic *F. vesiculosus*. These plants even died after 8 weeks. A lower degree of photosynthesis in *F. vesiculosus* after transfer from Atlantic salinity to lower salinity was found earlier by Russell (1987). The same has even been shown for other species, such as the red alga *Gracilaria verrucosa* (Koch & Lawrence, 1987). Bäck et al. (1992b) showed that Atlantic *F. vesiculosus* was able to grow and survive in a salinity of 6 psu, but with damaged apical meristems. In 1.5 psu their algae died within 7 weeks. Tropin et al. (2001) performed a transfer experiment where *F. vesiculosus* from 34 psu were transferred to 40, 20, 10 and 2 psu for 14 days. They found considerable cell damage and complete degradation of cell structures at 10 and 2 psu. Such a strong desalination obviously disturbs the cellular structures responsible for protein synthesis and energy metabolism. They drew the conclusion that the presence of *F. vesiculosus* would hardly be possible in the zones of strong and moderate desalination. But it is important to bear in mind that the acclimation/adaptation of *F. vesiculosus* to the Baltic conditions is something that has occurred during a long time (since the last glaciation). The fact that *F. vesiculosus* inhabits the brackish Baltic Sea indicates that this species has, over the years, been able to adapt to the low salinity. The transfer of Baltic *F. vesiculosus* to the higher salinity in the present study increased the P_{max} to such a value that after 1 week it equalled the P_{max} of Atlantic *F. vesiculosus*. This in turn indicates that these plants still possess the ability of a high P_{max}, but the demand for that is a more favourable condition with higher salinity. The results indicate that a new ecotypic form of *F. vesiculosus,* which is adapted to a low salinity, might be under evolution or even have evolved in the Bothnian Sea.

Other experiments with transfer of *F. vesiculosus* from Baltic to Atlantic water (Bäck et al., 1992a; Russell, 1988) have obtained opposite results, with decreased photosynthesis. It is important to note that in those cases the incubation times were only 24 or 48 h. It is most likely that the algae need a longer time to acclimatize to a changed salinity (as found by Tropin et al., 2001) and that other results might have been obtained if a longer exposure time had been used. The experimental time set-up of several weeks in the present study was chosen to allow the algae to become acclimatized to the changed salinity. It was earlier found that Baltic algae in media above 12 psu need 4–5 weeks to adjust turgor pressure by organic solutes and ions in a changed salinity (Bäck et al., 1992b).

A higher respiration rate might be a way for the alga to withstand a changed salinity. The respiration of Atlantic *F. vesiculosus* has been found to increase with decreased salinity, while the opposite have been found for Baltic *F. vesiculosus* (Bäck et al., 1992a). The reason for this might be that a higher metabolism is necessary to adjust the cells to the changed osmosis. In the present work the dark respiration of Atlantic *F. vesiculosus* increased when the plants were transferred to Baltic water, which supports earlier results by Bäck et al. (1992a). No such correlation could be found for the Baltic plants. It was shown earlier that Baltic *F. vesiculosus* responds to heavy metals by increasing respiration (Nygård & Ekelund, 1999). No such correlation was obtained here with UV-B radiation.

P_{max} shows that both Atlantic and Baltic *F. vesiculosus* are affected by the UV-B exposure. The reason for Baltic plants to be more sensitive to UV-B radiation than the Atlantic ones might be that they grow deeper in the water column, and are therefore normally only exposed to very low amounts of UV-B. These plants can therefore be predicted to be more sensitive to both high PAR and UV radiation. Typical negative effects of UV radiation in algae are often seen as destruction of D1 proteins in photosystem II (Campell et al., 1998) and effects on Rubisco (Bischof et al., 2000). Several reports have been presented regarding the relationship between tolerance to UV radiation and vertical distribution of macroalgae (Bischof et al., 1998a, 1998b; Karsten et al., 2001), but so far not much work has been done concerning the tolerance to UV radiation in different seawater salinities. Several marine macroalgae can be found even in estuaries and brackish waters, and therefore it is important to examine the influence of salinity on these species, especially when (as shown here) the P_{max} and tolerance to UV-B decrease with decreased salinity.

Pigments

The generally higher pigment content in the Atlantic *F. vesiculosus* is related to these plants' higher initial photosynthetic slopes at limiting irradiances (Figure 2). This higher content might be related to the more optimal salinity in the Atlantic. Part of the difference might be explained by the plants' different depth distribution and light regimes in the Baltic and the Atlantic. For example, the higher content of photoprotective pigments such as beta-carotene in the Atlantic plants might be a direct effect of their shallow depth distribution. *F. vesiculosus* in the Atlantic is found in the intertidal zone and is thereby exposed to high irradiance as well as UV radiation. During low tides these plants can be fully exposed to the strong light of the sun, but the Baltic plants always grow subtidally, due to absence of tides in the Baltic Sea, and are thereby submerged under a protective water column. The decrease of pigments which occurred when Atlantic *F. vesiculosus* was transferred to Baltic water, might be due to the general negative affects of the low salinity. These transferred plants died after 8 weeks in the low salinity. The results from P_{max} and pigment content analyses suggest that the optimal sea water salinity for Baltic *F. vesiculosus* is higher than the actual salinity in the northern Baltic. It is therefore likely that the plants of *F. vesiculosus* in northern Baltic do not experience an optimal situation, even when they are able to survive in the area.

Acknowledgments

The authors would like to thank Tomas Melander for diving assistance, Seved Granberg for valuable help with statistical analysis and Maurice O'Connor for comments on the language. We are also grateful to the Marine Station of Trondheim University for help with sea water supply.

References

Bäck S, Collins JC, Russel G (1992a) Comparative ecophysiology of Baltic and Atlantic *Fucus vesiculosus*. Mar. Ecol. Prog. Ser. 84: 71–82.

Bäck S, Collins JC, Russell G (1992b) Effects of salinity on growth of Baltic and Atlantic *Fucus vesiculosus*. Br. Phycol. J. 27: 39–47.

Bischof K, Hanelt D, Wiencke C (1998a) UV-radiation can affect depth-zonation of Antarctic macroalgae. Mar. Biol. 131: 597–605.

Bischof K, Hanelt D. Wiencke C (2000) Effects of ultraviolet radiation on photosynthesis and related enzyme reactions of marine macroalgae. Planta 211: 555–562.

Bischof K, Hanelt D, Tüg H, Karsten U, Brouwer PEM, Wiencke C (1998b) Acclimation of brown algal photosynthesis to altraviolet radiation in Arctic coastal waters (Spitsbergen, Norway). Polar Biology 20: 388–395.

Dawes CJ, Bird K, Hanisak MD (1988) Physiological responses of transplanted populations of *Sargassum pteropleuron* grown in Florida. Aquatic Bot. 31: 107–123.

Kalvas A, Kautsky L (1998) Morphological variation in *Fucus vesiculosus* populations along temperature and salinity gradients in Iceland. J. Mar. Biol. Assn. UK. 78: 985–1001.

Karsten U, Bischof K, Wiencke C (2001) Photosynthetic performance of Arctic macroalgae after transplantation from deep to shallow waters. Oecologia 127: 11–20.

Koch EW, Lawrence J (1987) Photosynthetic and respiratory responses to salinity changes in the red alga *Gracilaria verrucosa*. Bot. Mar. 30: 327–329.

Nygård CA, Ekelund NGA (1999) Effects of lead ($PbCl_2$) on photosynthesis and respiration of the bladder wrack, *Fucus vesiculosus*, in relation to different salinities. Wat. Air Soil Polln. 116: 549–565.

Russell G (1987) Spatial and environmental components of evolutionary change: Interactive effects of salinity and temperature on *Fucus vesiculosus* as an example. Helgol. Meeresunters. 41: 371–376.

Russell G (1988) The seaweed flora of a young semi-enclosed sea: The Baltic. Salinity as a possible agent of flora divergence. Helgol. Meeresunters. 42: 243–250.

Tropin IV, Radzinskaya NV, Voskoboinikov GM (2001) The influence of salinity on the rate of dark respiration and structure of the cells of brown algae thalli from the Barents Sea littoral. Biol. Bull. 30: 40–47.

Waern M (1952) Rocky shore algae in the Öregrund archipelago. Acta Phytogeogr. Sueca 30: 1–298.

Wallentinus I (1979) Environmental influences on benthic macrovegetation in the Trosa-Askö area, northern Baltic proper. II. The ecology of macroalgae and submerged phanerogams. Contrib. Askö Lab. Univ. Stockholm, 25: 1–210.

Wright SW, Jeffrey SW, Mantoura RFC, Llewellyn CA, Bjørnland T, Repeta D, Welschmeyer N (1991) Improved HPLC method for the analysis of chlorophylls and carotenoids from marine phytoplankton. Mar. Ecol. Prog. Ser. 77: 183–196.

Journal of Applied Phycology (2006) 18: 469–474
DOI: 10.1007/s10811-006-9047-5

Effects of environmental factors and metal ions on growth of the red alga *Gracilaria chorda* Holmes (Gracilariales, Rhodophyta)

H. Kakita* & H. Kamishima
Institute for Environmental Management Technology (Formerly Institute for Marine Resources and Environment), AIST Shikoku, National Institute of Advanced Industrial Science and Technology, Hayashi, Takamatsu, Kagawa 761-0395, Japan

*Author for correspondence: e-mail: h-kakita@aist.go.jp; fax: 81-878-693553

Key words: *Gracilaria*, growth, environmental factors, metal ions, ultra pure salt, artificial seawater

Abstract

Gracilaria is a potentially valuable source of marine biopolymers such as proteins and polysaccharides. In order to select suitable culture conditions, growth and tolerance of *Gracilaria chorda* Holmes from Shikoku Island in southwest Japan were investigated under variations of temperature (5–30 °C), photon irradiance (20–120 μmol photons $m^{-2}\,s^{-1}$), and photoperiod (12:12 h, 14:10 h light:dark regime) in a unialgal culture. *Gracilaria chorda* showed wide tolerances for all factors investigated, which is characteristic of eurythermal species. Maximum growth was observed at 18–24 °C. The optimum photon irradiance for algal growth was 60–120 μmol photons $m^{-2}s^{-1}$. Instead of using ordinary sea salt (NaCl) to prepare artificial seawater, ultra pure salt was adopted. *Gracilaria chorda* grew faster in artificial seawater made with ultra-pure salt than that made with ordinary sea salt, probably because the former medium was clear, while the latter was milky. Effects of some metal ions on the growth were tested with artificial seawater. Iron ions affected algal growth, but cobalt ions did not. This study enables us to determine suitable culture conditions for *G. chorda*. A scaled-up 30 l culture of *G. chorda* under such conditions was successful.

Introduction

The red algal genus *Gracilaria* is harvested and cultured on a commercial scale in many countries because it has considerable economic importance as an agarophyte. The total annual *Gracilaria* production in the world increased to more than 89,000 t, including 50,000 t of cultured production, in 1995. *Gracilaria* plants are also used as sources of traditional seaweed salad in Japan and feed for shellfish (abalone) in many countries. Recently some bioactive substances from *Gracilaria* spp. have been extracted and reported (Kakita et al., 2003).

Seasonality affects agar quality in various *Gracilaria* species (Oza, 1978; Hoyle, 1978, Whyte et al., 1981; Lahaye & Yaphe, 1988; Bird & Ryther, 1990, Luhan, 1992; Yenigul, 1993). For obtaining *Gracilaria* biopolymers of constant quality and quantity, a cultured strain is likely to be more suitable than a wild one.

Environmental factors including temperature, salinity and light play an important role in the growth, reproduction and distribution of marine algae (Gessner, 1970; Gessner & Schramm, 1971; Lüning, 1981; Lobban & Harrison, 1994). Temperature requirements for survival and growth of *Gracilaria* species have been extensively studied (Bird et al., 1979; Laing et al., 1989; Yokoya & Oliveira, 1992). Some *Gracilaria* species require less than 100 μmol photons $m^{-2}s^{-1}$ for optimal growth (Bird et al., 1979; Beer & Levy, 1983), while others require higher irradiance (Lapointe, 1981; Lapointe et al., 1984). However, few data on the effects of such environmental factors on the growth of Japanese *Gracilaria* species in controlled conditions are available (Orsco & Ohno, 1992; Chirapart et al., 1994; Yokoya et al., 1999). The aim of this study is to characterize the physiological responses of *G. chorda* to temperature, irradiance and photoperiod, assessing tolerance and optimal conditions for growth in unialgal cultures.

Several metals such as iron ions are regarded as essential components for algal growth (Matsunaga et al., 1998). Thus, effects of some metal ions on the growth were also tested in artificial seawater. We therefore tested and selected suitable culture conditions and prepared a scaled-up model of an artificial seawater system for *Gracilaria* cultivation.

Materials and methods

Stock unialgal culture strains

Stock unialgal cultures of the red alga, *G. chorda*, were started from spores released by fertile plants that were harvested in June 1998 from Seto Inland Sea off the coast of Tokushima city, Tokushima Pref., Japan. The establishment of unialgal strains followed the methods of Yamamoto and Sasaki (1987). Stock unialgal cultures were incubated with aeration at 20 °C, in a 14:10 h light:dark regime, in salinity of about 33‰, at photon irradiances of just 40 μmol photons $m^{-2}s^{-1}$ under cool-white lamps to inhibit growth in storage. Provasoli enriched seawater (PES) (Provasoli, 1968) was made using sterilized Yashima surface seawater (Yashima, Kagawa Pref., southwest Japan) without the addition of vitamins, and this medium was used for the stock unialgal culture (Yamamoto & Sasaki, 1987). Medium renewal was carried out bi-weekly.

Temperature, irradiance, and photoperiod

The growth of *G. chorda* was compared among various culture conditions. Variations of temperature (5–30 °C) and irradiance (20–120 μmol photons $m^{-2}s^{-1}$) were tested. Algal growth rates in long-day (14:10 h light:dark) regime were compared with short-day (12:12 h light:dark) regime at three different temperatures (10, 20, and 30 °C). Irradiances were measured with a LI-250 photometer equipped with a LI-193SA spherical quantum sensor (LI-COR, Inc). The controlled conditions were the same as described above for stock unialgal cultures except that irradiance was increased to 60 μmol photons $m^{-2}s^{-1}$ to promote growth.

For each experiment, five replicates of six apical segments (5 mm long and approximately 1.4 mg fresh weight) cut from the stock unialgal culture strains were inoculated into 200 mL conical flasks containing 200 mL of PES. The measurement of total fresh weight of the six apical fragments and the renewal of culture media were carried out weekly in a clean booth. The data on algal growth rate, which were measured from day 14 to day 21 of cultivation, were analyzed by one-way ANOVA (for temperature and irradiance experiments) followed by Tukey's multiple comparison test (Winer et al., 1991) or t-test (for photoperiod experiment). Relative growth rates *R* were calculated using the formula:

$$R = [\ln(W_t) - \ln(W_0)]t^{-1},$$

where W_0 is the initial fresh weight, W_t is the fresh weight after *t* days and t is the number of days (Kain, 1987). Growth rate (%) was defined as $R \times 100$.

Artificial seawater

Ordinary sea salt (sodium chloride; NaCl) contains magnesium and other ions as contaminants (Niino et al., 1993). Some adsorbents, such as chelate resins and zeolites, are known to adsorb magnesium ions. Thus, an ultra pure salt (NaCl) was purified from ordinary sea salt by passing a 5.844 % solution of ordinary sea salt through a column of $Na_6Al_6Si_{30}O_{72}\cdot 24H_2O$-type zeolite (clinoptilolite: Sun-Zeolite Co., Ltd, Akita Pref., Japan) at 27 °C. The solution passed was re-crystallized only once and dried to obtain an ultra pure salt as a white powder. Atomic absorption spectrochemical analysis of the ultra pure salt showed that it contained only about 0.00015 % (w/w) of magnesium ions on average. On the other hand, ordinary sea salt (before adsorption treatment) contained about 0.00753 % (W/W) of magnesium ions.

Artificial seawater solids, Sample A, were made up of 548 g of ultra pure salt (NaCl), 250 g of $MgCl_2\cdot 6H_2O$, 92.5 g of Na_2SO_4, 35.0 g of $CaCl_2\cdot 2H_2O$, 15.8 g of KCl, 4.5 g of $NaHCO_3$, 2.25 g of KBr, 0.75 g of H_3BO_3, 0.25 g of $SrCl_2$, 0.13 mg of $FeCl_3\cdot 6H_2O$, 8.75 mg of glycerophosphate disodium salt pentahydrate ($C_3H_7Na_2O_6P\cdot 5H_2O$), and 4.0 mg of $NaNO_3$. Artificial seawater solids, Sample B, were a similar composition to Sample A, with the exception of the substitution of ordinary sea salt (NaCl) for ultra pure salt. After mixing of the components mentioned above, several batches of each sample of artificial seawater solids were sealed in laminated bags and stored for 30 days at 20 °C.

After storage for 30 days at 20 °C, 40 g of each sample of artificial seawater solids were dissolved in 1 L of distilled water. The transparency of each artificial seawater solution was measured as absorbance at 660 nm. The average absorbances of artificial seawater Solution A and B were 0.0005 and 0.0028, respectively ($n = 6$). Artificial seawater Solution A was more transparent than Solution B ($t = -6.139$, $p < 0.001$). The artificial seawater Solution A was named AIST-01.

The growth of *G. chorda* in the artificial seawater Solutions A and B was compared. Each type of artificial seawater solids was dissolved in 25 L of distilled water and these were used as algal culture media. The culture conditions with aeration were set at a temperature of 20 °C, an irradiance of 80 μmol photons $m^{-2}s^{-1}$, a light:dark regime of 14:10 h, in a salinity of about 33‰. Medium renewal was carried out weekly. The data on algal growth rate, which were measured from day 14 to day 21 of cultivation, were analyzed by *t*-test ($n = 5$).

Metal ions

Fe-free artificial seawater solids, Sample C, was made of the same compounds as Sample A, but lacking $FeCl_3{\cdot}6H_2O$. Forty grams of the artificial seawater solids, Sample C, were dissolved in distilled water, to which was added a predetermined concentration of $FeCl_3{\cdot}6H_2O$ or $CoCl_3{\cdot}6H_2O$ solution, and volume was adjusted to 1 L with distilled water. Various concentrations of $FeCl_3{\cdot}6H_2O$ (0.5, 5, 50 and 500 μg L^{-1}) or $CoCl_3{\cdot}6H_2O$ (0.05, 0.5, 5, and 50 μg L^{-1}) were tested in artificial seawater Batch C. The culture conditions were the same as described above for the salts experiment. Medium renewal was carried out weekly. Data on algal growth rate, which was measured from day 14 to day 21 of cultivation, were analyzed by one-way ANOVA ($n = 3$).

Thirty liter scale cultivation

Artificial seawater Solution A (AIST-01) containing 50 μg L^{-1} of $FeCl_3{\cdot}6H_2O$ was named AIST-01-Fe50. Three 30 L tanks were prepared, each containing 30 L of AIST-01-Fe50 and one algal specimen (fresh weight 1.16 g, 1.30 g, and 1.41 g). Three other 30 L tanks were prepared, each containing 30 L of Yashima surface seawater and one algal specimen (fresh weight 1.10 g, 1.22 g, and 1.37 g). Thirty liter volume unialgal cultivations were maintained simultaneously in three growth chambers, each accommodating two tanks (Koito Seisakusho, Co., Tokyo, Japan, model SNIRI-100) with aeration. The cylindrical tanks used were transparent polycarbonate and had an inner diameter of 350 mm and a height of 460 mm. The culture conditions used were the same as described above for the salts experiment. Medium renewal was carried out weekly. Algal growth rates cultivated in AIST-01-Fe50 were compared with those in surface seawater. The data concerning algal growth rate, which were measured from day 14 to day 21 of cultivation, were analyzed by t-test ($n = 3$).

Results

Temperature had a significant effect on the growth rate of *G. chorda* over three weeks ($n = 5$, $F = 96.662$, $p < 0.001$). The optimum temperature for the growth of *G. chorda* was 18–24 °C. The growth rate of *G. chorda* ranged from 0.13% d^{-1} at 5 °C to 12.3% d^{-1} at 20 °C (Figure 1). The optimum irradiance for the growth of *G. chorda* was 60–120 μmol m^{-2} s^{-1} (Figure 2). Irradiance had a significant effect

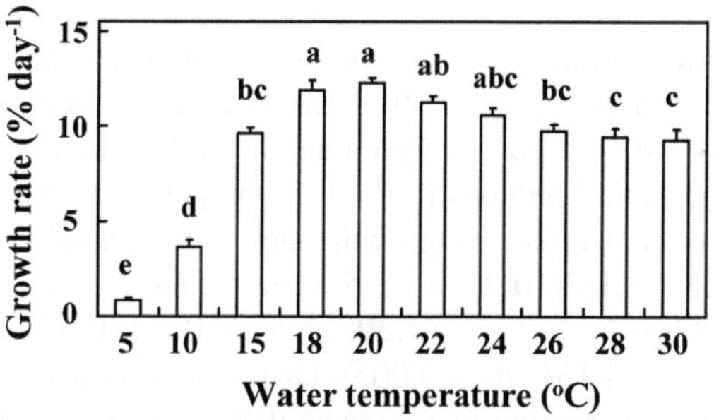

Figure 1. Effect of water temperature on algal growth. Growth rates of *G. chorda* cultivated for three weeks at different temperatures, constant photon irradiance (60 μmol photons m^{-2} s^{-1}) and photoperiod (14:10 h light:dark regime). Each data point is the mean of five replicates (means ± SE). Bars marked with the same letter are not significantly different according to Tukey's multiple comparison test ($p = 0.05$).

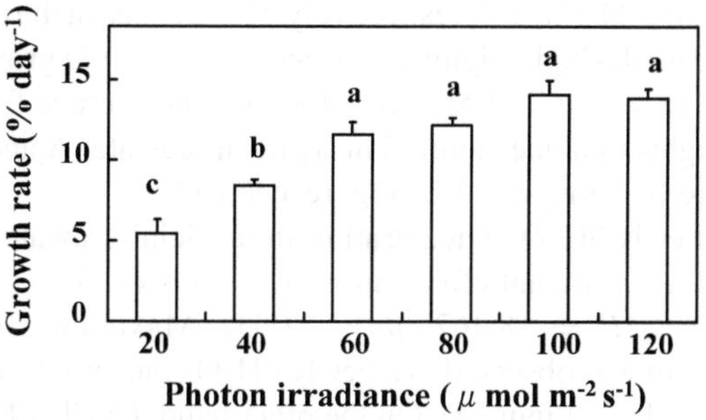

Figure 2. Effect of photon irradiance on algal growth. Growth rates of *G. chorda* cultivated for three weeks at different photon irradiances, constant temperature (20 °C) and photoperiod (14:10 h light:dark regime). Each data point is the mean of five replicates (means ± SE). Bars marked with the same letter are not significantly different according to Tukey's multiple comparison test ($p = 0.05$).

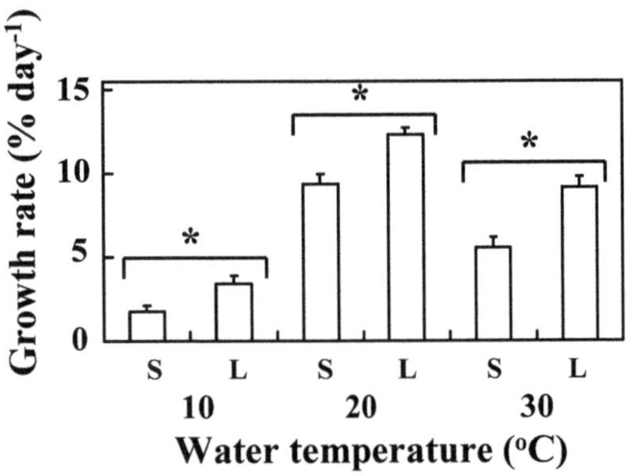

Figure 3. Effect of photoperiod on algal growth. Growth rates of *G. chorda* cultured for three weeks at different photoperiods and temperature (10, 20, or 30 °C) but constant photon irradiance (80 μmol photons $m^{-2}s^{-1}$). Each data point is the mean of five replicates (means ± SE). L: photoperiod is14:10 h light:dark regime. S: photoperiod is12:12 h light:dark regime. *Indicates that growth under a short-day (SD) was statistically different to growth under a long-day (LD) regime.

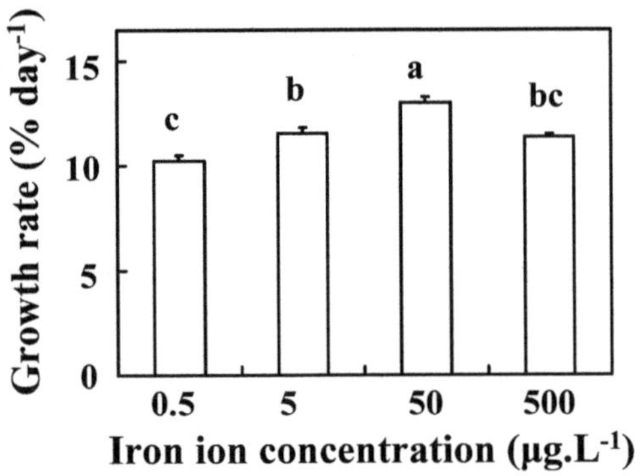

Figure 4. Effect of iron ions on algal growth. Growth rates of *G. chorda* cultivated for three weeks in artificial seawater Solution A containing different iron ion concentrations (0.5, 5, 50, and 500 μg l^{-1}), constant temperature (20 °C), irradiance (80 μmol photons $m^{-2}s^{-1}$) and photoperiod (14:10 h light:dark regime). Each data point is the mean of three replicates (means ± SE). Bars marked with the same letter are not significantly different according to Tukey's multiple comparison test ($p = 0.05$).

on growth rates of *G. chorda* over three weeks ($n = 5$, $F = 25.948$, $p < 0.001$). Growth rates varied from 5.44% d^{-1} at 20 μmol photons $m^{-2}s^{-1}$ to 14.1% d^{-1} at 100 μmol photons m^{-2} s^{-1}. Photoperiod also had a significant effect on growth rates of *G. chorda* over three weeks (10 °C: $n = 5, t = -2.648, p < 0.05$, 20 °C: $n = 5, t = -4.305, p < 0.005$, 30 °C: $n = 5, t = -4.131, p < 0.005$). Growth rate in a 14:10 h light:dark regime was greater than that in a 12:12 h light:dark regime (Figure 3). From the results of temperature and light experiments, suitable culture conditions were set at a temperature of 20 °C, an irradiance of 80 μmol $m^{-2}s^{-1}$, and a light:dark regime of 14:10 h.

The growth rates of the algae over three weeks of cultivation in artificial seawater A and B were 13.1% d^{-1} and 11.8% d^{-1}, respectively. Differences in the culture media had a significant effect on *G. chorda* growth rate ($n = 5$, $t = 4.555$, $p < 0.005$). The average fresh weights of algae cultured in artificial seawater A and B were 62.6 mg, and 52.1 mg, respectively.

$FeCl_3 \cdot 6H_2O$ concentration in artificial seawater A had a significant effect on growth rates of *G. chorda* ($n = 3$, $F = 18.467$, $p < 0.005$). Maximum algal growth was observed in a $FeCl_3 \cdot 6H_2O$ concentration of 50 μg L^{-1} (Figure 4). On the other hand, $CoCl_3 \cdot 6H_2O$ concentration did not affect the growth rate of the alga (data not shown).

On 30 L scale culture, the growth rates of algae over three weeks of cultivation in AIST-01-Fe50 medium and Yashima surface seawater were 9.2% d^{-1} and 1.6% d^{-1}, respectively. The artificial seawater solution (AIST-01-Fe50) accelerated algal growth of *G. chorda* over Yashima surface seawater ($n = 3$, $t = 29.303$, $p < 0.001$). After three weeks cultivation, the average fresh weights of algae cultured in AIST-01-Fe50 medium and Yashima surface seawater were 9.37 g/tank and 2.95 g/tank, respectively ($n = 3$).

Discussion

Gracilaria chorda tolerated a wide range of temperature variation, from 5 to 30 °C. The broad temperature tolerance of *G. chorda* is in accordance with observations that *Gracilaria* species from temperate waters tend to be eurythermal (Bird et al., 1979). Maximum growth of *G. chorda* was observed at 18–24 °C, and similar results have been observed in *G. tikvahiae* McLachlan (Bird et al., 1979), *G. chilensis* (Laing et al., 1989 as *G. sordida* Nelson); Yokoya and Oliveira, 1992, *Gracilaria* sp. (chorda-type) (Chirapart et al., 1994), and *G. vermiculophylla* (Yokoya et al., 1999). *Gracilaria chorda* has temperature responses similar to *G. tikvahiae* and *G. vermiculophylla*, growing well in temperatures as high as 30 °C, and tolerating low temperatures without necrosis of the thallus (Bird et al., 1979; Yokoya et al., 1999).

Maximum growth of *G. chorda* occurred at irradiances of 60–120 μmol photons $m^{-2}s^{-1}$, and these responses probably influence its intertidal distribution along the Japanese coast. These irradiances are higher than those observed for *G. tikvahiae*, which was light-saturated for growth at less than 50 μmol photons $m^{-2}s^{-1}$, and became necrotic at about 65 μmol photons $m^{-2}s^{-1}$ (Bird et al., 1979). On the other hand, optimum growth in higher light levels was reported in *G. foliifera v. angustissima* (Harvey) Taylor (Lapointe, 1981) and *G. chilensis* (Laing et al., 1989, as *G. sordida*). The results of temperature and irradiance experiments show that *G. chorda* is tolerant of wide variations in temperature and irradiance. The result of the photoperiod experiment indicates that *G. chorda* grows well under a long-day photoperiod, similar to that in their natural environment. These findings show that *G. chorda* could be cultivated economically in temperate brackish regions.

In Japan, most salt (NaCl) is purified from seawater. The final step in Japanese salt manufacturing is drying of recrystallized salt slurry at 130 °C. $MgCl_2 \cdot 6H_2O$ (magnesium chloride hexahydrate), which is adsorbed to the surface of NaCl crystals, changes to MgOHCl (basic magnesium chloride: magnesium hydroxide chloride) during heat treatment at temperatures higher than 110 °C. (Niino et al., 1992). After heat treatment, MgOHCl is resolved with moisture in the air, changing to $Mg(OH)_2$ (magnesium hydroxide), and finally to basic magnesium chloride [$Mg_2(OH)_3Cl.4H_2O$ (dimagnesium trihydroxide chloride tetrahydrate)] or [$Mg_3(OH)_5Cl \cdot 4H_2O$ (trimagnesium pentahydroxide chloride tetrahydrate)]. Basic magnesium chloride induces production of $CaCO_3$ (Niino et al., 1993). $CaCO_3$ is insoluble and reduces the transparency of a solution. Artificial seawater containing $CaCO_3$ has disadvantages of low transparency and the promotion of adhesion of insoluble $CaCO_3$ to the algal surface. These two disadvantages inhibit algal growth.

Because the ultra pure salt lacks magnesium ions, the artificial seawater Sample A (ultra pure salt-based) was more transparent than artificial seawater Sample B (ordinary sea salt-based): this may explain the faster growth in A. The high transmission of the seawater prepared with ultra pure salts seems to be effective for acceleration of algal growth. Because low light transmission in the medium results in a wide distribution of irradiation in a culture tank in seawater prepared with non-purified salts, algal growth rates are likely to remain lower than those in seawater prepared with ultra pure salts, even if the light intensity is increased. It may be more effective with large scale tanks.

Commercial scale cultivation of algae in tanks has the advantage of only minor biological contaminants compared with those in the field. Thus, high quality algae are obtained from tank culture. One of the biggest problems in commercial scale culture is that algal growth rate declines when algal density increases. Most commercial scale cultures aim to obtain a lot of cultured algae rapidly. Optimum irradiance and nutrients for algal growth throughout the tank are necessary for rapid growth and high density cultivation. Artificial seawater Solution A (ultra pure salt-based) was more transparent than Solution B (common salt-based). Components of artificial seawater can be manipulated to provide a suitable medium for each algal species. Although we have completed only a laboratory-scale manufacturing process for ultra pure salt, with a yield of 20 kg, the establishment of a large manufacturing process for ultra pure salt would be essential for its application to commercial scale culture. This would enable production of an artificial seawater solution useful for algal cultivation, though some limitations may remain to be solved.

Conclusions

This study enabled us to identify the optimum temperature and light conditions for cultivating *G. chorda*. Using artificial seawater also enabled the study of the effects of Fe and Co ions (micro-nutrients) on algal growth. Artificial seawater is useful for (1) obtaining growth responses to micro-nutrients, (2) maintaining constant quality, (3) growing axenic cultures, (4) cultivating algal strains that are susceptible to microorganisms. A 30 L scale culture of *G. chorda* under such conditions was successful. Ultra pure salt was superior to ordinary sea salt as a component of artificial seawater solids for algal growth.

Acknowledgements

We are grateful to Dr Hirotoshi Yamamoto of Hokkaido University and Dr Ryuta Terada of Kagoshima University for identifying *Gracilaria* spp. We are also grateful to Dr Masao Ohno of Kochi University and Dr Nair S. Yokoya of Instituto de Botanica (Brazil) for much helpful advice. We thank Ms. Satoko Tsukuda for her technical assistance.

References

Beer S, Levy I (1983) Effects of photon fluence rate and light spectrum composition on growth, photosynthesis and pigment relations in *Gracilaria* sp. J. Phycol. 19: 516–522.

Bird NL, Chen LCM, McLachlan J (1979) Effects of temperature, light and salinity on growth in culture of *Chondrus crispus*, *Furcellaria lumbricalis*, *Gracilaria tikvahiae* (Gigartinales, Rhodophyta) and *Fucus serratus* (Fucales, Phaeophyta). Bot. Mar. 22: 521–527.

Bird KT, Ryther JH (1990) Cultivation of *Gracilaria verrucosa* (Gracilariales, Rhodophyta) strain G-16 for agar. Hydrobiologia 204/205: 347–351.

Chirapart A, Ohno M, Sawamura M, Kusunose H (1994) Effect of temperature on growth rate and agar quality of a new member of Japanese *Gracilaria* in Tosa Bay, southern Japan. Jpn. J. Phycol. 42: 325–329.

Gessner F (1970) Temperature. Plants. In Kinne, O. (ed.), Marine Ecology, Wiley-Interscience, London: pp. 363–406.

Gessner F, Schramm W (1971) Salinity. Plants. In Kinne, O. (ed.), Marine Ecology, Wiley-Interscience, London: pp. 705–820.

Hoyle MD (1978) Agar studies in two *Gracilaria* species (*G. bursapastoris* (Gmelin) Silva and *G. coronopifolia* J. Ag.) from Hawaii. II Seasonal aspects. Bot. Mar. 21: 347–352.

Kakita H, Kamishima H, Ohno M, Chirapart A (2003) Marine biopolymers from the red algae, *Gracilaria* spp. In Fingerman M., Nagabhushanam R. (eds), Recent Advances in Marine Biotechnology. Volume 9. Biomaterials and bioprocessing. Science Publishers Inc., Enfield, NH: pp. 79–109.

Kain JM (1987) Seasonal growth and photoinhibition in *Plocamium cartilagineum* (Rhodophyta) off the Isle of Man. Phycologia 26: 88–99.

Lahaye M, Yaphe W (1988) Effect of seasons on the chemical structure and gel strength of *Gracilaria pseudoverrucosa* agar (Gracilariaceae, Rhodophyta). Carboh. Polym. 8: 285–301.

Laing WA, Christeller JT, Terzaghi BE (1989) The effect of temperature, photon flux density and nitrogen on growth of *Gracilaria sordida* Nelson (Rhodophyta). Bot. Mar. 32: 439–445.

Lapointe BE (1981) The effects of light and nitrogen on growth, pigment content and biological composition of *Gracilaria foliifera* v. *angustissima* (Gigartinales, Rhodophyta). J. Phycol. 17: 90–95.

Lapointe BE, Tenure KR, Dawes CJ (1984) Interactions between light and temperature on the physiological ecology of *Gracilaria tikvahiae* (Gigartinales: Rhodophyta). Mar. Biol. 80: 161–170.

Lobban CS, Harrison PJ (1994) Seaweed Ecology and Physiology, Cambridge University Press, Cambridge: 366 pp.

Luhan MRJ (1992) Agar yield and gel strength of *Gracilaria heteroclada* collected from Iloilo, central Philippines. Bot. Mar. 35: 169–172.

Lüning K (1981) Light. In Lobban, CS, Wynne MJ (eds), The Biology of Seaweeds, Blackwell Scientific Publications, Oxford: pp. 326–355.

Matsunaga K, Nigi G, Suzuki Y, Yasui H, Deein G (1998) Effect of fulvic acid-Fe derived from the forest on the growth of *Laminaria religiosa* Miyabe and *Undaria pinnatifida* Suringar. Bulletin of Society of Sea Water Science Japan 52: 315–318.

Niino Y, Nishimura H, Arita M (1992) Studies on drying temperature conditions for measuring loss of mass on drying of common salt transition temperature of salt hydrates. Bulletin of the Society of Sea Water Science Japan 46: 150–157 (in Japanese with English abstract).

Niino Y, Nishimura H, Arita M (1993) The formation of hard-to-dissolve matter in dry salt. Bulletin of the Society of Sea Water Science Japan 47: 74–80 (in Japanese with English abstract).

Orsco CA, Ohno M (1992) Growth rates of *Gracilaria* species (Gracilariales, Rhodophyta) from Tosa Bay, southern Japan. Jpn. J. Phycol. 40: 239–244.

Oza RM (1978) Studies on Indian *Gracilaria*. IV. Seasonal variation in agar and gel strength of *Gracilaria corticata* J. Ag. occurring on the coast of Veraval. Bot. Mar. 21: 165–167.

Provasoli L (1968) Media and prospects for the cultivation of marine algae. In Watanabe A, Hattori A. (eds.), Cultures and collections of Algae, Japanese Society of Plant Physiology, Tokyo: pp. 63–75.

Whyte JNC, Englar JR, Saunder RG, Lindsay JC (1981) Seasonal variations in the biomass, quantity and quality of agar, from the reproductive and vegetative stages of *Gracilaria* (*verrucosa* type). Bot. Mar. 24: 493–501.

Winer BJ, Brown DR, Michels KM (1991) Statistical principles in experimental design, third edition, McGraw-Hill, New York: pp. 153–198.

Yamamoto H, Sasaki J (1987) Cross experiment between so-called *Gracilaria verrucosa* (Huds.) Papenfuss from two localities, Shinori and Kikonai in Hokkaido. Bull. Fac. Fish. Hokkaido 38: 1–4.

Yenigul M (1993) Seasonal changes in the chemical and gelling characteristics of agar from *Gracilaria verrucosa* collected in Turkey. Hydrobiologia 260/261: 627–631.

Yokoya NS, Kakita H, Obika H, Kitamura T (1999) Effects of environmental factors and plant growth regulators on growth of the red alga *Gracilaria vermiculopylla* from Shikoku Island, Japan. Hydrobiologia 398/399: 339–347.

Yokoya NS, Oliveira EC (1992) Temperature responses of economically important red algae and their potential for mariculture in Brazilian waters. J. Appl. Phycol. 4: 339–345.

Journal of Applied Phycology (2006) 18: 475–481
DOI: 10.1007/s10811-006-9054-6

A genomic and phylogenetic perspective on endosymbiosis and algal origin

Hwan Su Yoon, Jeremiah D. Hackett & Debashish Bhattacharya*
Department of Biological Sciences and Roy J. Carver Center for Comparative Genomics, University of Iowa, 446 Biology Building, Iowa City, Iowa 52242-1324

*Author for correspondence: e-mail: debashi-bhattacharya@uiowa.edu; fax: (319) 335-1069

Received 4 August 2004; accepted 17 November 2004

Key words: algal evolution, chromalveolates, endosymbiosis, gene transfer, plastid

Abstract

Accounting for the diversity of photosynthetic eukaryotes is an important challenge in microbial biology. It has now become clear that endosymbiosis explains the origin of the photosynthetic organelle (plastid) in different algal groups. The first plastid originated from a primary endosymbiosis, whereby a previously non-photosynthetic protist engulfed and enslaved a cyanobacterium. This alga then gave rise to the red, green, and glaucophyte lineages. Algae such as the chlorophyll *c*-containing chromists gained their plastid through secondary endosymbiosis, in which an existing eukaryotic alga (in this case, a rhodophyte) was engulfed. Another chlorophyll *c*-containing algal group, the dinoflagellates, is a member of the alveolates that is postulated to be sister to chromists. The plastid in these algae has followed a radically different path of evolution. The peridinin-containing dinoflagellates underwent an unprecedented level of plastid genome reduction with the ca. 16 remaining genes encoded on 1–3 gene minicircles. In this short review, we examine algal plastid diversity using phylogenetic and genomic methods and show endosymbiosis to be a major force in algal evolution. In particular, we focus on the evolution of targeting signals that facilitate the import of nuclear-encoded photosynthetic proteins into the plastid.

Introduction

The eukaryotic photosynthetic organelle (plastid) is critical to life on our planet because of its contribution to global primary production. Ten different types of plastids are known and are found in evolutionarily divergent eukaryotic clades (Baldaif et al., 2000; Bhattacharya et al., 2004). The endosymbiosis hypothesis was put forth to explain the origin of plastids and mitochondria (Margulis, 1970; Mereschkowsky, 1905) and has been extensively supported with modern molecular evolutionary analyses. In plastid primary endosymbiosis, a non-photosynthetic protist engulfed a cyanobacterium and converted it into a permanent photosynthetic organelle. This photosynthetic eukaryote gave rise to the red, green, and glaucophyte algae that have a plastid bound by two membranes. Thereafter, plastids were horizontally spread into the remaining photosynthetic protist groups through secondary endosymbiosis, in which non-photosynthetic cells engulfed an existing (red or green) alga. This proces resulted in the plastid of cryptophytes, haptophytes, stramenopiles, apicomplexans, dinoflagellates (red algal endosymbiont), euglenophytes, and chlorarachniophytes (green algal endosymbiont, Bhattacharya & Medlin, 1995; Cavalier-Smith, 1986; Douglas, 1998; Douglas et al., 1991; Gibbs, 1978; McFadden et al., 1994; Yoon et al., 2002b; Zhang et al., 1999). However, endosymbiosis did not stop there because in dinoflagellates the existing plastid of red algal origin was replaced on multiple independent occasions with this organelle from an alga containing a secondary plastid (a cryptophyte, haptophyte or stramenopile: tertiary endosymbiosis) or a primary plastid (a green alga) (Chesnick et al., 1997; Hackett et al., 2003; Ishida & Green, 2002; Tengs et al., 2000; Watanabe et al., 1990; Yoon et al., 2002a).

The development of large scale sequencing and genomic approaches has greatly augmented our understanding of algal evolution. These methods have been applied to generate complete genome or expressed sequence tag (EST) databases of model algae or protists as well as to generate broadly sampled multi-gene phylogenies. In this paper, we discuss algal diversity from the perspective of plastid endosymbiosis, and present a brief summary of recent findings from genomic and phylogenetic approaches. In addition, we examine the leader sequences of nuclear-encoded plastid genes that have resulted from intracellular gene transfer and that make possible plastid targeting.

Endosymbiosis is an important driving force in algal evolution

Primary endosymbiosis

Rhodophyta, Viridiplantae (green algae and land plants), and Glauco(cysto)phyta contain plastids surrounded by a double membrane that very likely originated through a single primary endosymbiosis in the common ancestor of these taxa (Bhattacharya & Medlin, 1995; Delwiche et al., 1995; Gray, 1992; McFadden, 2001; Moreira et al., 2000; Matsuzaki et al., 2004; McFadden & van Dooren, 2004). Molecular clock analysis using a concatenated data set of six plastid genes and multi-fossil calibrations suggest that the primary endosymbiosis occurred around 1.6 billion years ago (Yoon et al., 2004). This estimate has been independently confirmed by multi-protein analyses of nuclear loci that suggest a date of 1.6–1.5 BY for primary plastid origin (Hedges et al., 2004; Hackett et al., 2006). Despite their ancient origin, the monophyly of Plantae is moderately supported by recent molecular phylogenetic studies using nuclear and mitochondrial genes (Baldauf et al., 2000; Moreira et al., 2000; Palmer, 2003; Rodriguez-Ezpeleta et al., 2005). A broadly sampled tree of microbial eukaryotes is urgently needed to test the monophyly of Plantae (and other groups – see below).

Secondary endosymbiosis

The putative lineage Chromista, which comprises the cryptophytes, haptophytes, and stramenopiles, contain chlorophyll *c* in their 4-membrane bound plastid (Cavalier-Smith, 1986). The chromist plastid is not located in the cytosol but rather within the rough endoplasmic reticulum (RER), which is connected to the outermost membrane of the plastid and is referred to as the chloroplast endoplasmic reticulum (CER). Secondary endosymbiosis, in which the non-photosynthetic ancestor of chromists engulfed an existing red alga, explains plastid origin in this group (Bhattacharya & Medlin, 1995; Douglas et al., 1991; Fast et al., 2001; Gibbs, 1981; Harper & Keeling, 2003). Evidence for this secondary endosymbiosis comes from the cryptophytes that retain the remnant nucleus of the red algal endosymbiont, the nucleomorph, between the two inner and two outer plastid membranes. The haptophytes and stramenopiles have presumably lost the nucleomorph after their divergence from the cryptophytes. Our molecular clock analysis suggests a minimum age of 1.3 BY for this secondary endosymbiotic event and around 1.2 BY for the divergence of cryptophytes from the other chromists and 1 BA for the split of haptophytes and stramenopiles (Yoon et al., 2004).

Alveolates, which comprise the dinoflagellates, apicomplexans, and ciliates, are postulated to be sister to the chromists (together, the chromalveolates; Cavalier-Smith, 1999 [see Fast et al., 2001; Harper & Keeling, 2003; Bhattacharya et al., 2004]). The chromalveolate common ancestor most likely contained a red algal secondary endosymbiont (Cavalier-Smith, 1999) that was apparently lost in the ciliates. In the apicomplexans, such as the well-known human parasite *Plasmodium falciparum* Welch, the remnant plastid (called the apicoplast) genome was reduced to a 35 Kb circle (Williamson et al., 1994). However, the phylogenetic history of apicoplasts remains unclear because of the high divergence of the encoded sequences that usually results in long branch artifacts in trees (Funes et al., 2002; Waller et al., 2003; Zhang et al., 1999, 2000). An important data set that supports chromalveolate monophyly is the presence of a unique glyceraldehyde-3-phosphate dehydrogenase (GAPDH) gene replacement shared by these taxa (Fast et al., 2001; Harper & Keeling, 2003). In addition, because the plastid genes of dinoflagellates were re-organized into mini-circles and most of the plastid genes were transferred to the nucleus (followed by high sequence divergence rates in many of these coding regions), it is difficult to accurately infer the phylogeny of dinoflagellate plastids (Bachvaroff et al., 2004; Hackett et al., 2004; Zhang et al., 1999). To resolve this issue, we sequenced five minicircle-encoded plastid proteins from a handful of peridinin dinoflagellates and from fucoxanthin-containing taxa. These latter taxa presumably gained their plastid through a haptophyte tertiary endosymbisois (Ishida & Green, 2002; Tengs et al., 2000; Yoon et al., 2005). In the resulting trees, the evolutionary origin of the peridinin plastids remains unclear but this clade is clearly positioned as a monophyletic lineage within the red algae with a weak sister group relationship to the stramenopiles. This result is consistent with the chromalveolate hypothesis and potentially explains the monophyletic origin of the plastid (and by extension, the host cells; Cavalier-Smith, 1999; Fast et al., 2001; Harper & Keeling, 2003). However, it is critical to verify this hypothesis with a broad taxon sampling and a multi-gene analysis of the host cells.

Euglenophytes and chlorarachniophytes acquired their green algal plastid through independent secondary endosymbioses (Archibald & Keeling, 2002; Baldauf, 2003; Bhattacharya et al., 2004; Palmer, 2003). However, if secondary plastid loss has occurred more frequently then we postulate, genomic analysis may be the best, and perhaps only, approach to identify the number of plastid endosymbioses that have occurred during eukaryotic evolution. The finding of nuclear-encoded genes of photosynthetic function in presently aplastidial cells will, for example, allow us to more accurately map the ancestral plastid distribution on the host tree.

Genomic approaches for clarifying algal evolution

Genomic methods such as whole genome random shotgun and EST approaches have recently been used to significantly improve our understanding of algal evolution. In one study Martin et al. (2002) found that ca. 18% of the nuclear genes in *Arabidopsis thaliana* (L.) Heynh. (of both photosynthetic and non-photosynthetic function) originated through intracellular gene transfer from the original cyanobacterial primary endosymbiont. This suggests that primary endosymbiosis resulted in massive lateral gene transfer from the endosymbiont to the nucleus and subsequently resulted in the enrichment and potential re-organization of the nuclear genome.

Three EST studies have thus far been done with dinoflagellates (i.e., *Alexandrium tamarense* (Lebour.) E. Balech, *Amphidinium carterae* Hulburth, and *Lingulodinium polyedrum* (F. Stein) J. D. Dodge)) and many others are underway. These studies have identified a massive transfer of plastid genes to the nucleus (Bachvaroff et al., 2004; Hackett et al., 2004) and intriguingly some of these genes are of green algal origin. This suggests that that there has either been multiple lateral gene transfers from green algal sources or, less parsimoniously, an as yet unsubstantiated green algal endosymbiosis (Hackett et al., 2004). Other examples of gene transfer have been reported in the chlorarachniophyte alga, *Bigelowiella natans* Moestrup which gained its plastid through a green algal secondary endosymbiosis (Archibald et al., 2003). EST analysis of this taxon showed numerous lateral transfers of genes from streptophyte, stramenopile, red algal, and bacterial sources. Furthermore, the whole genome sequence from the apicomplexan parasite *P. falciparum* reveals that 11% (581/5268 proteins) of genes of plastid function are still maintained in the nucleus (Gardner et al., 2002). It is interesting that these genes are absent from another apicomplexan, *Cryptosporidium parvum* Tyzzer, that apparently lacks an apicoplast (Abrahamsen et al., 2004). Taken together, genomic approaches are becoming ever more popular and result in an unprecedented quantity and quality of data that provide critical evolutionary information. However, it is important to keep in mind that single sequenced taxa (model or otherwise) do not adequately represent the evolutionary diversity of eukaryotes and taxonomically broadly sampled genomics projects hold the greatest promise for clarifying algal evolution.

Protein import system: Leader sequence

Following transfer to the nucleus, the proper function of the proteins involved in plastid function relies on their successful import into this organelle. This process occurs because of the presence of a N-terminal extension on each protein that specifies organellar import (Martin & Herrmann, 1998). The nature of these "leader" sequences depends on the ultrastructure of the plastid, such as the number of bounding membranes, and the location of the organelle in the cytoplasm (for detailed review, see Kilian & Kroth, 2003).

Two-membrane plastids located in the cytosol

The cyanobacterial origin of plastids bound by two membranes is strongly supported by the finding of homology of the protein import channel of the translocon of the inner/outer envelope of the plastid (i.e., Tic20, Tic55, and Toc75) among plants and cyanobacteria (Eckart et al., 2002; Heins et al., 1998; Kilian & Kroth, 2003). The green, red, and glaucophyte algae contain leader sequences (Figure 1A) to target proteins to the plastidial Tic/Toc system (see Matsuzaki et al., 2004; McFadden & van Dooren, 2004). These residues are encoded on the 5′-terminus of the open reading frame and are of length 25–125 amino acids (Cavalier-Smith, 2000; McFadden, 1999; Nassoury et al., 2003; Waller et al., 1998). Within the plastid, an endopeptidase

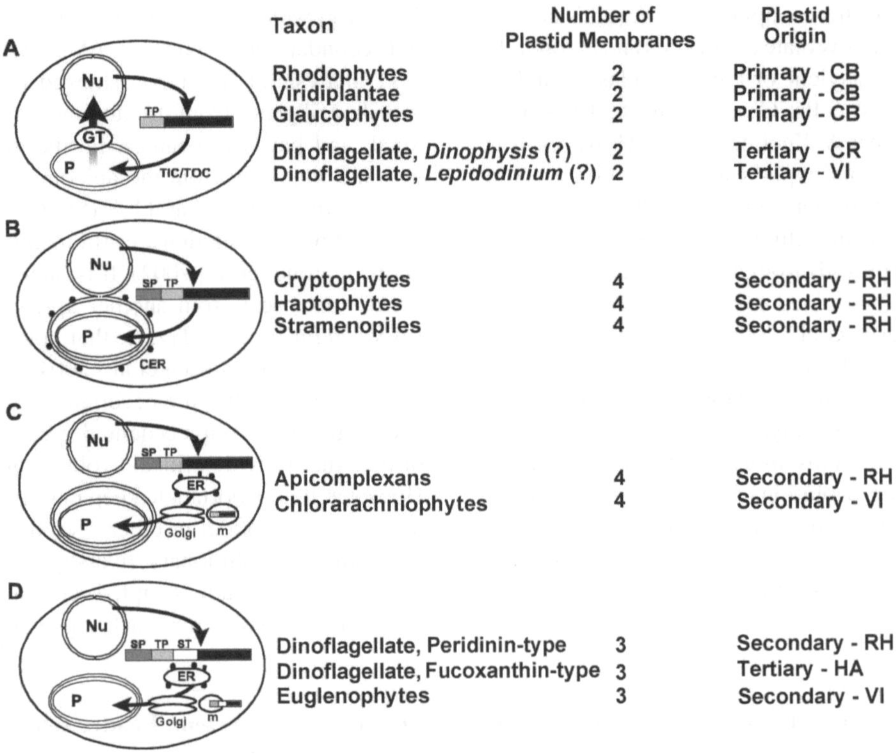

Figure 1. Plastid origin and the protein import system in photosynthetic eukaryotes. The protein import system reflects the plastid ultrastructure rather than the source of the organelle. (A) After primary endosymbiosis, massive plastid gene transfer (GT) occurred to the nucleus (Nu). A transit peptide (TP) that modified the N-terminus targets the functional proteins to the 2-membrane bound plastid (P) through the translocon of the inner/outer plastid envelope (TIC/TOC). This import system is found in algae containing primary endosymbionts, however, it may also be present in the 2-membrane bound plastid of the dinoflagellates *Dinophysis* and *Lepidodinium.* (B) Chromista that contain 4-membrane plastids with a chloroplast endoplasmic reticulum (CER) have a modified bipartite leader sequence, which targets the proteins to the CER with the signal peptide (SP) in addition to typical transit peptide. (C) The secretory pathway (ER and Golgi apparatus) is involved in the protein import system in the apicomplexa and chlorarachniophytes. The small circle represents microsomes (m) that contain the transit peptide and the functional protein via the secretory pathway. (D) The 3-membrane bound plastid in the dinoflagellates and euglenophytes contain a tripartite leader sequence. The second hydrophobic region (ST) acts as a stop transfer signal that generates a functional protein in the cytoplasm (m). CB, cyanobacterium; CR, cryptophyte; HA, haptophyte; RH, rhodophyte; VI, Viridiplantae.

cleaves the leader sequence (transit peptide). The two-membrane bound plastid in the dinoflagellates, *Dinophysis* spp. and *Lepidodinium viride* M. Watanabe, S. Suda, I. Inoye, T. Sawaguchi and M. Chihara which did not originate through primary endosymbiosis (rather via cryptophyte and green algal plastid replacements, respectively), most likely use the Tic import pathway of the inner plastid membrane that has been found in all algae and plants (McFadden & van Dooren, 2004).

Four-membrane bound plastids located in the CER of the lumen

In addition to the two inner membranes, chromists contain an additional two membranes that necessitate a more complex targeting signal (Figure 1B). Because the plastid of chromists is located in the CER, targeting into this membrane requires a classic signal peptide that has a hydrophobic region (Apt et al., 2002). This bipartite leader sequence subsequently targets the protein across the inner two membranes with a downstream transit peptide.

Four-membrane plastids located in the cytosol

Although the plastids of apicomplexans and chlorarachniophytes are of independent origins (from a red and a green alga, respectively), they both contain a four-membrane bound plastid located in the cytosol (unlike the chromists, Figure 1C). The bi-partite leader sequence, that specifies a signal peptide and a transit peptide, target the proteins into the cytosolic RER where the signal peptide is cleaved (McFadden, 1999;

Waller et al., 1998). During passage through the Golgi system, the secretory vesicles fuse with the outermost membrane of the plastid and thereafter, the transit peptide directs the protein through the Tic/Toc system. A secretory system-dependent transport system has been found in the four-membrane bound apicoplast in *P. falciparum* and surprisingly, as well as in the three-membrane bound plastid in *Euglena gracilis* G. A. Klebs and *Gonyaulax polyedra* F. Stein that do not contain a CER (Nassoury et al., 2003; Sulli et al., 1999; Waller et al., 2000).

Three membrane-bound plastids located in the cytosol

Dinoflagellates and euglenophytes contain plastids bound by three membranes that do not have a connection between the outer plastid membrane and the endomembranes (Figure 1D). Tripartite leader sequences, which consist of a hydrophobic signal peptide, a transit peptide, and a second hydrophobic region, were found in these taxa (Hackett et al., 2004; Nassoury et al., 2003; Sulli et al., 1999). The second hydrophobic region acts as a "stop transfer signal" and the functional protein is located in the cytoplasmic side of the ER after cleavage of the signal peptide (Nassoury et al., 2003). Microsomes that contain the functional transit peptide pass through the membranes via a secretory pathway (Golgi apparatus) followed by subsequent vesicular transport across the cytoplasm before entering the plastids (Nassoury et al., 2003; Sulli et al., 1999). A tripartite leader sequence in the *psb*O gene has been identified from the fucoxanthin-containing dinoflagellate, *Karenia brevis* (C. C. Davis) G. Hansen & Ø. Moestrup (Ishida & Green, 2002).

Conclusions

Great progress has recently been made in generating the outline of the eukaryotic tree of life using molecular phylogenies (e.g., Baldauf, 2003). The deep branches of the tree remain however unsubstantiated and await a rigorous multi-gene approach with a broad taxonomic sampling. This type of analysis will ultimately resolve the main splits in the algal tree and establish the timing of algal origins. The methods of phylogenetics and genomics provide significant data but the challenge remains to extensively sample representatives of all the major algal groups and relevant non-algal groups. Our understanding of algal biodiversity has also been significantly changed by environmental PCR analyses of extreme environments such as the deep sea and hot springs (Lopez-Garcia et al., 2001; Moon-van der Staay et al., 2001; Ciniglia et al., 2004). Furthermore, the picoplankton in both coastal and open ocean environments promises to be a potentially endless source of novel taxa. Because many of these lineages are positioned basal in trees, the environmental PCR method provides a powerful tool for understanding early algal evolution. In this regard, the finding that many basal lineages such as the marine stramenopiles and alveolates are heterotrophic, forces us to postulate multiple secondary plastid losses in chromalveolates to be consistent with the ideas presented here.

Recently, an environmental meta-genomic approach using shotgun whole genome sequencing was used to study bacterial diversity in the oligotrophic Sargasso Sea (Venter et al., 2004). This breakthrough work identified at least 1800 genomic species including 148 previously unknown bacterial phylotypes and over 1.2 million previously unknown genes. Although an imprecise and incomplete approach to generating complete genome sequences, this remarkable body of data challenges all biologists to account more rigorously for microbial diversity when studying eukaryotic and in particular, algal evolution. Clearly, we are just beginning to understand the complex history of microbial eukaryotes and the future holds great promise in clarifying the phylogeny of algae and their place in the tree of life.

Acknowledgements

This work was primarily supported by grants from the United States National Science Foundation to D.B. (grants DEB 01-07754, MCB 02-36631).

References

Abrahamsen MS, Templeton TJ, Enomoto S, Abrahante JE, Zhu G, Lancto CA, Deng M, Liu C, Widmer G, Tzipori S, Buck GA, Xu P, Bankier AT, Dear PH, Konfortov BA, Spriggs HF, Iyer L, Anantharaman V, Aravind L, Kapur V (2004) Complete genome sequence of the apicomplexan, *Cryptosporidium parvum*. Science 304: 441–445.

Apt KE, Zaslavkaia L, Lippmeier JC, Lang M, Kilian O, Wetherbee R, Grossman AR, Kroth PG (2002) *In vivo* characterization of diatom multipartite plastid targeting signals. J. Cell Sci. 115: 4061–4069.

Archibald JM, Keeling PJ (2002) Recycled plastids: A 'green movement' in eukaryotic evolution. Trends in Genet. 18: 577–584.

Archibald JM, Rogers MB, Toop M, Ishida K, Keeling PJ (2003) Lateral gene transfer and the evolution of plastid-targeted proteins in the secondary plastid-containing alga *Bigelowiella natans*. PNAS 100: 7678–7683.

Bachvaroff TR, Concepcion GT, Rogers CR, Herman EM, Delwiche CF (2004) Dinoflagellate expressed sequence tag data indicate massive transfer of chloroplast genes to the nuclear genome. Protist 155: 65–78.

Baldauf SL (2003) The deep roots of eukaryotes. Science 300: 1703–1706.

Baldauf SL, Roger AJ, Wenk-Siefert I, Doolittle WF (2000) A kingdom-level phylogeny of eukaryotes based on combined protein data. Science 290: 972–977.

Bhattacharya D, Medlin L (1995) The phylogeny of plastids: A review based on comparisons of small-subunit ribosomal RNA coding regions. J. Phycol. 31: 489–498.

Bhattacharya D, Yoon HS, Hackett JD (2004) Photosynthetic eukaryotes unite: Endosymbiosis connects the dots. BioEssays 26: 50–60.

Cavalier-Smith T (1986) The kingdon chromista: Origin and systematics. In: Round FE, Chapman DJ (eds.) Progress in Phycological Research No. 4, Biopress, Bristol, pp 309–347.

Cavalier-Smith T (1998) A revised six-kingdom system of life. Biol. Rev. 73: 203–266.

Cavalier-Smith T (1999) Principles of protein and lipid targeting in secondary symbiogenesis: Euglenoid, dinoflagellate, and sporozoan plastid origins and the eukaryote family tree. J. Eukaryot. Microbiol. 46: 347–366.

Cavalier-Smith T (2000) Membrane heredity and early chloroplast evolution. Trends in Plant Sci. 5: 174–182.

Chesnick JM, Kooistra WH, Wellbrock U, Medlin LK (1997) Ribosomal RNA analysis indicates a benthic pennate diatom ancestry for the endosymbionts of the dinoflagellates *Peridinium foliaceum* and *Peridinium balticum* (Pyrrhophyta). J. Eukaryot. Microbiol. 44: 314–320.

Ciniglia C, Yoon HS, Pollio A, Pinto G, Bhattacharya D (2004) Hidden biodiodiversity of the extremophilic Cyanidiales red algae. Mol. Ecol. 13: 1827–1838.

Delwiche CF, Kuhsel M, Palmer J. D (1995) Phylogenetic analysis of *tuf*A sequences indicates a cyanobacterial origin of all plastids. Mol. Phylogenet. Evol. 4: 110–128.

Douglas SE (1998) Plastid evolution: Origins, diversity, trends. Curr. Opin. Genet. Dev. 8: 655–661.

Douglas SE, Murphy CA, Spencer DF, Gray MW (1991) Cryptomonad algae are evolutionary chimaeras of two phylogenetically distinct unicellular eukaryotes. Nature 350: 148–151.

Eckart K, Eichacker L, Sohrt K, Schleiff E, Heins L, Soll J (2002) A toc75-like protein import channel is abundant in chloroplasts. EMBO Reports 3: 557–562.

Fast NM, Kissinger JC, Roos DS, Keeling PJ (2001) Nuclear-encoded, plastid-targeted genes suggest a single common origin for apicomplexan and dinoflagellate plastids. Mol. Biol. Evol. 18: 418–426.

Funes S, Davidson E, Reyes-Prieto A, Magallon S, Herion P, King MP, Gonzalez-Halphen D (2002) A green algal apicoplast ancestor. Science 298: 2155.

Gardner MJ, Hall N, Fung E, White O, Berriman M, Hyman RW, Carlton JM, Pain A, Nelson KE, Bowman S, Paulsen IT, James K, Eisen JA, Rutherford K, Salzberg SL, Craig A, Kyes S, Chan MS, Nene V, Shallom SJ, Suh B, Peterson J, Angiuoli S, Pertea M, Allen J, Selengut J, Haft D, Mather MW, Vaidya AB, Martin DM, Fairlamb AH, Fraunholz MJ, Roos DS, Ralph SA, McFadden GI, Cummings LM, Subramanian GM, Mungall C, Venter JC, Carucci DJ, Hoffman SL, Newbold C, Davis RW, Fraser CM, Barrell B (2002) Genome sequence of the human malaria parasite *Plasmodium falciparum*. Nature 419: 498–511.

Gibbs SP (1978) The chloroplasts of *Euglena* may have evolved from symbiotic green algae. Can. J. Bot. 56: 2883–2889.

Gibbs SP (1981) The chloroplasts of some algal groups may have evolved from endosymbiotic eukaryotic algae. Annal. NY Acad. Sci. 361: 193–208.

Gray MW (1992) The endosymbiont hypothesis revisited. Int. Rev. Cytol. 141: 233–357.

Hackett JD, Maranda L, Yoon HS, Bhattacharya D (2003) Phylogenetic evidence for the cryptophyte origin of the plastid of *Dinophysis* (dinophysiales, dinophyceae). J. Phycol. 39: 440–448.

Hackett JD, Yoon HS, Butterfield NJ, Sanderson MJ, Bhattacharya D (2006) Plastid endosymbiosis: Origins and timing of events. In: Falkowski P, Knoll A (eds.) Evolution of Aquatic Photoautotrophs, Academic Press, in press.

Hackett JD, Yoon HS, Soares MB, Bonaldo MF, Casavant TL, Scheetz TE, Nosenko T, Bhattacharya D (2004) Migration of the plastid genome to the nucleus in a peridinin dinoflagellate. Curr. Biol. 14: 213–218.

Harper JT, Keeling PJ (2003) Nucleus-encoded, plastid-targeted glyceraldehyde-3-phosphate dehydrogenase (GAPDH) indicates a single origin for chromalveolate plastids. Mol. Biol. Evol. 20: 1730–1735.

Hedges SB, Blair JE, Venturi ML, Shoe JL (2004) A molecular timescale of eukaryote evolution and the rise of complex multicellular life. BMC Evol. Biol. 4: 2.

Heins L, Soll J, Collinson I (1998) The protein translocation apparatus of chloroplast envelopes. Trends in Plant Sci. 3: 56–61.

Ishida K, Green BR (2002) Second- and third-hand chloroplasts in dinoflagellates: Phylogeny of oxygen-evolving enhancer 1 (psbO) protein reveals replacement of a nuclear-encoded plastid gene by that of a haptophyte tertiary endosymbiont. PNAS 99: 9294–9299.

Kilian O, Kroth PG (2003) Evolution of protein targeting into "complex" plastids: The "secretory transport hypothesis". Plant Biol. 5: 350–358.

Lopez-Garcia P, Rodriguez-Valera F, Pedros-Alio C, Moreira D (2001) Unexpected diversity of small eukaryotes in deep-sea Antarctic plankton. Nature 409: 603–607.

Margulis L (1970) Origin of eukaryotic cells. Yale University Press, New Haven, CT.

Martin W, Herrmann RG (1998) Gene transfer from organelles to the nucleus: How much, what happens, and why? Plant Physiol. 118: 9–17.

Martin W, Rujan T, Richly E, Hansen A, Cornelsen S, Lins T, Leister D, Stoebe B, Hasegawa M, Penny D (2002) Evolutionary analysis of *Aarabidopsis*, cyanobacterial, and chloroplast genomes reveals plastid phylogeny and thousands of cyanobacterial genes in the nucleus. PNAS 99: 12246–12251.

Matsuzaki M, Misumi O, Shin-i T, Maruyama S, Takahara M, Miyagishima S, Mori T, Nishida K, Yagisawa F, Nishida K, Yoshida Y, Nishimura Y, Nakao S, Kobayashi T, Momoyama Y, Higashiyama T, Minoda A, Sano M, Nomoto H, Oishi K, Hayashi H, Ohta F, Nishizaka S, Haga S, Miura S, Morishita T, Kabeya Y, Terasawa K, Suzuki Y, Ishii Y, Asakawa S, Takano H, Ohta N, Kuroiwa H, Tanaka K, Shimizu N, Sugano S, Sato N, Nozaki H, Ogasawara N, Kohara Y, Kuroiwa T (2004) Genome sequence of the ultrasmall unicellular red alga *Cyanidioschyzon merolae* 10D. Nature 428: 653–657.

McFadden GI (1999). Plastids and protein targeting. J. Eukaryot. Microbiol. 46: 339–346.

McFadden GI (2001) Primary and secondary endosymbiosis and the origin of plastids. J. Phycol. 37: 951–959.

McFadden GI, van Dooren GG (2004) Evolution: Red algal genome affirms a common origin of all plastids. Curr. Biol. 14: R514–R516.

McFadden GI, Gilson PR, Hofmann CJ, Adcock GJ, Maier UG (1994) Evidence that an amoeba acquired a chloroplast by retaining part of an engulfed eukaryotic alga. PNAS 91: 3690–3694.

Mereschkowsky C (1905) Über Natur und Ursprung der Chromatophoren im Pflanzenreiche. Biol. Centralbl. 25: 593–604.

Moon-van der Staay SY, Wachter RDe, Vaulot D (2001) Oceanic 18s rDNA sequences from picoplankton reveal unsuspected eukaryotic diversity. Nature 409: 607–610.

Moreira D, Le Guyader H, Phillippe H (2000) The origin of red algae and the evolution of chloroplasts. Nature 405: 69–72.

Nassoury N, Cappadocia M, Morse D (2003) Plastid ultrastructure defines the protein import pathway in dinoflagellates. J. Cell Sci. 116: 2867–2874.

Palmer JD (2003) The symbiotic birth and spread of plastids: How many times and whodunit. J. Phycol. 39: 4–12.

Rodriguez-Ezpeleta N, Brinkmann H, Burey SC, Roure B, Burger G, Loffelhardt W, Bohert HJ, Philippe H, Lang BF (2005) Monophyly of primary photosynthetic eukaryotes: green plants, red algae, and glaucophytes. Curr. Biol. 15: 1325–1330.

Sulli C, Fang Z, Muchhal U, Schwartzbach SD (1999) Topology of *Euglena* chloroplast protein precursors within endoplasmic reticulum to Golgi to chloroplast transport vesicles. J. Biol. Chem. 274: 457–463.

Tengs T, Dahlberg OJ, Shalchian-Tabrizi K, Klaveness D, Rudi K, Delwiche CF, Jakobsen KS (2000) Phylogenetic analyses indicate that the 19′hexanoyloxy-fucoxanthin-containing dinoflagellates have tertiary plastids of haptophyte origin. Mol. Biol. Evol. 17: 718–729.

Venter JC, Remington K, Heidelberg JF, Halpern AL, Rusch D, Eisen JA, Wu D, Paulsen I, Nelson KE, Nelson W, Fouts DE, Levy S, Knap AH, Lomas MW, Nealson K, White O, Peterson J, Hoffman J, Parsons R, Baden-Tillson H, Pfannkoch C, Rogers Y-H, Smith HO (2004) Environmental genome shotgun sequencing of the Sargasso Sea. Science 304: 66–74.

Waller RF, Keeling PJ, Donald RG, Striepen B, Handman E, Lang-Unnasch N, Cowman AF, Besra GS, Roos DS, McFadden GI (1998) Nuclear-encoded proteins target to the plastid in *Toxoplasma gondii* and *Plasmodium falciparum*. PNAS 95: 12352–12357.

Waller RF, Keeling PJ, van Dooren GG, McFadden GI (2003) Comment on "a green algal apicoplast ancestor". Science 301: 49.

Waller RF, Reed MB, Cowman AF, McFadden GI (2000) Protein trafficking to the plastid of *Plasmodium falciparum* is via the secretory pathway. EMBO J. 19: 1794–1802.

Watanabe MM, Suda S, Inouye I, Sawaguchi T, Chihara M (1990) *Lepidodinium viride* gen. et sp. nov. (Gymnodiniales, Dinophyta), a green dinoflagellate with chlorophyll *a*- and *b*-containing endosymbiont. J. Phycol. 26: 741–751.

Williamson DH, Gardner MJ, Preiser P, Moore DJ, Rangachari K, Wilson RJ (1994) The evolutionary origin of the 35 kb circular DNA of *Plasmodium falciparum*: New evidence supports a possible rhodophyte ancestry. Mol. Gen. Genet. 243: 249–252.

Yoon HS, Hackett JD, Bhattacharya D (2002a) A single origin of the peridinin- and fucoxanthin-containing plastids in dinoflagellates through tertiary endosymbiosis. PNAS 99: 11724–11729.

Yoon HS, Hackett JD, Ciniglia C, Pinto G, Bhattacharya D (2004) A molecular timeline for the origin of photosynthetic eukaryotes. Mol. Biol. Evol. 21: 809–818.

Yoon HS, Hackett JD, Pinto G, Bhattacharya D (2002b) The single, ancient origin of chromist plastids. PNAS 99: 15507–15512.

Yoon HS, Hackett JD, Van Dolah FM, Nosenko T, Lidie KL, Bhattacharya D (2005) Tertiary endosymbiosis driven genome evolution in dinoflagellate algae. Mol. Biol. Evol. 22: 1299–1308.

Zhang Z, Green BR, Cavalier-Smith T (1999) Single gene circles in dinoflagellate chloroplast genomes. Nature 400: 155–159.

Zhang Z, Green BR, Cavalier-Smith T (2000) Phylogeny of ultra-rapidly evolving dinoflagellate chloroplast genes: A possible common origin for sporozoan and dinoflagellate plastids. J. Mol. Evol. 51: 26–40.

Journal of Applied Phycology (2006) 18: 483–487
DOI: 10.1007/s10811-006-9049-3

Isolation of pollutant (pine needle ash)-responding genes from tissues of the seaweed *Ulva pertusa*

Se-Eun Kang[1], Long-Guo Jin[1], Jae-Suk Choi[1], Ji-Young Cho[1], Hyun-Woung Shin[2] & Yong-Ki Hong[1,*]
[1]*Department of Biotechnology, Pukyong National University, Namku, Busan 608-737, Korea;* [2]*Department of Marine Biotechnology, Soonchunhyang University, Asan 336-900, Korea*
*Author for correspondence: e-mail: ykhong@pknu.ac.kr

Key words: ash, differential display, pine needle, *Ulva pertusa*, seaweed

Abstract

Genetic responses of the seaweed *Ulva pertusa* to pine needle ash have been compared using differential display technique. The tissue viability was assessed to evaluate the stress level with triphenyltetrazolium chloride. Total RNA, from tissues treated in seawater containing ash, was reverse transcribed and amplified by PCR with arbitrary primers. The genetic fragments responding to the stress were selectively isolated from agarose gel and sequenced with a DNA auto sequencer. According to sequence analysis, an ash-inducible gene (342 bp) and an ash-suppressed gene (1690 bp) were identified as hypothetical proteins.

Introduction

During the winter and spring, forest fires often occur in Korea (www.foa.go.kr). In the forest area, the pine tree *Pinus densiflora* is the most dominant tree. The ash derived from the fire may be one of the natural hazards that cause damage to the seaweeds near a river mouth. Especially through the short rivers, the ash reaches the coast without self-purification, and affects marine organisms, including useful seaweeds. *Ulva pertusa* Kjellman (sea lettuce) is one of the most widely distributed species in the coastal area near river mouths, and also commonly found occupying a range of shores and habitats, polluted and unpolluted. *U. pertusa* has thick, stiff foliose, perforated blades with two cell layers. It occurs throughout the year, propagating mostly in the winter and spring periods. It is simple in structure and very easy to handle in culture and seems to be consistent in its behavior when collected from different habitat sources (Burrows, 1971). The growth of tissue was accelerated in seawater containing humic-like substances (Asahina et al., 1999). Moreover, *Ulva* has been well investigated with respect to cellular developmental properties (Reddy et al., 1992; Nakanishi et al., 1996), physiology (Floreto et al., 1993), biochemical constituents (Okano & Aratani, 1979), and molecular genetics (Lim et al., 1983). Thus, it is ideally suited as a marine bioassay test organism. With the exception of obvious lethal damage associated with extreme environmental conditions, it is difficult to evaluate the occurrence and severity of stress in natural populations of seaweed. There is a need to develop molecular and biochemical markers specific for an individual stress or group of stresses to allow the prompt, unambiguous and direct determination of stress. We measured the viability against pine needle ash as a pollutant stress source and screened the differential display of gene expression for detection of stress markers at the RNA expression level.

Materials and methods

Thallus

Fresh thalli of the green alga *Ulva pertusa* Kjellman were collected from the Chongsapo area in Busan, Korea. Tissues were cleaned by brushing thoroughly and sonicating (47 kHz) twice for 1 min in autoclaved seawater, and immersed in 1% Betadine for 2 min to eliminate epiphytes (Jin et al., 1997). They were then rehabilitated at 18 °C in PES (Provasoli, 1968) for a day before use.

Pollutant treatment

Ash from pine (*Pinus densiflora*) needles was prepared by burning in an oven at 200 °C for 1 h. After this preparation, 1 g of fresh needles resulted in 0.5 g dry ash weight, 25 mg water-soluble extract (1d, RT), and an absorbance of 0.61 at 230 nm for the 10-fold diluted extract. To measure the stress level induced by the ash, 0.1 g of the *U. pertusa* tissue was incubated under 50 μmol photons $m^{-2}\,s^{-1}$ at 18 °C for 1 d in 100 ml PES containing various concentrations of ash. Viability of the tissue was measured by the TTC method that is based on the enzymatic reduction of colorless 2,3,5-triphenyltetrazolium chloride to a red-colored triphenylformazan (Nam et al., 1998). The relative viability (%) for each ash concentration was expressed on a dose-response curve. From the curves, we determined the MNLC (maximum non-lethal concentration), LC_{50} (lethal concentration 50), and MLC (minimum lethal concentration).

RNA extraction

For the RNA extraction, pollutant-treated and control tissues were processed by the LiCl-guanidinium method (Hong et al., 1995). Briefly, 0.6 g of wet tissues was powdered in liquid nitrogen and heated in 4 ml of RNA extraction solution at 55 °C for 10 min. Total nucleic acid was precipitated after addition of the same volume of 4 M LiCl at 4 °C for 1 h. To remove DNA, a 100 μl aliquot was incubated with 4 μL of RNase-free DNase I (1 unit μL^{-1}). The total RNA was adjusted to 0.5 $\mu g\,\mu L^{-1}$ with RNasin (0.5 unit μL^{-1}) in DEPC water for the cDNA synthesis.

cDNA synthesis

Five μL of total RNA (2.5 μg) was used as a template in 20 μL of reaction mixture, according to the Invitrogen cDNA synthesis protocol. To prime total RNA, 1 μL of random hexamers (1 $\mu g\,\mu L^{-1}$) was added. cDNA synthesis was carried out at 42 °C for 2 h by avian virus reverse transcriptase (10 unit μL^{-1}).

Differential display

PCR amplification was carried out using a DNA thermal cycler (Perkin-Elmer, Norwalk, CT). Arbitrary primers of 10-base oligonucleotides were purchased from Operon Technologies, Inc. (Alameda, CA). A 25 μL PCR reaction mixture contained 1 μL of the cDNA, 5 pM of arbitrary primer, 10 mM Tris-HCl, pH 8.3, 50 mM KCl, 2 mM $MgCl_2$, 0.001% gelatin, 0.5% Tween 20, 0.1 mM of each dNTP, and 1 unit of *Taq* DNA polymerase. The cycling parameters included an initial incubation at 94 °C for 5 min followed by 45 cycles of 5 sec denaturation at 94 °C, 1 min annealing at 36 °C, and 2-min extension at 72 °C (Yu & Pauls, 1992). A 10 μL sample of PCR product was loaded on a 3% agarose gel containing 0.5 $\mu g\,mL^{-1}$ ethidium bromide and run for 1.5 h at 5 V cm^{-1} in 0.5 × TAE buffer (20 mM Tris-acetate, 1 mM EDTA, pH 8.0). Amplified cDNA fragments induced by the stress were selectively isolated from the gel.

Sequence analysis

The cDNA fragments were cloned into pCR2.1 using the TA cloning kit (Invitrogen). Both strands of the cloned cDNA fragments were sequenced with a DNA auto sequencer (Perkin-Elmer ABI Prism 377). Sequences were analyzed for open reading frames (ORFs) using the ORF Finder option from NCBI (www.ncbi.nlm.nih.gov), and also aligned using the BLASTX option to detect GenBank proteins that are most likely to be related. The G + C content and molecular weight were determined using the DNAsis and ExPASY Molecular Biology programs (www.expasy.org), respectively.

Results

Ulva pertusa tissues were treated with various concentrations of ash to identify the concentration at which a stress response was induced. From the dose-response curve, the MNLC, LC_{50} and MLC induced by the ash were determined to be 60, 350 and 550 mg mL^{-1}, respectively (Figure 1). For the differential display screening of ash stress responsive genes, the tissues were treated in PES at MNLC for 24 h, LC_{50} for 1 h, LC_{50} for 6 h, and LC_{50} for 24 h. Total RNA was extracted from each stressed tissue using the LiCl-guanidinium method. The RNA yield after removing DNA was approximately 169 μg from 1 g of wet tissue.

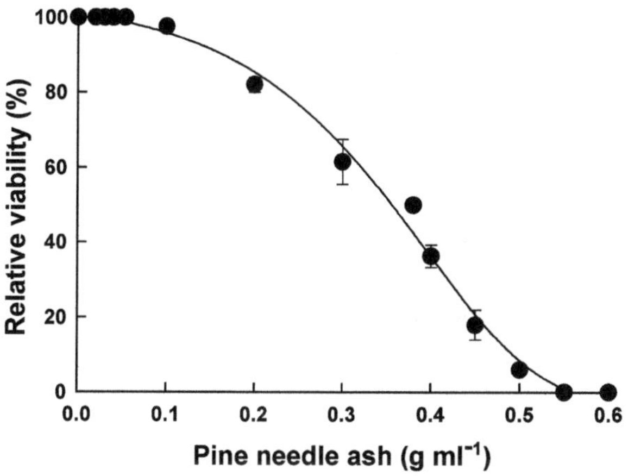

Figure 1. Effect of pine needle ash on viability of tissues of the seaweed *U. pertusa*. Viability was measured by the TTC method, and expressed as a relative percentage.

The total RNA showed ribosomal RNAs bands, and the ratio of A_{260} $A_{280}{}^{-1}$, generally indicative of protein contamination, was close to the 2.0 value expected for pure RNA. Thus, the RNA obtained was suitable for use as a cDNA template. Ash-responsive transcripts were identified by comparing the ash-treated tissues to the control tissue. Sixty different arbitrary primers were screened for primers to amplify cDNA from ash-stressed tissues. Among them, ash-resistant inducible genes, showing no band in the control but appearing in the treated lanes, were amplified by 2 primers. Ash-suppressed genes, showing bands in the control but lacking in the treated lanes, were amplified by 7 different primers. Especially with the primer OPA 4 (AATCGGGCTG) as a single primer, the differential display pattern showed an intense band (marked as an arrow on lane 6) from the strongly stressed tissues under the LC_{50} condition (Figure 2A). With the primer OPA 18 (AGGTGACCGT) as a single primer, differential display showed a suppressed gene from the mild-stressed tissues even at the sub-lethal MNLC condition (Figure 2B). The above two ash-responsive genes have been consistently isolated and showed the greatest difference in intensity in the control relative to the treated lanes. First, the nucleotide sequence of the ash-inducible cDNA fragment of 342 bp (AI342), transiently up-regulated by ash treatment, was determined. Sequence analysis shows two direct repeats of AGTTTTTCT and the G + C content of 52%. It encodes a protein of at least 107 amino acids (Figure 3), with a predicted molecular weight of more than 12 kD. Five amino acids (Ala, Arg, Leu, Ser, Val) formed more than 46% of the protein. The most common amino acid, serine, comprised 11% of the total amino acids. A BLAST search indicated that the deduced amino acid sequence has no conserved domain and no significant similarity to known proteins from all GenBank sources. Thus, we assume this cDNA corresponds to a hypothetical protein.

Next, an ash-suppressed gene fragment of 1690 bp (AS1690), that was transiently down-regulated by ash treatment, was sequenced to identify the cDNA fragment. The fragment included one ORF of 678 bp (22–699). The ORF shows two direct repeats of

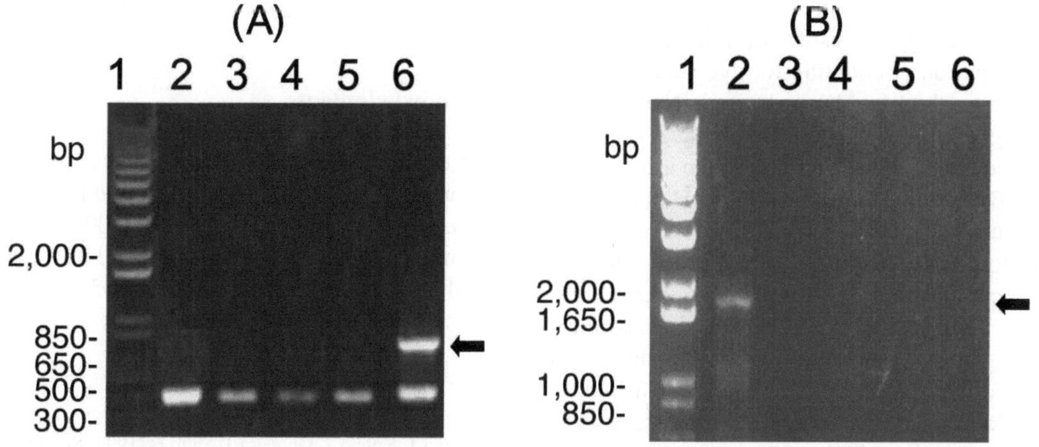

Figure 2. Patterns of differential display detected in cDNAs from *U. pertusa* tissues treated by pine needle ash. Reverse transcribed cDNAs from the tissues, treated at MNLC for 24 h (lane 3), LC_{50} for 1 h (lane 4), 6 h (lane 5) and 24 h (lane 6), were amplified using the arbitrary primers OPA 4 (A) and OPA 18 (B), respectively. The control reaction (lane 2) followed the same procedure without ash treatment. The molecular size marker (lane 1) is the 1 kb DNA ladder from BRL/Gibco.

1RSFSESRPVTVQDLIITAPLARSRCVYGCQFFSLQTSSAHSAAYRMEGCLHLLQNPNR
VCLWAHSRYWWGRYTPVLVVQCVPHPAVVLASALSSPIKGEFQHTGGRY*WIRAR 113

Figure 3. Deduced amino acid sequence of the ash-inducible cDNA fragment of 342 bp (AI342). Symbol * indicates the termination codon TAG.

1MRRLPIGELAGRFGNLGRRIWYMAQGQDPEPVKQSVAAPKSIGHGKVMPPNTKNLTTV
TMYFHHMAEKVAARLRKHEFEATIYFIGFLTDQEWVARKMQTEMPTNDSQQIKKLVNIF
LRDYWHKQGSFQVQITALDPKDQNHQMSLFGERDIQREQLNLAMDKINERYGEFTLAPA
NCCIVQICQMLLLHGSLLVIVKQYKTDQAKSAGLIMLTYLRMCQELLR* 226

Figure 4. Deduced amino acid sequence of an ORF from the ash-suppressed cDNA fragment of 1690 bp (AS1690). Symbol * indicates the termination codon TAA.

ATATATTTT and the G + C content of 41%. It encodes a 226 amino acid protein (Figure 4) with a predicted molecular weight of 26 kD. Six amino acids (Ala, Arg, Gln, Ile, Leu, Lys) formed more than 44% of the protein. The amino acid leucine comprised 11% of the total amino acids. The deduced amino acid sequence shows homology to a hypothetical protein (Accession No. ZP00060547) from *Clostridium thermocellum* with a score of 153 and an E value of 2e-09 (29% identity and 50% positive). Thus, we assume this cDNA is also for a hypothetical protein.

Discussion

For seaweed near a river mouth, one of the possible natural hazards could be caused by ash released from forest fires during the winter and spring. Most seaweed reproduces and grows during this season. Seaweed tissues have been used as indicators for environmental assessment (Fletcher, 1991). Instead of waiting for a seaweed growth period or counting for dye-stained cells, a quantitative method has been devised (Nam et al., 1998). Using this quantitative assay, tissue viability was measured for the stress strength in the ash-treated *U. pertusa*. By addition of ash at LC_{50}, pH change of the PES culture was minimal, from pH 8.10 to pH 8.17. The extent to which plant material is burned will determine the composition of chemical compounds that are released. Combustion of plant matter is known to release nitrogenous compounds (Perez-Fernandez & Rodriguez-Echeverria, 2003). However, pine ash may also release lots of phenolic-related substances, as seen in the absorbance value of 6.1 at 230 nm from 0.5 g ash. Thus, the stress to seaweed tissues could be caused by a heterogeneous mixture of substances rather than by a mere change of pH or a simple compound.

Biological reactions under the burden of stress can be understood at a gene expression level by means of RT-PCR, especially by the differential display technique (Liang & Pardee, 1992; Hong et al., 1995). Differential display provides the means for the detection and isolation of the responding genes that are uniquely expressed by the environmental stress (Rubinelli et al., 2002). Sequence analysis in this study revealed that a differentially-expressed cDNA fragment of AI342 showed no significant homology to known proteins. The other cDNA of AS1690 appears to be for a hypothetical protein loosely related to one found in a bacterium. The lack of sequence identity might be because few seaweed genes are found in databases, and large evolutionary distances prevent matches with known land plants with highly divergent sequences. Our results illustrate the utility of the differential display technique for isolating and identifying genes that are uniquely expressed by stress in seaweed tissues. The role of the gene products, as yet uncharacterized stress proteins, requires further investigation.

Acknowledgments

We thank Dr D. Coury for his critical reading of the manuscript. This work was supported by the National R & D Program, National Hazard 00-J-ND-01-B-26 from Korean Ministry of Science and Technology and by the Brain Korea 21 Project in 2005.

References

Asahina T, Kanno T, Onodera T, Saito H (1999) Substances for hastening the growth of *Ulva pertusa* in treated wastewater. Bull. Jpn. Soc. Sci. Fish. 65: 810–817.

Burrows EM (1971) Assessment of pollution effects by the use of algae. Proc. Roy. Soc. London, Ser. B, 177: 295–306.

Fletcher RL (1991) Marine macroalgae as bioassay test organisms. In Abel PD, Axiak V (eds), Ectoxicology and the Marine Environment. Ellis Horwood Ltd., New York, pp. 111–131.

Floreto EAT, Hirata H, Ando S, Yamasaki S (1993) Effects of temperature, light intensity, salinity and source of nitrogen on the growth, total lipid and fatty acid composition of *Ulva pertusa* Kjellman (Chlorophyta). Bot. Mar. 36: 149–158.

Hong YK, Sohn CH, Polne-Fuller M, Gibor A (1995) Differential display of tissue-specific messenger RNAs in *Porphyra perforata* (Rhodophyta) thallus. J. Phycol. 31: 640–643.

Jin HJ, Seo GM, Cho YC, Hwang EK, Sohn CH, Hong YK (1997) Gelling agents for tissue culture of the seaweed *Hizikia fusiformis*. J. Appl. Phycol. 9: 489–493.

Liang P, Pardee AB (1992) Differential display of eukaryotic messenger RNA by means of the polymerase chain reaction. Science 257: 967–971.

Lim BL, Hori H, Osawa S (1983) The nucleotide sequences of 5S rRNAs from a multicellular green alga, *Ulva pertusa*, and two brown algae, *Eisenia bicyclis* and *Sargassum fulvellum*. Nucl. Acids Res. 11: 1909–1912.

Nakanishi K, Nishijima M, Nishimura M, Kuwano K, Saga N (1996) Bacteria that induce morphogenesis in *Ulva pertusa* (Chlorophyta) grown under axenic conditions. J. Phycol. 32: 479–482.

Nam BH, Jin HJ, Kim SK, Hong YK (1998) Quantitative viability of seaweed tissues assessed with 2,3,5-triphenyltetrazolium chloride. J. Appl. Phycol. 10: 31–36.

Okano M, Aratani T (1979) Constituents in marine algae. 1. Seasonal variation of sterol, hydrocarbon, fatty acid, and phytol fractions in *Ulva pertusa*. Bull. Jpn. Soc. Sci. Fish. 45: 389–393.

Perez-Fernandez MA, Rodriguez-Echeverria S (2003) Effect of smoke, charred wood, and nitrogenous compounds on seed germination of ten species from woodland in central-western Spain. J. Chem. Ecol. 29: 237–251.

Provasoli L (1968) Media and prospects for the cultivation of marine algae. In Watanabe A, Hattori A (eds), Cultures and Collections of Algae. The Japanese Society of Plant Physiologists, Tokyo, pp. 63–75.

Reddy CRK, Iima M, Fujita Y (1992) Induction of fast-growing and morphologically different strains through intergeneric protoplast fusions of *Ulva* and *Enteromorpha* (Ulvales, Chlorophyta). J. Appl. Phycol. 4: 57–65.

Rubinelli P, Siripornadulsil S, Gao-Rubinelli F, Sayre RT (2002) Cadmium- and iron-stress-inducible gene expression in the green alga *Chlamydomonas reinhardtii*: Evidence for H43 protein function in iron assimilation. Planta 215: 1–13.

Yu K, Pauls KP (1992) Optimization of the PCR program for RAPD analysis. Nucl. Acids Res. 20: 2606.

Journal of Applied Phycology (2006) 18: 489–496
DOI: 10.1007/s10811-006-9052-8

Isolation and identification of gametogenesis-related genes in *Porphyra yezoensis* (Rhodophyta) using subtracted cDNA libraries

Makoto Kakinuma*, Izumi Kaneko, Daniel A. Coury, Takuya Suzuki & Hideomi Amano
Laboratory of Marine Biochemistry, Faculty of Bioresources, Mie University, 1515 Kamihama, Tsu, Mie 514-8507, Japan

*Author for correspondence: e-mail:kakinuma@bio.mie-u.ac.jp; fax: +81 59 231 9557

Key words: cDNA, gametogenesis, gene expression, *Porphyra yezoensis*, Rhodophyta, subtracted cDNA library

Abstract

Gametogenesis of *Porphyra yezoensis* thalli is induced by ageing as well as by changing water temperature and photoperiod. Under laboratory conditions, thalli cultivated at 10 °C with a 10:14 h (light: dark) cycle develop vegetatively to adult form without gametogenesis. On the other hand, sexual reproduction, which involves differentiation of vegetative cells and subsequent gametogenesis, is induced by culturing at 15 °C with a 16: 8 h (light: dark) cycle. We have constructed subtracted cDNA libraries enriched for differentially expressed transcripts in vegetative and reproductive thalli, and randomly selected 1,152 cDNAs from each subtracted library. Results of the dot blot analyses used for identification of differentially expressed cDNAs indicated that mRNA levels of 176 and 138 cDNAs tended to increase in the vegetative and reproductive thalli, respectively. BLAST analysis of nucleotide and deduced amino acid sequences showed that the cDNAs represented 63 and 59 unique clones for the vegetative and reproductive cDNA enriched subtracted libraries, respectively. Interestingly, some of the cDNAs isolated from the reproductive subtracted library were homologous to genes encoding protein kinases, GTP-binding protein, and heat shock proteins involved in signal transduction and the molecular chaperon system.

Introduction

The genus *Porphyra* contains several species that include the edible laver, and is of considerable economic importance in many places of the world, especially in Asia (Zemke-White & Ohno, 1999). In Japan, approximately 60,000 tonnes (dry weight) were produced in cultivation farms each year (Zemke-White & Ohno, 1999). Thus, *Porphyra* is one of the most extensively cultivated seaweeds used as food in Japan. *Porphyra* displays a unique heteromorphic, digenetic life cycle that consists of a leafy gametophyte and a filamentous sporophyte. The difference between these two developmental phases is usually associated with different chromosome ploidy level. In addition, these two generations show many different structural features, such as chloroplast number and cell wall composition (Cole and Conway, 1975; Mukai et al., 1981). Therefore, in order to elucidate molecular mechanisms underlying these differences between the gametophytic and sporophytic generations, analyses of differentially expressed genes have been performed using subtracted cDNA libraries (Liu et al., 1994a) and expressed sequence tags (ESTs) (Nikaido et al., 2000; Asamizu et al., 2003) for both generations.

Some *Porphyra* species also have properties that make them a suitable system for the study of cellular differentiation, since the thallus is formed by a single cell layer. In the leafy gametophyte of *Porphyra*, vegetative cells differentiate into sexually mature male and female cells. In order to identify tissue-specific genetic markers of differentiation in *Porphyra*, RNA transcripts among morphologically distinct regions of the differentiated tissue have been compared using differential display, and a few genetic markers specific to each tissue have been isolated (Hong et al., 1995). However, the molecular mechanisms underlying and controlling the differentiation of vegetative cells into reproductive cells are still poorly understood.

Among *Porphyra* species, *P. yezoensis* has recently been recognized as a useful model organism for fundamental and applied studies of marine algae (Waaland et al., 2004) since the life cycle can be completed within a few months in laboratory culture (Kuwano et al., 1996), the genome size is similar to other higher plant model organisms such as *Arabidopsis* and rice (Kapraun et al., 1991; Le Gall et al., 1993), and a public EST database exists (Nikaido et al., 2000; Asamizu et al., 2003). In addition, gamete formation (sexual differentiation) of vegetative cells within the leafy gametophyte can be easily induced by changing two extrinsic signals, photoperiod and water temperature, in laboratory culture (Iwasaki, 1979). Therefore, this alga is an ideal research tool to investigate the molecular mechanisms related to differentiation of vegetative cells, and to identify genes that regulate gametogenesis of *P. yezoensis* thalli that are induced or repressed in response to these changing conditions.

As a first step toward understanding the molecular mechanisms for reproductive cell differentiation induced by changing cultivation conditions, the present study was undertaken to construct subtracted cDNA libraries enriched for differentially expressed messages in vegetative and reproductive thalli and to investigate the levels of transcript accumulation for each subtracted cDNA by dot blot analysis. In addition, the cDNAs corresponding to putative differentially expressed transcripts in the vegetative and reproductive thalli were sequenced to identify candidate genes important for gametogenesis.

Materials and methods

Materials and cultivation conditions

Leafy gametophytes of *P. yezoensis* (strain FA-89) were grown in one-fifth strength Provasoli's enriched seawater (1/5 PES) medium (Provasoli, 1968). The culture medium was changed every three days throughout the experiments. The culture was aerated with filtrated air, irradiated with 40 μmol photons $m^{-2}\,s^{-1}$ light on a 10:14 h (light: dark) cycle (10L/14D), and maintained at 10 °C. When the thalli reached an average length of 5 cm, the thalli were sampled as "vegetative thalli" after microscopic confirmation that only vegetative cells were present.

Gametogenesis of *Porphyra* thalli was induced by changing the photoperiod and water temperature. Thalli grown to an average length of 5 cm were inoculated into 1/5 PES medium and maintained on a 16 L/8D photoperiod at 15 °C. The morphology of the thallus cells was examined daily under a microscope. Thalli cultivated for three or seven days under the induction conditions were collected as "induced thalli" and "reproductive thalli", respectively. The cells in the induced thalli showed no differentiation, whereas in the reproductive thalli differentiation was limited to marginal files of cells in the upper one-third of the thallus, representing approximately 5% of the total surface area.

Construction of subtracted cDNA libraries

Poly(A)$^+$ mRNA was isolated directly from fresh thalli using a QuickPrep *Micro* mRNA Purification Kit (Amersham Biosciences). Two micrograms of mRNA from the vegetative and reproductive thalli were used for double-stranded cDNA synthesis and then for subtracted cDNA library construction using the PCR-Select^{TM} cDNA Subtraction Kit (Clontech), according to the manufacturer's instructions. In the case of preparation of the forward subtracted (FS) cDNA library that is enriched for differentially expressed transcripts in vegetative thalli, cDNAs from the vegetative and reproductive thalli were used as tester and driver cDNAs, respectively. The reverse subtracted (RS) cDNA library, enriched for differentially expressed transcripts in reproductive thalli, was prepared by using cDNAs from the reproductive and vegetative thalli as tester and driver cDNAs, respectively. Before subtractive hybridizations, sub-pools of each tester cDNA ligated to a different adaptor supplied in the kit were pooled as forward unsubtracted (FU) cDNA from the vegetative thalli and reverse unsubtracted (RU) cDNA from the reproductive thalli. The FS and RS cDNAs obtained were subcloned into a pT7Blue *T*-Vector (Novagen) and subsequently used to construct FS and RS cDNA libraries in *E. coli* JM109. After plating on LB-agar plates supplemented with ampicillin, X-gal, and ITPG, 1152 white colonies from each transformation were picked into twelve 96 well microtiter plates containing LB and ampicillin. The cultures were grown overnight with shaking and stored at −80 °C after the addition of an equal volume of 50% glycerol.

Screening of subtracted cDNA libraries

In order to identify differentially expressed transcripts in the vegetative and reproductive thalli, dot blot analyses were performed on the FS and RS libraries. For each library, cDNA inserts were amplified directly from the 1152 liquid cultures in twelve 96 well PCR plates using the nested adaptor primers from the CLONTECH cDNA Subtraction Kit. The twelve plates from each library were divided into four sets of three plates each. PCR products from each set of three plates (288 independent clones) were arrayed in triplicate onto a Hybond^{TM}-N^{+} nylon membrane (Amersham Biosciences). Four replicate membranes were prepared from each set of three plates. In total, four sets of four replicate membranes were prepared for each library, and each replicate membrane within a set was hybridized separately with either the FS, RS, FU, or RU cDNA probe.

For preparation of cDNA probes, FS, RS, FU, and RU cDNAs were random DIG-labeled using DIG-High Prime (Roche Diagnostics). Membranes were prehybridized at 42 °C for 1 h in DIG Easy Hyb (Roche Diagnostics) and hybridized at 42 °C for 16 h in the same solution after adding 200 ng of the DIG-labeled probe. After hybridization, the membrane was washed twice with 2 X SSC containing 0.1% SDS at room temperature for 10 min and twice with 0.2 X SSC containing 0.1% SDS at 68 °C for 15 min. Signal generation of the DIG-labeled probe was performed using the ECF substrate (Amersham Biosciences) according to the manufacturer's instructions. The hybridized membranes were scanned using a STORM^{TM} 860 and quantified using ImageQuant V1.2 (Amersham Biosciences).

Sequence analysis

Purified plasmid DNA isolated using the QIAprep Spin Miniprep Kit (QIAGEN) was sequenced with the aid of a DYEnamic^{TM} ET Terminator Cycle Sequencing Kit (Amersham Biosciences) and an ABI PRISM Model 373A Sequencer (PE Biosystems). Database searches and similarity analyses of cDNA nucleotide sequences were carried out with the BLASTN and BLASTX programs (Altschul et al., 1990, 1997) against public nucleotide, EST, and protein databases.

Northern blot analysis

For the preparation of a cDNA probe, 10 pg of the cDNA fragment was DIG-labeled using the PCR DIG Probe Synthesis Kit (Roche Diagnostics). Ten micrograms of total RNA for each thallus type (vegetative, induced, and reproductive) were size-fractionated by electrophoresis on a 1.0% agarose-formaldehyde gel and capillary-transferred to a Hybond^{TM}-N^{+} nylon membrane. The RNA blot was pre-hybridized at 50 °C for 1 h in DIG Easy Hyb (Roche Diagnostics) and hybridized at 50 °C for 16 h in the same solution after adding the DIG-labeled cDNA probe. After hybridization, stringency washes and signal generation and detection of the DIG-labeled probe on the Northern blot were performed as for the dot blot analysis.

Results

Isolation of cDNAs for differentially expressed transcripts in the vegetative and reproductive thalli

Porphyra thalli grown to the average length of 5 cm in 1/5 PES medium under a 10L/14D photoperiod at 10 °C (vegetative thalli) contained only vegetative cells by microscopic examination. When these thalli were transferred to fresh medium and cultivated on a 16L/8D photoperiod at 15 °C for three days (induced thalli), differentiation of vegetative cells and subsequent gametogenesis was not observed. After four days, differentiated cells (sexually mature male cells) were clearly observed in the tips of thalli, and after seven days, the marginal cells in the thalli (reproductive thalli) had almost completely differentiated into reproductive cells.

Messenger RNA extracted from the vegetative and reproductive thalli was subsequently used for cDNA synthesis and construction of the FS and RS cDNA libraries enriched for differentially expressed transcripts in vegetative and reproductive thalli, respectively. After transformation of *E. coli* with the FS and RS cDNA libraries, 1152 recombinant (white) colonies obtained from each subtracted library were randomly selected and those inserted cDNAs were amplified by PCR. When 192 PCR products (8 per set of 96 colonies) were examined, the average size of cDNA inserts was approximately 400 bp.

Individual clones of the subtracted cDNA libraries were screened for differential expression by hybridizing with either FS, RS, FU, or RU cDNA probes. The distribution of signal intensity ratios (FS/RS and RS/FS for the screening of the FS and RS cDNAs, respectively) in dot blot analyses for the 1152 randomly selected cDNAs from each subtracted

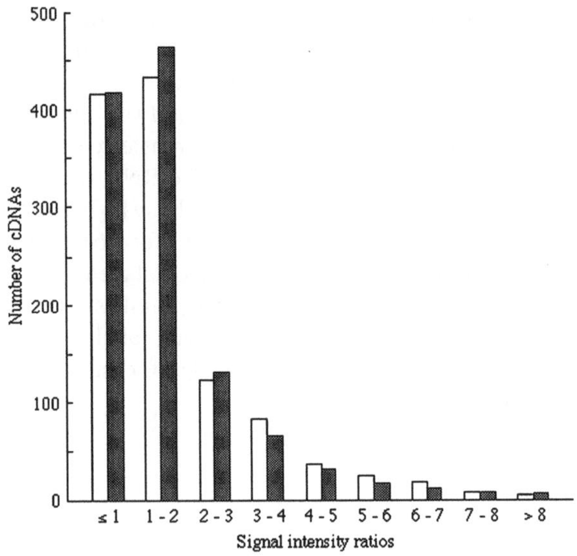

Figure 1. Distribution of signal intensity ratios in dot blot analyses for 1,152 randomly selected cDNAs from each subtracted library. Open and gray bars represent the number of cDNAs from the forward subtracted (FS) and reverse subtracted (RS) cDNAs, respectively. The signal intensity ratios for the FS and RS cDNAs were calculated from the signal intensity of FS to RS cDNA probes and of RS to FS cDNA probes, respectively. cDNA clones which showed a difference in signal intensity ratio of more than 3-fold in dot blot analyses using FS and RS cDNA probes were selected for further studies.

library is shown in Figure 1. Quantitative analysis of signal intensity for the dot blots showed that approximately three-quarters of the recombinant colonies in the FS and RS cDNA libraries had less than a two-fold difference in signal intensity. On the other hand, 176 of the clones screened from the FS library and 138 of the clones screened from the RS library, had greater than a three-fold difference in signal intensity. Therefore, these clones in the FS and RS cDNA libraries were selected for further investigation.

Identification of differentially expressed transcripts in the vegetative and reproductive thalli

Positive clones selected by dot blot analyses were sequenced to identify the corresponding genes. The 176 forward subtracted and 138 reverse subtracted clones were found to represent 63 and 59 unique sequences, respectively. Approximately 84% (53/63) of the forward subtracted clones showed similarity ($E < 0.005$, score >90) to genes and *Porphyra* ESTs registered in the public databases, while 73% (43/59) of the reverse subtracted clones also had significant matches in the public databases (Table 1). Therefore, a significant number of the clones remain to be categorized.

Table 1. BLAST results of similarity searches against the public nucleotide and protein databases.

Similarity	Number of unique cDNAs: Forward subtracted	Reverse subtracted
Genes of known or putative function[a]	33 (12)	26 (15)
Porphyra ESTs of unknown function[b]	20	17
No similarity[c]	10	16
Total	63	59

[a] Similarity threshold set at $E < 0.005$ and score >90 for genes of known or putative function. Number of cDNAs in parentheses also showed significant similarity ($E <$ 1e-60 and score >120) to *Porphyra* ESTs.
[b] Similarity threshold set at $E <$ 1e-40 and score >90.
[c] Number of cDNAs that showed no similarity ($E \geq 0.005$ and score ≤90) to genes or ESTs of known or putative function.

Thirty-three of the forward subtracted clones and 26 of the reverse subtracted clones that showed sequence similarity to genes of known or putative functions were classified according to their putative biological roles and biochemical functions (Table 2). The largest category (12/33) was forward subtracted clones with similarities to genes that function in protein synthesis, mainly corresponding to genes encoding various ribosomal proteins ($E <$ 1e-23, score >270). Interestingly, several homologues of genes associated with signal transduction (e.g. a small GTP-binding protein ($E =$ 2e-24, score = 281), a MAP kinase ($E =$ 1e-22, score

Table 2. Functional classification of the subtracted cDNA clones based on similarity to known protein genes.

Functional categories	Number of protein genes: Forward subtracted	Reverse subtracted
Energy metabolism	5	5
Protein fate	5	6
Protein synthesis	12	1
Transport and binding proteins	2	1
Transcription and regulation	1	4
Signal transduction	0	3
Structure and membrane proteins	0	2
Fatty acid metabolism	3	0
Intermediary metabolism	5	4
Total	33	26

= 265), and a SNF1/AMP-activated protein kinase (E = 5e-12, score = 174)) and protein fate (e.g. HSP90 (E = 4e-18, score = 151), HSP70 (E < 1e-46, score >470), ubiquitin (E = 3e-30, score = 331), and some proteasome subunits (E < 1e-16, score >210)) were also identified.

Expression of a small GTP-binding protein in different growth phases

We chose a 368 bp cDNA putatively encoding a small GTP-binding protein to confirm that the isolated cDNA clones truly represent mRNAs differentially expressed in vegetative and reproductive thalli. The clone was hybridized to mRNA from the three phases (vegetative, induced, and reproductive) in Northern hybridization experiments (Figure 2). The apparent larger size of the band in the reproductive thalli total RNA lane is most likely due to polysaccharides inhibiting the rate of RNA migration in the gel, since the ethidium bromide-stained gel also showed the same shift of rRNA in this lane relative to the other two (data not shown). The relative mRNA level of the small GTP-binding protein gene was almost the same in the vegetative and induced thalli, but the level was 2.6 times higher in the reproductive thalli than in the vegetative thalli.

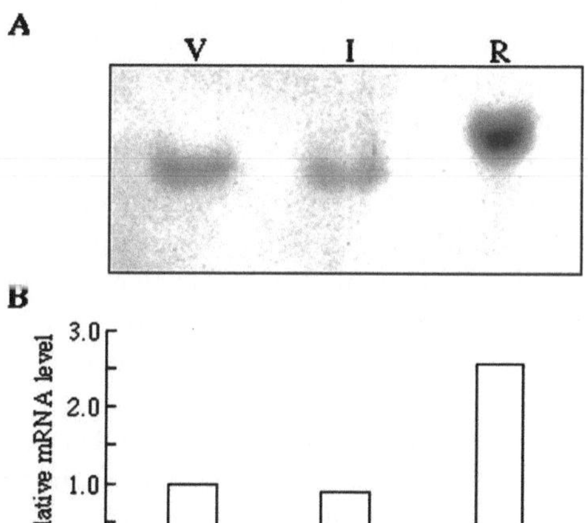

Figure 2. Northern blot analysis (A) and relative expression levels (B) of mRNA encoding a small GTP-binding protein in *P. yezoensis* thalli. In panel A, total RNA (10 μg per lane) from vegetative (V), induced (I), and reproductive thalli (R) were electrophoretically separated on a 1.0% agarose-formaldehyde gel, blotted onto a nylon membrane, and hybridized with DIG-labeled cDNA probe from the insert of clone pT7B-RS-H3-8. In panel B, expression levels in the induced (I) and reproductive thalli (R) represent transcript abundance relative to the vegetative thalli (V). The rRNA band stained with ethidium bromide was used to adjust values for equal loading (data not shown).

Discussion

Many studies have been performed that focused on morphological and physiological differences between the leafy gametophyte and the filamentous sporophyte in the life cycle of *Porphyra* species. Gametophyte- and sporophyte-specific cDNAs that encode proteins such as elongation factors, serine protease-like proteins, polysaccharide-binding proteins, and lipoxygenases have been isolated by differential screening and subtraction of phase-specific cDNA libraries (Liu et al., 1994a,b, 1996a,b,c). Recently, EST analysis has been performed to identify candidate genes related to the morphological and physiological differences between the gametophytic and sporophytic generations, and large numbers of cDNAs have been identified (Nikaido et al., 2000; Asamizu et al., 2003). Thus, potential genetic markers specific to two generations are well known. However, because these are generation-specific cDNAs rather than cell differentiation-specific genes, the genes regulating the maturation process of *Porphyra* thalli remain poorly understood. In this paper, we describe initial results examining the molecular mechanisms regulating the reproductive maturation process in *Porphyra* thalli.

We constructed two subtracted cDNA libraries, enriched for differentially expressed transcripts in vegetative thalli and reproductive thalli that were artificially induced to undergo differentiation in laboratory culture by changing the photoperiod and water temperature. In order to remove false positives, 1152 recombinant clones from each subtracted cDNA library were screened by dot blot hybridization with four DIG-labeled cDNA probes corresponding to forward and reverse, subtracted and unsubtracted cDNAs. cDNAs that showed greater than a three-fold difference in signal intensity between the forward and reverse subtractions were sequenced, and the sequence data subjected to clustering by BLAST analysis. These cDNAs were found to represent 63 and 59 unique clones from the forward and reverse subtracted libraries, respectively (Table 1), with average size of approximately 370 bp. This insert size is shorter than those reported previously (Diatchenko et al., 1996), which may reflect an increased frequency of *Rsa*I restriction sites in the *P. yezoensis* genome compared to the human genome.

The size also is shorter than the average insert and EST sequences in the *Porphyra* normalized library (approximately 970 and 470 bp, respectively) (Nikaido et al., 2000). Therefore, in order to generate more sequence information for each clone, one possibility is to use a six-base recognition enzyme rather than the four-base recognition enzyme *Rsa*I used in the CLONTECH PCR-Select™ cDNA Subtraction Kit.

Results of similarity searches for each unique cDNA using the BLASTX program showed that 33 forward subtracted and 26 reverse subtracted cDNA clones were putative homologues of known functional genes registered in public databases (Table 1). The largest functional category, genes involved in protein synthesis, appeared to be down regulated in reproductive thalli compared to vegetative thalli (Table 2). This apparent down regulation does not result in a decreased growth rate in reproductive thalli (I. Kaneko & M. Kakinuma, unpublished data). However, at approximately seven days of induction (i.e., the time point at which "reproductive" mRNA was sampled), the pace of differentiation picks up rapidly, suggesting that the protein synthesis machinery already present may be sufficient to carry the thallus through the rest of the developmental program with reduced expression of protein synthesis genes. As is the case with many of the genes identified in this screen, further confirmation of differential expression is required.

Since, in this study, thallus maturation was artificially induced by changing photoperiod and water temperature, we are largely interested in the cDNAs involved in signal transduction (e.g. small GTP-binding protein and protein kinases) and protein fate (e.g. HSP90). Small GTP-binding proteins have been shown to participate in signal transduction, cell proliferation and differentiation, and membrane vesicle transport (Balch, 1990; Bourne et al., 1990; Hall, 1990). In higher plants, it has been reported that light regulates the changes in steady-state levels of several small GTP-binding protein mRNAs, and that phytochrome mediates the changes in a negative manner (Yoshida et al., 1993; Inaba et al., 1999). On the other hand, some small GTP-binding proteins in *Chlamydomonas* are used for household functions responsible for vesicle transport rather than for cell differentiation (Dietmaier et al., 1995). In the case of maturation process of *P. yezoensis*, expression of the small GTP-binding protein was dramatically increased in the reproductive thalli (Figure 2). This result suggests a possibility that the small GTP-binding protein plays an important, yet unknown role in cell differentiation in the thallus maturation process.

For other interesting genes such as protein kinases and HSP90, expression profiles during the maturation process have not yet been investigated by Northern blot analysis. However, it is well known that the MAP kinases mediate intracellular phosphorylation events linking receptor activation to the control of cell proliferation, chemotaxis, differentiation, and stress responses (Schaeffer & Weber, 1999). Also, the SNF1/AMP-activated protein kinase is commonly activated in response to cellular and environmental stress responses (Hardie et al., 1998). The HSP90 family in most eukaryotic cells binds to and regulates the activity of functionally important proteins such as steroid hormone receptors and protein kinases (Pratt et al., 2001). In addition, it has been reported that HSP90 function is linked to the development of pollen in higher plants and of the female gametophytes in algae (Marrs et al., 1993; Yabe et al., 1994; Lee et al., 1998). Cells within *P. yezoensis* thalli have differentiated specific signal transduction pathways for response to and integration of extracellular stimuli. Therefore, it is possible that protein kinases and HSP90 in *P. yezoensis* cells also link perception of extracellular stimuli and sexual differentiation.

One of the surprising results of the similarity search using isolated cDNAs from FS and RS libraries is that many of cDNAs isolated by subtractive hybridization have no known homologues, even among the available *Porphyra* ESTs (Nikaido et al., 2000; Asamizu et al., 2003). Because the *Porphyra* ESTs were derived from vegetative (non-induced or reproductive) gametophyte tissue, it is possible that these clones might correspond to rare genes related to the maturation processes of *Porphyra* thalli. Northern blot analyses using these clones probed against RNA from different phases of the maturation process are currently being carried out in our laboratory.

In this paper we have identified a number of interesting candidate genes that might play an important role in gametogenesis in *P. yezoensis*. However, it is possible that changes in expression of these genes may only be a response to the change in temperature and photoperiod and are unrelated to gametogenesis. We have observed that under our laboratory conditions gametogenesis is induced even in young thalli when either or both conditions are changed, suggesting a direct link between the increase in temperature and photoperiod and induction of gametogenesis. We are currently developing an *in situ* hybrizidation protocol to determine if changes in expression of the candidate genes are focused in the area of sexual differentiation. We are also constructing subtracted libraries using induced and reproductive tissue. Theoretically, since both of these phases will have been grown under the increased temperature and photoperiod conditions, differences in expression of temperature and light response genes will be minimized. Ultimately, we would like to be able to knock out expression of these genes and look for the effect on gametogenesis, although this approach is not yet technically feasible.

Acknowledgements

This study was supported by the National Research Institute of Fisheries Science, Fisheries Research Agency, Japan, and was funded in part by a Grant-in-Aid from the Fisheries Agency, Government of Japan. We thank Dr. M. Iwabuchi of the Fukuoka Fisheries and Marine Technology Research Center for supplying *P. yezoensis* strain FA-89, and for help with its culture. We also thank Dr. T. Morita of the Marine Productivity Division of the National Research Institute of Fisheries Science, Japan, for help with the experiments.

References

Altschul SF, Gish W, Miller W, Myers EW, Lipman DJ (1990) Basic local alignment search tool. J. Mol. Biol. 215: 403–410.

Altschul SF, Madden TL, Schäffer AA, Zhang J, Miller W, Lipman DJ (1997) Gapped BLAST and PSI-BLAST: A new generation of protein database search programs. Nucleic Acids Res. 25: 3389–3402.

Asamizu E, Nakajima M, Kitade Y, Saga N, Nakamura Y, Tabata S (2003) Comparison of RNA expression profiles between the two generations of *Porphyra yezoensis* (Rhodophyta), based on expressed sequence tag frequency analysis. J. Phycol. 39: 923–930.

Balch WE (1990) Small GTP-binding proteins in vesicular transport. Trends in Biochem. Sci. 15: 473–477.

Bourne HR, Sanders DA, McCormick F (1990) The GTPase superfamily: A conserved switch for diverse cell functions. Nature 348: 125–132.

Cole KM, Conway E (1975) Phenetic implications of structural features of the perennating phase in the life history of *Porphyra* and *Bangia* (Bangiophyceae, Rhodophyta). Phycologia 16: 205–216.

Diatchenko L, Lau YFC, Campbell AP, Chenchik A, Moqadam F, Huang B, Lukyanov S, Lukyanov K, Gurskaya N, Sverdlov ED, Siebert PD (1996) Suppression subtractive hybridization: A method for generating differentially regulated or tissue-specific cDNA probes and libraries. PNAS 93: 6025–6030.

Dietmaier W, Fabry S, Huber H, Schmitt R (1995) Analysis of a family of *ypt* genes and their products from *Chlamydomonas reinhardtii*. Gene 158: 41–50.

Hall, A (1990) The cellular functions of small GTP-binding proteins. Science 249: 635–640.

Hardie GD, Carling D, Carlson M (1998) The AMP-activated/SNF1 protein kinase subfamily: Metabolic sensors of the eukaryotic cell? Annu. Rev. Biochem. 67: 821–855.

Hong YK, Sohn CH, Polne-Fuller M, Gibor A (1995) Differential display of tissue-specific messenger RNAs in *Porphyra perforata* (Rhodophyta) thallus. J. Phycol. 31: 640–643.

Inaba T, Nagano Y, Sakakibara T, Sasaki Y (1999) Identification of a *cis*-regulatory element involved in phytochrome down-regulated expression of the pea small GTPase gene *pra2*. Plant Physiol. 120: 491–499.

Iwasaki, H (1979) 2.2.E(IV) *Porphyra tenera* (Rhodophyta). In Nishizawa, K., Chihara, M. (eds), *Methods in Phycological Studies*, Kyoritsu Shuppan Co., Ltd., Tokyo, pp. 118–126.

Kapraun DF, Hinson TK, Lemus AJ (1991) Karyology and cytophotometric estimation of inter- and intraspecific nuclear DNA variation in four species of *Porphyra* (Rhodophyta). Phycologia 30: 458–466.

Kuwano K, Aruga Y, Saga N (1996) Cryopreservation of clonal gametophytic thalli of *Porphyra* (Rhodophyta). Plant Sci. 116: 117–124.

Lee YK, Kim SH, Hong CB, Chah OK, Kim GH, Lee IK (1998) Heat shock protein 90 may be involved in differentiation of the female gametophytes in *Griffithsia japonica* (Ceramiales, Rhodophyta). J. Phycol. 34: 1017–1023.

Le Gall Y, Brown S, Marie D, Mejjad M, Kloareg B (1993) Quantification of nuclear DNA and GC content in marine microalgae by flow cytometry of isolated nuclei. Protoplasma 173: 123–132.

Liu QY, Meer JP, Reith ME (1994a) Isolation and characterization of phase-specific complementary DNAs from sporophytes and gametophytes of *Porphyra purpurea* (Rhodophyta) using subtracted complementary DNA libraries. J. Phycol. 30: 513–520.

Liu QY, Reith ME (1994b) Isolation of a gametophyte-specific cDNA encoding a lipoxygenase from the red alga *Porphyra purpurea*. Mol. Mar. Biol. Biotechnol. 3: 206–209.

Liu QY, Baldauf SL, Reith ME (1996a) Elongation factor 1α genes of the red alga *Porphyra purpurea* include a novel, developmentally specialized variant. Plant Mol. Biol. 31: 77–85.

Liu QY, Ross N, Lanthier P, Reith M (1996b) A gametophyte cell wall protein of the red alga *Porphyra purpurea* (Rhodophyta) contains four apparent polysaccharide-binding domains. J. Phycol. 32: 995–1003.

Liu QY, Reith M (1996c) A sporophyte cell wall protein of the red alga *Porphyra purpurea* (Rhodophyta) is a novel member of the chymotrypsin family of serine proteases. J. Phycol. 32: 1003–1009.

Marrs KAE, Casey ES, Capitant SA, Bouchard RA, Dietrich PS, Mettler IK, Sinibaldi RM (1993) Characterization of two maize hsp90 heat shock protein genes and expression during heat shock, embryogenesis, and pollen development. Dev. Genet. 14: 27–41.

Mukai LS, Craigie JS, Brown RG (1981) Chemical composition and structure of the cell walls of the conchocelis and thallus phases of *Porphyra tenera* (Rhodophyceae). J. Phycol. 17: 192–198.

Nikaido I, Asazumi E, Nakajima M, Nakamura Y, Saga N, Tabata S (2000) Generation of 10,154 expressed sequence tags from a leafy gametophyte of a marine red alga, *Porphyra yezoensis*. DNA Res. 7: 223–227.

Pratt WB, Krishna P, Olsen LJ (2001) Hsp90-binding immunophilins in plants: the protein movers. Trends Plant Sci. 6: 54–58.

Provasoli, L (1968) Media and prospects for the culltivation of marine algae. In Watanabe, A., Hattori, A. (eds), *Culture and Collection of Algae: Proceedings of US-Japan Conference in Hakone*, Japanese Society of Plant Physiologists, Tokyo, pp. 63–75.

Schaeffer HJ, Weber MJ (1999) Mitogen-activated protein kinases: Specific messages from ubiquitous messengers. Mol. Cell. Biol. 19: 2435–2444.

Waaland JR, Stiller JW, Cheney DP (2004) Macroalgal candidates for genomics. J. Phycol. 40: 26–33.

Yabe N, Takahashi T, Komeda Y (1994) Analysis of tissue-specific expression of *Arabidopsis thaliana* hsp90-family gene *HSP81*. Plant Cell Physiol. 35: 1207–1219.

Yoshida K, Nagano Y, Murai N, Sasaki Y (1993) Phytochrome-regulated expression of the genes encoding the small GTP-binding proteins in peas. PNAS 90: 6636–6640.

Zemke-White WL, Ohno M (1999) World seaweed utilization: An end-of-century summary. J. Appl. Phycol. 11: 369–376.

Journal of Applied Phycology (2006) 18: 497–504
DOI: 10.1007/s10811-006-9055-5

Molecular systematics and phylogenetics of Gracilariacean species from the Mediterranean Sea

G.M. Gargiulo*, M. Morabito, G. Genovese & F. De Masi
Department of Botanical Sciences, University of Messina, Salita Sperone, 31, S. Agata, 98166 Messina, Italy
*Author for correspondence: e-mail: ggargiulo@unime.it

Key words: Gracilariales, *rbc*L gene, molecular systematics, Mediterranean Sea, Rhodophyta, phylogeny

Abstract

Mediterranean reports of Gracilariaceae species, in particular those assigned to the *G. verrucosa* complex, were re-examined with the use of molecular tools, in order to verify their systematic position and better understand their distribution. Within this complex, we recognized four distinct taxa: *Gracilariopsis longissima, Gracilaria gracilis, Gracilaria longa* and a possible new species. The *rbc*L gene sequences, together with those of other terete Mediterranean entities, were included in a broad molecular phylogeny of the family. The reproductive characters of the studied taxa do not fit completely with published hypothesis on the generic and intrageneric relationships, suggesting that the anatomy of some subgroups should be better characterized.

Introduction

The order Gracilariales is well defined from both an anatomical (Fredericq & Hommersand, 1989a,b; Fredericq & Hommersand, 1990) and a molecular point of view (Freshwater et al., 1994; Saunders & Kraft, 1997; Harper & Saunders, 2001) within the Rhodophyta, but the intergeneric taxonomy had a somewhat more complex history (Bird et al., 1992; Bird et al., 1994; Bird, 1995; Bellorin et al., 2002). Recently, clarifying insights have come from systematic studies of the order that consider the ontogenesis of reproductive structures and *rbc*L gene phylogeny (Liao & Hommersand, 2003; Gurgel & Fredericq, 2004), nevertheless many problems remain at a lower taxonomic level, such as the *G. verrucosa*-complex. Populations previously referred to this taxon are now considered either *Gracilariopsis longissima* (S. Gmelin) Steentoft, Irvine et Farnham or *Gracilaria gracilis* (Stackhouse) Steentoft, Irvine et Farnham, or described as new species (Abbott, 1985; Zhang & Xia, 1985; Bird et al., 1986; Gargiulo et al., 1987; Steentoft et al., 1995). However, the name *Gracilaria verrucosa* (Hudson) Papenfuss, despite having been rejected (Irvine & Steentoft, 1995), is still used to identify economically important entities (Skiptsova, 2000; Imbs et al., 2001; Mancinelli & Rossi, 2001; Rath & Adhikary, 2002; Wang, 2002). Mediterranean reports of Gracilariacean species need re-examination after these systematic revisions. These taxa are frequently included in Mediterranean check-lists because of mere nomenclature changes rather than from actual verifications of specimens (Furnari et al., 2003). Other Mediterranean entities, e.g. *Gracilaria longa* Gargiulo, De Masi et Tripodi, share a similar morphology, making the segregation of these taxa difficult (Bird & Rice, 1990; Steentoft et al., 1995).

The aim of our study is to verify the systematic position and the Mediterranean distribution of some *G. verrucosa*-like populations, using *rbc*L gene sequence analysis. Moreover, sequence data set includes other terete Mediterranean species and those available for *Gracilaria* species from other geographical areas, in order to test how Mediterranean taxa match with proposed phylogenetic hypotheses.

Materials and methods

Sequence data generated for *rbc*L gene were submitted to GenBank and accession numbers together with collection information are given in Table 1. DNA was isolated from freshly collected or dried thalli (both silica gel preserved or recovered from herbarium sheets) with a modified CTAB protocol (Doyle & Doyle, 1987). Ground material was incubated in 2× CTAB buffer (0.1 M Tris-HCl, 0.05 M Na_2EDTA, 1.4 M NaCl, 2% CTAB, 1% PVP, 0.5% (v/v) β-mercaptoethanol) for 120–180 min at room temperature under constant agitation. Polysaccharides were precipitated with incubation with 2.5 M K acetate on ice for 20 (Saunders, 1993). Nucleic acids were extracted three times with 1 volume of phenol-chlorophorm-isoamilic alcohol (25:24:1) and twice with 1 volume of chlorophorm-isoamilic alcohol (24:1), precipitated with isopropanol and redissolved in 0.01 M Tris-HCl-0.001 M EDTA pH 8.0 (Sambrook et al., 1989). If consistent amounts of RNA were detected a digestion with RNase was performed (Sambrook et al., 1989). Voucher specimens were preserved in 4% formalin in seawater, or dried in silica gel, or pressed as herbarium sheets and deposited in the Phycological Herbarium of the Department of Botanical Sciences of the University of Messina (MS). In order to prevent errors in sorting of samples, each DNA isolation was performed from a single individual, a fragment of which was kept as voucher formalin preserved and/or pressed for further inspections.

The *rbc*L gene was PCR amplified using primers listed in Freshwater and Rueness (1994). Sequencing reactions were performed by an external company (MWG Biotech AG, Ebersberg, Germany).

Nucleotide sequences were aligned by eye unambiguously due to the absence of insertion or deletion mutations. Additional published *rbc*L gene sequences from species of Gracilariaceae (Gurgel et al., 2003a; Gurgel & Fredericq, 2004) were added to the alignment. A data set of 52 *rbc*L gene sequences of Gracilariaceae was used for phylogenetic analyses. When more sequences were available for each species, just one was used, with the exception of *Gracilariopsis longissima* for which three sequences were used due to the higher sequence divergence and related taxonomic implications (see discussion). Three representatives, respectively from Halymeniales, Rhodymeniales and Plocamiales, were selected as outgroup taxa (Saunders & Kraft, 1997) (Table 1). The final alignment included 55 taxa of 1234 characters. When presented, sequence divergence is expressed as uncorrected nucleotide substitutions percentage.

All phylogenetic analyses were performed in PAUP* 4b10 for the Macintosh (Swofford, 2002). The model of sequence evolution was selected according to a hierarchical likelihood ratio test as implemented in Modeltest 3.06 (Posada & Crandall, 1998). The model selected (a general time reversible model with invariable sites and gamma distribution, GTR+I+G; Lanave et al., 1984), and associated parameters (base frequencies: $A = 0.3476$, $C = 0.1109$, $G = 0.1601$, $T = 0.3815$; substitution rate matrix: $A-C = 1.1547$, $A-G = 6.5514$, $A-T = 0.9262$, $C-G = 1.9585$, $C-T = 12.7219$, $G-T = 1.0000$; proportion of invariable sites = 0.5250, gamma parameter = 1.1055) were used in distance and maximum likelihood (ML) analyses.

Distance phylogenies were constructed with a neighbor joining (NJ) algorithm and with a heuristic search under the criterion of minimum evolution (ME), with 1000 random addition sequence replicates, holding 20 trees at each step, tree bisection and reconnection (TBR) as branch-swapping algorithm, saving all minimal trees (MulTrees). The steepest descent option in the branch swapping procedure was not used because of an unfixed bug in the current beta version of PAUP* (http://paup.csit.fsu.edu/problems.html).

Parsimony analysis was conducted under a heuristic search similarly to ME analysis. ML analysis was performed under a heuristic search, with 10 random addition sequence replicates, holding 1 tree at each step, with TBR branch-swapping algorithm and MulTrees option in effect. Distance and parsimony analyses were subjected to bootstrap re-samplings to estimate robustness of the internal nodes (Felsenstein, 1985), basing on 1000 replicates, but with 10 random addition sequence replicates, holding 1 tree at each step, in the heuristic searches. Bootstrap resampling was not performed on maximum likelihood analysis, due to computational limitations. In all analyses unrooted trees were constructed, and subsequently rooted with reference to the outgroup taxa.

Results

Among the 1234 bp analysed (positions 117–1350, 84.12% of the entire length of the gene), 449 were parsimony informative. Parsimony analyses resulted in 25 most parsimonious (MP) trees (tree length 2366, consistency index 0.3407, retention index = 0.5917), not shown. Distance analyses resulted in a NJ tree and a ME tree (ME score = 2.57491), not shown, similar to the MP trees. ML analysis resulted in a phylogenetic tree (ln likelihood = −12762.10785, topology recovered 9 times out of 10 replicates), presented in Figure 1.

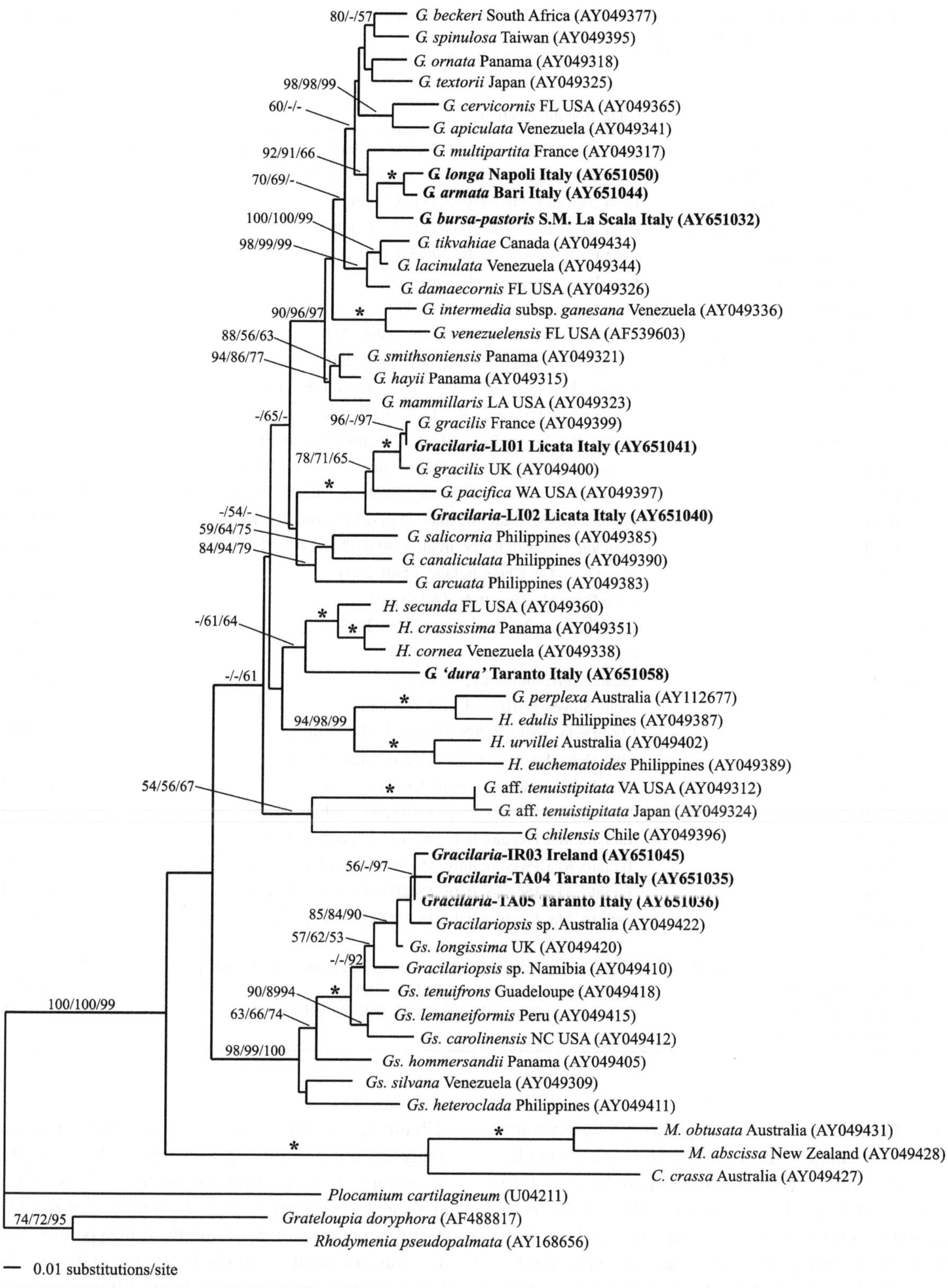

Figure 1. ML phylogram (ln L= −12762.10785), with bootstrap values inferred from respectively NJ, ME, MP analyses; branches with 100% support in all analyses are marked with an asterisk. GenBank accession numbers of the *rbc*L gene sequences are reported in brackets; sequences generated in the present study are indicated in bold.

Table 1. List of specimens sequenced in this study, with the GenBank accession numbers of the relative *rbc*L gene sequences. Sequences used in phylogenetic analyses are marked with an asterisk

Species	Collection information	GenBank accession number
Gracilaria armata (C.A. Agardh) Greville	Grotta della Regina, Bari, Italy; 01/06/2002; coll. C. Perrone & G. Felicini	AY651044*
Gracilaria bursa-pastoris (Gmelin) Silva	Clew Bay, Co. Mayo, Ireland; 28/08/2002; coll. R.J. Wilkes	AY651049
Gracilaria bursa-pastoris (Gmelin) Silva	Izola, Slovenia; 06/07/2003; coll. C. Battelli	AY651056
Gracilaria bursa-pastoris (Gmelin) Silva	Izola, Slovenia; 08/06/2003; coll. C. Battelli	AY651057
Gracilaria bursa-pastoris (Gmelin) Silva	Lake Faro, Messina, Italy; 20/05/2002	AY651038
Gracilaria bursa-pastoris (Gmelin) Silva	Lake Faro, Messina, Italy; 23/10/2001	AY651031
Gracilaria bursa-pastoris (Gmelin) Silva	Posillipo, Napoli, Italy; 23/06/2002	AY651047
Gracilaria bursa-pastoris (Gmelin) Silva	S. Maria La Scala, Catania, Italy; 06/06/2001	AY651032*
Gracilaria bursa-pastoris (Gmelin) Silva	S. Maria La Scala, Catania, Italy; 13/05/2002	AY651039
Gracilaria bursa-pastoris (Gmelin) Silva	Taranto, Mar Piccolo, Italy; 13/02/2002; coll. E. Cecere	AY651033
Gracilaria 'dura'	Taranto, Mar Piccolo, Italy; 13/02/2002; coll. E. Cecere	AY651058*
Gracilaria longa Gargiulo, De Masi et Tripodi	Napoli, Rivafiorita, Italy; 01/11/1981, holotype specimen	AY651050*
Gracilaria-EN01	Daynee Bay, North Cornwall, England, UK; coll. M.J. Homes & J.A. Brodie	AY651037
Gracilaria-EN02	St. Just in Roseland, Fal Estuary, England, UK; 23/01/2000; coll. Tean Mitchell	AY651046
Gracilaria-IR01	Finnavarra, Co. Clare, Ireland; 22/07/2003; coll. R.J. Wilkes	AY651042
Gracilaria-IR02	Kenmare Bay, Co. Kerry, Ireland; 30/07/2002; coll. R.J. Wilkes	AY651043
Gracilaria-IR03	New Quay, Co. Clare, Ireland; 05/01/2003	AY651045*
Gracilaria-LF01	Lake Faro, Messina, Italy; 08/05/2001	AY651030
Gracilaria-LG01	Lake Ganzirri, Messina, Italy; 28/10/2003	AY651053
Gracilaria-LI01	Licata, Agrigento, Italy; 17/06/2002	AY651041*
Gracilaria-LI02	Licata, Agrigento, Italy; 17/06/2002	AY651040*
Gracilaria-SL01	Izola, Slovenia; 01/2003; coll. C. Battelli	AY651054
Gracilaria-SP01	Ebro Delta, Spain; 29/05/2003; coll. M.A. Ribera	AY651055
Gracilaria-SP02	Ebro Delta, Spain; 29/05/2003; coll. M.A. Ribera	AY651048
Gracilaria-SP03	Ebro Delta, Spain; 29/05/2003; coll. M.A. Ribera	AY651059
Gracilaria-TA01	Taranto, Mar Piccolo, Italy; 14/10/2003; coll. E. Cecere	AY651051
Gracilaria-TA02	Taranto, Mar Piccolo, Italy; 06/06/2002; coll. E. Cecere	AY651052
Gracilaria-TA03	Taranto, Mar Piccolo, Italy; 15/02/2002; coll. E. Cecere	AY651034
Gracilaria-TA04	Taranto, Mar Piccolo, Italy; 15/02/2002; coll. E. Cecere	AY651035*
Gracilaria-TA05	Taranto, Mar Piccolo, Italy; 15/02/2002; coll. E. Cecere	AY651036*
Outgroup taxa:		
Grateloupia doryphora (Montagne) Howe	Playa de San Francisco, Bahia de Ancon, Peru; coll. P. Carbájal, 15.ix.01	AF488817
Plocamium cartilagineum	Pigeon Point, San Mateo Co., CA, USA	U04211
Rhodymenia pseudopalmata (J.V. Lamouroux) Silva	Port Aransas Jetty, TX, USA; coll. C.F. Gurgel 17.v.98	AY168656

On a basal level three main phylogenetic lineages were evident, namely the *Curdiea-Melanthalia* clade, the *Gracilariopsis* clade and the *Gracilaria sensu lato* clade (Gurgel & Fredericq, 2004), including *Hydropuntia* and *G. chilensis* group, with variable bootstrap support. In all phylogenetic analyses, *G. longa* and *G. armata* (C. Agardh) J. Agardh were sister taxa with full support, allied within a clade encompassing *G. bursa-pastoris* (S.G. Gmelin) Silva and *G. multipartita* (Clemente) Harvey, strongly supported in distance analyses, but poorly in parsimony analysis (92/91/66 bootstrap percentage values respectively in NJ/ME/MP analyses). The sequence of *Gracilaria*-SP01 did not differ from the sequence from the holotype of *G. longa* and was not included in phylogenetic analyses. All *G. bursa-pastoris* isolates, both from different collection sites in the Mediterranean and from Ireland, showed a moderate sequence divergence (0.00–1.22%). Only the sequence of the sample from S. Maria La Scala (Sicily, Italy) was included in the final alignment.

Gracilaria-SP02, *Gracilaria*-IR01, *Gracilaria*-LF01, *Gracilaria*-LG01, *Gracilaria*-LI01, *Gracilaria*-TA01 clustered within a clade including *G. gracilis* from Wales, the type area (not shown). All specimens from Mediterranean sites showed a low sequence divergence (0.00–0.41%), that increased (up to 1.22%) when North Atlantic samples were included. *Gracilaria*-LI02 clustered basally within a clade including *G. gracilis* and *G. pacifica* I.A. Abbott. Its sequence showed a consistent divergence from *G. gracilis* specimens, ranging from 4.78 to 5.11%, and from *G. pacifica*, 5.59%.

G. 'dura' grouped in all analyses with the clade containing *H. crassissima* (Crouan et Croaun in Mazé et Schramm) M.J. Wynne and related Atlantic *Hydropuntia* species, though with no to low bootstrap support (–/61/64).

In all analyses, *Gracilaria*-EN01, *Gracilaria*-SP03, *Gracilaria*-SL01, *Gracilaria*-IR02, *Gracilaria*-IR03, *Gracilaria*-EN02, *Gracilaria*-TA02, *Gracilaria*-TA03, *Gracilaria*-TA04, *Gracilaria*-TA05 clustered together with a topotype specimen of *Gracilariopsis longissima* and an isolate of *Gracilariopsis* sp. from Australia, with high bootstrap support (85/84/90). Within this clade, all specimens, both from the Mediterranean and north eastern Atlantic, always formed a separate group from the topotype specimen (up to 1.93% sequence divergence) with variable bootstrap support (56/–/97).

Discussion

This study represents the first molecular contribution to the taxonomy of gracilarioid algae from the Mediterranean Sea. The sequence data obtained permitted us to solve the taxonomic position of some populations in the *G. verrucosa* complex, which had proven difficult using a classic morpho-anatomical approach.

Specimens *Gracilaria*-LI01, *Gracilaria*-LF01, *Gracilaria*-LG01 and *Gracilaria*-SP02 grouped with *G. gracilis* from the type area. Even if they appear a more genetically homogeneous group than Atlantic specimens, their sequence divergence from the latter is not large enough (up to 1.22%) to suggest that they are a distinct species. In the Mediterranean Sea, the only report of *G. gracilis*, verified on the basis of molecular investigations, was that for the Gulf of Taranto (Morabito et al., 2003b). Its geographical distribution is now expanded also to Sicily, in Licata and in two brackish lakes in the coasts of the Straits of Messina, and to Spain, at the delta of the river Ebro.

The isolate *Gracilaria*-LI02 from southern Sicily is strongly allied as a sister species within the clade comprising *G. gracilis* and *G. pacifica*. It diverges from both species with a distance ranging from 4.78 to 5.59%. According to Gurgel et al. (2001), this difference is much larger than the intraspecific genetic distance within *Gracilaria sensu stricto* (0.00–1.89%), permitting us to considering it a separate taxon. At present, it does not fit any species reported for the Mediterranean; nevertheless, a critical study on its anatomical and reproductive features is needed before giving it a formal taxonomic status.

Gracilaria-SP01 has an identical sequence of the holotype of *G. longa*, clearly a distinct taxon from *G. gracilis* (Morabito et al., 2003a). This species is actually present in north eastern Spain (Ebro delta) other than in the type locality, Naples.

Gracilaria-SP03, *Gracilaria*-SL01, *Gracilaria*-TA02, *Gracilaria*-TA03, *Gracilaria*-TA04 and *Gracilaria*-TA05, together with some North Atlantic specimens (*Gracilaria*-IR02, *Gracilaria*-IR03, *Gracilaria*-EN 01, *Gracilaria*-EN02), clustered separately from the topotype of *Gracilariopsis longissima*, within a clade including *Gracilariopsis* sp. from Australia, even if without bootstrap support.

Sequence divergence among the studied specimens and *Gs. longissima* from Wales is very high (up to 1.93 %), and near to the value corresponding to separate species in *Gracilariopsis* (Gurgel et al., 2003b). However, the relationships among specimens currently included in *Gs. longissima* and *Gracilariopsis* sp. from Australia are not supported in the *rbc*L phylogeny. A different more rapidly evolving molecular marker, e.g. the internal transcribed spacer region (Goff et al., 1994), might be useful in clarifying this topic, as previously noted by Gurgel et al. (2003b). At the moment, we prefer to maintain this entity within *Gs. longissima*. Consequently, this species, previously reported for the Venice lagoon (Gurgel & Fredericq, 2004) on the basis of molecular evidences, is also found in Spain, Slovenia and the Gulf of Taranto.

According to molecular phylogenies (Bird et al., 1992; Bellorin et al., 2002; Gurgel & Fredericq, 2004) three lineages are present in the order Gracilariales: the *Curdiea-Melanthalia* clade, the *Gracilariopsis* clade and a *Gracilaria* clade. Only the last two are represented by the Mediterranean specimens included in our study. Within the *Gracilaria* clade, Gurgel and Fredericq (2004) recognized nine distinct evolutionary lineages, encompassing a new genus, based on *G. chilensis*, two subgroups joined in the genus *Hydropuntia*, and the remaining subgroups united in the genus *Gracilaria sensu stricto*. The last is characterized by a *Gracilaria sensu stricto* type cystocarp (Fredericq & Hommersand, 1990) and by spermatangial conceptacles in pits ranging from *verrucosa*-type, present in the most basal subgroups, to *textorii*-type in the most derived (Liao & Hommersand, 2003). Among Mediterranean species analyzed, *G.longa*, *G. armata* and *G. bursa-pastoris* clustered in a clade together with *G. multipartita*, a north eastern Atlantic species, with high bootstrap support in NJ and ME analyses and low support in MP (92/91/66). Within this clade *G. longa* and *G. armata* are sister taxa with full support. It is noteworthy that a species, *G. longa*, with *verrucosa*-type spermatangial conceptacles (Gargiulo et al., 1987), falls within a clade defined by a *textorii*-type configuration (Gargiulo et al., 1992).

The studied specimen of *G. 'dura'* fits the concept considered typical for this species in the Mediterranean Sea; nevertheless, a comparison with samples collected from the type locality would be necessary. This species has *verrucosa*-type spermatangia that never become confluent (Gargiulo et al., 1992), but its sequence is allied with *Hydropuntia* species and clustered basally within the Atlantic *Hydropuntia* subgroup even if with low bootstrap support.

The three genera recognized by Gurgel and Fredericq (2004), the new one based on *G. chilensis*, *Hydropuntia* and *Gracilaria*, received no to low bootstrap support. The first had low bootstrap support in all analyses (54/56/67), while no support was gained by *Hydropuntia*. Within the latter, the Pacific *Hydropuntia* group received very high bootstrap support (94/98/99), while the Atlantic *Hydropuntia* group had full support without the inclusion of *G. 'dura'*, considering it low (-/61/64). At present we do not transfer *G. 'dura'* to the genus *Hydropuntia*, because of the unsolved relationships, both anatomical and molecular, among the taxa included in this complex, as pointed out also by Bellorin et al. (2002; 2004). *Gracilaria sensu stricto* received nearly no bootstrap support (54% in ME analysis), encompassing the six evolutionary subgroups, all with high bootstrap support but the *G. bursa-pastoris* subgroup, which had moderate support (70/69/-).

Conclusions

The comparison of *rbc*L gene sequences proved effective in our investigation to clarify relationships among species within the *G. verrucosa* complex. On the other hand, among very close entities, namely *Gs. longissima* isolates, the resolution provided by this marker is not sufficient to achieve a conclusive taxonomic assessment. A comprehensive study of different populations within this complex with a comparative molecular and morpho-anatomical approach is needed to solve their relationships.

The most basal nodes of the proposed phylogeny are not well supported, suggesting that a more conserved molecular marker is needed to better resolve problems at the generic and intrageneric levels. SSU rRNA gene phylogenies (Bird et al., 1992; Bird et al., 1994; Bellorin et al., 2002) are promising in such a direction, but in these analyses critical taxa, especially those included in the *Hydropuntia* concept from different geographical regions, were poorly represented. In addition, reproductive characters of the studied Mediterranean taxa do not fit completely with the *rbc*L gene based phylogenetic hypothesis (Gurgel & Fredericq, 2004), suggesting that the anatomy of some subgroups should be better characterized (see also Bellorin et al., 2004). Gargiulo et al. (1992) observed that some other Mediterranean species had mixed characters regarding the generic lineages within the Gracilariaceae. Therefore, the addition of such entities might be useful to a global systematic scheme of such a challenging family.

Acknowledgments

The authors wish to thank the many collectors who provided samples for analyses, C. Battelli, J.A. Brodie, E. Cecere, G. Felicini, M.J. Homes, T. Mitchell, C. Perrone, M.A. Ribera, R.J. Wilkes. A special acknowledgement is due to Suzanne Fredericq, who permitted us to use many *rbc*L sequence data from Gracilariaceae before being released on GenBank. We are also grateful to two anonymous referees for their careful revisions. This study was supported by grants from the University of Messina, Italy to G.M. Gargiulo (PRA 2002) and to M. Morabito (PRA-GR 2002).

References

Abbott IA (1985) *Gracilaria*. Part 2: Taxonomic and morphological studies. New species of *Gracilaria* Grev. (Gracilariaceae,

Rhodophyta) from California and Hawaii. In Abbott IA, Norris JN (eds), Taxonomy of economic seaweeds: With reference to some Pacific and Caribbean species. University of California Sea Grant College Program, La Jolla, CA (USA): 115–122.

Bellorin AM, Oliveira MC, Oliveira EC (2002) Phylogeny and systematics of the marine algal family Gracilariaceae (Gracilariales, Rhodophyta) based on small subunit rDNA and ITS sequences of Atlantic and Pacific species. J. Phycol. 38: 551–563.

Bellorin AM, Oliveira MC, Oliveira EC (2004) *Gracilaria vermiculophylla*: A western Pacific species of Gracilariaceae (Rhodophyta) first recorded from the eastern Pacific. Phycol. Res. 52: 69–79.

Bird CJ (1995) A review of recent taxonomic concepts and developments in the Gracilariaceae (Rhodophyta). J. Appl. Phycol. 7: 255–267.

Bird CJ, Rice EL (1990) Recent approaches to the taxonomy of the Gracilariaceae (Gracilariales, Rhodophyta) and the *Gracilaria verrucosa* problem. Hydrobiologia 204/205: 111–118.

Bird CJ, McLachlan J, De Oliveira EC (1986) *Gracilaria chilensis* sp. nov. (Rhodophyta, Gigartinales) from Pacific South America. Can. J. Bot. 64: 2928–2934.

Bird CJ, Rice EL, Murphy CA, Ragan MA (1992) Phylogenetic relationships in the Gracilariales (Rhodophyta) as determined by 18S rDNA sequences. Phycologia 31: 510–522.

Bird CJ, Ragan MA, Critchley AT, Rice EL, Gutell RR (1994) Molecular relationships among the Gracilariaceae (Rhodophyta): further observations on some undetermined species. Europ. J. Phycol. 29: 195–202.

Doyle JJ, Doyle JL (1987) A rapid isolation procedure for small quantities of fresh leaf tissue. Phytochem. Bull. 19: 11–15.

Felsenstein J (1985) Confidence limits in phylogenies: an approach using the bootstrap. Evolution 39: 783–791.

Fredericq S, Hommersand MH (1989a) Comparative morphology and taxonomic status of *Gracilariopsis* (Gracilariales, Rhodophyta). J. Phycol. 25: 228–241.

Fredericq S, Hommersand MH (1989b) Proposal of the Gracilariales ord. nov. (Rhodophyta) based on an analysis of the reproductive development of *Gracilaria verrucosa*. J. Phycol. 25: 213–227.

Fredericq S, Hommersand MH (1990) Diagnoses and key to the genera of the Gracilariaceae (Gracilariales, Rhodophyta). Hydrobiologia 204/205: 173–178.

Freshwater DW, Rueness J (1994). Phylogenetic relationships of some European *Gelidium* (Gelidiales, Rhodophyta) species based on *rbc*L nucleotide sequences analysis. Phycologia 33: 187–194.

Freshwater DW, Fredericq S, Butler BS, Hommersand MH, Chase MW (1994) A gene phylogeny of the red algae (Rhodophyta) based on plastid *rbc*L. PNAS 91: 7281–7285.

Furnari G, Giaccone G, Cormaci M, Alongi G, Serio D (2003) Biodiversità marina delle coste italiane: catalogo del macrofitobenthos. Biol. Mar. Med. 10: 3–482.

Gargiulo GM, De Masi F, Tripodi G (1987) Structure and reproduction of *Gracilaria longa* sp. nov. (Rhodophyta, Gigartinales) from the Mediterranean Sea. Gior. Bot. Ital. 121: 247–257.

Gargiulo GM, De Masi F, Tripodi G (1992) Morphology, reproduction and taxonomy of the Mediterranean species of *Gracilaria* (Gracilariales, Rhodophyta). Phycologia 31: 53–80.

Goff LJ, Moon DA, Coleman AW (1994) Molecular delineation of species and species relationships in the red algal agarophytes *Gracilariopsis* and *Gracilaria* (Gracilariales). J. Phycol. 30: 521–537.

Gurgel CFD, Fredericq S (2004) Systematics of the Gracilariaceae (Gracilariales, Rhodophyta): a critical assessment based on *rbc*L sequence analyses. J. Phycol. 40: 138–159.

Gurgel CFD, Fredericq S, Norris JN (2001) Phylogeny, taxonomy and biogeography of the Gracilariaceae (Gracilariales, Rhodophyta) with emphasis on the northwestern Atlantic. Phycologia 40: 21.

Gurgel CFD, Fredericq S, Norris JN (2003a) *Gracilariopsis silvana* sp. nov., *G. hommersandii* sp. nov., and *G. cata-luziana* sp. nov., three new species of Gracilariaceae (Gracilariales, Rhodophyta) from the Western Atlantic. Hidrobiológica 13: 57–68.

Gurgel CFD, Liao LM, Fredericq S, Hommersand MH (2003b) Systematics of *Gracilariopsis* (Gracilariales, Rhodophyta) based on *rbc*L sequence analyses and morphological evidence. J. Phycol. 39: 154–171.

Harper JT, Saunders GW (2001) Molecular systematics of the Florideophyceae (Rhodophyta) using nuclear large and small subunit rDNA sequence data. J. Phycol. 37: 1073–1082.

Imbs AB, Vologodskaya AV, Nevshupova NV, Khotimchenko SV, Titlyanov EA (2001) Response of prostaglandin content in the red alga *Gracilaria verrucosa* to season and solar irradiance. Phytochem. 58: 1067–1072.

Irvine L, Steentoft M (1995) Proposal to reject the name *Fucus verrucosus* Huds. (Rhodophyta). Taxon 44: 223–224.

Lanave C, Preparata G, Saccone C, Serio G (1984) A new method for calculating evolutionary substitution rates. J. Mol. Evol. 20: 86–93.

Liao LM, Hommersand MH (2003) A morphological study and taxonomic reassessment or the generitype species in the Gracilariaceae. J. Phycol. 39: 1207–1232.

Mancinelli G, Rossi L (2001) Indirect, size-dependent effects of crustacean mesograzers on the Rhodophyta *Gracilaria verrucosa* (Hudson) Papenfuss: Evidence from a short-term study in the Lesina Lagoon (Italy). Mar. Biol. 138: 1163–1173.

Morabito M, Genovese G, De Masi F, Gargiulo GM (2003a) Molecular differentiation of two species of *Gracilaria* (Gracilariales, Rhodophyta). 51st Winter Meeting of the British Phycological Society, National University of Ireland, Galway (Ireland). The Phycologist 64: 33.

Morabito M, Genovese G, Cecere E, Gargiulo GM (2003b) Gracilariaceae (Rhodophyta) from the Gulf of Taranto (Italy). 3rd European Phycological Congress, Queen's University, Belfast (UK), Programme and book of abstracts: 88.

Posada D, Crandall KA (1998) MODELTEST: testing the model of DNA substitution. Bioinformatics Application Note 14: 1998.

Rath J, Adhikary S (2002) Growth and spectral characteristics of four algae and agar yield of *Gracilaria verrucosa* at different salinity gradients of Chilika lake. Seaweed Research and Utilization 24: 41–46.

Sambrook J, Fritch EF, Maniatis T (1989) Molecular cloning: a laboratory manual. 2nd edition. Cold Spring Harbor Laboratory Press, New York.

Saunders GW (1993) Gel purification of red algal genomic DNA: an inexpensive and rapid method for the isolation of polymerase chain reaction-friendly DNA. J. Phycol. 29: 251–254.

Saunders GW, Kraft GT (1997) A molecular perspective on red algal evolution: focus on the Florideophycidae. In Bhattacharya D.

(ed.), Origins of algae and their plastids. Springer-Verlag Vienna 115–138.

Skiptsova A (2000) The influence of rope cultivation of *Gracilaria verrucosa* on wild community of *Zostera japonica* in the lagoons of southern Primorye, Sea of Japan. Biologiya Morya 26: 40–44.

Steentoft M, Irvine LM, Farnham WF (1995) Two terete species of *Gracilaria* and *Gracilariopsis* (Gracilariales, Rhodophyta) in Britain. Phycologia 34: 113–127.

Swofford DL (2002) PAUP*. Phylogenetic Analysis Using Parsimony (*and Other Methods). Version 4.0b10. Sinauer Associates, Sunderland, Massachusetts.

Wang G., 2002. Isolation and purification of phycoerythrin from red alga *Gracilaria verrucosa* by expanded-bed-adsorption and ion-exchange chromatogaphy. Chromatographia 56: 509–513.

Zhang JF, Xia BM (1985) On *Gracilaria asiatica* sp. nov. and *G. verrucosa* (Huds.) Papenfuss. Oceanol. Limnol. Sinica 16: 175–180.

Journal of Applied Phycology (2006) 18: 505–519
DOI: 10.1007/s10811-006-9056-4

Long term variability in the structure of kelp communities in northern Chile and the 1997–98 ENSO

Julio A. Vásquez[1,2,*], J. M. Alonso Vega[1,2] & Alejandro H. Buschmann[3]
[1]*Departamento de Biología Marina, Facultad de Ciencias del Mar, Universidad Católica del Norte. Casilla 117. Coquimbo. Chile;* [2]*Centro de Estudios Avanzados de Zonas Aridas. CEAZA-Coquimbo Chile;* [3]*I-MAR, Universidad de Los Lagos, Casilla 557, Puerto Montt, Chile*

*Author for correspondence: e-mail: jvasquez@ucn.cl

Key words: *Lessonia*, *Macrocystis*, southern hemisphere, subtidal rocky shore

Abstract

This is the first study on the south eastern Pacific coast of South America which details long term, interannual variability in the structure of subtidal rocky-bottom kelp-dominated communities before, during, and after the El Niño Southern Oscillation (ENSO) event of 1997–1998 in northern Chile (23°S). The temporal patterns of the main components of these ecosystems, which included *Macrocystis integrifolia*, *Lessonia trabeculata*, echinoids and asteroids, were evaluated seasonally between 1996 and 2004. *M. integrifolia* demonstrated high interannual variability in temporal patterns of abundance. The 1997–1998 ENSO did not significantly modify the temporal patterns of *Macrocystis*, although local extinction of *M. integrifolia* beds occurred during negative thermal anomalies in 1999–2000 (La Niña event), facilitating the establishment of urchin dominated "barren grounds". The abundance of *Lessonia trabeculata* showed little temporal variability, and this species dominated the deeper regions of the kelp assemblage (8–13 m depth).

The structure of the kelp communities in the study area is regulated by a trophic cascade which modulates alternation between kelp dominated areas and sea urchin barrens. In this context, frequent and intense upwelling of cold water high in nutrients favors the establishment and persistence of kelp assemblages. During ENSO, coastal upwellings can mitigate superficial warming of coastal water and increase the nutrient concentration in the water column. Superficial warming during the 1997–1998 ENSO induced spawning by different species of echinoderms, which resulted in major recruitment of these species during 1999. Top down events, such as the decrease in densities of the asteroids after the 1997–1998 ENSO event, favored increases in densities of benthic grazers, which caused significant decreases in abundance of *M. integrifolia*. The re-establishment of the adult fraction of the carnivore (starfish) guild coincided with a decrease in the density of sea urchins and thus re-establishment of the kelp. In the temperate south eastern Pacific, oceanographic events, which act on different spatial-temporal scales, trigger trophic cascades that act at local levels, producing interannual variability in the structure of kelp communities. On the other hand, considering the high macroinvertebrate diversity associated with kelp assemblages, the transitions between kelp-dominated areas and sea urchin barrens do not appear to significantly affect the biodiversity of these assemblages of benthic invertebrates.

Introduction

El Niño Southern Oscillation (ENSO) is an irregular fluctuation involving the entire tropical Pacific Ocean and global atmosphere (Fiedler, 2002). ENSO itself consists of an unstable interaction between sea surface temperature (SST) and atmospheric pressure. ENSO produces interannual variability in the oceanographic climate (Dayton et al., 1999), with alternating warm and cold periods resulting from positive (El Niño), and negative (La Niña) thermal anomalies of the SST, in 2 to 7 year feedback cycles (Fiedler,

2002). Differences in frequency, intensity, and magnitude of ENSO events have been associated with ocean regime-shifts caused by the Pacific Decadal Oscillation (PDO) and global warming (Steneck et al., 2002). During high-intensity, high-magnitude ENSO events (eg. 1982–1983 El Niño, 1997–1998 El Niño), Kelvin waves are propagated from the tropics both northward (North America) and southward (South America) along the eastern Pacific coastline. Although their manifestation decreases with increase in latitude, they may be detected beyond 35° (Halpin et al., 2004). The advance of these waves impinging on the coastline lowers the thermocline, increases sea level, modifies the direction and velocities of currents, and decreases or prevents coastal upwelling (Takesue et al., 2004). Changes in the oceanographic climate caused by high-intensity ENSO events have an important role as a disturbing process at temperate latitudes along the eastern Pacific coastline, producing bathymetric migrations of organisms, invasions of exotic species, behavioral alterations, and positive or negative changes in abundance, the latter of which may reduce population densities to local extinction (see Tegner & Dayton, 1987; Glynn, 1988; Dayton et al., 1999). Modifications of the coastal biota may be observed on both local and regional geographic scales (Camus, 2001; Edwards, 2004). Reductions in populations or local extinction processes generated by ENSO events are very important for "engineer species" in ecosystems (*sensu* Jones et al., 1994) such as kelp. The presence of these species determines the diversity, complexity, structure, and functioning of their associated communities (Graham, 2004).

Long-term studies on the North America west coast have shown that ENSO events alter the structure and organization of subtidal kelp communities in temperate latitudes, modifying patterns of persistence, stability, succession, species diversity, and abundance (Dayton et al., 1992, 1999; Tegner et al., 1997). Moreover, ENSO events have been considered as large-scale disturbances, which produce phase shifts between, kelp-dominated to sea urchin-dominated states (Tegner & Dayton, 1991; Steneck et al., 2002). In kelp forests, population changes in top predators commonly drive these shifts through top-down forcing processes (Estes et al., 2004). However, in California kelp forests, factors connected with anthropogenic impacts (see Tegner & Dayton, 1991; Dayton et al., 1998), may have buffered the phase shift to sea urchin-dominated states and facilitated recovery from ENSO disturbances (Steneck et al., 2002). In contrast, most studies of subtidal kelp communities in the Southern Hemisphere are short-term (one–two years), or are limited to high latitudes (≥40°S) where the influence of ENSO is minimal (Halpin et al., 2004). As such, there are no long-term data concerning the effects of large-scale, low frequency ENSO events on the structure of South American kelp communities.

In northern Chile and southern Peru (10°–30°S), protected and semi-exposed shallow subtidal hard-bottom environments (*ca.* 20 m depth) are dominated by two kelp species from the Order Laminariales, including *Lessonia trabeculata* Villouta & Santelices and *Macrocystis integrifolia* Bory. Although there are some reports in the literature on the ecology of *Lessonia trabeculata* (see Vásquez, 1992; Tala et al., 2004), data are scarce on the population biology of *M. integrifolia* and the *Macrocystis-Lessonia* assemblage in northern Chile. Available information is restricted only to standing stock evaluations and observations on reproductive activity in controlled environments and in the field (Buschmann et al., 2004; Vega et al., 2004).

The subtidal kelp ecosystems on the South American west coast are highly productive, hosting diverse and abundant macroinvertebrates and fishes (Vasquez et al., 2001a). *M. integrifolia* and *L. trabeculata*, are highly sensitive to positive SST anomalies and low nutrient concentrations on the coast caused by ENSO events, and experienced high mortalities during the 1982–83 (Tomicic, 1985; Soto, 1985; Glynn, 1988) and 1997–98 (Godoy, 2000; Llellish et al., 2001) ENSO events. Without kelp assemblages, subtidal rocky reefs form alternative states, the most common of which is the "barren-ground" associated with sea urchins (Vásquez, 1992). The most important grazers in such systems are the sea urchins *Tetrapygus niger* (Molina) and the sympatric but less common species *Loxechinus albus* (Molina) (Rodriguez & Ojeda, 1993). *T. niger* is an omnivore, while *L. albus* is an herbivore and feeds on foliose algae and drifting algal rafts (Contreras & Castilla, 1987). Although both species can completely destroy kelp beds on a local scale (see Dayton, 1985; Buschmann et al., 2003), *T. niger* is primarily responsible for generation and maintenance of the barren grounds typically observed in northern Chile (Vásquez & Buschmann, 1997).

A guild of carnivores (starfish), regulates spatial and temporal patterns of abundance and diversity of the benthic grazers (Viviani, 1978; Vásquez & Buschmann, 1997). The fishes associated with the kelp communities have a broad trophic spectrum and eat few sea urchins (Medina et al., 2004). In northern Chile ENSO produces trophic cascades by top-down processes (starfish↔sea urchins↔kelp), which modify the patterns of biodiversity, stability and persistence of subtidal kelp communities (Vásquez & Vega, 2004). In this context long-term monitoring has permitted postulation that the 1997–1998 ENSO event generated: (1) interannual variability in the abundance of the main functional groups associated with the kelp, (2) differential responses of the species or functional groups, and (3) changes in the structure and organization of the kelp communities.

The present study evaluates long term patterns of abundance in key species which regulate subtidal communities on rocky bottoms dominated by kelp in northern Chile, and incorporates the effects of the 1997–1998 ENSO event. Data were obtained from before, during, and after this ENSO event. Data are presented on the effects of different groups of key benthic species on the structure and organization of these subtidal kelp communities, including the kelp species, grazers (sea urchins), predators (sea stars), and macroinvertebrate species forming the more common assemblages in this region.

Materials and methods

Study area

Shallow, subtidal, rocky-bottom communities dominated by kelp were evaluated seasonally between July 1996 and August 2004 at Caleta Constitución (23°26′S, 70°36′W). This bay, on the southern end of the Mejillones Peninsula in northern Chile (Figure 1) is in a region of permanent upwelling (Takesue et al., 2004), semi-protected from prevailing winds by Santa Maria Island. The bottom substrate consists of rocky platforms, which drop to depth, plus scattered boulders separated by channels floored with coarse sand. Kelp beds are widely distributed throughout the bay; the sub-canopy undergrowth comprises various foliose, turf, and crustose macroalgae. These include crustose Corallinales, turfs of Gelidiales and/or Ceramiales, and often patches of *Halopteris spp.*, *Glossophora kunthii* (C. Ag.) J. Ag., *Asparagopsis armata* Harley and *Rhodymenia spp.* and *Chondrus canaliculatus* (C. Ag.) Grev. (Vásquez et al., 2001b). Descriptions of the study site and marine ecosystem are given by Vásquez et al. (1998).

Oceanographic conditions

Mean daily *in situ* water temperatures were measured on continuous-register thermographs (Onset Computer Corp., MA, USA) placed at 3 m depths along the shallow limits of the kelp. When *in situ* records of oceanographic variables were discontinued, large-scale climatic indexes were used, which permitted description of oceanographic conditions, and for making approximations of ecological processes that acted on smaller scales (Stenseth et al., 2003). Warm and cool phases of the ENSO were determined using monthly averages of the Southern Oscillation Index (SOI) and the Multivariate El Niño Index (MEI) for the period (1996–2004), from the Bureau of Metereology, Australia (www.bom.gov.au/climate/current/) and Climate Diagnostic Center of NOAA (www.cdc.noaa.gov/~kew/MEI/mei.html), respectively. Information on the temporal variability of upwelling events in the region (23–25°S) was obtained from the monthly average index of upwelling (Offshore Eckman Transport, OET) between 1996 and 2001, from the Pacific Environmental Laboratory (PFEL, (www.pefg.noaa.gov/products/PFELindices.html). A detailed description of this calculation, and characteristics of the area of influence of the SOI, MEI and OET indexes is presented by Navarrete et al. (2002).

Temporal patterns of kelp abundance

Temporal patterns of abundance of *M. integrifolia* and *L. trabeculata* were evaluated seasonally on four haphazardly chosen transects which were established perpendicular to the coastline from the intertidal to 15 m depth. Each transect was 160 m long and 1 m wide. Two SCUBA divers swam each transect, counting juvenile and adult sporophytes within 0.5 m on each side of it's axis. Kelp juveniles were sporophytes with up to two lanceolate and laminar fronds, without reproductive structures, and with maximum holdfast diameters of $\leq$1 cm. Densities of juveniles and adults were expressed as the number of plants per m^2 ($N = 4$).

Temporal patterns of grazer (sea urchins) abundance

Temporal changes in the density of grazers were determined by seasonal evaluation of 34 steel-frame quadrats of 0.25 m^2 each, haphazardly tossed from a boat between the perpendicular transects described above. Densities of the sea urchins were expressed as the number of individuals per 0.25 m^2.

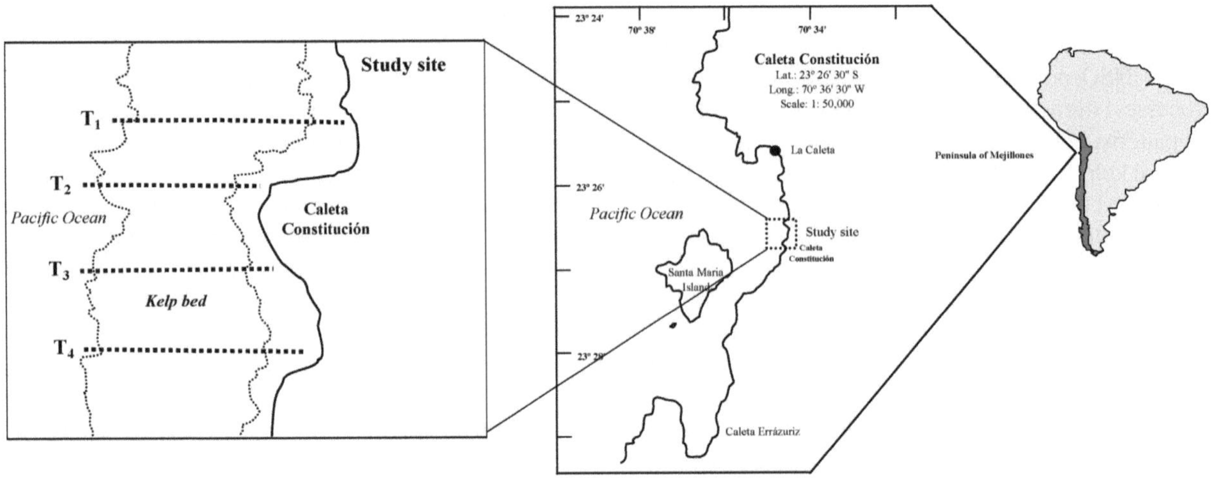

Figure 1. Geographic location of the study area, showing sampling sites and positions of transects.

Temporal patterns of carnivore (sea stars) abundance

To evaluate the temporal distribution of the sea star carnivorous guild, we used the same methods and sampling units as for the kelp assemblage. The number of asteroids per species was counted for each sampling unit ($N = 4$, transect of 160 m^2), with individual densities expressed per m^2.

Temporal patterns of macroinvertebrate assemblages in subtidal kelp beds

Changes in hard-bottom community structure before (1996), during (1997–98), and after (1999–2000) the 1997–1998 ENSO were evaluated seasonally by means of analysis of benthic macroinvertebrate assemblages associated with subtidal kelp beds. The composition and abundance of the benthic macroinvertebrates species was evaluated using destructive sampling with 0.25 m^2 quadrats. Twenty steel-frame quadrats were thrown haphazardly from a boat over depths of between two and 15 m, covering the entire range of the kelp beds (2–15 m depth). Divers, using numbered, 1-mm mesh collecting bags, recovered all the fauna occurring within the quadrats. Collected material was transferred to numbered plastic bags onshore, fixed in 8% formalin dissolved in seawater and later preserved in 70% alcohol. In the laboratory, the invertebrates were sorted and identified to the lowest taxonomic level possible using literature listed by Lancelloti and Vásquez (1999, 2000). The number of individuals of each macroinvertebrate species was counted in each sample unit. Temporal patterns of invertebrate assemblages associated with subtidal kelp beds was analyzed using univariate biodiversity indexes (species richness (S')) and biodiversity index of Shannon Weiner [H', J']).

Statistical analyses

A multifactorial analysis of variance (ANOVA) using the species, years and seasons as the main variables was used to evaluate the hypothesis that ENSO events generated long-term variability in patterns of abundance of key species (kelp, sea urchins and starfish), which structured subtidal kelp communities at the study site. The multifactorial analysis of variance (ANOVA) was done after visual determination of normality of the data and homoscedasticity of variances by means of a Bartlett test (Sokal & Rohlf, 1981), using SYSTAT 8.0® computational software for Windows; transformations (root abundance + 1) were applied when necessary to improve homoscedasticity (Sokal & Rohlf, 1981). An *a posteriori* Tukey test was used in order to determine which groups differed from others (Sokal & Rohlf, 1981). The relationship between mean abundance of kelp, sea urchins and starfish was determined using a Pearson Correlation Analysis (Sokal & Rohlf, 1981).

Results

Oceanographic conditions

In situ sea temperature showed a seasonal pattern, with warm water between December and March (summer)

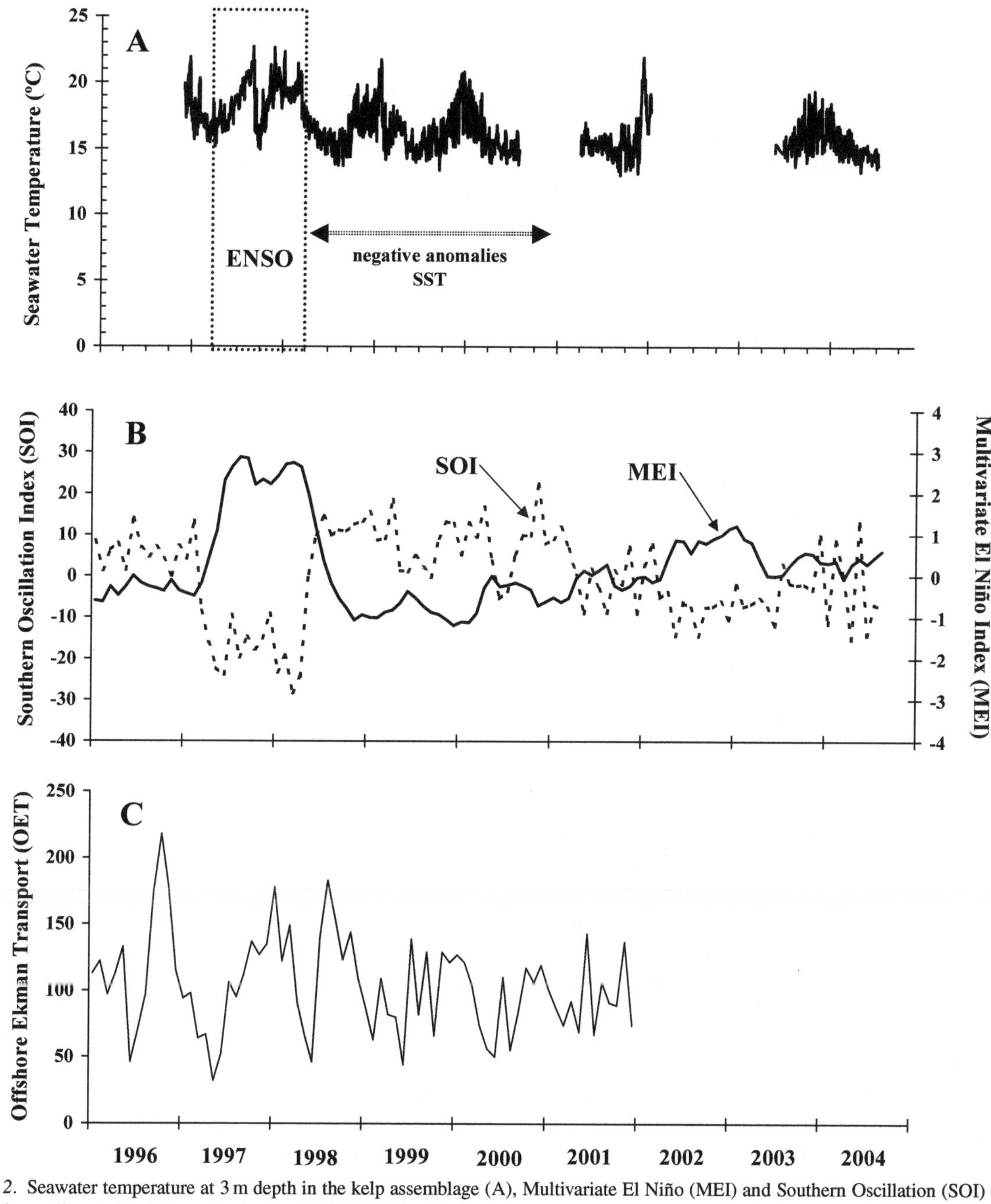

Figure 2. Seawater temperature at 3 m depth in the kelp assemblage (A), Multivariate El Niño (MEI) and Southern Oscillation (SOI) indexes (B), and upwelling index (Eckman transport, OET) (C) during the study period.

and cool water between June and September (winter) (Figure 2A). Between April 1997 and March 1998 the water was unusually warm, with maximum positive thermal anomalies fluctuating between +2° and +2.5°C; an exception occurred between August and November 1997 when upwelling lowered the seawater temperatures, interrupting the continuity of the anomalous warm period. Beginning in April 1998, cooling of the water began with weak, moderate, and strong levels when the anomaly ranged between −0.5° and −1.5 °C until the end of 2000 (La Niña event; Figure 2A). The Southern Oscillation Index (SOI, Figure 2B) and the Multivariate El Niño Index (MEI, Figure 2B) detected normal conditions in 1996 lasting until summer 1997. An ENSO event was recorded between May 1997 and March 1998, which was of high intensity and magnitude, coinciding with the thermal anomaly detected by the *in situ* temperature records (Figure 2A). Following the normal-cold period of 1998–2001, a new ENSO manifestation was detected which was of low to moderate intensity between April 2002 and April 2003 (Figure 2B), with a positive thermal anomaly of +1 °C. The mean values for the upwelling index (OET) were always positive during the study period, and represented continuity over time of the Ekman transport in the region (Figure 2C). The upwelling index showed a greater offshore transport between September and December (spring), and lower intensities of upwelling between April and July of each year (Figure 2C). The highest upwelling activity occurred during the spring of 1996, decreasing significantly in May 1997 at the beginning of the 1997–1998 ENSO event. Nevertheless, the Ekman transport remained active, constant, and intense between July 1997 and February 1998 (Figure 2C) during the maximum positive thermal anomalies of the 1997–1998 ENSO.

Temporal patterns of kelp abundance

The temporal patterns of abundance of *M. integrifolia* differed significantly from those of *L. trabeculata* (Table 1A). *M. integrifolia* showed marked annual changes, with maximum abundances of adult sporophytes during 1997–1998 (1997–1998 ENSO period) and minima during 2000–2002 (period of negative SST anomalies), reaching critical levels of abundance (0.1 to 0.6 sporophytes/m^2) in 2000 (Figures 3A–B). In the fall of 2001, the population of *M. integrifolia* started to become re-established, reaching its maximum density during 2003 (Figure 3A). In contrast, the average abundance of *L. trabeculata* during the study period was 0.5 ± 0.9 sporophytes/m^2, with stochastic changes (Figures 3 C–D). The temporal patterns of juveniles of *M. integrifolia* were significantly different from those of *L. trabeculata* (Table 1B, Figures 3 B–D). The abundances of juveniles of *M. integrifolia* increased during the 1997–1998 ENSO, and during the re-establishment of kelp bed from 2001 to 2003 (Figure 3C). There was, however, a decrease due to a failure in recruitment in 1999–2000, which helped cause the disappearance of the *M. integrifolia* bed during the negative SST anomalies (La Niña event). In contrast, the abundance of *L. trabeculata* juveniles increased mainly in the spring (Figure 3D), even though juveniles of this species can be present throughout the entire year (i.e. 2003; Figure 3D).

Table 1. Multifactorial Analysis of Variance (ANOVA) using species, year and season as main variables to evaluate the hypothesis that ENSO event generate long-term variability in abundance of key species (kelp, sea urchins and starfish).

Factor	df	MS	F	*p*-level
(A) Kelp adults				
Species	1	0.181	4.797	0.116
Year	6	1.408	8.821	0.001
Season	3	0.014	5.542	0.001
Species vs Year	6	1.185	10.685	0.001
Species vs Season	3	0.038	14.486	0.001
Year vs Season	18	0.160	61.231	0.001
Species vs Year vs Season	18	0.111	42.539	0.001
Residuals	168	0.003		
(B) Kelp juveniles				
Species	1	4.432	4.777	0.117
Year	6	3.554	2.196	0.092
Season	3	0.562	17.701	0.001
Species vs Year	6	3.932	2.144	0.098
Species vs Season	3	0.927	29.231	0.001
Year vs Season	18	1.618	50.970	0.001
Species vs Year vs Season	18	1.833	57.730	0.001
Residuals	168	0.032		
(C) Sea Urchins				
Species	1	363.45	274.92	0.001
Year	6	36.66	23.39	0.001
Season	3	3.61	3.49	0.015
Species vs Year	6	29.18	12.63	0.001
Species vs Season	3	1.32	1.28	0.280
Year vs Season	18	1.57	1.52	0.075
Species vs Year vs Season	18	2.31	2.23	0.002
Residuals	1848	1.03		
(D) Sea Star				
Species	3	0.1577	16.792	0.001
Year	6	0.0746	3.535	0.017
Season	3	0.0267	137.647	0.001
Species vs Year	18	0.0222	3.075	0.001
Species vs Season	9	0.0094	48.510	0.001
Year vs Season	18	0.0211	109.032	0.001
Species vs Year vs Season	54	0.0072	37.273	0.001
Residuals	336	0.0002		

Significant differences $p < 0.05$.

Temporal patterns of grazer (sea urchin) abundance

The black sea urchin (*T. niger*) was the most conspicuous herbivore at Caleta Constitución, coexisting with the significantly less abundant sea urchin *L. albus* (Figure 4). The benthic grazer abundances varied significantly between years (Table 1C), showing

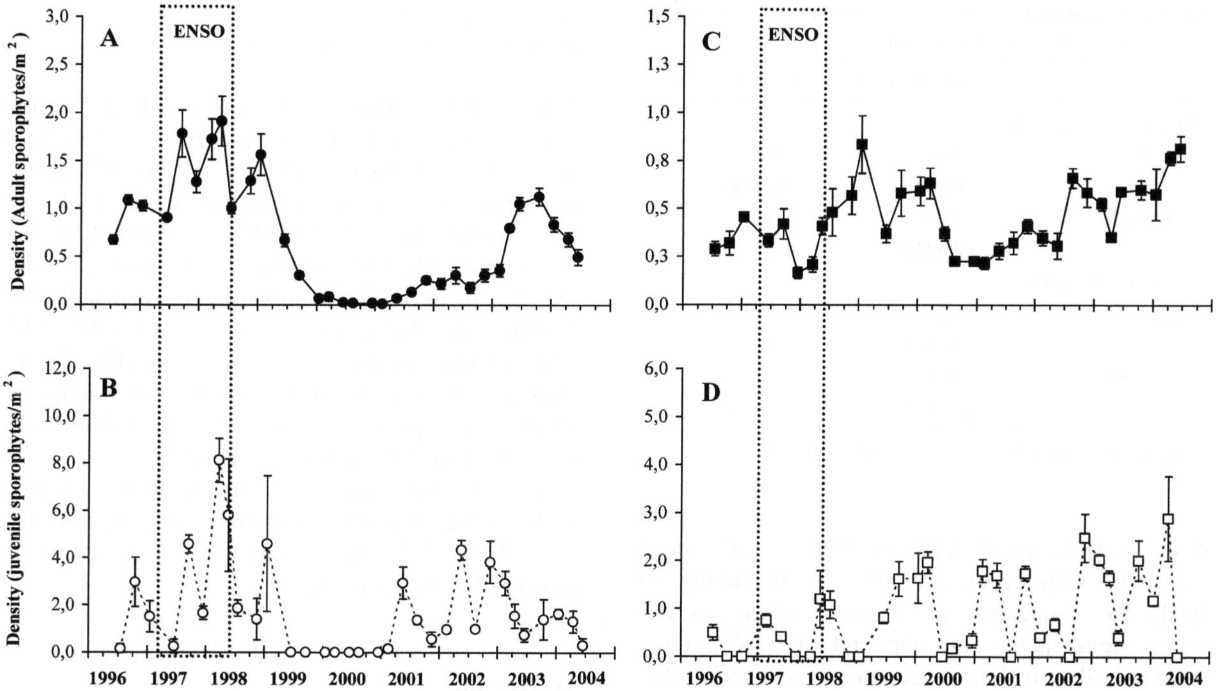

Figure 3. Long-term variability of kelp densities (1996–2004): Density of adult (A) and juvenile (B) sporophytes of *Macrocystis integrifolia*, and adults (C) and juveniles (D) of *Lessonia trabeculata* in subtidal habitats at Caleta Constitución, Antofagasta, Chile.

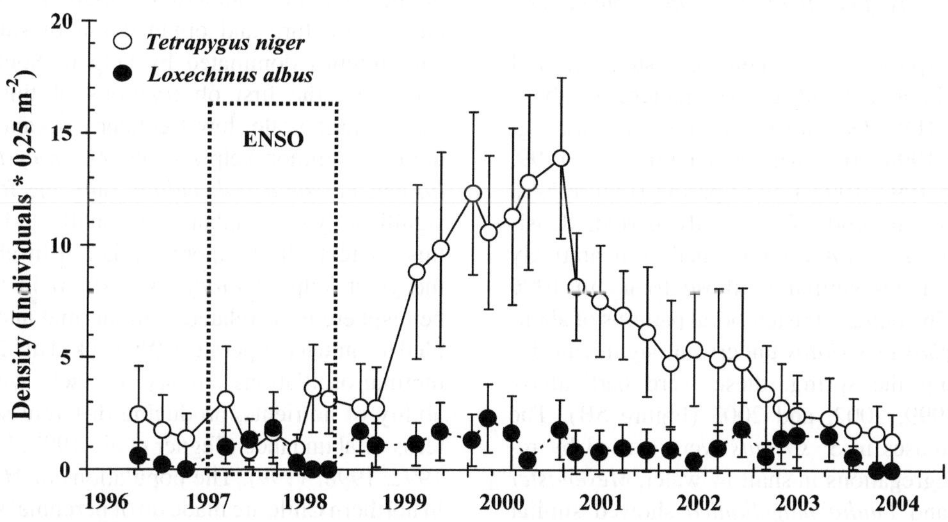

Figure 4. Long-term variability of grazer densities (1996–2004): *Tetrapygus niger* and *Loxechinus albus* in rocky subtidal habitats at Caleta Constitución, Antofagasta, Chile.

three different population levels throughout the study period. Sea urchins were less abundant between 1996 and 1998, including the 1997–1998 ENSO (Figure 4). During periods of negative SST anomalies (1999–2000), the mean density of grazers increased, tripling its mean density between 1996–1998 (Figure 4). This change in temporal patterns of abundance of *T. niger* coincided with the local extinction of the *M. integrifolia* population. An inverse and significant correlation suggested that the density of juvenile and adult *M. integrifolia* sporophytes decreased with increasing numbers of *T. niger* (Table 2). In contrast, there were no significant correlations between *T.*

Table 2. Pearson correlation coefficient (probability in parentesis) between sea urchins and kelp abundance.

	Tetrapygus niger	*Loxechinus albus*
Macrocystis integrifolia		
Adults	−0,67	−0,17
	(0,0001)	(0,3631)
Juveniles	−0,51	0.06
	(0,0036)	(0,7374)
Lessonia trabeculata		
Adults	−0,22	0,30
	(0,2343)	(0,1075)
Juveniles	−0,36	0,25
	(0,0507)	(0,1868)

Boldface indicates significant association, alpha = 0.05.

niger and *L. trabeculata* or between *L. albus* and both kelp species (Table 2). Beginning in 2001, the abundance of *T. niger* began to decrease significantly until the end of 2003, giving values similar to those encountered between 1996–1998 (Figure 4).

Temporal patterns of carnivore (sea star) abundance

The asteroid species of this benthic system differed significantly in annual and seasonal patterns of abundance (Table 1D). *Heliaster helianthus* increased significantly in 1998–2000 with a maximum in 1999, including the 1997–1998 ENSO period (Figure 5A). From 2001 to the end of the study period, densities of *Heliaster helianthus* remained comparatively low, with averages similar to those found in 1996 (Figure 5A). In contrast, the temporal patterns of abundance of *Stichaster striatus* underwent significant increases during the spring: these were particularly notable in 1999, 2002, and 2003 (Figure 5B). The seasonal increases in *S. striatus* were caused by reproductive aggregations in shallow water. *Meyenaster gelatinosus* and *Luidia magellanica* showed similar tendencies in their temporal patterns (Figures 5C y D): both decreased significantly in 1998, immediately after the 1997–1998 ENSO, and re-established their densities during the cool period of 1999–2000. There were positive and significant correlations among the different species of asteroids as well as among the asteroids and echinoids (Table 3), suggesting common populational responses to interannual variations in the oceanographic climate within this subtidal ecosystem.

Temporal patterns of benthic macroinvertebrate assemblages

The temporal variation in species richness of the macroinvertebrate assemblage associated with the kelp assemblages fluctuated between 50 and 80 species during the study period, without detection of relevant breaks during the 1997–1998 ENSO (Figure 6A). The biodiversity (H') and uniformity (J') indexes only detected a break in the temporal patterns in fall 1998 (Figure 6B), during the decline of the 1997–1998 ENSO event. There was a decrease in the diversity indices as a consequence of the numerical dominance of filter-feeding species (i.e. tunicates, mussels) that cover rocky bottom. The total macroinvertebrate density, in contrast to the other community variables, showed high temporal variability before (1996), and after the 1997–1998 ENSO, in contrast with the low densities during ENSO (Figure 6C).

Discussion

The present study demonstrates some of the effects of the El Niño Southern Oscillation (ENSO) event on the structure and organization of subtidal rocky communities dominated by kelp in South America. These are the first observations of this type on a such a large-scale, low frequency oceanographic. Of the two common kelp species, *Macrocystis integrifolia* and *Lessonia trabeculata*, only the former varies significantly in abundance seasonally and annually. In this context, the temporal variation in abundance of the giant kelp M*acrocystis pyrifera* in the northern hemisphere is correlated with thermal anomalies coupled to annual temperature (Steneck et al., 2002). These thermal oscillations co-vary inversely with the availability of nutrients, producing different seasonal patterns of abundance (Tegner et al., 1997; Dayton et al., 1992, 1998, 1999). The populations of *M. integrifolia* in northern Chile are made up of perennial sporophytes, which maintain average abundances throughout the year, with seasonal variability only in growth and reproduction (Buschmann et al., 2004). As in populations of *M. pyrifera* in California and Mexico (Ladah et al., 1999; Edwards, 2004), it may be predicted that the temporal stability of South American populations of *M. integrifolia* could be interrupted by (1) positive thermal anomalies generated by ENSO events which produce local mortalities with highest intensity at the lower latitudes, and (2) the rate of post-ENSO

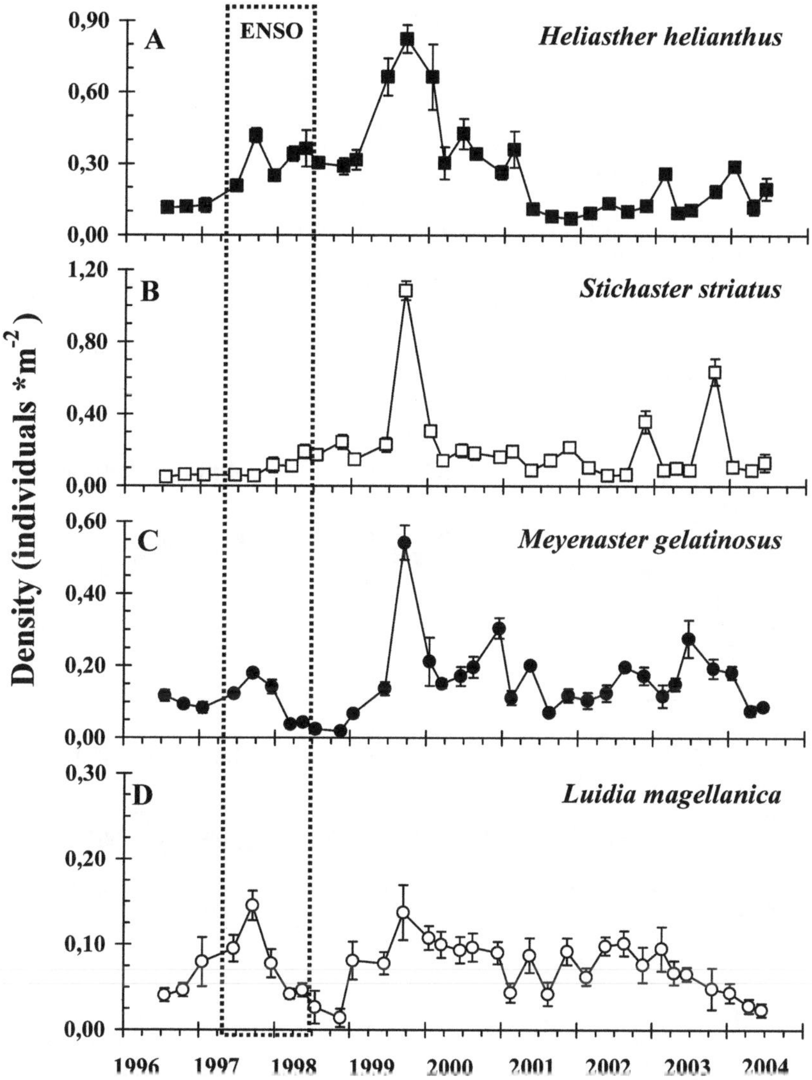

Figure 5. Long term variability of sea star densities (1996–2004): (A) *Heliasther helianthus*, (B) *Stichaster striatus*, (C) *Meyenaster gelatinosus* and (D) *Luidia magellanica* at Caleta Constitución, Antofagasta, Chile.

recovery, that may depend on the intensity of negative thermal anomalies (La Niña). Both factors would be expected to generate interannual variability that have not always been taken into account in understanding the functioning of these communities. Our observations of the structure and organization of the kelp assemblages in northern Chile, made during the study period which included the 1997–1998 ENSO event, nevertheless were an exception to the above two possibilities, since the abundance of *M. integrifolia;* (1) increased significantly during the 1997–1998 ENSO event, (2) decreased during the 1999–2001 La Niña event to levels near zero in 2000, and (3) became re-established during a period with a positive thermal anomaly in 2002–2003. In this context, a few fertile sporophytes survived the local disappearance of *M. integrifolia*, generating reproductive propagules for the re-establishment of the population (Vega et al., 2004). Also, drifting rafts and seed banks of microscopic dormant stages (gametophytes) may be included in possible complementary strategies for repopulation of this kelp species (Ladah et al., 1999; Buschmann et al., 2004; Vega et al., 2004).

Populations of *L. trabeculata* in northern Chile are made up of perennial and long-lived sporophytes (Vásquez, 1992; Tala et al., 2004), partially

Table 3. Pearson correlation coefficient (probability in parentesis) between Echinoids and Aesteroids abundance.

	Echinoids		Asteroids		
Asteroideos	*T. niger*	*L. albus*	*H. helianthus*	*S. striatus*	*M. gelatinosus*
Heliasther helianthus	0,45 **(0,0096)**	0,23 (0,2040)	–	–	–
Stichaster striatus	0,30 **(0,0992)**	0,37 **(0,0387)**	0,57 **(0,0006)**	–	–
Megenasster gelatinosus	0,47 **(0,0063)**	0,49 **(0,0043)**	0,43 **(0,0148)**	0,65 **(0,0001)**	–
Luidia magellanica	0,42 **(0,0163)**	0,38 **(0,0317)**	0,38 **(0,0320)**	0,23 (0,2124)	0,62 **(0,0002)**

Boldface indicates significant association, alpha = 0.05.

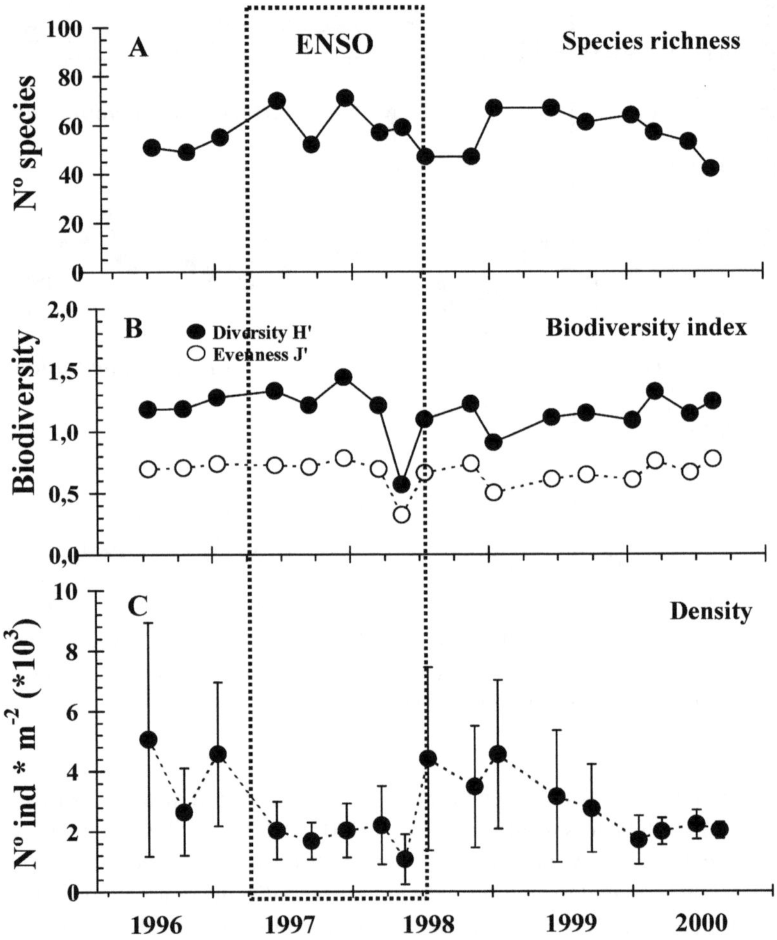

Figure 6. Long term variability of biodiversity indexes (1996–2000) in subtidal kelp communities at Caleta Constitución, Antofagasta, Chile: (A) Species richness, (B) Diversity (H'), Evenness (J'), and (C) Total density of macroinvertebrates.

explaining the temporal patterns of bathymetric distribution of *L. trabeculata* between 1996 and 2004. On the Peruvian coast, the mortality rates of *L. trabeculata* sporophytes during the 1997–1998 ENSO were inversely correlated with depth, with highest survival between 12 and 15 depths (Fernández et al., 1999). *M. integrifolia* dominated rocky bottoms at 5–8 m in the area of the present study, while *L. trabeculata* was dominant at greater depths (8–13 m; Vega et al., 2004).

The temporal pattern of abundances of juveniles differed between the two kelp species over long-term periods. Whereas *M. integrifolia* recruits throughout the year (as in other wave-protected environments; see Graham et al., 1997), *L. trabeculata* recruits during the winter months, thus producing a greater abundance of juveniles during the spring. These differences in recruitment patterns (annual *vs* seasonal) suppose different reproductive strategies that may in part explain the temporal dynamics and longevity of the assemblage. These hypotheses need to be studied, experimentally, in the future.

Although the 1997–1998 ENSO event was a catastrophic occurrence which produced local kelp extinctions at low latitudes on the coasts of both Chile and Peru in the Southern Hemisphere (Fernández, et al., 1999; Godoy, 2000; Llellish et al., 2001; Martínez et al., 2003), and in California and Mexico in the Northern Hemisphere (Ladah et al., 1999; Edwards, 2004), local conditions permitted persistence of the kelp assemblages (Martínez et al., 2003; Vega et al., 2004). Here, the maintenance of temporal patterns of *M. integrifolia* and *L. trabeculata* during the 1997–1998 ENSO event in northern Chile, may be explained by the frequency and intensity of coastal upwelling (Lagos et al., 2002), which minimized the warming effects at the SST, maintaining high concentrations of nutrients within coastal environments (Takesue et al., 2004).

On the California coast, the recovery of *Macrocystis* post 1997–1998 ENSO was favored by the rapid establishment of a cold period (1998–2000 La Niña; Edwards, 2004) and the survival of sporophytes in deep environments (Ladah et al., 1999). In the Southern Hemisphere the re-colonization rate of the kelp assemblages occurred comparatively slowly (Martínez et al., 2003), although cool conditions of 1998–2000 added to the effects of upwelling. The abundance of *M. integrifolia* in the study area was modified by a significant reduction in the adult population and lack of recruitment of juvenile sporophytes. Thus, the disappearance of the *M. integrifolia* population occurred two years after the 1997–1998 ENSO event, and was significantly, inversely correlated with the increase in the population of the black sea urchin *Tetrapygus niger*. This contrasts with information from other areas of the south eastern Pacific during the 1997–1998 ENSO event, where superficial warming decreased the abundance of kelp on shallow bottoms, inducing migrations of grazers to deeper zones in search of food (Vásquez & Buschmann, 1997; Fernández et al., 1999; Godoy, 2000; Llellish et al., 2001). This type of migratory behavior of benthic grazers such as sea urchins and gastropods on hard bottoms produces communities dominated by crustose calcareous algae ("barren ground", *sensu* Lawrence, 1975). In the Northern Hemisphere, events that impact the abundances of high-level predators, and low levels of availability of drift algae promote the formation of barren ground; this occurrence is not necessarily linked to ENSO (Tegner & Dayton, 1991; Steneck et al., 2002; Estes et al., 2004). The urchin-crustose algae association persists until the sea urchin population is decimated by disease, migration, or predator pressure, which act together to promote re-establishment of the kelp (Dayton et al., 1992; Estes et al., 2004; Graham, 2004).

It has been noted that areas with intense and permanent offshore transport, such as the study area on the Mejillones Peninsula, are typified by high survival, retention, and transport of echinoderm larvae toward the coast (Ebert & Russell, 1988). During the ENSO cycle (1997–1998 El Niño and 1998–2000 La Niña), different events favored an increase in the urchin population during the cool phase, including (1) induction of spawning due to increases in SST and persistence of upwelling events, (2) significant reductions in densities of adult individuals of *M. gelatinosus* and *L. magellanica* (Vásquez et al., 1998; Vásquez & Vega, 2004) and (3) changes in the feeding behavior of *H. helianthus* (Tokeshi & Romero, 1995; Vásquez et al., 1998).

In the absence of other large predators on the south eastern Pacific coast, the sea stars form a predatory guild, which significantly lower the abundance of ben thic herbivores such as the sea urchins and gastropods (Vasquez & Buschmann, 1997). Although fishes such as *Graus nigra* and *Semicosyphus maculatus* may prey upon juvenile urchins, the sea urchins do not exceed 17% of their diets (Medina et al., 2004). The re-establishment of adult densities of *Meyenaster gelatinosus*, *Stichaster striatus*, *Heliater helianthus* and *Luidia magellanica* was associated with the temporal recovery of kelp assemblages in the study area. *Meyenaster* and *Luidia* also prey upon *H. helianthus and S. striatus* (Dayton et al., 1977; Viviani, 1978; Tokeshi & Romero, 1995). There is spatial segregation on a bathymetric gradient between the different species of sea stars (Vásquez & Vega, 2004). Encounters between these high level predators often result in autotomy of one or more of their rays (Lawrence & Vásquez, 1996). Here, sublethal predation between

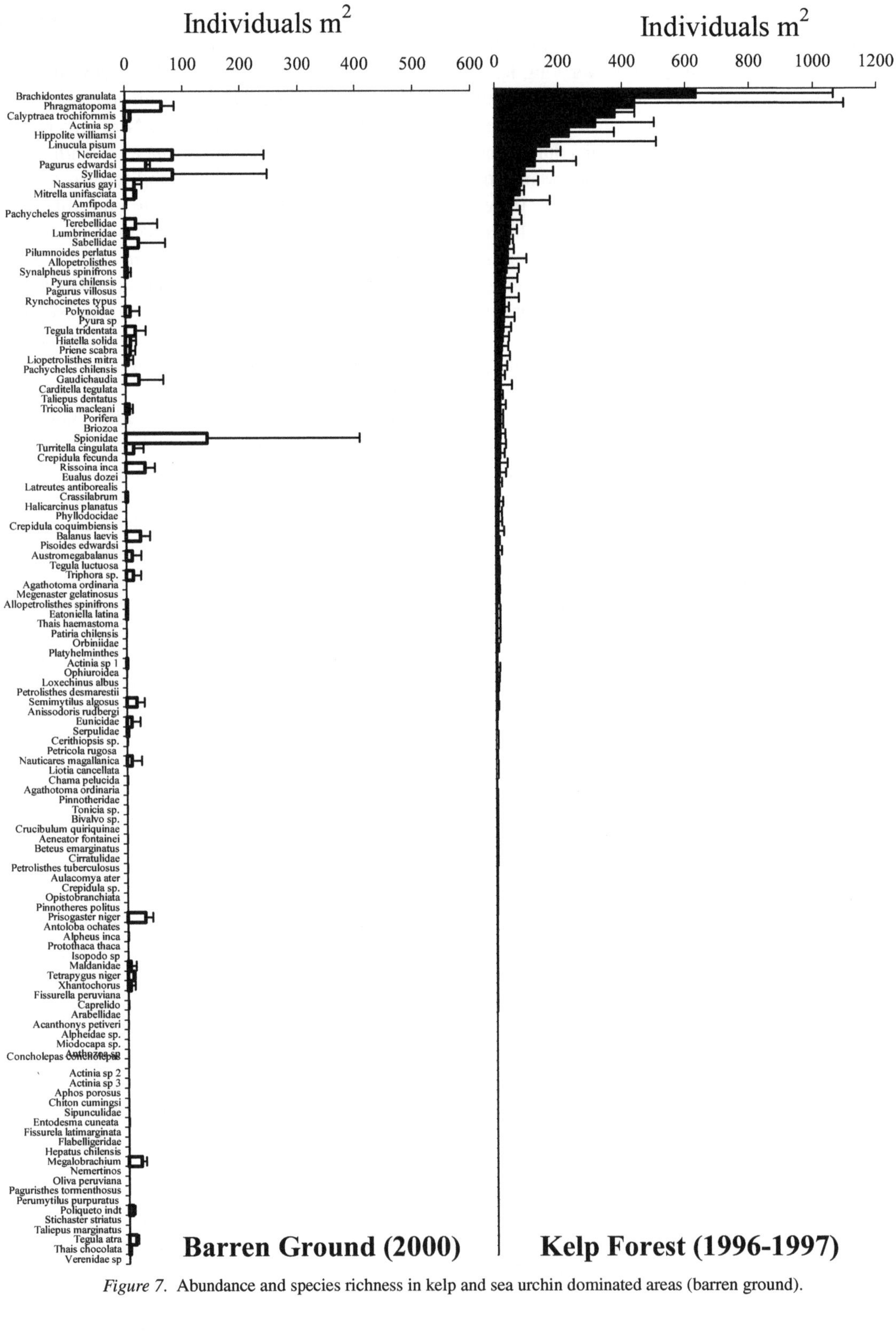

Figure 7. Abundance and species richness in kelp and sea urchin dominated areas (barren ground).

components of the asteroid guild has been suggested as evidence to explain the segregated patterns of distribution on the benthic gradient (Lawrence & Vásquez, 1996).

Different bottom-up and top-down events may regulate long-term ecosystem changes in northern Chile including: (**1**) The intensity and frequency of upwelling buffer the positive thermal anomalies in SST, maintaining high nutrient levels which favor the kelp during ENSO events. (**2**) Site-dependent oceanographic conditions may generate optimal conditions for spawning, larval development, and recruitment of echinoderms during and/or after ENSO event. (**3**) The population dynamics of adult starfish and sea urchins during ENSO events are essentially species-specific. (**4**) Species-specific population dynamics (i.e. *L. magellanica*) and changes in dietary composition (i.e. *H. helianthus*) during ENSO events, may promote population increases in *T. niger*. This seems to be a key factor in alternation of environments dominated by kelp beds and barren grounds. (**5**) The increase in population density of the adult fraction of the carnivore guild is correlated with the decline in densities and/or displacement to shallow bottoms of the more conspicuous herbivorous grazers.

Discrete and local oceanographic events (upwelling) as well as large-scale, low-frequency events (ENSO) generate interannual variability in species or groups of key species, which structure and organize subtidal, rocky reef communities in northern Chile. This translates to trophic cascades that modulate the temporal alternation of states dominated by kelp, and sea urchin barrens. These changes in submarine seascapes have been treated as "catastrophic" in the literature (Tomicic, 1985; Soto, 1985), although simple species diversity indexes revealed no significant changes with (1996–1998), or without (1999–2000) the presence of kelp assemblages (Figure 7). Here, the composition and species richness seem to indicate different organizational states of these communities. On a regional scale, this characteristic suggests the presence of a mosaic of subtidal seascapes in different seral stages of ecological succession (Tomicic, 1985; Vasquez et al., 1998; Camus, 2001, Dayton et al., 1998, 1999, Edwards 2004). Graham (2004), recently contrasting biodiversity and trophic complexity in sea urchin barrens and kelp-dominated habitats (*Macrocystis pyrifera*), did not find significant differences between alternate states. This also suggests a temporal resilience in the trophic web of the kelp forest over the long term, with few species exclusively associated with a determinate state.

Finally, the interannual variability in the structure and organization of subtidal kelp communities suggests the need for long term (8^+ yrs) monitoring programs. These would detect changes over time, which would not be evident in short or medium-term studies. This type of data would be useful in evaluating conservation and management of resources, and would broaden knowledge of the sustainability of the biological diversity of Chile's continental coastal marine ecosystem (Vasquez et al., 2001a). In this context, Chile's extensive coastline (18°–56°S) offers an ecological scenario that is unique in the world for evaluation of the effects of events that operate on different geographic scales.

Acknowledgements

This study was funded by grants from FONDECYT 5960001, 1000044, 1010706 and 1040425 to JAV.

References

Buschmann A, García C, Espinoza R, Filún L, Vásquez JA (2003) Sea urchins (*Loxechinus albus*) and kelp (*Macrocystis pyrifera*) in protected areas in southern Chile. In Lawrence J (ed.), Sea Urchins: Fisheries and Ecology. Proceeding of the International Conference on Sea-Urchin Fisheries and Aquaculture 2003: 120–130.

Buschmann A, Vásquez JA, Osorio P, Reyes E, Filun L, Hernández-González MC, Vega A (2004) The effect of water movement, temperature and salinity on abundance and reproductive patterns of *Macrocystis* spp. (Phaeophyta) at different latitudes in Chile. Marine Biology 145: 849–862.

Camus PA (2001) Biogeografía marina de Chile continental. Revista Chilena de Historia Natural 74: 587–617.

Contreras S, Castilla JC (1987) Feeding behavior and morphological adaptations in two sympatric sea urchins species in central Chile. Marine Ecology Progress Series 38: 217–224.

Dayton PK (1985) The structure and regulation of some South American kelp communities. Ecological Monographs 55: 447–468.

Dayton PK, Rosenthal RJ, Mahen LC, Antezana T (1977) Population structure and foraging biology of the predaceous Chilean asteroid *Meyenaster gelatinosus* and the scape biology of its prey. Marine Biology 39: 361–370.

Dayton PK, Tegner MJ, Parnell PE, Edwards PB (1992) Temporal and spatial patterns of disturbance and recovery in kelp forest community. Ecological Monographs 62: 421–445.

Dayton PK, Tegner MJ, Edwards PB, Riser KL (1998) Slinding, ghost, and reduced expectations in kelp forest communities. Ecological Applications 8: 309–322.

Dayton PK, Tegner MJ, Edwards PB, Riser KL (1999) Temporal and spatial of kelp demography: The role of oceanographic climate. Ecological Monographs 69 (2): 219–250.

Ebert TA, Rusell MP (1988) Latitudinal variation in size structure of the west coast purple sea urchins: A correlation with headlands. Limnology and Oceanography 33: 286–294.

Edwards MS (2004) Estimating scale-dependency in disturbance impacts: El Niños and giant kelp forests in the northeast Pacific. Oecologia 138: 436–447.

Estes JA, Danner EM, Doak DF, Konar B, Springer AM, Steinberg PD, Tinker MT, Williams TM (2004) Complex trophic interactions in kelp forest ecosystems. Bulletin of Marine Science 74: 621–638.

Fernandez E, Cordova C, Tarazona J (1999) Condiciones del bosque submareal de *Lessonia trabeculata* en la isla Independencia durante el evento El Niño 1997–1998. Revista Peruana de Biología, Volumen Extraordinario 47–59.

Fielder PC (2002) Environmental changes in the eastern tropical Pacific Ocean: Review of ENSO and decadal variability. Marine Ecology Progress Series 244: 265–283.

Glynn PW (1988) El Niño-Southern Oscillation 1982–1983: Nearshore population, community, and ecosystem responses. Annual Review of Ecology and Systematic 19: 309–345.

Godoy NE (2000) *Macrocystis integrifolia* (Laminariales, Phaeophyta) en el norte de Chile: Distribución espacial y fauna asociada. Tesis para optar al Titulo de Biólogo Marino. Departamento de Biología Marina, Facultad de Ciencias del Mar, Universidad Católica del Norte, Coquimbo: 58 pp.

Graham MH (2004) Effects of local deforestation on the diversity and structure of southern California giant kelp forest food webs. Ecosystems 7: 341–357.

Graham MH, Harrold C, Lisin S, Light K, Watanabe JM, Foster MS (1997) Population dynamics of giant kelp *Macrocystis pyrifera* along a wave exposure gradient. Marine Ecology Progress Series 148: 269–279.

Halpin PA, Strub PT, Peterson WT, Baumgartner TR (2004) An overview of interactions among oceanography, marine ecosystems, climatic and human disruptions along the eastern margins of the Pacific Ocean. Revista Chilena de Historia Natural 77: 371–410.

Jones CG, Lawton JH, Shachak M (1994) Organisms as ecosystem engineers. Oikos 69: 373–386.

Ladah LB, Zertuche-González JA, Hernández-Carmona G (1999) Giant kelp (*Macrocystis pyrifera*, Pheophyceae), recruitment nears its southern limit in Baja California after mass disappearance during ENSO 1997–1998. Journal of Phycology 35: 1106–1112.

Lagos NA, Barría ID, Paolini P (2002) Upwelling ecosystem of northern Chile: Integrating benthic ecology and coastal oceanography through remote sensing. In Castilla JC, Largier JL (eds.), The Oceanography and Ecology of the Nearshore and Bays in Chile, Ediciones Universidad Católica de Chile, Santiago, Chile: 117–141.

Lancelloti D, Vásquez JA (1999) Biogeographical patterns of benthic invertebrates in the southeastern Pacific litoral. Journal of Biogeography 26: 1001–1006.

Lancelotti D, Vásquez JA (2000) Zoogeografía de macroinvertebrados bentónicos de la costa de Chile: Contribución para la conservación marina. Revista Chilena de Historia Natural 73: 99–129.

Lawrence JM (1975) On the relationships between marine plants and sea urchins. Oceanography and Marine Biology Annual Review 13: 213–286.

Lawrence JM, Vásquez JA (1996) The effect of subletal predation on the biology of echinoderms. Oceanologica Acta 19: 1–10.

Lleellish J, Fernández E, Hooker Y (2001) Disturbancia del bosque submareal de *Macrocystis pyrifera* durante El Niño 1997–1998 en la Bahía de Pucusana. In Alveal K, Antezana T (eds.), Sustentabilidad de la Biodiversidad. Un Problema Actual: Bases Científico Técnicas, Teorizaciones y Proyecciones. Universidad de Concepción, Chile, 331–350.

Martínez EA, Cardenas L, Pinto R (2003) Recovery and genetic diversity of the intertidal kelp *Lessonia nigrescens* 20 years after El Niño 1982–83. Journal of Phycology 39: 504–508.

Medina M, Araya M, Vega C (2004) Alimentación y relaciones tróficas de peces costeros de la zona norte de Chile. Investigaciones Marinas (Chile) 32: 33–47.

Navarrete SA, Broitman B, Wieters EA, Finke GR, Venegas RM, Sotomayor A (2002) Recruitment of intertidal invertebrates in the southeast Pacific: Interannual variability and the 1997–1998 El Niño. Limnology and Oceanography 47: 791–802.

Rodríguez SR, Ojeda FP (1993) Distribution patterns of *Tetrapygus niger* (Echinodermata: Echinoidea) off the central Chilean coast. Marine Ecology Progress Series 101: 157–162.

Sokal RR, Rohlf FJ (1981) Biometry. The principles and practice of statistics in biological research, W. H. Freeman & Company, New York, 859 pp.

Soto R (1985) Efectos del fenomeno del El Niño 1982-83, en ecosistemas de la I Región. Investigaciones Pesqueras (Chile) 32: 199–206.

Steneck RS, Graham MH, Bourque BJ, Corbett D, Erlandson JM, Estes JA, Tegner MJ (2002) Kelp forest ecosystems: Biodiversity, stability, resilience and future. Environmental Conservation 29: 436–459.

Stenseth NC, Ottersen G, Hurrell JW, Mysterud A, Lima M, Chang K, Yoccoz NG, Adlandsvik B (2003) Studying climate effects on ecology through the use of climate indices: The North Atlanthic Oscillation, El Niño southern Oscillation and beyond. Proceeding of the Royal Society Biological Sciences 270: 2087–2096.

SYSTAT (1992) Statistics. Version 8.0 (ed.) SYSTAT Inc. Evanston. IL.

Takesue RK, van Geen A, Carriquiry JD, Ortiz E, Gonidez-Orta L, Granados I, Saldivar M, Ortlieb L, Guzman N, Castilla JC, Varas M, Salamanca M, Figueroa C (2004) Influence of coastal upwelling and ENSO on nearshore water along Baja California and Chile: Shore-based monitoring during 1997–2000. Journal of Geophysical Research 109: C03009, doi: 10.1029/2003JC001856.

Tala F, Edding M, Vasquez JA (2004) Aspect of the reproductive phenology of *Lessonia trabeculata* (Laminariales: Phaeophyceae) from three populations in northern Chile. New Zealand Journal of Marine and Freshwater Research 38: 255–266.

Tegner MJ, Dayton PK (1987) El Niño effects on southern California kelp forest communities. Advances in Ecological Research 17: 243–274.

Tegner MJ, Dayton PK (1991) Sea urchins, El Niño, and the long-term stability of southern California kelp forest communities. Marine Ecology Progress Series 77: 49–63.

Tegner MJ, Dayton PK, Edwards PB, Riser KL (1997) Large scale, low frequency oceanographic effects on the kelp forest succession: A tale of two cohorts. Marine Ecology Progress Series 146: 117–134.

Tokeshi M, Romero L (1995) Quantitative analysis of foraging behavior in a field population of the South American sun-star *Heliaster helianthus*. Marine Biology 122: 297–303.

Tomicic JJ (1985) Efectos del fenómeno del El Niño 1982-83, en las comunidades litorales de la Peninsula de Mejillones. Investigaciones Pesqueras (Chile) 32: 209–213.

Vásquez JA (1992) *Lessonia trabeculata*, a subtidal bottom kelp in northern Chile: A case study for a structural and geographical comparisons. In Seeliger U (ed.), Coastal Plants of Latin America, Academics Press, San Diego: 77–89.

Vásquez JA, Buschmann AH (1997) Herbivore-kelp interactions in Chilean subtidal communities: A review. Revista Chilena de Historia Natural 70: 41–52.

Vásquez JA, Vega JMA (2004) El Niño 1997–1998 en el norte de Chile: Efectos en la estructura y en la organización de comunidades submareales dominadas por algas pardas. In Avaria S, Carrasco J, Rutland J, Yañez E (eds.), El Niño-La Niña 1997-2000 y sus efectos en Chile Comité Oceanográfico Nacional, Chile, 119–135.

Vásquez JA, Camus PA, Ojeda FP (1998) Diversidad estructura y funcionamiento de ecosistemas costeros rocosos del norte de Chile. Revista Chilena de Historia Natural 71: 479–499.

Vásquez JA, Véliz D, Pardo LM (2001a) Vida bajo las grandes algas pardas. In Alveal K, Antezana T (eds.), Sustentabilidad de la Biodiversidad. Un Problema Actual, Bases Científico Técnicas, Teorizaciones y Perspectivas, Ediciones Universidad de Concepción, Concepción. Chile, 293–308.

Vásquez JA, Fonck E, Vega JMA (2001b) Diversidad, abundancia y variabilidad temporal de ensambles de macroalgas del submareal rocoso del norte de Chile. In Alveal K, Antezana T (eds.), Sustentabilidad de la Biodiversidad. Un Problema Actual, Bases Científico Técnicas, Teorizaciones y Perspectivas, Ediciones Universidad de Concepción, Concepción, Chile, 615–634.

Vega JMA, Vásquez JA, Buschmann A (2005) Population biology of the subtidal kelps *Macrocystis integrifolia* and *Lessonia trabeculata* (Laminariales, Phaeophyceae) in an upwelling ecosystem of northern Chile: Interannual variability and El Niño 1997–98. Revista Chilena de Historia Natural 78: in press.

Viviani C (1978) Predación interespecifica, canibalismo y autotomía como mecanismos de escape en las especies de Asteroidea (Echinodermata) en el litoral del Desierto del norte grande de Chile. Report. Laboratorio de Ecología Marina. Universidad del Norte, Iquique. 116 pp.

Journal of Applied Phycology (2006) 18: 521–527
DOI: 10.1007/s10811-006-9057-3

Recent fluctuations in distribution and biomass of cold and warm temperature species of Laminarialean algae at Cape Ohma, northern Honshu, Japan

Shinji Kirihara[1,*], Toshiki Nakamura[1], Naoto Kon[2], Daisuke Fujita[3] & Masahiro Notoya[3]
[1]*Aomori Prefectural Fisheries Research Center, Aquaculture Institute, 039-3381 Hiranai, Aomori, Japan;* [2]*MAC Co. Ltd, 039-3502 Aomori, Aomori, Japan;* [3]*Tokyo University of Marine Science and Technology, 108-8477 Minato-ku, Tokyo, Japan*

*Author for correspondence: e-mail: shinji_kirihara@ags.pref.aomori.jp

Key words: Laminariales, seaweed, changes of flora, long-term water temperature, warming, Cape Ohma, Japan

Abstract

The Cape Ohma region of Shimokita Peninsula, the northernmost point of Honshu Island, Japan, is subject to both the warm Tsugaru Current and the cold Kurile Current. As a result, the Laminarialean flora includes both cold temperature species (*Laminaria japonica* Areschoug, *Kjellmaniella crassifolia* Miyabe and *Costaria costata* (C. Agardh) Saunders) and warm temperature species (*Undaria peterseniana* (Kjellman) Okamura, *Ecklonia stolonifera* Okamura), as well as *Undaria pinnatifida* (Yendo) Okamura, which is distributed in both waters. The frequency of occurrence (as a measure of distribution) and the biomass of these species were recorded in June 1976 (at 50 points in depths between 8–30 m), July 1988 (192 points, 2.5–25 m) and July 2001 (78 points, 2.5–25 m). Comparison of these data revealed a decrease in cold temperature species and an increase in warm temperature species from 1976 or 1988 to 2001. Long-term data of seawater temperature measured at 5 m depth near the study site showed that mean temperatures in the middle of winter (late January to February) in 1989–2000 were 0.9–1.1 °C higher than those in 1980–1988. Higher seawater temperatures in the last decade appear to have affected the frequency of occurrence and biomass of the Laminarialean species along the coasts of Cape Ohma. This result supported our previous conclusion that 1 °C higher mean seawater temperature in late January caused a decrease in the biomass of *L. japonica* (by ca. 64%) along the same coast.

Introduction

The coasts around Cape Ohma, Shimokita Peninsula, the northernmost point of Honshu Island, Japan, are dominated by the warm Tsugaru Current from the Sea of Japan, but also affected by the cold Kurile Current (Oyashio) flowing down from the east coast of Hokkaido. Therefore, many phycologists have studied the seaweed flora around the cape (Yamada, 1928; Takamatsu, 1938; Kanda et al., 1950; Nanao, 1974; Notoya & Asuke, 1984; Notoya & Aruga, 1989), and a total of 10 Laminarialean species have been reported up to now, out of a total of 36 for Japan (Kawashima, 1989). These floristic data revealed that both cold temperature and warm temperature species occurred in a narrow area. However, the distribution and biomass of these Laminarialean species around the cape have never been studied in detail.

Around Cape Ohma, two cold temperature species, *Laminaria japonica* Areschoug and *Kjellmaniella crassifolia* Miyabe are economically valuable and are harvested at depths from 10 to 25 m, yielding hundreds of millions of yen a year (millions of US dollars per year). Unfortunately, the average yield per decade of these kelps has decreased from 3,304 t in the 1980s to 2,045 t in the 1990s. A previous study (Kirihara et al., 2003b) showed that natural growth of *L. japonica* at Shiriyazaki, the easternmost end of Shimokita Peninsula, was highly correlated with the water temperature from January to March. Around

Cape Ohma. However, most of the above floristic data were not quantitative and no attempt was made to analyze water temperatures in detail or to relate them to the Lmainarialean species present. We compare the distribution (as frequency of occurrence) and biomass of Laminarialean species around Cape Ohma in 1976, 1988 and 2001, in relation to long-term temperature records and known temperature responses of the species.

Material and methods

Analysis of water temperature

Water temperature data from 1980 to 2001 were obtained from the Abalone Culture Center at Sai located near Cape Ohma. Seawater was pumped from a depth of 5 m and the temperature was measured at 9 a.m. every day using an electronic recorder (RIGO Co. Ltd., accuracy ± 0.1°C). Data were then averaged for every ten days. The time series of ten-day averages was analyzed using a Trend Index for meteorological time series analysis (Suzuki, 1975) and Significance Probability. Long-term variation of water temperatures was examined using moving averages of twelve months. After plotting a linear regression, the year of the major discontinuity of the averages was tested according to the method of Tomosada (1994).

Distribution and biomass

Sampling was done in rocky subtidal areas (Figure 1) around Cape Ohma in June 1976, July 1988 and July 2001, by SCUBA diving. In June 1976, sampling was restricted to 50 points in the kelp harvesting zone (8 to 30 m depth); both destructive and nondestructive methods were used. In July 1988, sampling was increased to 192 points, at depths of 2.5 to 25 m; the destructive method was excluded. In July 2001, destructive and nondestructive sampling was done at 78 points ranging from 2.5 to 25 m depth. In the nondestructive sampling, a quadrat of 5 × 5 m was placed at each point and occurrences of Laminarialean species were recorded as presence or absence in each 25 m² quadrat. In the destructive sampling, a smaller quadrat of 50 × 50 cm was placed on a patch of each Laminarialean species within the large quadrat and the seaweeds were removed. The distribution of each species was described using percentage frequency of occurrence and biomass. Percentage frequency was calculated as

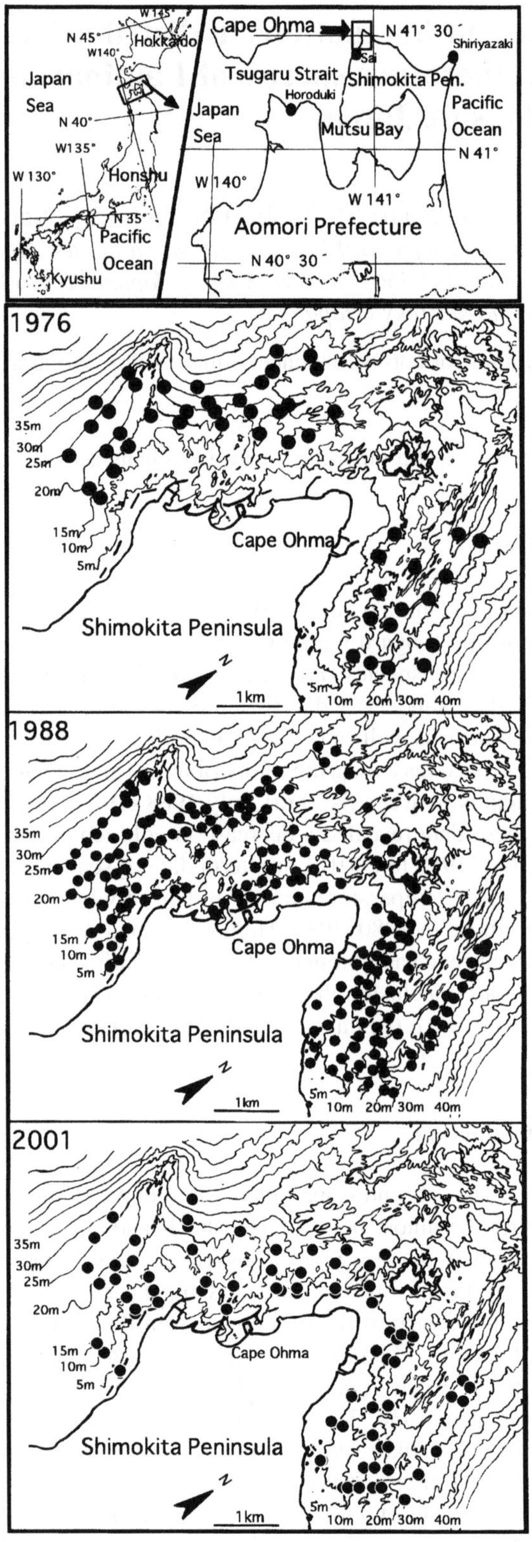

Figure 1. Map showing locations and contours around Cape Ohma. ●: sampling points in June 1976, July 1988 and July 2001.

the percentage of points (25 m^2 quadrats) in which a species was present out of the total number of points sampled. Biomass was calculated from the wet weight of each species in each 0.25 m^2 quadrat, collected and weighed in the laboratory.

Results

Water temperature

Mean monthly water temperatures at Sai from 1980 to 2000 are shown in Figure 2. The annual average water temperatures ranged from 13.2 °C (1984) to 15.3 °C (1990). The maximum water temperatures were recorded from August to early September, and ranged from 21.6 °C (1980) to 25.2 °C (1994). The minimum water temperatures were recorded from February to early April and ranged from 5.7 °C (1984) to 8.7 °C (1990). Trend Index and probability of significance examined from time series of the ten days averages of water temperature (Figure 2) showed similar changes. Trend Index and probability of significance were high from early spring to summer, but were low for the autumn to early winter. Water temperatures rose significantly (significance probability less than 5% and Trend Index 1.5 or more) in the periods from late January to late February, middle of March, late May and late August.

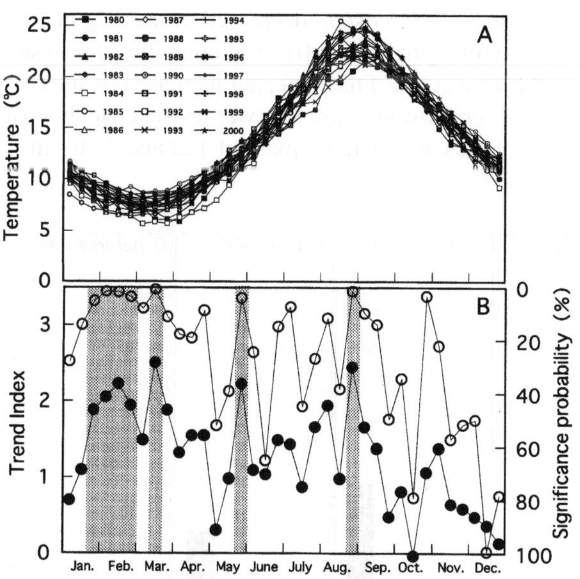

Figure 2. Water temperatures (average of every ten days) at Sai from 1980 to 2000 (A) and Trend Index (●) and significance probability (O) (B) obtained from time series analysis of the data in A. Shaded portions: significance probability less than 5% and trend index higher than 1.5.

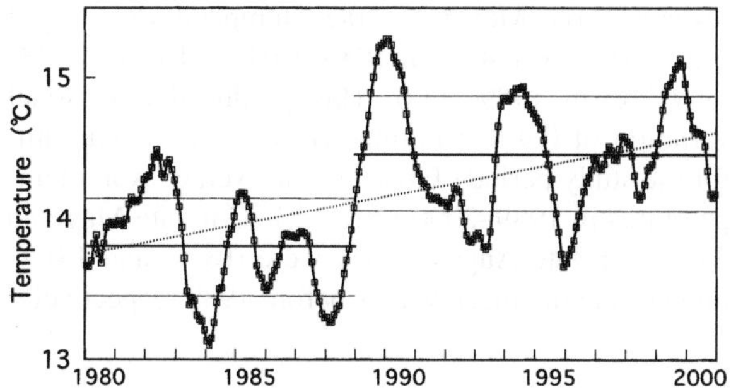

Figure 3. Water temperature (twelve month moving averages) at Sai. The dotted line shows the regression line of linear trend; the solid lines show averages and ranges of standard deviation in the periods 1980 to 1988 and 1989 to 2000.

Water temperatures (twelve months moving average) at Sai from 1980 to 2000 are shown in Figure 3. They showed large fluctuations every few years but also a tendency to rise over the whole period. The linear regression was: $Y = 0.00345\,X + 13.744$ $r = 0.499$ X, months; Y, temperature (°C). Since this indicates an increase in temperature with time, we then tested for the "jump year" or year of greatest increase in average water temperature. As a result, 1989 was found to be the most appropriate time to divide water temperatures (moving averages), in order that the standard deviations of both periods were the minimum and lower than the standard deviations of the linear trend.

Timings of significant elevations in water temperature from 1980–1988 to 1989–2000, selected from significance probability <5% and Trend Index >1.5, are shown in Table 1. From late January to middle March

Table 1. Timing of significant elevation in water temperature after 1989 (i.e., from 1980–1988 to 1989–2000), selected from significance probability <5% and Trend Index>1.5.

Month	Period	Average water temperatue (°C)		Difference (°C)
		1980–1988	1989–2000	
January	Late	8.4	9.2	0.9
February	Early	7.6	8.7	1.1
	Middle	7.4	8.3	0.9
	Late	7.1	8.0	0.9
March	Middle	7.0	7.9	0.9
May	Late	11.9	12.7	0.8
August	Late	22.4	23.3	0.9

(except early March) average temperatures ranged from 7.0 °C to 8.4 °C in 1980–1988 and from 7.9 °C to 9.2 °C in 1989–2000. This produced a seasonal variation of 0.9 °C (minimum) to 1.1 °C (maximum) for the study period (Table 1). The average water temperatures in 1989–2000 were 12.7 °C in late May and 23.3 °C in late August. They were 0.8 °C and 1.0 °C higher than the mean values before 1988 respectively.

Frequencies of Laminarialean species

Among the ten species of Laminariales reported around Cape Ohma, seven species (*L. japonica, K. crassifolia, Costaria costata* (C. Agardh) Saunders, *Agarum cribrosum* f. *rugosum* Yamada, *Undaria peterseniana* (Kjellman) Okamura, *Undaria pinnatifida* (Yendo) Okamura and *Ecklonia stolonifera* Okamura) were recognized in the present study. Among the remaining species, *Chorda filum* (Linnaeus) Stackhouse was an annual alga observed only in spring. *Alaria crassifolia* Kjellman occurred in the shallower zones. In addition, *Laminaria religiosa* Miyabe was not distinguished from *L. japonica* during the survey and included here as *L. japonica*.

Yearly changes in frequencies of the seven Laminarialean species are shown in Figure 4. Frequencies of the two valuable kelp-fishery species *L. japonica* and *K. crassifolia* were relatively high in 1976 and 1988, but decreased in 2001. These decreases were particularly marked in the kelp fishery depth zone (8–25 m), after 1988: Frequencies of *L. japonica* decreased from 85.7% in 1988 to 60.4% in 2001, and of *K. crassifolia* from 70.7% in 1988 to and 31.3% in 2001. Frequencies of *C. costata* decreased steadily from 44.0% in 1976 to 20.5% in 2001. Frequencies of *A. cribrosum* f. *rugosum* were always relatively low, but dropped from 14.0% in 1976 to 6.4% in 2001. The frequencies of three Laminarialean species increased between 1976 and 2001. *U. pinnatifida* was observed in 24.4% of survey points in 2001, though it was found only in 2.0% in 1976. *E. stolonifera* was not found in 1976, but was sampled in 35.9% of survey points in 2001. *U. peterseniana* was not detected in 1976 or 1988, but was recorded in 9.0% of the survey points in 2001.

Biomass of Laminarialean species

Biomasses, at the various sampling sites, of Laminarialean species recorded in June 1976 and in July 2001 are compared in Figure 5. In addition, the points where those species were observed in 1988 are indicated in the same figure. Biomass of *L. japonica* decreased from 1,906 g m^{-2} in 1976 to 1,614 g m^{-2} in 2001, but it was found widely around the cape during the survey. Biomass of *K. crassifolia* decreased from 583 g m^{-2} to 236 g m^{-2} during the same period. The quantities decreased greatly on the west coast of the cape compared with the east coast. Biomass of *L. japonica* and *K. crassifolia* were 1,811 g m^{-2} and 383 g m^{-2} respectively in the kelp fishery depth zone in 2001: both values were lower than in 1976.

Biomass of *C. costata* decreased from 220 g m^{-2} in 1976 to 99.3 g m^{-2} in 2001. The quantities decreased on the west coast of the cape, but increased on the east coast. *A. cribrosum* f. *rugosum* was restricted to the east coast of the cape, with biomass of 122 and 27.6 g m^{-2}

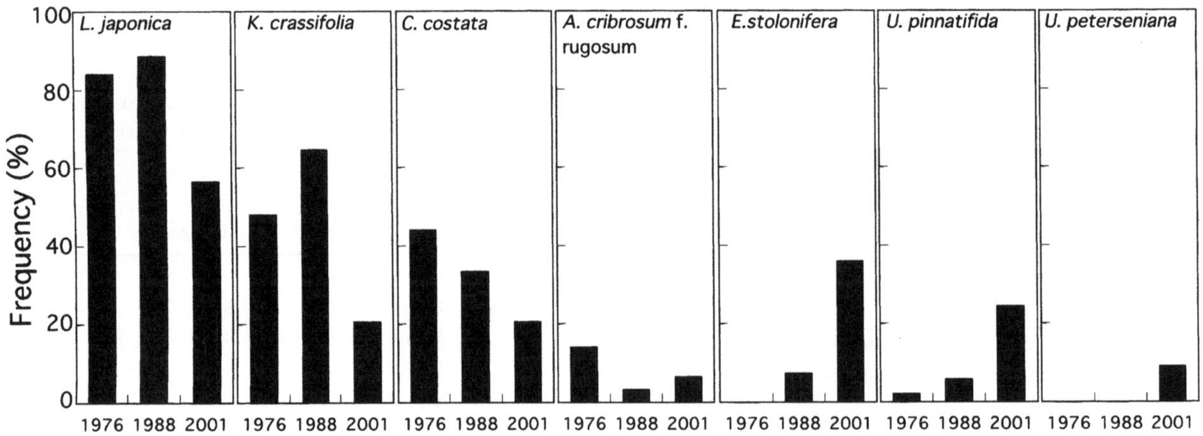

Figure 4. Frequencies of occurrence (present/total quadrats) of Laminarialean species around Cape Ohma in June 1976, July 1988 and July 2001.

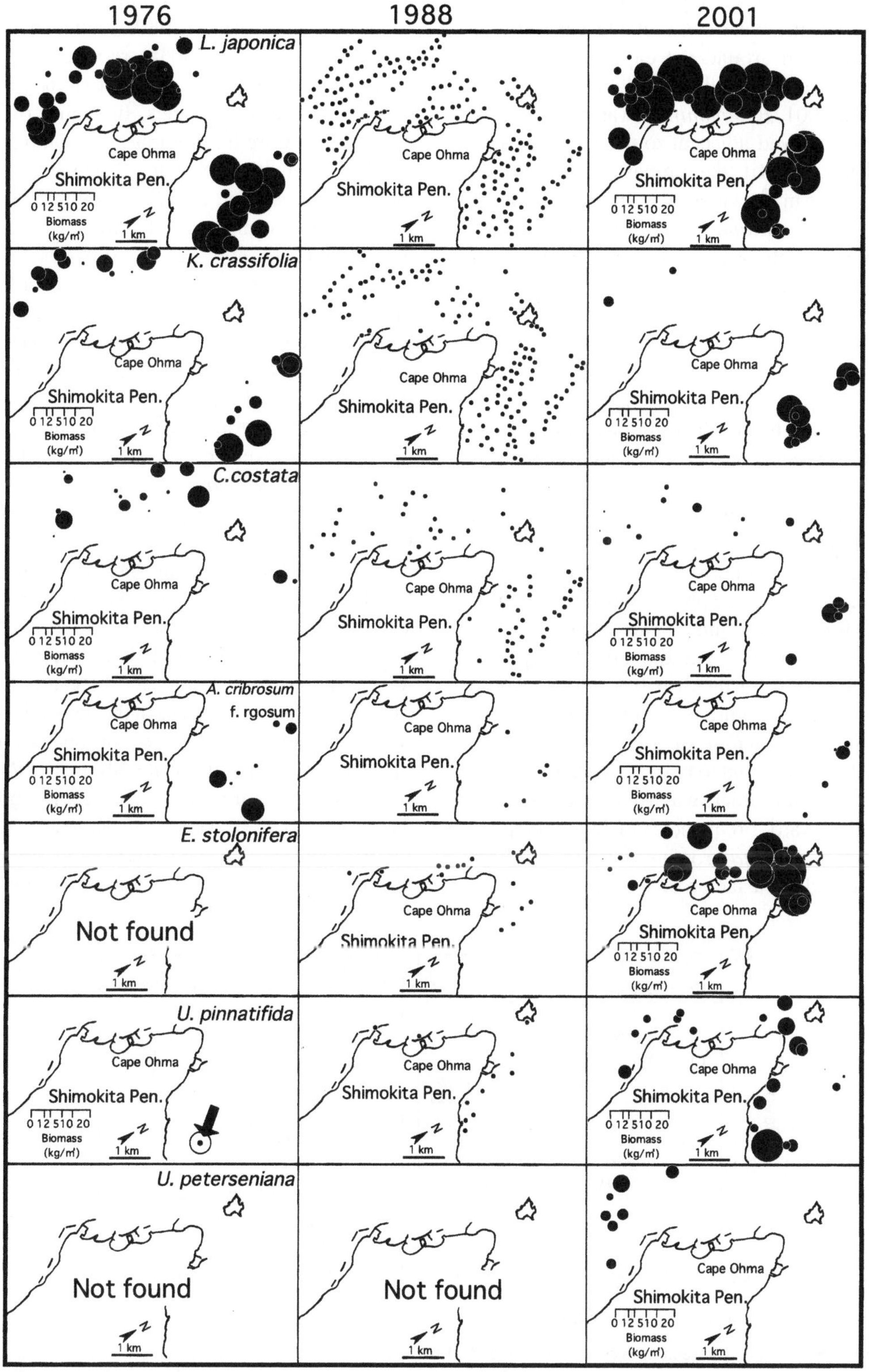

Figure 5. Distribution of Laminarialean species (biomass) around Cape Ohma in June 1976, July 1988 and July 2001. Bubble size shows relative biomass in 1976 and 2001.

in 1976 and 2001 respectively. *U. pinnatifida* was observed only on the west coast of the cape before 1988, but was found around the cape in 2001. The biomass increased dramatically from 4.4 g m^{-2} in 1976 to 248 g m^{-2} in 2001. *E. stolonifera* was not found in 1976, but was observed adjacent to the west coast of the cape in 1988. In 2001, the biomass of this species had reached, 783 g m^{-2}, which was similar to that of *L. japonica. U. peterseniana* was detected at comparatively deep points on the west coast of the cape only in 2001: biomass was 76.4 g m^{-2}.

Discussion

Frequencies of occurrence (distribution) and biomass of Laminariales species around Cape Ohma changed between 1988 and 2001, and mean water temperatures changed considerably around 1989. To compare increases and decreases in individual species, we examined the relationships between the trends we measured here and the biogographical and physiological data available for seven Laminarialean species. Data for this survey were compared with the latitudes reported by Kawashima (1989) for the distribution of Japanese Laminariales species (Figure 6 A). Frequencies of *U. pinnatifida* and *E. stolonifera* distributed south to the coast in Kyushu were found to have increased 4.7 to 12 times. On the other hand, there was a decrease in species distribution northward. Frequencies of *L. japonica, K. crassifolia and C. costata* growing north of northern Honshu decreased by a factor of 0.32 to 0.70. Saito (1956), Notoya and Asuke (1983), Nakahara (1984), Okada and Sanbonsuga (1980), Yamauchi et al. (2003), and Kirihara et al. (2003a) reported the maximum water temperatures for maturation in gametophytes of six Laminariales species. In Figure 6B the maximum temperature for maturation is compared with the trend in fluctuation. Species that mature at 22 °C or more (*U. pinnatifida* and *E. stolonifera*) increased in biomass, but most species that mature at temperatures below 20 °C (*L. japonica, K. crassifolia A. cribrosum* f. *rugosum* and *C. costata*) decreased in biomass. *U. peterseniana* is distributed on the coast of northern Kyushu (Kawashima, 1989) and matures at temperatures lower than 25 °C in gametophytes (Migita, 1963), but was excluded from the above discussion because of no records before 1988 at the cape. Comparison of these data revealed a decrease in cold temperature species and an increase of warm temperature species from 1976 or 1988 to 2001 along the coast of Cape Ohma.

In the long-term, mean seawater temperatures for late January to February for 1989 to 2000 were about 1 °C higher than those for 1980 to 1988. Kirihara et al. (2003b) reported that 1 °C higher mean seawater temperature in late January for the gametophytic stage (Tc_5 ,°C) caused a decrease in the densities of cold temperature species of *L. japonica* in June (L_1, individuals m^{-2}) by 64.1% at Shiriyazaki, Shimokita Peninsula, on the coast of Tsugaru Strait ($L_1 = \exp[11.600–1.024 \times Tc_5]$, $R = 0.957$). Moreover, Kinoshita and Shibata (1939) reported that the harvest of warm temperature species of *U. pinnatifida* increased because the water

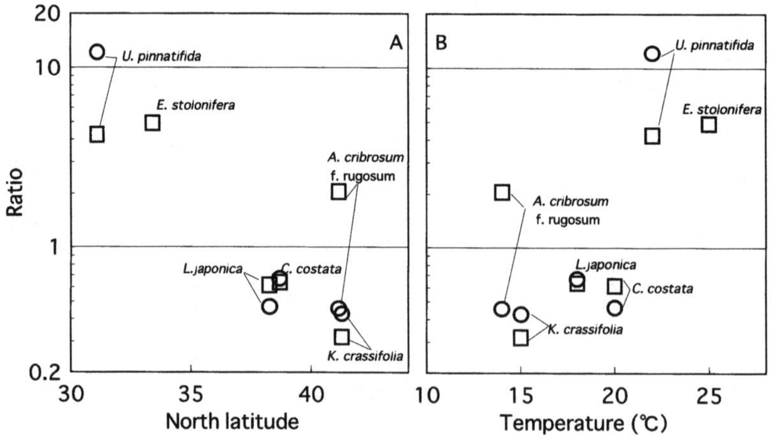

Figure 6. Relationship between the trend in fluctuation (ratio of 2001 to 1976 or 1988 frequencies, data from Figure 4) and biogeographic or physiological data for seven Laminarialean species. The latitude of the southern limit in Japan and the highest water temperature for maturation of gametophytes are shown in A and B, respectively. ○: ratio of 2001 to 1976; □: ratio of 2001 to 1988. (*U. peterseniana*, distributed on the coast of northern Kyushu and maturing at temperatures lower than 25 °C in gametophytes, was not shown here because of no record before 1988).

temperature from January to March was highst on the coast of Hokkaido. These reports provide further evidence that the higher seawater temperature could have affected frequencies and biomass of both the cold and warm temperature Laminariales species along the coast of Cape Ohma. *E. stolonifera* was distributed only from the west of Horoduki at Tsugaru Peninsula on the Tsugaru Strait coast to the Sea of Japan in 1948 (Kanda et al., 1950) as well as in 1960 through to 1971 (Nanao, 1974). Notoya and Asuke (1984) reported it was new to Cape Ohma in 1982. *E. stolonifera* is known as a unique Laminariales species which propagates vegetatively by stoloniferous rhizoids along the Japan coasts (Kawashima, 1989). Small holdfasts, of even 10 cm in diameter, seemed to be three years old, whereas subsequent propagation is reported to be fast (Notoya & Aruga, 1990). It appears that the warm temperature species, *E. stolonifera,* has rapidly expanded its distribution at Cape Ohma in the past two decades.

Frequency and biomass of two commercially valuable cold temperature species (*L. japonica, K. crassifolia*) also decreased from 1976 or 1988 to 2001 in the kelp fishery depth zone. A possible reason for the recent decrease in kelp yields around the cape could be floristic change as well as competition with warm temperature Laminarialean species that have increased due to warming of sea surface temperatures. On the west coast of the cape, the frequencies and biomass of two cold temperature species (*K. crassifolia* and *C. costata*) decreased and two warm temperature species that were not found in 1976 (*E. stolonifera, U. peterseniana*), increased after 1989. *A. cribrosum* f. *rugosum* was not found there in these surveys. The Tsugaru warm current flows onto the west coast of the cape directly, but a large, clockwise eddy goes westwards along the east coast. For this reason, compared to the east coast of the cape, the water temperature on the west coast is high all year round (Otani & Nakamura, 1985). Overall, it appears that changing water temperatures on both sides of the cape have led to changes in the composition of the Laminarialean flora.

References

Kanda G, Sakai Y, Mikami S (1950) Aomori-ken Yuyo Kaiso. Aomori Prefecture Fisheries Resources Research Report 1, Aomori Prefecture Fisheries Experimental Station 83–93 (in Japanese).

Kawashima S (1989) Nihon-san konbu zukan. Kitanihon Kaiyo Center, Sapporo, Japan, pp. 153–155, pp. 200–207 (in Japanese).

Kinoshita T, Shibata K (1939) Relation between the harvest of *Undaria pinnatifida* (Harv.) Sur. and sea temperature in Suttu region, Hokkaido. Bull. Jpn. Soc. Sci. Fish. 5(4): 191–193 (in Japanese).

Kirihara SY, Fujikawa MN (2003a) Effect of the temperature and light intensity on the growth of zoospore germling of *Kjellmaniella crassifolia* Miyabe (Laminariales, Phaeophyceae) in culture. Suisanzoshuku 51(3): 281–286 (in Japanese).

Kirihara ST, Nakamura MN (2003b) Effect of Water Temperature on the Growth of *Laminaria japonica* (Laminariales, Phaeophyta) at the Coast of Shiriyazaki, Shimokita Peninsula, Japan. Suisanzoshuku 51(3): 273–280 (in Japanese).

Nakahara H (1984) Alternation of generations of some brown algae in unialgal and axenic culture. Sci. Pap. Inst. Alg. Res., Hokkaido University 7: 77–194.

Nanao Y (1974) A distribution of the marine algae from the coast of Aomori Prefecture. Bull. Jpn. Soc. Phycol. 22(1): 29–38 (in Japanese).

Notoya M, Asuke M (1983) Influence of temperature on the zoospore germination of *Ecklonia stolonifera* Okamura (Phaeophyta, Laminariales) in culture. Jpn. J. Phycol. 31: 28–33 (in Japanese).

Notoya M, Asuke M (1984) Distribution of Laminariales plants along the coast of Aomori Prefecture. Scientific Reports Aquaculture Centre, Aomori Prefecture 3: 15–18 (in Japanese).

Notoya M, Aruga Y (1989) Vertical distribution of several species of macroalgae (Phaeophyta) along the coasts of Aomori Prefecture, Japan. Korean J. Phycol. 4(2): 165–170.

Notoya M, Aruga Y (1990) Relationship between size and holdfasts of *Ecklonia stolonifera* (Laminariales, Phaeophyta) in northern Honshu, Japan. Hydrobiologia 204/205: 241–246.

Migita S (1963) Studies on ecology and culture of *Undaria peterseniana* Bulletin of the Faculty of Fisheries, Nagasaki University 15:24–48 (in Japanese).

Okada Y, Sanbonsuga Y (1980) Effect of temperature on the growth and maturation of the female gametophytes of Laminariaceous plant. I. On *Laminaria japonica, L. ochotensis, L. diabolica, L. religiosa and L. angustata* var. *longissima* in culture. Hokkaido Regional Fishery Research Laboratory 50: 27–44 (in Japanese).

Otani K, Nakamura T (1985) Coastal Oceanography of Japanese Islands. Coastal Oceanography Research Committee, The Oceanographical Society of Japan (ed.), Tokai University Press, Tokyo, pp. 145–154 (in Japanese).

Saito Y (1956) An ecological study of *Undaria pinnatifida* Sur.-II. On the influence of the environmental factors upon the maturity of gametophytes and early development of sporophytes. Bull. Jpn. Soc. Sci. Fish. 22(4): 235–239 (in Japanese).

Suzuki E (1975) Kishou Toukei Gaku. Chijin Shokan, Tokyo, Japan, pp. 121–133 (in Japanese).

Takamatsu M (1938) Marine algae from Tsugaru Strait, northeastern Honshu, Japan. Saito-Ho-On Kai Museum Research Bulletin 17: 21–83.

Tomosada A (1994) Long-term variations of water temperature around Japan. Bull. Tohoku National Fisheries Research Institute 56: 1–45 (in Japanese).

Yamada Y (1928) Marine algae of Mutsu Bay and adjacent water. II. Scientific Reports Tohoku Imperial University 3: 497–534.

Yamauchi H, Takanashi K, Nakada K, Fujikawa Y, Aisaka K (2003) Seedling and cultivation experiments on local Laminariales seaweeds. Annual Reports Aquaculture Centre, Aomori Prefecture 32: 343–354 (in Japanese).

Journal of Applied Phycology (2006) 18: 529–541
DOI: 10.1007/s10811-006-9074-2

Introduced macroalgae – A growing concern

Britta Schaffelke[1,*], Jennifer E. Smith[2] & Chad L. Hewitt[3]
[1]*CRC Reef Research Centre and James Cook University, PO Box 772, Townsville QLD 4810, Australia; Current address: Australian Institute of Marine Science, PMB 3, Townsville MC QLD 4810, Australia;* [2]*Department of Botany, University of Hawaii Manoa, 3190 Maile Way, Honolulu, HI 96822, USA; Current address: University of California, 735 State St., Santa Barbara, CA 93101, USA;* [3]*Ministry of Agriculture and Forestry, Biosecurity New Zealand, PO Box 2526, Wellington 6001, New Zealand; Current address: National Centre for Marine and Coastal Conservation, Australian Maritime College, Private Bag 10, Rosebud VIC 3939, Australia*

*Author for correspondence: e-mail: b.schaffelke@aims.gov.au

Key words: introduced species, invasion biology, invasive macroalgae

Abstract

Introductions of non-indigenous species to new ecosystems are one of the major threats to biodiversity, ecosystem functions and services. Globally, species introductions may lead to biotic homogenisation, in synergy with other anthropogenic disturbances such as climate change and coastal pollution. Successful marine introductions depend on (1) presence of a transport vector, uptake of propagules and journey survival of the species; (2) suitable environmental conditions in the receiving habitat; and (3) biological traits of the invader to facilitate establishment. Knowledge has improved of the distribution, biology and ecology of high profile seaweed invaders, e.g. *Caulerpa taxifolia*, *Codium fragile* ssp. *tomentosoides*, *Sargassum muticum*, and *Undaria pinnatifida*. Limited, regional information is available for less conspicuous species. The mechanisms of seaweed introductions are little understood as research on introduced seaweeds has been mostly reactive, following discoveries of introductions. Sources of introductions mostly cannot be determined with certainty apart from those directly associated with aquaculture activities and few studies have addressed the sometimes serious ecological and economic impacts of seaweed introductions. Future research needs to elucidate the invasion process, interactions between invaders, and impacts of introductions to support prevention and management of seaweed introductions.

Introduction

Introduced species are considered to be one of the greatest threats to native marine biodiversity and resource values of the world's oceans (Norse, 1993; Vitousek et al., 1997; Carlton, 2000). Regional studies have identified hundreds of non-indigenous marine species (NIMS) introduced through human activities. These studies are, however, limited to a few countries or regions, i.e. Australia, Europe, New Zealand and the United States (e.g. Pollard & Hutchings, 1990; Cohen & Carlton, 1995; Cranfield et al., 1998; Coles et al., 1999; Ruiz et al., 2000; Hewitt et al., 2004). There is very little information on the status of NIMS in other regions (e.g. Williamson et al., 2002 for 20 member economies of the Asia-Pacific Economic Cooperation, APEC). The rate of introductions of NIMS has increased in the last 20 years, reflecting increased global trade but also more survey effort (Ruiz et al., 2000; Ribera Siguan, 2002; Hewitt, 2003a). Some NIMS have had catastrophic effects on the recipient ecosystem, e.g. the Asian clam (*Potamocorbula amurensis*) in San Francisco Bay (Nichols et al., 1994) and the comb jelly (*Mnemiopsis leidyi*) in the Black Sea (Kideys, 2002). The combined effects of global change and species introductions are believed to result in biotic homogenization (e.g., Olden et al., 2004; Olden and Poff, 2004; Wilkinson, 2004). Widespread generalists and opportunistic species will dominate ecosystems, a pattern already observed in locations affected by environmental

Table 1. Number of non-indigenous marine species (NIMS) introduced to various regions.

Location	Total extant NIMS	Macroalgal NIMS (no.)	Macroalgal NIMS (%.)	Reference
French Atlantic Coast	104	21	20	Goulletquer et al. (2002)
Italy	110	32	29	Occhipinti Ambrogi (2002)
North Sea coast	82	20	24	Reise et al. (2002)
Chile	32	12	38	Castilla et al. (in press)
Hawaii	89	21	24	Coles et al. (1999), Godwin (2001) and Smith et al. (2002)
New Zealand	109	19	17	Cranfield et al. (1998)
Port Phillip Bay, Australia	99	16	16	Hewitt et al. (2004)
United States (continental)	298	24	8	Ruiz et al. (2000)

degradation, and likely to be amplified by species introductions (McKinney & Lockwood, 1999).

Marine macroalgae are a significant component of introduced NIMS (Table 1). These include several high profile species that have caused significant ecological and economic impacts (e.g. *Caulerpa taxifolia* (Vahl) C. Agardh, *Codium fragile* (Suringar) Hariot ssp. *tomentosoides* (Van Goor) Silva, *Sargassum muticum* (Yendo) Fensholt and *Undaria pinnatifida* (Harvey) Suringar; e.g. Trowbridge, 1998; Boudouresque & Verlaque, 2002; Ribera Siguan, 2002, 2003; Wallentinus, 2002; Occhipinti-Ambrogi & Savini, 2003). Macroalgae are considered to be especially worrying NIMS as they may alter both ecosystem structure and function by monopolizing space, developing into ecosystem engineers, changing foodwebs, and spreading beyond their initial point of introduction through efficient dispersal capacities (Thresher, 2000).

The majority (80%) of marine macroalgal orders contain introduced species: 7 out of 9 orders in the phylum Chlorophyta, 16 out of 19 orders in the Rhodophyta, 8 out of 12 orders in the Phaeophyceae. The numbers of introduced species per order are highly correlated with total species number (Figure 1, Pearson-Product moment correlation: $r^2 = 0.91$, $p < 0.05$). However, some orders contain more, others less, introduced species than expected by chance alone; for example the Ectocarpales, Laminariales and Bonnemaisoniales have more, while the Chaetophorales, Fucales and Corallinales have less introduced species than expected (Smith et al., unpublished data).

Recent reviews of the status of introduced marine plants, both with a regional and global scope, include current inventories of introduced species as well as assessments of introduction vectors and mechanism that may influence invasion success (Wallentinus, 1999a, 2002; Verlaque, 2001; Ribera Siguan, 2002, 2003; Smith et al., 2002). Despite recent research, especially in the Pacific region and the Mediterranean Sea, we still have a limited understanding of the invasion process, the distribution and ecology of less conspicuous introduced macroalgae, and the ecological and economic impacts of marine invasions. In this review we will update current knowledge of seaweed introductions using recent case studies to illustrate the three main phases of the invasion process: uptake and transport, release and establishment, and spread and impact.

Uptake and transport

The first stage in the invasion process depends on the presence of a transport vector and the availability of suitable macroalgal life stages for uptake by this vector. The most important pathways for the transport of NIMS are associated with shipping vectors (ballast water and fouling of hulls), aquaculture and the aquarium trade (Ruiz et al., 2000; Carlton, 2001; Hewitt et al., 2004). It is often difficult to pinpoint a pathway for a specific introduction; it may differ between regions or the introduction may have occurred through multiple pathways.

Fouling of ships' hulls, structures or other surfaces and living epibiotically (e.g., on mollusks) or as boring organisms (e.g., the conchocelis phase of Porphyra species boring into mollusk shells) are considered to be the most important pathways for the unintentional introduction of macroalgae (Ribera Siguan, 2003). All macroalgae have the potential to colonise ships' hulls and other maritime structures, especially species that occur either within or in close proximity to port environments. In Port Phillip Bay, Australia, fouling of ships' hulls is considered to be the most

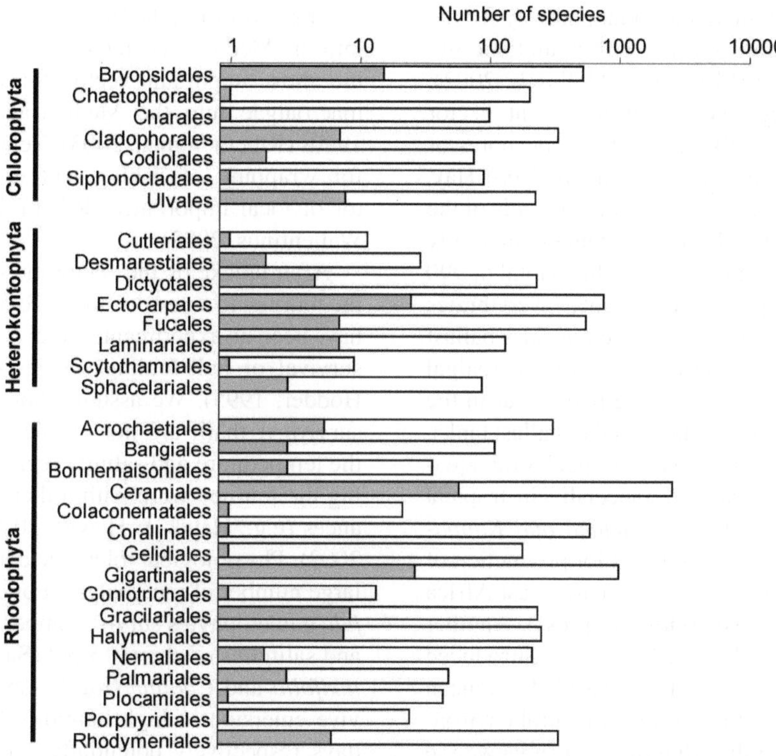

Figure 1. Proportion of number of introduced (grey bars) to total number of species (white bars) in macroalgal orders containing introduced species. Note logarithmic scale. Data are from a database with published records of introduced macroalgae (Smith et al., unpublished data) and AlgaeBase (http://www.algaebase.org, ©1996 – 2004 M.D. Guiry).

important vector for macroalgal introductions (Lewis, 1999; Hewitt et al., 2004). Availability of large numbers of propagules would facilitate colonisation of ships' hulls and other surfaces. For example, high-density populations of *U. pinnatifida* and *Laminaria japonica* J. E. Areschoug occur along the North West Pacific coasts, where the two species are widely cultivated. In these areas the probability is high for zoospores or gametophytes to settle on ships' hulls, aquaculture stock (e.g. oysters) and equipment. Introduced *Codium fragile* ssp. *tomentosoides* in Australia is also generally found in modified environments, often associated with shipping-related infrastructure such as marinas, wharfs, jetties, rip rap, and mooring sites.

In historical times, wooden ships carried vast amounts of fouling species, including macroalgae, on hulls and ballast rock (Carlton & Hodder, 1995; Carlton, 2003). This may explain the cosmopolitan distribution of many well-known fouling taxa such as members of the Ceramiaceae, Ectocarpaceae, Ulvaceae and Cladophoraceae. These are now considered to be 'cryptogenic' (of unknown origin, sensu Carlton, 1996a) in many locations, and include species that may have been introduced many centuries ago. The use of antifouling paint on modern vessels provides only partial protection. Even large vessels have areas on the hull and internal water intake structures (sea chests) that are not or incompletely antifouled and can be colonised by fouling species (Coutts et al., 2003). Smaller vessels, such as small commercial boats, private yachts and launches commonly used in coastal marine traffic may pose an even higher risk due to (i) their usual residence in coastal waters close to seaweed habitats, (ii) their frequently extensive mooring periods, (iii) their slow travel speed, and (iv) their highly variable hull maintenance patterns (Floerl & Inglis, in press; Floerl et al., in press). The incidence of hull fouling is likely to increase as the use of tributyltin (TBT), the main active ingredient in antifouling paints for commercial vessels, will be globally phased out by 2008 for environmental reasons (International Convention on the Control of Harmful Antifouling Systems on Ships 20012).[1]

[1] Adopted 5 October 2001, http://www.imo.org accessed 18 May

Ballast water is the most important pathway for the introduction of plankton, species with planktonic life history stages and fish (Minchin & Gollasch, 2002). Ballast water is suggested as an important vector for *U. pinnatifida*, possibly transporting zoospores or suspended gametophytes (Hay & Luckens, 1987; Hay, 1990). However, an extensive international study of the species composition in ballast water tanks found only fragments of four macroalgal taxa within a total of 990 taxa (bacteria, fungi, protozoans, algae, invertebrates and fishes; Gollasch et al., 2002). We consider ballast water to be a less important pathway for macroalgal introductions. However, macroalgae may occur in the much less studied sediments deposited in ballast tanks.

The direct introduction of seaweed species for aquaculture is an important vector, especially in tropical regions (Smith et al., 2002). *Eucheuma and Kappaphycus* species have been introduced for production of carrageenan to 26 countries in the Pacific, east Africa and the Caribbean (Zemke-White, in press). Another well-known example is the translocation of introduced *U. pinnatifida* for aquaculture from the Mediterranean to Brittany where it established in the natural environment and spread along the Atlantic coast (reviewed in Wallentinus, 1999b).

The transport and cultivation of NIMS in the domestic and international aquarium trade, including public, private and research aquarium facilities, are potential pathways for the introduction of macroalgae. Whole thalli, fragments or propagules can be released to waterways through untreated effluent or disposal of biomass. The best-known example is the introduction of *Caulerpa taxifolia* into the Mediterranean, presumably by accidental release from a public aquarium (Meinesz & Boudouresque, 1996). Eleven species of marine macroalgae are available through the aquarium trade (Wallentinus 2002), as well as 'live rock', natural substratum cultivated for its variety of attached epibionts including macroalgae (Wallentinus, 2002; Frisch & Murray, 2002).

Macroalgal introductions to Europe are dominated by associations with aquaculture vectors (Maggs & Stegenga, 1999; Reise et al., 1999; Ribera Siguan, 2002; Wallentinus, 2002). The large-scale import of the Pacific Oyster to Europe in the 20th century, typically without any quarantine measures (Wolff & Reise, 2002), may explain the high proportion of Pacific macroalgae in European waters (data in Wallentinus, 2002).

Transport from the Red Sea through the Suez Canal into the Mediterranean ('Lessepsian migrations') was the most important vector for the introduction of macroalgae into the Mediterranean until the early 1990s (Ribera Siguan, 2003). Release of seaweed used for wrapping of fishing bait or seafood is a vector of local importance (Ribera Siguan, 2002; 2003; Wallentinus, 2002).

After uptake by the vector the species must survive the journey to a new location. To our knowledge, there have been no experimental studies on the trans-oceanic survival of hull-fouling species (but see Carlton & Hodder, 1995). We assume, however, species would survive if thalli are not physically dislocated and if the temperature and salinity regimes encountered during the journey were within their physiological tolerances (e.g., Hayes & Hewitt, 2000; Hewitt & Hayes, 2002). Physiological tolerance data are available for a large number of macroalgae. For example, *U. pinnatifida* gametophytes survive temperatures of –1 to 30 °C and salinities of above 15 ppt (Saito, 1975). *Caulerpa taxifolia* and *Codium fragile* ssp. *tomentosoides* survive emersion in high humidity for up to 10 and 90 days, respectively, potentially enabling them to survive shipboard transport for extended periods, for example entangled in fishing nets (Sant et al., 1996; Schaffelke & Deane, in press).

Release and establishment

Following release, a successful invader must survive and establish itself in the receiving environment. This phase in the invasion process is least well known for seaweeds. The definition of establishment has been ambiguous in the literature (Hewitt et al., unpublished). Here we use the definition of established as forming 'a reproductive and self-sustaining population' (e.g., Case, 1996; Williamson & Fitter, 1996; Duncan et al., 2001).

Establishment success is mainly determined by a combination of the following three factors:

- Inoculum pressure (vector frequency and rate of vector infection);
- Abiotic and biological characteristics of the receiving environment; and
- Eco-physiological characteristics of the arriving species.

Embayments and estuaries appear to more prone to introductions than open coast habitats (Carlton, 1996b). These environments, however, also have high

2004.

inoculum pressure, i.e. one or more significant vectors are generally present in port environments and urbanised embayments (Ruiz et al., 2000). International ports and harbours are both primary points of inoculation and initial establishment but may also be source populations for secondary spread (e.g., Ruiz et al., 2000; Hewitt, 2002; Ruiz & Hewitt, 2002; Hewitt et al., 2004). The accidental release of aquarium species is also significantly correlated with urban centers. Similarly, aquaculture facilities are typically located in embayments, often immediately adjacent to port environments. These locations represent 'hot spots' of species introduction (Ruiz & Hewitt, 2002; Hewitt, 2003a), even though invasions do occur in a wide variety of marine habitats (Carlton, 2002).

Successful establishment of species after arrival is dependent on matching environmental conditions in the source and recipient environments (e.g., Hewitt & Hayes, 2002). For example, low winter temperatures seem to have prevented the establishment of *Porphyra yezoensis* Ueda, introduced for aquaculture to the east coast of the United States (Watson et al., 2000) and of *Caulerpa taxifolia* in Japan, where it escaped from an aquarium facility (Komatsu et al., 2003). Risk assessments for bioinvasions use environmental conditions to predict, for example, whether the ballast water taken up by a vessel is of high risk to the environment at the destination, and species-specific physiological tolerance data to identify the risk of inoculation of a specific site (Hewitt & Hayes, 2002).

Some recipient environment characteristics are associated with increased introduction incidence: low native biodiversity and anthropogenic disturbance (e.g., water and sediment pollution, structures providing artificial substrates and altered temperature regime due to effluents; reviewed in Carlton, 1996b and Gollasch & Leppäkoski, 1999). For example, the establishment of *U. pinnatifida* in Australia was facilitated by reduced native macroalgal cover (Valentine & Johnson, 2003; 2004). Habitat 'invasibility' is also dependent on functional diversity of macroalgal habitats, e.g. less diverse algal turf assemblages and seagrass meadows can promote the establishment of introduced *Caulerpa* species (Cecherelli & Cinelli, 1998; Ceccherelli et al., 2002). *C. taxifolia* establishment and proliferation has been linked to enrichment of substrata by urban wastewater and organic matter (Chisholm et al., 1997). Extensive blooms of non-indigenous *C. brachypus* Harvey, recently discovered in Florida, may also be linked to local nutrient enrichment (Jacoby et al., 2004).

There is evidence that changes caused by numerous introductions into one region can synergistically operate as a biological disturbance agent and pave the way for new introductions, which has been called "invasional meltdown" (Simberloff & von Holle, 1999).

The successful establishment of *Codium fragile* ssp. *tomentosoides* in the North West Atlantic (Nova Scotia, Canada) versus presence at only low abundances in the North East Atlantic (England) has been explained by biological differences of the native community, despite similar abiotic environmental characteristics (Chapman, 1999). In the northeast Atlantic benthic biodiversity and grazing rates are high, whereas in the northwest Atlantic periodic disturbance of native kelp beds by sea urchin grazing has opened a window suitable for *C. fragile* ssp. *tomentosoides* establishment. This has been facilitated by factors that disrupt the natural sea urchin/kelp dynamics: spreading of the introduced bryozoan *Membranipora membranacea* that overgrows kelp blades and of several introduced red seaweeds colonising vacant space created by urchin grazing (Levin et al., 2002). In addition, grazing pressure is reduced by the decimation of the sea urchin *Strongylocentrus droebachiensis* from an amoebic disease, which presumably is also introduced (Harris & Tyrell, 2001; Chapman et al., 2002), and by avoidance of *C. fragile* ssp. *tomentosoides* by grazers (Scheibling & Anthony, 2001).

Species traits may facilitate the establishment of NIMS. Applying the properties of successful invaders after Lodge (1993), several *r*-selected traits have been identified for *Codium fragile* ssp. *Tomentosoides,* such as high growth rate and reproductive output, vegetative and parthenogenetic reproduction, and broad environmental tolerances (Chapman, 1999). However, most of these characteristics also apply to non-invasive subspecies of *C. fragile* (Trowbridge, 1998). In contrast, a quantitative ranking of European introduced and native seaweed species (using categories of species traits such as dispersal capabilities, environmental tolerances, reproductive mode, and size) indicated that introduced species indeed have species traits that increase the likelihood of successful invasion (Nyberg & Wallentinus, in press). Species most likely to be successful are: *C. fragile* ssp. *tomentosoides, Caulerpa taxifolia, U. pinnatifida, Asparagopsis armata* Harvey and *Grateloupia doryphora* (the currently accepted name for this species is *G. turuturu* Yamada (Gavio & Fredericq, 2002), however, the identity of records from Sicily has recently been disputed (Wilkes et al., unpublished data)).

Spread and impact

From the initial incursion sites, for example close to international ports or aquaculture facilities, NIMS spread to other areas by natural dispersal or by domestic translocation. Vectors for domestic translocation are similar to those of the initial introduction, such as aquaculture stock movements, coastal and recreational shipping (Kinloch et al., 2003).

Caulerpa taxifolia in the Mediterranean has spread steadily since its introduction in 1984, with an estimated colonised area of 131 km^2 (Meinesz et al., 2001). However, the current distribution and local abundance is disputed and remote sensing results suggest that *C. taxifolia* cover along the south coast of France may have been overestimated by a factor of ten (Jaubert et al., 2003). Since the early 1990s a second *Caulerpa* species has been spreading in the Mediterranean Sea, now identified as the proposed combination *C. racemosa* var. *cylindracea* (Sonder) Verlaque, Huisman *et* Boudouresque (Verlaque et al., 2003). The rate of spread of *C. racemosa* var. *cylindracea* and the co-occuring *Womersleyella setacea* (Hollenberg) R. Norris (see below) is dramatic compared to other introduced macroalgae in Europe (Verlaque et al., 2004). *C. racemosa* var. *cylindracea* is competitively superior to *C. taxifolia*, where the two species co-occur (Piazzi et al., 2001a; Piazzi & Ceccherelli, 2002).

The understanding of one introduction often cannot predict other introductions of the same species, as the factors determining success of establishment and further spread are site- or time-specific (Grosholz, 1996; see above for *Codium fragile* ssp. *tomentosoides*). While Boudouresque and Verlaque (2002) do not consider *U. pinnatifida* as an invasive species in the Mediterranean Sea (defined as NIMS that spread from the point of introduction and become abundant; Kolar and Lodge, 2001) the species is invasive, indicated by the continuously expanding range, along the European Atlantic coast (Wallentinus, 1999b), the west coast of the United States and Mexico (Silva et al., 2002; L. Aguilar Rosas, pers. comm.) and in the southern hemisphere (Sinner et al., 2000; Casas et al., in press).

Studies of impacts of NIMS are often hampered by the lack of ecological baseline studies. Typically, studies are only initiated after the incursion has already occurred and use comparisons of sites colonised and un-colonised by NIMS (Hewitt, 2003b). In such a study on *U. pinnatifida* Forrest and Taylor (2002) found no differences in native species richness and abundance, but suggest that the lack of benthic community data before establishment of *U. pinnatifida* limits inferences. *U. pinnatifida* has caused changes to the composition of native macroalgal communities (Battershill et al., 1999; Sinner et al., 2000, Valentine and Johnson, 2003), as well as decreases in cover (Curiel et al. 1998, 2001) and diversity (Casas et al., in press). Short-term studies indicated that the presence of *Caulerpa taxifolia* had a negative effect on seagrass shoot density, especially under nutrient enrichment (Ceccherelli and Cinelli, 1997). In contrast, long-term experiments suggest that *C. taxifolia* and seagrass are likely to co-exist and that high nutrient availability will not change competitive relations (Ceccherelli and Sechi, 2002). Overgrowth by *C. racemosa* changed macroalgal community composition and seagrass shoot density (Piazzi et al., 2001b; Ceccherelli and Campo, 2002).

Impacts of NIMS may also change through time. NIMS often persist at low levels and later start to increase in abundance and spread, which Stockwell et al. (2003) attributes to either an initial period of adaptation or a change to previously functional environmental controls such as competition or grazing. In contrast, adaptations to NIMS may also occur by herbivores changing preferences from native species to NIMS (Stimson et al., 2001). In other cases NIMS are not preferred (Schaffelke et al., 1995), preferred by only a few grazers (Trowbridge, 1998; Thornber et al., 2004) or no change of grazer populations and feeding habits was observed (Francour et al., 1995).

Apart from the handful of high profile species, rhodophytes are the most prevalent group of introduced macroalgae (Ribera Siguan, 2003). It is likely that consequences of these introductions are underestimated because the taxa involved are often inconspicuous and difficult to identify to species level. This is further complicated by separate introductions of morphologically dissimilar generations (e.g. gametophytes vs. tetrasoporophytes of *Asparagopsis armata*, Maggs and Stegenga, 1999) or cryptic invasions of sibling species that are morphologically indistinguishable from native species (e.g. McIvor et al., 2001). The detection of cryptic invasions is much aided by molecular techniques, which can also assist in the assignment of source regions of introductions (see below).

At least 21 seaweed species have been introduced to the Hawaiian Islands, both accidentally and intentionally for seaweed aquaculture (Godwin, 2001; Smith et al., 2002). The islands represent one of the most heavily invaded tropical systems in the world. Several species (*Acanthophora* spicifera (M. Vahl) Børgesen, *Avrainvillea amadelpha* (Montagne) A. Gepp and E.S.

Gepp, *Gracilaria salicornia* (C. Agardh) E.Y. Dawson, *Hypnea musciformis* (Wulfen) J.V. Lamouroux, *Kappaphycus* spp. and *Eucheuma* spp.), predominantly Rhodophytes, are now established in high abundance and spreading (Smith et al., 2002; Conklin and Smith, in press; G. Zucarello, pers. comm.). Four of these species are overgrowing live hard corals, sometimes leading to coral mortality (Smith et al., 2002). Costs associated with *H. musciformis* blooms are ~US$55,000 per year for one town alone, for removal of rotting algal biomass washed up onto beaches (Van Beukering and Cesdar, 2004).

Womersleyella setacea is an Indo-Pacific species (e.g. Silva et al., 1996), recently introduced into the Mediterranean Sea (first report in Verlaque, 1989) where it is now widely distributed (Airoldi et al., 1995; Piazzi and Cinelli, 2001), and is also found in the Canary Islands (Haroun et al., 2002). The species has developed dense turf assemblages on rocky substratum and on seagrass rhizomes, with reduced biodiversity compared with unaffected sites (Piazzi et al., 2002; Piazzi and Cinelli, 2003).

Heterosiphonia japonica Yendo, a North Pacific species (e.g., Abbott and Hollenberg, 1976; Yoshida et al., 1990), was recently introduced to the East Atlantic and Mediterranean Sea, potentially by oyster imports, and has since spread along the Atlantic coasts of Spain, France, and Norway (Lein, 1999; Maggs and Stegenga, 1999; Verlaque, 2001). It is now the most common species in sheltered and semi-exposed subtidal locations along the south-west coast of Norway, overgrowing other benthos (Husa et al., 2004). Further north *H. japonica* is found mainly in or near harbours, indicating translocation by shipping and fishing activities (Husa et al., 2004).

Management options

Several steps have been identified as fundamental to the management of NIMS: prevention and monitoring; detection and rapid response; and long-term control. The development of awareness and understanding by public and political interests, appropriate research strategies, and information management and sharing, underpin these steps.

Prevention and monitoring

The most cost-effective management strategy in the marine environment will be to reduce the introduction risks through minimisation of inoculation frequency and propagule pressure. These options cannot be solely driven at a local or national level, but require significant international and regional cooperation (e.g., Bax et al., 2003; Hewitt, 2003a). Several international and multilateral regional actions have recently been enacted to reduce the rates of NIMS transfers from various vectors. Examples include: The International Convention for the Control and Management of Ship's Ballast Water and Sediments (http://www.imo.org; see also Hewitt, 2003a; McConnell, 2003) that now requires ratification. This convention will create a uniform standard for the regulation of ballast water management. The International Council for the Exploration of the Seas (ICES) developed a Code of Practice (CoP) for the Introductions and Transfers of Marine Organisms in 1994 (updated in 2003, available at http://www.ices.dk). This CoP aids the management of intentional introductions (e.g., mariculture and stocking) and accidental introductions associated with aquaculture species. However, most effective would be a preferential development of aquaculture of native species. APEC has undertaken an assessment of regulatory frameworks for NIMS management in member economies (Bax et al., 2003) to develop a common regional risk management framework for NIMS, primarily targeting ballast water and hull fouling. Altogether, these actions do not fully address the dominant pathways for macroalgal introductions, i.e., translocations for aquaculture and fouling of marine vessels and installations.

Detection, rapid response and long-term control

Most management plans for introduced species have elements of 'rapid response' for eradication action, identifying when and how to shift to long-term control (e.g., Wotton and Hewitt, 2004). Rapid response requires early detection, either through passive (e.g., public reporting) or active means (e.g., surveillance program) and an understanding of what is already present (e.g., baseline surveys). Australia and New Zealand have the established national systems of port baseline surveys using standardised collection methods (Hewitt and Martin, 2001; Ruiz and Hewitt, 2002).

The determination of source regions is fundamental to decisions on management action after discovery of an introduction. For example, *Caulerpa taxifolia* was discovered between 2000 and 2002 in both the USA (California) and in Australia (New South Wales and South Australia). There are no native *Caulerpa* species in California, making the determination as an introduced species unequivocal. The two populations were identified as genetically identical with the 'aquarium-Mediterranean strain' (Jousson et al., 2000). A campaign to eradicate the alga commenced in the same year. In Australia, however, *C. taxifolia* is native to the tropical and subtropical regions (reviewed in Phillips and Price, 2002). Several populations of *C. taxifolia* were discovered in the region around Sydney, more than 800 km south of previous records, and even further away around Adelaide. Using molecular markers, Schaffelke et al. (2002) and Murphy and Schaffelke (2003) ruled out that *C. taxifolia* was introduced from overseas, i.e. from the Mediterranean or overseas aquaria, with high confidence for three of six new locations. It is most likely that the new records are the result of domestic translocation(s) from Australian (sub)tropical populations, assisted by human activities such as boat traffic and fishing or through the domestic aquarium trade (Schaffelke et al., 2002). A number of molecular markers are identical between the 'aquarium-Mediterranean strain' and certain Australian populations, indicating that the latter may be the origin of the introduction into the international aquarium market and thence to the Mediterranean (Meusnier et al., 2002, 2004; Fama et al., 2002). Recent molecular research indicates that *C. taxifolia* consists of at least two incipient species, of which only one is known to be invasive, and that there is evidence for a second, previously unrecognised, introduction event into the Mediterranean Sea (Meusnier et al., 2002, 2004).

Rapid response activities entail a variety of methods (e.g., physical or chemical control) with the intent of containing and eradicating introduced species as soon after detection as possible. Benefits and hazards of eradication/control efforts need to be balanced against the benefits and hazards of doing nothing, specific to each species and location (Hewitt et al., in press). While physical removal/control of introduced macroalgae (especially *Caulerpa taxifolia*) has shown some promising results, other methods such as chemical and biological control had limited effect (reviewed in McEnnulty et al., 2001). Recent efforts to eradicate or control *C. taxifolia* using chemical methods are more encouraging. In Australia, application of sea salt at 50 kg m^{-2} was successful in reducing abundance of the target species by up to 95% whereas native seagrass and infauna were less affected and had largely recovered after 6 months (Glasby et al., unpub data). In California, chlorine was applied under black tarpaulins left in place for several months, after which no *C. taxifolia* was found in cores from the treated area (Anderson, 2002). Laboratory tests indicate that chlorine concentrations of 125 ppm for at least 30 min result in 100% mortality, and that treatment should preferably occur in winter when *C. taxifolia* grows slowly (Williams and Schroeder, 2004).

Manual removal of *U. pinnatifida* significantly reduces sporophyte numbers. However, in established populations, survival over more than 2.5 years occurs either through 'seed banks' of microscopic stages or selective gametophyte survival in microhabitats (Hewitt et al. in press). Zoospore release in *U. pinnatifida* in Australia is limited to larger sporophytes for most of the growing season, but late in the season small sporophytes form mature sporophylls within just one month (Schaffelke et al., in press). Hence, removal efforts need to be more frequent in the late season. Removal of *Kappaphycus* spp. in Hawai'i required $\sim$2 h m^{-2} and regrowth was rapid (Conklin and Smith, in press). Manual removal of introduced macroalgae is a long-term commitment, and needs to be coupled with vector management and education to reduce the chances of re-inoculation and spread, and with monitoring (and response) on a larger spatial scale for the early detection of new incursion sites.

The eradication, or even control, of marine invasive species is both technically difficult and costly (Meyerson & Reaser, 2002). Globally, few marine incursions have resulted in response actions and, of those, a limited number have succeeded (e.g., Bax, 1999; Culver & Kuris, 2000; Wotton et al., in press). The recent successful eradication of *U. pinnatifida,* from the Chatham Islands near New Zealand was achievable because it only occurred on a single sunken vessel hull on sandy substrate, limiting the likelihood of spread (Wotton et al., in press). The total cost of this eradication was in excess of $\sim$US$ 1.8 million (ibid.). The costs of the eradication campaign for *C. taxifolia* in California are to date $\sim$US$ 4 million (R. Woodfield, pers. comm.).

Lastly, the crux to any long-term control or eradication effort will be sustained public and political will. Biosecurity must identify impacts of introductions in economic and social terms, create gains that can be readily identified, but also communicate losses and identify the causes, and link biosecurity to tangible examples that remain of current interest.

Conclusion

Macroalgal introductions are increasing. Despite sometimes being a new resource for harvest and aquaculture, there is a growing understanding of the often significant ecological and economic impacts of these introductions. Only sustained biosecurity management, based on better knowledge of invasion ecology, will prevent these impacts from changing ecosystems as we know them.

Acknowledgements

We thank Mike Guiry for the provision of species numbers from AlgaeBase and Keith Hayes for information from the CSIRO NIMPIS database. We also thank the colleagues who made unpublished information available to us.

References

Abbott IA, Hollenberg GJ (1976) Marine Algae of California. Stanford University Press, Stanford, USA.

Airoldi L, Rindi F, Cinelli F (1995) Structure, seasonal dynamics and reproductive phenology of a filamentous turf assemblage on a sediment influenced, rocky subtidal shore. Bot. Mar. 38: 227–237.

Anderson LWJ (2002) *Caulerpa taxifolia* in the United States: rapid response and eradication program. International *Caulerpa taxifolia* Conference Proceedings, January 31-February 1, 2002. California Sea Grant College Program, La Jolla. Publication No. T-047 (CD-ROM).

Battershill C, Miller K, Cole R (1999) The understorey of marine invasions. Seafood NZ 6: 31–33.

Bax NJ (1999) Eradicating a dreissenid from Australia. Dreissena! 10: 1–5.

Bax N, Williamson A, Aguero M, Gonzalez E, Geeves W (2003) Marine invasive species: A threat to global biodiversity. Mar. Policy 27: 313–323.

Boudouresque CF, Verlaque M (2002) Biological pollution in the Mediterranean Sea: Invasive versus introduced macrophytes. Mar. Poll. Bull. 44: 32–38.

Carlton JT (1996a) Biological invasions and cryptogenic species. Ecology 77: 1653–1655.

Carlton JT (1996b) Pattern, process, and prediction in marine invasion ecology. Biol. Cons. 78: 97–106.

Carlton JT (2000) Global change and biological invasions in the Oceans. In Mooney HA, Hobbs RJ (eds), Invasive Species in a Changing World, Island Press, Washington DC, pp. 31–53.

Carlton JT (2001) Introduced species in U. S. coastal waters: Environmental impacts and management priorities. Pew Oceans Commission, Arlington, Virginia. 28 pp.

Carlton JT (2002) Bioinvasion ecology: Assessing invasion impact and scale. In Leppäkoski E, Gollasch S, Olenin S (eds), Invasive Aquatic Species of Europe. Distribution, Impacts and Management, Kluwer Academic Publishers, Dordrecht, pp. 7–19.

Carlton JT (2003) Community assembly and historical biogeography in the North Atlantic Ocean: The potential role of human-mediated dispersal vectors. Hydrobiologia 503: 1–8.

Carlton JT, Hodder J (1995) Biogeography and dispersal of coastal marine organisms: Experimental studies on a replica of a 16th-century sailing vessel. Mar. Biol. 121: 721–730.

Casas G, Scrosati R, Piriz ML, in press. The invasive kelp *Undaria pinnatifida* (Phaeophyceae, Laminariales) reduces native seaweed biodiversity in Nuevo Gulf (Patagonia, Argentina). Biological Invasions.

Case TJ (1996) Global patterns in the establishment and distribution of exotic birds. Biol. Cons. 78: 69–96.

Castilla JC, Uribe M, Bahamonde N, Clarke M, Desqueyroux-Faúndez R, Kong I, Moyano H, Rozbaczylo N, Santelices B, Valdovinos C, Zavala P, in press. Down under the southeastern Pacific: Marine non-indigenous species in Chile. Biological Invasions.

Ceccherelli G, Campo D (2002) Different effects of *Caulerpa racemosa* on two co-occurring seagrasses in the Mediterranean. Bot. Mar. 45: 71–76.

Ceccherelli G, Cinelli F (1997) Short-term effects of nutrient enrichment of the sediment and interactions between the seagrass *Cymodocea nodosa* and the introduced green alga *Caulerpa taxifolia* in a Mediterranean bay. J. Exp. Mar. Biol. Ecol. 217: 165–177.

Ceccherelli G, Cinelli F (1998) Habitat effect on spatio-temporal variability in size and density of the introduced alga *Caulerpa taxifolia*. Mar. Ecol. Prog. Ser. 163: 289–294.

Ceccherelli G, Sechi N (2002) Nutrient availability in the sediment and the reciprocal effects between the native seagrass *Cymodocea nodosa* and the introduced rhizophytic alga *Caulerpa taxifolia*. Hydrobiologia 474: 57–66.

Ceccherelli G, Piazzi L, Balata D (2002) Spread of introduced *Caulerpa* species in macroalgal habitats. J. Exp. Mar. Biol. Ecol. 280: 1–11.

Chapman AS (1999) From introduced species to invader: What determines variation in the success of *Codium fragile* ssp. *tomentosoides* (Chlorophyta) in the North Atlantic Ocean? Helgol. Meeresunters. 52: 277–289.

Chapman AS, Scheibling RE, Chapman ARO (2002) Species introductions and changes in the marine vegetation of Atlantic Canada. In Claudi R, Nantel P, Muckle-Jeffs E (eds), Alien Invaders in Canada's Waters, Wetlands and Forests, Natural Resources Canada, Canadian Forest Service, Science Branch, Ottawa, pp. 133–148.

Chisholm JRM, Fernex FE, Mathieu D, Jaubert JM (1997) Wastewater discharge, seagrass decline and algal proliferation on the Cote d'Azur. Mar. Poll. Bull. 34: 78–84.

Cohen AN, Carlton JT (1995) Non-indigenous aquatic species in a United States estuary: A case study of the biological invasions of the San Francisco Bay and Delta. Report to US Fish and Wildlife Service, Washington DC.

Coles SL, DeFelice LG, Carlton JT (1999) Historical and recent introductions of non-indigenous marine species into Pearl Harbour, Oahu, Hawaiian Islands. Mar. Biol. 135: 147–158.

Conklin EJ, Smith JE (2005) Abundance and spread of the invasive red alga, *Kappaphycus* spp., in Kane'ohe Bay, Hawai'i and an experimental assessment of management options. Biological Invasions 7: 1029–1039.

Coutts ADM, Moore KM, Hewitt CL (2003) Ships' sea chests: An overlooked transfer mechanism for non-indigenous marine species? Mar. Poll. Bull. 46: 1504–1515.

Cranfield HJ, Gordon DP, Willan RC, Marshall BA, Battershill CN, Francis MP, Nelson WA, Glasby CJ, Read GB (1998) Adventive species in New Zealand., NIWA Technical Report #34, Wellington, New Zealand.

Culver CS, Kuris AM (2000) The apparent eradication of a locally established introduced marine pest. Biological Invasions 2: 245–253.

Curiel D, Bellemo G, Marzocchi M, Scattolin M, Parisi G (1998) Distribution of introduced Japanese macroalgae *Undaria pinnatifida*, *Sargassum muticum* (Phaeophyta) and *Antithamnion pectinatum* (Rhodophyta) in the Lagoon of Venice. Hydrobiologia 385: 17–22.

Curiel D, Guidetti P, Bellemo G, Scattolin M, Marzocchi M (2001) The introduced alga *Undaria pinnatifida* (Laminariales, Alariaceae) in the Lagoon of Venice. Hydrobiologia 477: 209–219.

Duncan RP, Bomford M, Forsyth DM, Conibear L (2001) High predictability in introduction outcomes and the geographical range size of introduced Australian birds: A role for climate. J. Anim. Ecol. 70: 621–632.

Fama P, Jousson O, Zaninetti L, Meinesz A, Dini F, Giuseppe GDi, Millar AJK, Pawlowski J (2002) Genetic polymorphism in *Caulerpa taxifolia* (Ulvophyceae) chloroplast DNA revealed by a PCR-based assay of the invasive Mediterranean strain. J. Evol. Biol. 15: 618–624.

Floerl O, Inglis GJ (2005) Starting the invasion pathway: The interaction between source populations and human transport vectors. Biological Invasions 7: 589–606.

Floerl O, Inglis GJ, Marsh HM (2005) Selectivity in vector management: An investigation of the effectiveness of measures used to prevent transport of non-indigenous species. Biological Invasions 7: 459–475.

Forrest BM, Taylor MD (2002) Assessing invasion impact: Survey design considerations and implications for management of an invasive marine plant. Biological Invasions 4: 375–386.

Francour P, Harmelin-Vivien M, Harmelin JG, Duclerc J (1995) Impacts of *Caulerpa taxifolia* colonization on the littoral ichtyofauna of North-Western Mediterranean sea: Preliminary results. Hydrobiologia 300/301: 345–353.

Frisch S, Murray S (2002) The availability of *Caulerpa* spp. and "Live Rock" in retail aquarium outlets in southern California. International *Caulerpa taxfolia* Conference Proceedings, January 31-February 1, 2002. California Sea Grant College Program, La Jolla. Publication No. T-047 (CD-ROM).

Gavio B, Fredericq S (2002) *Grateloupia turuturu* (Halymeniaceae, Rhodophyta) is the correct name of the non-native species in the Atlantic known as *Grateloupia doryphora*. Eur. J. Phycol. 37: 349–359.

Godwin LS (2001) Hull fouling of maritime vessels as a pathway for marine species invasions to the Hawaiian Islands. In Proceedings of the 24th Meeting of the Marine Facilities Panel of the United States Japan Cooperative Program in Natural Resources, November 4–10, 2001: 1–10.

Gollasch S, Leppäkoski E (1999) Initial risk assessment of alien species in Nordic coastal waters. Nord 1999: 8. Nordic Council of Ministers, Copenhagen.

Gollasch S, Macdonald E, Belson S, Botnen H, Christensen JT, Hamer JP, Houvenaghel G, Jelmert A, Lucas I, Masson D, McCollin T, Olenin S, Persson A, Wallentinus I, Wetsteyn LPMJ, Wittling T (2002) Life in ballast tanks. In Leppäkoski E, Gollasch S, Olenin S (eds), Invasive Aquatic Species of Europe. Distribution, Impacts and Management, Kluwer Academic Publishers, Dordrecht: pp. 217–231.

Goulletquer P, Bachelet G, Guy Sauriau P, Noel P (2002) Open Atlantic coast of Europe-a century of in troduced species into French waters. In Leppäkoski E, Gollasch S, Olenin S (eds), Invasive Aquatic Species of Europe. Distribution, Impacts and Management, Kluwer Academic Publishers, Dordrecht, pp. 276–290.

Grosholz ED (1996) Contrasting rates of spread for introduced species in terrestrial and marine systems. Ecology 77: 1680–1686.

Haroun RJ, Gil-Rodriguez MC, de Castro JD, van Reine WFP (2002) A checklist of the marine plants from the Canary Islands (Central Eastern Atlantic Ocean). Bot. Mar. 45: 139–169.

Harris LG, Tyrrell MC (2001) Changing community states in the Gulf of Maine: synergism between invaders, overfishing and climate change. Biological Invasions 3: 9–21.

Hay CH (1990) The dispersal of sporophytes of *Undaria pinnatifida* by coastal shipping in New Zealand, and implications for further dispersal of *Undaria* in France. Br. Phycol. J. 25: 301–313.

Hay CH, Luckens PA (1987) The Asian kelp *Undaria pinnatifida* (Phaeophyta: Laminariales) found in a New Zealand harbour. NZ J. Bot. 25: 329–332.

Hayes K, Hewitt CL (2000) Risk assessment framework for ballast water introductions – Volume II. Centre for Research on Introduced Marine Pests Technical Report No. 21, CSIRO Marine Research, Hobart.

Hewitt CL (2002) The distribution and diversity of tropical Australian marine bio-invasions. Pacific Science 56: 213–222.

Hewitt CL (2003a) Marine biosecurity issues in the world oceans: Global activities and Australian directions. Ocean Yearbook 17: 193–212.

Hewitt CL (2003b) The diversity of likely impacts of introduced marine species in Australian waters. Records of the South Australian Museum Monographs Series 7: 3–10.

Hewitt CL, Hayes KR (2002) Risk assessment of marine biological invasions. In Leppäkoski E, Gollasch S, Olenin S (eds), Invasive Aquatic Species of Europe. Distribution, Impacts and Management, Kluwer Academic Publishers, Dordrecht: pp. 456–466.

Hewitt CL, Martin RB (2001) Revised protocols for baseline port surveys for introduced marine species – design considerations, sampling protocols and taxonomic sufficiency. Centre for Research on Introduced Marine Pests Technical Report No. 22. CSIRO Marine Research, Hobart.

Hewitt CL, Campbell ML, Thresher RE, Martin RB, Boyd S, Cohen BF, Currie DR, Gomon MF, Keough MJ, Lewis JA, Lockett MM, Mays N, McArthur MA, O'Hara TD, Poore GCB, Ross DJ, Storey MJ, Watson JE, Wilson RS (2004) Introduced and cryptogenic species in Port Phillip Bay, Victoria, Australia. Mar. Biol. 144: 183–202.

Hewitt CL, Campbell ML, McEnnulty F, Moore KM, Murfet NB, Robertson B, Schaffelke B (2005) Efficacy of physical removal of a marine pest: The introduced kelp *Undaria pinnatifida* in a Tasmanian Marine Reserve. Biological Invasions 7: 251–263.

Husa V, Sjøtun K, Lein TE (2004) The newly introduced species *Heterosiphonia japonica* Yendo (Dasyaceae, Rhodophyta): Geographical distribution and abundance at the Norwegian southwest coast. Sarsia 89: 211–217.

Jacoby C, Lapointe BE, Creswell L (2004) Are native and non-indigenous seaweeds overgrowing Florida's east coast reefs?

Florida Sea Grant College Program. SGEF-156. Available at: http://flseagrant.org/program_areas/coastal_habitats/publications/SGEF_156_web.pdf [15/09/2004].

Jaubert JM, Chisholm JRM, Minghelli-Roman A, Marchioretti M, Morrow JH, Ripley HT (2003) Re-evaluation of the extent of *Caulerpa taxifolia* development in the northern Mediterranean using airborne spectrographic sensing. Mar. Ecol. Prog. Ser. 263: 75–82.

Jousson O, Pawlowski J, Zaninetti L, Zechman FW, Dini F, Di Guiseppe G, Woodfield R, Millar A, Meinesz A (2000) Invasive alga reaches California. Nature 408: 157–158.

Kideys AE (2002) The comb jelly *Mnemiopsis leidyi* in the Black Sea. In Leppäkoski E, Gollasch S, Olenin S (eds), Invasive Aquatic Species of Europe. Distribution, Impacts and Management, Kluwer Academic Publishers, Dordrecht: pp. 56–61.

Kinloch M, Summerson R, Curran D (2003) Domestic vessel movements and the spread of marine pests. Department of Agriculture, Fisheries and Forestry, Commonwealth of Australia, Canberra.

Kolar CS, Lodge DM (2001) Progress in invasion biology: Predicting invaders. Trends Ecol. Evol. 16: 199–204.

Komatsu T, Ishikawa T, Yamaguchi N, Hori Y, Ohba H (2003) But next time? Unsuccessful establishment of the Mediterranean strain of the green seaweed *Caulerpa taxifolia* in the Sea of Japan. Biological Invasions 5: 275–277.

Lein TE (1999) A newly immigrated red alga ('*Dasysiphonia*', Dasyaceae, Rhodophyta) to the Norwegian coast. Sarsia 84: 85–88.

Levin PS, Coyer JA, Petrik R, Good TP (2002) Community-wide effects of nonindigenous species on temperate rocky reefs. Ecology 83: 3182–3193.

Lewis JA (1999) A review of the occurrence of exotic macroalgae in Southern Australia, with emphasis on Port Phillip Bay, Victoria. In Hewitt, CL, ML Campbell, RE Tresher & RB Martin (eds), Marine Biological Invasions of Port Phillip Bay, Victoria. Centre for Research on Introduced Marine Pests Technical report No. 20. CSIRO Marine Research, Hobart: pp. 61–87.

Lodge DM (1993) Biological invasions: Lessons for ecology. Trends Ecol. Evol. 8: 133–137.

Maggs CA, Stegenga H (1999) Red algal exotics on the North Sea coasts. Helgol. Meeresunters. 52: 243–258.

McConnell ML (2003) Ballast and biosecurity: the legal, economic and safety implications of the developing international regime to prevent the spread of harmful aquatic organisms and pathogens in ship's ballast water. Ocean Yearbook 17: 213–255.

McEnnulty F, Bax NJ, Schaffelke B, Campbell ML (2001) A review of rapid response options for the control of ABWMAC listed introduced species and related taxa in Australian waters. Centre for Research on Introduced Marine Pests Technical Report No. 23. CSIRO Marine Research, Hobart.

McIvor L, Maggs CA, Provan J, Stanhope MJ (2001) RbcL sequences reveal multiple cryptic introductions of the Japanese red alga *Polysiphonia harveyi*. Mol. Ecol. 10: 911–919.

McKinney ML, Lockwood JL (1999) Biotic homogenization: A few winners replacing many losers in the next mass extinction. Trends Ecol. Evol. 14: 450–453.

Meinesz A, Boudouresque CF (1996) On the origin of *Caulerpa taxifolia* in the Mediterranean Sea. Science et Vie 319: 603–613.

Meinesz A, Belsher T, Thibaut T, Antolic B, Ben Mustapha K, Boudouresque CF, Chiaverini D, Cinelli F, Cottalorda J-M, Djellouli A, Abed A El, Orestano C, Grau AMM, Ivesa L, Jaklin A, Langar H, Massuti-Pascual E, Peirano A, Tunesi L, de Vaugelas J, Zavodnik N, Zuljevic A (2001) The introduced green alga *Caulerpa taxifolia* continues to spread in the Mediterranean. Biological Invasions 3: 201–210.

Meusnier I, Valero M, Destombe C, Gode C, Desmarais E, Bonhomme F, Stam WT, Olsen JL (2002) Polymerase chain reaction-single strand conformation polymorphism analyses of nuclear and chloroplast DNA provide evidence for recombination, multiple introductions and nascent speciation in the *Caulerpa taxifolia* complex. Mol. Ecol. 11: 2317–2325.

Meusnier I, Valero M, Olsen JL, Stam WT (2004) Analysis of rDNA ITS1 indels in *Caulerpa taxifolia* (Chlorophyta) supports a derived, incipient species status for the invasive strain. Eur. J. Phycol. 39: 83–92.

Meyerson LA, Reaser JK (2002) Biosecurity: Moving toward a comprehensive approach. BioScience 52: 593–600.

Minchin D, Gollasch S (2002) Vectors- how exotics get around. In Leppäkoski E, Gollasch S, Olenin S (eds), Invasive Aquatic Species of Europe. Distribution, Impacts and Management Kluwer Academic Publishers, Dordrecht, pp. 183–192.

Murphy NE, Schaffelke B (2003) Use of amplified fragment length polymorphism (AFLP) as a new tool to explore the invasive green alga *Caulerpa taxifolia* in Australia. Mar. Ecol. Prog. Ser. 246: 307–310.

Nichols FH, Thompson JK, Schemel LE (1994) Remarkable invasion of San Francisco Bay (California, USA) by the Asian clam *Potamocorbula amurensis*. II. Displacement of a former community. Mar. Ecol. Prog. Ser. 66: 95–101.

Norse EA (1993) Global Marine Biological Diversity. A Strategy for Building Conservation into Decision Making, Island Press, Washington DC.

Nyberg CD, Wallentinus I (2005) Can species traits be used to predict marine macroalgal introductions? Biological Invasions 7: 265–279.

Occhipinti Ambrogi A (2002) Current status of aquatic introductions in Italy. In Leppäkoski E, Gollasch S, Olenin S (eds), Invasive Aquatic Species of Europe. Distribution, Impacts and Management, Kluwer Academic Publishers, Dordrecht, pp. 311–324.

Occhipinti Ambrogi A, Savini D (2003) Biological invasions as a component of global change in stressed marine ecosystems. Mar. Poll. Bull. 46: 542–551.

Olden JD, Poff NL (2004) Clarifying biotic homogenization. Trends Ecol. Evol. 19: 282–283.

Olden JD, Poff NL, Douglas MR, Douglas ME, Fausch KD (2004) Ecological and evolutionary consequences of biotic homogenization. Trends Ecol. Evol. 19: 18–24.

Phillips JA, Price IR (2002) How different is Mediterranean *Caulerpa taxifolia* (Caulerpales: Chlorophyta) to other populations of the species? Mar. Ecol. Prog. Ser. 238: 61–71.

Piazzi L, Ceccherelli G (2002) Effects of competition between two introduced *Caulerpa*. Mar. Ecol. Prog. Ser. 225: 189–195.

Piazzi L, Cinelli F (2001) Distribution and dominance of two introduced turf-forming macroalgae on the oast of Tuscany, Italy, northwestern Mediterranean Sea in relation to different habitats and sedimentation. Bot. Mar. 44: 509–520.

Piazzi L, Cinelli F (2003) Evaluation of benthic macroalgal invasion in a harbour area of the western Mediterranean Sea. Eur. J. Phycol. 38: 223–231.

Piazzi L, Balata D, Cinelli F (2002) Epiphytic macroalgal assemblages of *Posidonia oceanica* rhizomes in the western Mediterranean. Eur. J. Phycol. 37: 69–76.

Piazzi L, Balata D, Ceccherelli G, Cinelli F (2001a) Comparative study of the growth of the two-occurring introduced green algae

Caulerpa taxifolia and *Caulerpa racemosa* along the Tuscan coast (Italy, western Mediterranean). Cryptogamie, Algologie 22: 459–466.

Piazzi L, Ceccherelli G, Cinelli F (2001b) Threat to macroalgal diversity: effects of the introduced green alga *Caulerpa racemosa* in the Mediterranean. Mar. Ecol. Prog. Ser. 210: 149–159.

Pollard DA, Hutchings PA (1990) A review of exotic marine organisms introduced to the Australian Region. II. Invertebrates and algae. Asian Fish. Sci. 3: 223–250.

Reise K, Gollasch S, Wolff WJ (1999) Introduced marine species of the North Sea coasts. Helgol. Meeresunters. 52: 219–234.

Reise K, Gollasch S, Wolff WJ (2002) Introduced marine species of the North Sea coasts. In Leppäkoski E, Gollasch S, Olenin S (eds), Invasive Aquatic Species of Europe. Distribution, Impacts and Management, Kluwer Academic Publishers, Dordrecht, pp. 260–266.

Ribera Siguan MA (2002) Review of non-native marine plants in the Mediterranean Sea. In Leppäkoski E, Gollasch S, Olenin S (eds), Invasive Aquatic Species of Europe. Distribution, Impacts and Management, Kluwer Academic Publishers, Dordrecht, pp. 291–310.

Ribera Siguan MA (2003) Pathways of biological invasions of marine plants., In Ruiz GM, Carlton JT (eds), Invasive Species: Vectors and Management Strategies, Island Press, Washington, pp. 183–226.

Ruiz GM, Hewitt CL (2002) Toward understanding patterns of coastal marine invasions: A prospectus. In Leppäkoski E, Gollasch S, Olenin S (eds), Invasive Aquatic Species of Europe. Distribution, Impacts and Management, Kluwer Academic Publishers, Dordrecht, pp. 529–547.

Ruiz GM, Fofonoff PW, Carlton JT, Wonham MJ, Hines AH (2000) Invasion of coastal marine communities in North America: apparent patterns, processes, and biases. Annu. Rev. Ecol. Syst. 31: 481–531.

Saito Y (1975) Undaria. In Tokida J, Hirose H (eds), Advance of Phycology in Japan, Junk Publishers, The Hague, pp. 304–320.

Sant N, Delgado O, Rodriguez-Prieto C, Ballesteros E (1996) The spreading of the introduced seaweed *Caulerpa taxifolia* (Vahl) C. Agardh in the Mediterranean Sea: Testing the boat transportation hypothesis. Bot. Mar. 39: 427–430.

Schaffelke B, Deane D (2005) Desiccation tolerance of the introduced marine green alga *Codium fragile* ssp. *tomentosoides*-Clues for likely transport vectors? Biological Invasions 7: 557–565.

Schaffelke B, Evers D, Walhorn A (1995) Selective grazing of the isopod *Idotea baltica* between *Fucus evanescens* and *F. vesiculosus* from Kiel Fjord (Western Baltic). Mar. Biol. 124: 215–218.

Schaffelke B, Murphy N, Uthicke S (2002) Using genetic techniques to investigate the sources of the invasive alga *Caulerpa taxifolia* in three new locations in Australia. Mar. Poll. Bull. 44: 204–211.

Schaffelke B, Campbell ML, Hewitt CL (2005) Reproductive phenology of the introduced kelp *Undaria pinnatifida* (Phaeophyta, Laminariales) in Tasmania, Australia. Phycologia 44: 84–94.

Scheibling RE, Anthony SX (2001) Feeding, growth and reproduction of sea urchins (*Strongylocentrotus droebachiensis*) on single and mixed diets of kelp (*Laminaria* spp.) and the invasive alga *Codium fragile* ssp. *tomentosoides*. Mar. Biol. 139: 139–146.

Silva PC, Basson PW, Moe RL (1996) Catalogue of the Benthic Marine Algae of the Indian Ocean. University of California Publications in Botany 79. University of California Press, Berkeley.

Silva PC, Woodfield RA, Cohen AN, Harris LH, Goddard JHR (2002) First report of the Asian kelp *Undaria pinnatifida* in the northeastern Pacific Ocean. Biological Invasions 4: 333–338.

Simberloff D, von Holle B (1999) Positive interactions of nonindigenous species: invasional meltdown? Biological Invasions 1: 21–32.

Sinner J, Forrest B, Taylor M (2000) A strategy for managing the Asian kelp *Undaria*: Final report. Report to Ministry of Fisheries, New Zealand. Cawthron Report No. 578. Cawthron Institute, Nelson.

Smith JE, Hunter CL, Smith CM (2002) Distribution and reproductive characteristics of nonindigenous and invasive marine algae in the Hawaiian Islands. Pac. Sci. 56: 299–315.

Stimson J, Larned ST, Conklin E (2001) Effects of herbivory, nutrient levels, and introduced algae on the distribution and abundance of the invasive macroalga *Dictyosphaeria cavernosa* in Kaneohe Bay, Hawaii. Coral Reefs 19: 343–357.

Stockwell CA, Hendry AP, Kinnison MT (2003) Contemporary evolution meets conservation biology. Trends Ecol. Evol. 18: 94–101.

Thornber CS, Kinlan BP, Graham MH, Stachowicz JJ (2004) Population ecology of the invasive kelp *Undaria pinnatifida* in California: Environmental and biological controls on demography. Mar. Ecol. Prog. Ser. 268: 69–80.

Thresher RE (2000) Key threats from marine bioinvasions: a review of current and future issues. In Pederson J (ed.), Marine Bioinvasions, Proceedings of the First National Conference, January 24–27, 1999. Massachusetts Institute of Technology, Sea Grant College Program, Boston: pp. 24–36.

Trowbridge CD (1998) Ecology of the green macroalga *Codium fragile* (Suringar) Hariot 1889: Invasive and non-invasive subspecies. Oceanogr. Mar. Biol. Ann. Rev. 36: 1–64.

Valentine JP, Johnson CR (2003) Establishment of the introduced kelp *Undaria pinnatifida* in Tasmania depends on disturbance to native algal assemblages. J. Exp. Mar. Biol. Ecol. 295: 63–90.

Valentine JP, Johnson CR (2004) Establishment of the introduced kelp *Undaria pinnatifida* following dieback of the native macroalga *Phyllospora comosa* in Tasmania, Australia. Mar. Freshw. Res. 55: 223–230.

Van Beukering P, Cesar HSJ (2004) Ecological economic modeling of coral reefs: evaluating tourist overuse at Hanauma Bay and algae blooms at the Kihei Coast, Hawai'i. Pac. Sci. 58: 243–260.

Verlaque M (1989) Contribution a la flore des algues marines de Méditerranée: Especes rares ou nouvelles pour les côtes Françaises. Bot. Mar. 32: 101–113.

Verlaque M (2001) Checklist of the macroalgae of Thau Lagoon (Herault, France), a hot spot of marine species introduction in Europe. Oceanol. Acta 24: 29–49.

Verlaque M, Durand C, Huisman JM, Boudouresque C-F, Le Parco Y (2003) On the identity and origin of the Mediterranean invasive *Caulerpa racemosa* (Caulerpales, Chlorophyta). Eur. J. Phycol. 38: 325–339.

Verlaque M, Afonso-Carrillo J, Gil-Rodriguez M Candelaria, Durand C, Boudouresque CF, La Parco Y (2004) Blitzkrieg in a marine invasion: *Caulerpa racemosa* var. *cylindracea* (Bryopsidales, Chlorophyta) reaches the Canary Islands (NE Atlantic). Biological Invasions 6: 269–281.

Vitousek PM, Mooney HA, Lubchenco J, Melillo JM (1997) Human dominition of Earth's ecosystems. Science 277: 494–499.

Wallentinus I (1999a) Introduction and transfer of plants. In ALS Munro, SD Utting, I Wallentinus (eds), Status of introductions of non-indigenous marine species to north Atlantic waters 1981–1991. ICES Cooperative Research Report No. 231. International Council for the Exploration of the Sea, Copenhagen: pp. 1–32.

Wallentinus I (1999b) Undaria pinnatifida. In Gollasch S, Minchin D, Rosenthal H, Voigt M (eds). Case Histories on Introduced Species: Their General Biology, Distribution, Range Expansion and Impact, Department of Fishery Biology, Institut for Marine Science, University of Kiel, Germany: pp. 11–19.

Wallentinus I (2002) Introduced marine algae and vascular plants in European aquatic environments. In Leppäkoski E, Gollasch S, Olenin S (eds), Invasive Aquatic Species of Europe. Distribution, Impacts and Management, Kluwer Academic Publishers, Dordrecht, pp. 27–52.

Watson KL, Cheney, DP Levine I (2000) Biomonitoring of an aquacultured introduced seaweed, *Porphyra yezoensis* (Rhodophyta, Bangiophycidae) in Cobscook Bay, Maine, USA. Marine Bioinvasions, Proceedings of the First National Conference, January 24–27, 1999. Massachusetts Institute of Technology, Sea Grant College Program, Boston: pp. 260–264.

Wilkinson DM (2004) The long history of biotic homogenization. Trends Ecol. Evol. 19: 283–284.

Williams SL, Schroeder SL (2004) Eradication of the invasive seaweed *Caulerpa taxifolia* by chlorine bleach. Mar. Ecol. Prog. Ser. 272: 69–76.

Williamson AT, Bax NC, Gonzalez E, Geeves W (2002) Development of a regional risk management framework for APEC economies for use in the control and prevention of introduced marine pests. APEC Secretariat, Technical Reports of the Marine Resource Conservation Working Group. CSIRO Centre for Research on Introduced Marine Pests, Hobart and Inter-American Centre for Sustainable Ecosystems Development, Santiago.

Williamson MH, Fitter A (1996) The characteristics of successful invaders. Biol. Cons. 78: 163–170.

Wolff WJ, Reise K (2002) Oyster imports as a vector for the introduction of alien species into northern and western European coastal waters. In Leppäkoski E, Gollasch S, Olenin S (eds), Invasive Aquatic Species of Europe. Distribution, Impacts and Management, Kluwer Academic Publishers, Dordrecht, pp. 193–205.

Wotton DM, Hewitt CL (2004) Marine biosecurity post-border management: Developing incursion response systems for New Zealand. NZ J. Mar. Freshw. Res. 38: 553–559.

Wotton DM, O'Brien C, Stuart MD, Fergus DJ (2004) Eradication success Down Under: Heat treatment of a sunken trawler to kill the invasive seaweed *Undaria pinnatifida.* Mar. Poll. Bull. 49: 844–849.

Yoshida T, Nakajima Y, Nakata Y (1990) Check-list of marine algae of Japan (revised in 1990). Jpn. J. Phycol. 38: 269–320.

Zemke-White WL (2004) Assessment of the current knowledge on the environmental impacts of seaweed farming in the tropics. In: Phang et al. (eds), Marine Science into the New Millennium: New Perspectives and Challenges. Proceedings of the Asia-Pacific Marine Science and Technology Conference, 12–16 May 2002, Kuala Lumpur, Malaysia, 465–476.

Journal of Applied Phycology (2006) 18: 543–550
DOI: 10.1007/s10811-006-9061-7

Time-space characterization of commercial seaweed species from the Gulf of California using a geographical information system

J.A. Zertuche-González*, L.A. Galindo-Bect, I. Pacheco-Ruíz & A. Galvez-Telles
Instituto de Investigaciones Oceanológicas, Universidad Autónoma de Baja California. Km 106 Carr. Tijuana-Ensenada Ensenada. B.C. 22860, Mexico

*Author for correspondence: e-mail: zertuche@uabc.mx

Key words: Geographical Information System, GIS, Gulf of California, commercial seaweeds

Abstract

The Gulf of California, considered one of the most pristine areas of the world, hosts more than 50 seaweed species that have commercial applications. Only one species, however, is presently harvested commercially. In order to establish potential areas for seaweed use, a Geographical Information System (GIS) was used to determine areas for potential seaweed exploitation based on more than 9000 literature records. The system allows for the determination of sites, areas and times of the year when commercial species may be available. This information is being considered in a zoning program that would determine the areas of the Gulf sustainable for use or conservation. Temperature data were also included in order to determine potential areas for seaweed cultivation. GIS proves to be a powerful tool for large-scale management of seaweed resources.

Introduction

The Gulf of California, the youngest sea on Earth, is considered one of the most productive and pristine marine areas of the world (Alvarez-Borrego, 1983). It is responsible for 50% of the fisheries production in Mexico and 90% of the cultivated shrimp. It consists of a semi-enclosed basin of rectangular shape, approximately 1500 km long and 150 km wide, on average, with a mouth in the south connected to the Central Eastern Pacific. It covers an area of 260,000 km^2, comparable to the Red Sea, and encloses more than 900 islands and islets with many marine endemic species.

The Gulf of California is rich in all kinds of marine species including seaweeds. This has motivated many efforts by non-governmental organizations and government agencies to promote areas of the Gulf for conservation. At the same time, however, because it is one of the less developed areas in Mexico, the Gulf is currently being considered for many large-scale development projects including the construction of marinas, hotels and aquaculture farms.

The Gulf has also been recognized for its rich seaweed flora (Dawson, 1944; Norris, 1975). Of the 580 species mentioned by Espinoza-Avalos (1993) as reported for the Gulf, Pacheco-Ruíz and Zertuche-González (1996) have recognized at least 55 species that could have commercial applications, but only one is presently exploited (Pacheco Ruíz et al., 2003). Nevertheless, in the last decade, several of these species have been found to exist in sufficient amounts to be harvested commercially (Barilotti and Zertuche-González, 1990; Casas-Valdez et al., 1993; Hernández-Carmona et al., 1990; Pacheco-Ruíz & Zertuche-González, 1999; Pacheco-Ruíz et al., 1998, 2002).

The information currently available indicates that the Gulf of California may be an important source of commercial seaweed. This information, however, is not sufficient to define priority areas, their location and extent, which could be earmarked for seaweed harvesting or cultivation, nor the potential conflict with zones considered for conservation. A survey to determine the distribution of commercial species and the feasibility for exploitation in terms of abundance in space and time would be costly and lengthy for an area the size of the Gulf of California.

In this study, we use a Geographical Information System (GIS) to characterize the Gulf of California in terms of the commercially valuable species reported in the literature. We test the possibility of defining areas of importance for commercial seaweeds by providing geographic references to all species reported in the literature and plotting the records on a map with the help of GIS. The records were superposed on a map of surface water temperature in order to define oceanographic regions of importance. Special attention was giving to those commercial species for which abundance data are reported in the literature.

With this study we expect to provide an inexpensive alternative for characterizing areas of importance for seaweed exploitation or cultivation, as well as to identify areas or species that require more studies, particularly in broad-scale regions. To our knowledge, this is the first study considering the use of GIS on the characterization of seaweed resources on broad areas.

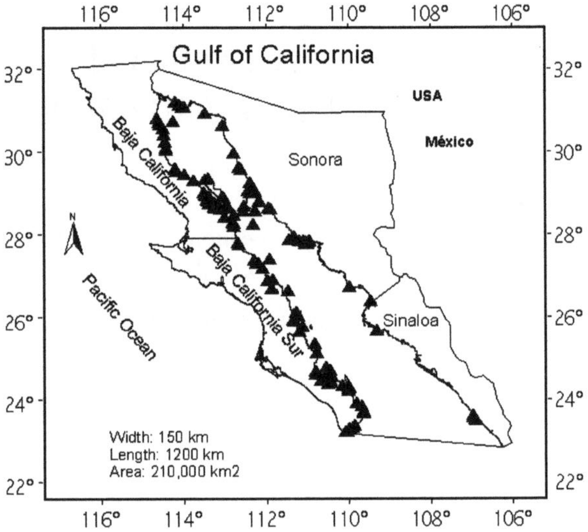

Figure 1. Sites with commercially valuable seaweed species reported from the Gulf of California.

Materials and methods

The extent of the Gulf of California varies depending on the criteria used to define its southern limit (Alvarez-Borrego, 1983). For the purpose of this work, the southern limit is established at 23°N latitude, which coincides with the southern end of the peninsula on the west coast and the southern limit of the state of Sonora on the east coast. Thus, four states border the Gulf of California: Baja California and Baja California Sur on the west coast, and Sonora and Sinaloa on the east coast (Figure1).

From a data base containing 9481 records of seaweeds reported for the Gulf of California from 1911 to 2003 (58 papers), the information relating to 55 species regarded as commercially valuable by Pacheco-Ruíz & Zertuche-González (1996), was displayed on a map to characterize their distribution within the Gulf in time and space. From these species, we selected those for which biomass studies were available to be analyzed in terms of the geographical and seasonal presence. The information was displayed and analyzed using Arcview 3.2a GIS. The data base considers only records published in refereed journals. Herbarium records, varieties, or species reported as doubtful were not included. Taxonomic and geographic attributes were considered. Seasonality was included when reported or when it was possible to deduce it from the manuscript. Most papers did not include geographic references. Therefore, based on the name of the localities, geographic coordinates were obtained from maps of the Mexican National Institute of Statistics, Geography and Informatics – INEGI – (Instituto Nacional de Estadística, Geografia e Informática). Records were grouped into "sites" when they were near an officially recognized map location. For instance, all records for a specific Bay or Point were grouped together.

Monthly mean surface temperatures were obtained from the PROMETEO data base (WWF, Program Mexico 2001). Seasonal temperature maps were derived from the average of three months. Data from January to March were used for winter, from April to June for spring, from July to September for summer and from October to December for fall. Site records for selected commercial species were superimposed on the temperature maps to determine their seasonal geographical distribution.

Results

Of the 160 sites recognized in the data base, 137 included commercially valuable seaweeds; commercial species occur throughout the Gulf of California, except for the southern east coast (state of Sinaloa), where the lack of studies was made evident by the low number of records (Figure 1).

Table 1. Seaweed biomass and number of sites evaluated versus the number of sites where these species have been reported

Species	Biomass (dry tons)	# of sites evaluated	# of sites present	References for biomass
Ulva lactuca	350	4	28	Pacheco-Ruíz et al. (2002)
Eucheuma uncinatum	165	1	27	Guzmán del Proo et al. (1986)
Chondracantus squarrulosus	160	1	19	Pacheco-Ruiz et al. (2000)
Gracilariopsis lemaneiformis	4060–5751	5	35	Pacheco-Ruiz et al. (1999)
Sargassum ssp	31,000	9	69	Pacheco-Ruíz et al. (1998)
Sargassum ssp	18,901(*)	1	69	Hernández-Carmona et al. (1993)
Sargassum ssp	7,250 (*)	1	69	Casas-Valez et al. (1993)

(*) Biomass reported as wet weight.

The species *Gracilariopsis lemaneiformis* (Bory de Saint-Vincent) E.Y. Dawson, Acleto & Foldvik (Pacheco-Ruíz & Zertuche-González, 1999) , *Ulva lactuca* Linnaeus (Pacheco-Ruíz et al., 2002), *Eucheuma uncinatum* Setchell & Gardner (Barilotti and Zertuche-González, 1990) and *Chondracantus squarrulosus* (Setchell & Gardner) Hughey, P.C. Silva & Hommersand (Pacheco-Ruíz et al., 2001), and the *Sargassum* species *Sargassum johnstonii* Setchell & Gardner, *Sargassum sinicola* Setchell & Gardner, *Sargassum herporhizum* Setchell & Gardner and *Sargassum lapazeanum* Setchell & Gardner were reported in the literature as present in large quantities (Hernández-Carmona et al., 1990; Casas-Valdez et al., 1993; Pacheco-Ruíz et al., 1998).

Seaweed biomass data is only available for species on the west coast and for few sites from the many that are present (Table 1). *Sargassum* species are by far the most abundant flora. These studies, however, do not always differentiate among the different species of *Sargassum*. Pacheco-Ruíz et al. (1998) reports *S. johnstonii* as the most abundant.

Species reported to occur in large quantities showed a strong seasonal variation. They all showed a maximum distribution in winter-spring and minimum in the summer-fall, except for *Sargassum lapazeanum,* with maximum distribution in spring-summer. However, their maximum distribution varies geographically.

Surface water temperature in the Gulf ranges from 15 to 30 °C throughout the year. The largest distribution

Table 2. Number of sites (#s), temperature range (T °C) and latitude range (°) of commercial species per season

	Spring			Summer			Fall			Winter			
Species	# s	T °	L.	# s	T °	L.	# s	T °	L.	# s	T °	L.	Lat
Ulva lactuca	16	18	23	7	26	23	2	25	23	11	15	23	23–29
		25	29		30	29		26	24		22	29	
Eucheuma uncinatum	18	18	27	4	26	27	2	21	27	9	15	27	27–31
		24	31		30	29		24	28		18	30	
Chondracantus squarrulosus	13	18	27	2	26	28	3	21	28	6	15	27	27–31
		24	31		30	31		25	31		18	31	
Gracilariopsis lemaneiformis	21	18	24	6	26	27	11	21	24	14	15	24	24–31
		24	31		30	30		26	31		21	31	
Sargassum johnstonii	19	18	24	5	26	27	5	21	28	12	15	25	24–31
		23	31		30	31		24	31		20	31	
Sargassum lapazeanum	10	18	22	10	28	24	6	21	24	3	16	24	22–30
		24	28		30	28		26	29		21	30	
Sargassum sinicola	23	18	22	23	26	23	20	21	23	25	15	23	22–31
		25	31		30	31		26	31		22	31	

Lat is the latitudinal distribution throughout the year.

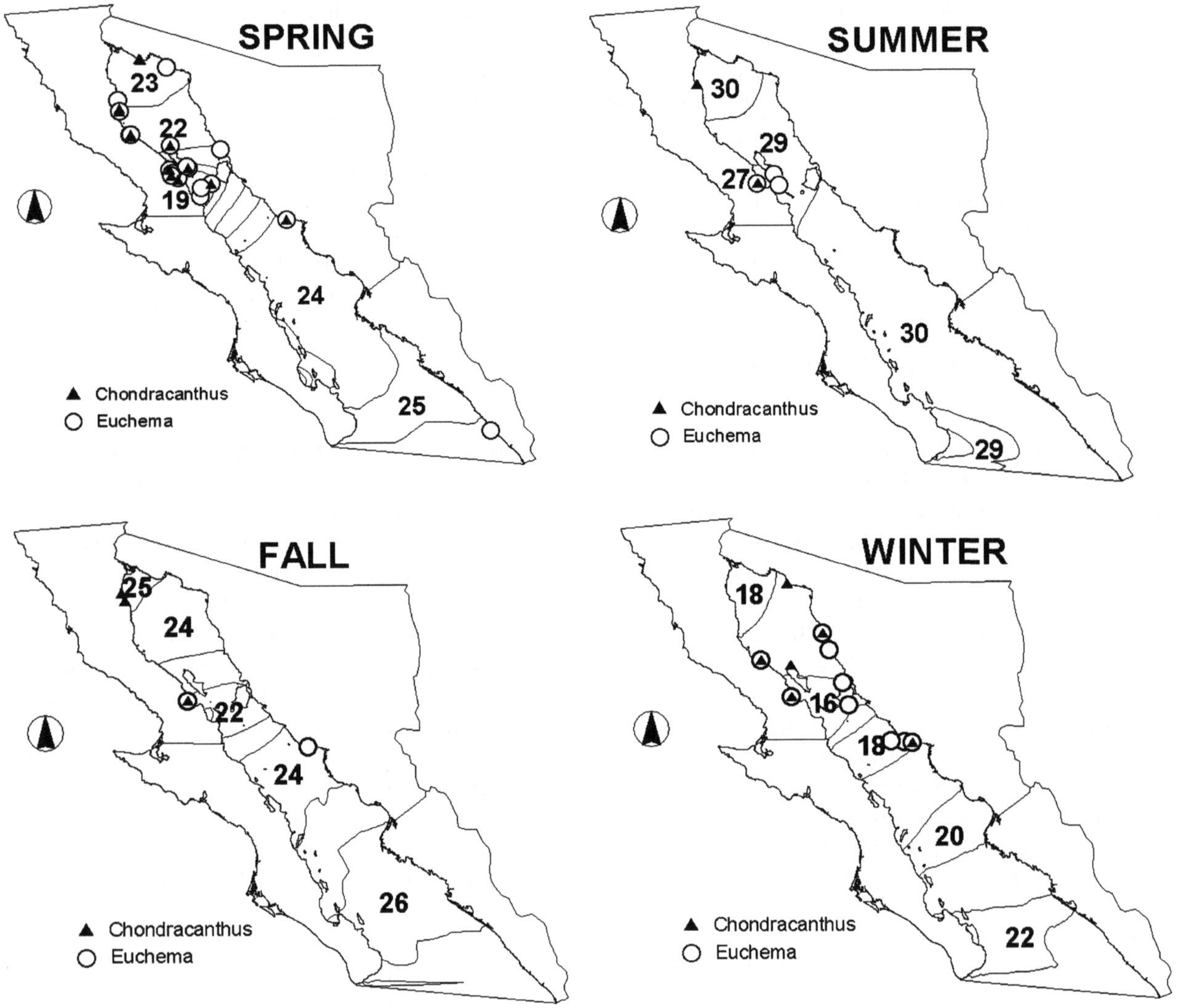

Figure 2. Seasonal distribution of sites of the carrageenophytes *Eucheuma uncinatum* and *Chondracantus squarrulosus* in the Gulf of California.

of commercial species is limited to areas and times when water temperature varies between 16 and 24 °C. The large temperature gradient in the Gulf of California makes it a sea of extreme climates when compared with the Pacific Ocean at similar latitudes (Table 2). An important anomaly is the fact that the coolest area is not the northern-most but the region of the large islands (28 to 30 °N), and that the largest annual temperature gradient is present in this region and to the north. Most of the seaweed records are, in fact, between 28–30 °N. Thus, while the annual temperature range at the mouth is 7 °C (from 22 to 29 °C), the gradient in the large islands region is 11 °C (from 16 to 27 °C) and in the north region 14 °C (from 18 to 30 °C). Surface temperatures showed strong seasonal variation in the Gulf, ranging from 18 to 25 °C in spring, from 26 to 30 °C in summer, from 20 to 27 °C in fall and from 15 to 23 °C in winter. Minimum temperatures, however, are present in the north-central region, from 28 to 30 °N, increasing to the north and south (Table 2).

The distribution of the carrageenophytes *E. uncinatum* and *C. scuarrulosus* is limited to the north of 28 °N, approximately. Both species tend to disappear in fall. Large beds are reported only in the north central region (between 28 °N and 29 °N) of the west coast of the Gulf (Figure 2).

Sargassum spp are distributed throughout the Gulf. The records for *S. johnstonii*, the only species reported to be present in large quantities, are concentrated in the central and northern region (north of 28 °N), while the

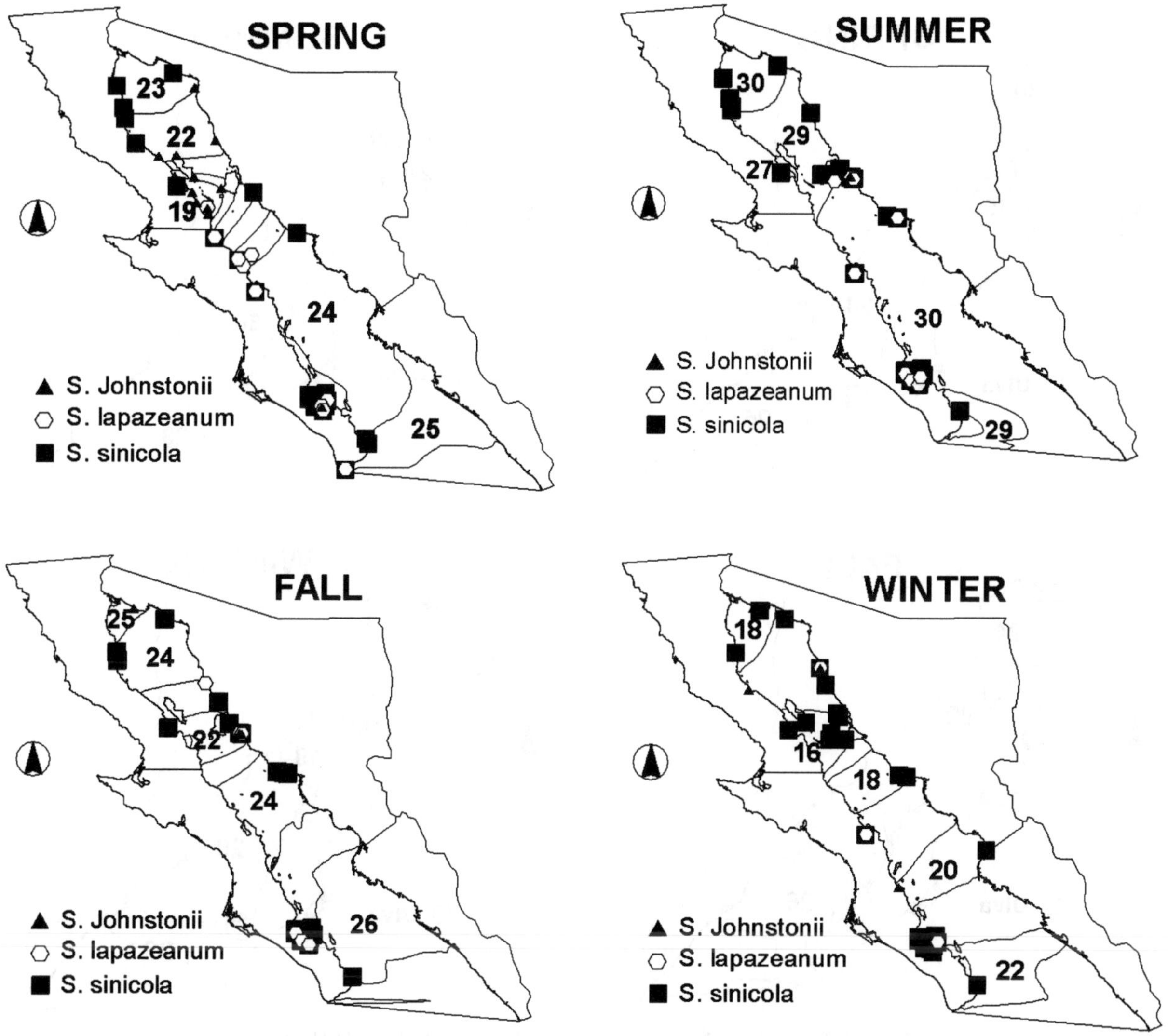

Figure 3. Seasonal distribution of sites of *Sargassum johnstonii, Sargassum lapazeanum* and *Sargassum herporizum* in the Gulf of California.

records for *S. lapazeanum* are restricted to the southern region (south of 28 °N) on the west coast. *S. sinicola* is reported throughout the Gulf, with the exception of the southern east coast (south of 27 °N), for which there are no records of *Sargassum* species (Figure 3).

The distribution of *Ulva lactuca* is limited to the central and southern region, below 30 °N. The records showed two distinctive distributions, one between 28 °N and 29 °N, which is absent in fall, and another in the southern region, concentrated between 23 °N and 26 °N, present all year around (Figure 4).

Gracilariopsis lemaneiformis occurs in practically all of the Gulf but in greater abundance during spring and winter. In fall, it is absent from most of the west coast, except in the northern and southern regions, where temperatures are warmer than the region of the islands; however, this pattern of distribution does not replicate on the east coast (Figure 5). *Gracilariopsis lemaneiformis* is the only species currently harvested in the Gulf of California (Pacheco-Ruíz et al., 2003). Commercial harvesting occurs between 28 °N and 29 °N, where Pacheco-Ruiz et al. (1999) estimated more than 4000 harvestable dry tons. This study shows many others sites where this species is present, particularly around 28° and 29 °N on the east coast of the Gulf and 24 °S on the west coast (Figure 5).

In the case of *Sargassum,* distinct areas were defined for *S. johnstonii* and *S. lapazeanum*, the former being

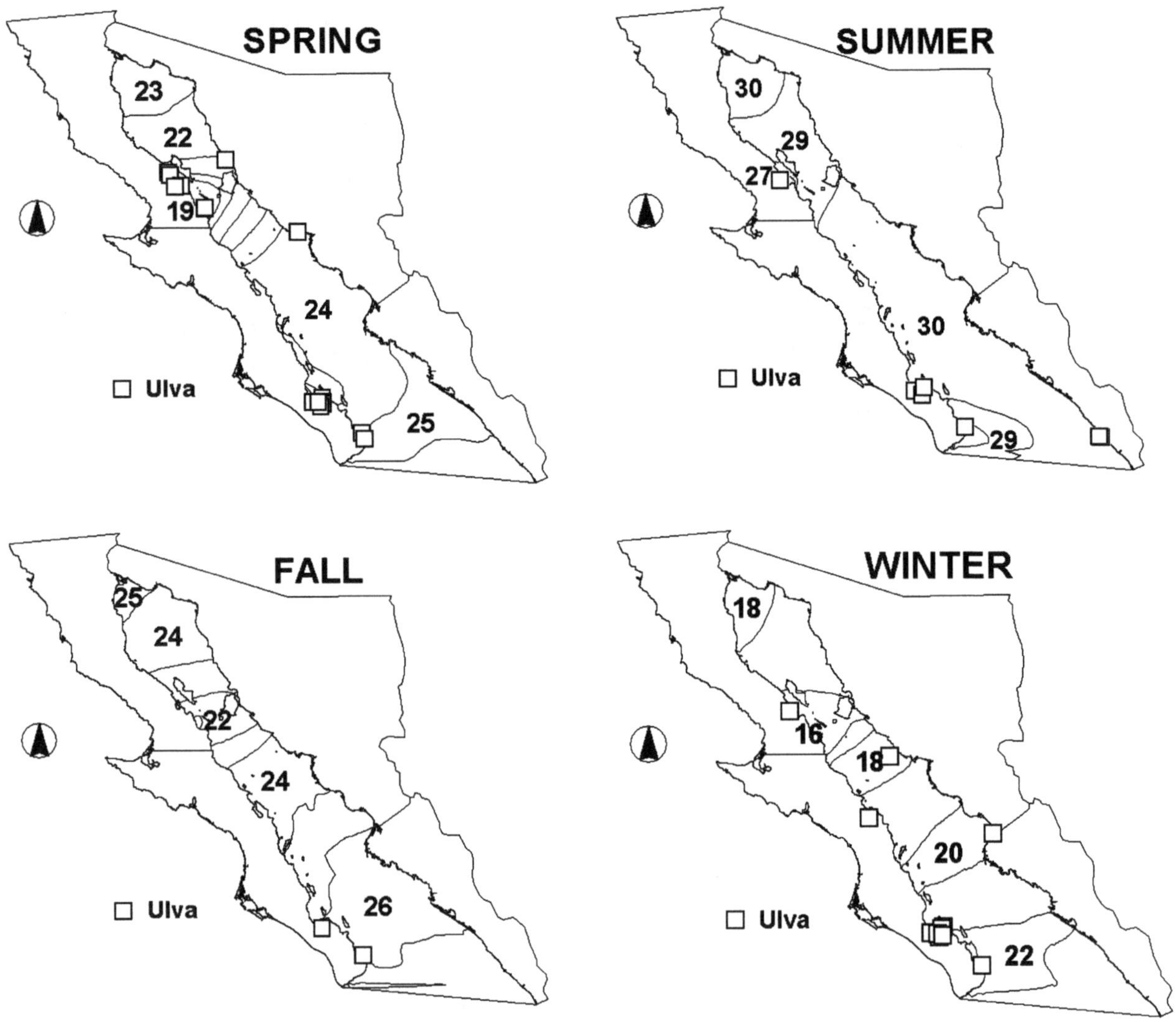

Figure 4. Seasonal distribution of sites of *Ulva lactuca* in the Gulf of California.

more abundant to the north of 27 °N and the latter to the south of 26 °S (Figure 3, Table 2). Similarly, large beds of *U. lactuca* do not occur north of 29 °30′N, but distinct populations are present (Figure 4). This distribution suggests a greater resistance for the southern population to warm temperatures, which may translate into longer harvesting periods.

Of the seven species with commercial application analyzed in this study, three are considered endemic: *E. uncinatum*, *C. squarrulosus* and *S. johnstonii.*

Discussion

Most of the few pristine coastal environments in the world are located in developing countries, where social pressure for development is challenging the conservation of marine resources. Governments are faced with the difficult task of designating areas for conservation or exploitation, and there is usually insufficient information on which to base their decision. The characterization of marine environments on a meso-scale represents a major challenge. GIS can be a powerful tool for analyzing databases of biological species and other information useful for management. However, its use as a tool to analyze information depends on the reliability of the database. Wrong interpretations may be made if insufficient data are available. On the other hand, GIS is a useful tool to identify geographical areas where there have been few or no studies. It is obvious from the data used in this study, taken from more than

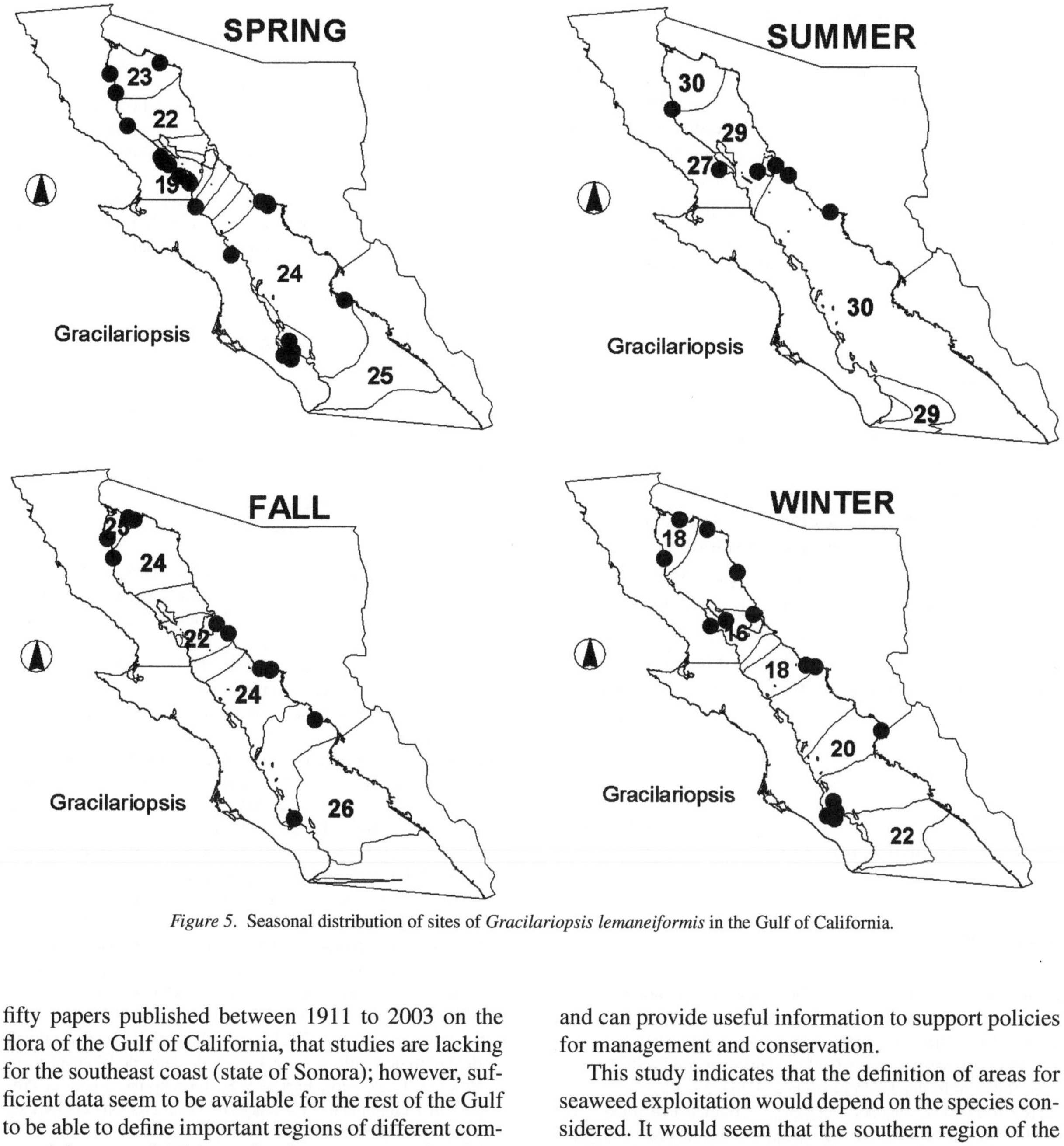

Figure 5. Seasonal distribution of sites of *Gracilariopsis lemaneiformis* in the Gulf of California.

fifty papers published between 1911 to 2003 on the flora of the Gulf of California, that studies are lacking for the southeast coast (state of Sonora); however, sufficient data seem to be available for the rest of the Gulf to be able to define important regions of different commercial seaweed. This study gives an example of the utility of the use of a GIS as a tool for the preliminary assessment for a broad area of the potential of seaweeds as a resource. The study identifies the location of the species in time and if there are biomass estimates or not. Furthermore, by comparing the number of sites in which the biomass has been determined against the number of sites where the seaweed is present, a potential for future surveys can be obtained. This analysis could be obtained for any species present in the region and can provide useful information to support policies for management and conservation.

This study indicates that the definition of areas for seaweed exploitation would depend on the species considered. It would seem that the southern region of the Gulf would be more suitable for tropical species suited to grow in warmer and relatively stable temperatures, while the islands region would be favorable for temperate species, although for just part of the year when relatively low temperatures prevail (winter and spring, 16 to 23 °C). This characteristic would be fundamental for establishing different zones in the Gulf for seaweed exploitation or cultivation.

Based on the fact that 85% of the sites studied in this region contain valuable species and that at least seven of them are reported to exist in large quantities even though few sites biomass has been evaluated, the Gulf of California could be an important source of seaweeds with commercial applications. The unique oceanographic conditions define different regions in the Gulf where different species predominate. Nevertheless, as a whole, seaweed availability shows a strong seasonality that would limit the possibility of seaweed utilization to only three to six months of the year. It is clear from this study that studies are lacking on the southeast coast of the Gulf.

Conclusions

GIS can be a powerful tool to support research, management or planning programs where special analysis of broad areas is necessary. In this study, the application of the use of a GIS to determine the potential availability of commercial seaweeds from the Gulf of California proved useful to identify valuable species, their location in time and space and the degree of knowledge about their biomass estimates. This approach can be used as an example to be applied in other areas of the world for the characterization of seaweed resources.

Acknowledgments

We thank the Universidad Autónoma de Baja California for funding this study.

References

Alvarez-Borrego S (1983) Gulf of California. In Ketchum BH (ed.), Estuaries and Enclosed Seas, Elsevier, Amsterdam: 427–449.

Barilotti DC, Zertuche-González JA (1990) Ecological effects of seaweed harvesting in the Gulf of California and Pacific Ocean of Baja California. Hydrobiologia 204/205: 35–40.

Casas-Valdez MM, Sánchez-Rodríguez I, Hernández-Carmona G (1993) Evaluación de *Sargassum* spp en la costa oeste de Bahía Concepción, B.C.S., Mexico. Investigaciones Marinas. CICIMAR 8: 61–69.

Dawson EY (1944) The marine algae of the Gulf of California. Allan Hancock Pacific Expeditions 3: 189–453.

Espinoza-Avalos J (1993) Macroalgas Marinas del Golfo de California. In Salazar-Vallejo SI and González NE eds. Biodiversidad Marina costera de México, Comisión Nacional de Biodiversidad y CIQRO, Mexico, 865 pp.

Guzmán-del-Proo SA, Casas-Valdez M, Díaz-Carrillo A, Díaz-López ML, Pineda Barrera J, Sánchez-Rodríguez MA (1986) Diagnóstico sobre las investigaciones y explotación de las algas marinas en México. Investigaciones Marinas. CICIMAR 3: 1–63.

Hernández-Carmona G, Casas-Valdez MM, Fajardo-León C, Sánchez-Rodríguez I, Rodríguez-Montesinos E (1990) Evaluación de *Sargassum* spp en la Bahía de la Paz, B.C.S., México. Investigaciones Marinas. CICIMAR 5: 11–18.

INEGI. Instituto Nacional de Estadística, Geografia e Informática. México. http://www.inegi.gob.mx.

Norris JN (1975) Marine algae of the northern Gulf of California. Dissertation. University of California. Santa Barbara. 575 pp.

Pacheco-Ruíz I, Zertuche-González JA (1996) The commercially valuable seaweeds of the Gulf of California. Bot. Mar. 35: 201–206.

Pacheco-Ruíz I, Zertuche-González JA, Chee-Barragán A, Blanco-Betancourt R (1998) Distribution and Quantification of *Sargassum* Beds Along the West Coast of the Gulf of California, Mexico. Bot. Mar. 41: 203–208.

Pacheco-Ruíz I, Zertuche-González JA (1999) Population structure and reproduction of the carrageenophyte *Chondracantus pectinatus* Daw. from the Gulf of California. Hydrobiologia 398/399: 159–165.

Pacheco-Ruíz I, Zertuche-González JA, Correa-Díaz F, Arellano-Carbajal Fausto (1999) *Gracilariopsis lemaneiformis* (Bory) Dawson, Acleto et Foldvik Beds Along the West Coast of the Gulf of California, Mexico. Hydrobiologia 398/399: 509–514

Pacheco-Ruíz I, Zertuche-González JA, Chee-Barragán A, Arroyo-Ortega E (2002) Biomass and potential comercial utilization of *Ulva lactuca* (Chlorophyta, Ulvaceae) beds along the north-west coast of the Gulf of California. Phycologia 41: 199–201.

Pacheco Ruíz I, Zertuche-González JA, Che-Barragan A (2003) Commercial exploitation of *Gracilariopsis lemaneiformis* in the Gulf of California. Proceedings of the XVII International Seaweed Symposium. Oxford University Press, pp. 101–105.

WWF Programa México 2001 Coalición para la Sustentabilidad del Golfo de California. 2001. Base de datos de Biodiversidad, Procesos Ecológicos, Físicos y Socioeconómicos para el Proceso de Definición de Prioridades de Conservación del Golfo de California, México. México, D.F.

Journal of Applied Phycology (2006) 18: 551–556
DOI: 10.1007/s10811-006-9070-6

Phenology of *Chondrus ocellatus* in Cheongsapo near Busan, Korea

Y.S. Kim [1,*], H.G. Choi[2] & K.W. Nam[3,*]
[1]*School of Marine Life Science, Kunsan National University, Kusan, Jeonbuk 573-701, Korea;* [2]*Faculty of Life Science, Wonkwang University, Iksan, Jeonbuk 570-749, Korea;* [3]*Department of Marine Biology, Pukyong National University, Nam-gu, Busan, 608-737, Korea*

*Author for correspondence: e-mail: kimys@kunsan.ac.kr, kwnam@pknu.ac.kr

Key words: *Chondrus ocellatus*, reproductive phenology, Korea, growth

Abstract

The reproductive phenology of *Chondrus ocellatus* and the effects of temperature and light on its growth were examined in Cheongsapo near Busan, Korea, from September 1994 to August 1995. The vegetative plants dominated over the year, with a peak occurrence in January. Gameto- and tetrasporophytes were most abundant in November and August. All vegetative and reproductive plants had a peak both in length and weight in October, when seawater temperature was highest (24 °C). In laboratory culture, the maximum relative growth rate (RGR) of 2.94% day^{-1} was obtained at 20 °C and 100 μmol photons $m^{-2}\,s^{-1}$, whereas the lowest value was recorded at 25 °C and 100 μmol photons $m^{-2}\,s^{-1}$ in a 12: 12 h LD photoperiod regime. Among the three photoperiod regimes (8:16 h, 12:12 h, 16:8 h LD) tested, there was evidence of a higher RGR in the 12:12 h LD cycle. This result suggests that the growth and reproduction of *C. ocellatus* are correlated with the seawater temperature based on laboratory culture and field observations.

Introduction

The genus *Chondrus* Stackhouse, which is widely distributed in temperate and cold-temperate waters (Lüning, 1985), has long been used as a source of gelling and stabilizing agent in foods (Taylor & Chen, 1994). Because of its commercial importance, there have been numerous studies on the genus to elucidate its biology and ecology (see Taylor & Chen, 1994). However, most studies have been based on Irish moss, *Chondrus crispus* Stackhouse. *Chondrus ocellatus* Holmes is mainly distributed on the coasts of Korea, Japan, China and Taiwan (Taylor & Chen, 1994). Although there are several reports on the life history, morphology and growth of this species (Ji & Guo, 1992; Brodie et al., 1993; Li et al., 1994), little is known about its phenology of growth and reproduction. In Korea, *C. ocellatus* inhabits the lower intertidal zone and is abundant on moderately exposed rocky shores (Kang, 1968). Recently, Choi and Kim (1999) reported that carrageenans from the Korean species have valuable anticoagulant properties.

Thus, the aim of this study was to investigate the reproductive phenology of *C. ocellatus,* in Cheongsapo near Busan in Korea, as baseline information for its management in the future. The present work provides new information on seasonal patterns in the proportions of reproductive phases and the sizes of individual thalli, in *C. ocellatus.* In addition, the effects of temperature and light on its growth are also examined in laboratory culture to compare with field observations.

Materials and methods

Cheongsapo is on the south eastern coast of Busan (129°12′E, 35°9′N), Korea. This site has a shallow and gently sloping intertidal zone. *Chondrus ocellatus* was sampled monthly from September 1994 to August 1995. Fifty fronds were randomly collected on the rocks of sampling areas, including rock pools, at each time. Plants were transported to the laboratory, then sorted by reproductive status (i.e. vegetative, gametophytic or tetrasporic), using visual examination and confirmation under a microscope. The occurrence of each reproductive stage was expressed as a percentage of the total number of plants analyzed. The length and fresh weight of each plant was measured after it had been rinsed in tap water, drained, and blotted.

Single apices of 5 mm length were excised from the plants and their growth evaluated in an experimental matrix (after a 24 h acclimatization period) under temperatures of 15, 20 and 25 °C and 40, 60 and 100 μmol photons m^{-2} s^{-1} at 12:12 LD photoperiod regime. Apical segments were cultured at 15 °C and 60 μmol photons m^{-2} s^{-1} at three photoperiod regimes (8:16 LD, 12:12 LD and 16:8 LD) for 60 days. Irradiance was measured using a Li-Cor Model Li-1400 quantum meter. For each condition, 30 segments were individually weighed and inoculated into a culture vessel containing 200 ml of culture medium (PES) (Provasoli, 1968). Media were changed every seven days. Each treatment was replicated 3 times. The relative growth rate (RGR) using fresh weight data was calculated for each replicate according to the following formula:

$$\mathrm{RGR} = \ln(W_t / W_0)t^{-1} \times 100$$

where W_0 is initial wet weight and W_t is the wet weight after t days. The seawater temperature data were obtained from NFRDI (National Fisheries Research and Development Institute).

Statistical analyses were performed using STATISTICA v. 5.0. A two-way ANOVA was used to test the effects of temperature and irradiance on the RGR of *Chondrus ocellatus*. A one-way ANOVA was applied to examine the effect of photoperiod in the RGR of the species. When significant differences between treatments were detected, the Tukey test was applied (Sokal & Rohlf, 1995).

Results

Monthly seawater temperature varied from 11 °C in March to 24 °C in September during the study period (Figure 1). Vegetative, gametophytic and tetrasporic plants were found in fluctuating ratios throughout the year (Figure 2). Vegetative plants were relatively abundant compare to reproductive plants during the entire study (above 36%) peaking in January (60%) with a minimum in November and August (36%). Maximum abundance of reproductive plants including both gameto- and tetrasporophytes was observed in November and August (64%) while minimum values were found in January (40%). Gametophytes were most abundant from September to December (30–40%), when temperature and daylength decreased. By contrast, tetrasporophytes were relatively abundant in other months, particularly, from April to August (36–44%), when temperature and daylength increased.

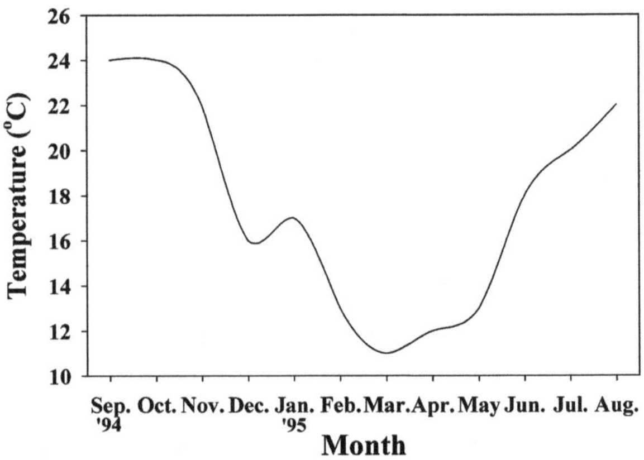

Figure 1. Monthly variations of seawater temperature at Cheongsapo in Korea.

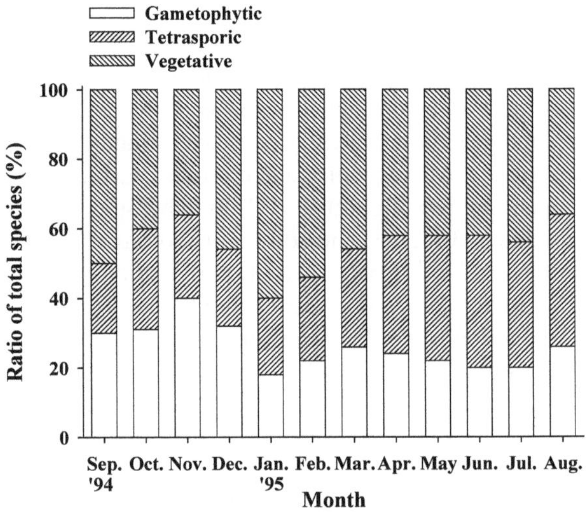

Figure 2. Monthly proportions of reproductive phases of *Chondrus ocellatus* during the sampling period.

The average length per plant was 6.99 ± 3.29 cm (mean ± SD) for gametophytic plants, 7.93 ± 3.61 cm for tetrasporic plants and 5.99 ± 2.89 cm for vegetative plants (Figure 3). Vegetative plants were smaller than reproductive plants. Growth in thallus length of vegetative and reproductive plants was highest in October, May and August whereas it was lowest in February and March (Figure 3). Monthly size distributions of thalli including all reproductive phases varied during the study period (Figure 4). From December to February

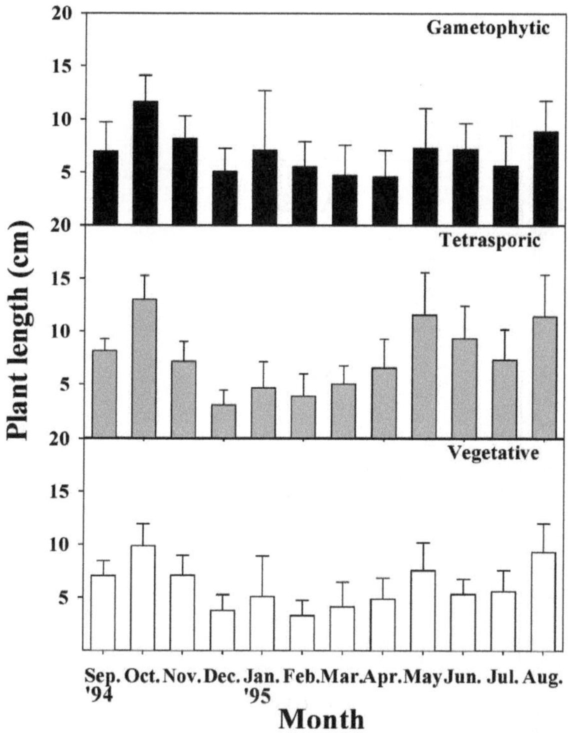

Figure 3. Mean plant length (± SD) for each reproductive phase of *Chondrus ocellatus* during sampling period.

(winter), small plants (below 4 cm) constituted about 60% of the population. Medium-sized plants (4–12 cm) comprised above 50% of the population from March to November except October.

In terms of thallus weight (Figure 5), the average wet weight per plant was 6.92 ± 3.47 g (mean ± SD) for gametophytes, 7.28 ± 4.29 g for tetrasporophytes and 4.22 ± 3.32 g for vegetative plants. Seasonal variations of wet weight of thallus showed a trend towards heavier plants in spring and autumn and lighter plants in winter.

In laboratory culture, a maximum RGR of 2.94% day^{-1} was obtained at 20 °C and 100 μmol photons m^{-2} s^{-1}, whereas the lowest value was recorded at 25 °C and 100 μmol photons m^{-2} s^{-1} (Figure 6). RGR of *Chondrus ocellatus* was the highest in 20 °C and there were significant differences in RGR among the examined temperatures of 15, 20 and 25 °C (ANOVA, $p < 0.05$). A Tukey test revealed the RGR of *C. ocellatus* was significantly different between 20 and 25 °C. With respect to photoperiod, although the highest RGR was apparently at 12:12 h LD, followed by 16:8 h LD and 8:16 h LD, there were no significant differences in RGR among them (ANOVA, $p > 0.05$, Figure 7).

Discussion

Vegetative, gametophytic and tetrasporic plants were found throughout the year. Reproductive plants including both gameto- and tetrasporophytes were most abundant in November and August when the temperature was 20 °C. However, sexual and asexual plants showed different patterns. Gametophytes were most abundant from September to December, when temperature and daylength decrease. In contrast, tetrasporophytes exceeded gametophytes in other months, particularly from April to August, when temperature and daylength increase. This pattern of association of reproductive plants with lower seawater temperatures and solar radiation is also found in *Gracilaria* (Piriz, 1996) and may result from different in physiological responses of gametophyte and tetrasporophyte phases to temperature and irradiance. Hannach and Santelices (1985) reported that gametophytic plants of *Iridaea* had a higher growth rate than tetrasporic plants under the same experimental conditions.

The growth of vegetative and reproductive plants (measured as thallus length) was highest in October, May and August when temperatures were highest at around 22–24 °C. In contrast, thalli were shortest in February and March, when temperature decreased from 15 °C to the lowest recorded value of 11 °C. This seasonal pattern in size distribution may be related not only to seawater temperature but also to reproduction and wave action. Generally, in July and August, several typhoons strike the Korean Peninsula, causing many seaweeds to become detached from the substratum. Also, the fronds of *C. ocellatus* die back after reproduction, as found in *Gracilaria* and *Eucheuma* (Dawes et al., 1974; Destombe et al., 1988) and it affects the size distribution of the species. In the study area, reproductive thalli are mainly seen from spring to autumn. The empty reproductive structures (after releasing spores) probably weaken the fronds, leading to decay of thalli which is accelerated by tyhoons tearing the weak fronds. Therefore it is not easy to determine the relative importance, to size distribution, of typhoons and reproduction in *C. ocellatus*. The rapid increase in the frequency of small plants (below 4 cm) from December to February may reflect the decay of larger plants and recruitment of new individuals after reproduction. High frequency of smaller plants during the period mainly results from dying back caused by reproduction and typhoons rather than new recruitments because we observed many smaller fronds have been torn. In the field population, larger plants become more

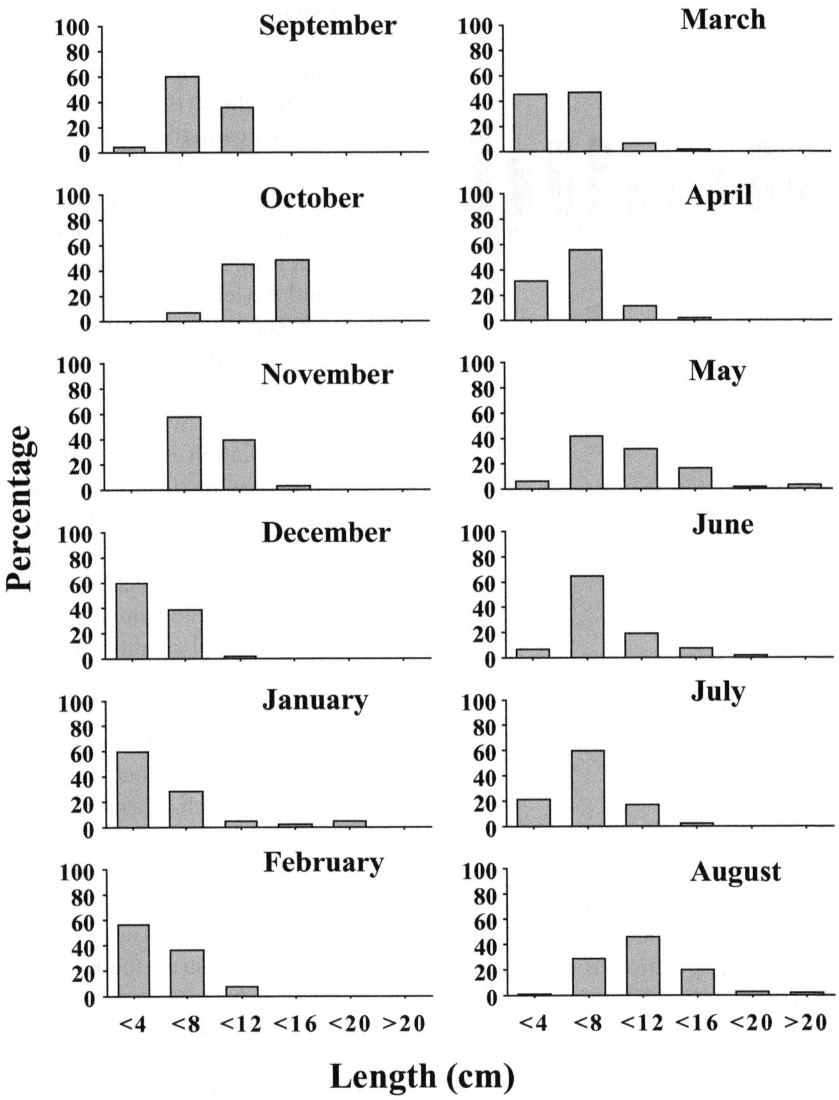

Figure 4. Size frequencies of *Chondrus ocellatus* collected over all the sampling period.

abundant from March to October and they were maximal in October, when seawater temperature is about 24 °C. In culture, however, the RGR of *C. ocellatus* was maximal at 20 °C and lowest at 25 °C. The large size of the plants in October is probably the result of growth in the previous months (June–August), when mean seawater temperature is about 20 °C. A similar pattern is also found in the weight of plants. These results suggest that the growth and reproduction of *C. ocellatus* are related to water temperature.

With respect to photoperiod, the highest RGR was found in 12:12 h LD, followed by 16:8 h LD and 8:16 h LD, indicating that this factor also affects the growth of *C. ocellatus*.

RGR is an important element in the evaluation of potential biomass production. In commercial algae, high RGR is essential for mass production. The RGR of *C. ocellatus* in culture ranged from 1.33 to 2.94% day^{-1}. These values are lower than those reported for *Chondrus crispus* (2–4% day^{-1}: Chopin et al., 1999) and for the other carrageenophytes *Eucheuma* and *Kappaphycus* (2–6% day^{-1} : Braud & Perez, 1979; Ohno et al., 1994). However, the RGR of *C. ocellatus* could be enhanced by culture in better designed growth conditions, with optimal temperature, light, salinity and nutrients. Thus, future studies on the growth of *C. ocellatus* should be carried out to determine these optimal conditions for commercial cultivation.

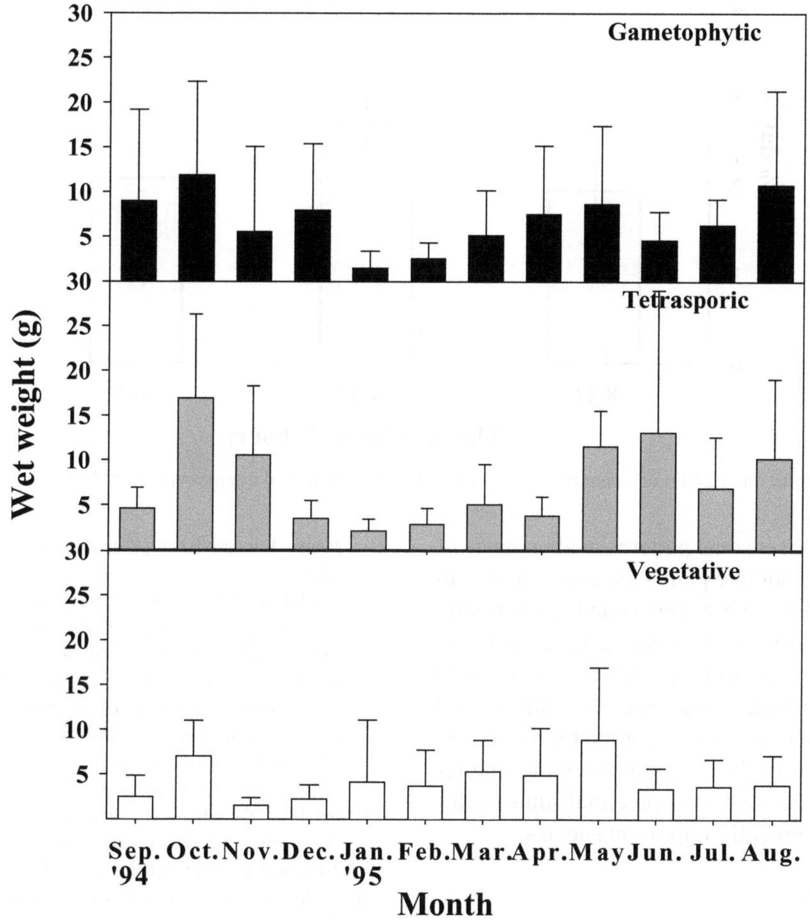

Figure 5. Mean wet weight (± SD) of plants for each reproductive phase of *Chondrus ocellatus* during sampling period.

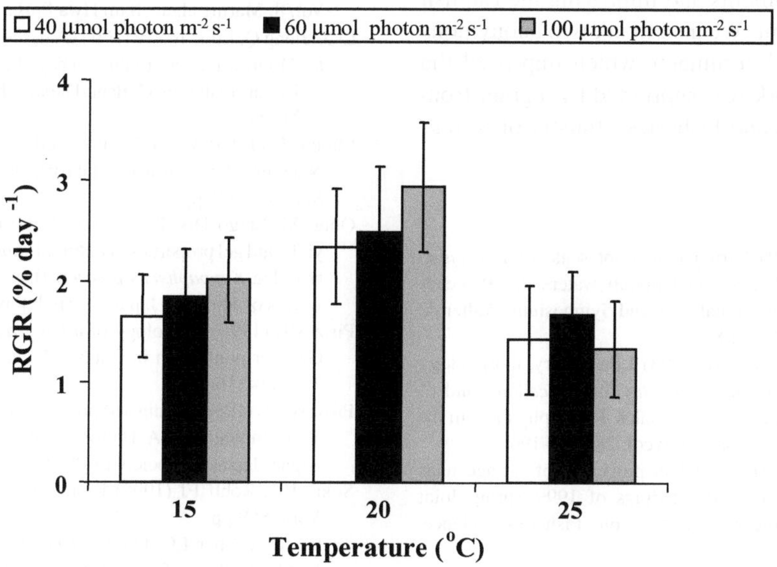

Figure 6. Relative growth rate (± SD) of *Chondrus ocellatus* at different temperatures and light intensities under 12 L:12 D photoperiod regime.

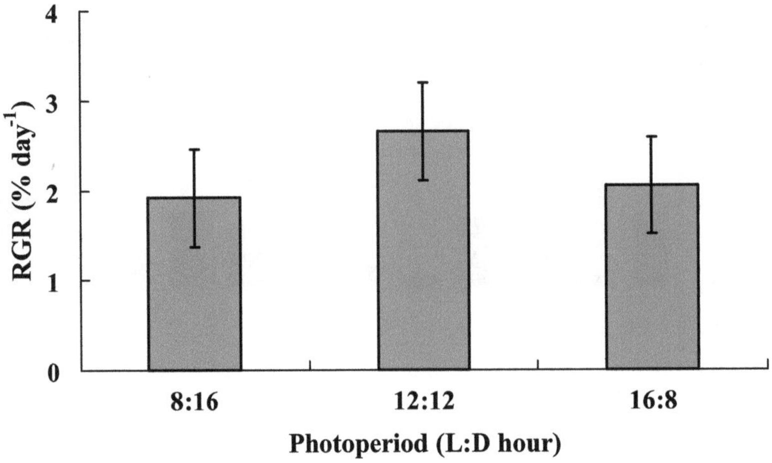

Figure 7. Relative growth rate (± SD) of *Chondrus ocellatus* at different photoperiod regimes under 15 °C and 60 μmol photons m^{-2} s^{-1}.

In conclusion, the seasonal patterns in abundance of vegetative and reproductive plants are associated with the combined factors of photoperiod and temperature. Sexual plants are most abundant near autumn and winter, and tetrasporophytes in spring and summer. Growth is greatest when temperatures are highest (summer and autumn). These results are useful for the timing of harvests of *C. ocellatus* in the field. Furthermore, the results of this study are useful for potential future cultivation of this economically important species.

Acknowledgments

We wish to express our sincere gratitude to Dr. Anderson for his kind comments and improving the English text of this manuscript. We also thank two anonymous reviewers for helpful comments which improved the manuscript. This work was supported by a grant from the Maritime Affairs and Fisheries Ministry of Korea.

References

Braud JP, Perez R (1979) Farming on pilot scale of *Eucheuma spinosum* (Florideophyceae) in Djibouti waters. In: Proceedings of the Tenth International Seaweed Symposium. Walter de Gruyter, Berlin, pp. 553–558.

Brodie J, Guiry MD, Masuda M (1993) Life history, morphology and crossability of *Chondrus ocellatus* forma *ocellatus* and *C. ocellatus* forma *crispoides* (Gigartinales, Rhodophyta) from the north-western Pacific. Europ. J. Phycol. 28: 183–196.

Choi Y, Kim SK (1999) Anticoagulant properties of carrageenans from *Chondrus ocellatus*. Proceedings of 1999 Spring Joint Meeting of the Korean Societies on Fisheries Science, pp. 151–152.

Chopin T, Sharp G, Belyea E, Semple R, Jones D (1999) Open-water aquaculture of the red alga *Chondrus crispus* in Prince Edward Island, Canada. Hydrobiologia 398/399: 417–425.

Dawes CJ, Mathieson AC, Cheney DP (1974) Ecological studies of floridiean *Eucheuma* (Rhodophyta, Gigartinales). I. Seasonal growth and reproduction. Bull. Mar. Sci. 24: 235–273.

Destombe C, Godin J, Bodard M (1988) The decay phase in the life history of *Gracilaria verrucosa*: The consequences in intensive cultivation. In Stadler T, Molion J, Verdus MC, Karamanos Y, Morvan H, Christiaen D (eds). Algal Biotechnology. Elsevier Applied Science, London, pp. 287–303.

Hannach G, Santelices B (1985) Ecological differences between the isomorphic reproductive phase of two species of *Iridaea* (Rhodophyta; Gigartinales). Mar. Ecol. Prog. Ser. 22: 291–303.

Ji Y, Guo J (1992) The effect of temperature on the growth and development of *Chondrus ocellatus*. Journal of Dalian Fisheries College/Dalian Shuichan Xueyuan Xuebao 7: 32–37.

Kang JW (1968) Illustrated Encyclopedia of Fauna & Flora of Korea Vol. 8. Marine algae. Sam Hwa Press, Seoul 465 pp.

Li X, Jiang Q, Lu J, Tao W (1994) A description of *Chondrus ocellatus* Holmes and its variation in bay of Liadong Peninsula. Journal of Dalian Fisheries College/Dalian Shuichan Xueyuan Xuebao 9: 21–25.

Lüning K (1985) Meeresbotank: Verbreitung, Ökophysiologie und Nutzung der marinen Makroalgen. Georg Thieme Verlag, Stuttgart. 375 pp.

Ohno M, Largo DB, Ikumoto T (1994) Growth rate, carrageenan yield and gel properties of cultured *kappa*-carrageenan producing red alga *Kappaphycus alvarezi* (Doty) Doty in the subtropical waters of Shikoku, Japan. J. Appl. Phycol. 6: 1–5.

Piriz ML (1996) Phenology of a *Gigartina skottsbergii* Setchell et Gardner population in Chubut Province (Argentina). Bot. Mar. 39: 311–316.

Provasoli L (1968) Media and prospects for the cultivation of marine algae. In Watanabe A, Hattori A (eds.), Cultures and Collection of Algae. Japanese Society of Plant Physiology, Tokyo, pp. 63–77.

Sokal RR, Rohlf FJ (1995) Biometry, 3rd edition. Freeman, New York. 859 pp.

Taylor ARA, Chen LCM (1994) *Chondrus* Stackhouse. In Akatsuka I. (ed.), Biology of Economic Algae, SPB Academic Publishing, Hague, pp. 35–76.

Journal of Applied Phycology (2006) 18: 557–563
DOI: 10.1007/s10811-006-9065-3

Seasonality pattern of biomass accumulation in a drifting *Furcellaria lumbricalis* community in the waters of the West Estonian Archipelago, Baltic Sea

Georg Martin*, Tiina Paalme & Kaire Torn
Estonian Marine Institute, University of Tartu, Mäealuse 10 a, 12618 Tallinn, Estonia
*Author for correspondence: e-mail georg.martin@ut.ee

Key words: loose-lying *Furcellaria lumbricalis*, *Coccotylus truncatus*, growth rate

Abstract

A free-floating, loose form of *Furcellaria lumbricalis* (Huds.) Lamour is rare in the Baltic Sea area. Kassari Bay, situated in the West Estonian Archipelago Sea area contains the largest known community of this kind. Here the free-floating mixed *Furcellaria lumbricalis-Coccotylus truncatus* (Paela) M. J. Wynne et J. N. Heine community inhabits sandy bottom, covering up to 120 km^2. Commercial exploitation of the community started in 1966 and has led to regular monitoring surveys for the quantification of the commercial resource. The aim of the present study was to determine the potential growth rates of the two community-forming species as well as to test different environmental factors affecting their growth. Results showed that the highest growth rates were measured in shallower depths (4 m) for both species. The seasonal growth pattern was also very similar for both species, showing the highest growth rates during the beginning of summer. Incubation of both species in another sea area with apparently similar basic environmental conditions (the northern part of the Gulf of Riga, Kõiguste Bay) resulted in significantly lower growth rates during the whole incubation period.

Introduction

In the Baltic Sea, at least two ecologically distinct forms of the red algal species *Furcellaria lumbricalis* are found. The attached form of this species is very common on the hard bottoms of the lower part of the phytobenthic zone of the Baltic Sea (Nielsen et al., 1995). Loose-lying *F. lumbricalis* is, on the contrary, very unique. Only three localities have been described as having large communities of this form in the Baltic Sea. One of these (Puck Bay) has already lost the population due to eutrophication and pollution problems (Martin et al., 1996; Kruk-Dowgiallo & Ciszewski, 1994). Austin (1959) described a similar agglomeration of loose *Furcellaria* in the central Kattegat area. The sea area of the West Estonian Archipelago hosts the largest known community of this kind, where a mixed community of loose-lying *Furcellaria lumbricalis* and *Coccotylus truncatus* covers up to 120 km^2 of sea bottom with more than 140 000 tons of wet biomass in Kassari Bay. The community was described for the first time by Kireeva (1961, 1964). The mean biomass of this community varied between 500 and 1000 g of wet weight m^{-2} and occasionally reached a maximum of 2.1 kg wet weight m^{-2} (Trei, 1975, 1976; Martin et al., 1996). The community was found on sandy substrates at depths between 5 and 9 m, where it formed a 0.15 to 0.3 m thick carpet on the seafloor. The proportion of the two dominant species differed slightly depending on locality but usually 60–70% of the biomass was *F. lumbricalis,* while *C. truncatus* accounted for 30–35%, on average. The proportion of other species was usually low, less than 5% (Trei, 1975, 1976; Martin et al., 1996).

The loose red algal community has been used as raw material for agar production since 1966 and annual yields have been estimated to be near 1000 t wet weight. The status of the community has been

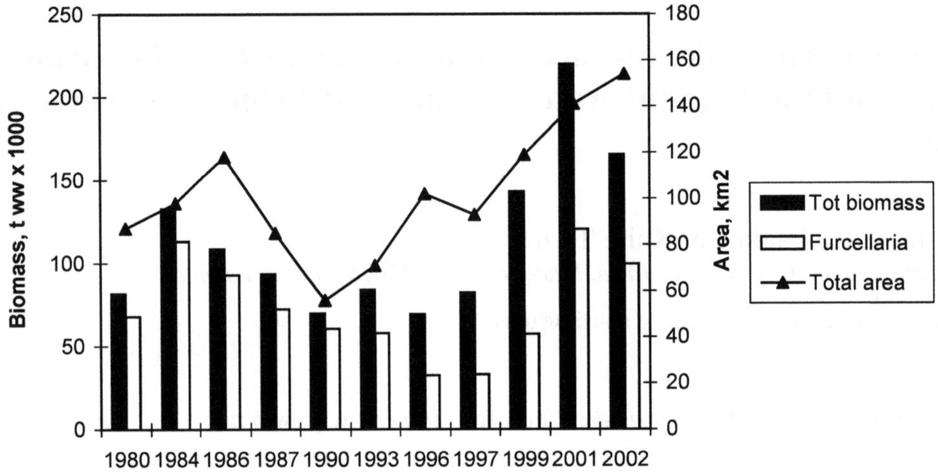

Figure 1. Dynamics of different characteristics of loose red algae community in Kassari Bay according to the results of commercial resource monitoring studies. Parameters shown are total biomass of the community, biomass of the species *Furcellaria lumbricalis* in the community and total area of the community.

monitored regularly and a decline in the loose *Furcellaria lumbricalis – Coccotylus truncatus* community was recorded in the Kassari Bay of the Väinameri area during 1996–97. Since then, both the total area of the community and total biomass have been steadily increasing (Figure 1). The decline in the loose red algal community was due to the extensive overgrowing of filamentous brown algae, which fixed the algal carpet and caused oxygen deficiency in the near-bottom layer (Martin & Kukk, 1997a,b, 1998, 1999).

The aim of this study was (1) to experimentally compare the growth rates of loose-lying form of *F. lumbricalis* with the accompanying species *C. truncatus,* (2) to compare the growth rates of these species in different ecological conditions (two different experimental sites, three different incubation depths) and (3) to follow seasonal changes in production.

Material and methods

Study area

Väinameri (inner sea of West-Estonian Archipelago) is formed by a system of straits connecting the waters of the Gulf of Riga to the Baltic Proper and the entrance to the Gulf of Finland (Figure 2). The total surface area of the system is 2243 km^2 (Suursaar et al., 1998). The mean depth of the whole system is less than 10 m. Kassari Bay, in the western part of the area, is connected to the Baltic Proper through the narrow Soela Strait and separated from the eastern part by a grid of islets. Hydrologically, this area behaves differently from the other parts of the Väinameri as it is more influenced by the saline waters of the Baltic Proper (Suursaar et al., 1998). The impact of the riverine inflow on the system is very small; the amount of fresh water entering the system reaches only 1 km^3 yr (Astok et al., 1999). The sea-floor is mainly of soft sediments, including fine mud and sand fractions. Harder substrata such as gravel or boulders can be found only in the most shallow and wave exposed areas. Due to the shallowness and the substrate being dominated by fine sediment fractions on the bottom, the water transparency is often very poor. After storms the Secchi depth may decrease to 0.5 m, while in the case of prolonged calm weather conditions the photic zone reaches the bottom in about 90% of this area.

The Gulf of Riga has a surface area of 16 330 km^2 with a water volume of 424 km^3, which makes up 3.9% of the total area of the Baltic Sea and 2.1% of its volume (Berzinsh, 1995). The volume of the annual freshwater input to the system is estimated to about 31 km^3 (Yurkovskis et al., 1993). An important feature of the basin is the lack of permanent stratification which enables intensive water exchange processes between the deep and surface layers. The nutrient regime of the Gulf of Riga basin differs greatly from that of the other parts of the Baltic, having several time higher nutrient concentrations compared to adjacent basins (Astok et al., 1999, Yurkovskis et al., 1993).

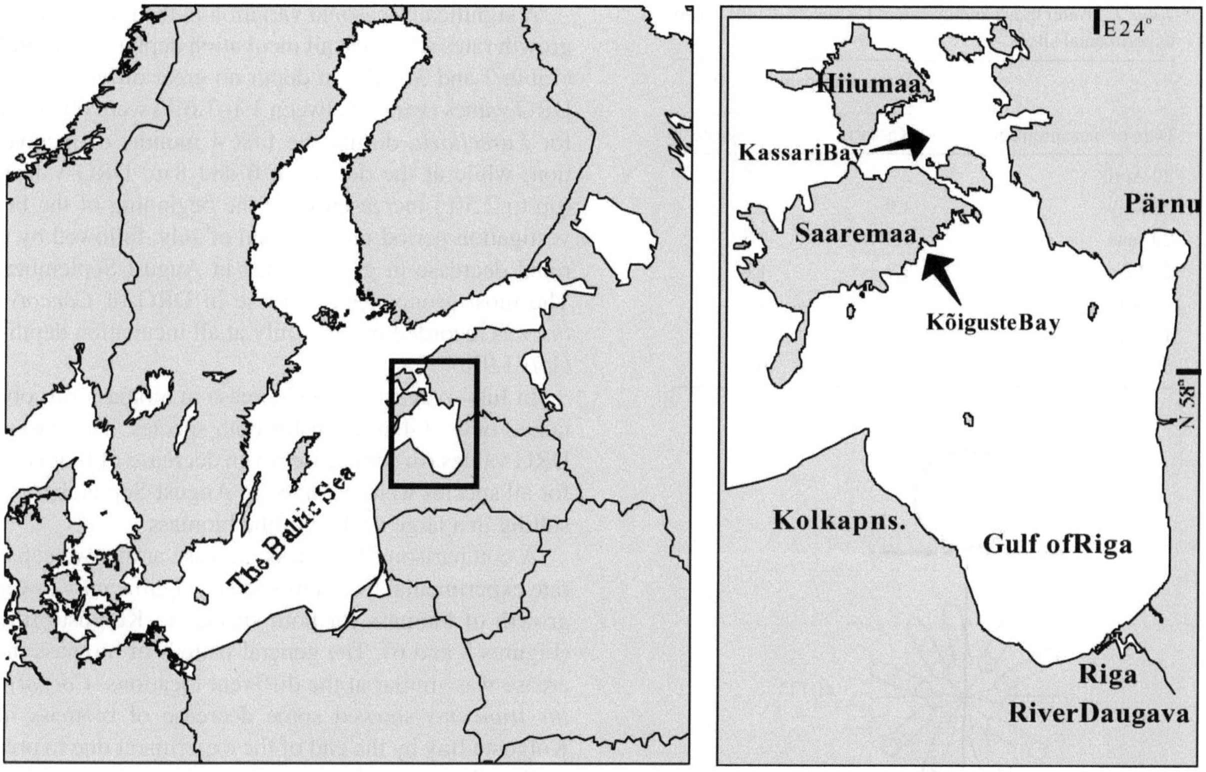

Figure 2. Study area. Location of two experimental sites are shown by arrows.

Field experiments

Field experiments for estimating the growth rate of loose-lying *F. lumbricalis* and the accompanying species *C. truncatus* were carried out in the period 20 April 2002–21 October 2002 in Kassari Bay (West-Estonian Archipelago Sea) and Kõiguste Bay (northern part of the Gulf of Riga) (Figure 2). *In situ* incubations of algal material were performed in special nylon mesh bags (with plastic frame inside; diameter 5.5 cm, height 20 cm) of 1 mm mesh size. Mesh bags with freshly collected algae (about 2–5 g wet weight per bag) which were free of macroepiphytes were incubated at depths of 4, 6 and 8 meters (Figure 3). Once a month 5 replicates from each depth of each species were collected for determination of growth rate. The changes in algal biomasses are presented as a percentage of the initial value (mean ± SE). Relative daily growth rates (DGR) were calculated for six experimental periods: 20 April–22 May (I); 22 May–19 June (II); 19 June–20 July (III); 20 July–22 August (IV); 22 August–16 September (V) and 16 September–21 October (VI), using the equation:

$$DRG(\%) = [(ln W_1 - \ln W_o)/n - 1]100$$

where n is the number of days of the incubation period, W_1 and W_o are the final and the initial weight of the algal material. Water temperature at incubation depths was measured as single measurements at the moment of sampling. Water transparency was estimated by Secchi disc.

Results

Environmental parameters

Water transparency was low during the entire observation period, generally not exceeding 2 m (Table 1). The water temperature at incubation depths varied between 0.5 and 22.8°C, measured between April and October 2002 (Figure 4). There were no major differences in the measured parameters for the two experimental sites.

Growth rates

The highest growth rates during the investigation period, in both experimental localities, for the loose-lying

Table 1. Water transparency values (as Secchi depth) in two experimental sites

	Water transparency (m)	
Date of measurement	Kassari	Kõiguste
20 April	2	1.5
23 May	1.9	2.1
20 June	2.3	2.3
21 July	1.8	1.8
23 Aug.	1.8	1.7
17 Sept.	3.5	3.3
22 Oct.	1	0.7

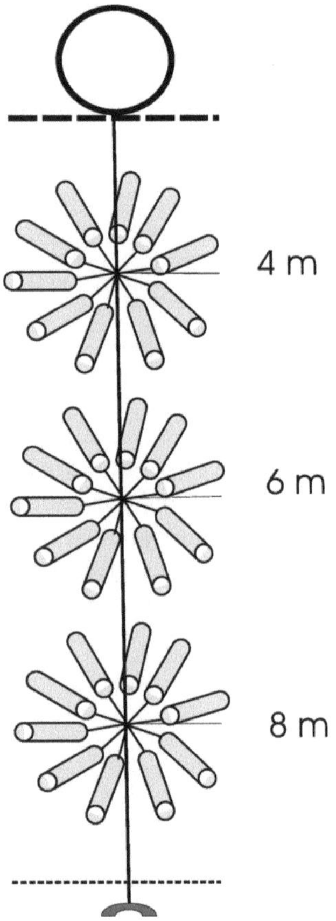

Figure 3. Illustration of experimental setup.

form of *Furcellaria lumbricalis* as well for *Coccotylus truncatus* were obtained at the incubation depth of 4 m, resulting in an increase in initial biomasses (on dry weight bases) of 268% and 238% respectively at the end of the incubation period. At the depths of 6 and 8 m the biomass increment was significantly lower (Figures 5 and 6).

A significant seasonal variation in the relative daily growth rates (DRG) at all incubation depths was found (Table 2 and 3). At 4 m depth no great differences in DRG values (varied between 1.1–1.6%) were obtained for *Furcellaria* during the first 4 months of incubation, while at the depths of 6 and 8 m, DRG values (up to 2.5%) increased from the beginning of the investigation period until the end of July, followed by a rapid decrease in growth rates in August-September. The most pronounced increase in DRG of *Coccotylus* was recorded in June–July at all incubation depths (1.7–1.9%).

In July–August a large increase in biomass was obtained only at 4 m depth for both species. The lowest DRG values (in some cases even decrease of biomass) for all species were obtained in August-September resulting in a large decline of the biomass.

A comparison of biomass increase at the two separate experimental locations showed significantly lower growth of biomass for both species in Kõiguste Bay (Figures 5 and 6). The general pattern of biomass increase was similar at the different locations. *Coccotylus truncatus* showed some decrease of biomass in Kõiguste Bay by the end of the experiment due to prevailing decomposition processses (Figures 5 and 6). In our study, both incubation period and incubation depth significantly affected the growth rates of loose-lying form of *Furcellaria* as well *Coccotylus truncatus* (2-way-ANOVA; $p < 0.001$). At a depth of 4 m the gain of biomass, in both the species studied, was significantly higher compared to 6 m and 8 m, indicating the more favourable growing conditions at shallower depths.

Discussion

In natural conditions, most of the loose *Furcellaria-Coccotylus* community is found at depths of 6–9 m in Kassari Bay (Martin et al., 1996). In our experiment the highest growth was observed much shallower than that. This contradiction can be explained by several other limiting environmental factors such as wave action, biological interactions, keeping most of the loose algae community in deeper water. Most of the added biomass (compared to the initial condition) was gained during the first half of the observation period and the pattern was similar in both studied species and experimental sites. This indicates that most of the net annual production of the natural algal communities is achieved during the spring and beginning of summer. According to

Table 2. Summary of two-way ANOVAs for *Furcellaria lumbricalis* growth parameters measured in Kassari (a) and Kõiguste (b) Bays over 6 experimental periods at the depths of 4, 6 and 8 m

			Biomass increment per experimental period			DGR	
Source of variation	*df*		*F*	*P*		*F*	*P*
Experimental	5	a)	430.30	<0.001	a)	203.54	<0.001
Period		b)	93.90	<0.001	b)	13.03	<0.001
Incubation depth	2	a)	258.46	<0.001	a)	860.39	<0.001
		b)	45.15	<0.001	b)	46.24	<0.001
Experimental	10	a)	27.59	<0.001	a)	26.59	<0.001
Period × incubation depth		b)	3.39	0.001	b)	2.93	0.004

Table 3. Summary of two-way ANOVAs for *Coccotylus truncatus* growth parameters measured in Kassari (a) and Kõiguste (b) Bays over 6 experimental periods at the depths of 4, 6 and 8 m

			Biomass increment per experimental period			DGR	
Source of variation	*df*		*F*	*P*		*F*	*P*
Experimental	5	a)	147.72	<0.001	a)	79.19	<0.001
Period		b)	26.75	<0.001	b)	7.37	0.001
Incubation depth	2	a)	119.04	<0.001	a)	467.04	<0.001
Experimental	10	b)	49.69	<0.001	b)	88.70	<0.001
Period × incubation depth		a)	9.21	<0.001	a)	12.91	<0.001
		b)	2.12	0.033	b)	1.38	0.206

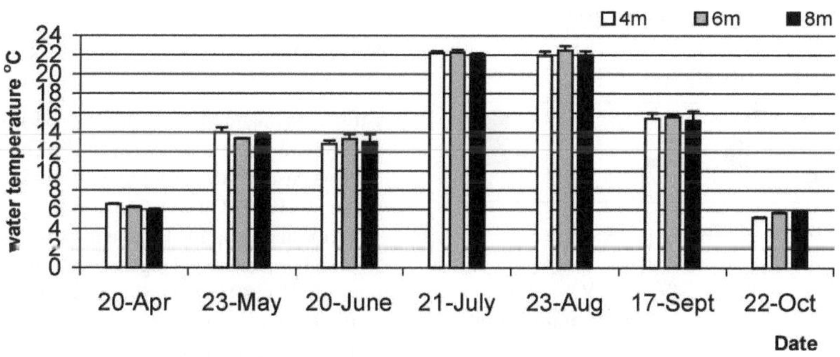

Figure 4. Water temperature measured in Kassari Bay at incubation depths.

literature data this is also true for many other macroalgal species (Kiirikki, 1996).

Biomass dynamics of loose-lying *Furcellaria* and *Coccotylus,* during the entire experimental period, followed a similar seasonal pattern at all incubation depths and both experimental sites, reaching the highest biomass in August. Loose-lying *F. lumbricalis* and *C. truncatus* together form a unique mixed community in Kassari Bay and thus are most probably adapted to similar environmental conditions, which was reflected in their similar growth in the experiment.

In the second half of the investigation period, the observed slight decline in biomass of both species was probably due to the onset of decomposition processes in the algal material. The latter would have been favoured by relatively high water temperature: as a rule decomposition processes depend on temperature (Carpenter & Adams, 1979; Birch et al., 1983). Besides high water temperature, losses of algal biomass could also be attributed to low photosynthetic activity of the algae caused by low water transparency due to increased pelagic productivity during this warm period. It was

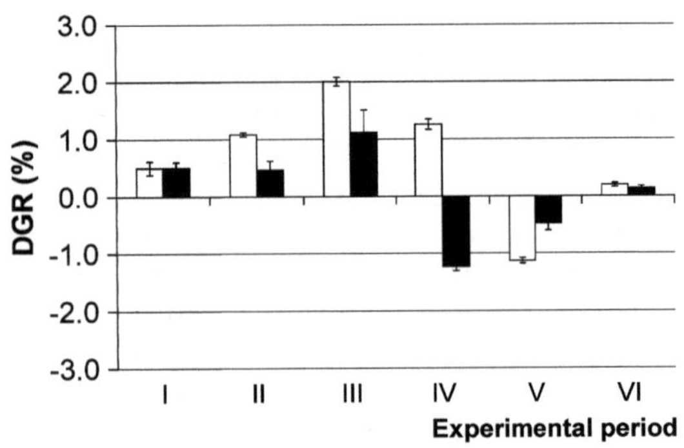

Figure 5. Relative daily growth rates (DGR) ± standard error ($n = 5$) of *Furcellaria lumbricalis* calculated for six experimental periods: 20 Apr.–22 May (I); 22 May–19 Jun. (II); 19 Jun.–20 Jul. (III); 20 Jul.–22 Aug. (IV); 22 Aug.–16 Sept. (V) and 16 Sept.–21 Oct. (VI) in Kassari Bay (white bars) and Kõiguste Bay (black bars) at the depths of 4 m (A), 6 m (B) and 8 m (C).

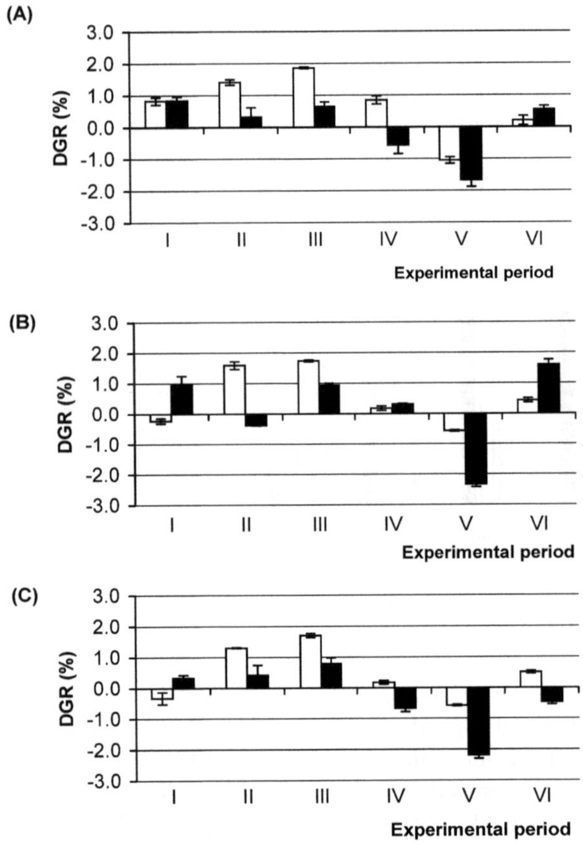

Figure 6. Relative daily growth rates (DGR) ± standard error ($n = 5$) of *Coccotylus truncatus* calculated for six experimental periods: 20 Apr.–22 May (I); 22 May–19 Jun. (II); 19 Jun.–20 Jul. (III); 20 Jul.–22 Aug. (IV); 22 Aug.–16 Sept. (V) and 16 Sept.—21 Oct. (VI) in Kassari Bay (white bars) and Kõiguste Bay (black bars) at the depths of 4 m (A), 6 m (B) and 8 m (C).

remarkable that the high water temperature values had similar effects on both the algal species studied.

The relatively low increases in biomass that were obtained at the Kõiguste experimental site, for both tested species, cannot be explained in terms of environmental variables measured during the experiment, because both transparency and water temperature values stayed similar at both sites. However, there may have been differences in nutrients in the two areas. It is stated that the waters of northern part of Gulf of Riga have higher nutrient concentrations compared to other parts of the Baltic Sea and also lower water transparency in general (HELCOM, 2002). In our case the exact mechanism remained unclear but the difference between two experimental sites was obvious.

Acknowledgements

The present study was conducted under the framework of the Estonian governmental programme no. 0182578s03 and publication made possible with support from the grant no. 5103 of Estonian Science Foundation. The authors thank two anonymous referees for valuable comments on the manuscript.

References

Astok V, Otsmann M, Suursaar Ü (1999) Water exchange as the main physical process in semi-enclosed marine systems: The Gulf of Riga case. Hydrobiologia 393: 11–18.

Austin A (1959) Observations on *Furcellaria fastigiata* (L.) Lam. forma aegagrophila Reinke in Danish waters together with a note on other unattached algae. Hydrobiologia 14(3–4): 255–277.

Berzinsh V (1995) Hydrology. In Ojaveer E (ed.), Ecosystem of the Gulf of Riga between 1920 and 1990. Estonian Academy Publishers. Tallinn 7–31.

Birch PB, Gabrielson JO, Hamel KS (1983) Decomposition of *Cladophora* I. Field studies in the Peel-Harvey estuarine system, western Australia. Bot. Mar. 26: 165–171.

Carpenter SR, Adam MS (1979) Effects of nutrients and temperature on decomposition of *Myriophyllum spicatum* L. in a hard-water eutrophic lake. Limnol. Oceanogr. 24: 520–528.

HELCOM (2002) Environment of the Baltic Sea Area 1994–1998. Baltic Sea Environmental Proceedings 82B: 214.

Johansson G, Snoeijs P (2002) Macroalgal photosynthetic responses to light in relation to thallus morphology and depth zonation. Mar. Ecol. Prog. Ser. 244: 63–72.

Kiirikki M (1996) Mechanisms affecting macroalgal zonation in the northern Baltic Sea. Europ. J. Phycol. 31: 225–232.

Kireeva MS (1961) Amount of *Furcellaria fastigiata* (Huds.) Lamour. in the Baltic Sea. (Area of Saaremaa and Hiiumaa Islands). Trudy NIIRH SNH Latv. SSR 3: 411–417 (in Russian).

Kireeva MS (1964) Aggregations of unattached red algae in the sea areas of Soviet Union. In: Resources of marine algae and their use. Moscow pp. 1–25 (in Russian).

Kruk-Dowgiałło L, Cisewski P (1994) Zatoka Pucka. Możliwości rewaloryzacji. Institut Ochrony Środowiska, Warsawa, 178.

Martin G, Paalme T, Kukk H (1996) Long-term dynamics of the commercial useable *Furcellaria lumbricalis-Phyllophora truncata* community in Kassari Bay, West Estonian Archipelago, the Baltic Sea. Proceedings of Polish-Swedish Symposium on Baltic coastal fisheries Resources and Management, 2-3 April 1996, Gdynia, Poland, pp. 121–129.

Martin G, Kukk H (1997a) Environmental factors limiting phytobenthos communities in the Gulf of Riga and West-Estonian Archipelago Sea. Phycologia 36: 70.

Martin G Kukk H (1997b) Unattached macroalgal communities – adaptation to unfavourable environmental conditions. In, Recruitment Dynamics of Exploited Marine Populations: Physical-Biological Interactions, ICES International Symposium, Book of Abstracts: 94.

Martin G, Kukk H (1998) The structure of benthic littoral communities of the West-Estonian Archipelago area asa reflection of unique hydrodynamical conditions. In Brackish Water Ecosystems, ICES International Symposium, Book of Abstracts: 20.

Martin G, Kukk H (1999) Environmental factors forcing the dynamics and the structure of loose *Furcellaria lumbricalis-Coccotylus truncatus* community in Kassari Bay, the inner sea of West-Estonian Archipelago, NE Baltic Sea. In Abstracts of 34th European Marine Biology Symposium: 6.

Nielsen R, Kristiansen A, Mathiesen L, Mathiesen H (1995) Distribution index of the benthic macroalgae of the Baltic Sea area. Act. Bot. Fennica 155: 55.

Suursaar Ü, Astok V, Otsmann M (1998) The front of Väinameri. In EMI Report Series 9: 23–33.

Trei T (1975) Flora and vegetation in the coastal waters of Western Estonia. Merentutkimuslait Julk 239: 348–351.

Trei T (1976) Brown and red algae in the coastal waters of western Estonia. Zvaigzne, Riga pp. 1–87 (in Russian).

Yurkovskis A, Wulff F, Rahm L, Andrushaitis A, Rodriguez-Medina M (1993) A nutrient budget of the Gulf of Riga; Baltic Sea. Estuar. Coast. Shelf. Sci. 37: 113–127.

Journal of Applied Phycology (2006) 18: 565–573
DOI: 10.1007/s10811-006-9058-2

Factors influencing the growth rates of three commercial eucheumoids at coastal sites in southern Kenya

J.G. Wakibia[1,3,*], J.J. Bolton[2], D.W. Keats[1] & L.M. Raitt[1]
[1]*Department of Biodiversity and Conservation Biology, University of the Western Cape, Private Bag X17, Bellville 7535, South Africa;* [2]*Department of Botany, University of Cape Town, 7701 Rondebosch, South Africa;* [3]*Kenya Marine and Fisheries Research Institute. P.O Box 81651, Mombasa, Kenya*

*Author for correspondence: e-mail: jwakibia@kmfri.co.ke; jgwakibia@yahoo.com

Key words: Eucheuma denticulatum, Kappaphycus alvarezii, growth, commercial eucheumoids, morphotypes, mariculture, Kenya

Abstract

As a possible means of improving the livelihoods of local villagers, off-bottom rope cultivation of commercial eucheumoids was investigated on the southern Kenyan coast at three sites, representative of the variety of environments. The morphotypes used were brown *Eucheuma denticulatum* and green and brown *Kappaphycus alvarezii*. The study was carried out over a 15 month period from August 2001 until October 2002. Relative growth rates were highest at a sandy flat in a mangrove system (Gazi; 5.6% d^{-1}), and lowest in an intertidal reef flat (Kibuyuni; 3.2% d^{-1}) with a lagoon being intermediate (Mkwiro; 4.8% d^{-1}). The brown *E. denticulatum* had the highest growth rate of 4.7% d^{-1} compared to the green and brown *K. alvarezii* which were 4.3% d^{-1} and 4.2% d^{-1} , respectively. Growth was more variable at Kibuyuni and Mkwiro. The growth was higher during the southeast monsoon (4.7% d^{-1}) than during the northeast monsoon (4.0% d^{-1}). This is part of a larger study and the effects of water motion, salinity, temperature, thallus nitrogen, and 'ice-ice' syndrome on growth of morphotypes is discussed. The water motion was observed to increase thallus nitrogen and hence the growth of eucheumoids. The 'ice-ice' condition affected both brown *E. denticulatum* and brown *K. alvarezii* but not green *K. alvarezii*. The results suggest that commercial cultivation of eucheumoids in Kenya will be feasible.

Introduction

The marine algae, *Eucheuma denticulatum* (Burman) Collins and Hervey and *Kappaphycus alvarezii* (Doty) Doty ex P.C. Silva are the most important carrageenophytes commercially (McHugh, 2003). However, several morphological and pigment forms of each species are used in farming, which makes the taxonomy and naming of commercial *Kappaphycus* and *Eucheuma* species problematic (Doty, 1987). The term "eucheumoid" (by adding "oid" to the genus *Eucheuma* as in gracilarioid) is used here to refer to both *E. denticulatum* and *K. alvarezii* and their associated forms. The nomenclature of *Eucheuma* and *Kappaphycus* species is according to Silva et al. (1996) and Prud'homme van Reine and Trono (2001).

The demand for carrageenan has been increasing over the years, and to meet this requirement (and possibly at a cheaper cost), carrageenan producers such as CP Kelco and FMC BioPolymer have been promoting the geographic spread of eucheumoids to non-endemic locations such as Indonesia, Tanzania, Madagascar, Fiji and Kiribati (Ask et al., 2003). In addition, the mariculture of eucheumoids is also gaining popularity in developing countries as a source of foreign exchange and a means of broadening the livelihoods for coastal communities (Ask & Azanza, 2002; Ask et al., 2003). For example, about 180 000 families in the Philippines (Hurtado & Cheney, 2003) and over 20 000 people in Tanzania (Mtolera, 1996) benefit from seaweed cultivation. The effects of the introductions on the local environment have not yet been determined.

Thus these species are being cultivated in East African countries such as Djibouti (Braud & Perez, 1978), Madagascar (Mollion & Braud, 1993), Zanzibar (Lirasan & Twide, 1993) and recently in Mozambique (R. Piezas, pers. comm.). Although Kenya has similar environmental conditions to Tanzania (McClanahan, 1988), there is as yet no commercial exploitation or farming of seaweeds. Coppejans (1989) reported that there is little suitable shallow water area (compared to Tanzania) for seaweed cultivation in Kenya. He also suggested that there might be conflicts for water space with other reef users. However, there are no reports of the suitable area for seaweed cultivation in Kenya. Furthermore, some farming systems such as the floating raft and long line methods are used in deep waters (Hurtado-Ponce et al., 2001; Paula & Pereira, 2003). Yarish and Wamukoya (1990) found no suitable natural sources and concluded that utilisation of eucheumoids in Kenya could only be realised by mariculture, which is supported by other reports (Lirasan & Twide, 1993; Oyieke, 1998; UNEP, 1998). However, McHugh (2002) did not consider that Kenya had good prospects for a seaweed industry. He felt that the pilot studies were not promising but he did not reference these studies. The purpose of this study was to evaluate the feasibility of growing three commercial eucheumoids under field conditions in Kenya.

Materials and methods

Study sites

Based on a preliminary survey of potential study sites along the southern Kenya coast, three areas, Gazi Bay, Kibuyuni and Mkwiro were selected because they are sheltered from wave action, accessible, and represent a range of environmental conditions in Kenya. Gazi Bay (4°25′S, 39°30′E) is a shallow mangrove system which receives freshwater from rivers. However, both a shoreward wind and tidal currents mix the water in the Bay, leading to seawater with near oceanic salinity (Kitheka, 1996). A prominent seagrass bed of *Thalassia hemprichii* (Ehrenberg) Ascherson exists at the centre of the bay. However, at the cultivation sandy flat, only *Halodule*, *Halophila* and *Cymodocea* were common. The major substratum consisted of a mixture of sand and silt. The sandy flat was covered with approximately 20–30 cm of water at the lowest tide and 3.8 m at the highest tide.

Kibuyuni (4°38′S, 39°20′E) is a large, intertidal reef flat covered by a belt of *Thalassodendron ciliatum* (Forsskal) den Hartog. Patches of *E. denticulatum* and *Kappaphycus striatus* (M.F. Schmitz) Doty ex P. Silva were also common in this area. There was however, insufficient and unhealthy material for the study. The substratum at the study site consisted of coral rubble and small pockets of sand. The reef-flat was covered with 10 cm of seawater at the lowest tide and 3.2 m at the highest tide. The dominant animals were soft corals, large sponges, starfishes, brittle stars, sea urchins and rabbit fishes.

Mkwiro (4°40′S, 39°23′E) is located on the eastern side of Wasini Island, about 2 km from the mainland. It is a lagoon characterized by sandy substratum where *Thalassia hemprichii*, *Syringodium isoetifolium* (Ashers.) Dandy., *Turbinaria*, and *Sargassum* predominate, with occasional coral heads. *Eucheuma platycladum* F. Schmitz was found growing in patches. Numerous sea urchins, starfishes, and soft corals were also evident at the site.

The coastal belt of Kenya experiences a tropical monsoon climate dominated by two seasons, the southeast monsoon (SEM) prevailing from April to September and the northeast monsoon (NEM) from October to March. The two seasons are characterised by distinct differences in physical and chemical conditions of the coastal waters (McClanahan, 1988). The SEM is associated with strong winds, low air and water temperatures, low solar radiation and heavy rains, with the lowest tides occurring during the night. During the NEM, these conditions are reversed with the lowest tide occurring during the day. The tides are mixed semidiurnal, with tidal ranges of about 4.0 m (McClanahan, 1988).

Plant materials and cultivation methods

Three morphotypes from two species (brown *Eucheuma denticulatum*, and green and brown *Kappaphycus alvarezii*), were collected from a seaweed farm in Zanzibar, although they originally came from Bohol, Philippines. The plant material (2.5 kg for each morphotype) were washed clean of silt, associated animals and plants, and transported to Mombasa, Kenya where the materials were placed in outdoor culture tanks at Kenya Marine and Fisheries Research Institute (KMFRI) for 4 weeks, inspected and cleaned of any foreign organisms before use.

The fixed, off-bottom farming technique as described by Lirasan and Twide (1993) was used. In order to compare the growth rates of the three morphotypes, 4 plots (5 × 1.5 m) were set up at each site, each containing four polypropylene ropes (one for each morphotype and an empty rope) stocked with twenty healthy cuttings, each weighing between 80 and 100 g. The empty rope was used as a control for weights of sand and filamentous epiphytes. At the end of each culture period (about 30 d), the weight of each rope with its cuttings, less the weight of the empty rope, was averaged to the number of cuttings remaining on the rope. The number of missing cuttings and those with 'ice-ice' syndrome (white, soft and partly dissolved thallus) on every rope were recorded. Healthy cuttings from each plot were used for each succeeding culture period.

The growth study was performed from August 2001 to October 2002. However, due to logistical problems it was not possible to collect data at Mkwiro from May to July 2002. The relative growth rate (RGR), expressed as percent increase in wet weight per day, was determined for each rope according to the following formula:

$$\text{RGR} = [(Wt/Wo)^{1/t}-1] \times 100$$

where W_o = average cutting wet weight at start, W_t = average cutting wet weight at time t and t = time intervals (days). Number of plants lost and the occurrence of 'ice-ice' were expressed as a percentage. Grazing damage was not measured.

Plant tissue analysis

The harvested material was oven dried at 45 °C for 72 h and stored in polythene bags for carrageenan extraction (data not presented here) and nitrogen and phosphorus tissue analysis. The material for N and P analysis was rinsed in distilled water, oven dried at 60 °C for 24 h, and ground in a Wiley mill (40-mesh size). Prior to analysis, the milled powders were redried to constant weight at 60 °C and cooled in a desiccator over silica gel. Total N was determined with a LECO FP528 Nitrogen Analyzer. For total P, 0.5 g of seaweed sample was dry ashed at 450 °C. The ash was wetted with deionised water, dissolved in 1:1 HCl (5 ml) and diluted to 50 ml. The total P concentration was determined with a Varian Vista-MPX ICP Spectrophometer. The N and P analyses were carried out by Bemlab (Cape Town, South Africa).

Environmental factors

Various environmental factors were recorded from each site once every two weeks both during the flood and ebb tides. These factors were measured at the depth where the plants were growing. Water temperature and salinity readings were determined using a mercury thermometer and a refractometer (Atago, Japan), respectively. Photon fluency was measured using a LI-193SA Spherical Quantum Sensor and LI-1400 Datalogger (Li-Cor, USA). Water pH and dissolved oxygen were measured using a portable pH meter (Beckman Instruments, USA) and WTW Multline P4 Universal meter (Germany), respectively. The water motion was determined by the dissolution of spherical plaster of Paris balls according to a modified clod card method of Doty (1971). The balls were made by pouring a mixture of water and plaster of Paris (1:1) into tennis ball moulds (about 5 cm diameter) and then inserting a stick (15 cm long, 0.5 cm diameter) into each ball. The moulds were removed after the hardening of plaster balls and the rough surfaces smoothed by filing with sand paper. Duplicate balls (each weighing about 30 g) were mounted at each site by threading the stick through the lay of 4 mm polypropylene ropes. They were collected after 24 h, rinsed in fresh water, dried and weighed. The diffusion factor (DF) was determined by the ratio of weight lost in the field compared to duplicate plaster balls that had been kept in 20 l of still seawater for four days.

Two surface seawater samples for the determination of nitrate and phosphate levels were collected at each site every two weeks, stored in a cool box at 4 °C and transported to the laboratory. Water samples were filtered through Whatman GF/C filters and analysed for nitrate and phosphate by the modified automated method of Parsons et al. (1984) using the continuous flow analyzer as applied in the Technicon Auto Analyzer II system. Two similar water samples were also collected to determine total suspended matter (TSM). The water was filtered through previously dried and weighed Whatman GF/C filter papers which were then dried and re-weighed to measure the TSM content of the water.

Data on total rainfall for Gazi Bay were supplied by the local Agriculture office while maximum and minimum air temperature measurements were obtained from the Kenya Meteorological weather station at Matuga, about 40 km from the study sites. Data analyses were made by General Linear Model Procedures followed by determining differences among individual mean values by Least Significant Difference test at $p < 0.05$. Pearson's product moment correlation test was used to determine the linear relationship between treatments. Statistical analyses were performed using the SAS Program, Version 8.2.

Results

Growth rate

The monthly relative growth rates (RGRs) are presented in Figure 1. There is a general pattern of initial high growth followed by a decline. There was a general increased growth rate in most of the morphotypes from March to August/September, followed by a decline again. The low growth rate at Kibuyuni from July to September was probably due to grazing by fish and the sea urchin, *Tripneustes gratilla*. The growth rate was more variable at both Mkwiro and Kibuyuni than at Gazi, where grazers were not abundant. The RGRs were highest at Gazi and lowest at Kibuyuni with Mkwiro being intermediate (Table 1).

Brown *E. denticulatum* had a significantly higher RGR than the green and brown *K. alvarezii* which were not significantly different from each other in RGR (Table 2).

The growth rates were higher ($p < 0.01$) during the southeast monsoon ($4.7 \pm 1.7\%\ d^{-1}$) than during the northeast monsoon ($4.0 \pm 1.8\%\ d^{-1}$). The month of August 2001 ($6.0 \pm 0.3\%\ d^{-1}$) had the highest RGR whereas the lowest growth was recorded in January 2002 ($3.1 \pm 0.3\%\ d^{-1}$). Growth of eucheumoids varied from one year to another as the RGR values for August to October in 2001 were significantly higher ($p < 0.01$) than those in August to October 2002.

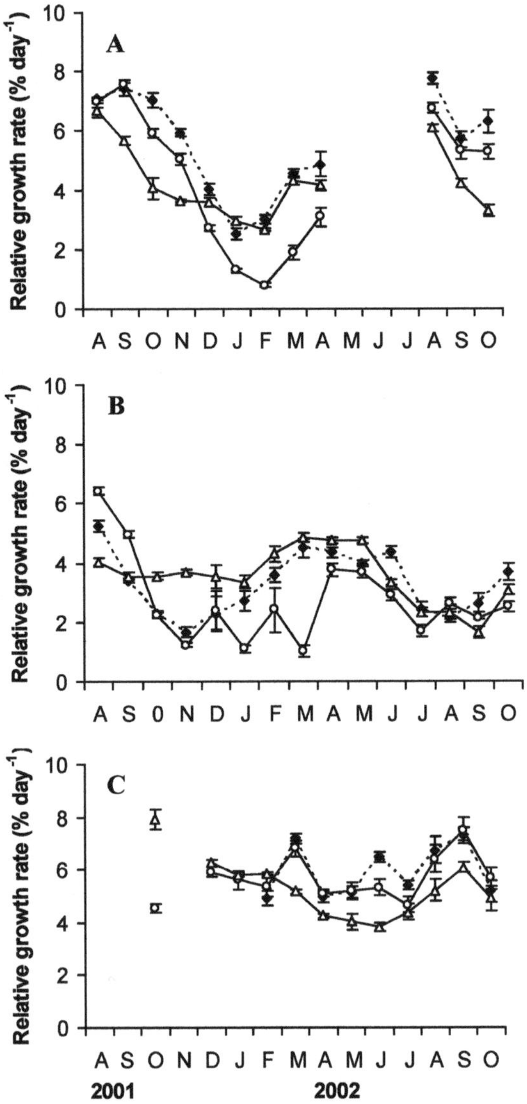

Figure 1. Monthly relative growth rates (RGRs) of three eucheumoid morphotypes, brown *E. denticulatum* (-♦-♦-), green (-△-△) and brown (-○-○-) *K. alvarezii* grown at three sites (*A* = Mkwiro; *B* = Kibuyuni; *C* = Gazi) in southern Kenya (mean ± SE, $n = 10$–15).

Tissue analysis and seaweed factors

Analyses are presented by morphotypes in Table 2. The highest thallus P values were obtained for brown *E. denticulatum* while both brown and green *K. alvarezii* had the same levels. Thallus N levels were similar for all the three morphotypes. However, significant differences in thallus N and P contents were observed between sites (Table 1). Thallus N levels were highest in Gazi plants, and lowest in Kibuyuni plants with those at Mkwiro having an intermediate content. Thallus P content was higher in Mkwiro plants than in those at Kibuyuni which in turn was higher than in the Gazi plants (Table 1). Highest thallus N ($1.10 \pm 0.04\%$ dry wt) and P ($0.087 \pm 0.003\%$ dry wt) levels were observed during the southeast monsoon while lowest values of 0.94 ± 0.02 and $0.079 \pm 0.003\%$ dry wt were recorded during the northeast monsoon for thallus N and P, in that order. There were no differences in plant loss among the morphotypes (Table 2) but there were between the sites (Table 1). The plant loss and 'ice-ice' occurrence at Kibuyuni was higher than at Mkwiro, whereas there was no incidence of 'ice-ice' at Gazi.

Table 1. N and P thallus content (% of dry wt) and seaweed parameters of eucheumoids grown at three sites in southern Kenya (mean ± SE, $n = 39$–117)

Parameter	Gazi	Kibuyuni	Mkwiro
Thallus N	1.22 ± 0.03[a]	0.83 ± 0.02[c]	1.02 ± 0.03[b]
Thallus P	0.077 ± 0.004[c]	0.082 ± 0.003[b]	0.090 ± 0.005[a]
% Plant loss	2.4 ± 0.7[b]	5.4 ± 0.6[a]	3.5 ± 0.6[b]
% Ice-ice syndrome	0.0[c]	2.5 ± 0.5[a]	1.4 ± 0.4[b]
Relative growth rate (% d^{-1})	5.6 ± 0.1[a]	3.2 ± 0.1[c]	4.8 ± 0.2[b]

Means with the same letter in each row are not significantly different at $p < 0.05$.

Table 2. N and P thallus content (% of dry wt) and seaweed parameters of three eucheumoids grown at three sites in southern Kenya (mean ± SE, $n = 34$–133)

Parameter	Brown *E. denticulatum*	Brown *K. alvarezii*	Green *K. alvarezii*
Thallus P	0.101 ± 0.003[a]	0.074 ± 0.002[b]	0.068 ± 0.002[b]
Thallus N	1.00 ± 0.03[a]	1.01 ± 0.04[a]	1.02 ± 0.04[a]
% Plant loss	3.7 ± 0.5[a]	4.7 ± 0.7[a]	3.5 ± 0.7[a]
% Ice-ice syndrome	0.8 ± 0.2[b]	3.4 ± 0.7[a]	0.1 ± 0.1[b]
Relative growth rate (% d^{-1})	4.7 ± 0.2[a]	4.2 ± 0.1[b]	4.3 ± 0.2[b]

Means with the same letter in each row are not significantly different at $p < 0.05$.

Environmental parameters

Table 3 shows the summary of environmental factors determined during the study period. Phosphate and nitrate levels were both higher at Kibuyuni and Mkwiro than at Gazi. The high nutrient levels at both sites were presumably due to nutrient excretions by the numerous invertebrates at both sites. Although Gazi had slightly lower levels of nutrients, the greater water motion probably made more nutrients available to the plants. Salinity varied slightly over the study period at each site (Table 3). Relatively higher salinities were recorded at both Kibuyuni and Mkwiro than at Gazi. The relatively low salinity at Gazi was presumably due to dilution with fresh water. The total annual rainfall data recorded at Gazi were 1578 and 1661 mm for 2001 and 2002, respectively. The diffusion factors calculated for Gazi were significantly higher than those for Mkwiro and Kibuyuni (Table 3). The high total suspended matter and low pH at Gazi were probably due to the influence of incoming run-off and sediments that were brought to the Bay during the rain. Water temperature (28.5–29.3° C), photon fluency (1042–1317 μmol photons m^{-2} s^{-1}) and dissolved oxygen (5.48-6.56 mg l^{-1}) were all similar at all the study sites.

Several epiphytes and grazers were observed at the study sites. The most common epiphytes were *Enteromorpha ramulosa* (J. E. Smith), particularly during the SEM, and the blue-green alga, *Lyngbya majuscula* Harvey ex Gomont (NEM). The most destructive grazers were the rabbit fishes (Siganidae) and an echinoderm, Tripneustes *gratilla*. The experimental plants at Kibuyuni were heavily grazed in July–Sept 2002, resulting in low net growth rates during this period.

Correlation analyses

Table 4 shows positive correlations between growth rates of the three morphotypes and both water motion (diffusion factor) and thallus nitrogen content. Surprisingly all other significant correlations were negative where they occurred: thus growth was reduced at both temperature extremes, inversely proportional to water salinity and as could be expected, negative correlations were obtained between the growth rates of both the brown eucheumoids and percent 'ice-ice'. Thallus N contents of brown *E. denticulatum* and brown *K. alvarezii* were inversely correlated with 'ice-ice' ($r = -0.361$, $p < 0.05$ and $r = -0.523$, $p < 0.01$, respectively), whereas very little 'ice-ice' was observed for green *K. alvarezii* ($r = -0.187$, $p > 0.05$).

Discussion

The growth rates of eucheumoids have been observed to vary with species and morphotypes (Doty, 1987). The growth rates of all morphotypes (Table 2) were

Table 3. Environmental factors measured at three study sites in southern Kenya (mean ± SE, $n = 9$–15)

Environmental parameter	Sites		
	Gazi	Kibuyuni	Mkwiro
Nitrate (μM)	1.39 ± 0.27^b	2.76 ± 0.57^a	2.43 ± 0.43^a
Phosphate (μM)	0.77 ± 0.14^c	1.33 ± 0.18^b	1.49 ± 0.25^a
Salinity (‰)	34.8 ± 0.5^b	35.6 ± 0.3^a	35.6 ± 0.3^a
Water motion (diffusion factor)	6.16 ± 0.31^a	4.72 ± 0.21^c	5.54 ± 0.28^b
pH	7.74 ± 0.05^b	8.19 ± 0.11^a	8.15 ± 0.08^a
Total suspended matter (mg l^{-1})	13.2 ± 0.7^a	12.8 ± 2.0^a	8.7 ± 1.0^b

Means with the same letter in each row are not significantly different at $p < 0.05$.

Table 4. Correlation coefficients of three eucheumoid growth rates with thallus N and P (% dry wt) and environmental factors measured in southern Kenya

Parameter	Brown *E. denticulatum*	Brown *K. alvarezii*	Green *K. alvarezii*
Water nitrate	−0.138	−0.209	0.005
Water phosphate	0.055	0.101	0.073
Maximum air temperature	−0.333*	−0.413**	0.021
Minimum air temperature	−0.246	−0.393*	0.022
Diffusion factor	0.518**	0.591**	0.489**
Salinity	−0.391*	−0.363*	−0.180
Thallus N	0.623**	0.790**	0.534**
Thallus P	−0.023	0.242	−0.121
% Ice-ice syndrome	−0.472**	−0.611**	−0.098

*Significant at $p < 0.05$, **Significant at $p < 0.01$.

above the recommended commercial value of 3.5% d^{-1} for eucheumoid farming (Doty, 1987) and similar to those observed for eucheumoids by other authors (Braud & Perez, 1978; Glenn & Doty, 1990; Lirasan & Twide, 1993; Dawes et al., 1994; Ohno et al., 1996; Hurtado-Ponce et al., 2001). The maximum growth rates (6.9 to 7.8% d^{-1}) reported in this study are lower than the maximum value of 10.6% d^{-1} recorded for *K. alvarezii* (Azanza-Corrales & Aliaza, 1999). On the other hand, the growth rates found were higher than those for both *E. denticulatum* and *K. striatus* (mean of 1.4 % d^{-1}) grown in Madagascar (Mollion & Braud, 1993).

The growth rates of eucheumoids vary with species, as observed in this study, where higher growth was obtained for brown *E. denticulatum* than for both brown and green *K. alvarezii* (Table 2). A similar observation has been reported by Russell (1982) in Fiji waters where the growth rate of *E. denticulatum* was higher (7.5 % d^{-1}) than that of *K. striatus* (6.1% d^{-1}). However, several authors have reported higher growth rates for *K. alvarezii* than in *E. denticulatum*. They attributed high growth of *K. alvarezii* to its wide range of tolerances to ecological factors (Dawes et al., 1994), greater response to water motion (Glenn & Doty, 1992), and morphological variability (Doty, 1987). On the other hand, no significant difference in growth rate was found between *E. denticulatum* and *K. striatus* by Mollion and Braud (1993). The higher growth rates for *E. denticulatum* than the two forms of *K. alvarezii* in this study may partly be due to the great water motion and the tidal flushing as was observed in Zanzibar (Mshigeni, 1994) and its better tolerance to the adverse conditions, as well as other factors. The high tidal amplitude (4 m) in Kenya (McClanahan, 1988) may have enhanced diffusion of nutrients in the study areas. It has been suggested that the algal responses to different water motion levels and other relative factors may vary between species (Glenn & Doty, 1990; Hurd, 2000). The brown and green morphotypes of *K. alvarezii* had similar growth rates in this study, as observed by other authors (Dawes et al., 1994; Ohno et al., 1994)

There were distinct differences in growth of eucheumoids at the three sites (Table 1). The variation in growth rates obtained for the morphotypes can possibly be explained by differences in water motion at the three sites (Table 4). Similar results of growth differences at different sites due to water motion have been reported for *E. denticulatum*, *K. alvarezii* and *K. striatus* in Kaneohe Bay, Hawaii (Glenn & Doty, 1992). It has been suggested that water motion enhances growth rates of seaweeds by reducing the thickness of the diffusion boundary layer around the algal surface (Wheeler, 1988). The decreased boundary layer then increases the supply of inorganic carbon, phosphate, nitrate and other micronutrients to the thallus (Wheeler, 1988; Hurd, 2000), and facilitates the removal of algal metabolic products such as O_2 and OH^- ions (Gonen et al., 1995), excess hydrogen peroxide and halogenated organic compounds (Mtolera, 1996) away from the plant surface. In addition, at Gazi Bay, there were strong tidal currents reaching velocities up to 0.6 m s^{-1} with far-reaching effects on nutrient and material exchanges (Kitheka, 1996), and hence, probably, the high growth.

The high growth rates at Gazi may be attributed to an increased supply of inorganic nitrogen due to high water motion as shown by the high thallus N level (Tables 1 & 4). A critical level of nitrogen and phosphorus for maximum growth has been observed for eucheumoids: 1.00% dry wt N has been suggested for *K. alvarezii* (Li et al., 1990) and between 1.19 and 2.53% dry wt N for *K. striatus* (Mairh et al., 1999). From these two studies it appears that the critical level of nitrogen for eucheumoids ranges from 1.00 to 2.53% dry wt and consequently, slower growing plants at Kibuyuni with thallus N value of 0.83% dry wt were presumably N-limited. N-limited eucheumoids have also been reported by Hurtado-Ponce (1995) and Mollion and Braud (1993). Thallus P content was not correlated with plant growth at any site, suggesting that the plants were probably not P-limited. Lower thallus P values of 0.04 and 0.03% dry wt were reported, respectively, for *E. denticulatum* and *Kappaphycus* from thalli that had been rinsed to remove surface salt (Doty, 1987). However, the critical level of N and P varies with photon fluency, nutrients and temperature, as well as other factors (Lobban & Harrison, 1994).

The highest water (30.6 ° C) and air maximum temperatures (33.0 °C) were measured in February, coinciding with periods of low growth rates. During this time the lowest tides occurred around the mid-day, maximising the stress. On the other hand, high growth values were obtained in August/September when the average monthly water and maximum temperature readings were 28.4 and 29.1 °C, respectively. It appears that temperatures between 28 and 30° C were favourable for the three morphophytes in this study, as has been suggested by Hurtado-Ponce (1992) and Ohno et al. (1994). Paula and Pereira (2003) showed that temperature was the main factor influencing growth of brown *K. alvarezii* grown on a raft in Brazil. High temperatures (>31°C) have been observed to reduce the growth rates of eucheumoids in this study, as well as in other tropical areas, such as Vietnam (Ohno et al., 1996) and Madagascar (Mollion & Braud, 1993).

Monthly average salinities at the three sites ranged from 34.3–37.4‰. The salinity range was generally oceanic (Lobban & Harrison, 1994) and within the required levels for eucheumoid farming (Hurtado-Ponce, 1992; Ohno et al., 1994). However, the results in this study showed a significant negative correlation between salinity and the growth rate of both brown morphotypes (Table 4). The salinity pattern seemed to follow that of temperature where higher values were recorded during NEM than during the SEM, whereas the growth rates of both brown morphophytes were higher during SEM than in NEM. The growth rates of green *K. alvarezii* were similar in both SEM and NEM. It appeared probable that salinity and temperature affected the growth interactively. Synergistic effects of salinity and temperature have been observed to affect the rate of photosynthesis in *Eucheuma isiforme* (Mathieson & Dawes, 1974). The growth of green *K. alvarezii* was not significantly correlated with both salinity and temperature as the other morphotypes; presumably due to greater tolerance.

Low growth rates of eucheumoids were recorded at Kibuyuni where a high 'ice-ice' occurrence was also observed. The cause of 'ice-ice' is not well understood but there has been a general consensus that 'ice-ice' is due to physical and chemical stresses (Uyenco et al., 1981; Largo et al., 1995; Mtolera, 1996). High 'ice-ice' occurrence was observed at Kibuyuni from November to January, coinciding with the period of high temperature and salinity. Grazing of the plants and presence of *L. majuscula* forming mats on thalli were also observed during this time. It appeared that the high 'ice-ice' at the site was promoted by a combination of environmental (high temperature/salinity and low water motion) and biological factors (grazers and epiphytes). Grazing by tip-nipping and removing of cortical layers from thalli, especially by sea urchins, probably reduced the photosynthetic capacity of the plants and left wounds that were potential sites for pathogen infection. The covering of plants by epiphytes could also have limited the amount of photon fluency for the seaweeds, hence low growth. An inverse correlation between percent 'ice-ice' and growth of both brown *E. denticulatum* and brown *K. alvarezii* was observed but none for the green *K. alvarezii* (Table 4), implying that the green morphotype was possibly 'ice-ice' resistant. A significant correlation between percent 'ice-ice' occurrence and thallus nitrogen content of both brown *E. denticulatum* and brown *K. alvarezii* was also obtained (see results) but none for the green morphophyte. Low thallus N content was reputed to promote 'ice-ice' in Madagascan eucheumoids when thallus N content was 0.88% of dry wt (Mollion & Braud, 1993), comparable to a mean value of 0.83% of dry wt for morphotypes at Kibuyuni (Table 1). Growth of eucheumoids may also have been affected by the presence of seagrasses, which are competitors for nutrients and space, as observed for *E. denticulatum* in Zanzibar (Mtolera, 1996). High levels of water nitrate and phosphate were measured at Kibuyuni but the thallus N and P contents were low, suggesting that other plants, including the phytoplankton, were competing for nutrients. The growth patterns of eucheumoids at the three sites support the complex multifactorial theory of Doty (1987) which suggests that light, water motion, temperature and water quality determines the fertility of a site for seaweed growth. However, other factors, such as epiphytes and grazers should also be incorporated in the model as observed in this study.

In conclusion, results of the present study indicate good prospects for commercial cultivation of eucheumoids at all three sites in southern Kenya, including a mangrove bay environment. However, it is apparent that attention should be given to a survey of potential sites for commercial eucheumoid farming along the entire Kenyan coast, with water motion, temperature and salinity, among other criteria being considered. Selection of local eucheumoid strains with good growth and carrageenan properties should also be given priority.

Acknowledgements

Financial support was provided by International Ocean Institute (IOI-HQ, IOI-Southern Africa & IOI-Eastern Africa) under the auspices of the Women and Sea Programme, International Foundation for Science, National Research Foundation, Department of Environment and Tourism, Kenya Marine and Fisheries Research Institute, Western Indian Ocean Marine Science Association and University of the Western Cape. Kimathi is thanked for excellent technical assistance; Mitto for nutrient analyses; Gaya and Omondi for piloting boats; Frikkie for statistical analyses; Muhoro and Kibue for availing meteorological data; and Dr. Kazungu, Director of KMFRI and Dr. Prochazka, Director of IOI-SA for their support.

References

Ask EI, Azanza RV (2002) Advances in cultivation technology of commercial eucheumatoid species: a review with suggestions for future research. Aquaculture 206: 257–277.

Ask EI, Batibasaga A, Zertuche-Gonzalez JA, de San M (2003) Three decades of *Kappaphycus alvarezii* (Rhodophyta) introduction to non-endemic locations. In Chapman ARO, Anderson RJ, Vreeland VJ, Davison IR (eds), Proceedings of the 17th International Seaweed Symposium, Oxford University Press, Oxford: 49–57.

Azanza – Corrales RV, Aliaza TT (1999) In vitro carpospore release and germination in *Kappaphycus alvarezii* (Doty) Doty from Tawi-Tawi, Philippines. Botanica Marina 42: 281–284.

Braud JP, Perez R (1978) Farming on pilot scale of *Eucheuma spinosum* (Florideophyceae) in Djibouti waters. In Jensen A, Stein J (eds), Proceedings of the 9th International Seaweed Symposium. Science Press, Princeton 533–539.

Coppejans E (1989) Preliminary study of *Eucheuma* and *Gracilaria* culture off Kenya. In Kain JM, Andrews JW, McGregor BJ, (eds), Outdoor seaweed cultivation. Commission of the European Communities, Brussels 81–83.

Dawes CJ, Lluisma AO, Trono CG (1994) Laboratory and field growth studies of commercial strains of *Eucheuma denticulatum* and *Kappaphycus alvarezii* in the Philippines. Journal of Applied Phycology 6: 21–24.

Doty MS (1971) Measurement of water movement in reference to benthic algal growth. Botanica Marina 14: 32–35.

Doty MS (1987) The production and use of *Eucheuma*. FAO Fisheries Technical Paper No. 281, Rome: 123–164.

Glenn EP, Doty MS (1990) Growth of seaweeds *Kappaphycus alvarezii, K. striatus* and *Eucheuma denticulatum* as affected by environment in Hawaii. Aquaculture 84: 245–255.

Glenn EP, Doty MS (1992) Water motion affects the growth rates of *Kappaphycus alvarezii* and related red seaweeds. Aquaculture 108: 233–246.

Gonen, Y, Kimmel E, Friedlander M (1995) Diffusion boundary layer transport in *Gracilaria conferta* (Rhodophyta). Journal of Phycology 31: 768–773.

Hurd CL (2000) Water motion, marine macroalgal physiology, and production. Journal of Phycology 36: 453–472.

Hurtado AQ, Cheney DP (2003) Propagule production of *Eucheuma denticulatum* (Burman) Collins et Harvey by tissue culture. Botanica Marina 46: 338–341.

Hurtado-Ponce AQ (1992) Cage culture of *Kappaphycus alvarezii* var. *tambalang* (Gigartinales, Rhodophyceae). Journal of Applied Phycology 4: 311–313.

Hurtado-Ponce AQ (1995) Carrageenan properties and proximate composition of three morphotypes of *Kappaphycus alvarezii* Doty (Gigartinales, Rhodophyta) grown at two depths. Botanica Marina 38: 215–219.

Hurtado-Ponce AQ, Agbayani RF, Sanares R, de Castro-Mallare TR (2001) The seasonality and economic feasibility of cultivating *Kappaphycus alvarezii* in Panagatan Cays, Caluya, Antique, Philippines. Aquaculture 199: 295–310.

Kitheka JU (1996) Water circulation and coastal trapping of brackish water in a tropical mangrove-dominated bay in Kenya. Limnology and Oceanography 41: 169–176.

Largo DB, Fukami K, Nishijima T, Ohno M (1995) Laboratory-induced development of the ice-ice disease of the farmed red algae *Kappaphycus alvarezii* and *Eucheuma denticulatum* (Solieriaceae, Gigartinales, Rhodophyta). Journal of Applied Phycology 7: 539–543.

Li R, Li JJ, Wu CY (1990) Effect of ammonium on growth and carrageenan content in *Kappaphycus alvarezii* (Gigartinales, Rhodophyta). Hydrobiologia 204/205: 499–503.

Lirasan T, Twide P (1993) Farming *Eucheuma* in Zanzibar, Tanzania. Hydrobiologia 260/261: 353–355.

Lobban CS, Harrison PJ (1994) Seaweed ecology and physiology, Cambridge University Press, Cambridge, 365 pp.

Mairh OP, Zodape ST, Tewari A, Mishra JP (1999) Effect of nitrogen sources on the growth and bioaccumulation of nitrogen in marine red alga *Kappaphycus striatus* (Rhodophyta, Solieriaceae) in culture. Indian Journal of Marine Sciences 28: 55–59.

Mathieson AC, Dawes CJ (1974) Ecological studies of Floridian *Eucheuma* (Rhodophyta, Gigartinales). II. Photosynthesis and respiration. Bulletin Marine Science 24: 274–285.

McClanahan TR (1988) Seasonality in East Africa's coastal waters. Marine Ecology Progress Series 44: 191–199.

McHugh DJ (2002) Prospects for seaweed production in developing countries. FAO Fisheries Circular No. 968, Rome, 28 pp.

McHugh DJ (2003) A guide to the seaweed industry. FAO Fisheries Technical Paper No. 441, Rome 105 pp.

Mollion J, Braud JP (1993) A *Eucheuma* (Solieriaceae, Rhodophyta) cultivation test on the south-west coast of Madagascar. Hydrobiologia 260/261: 373–378.

Mshigeni KE (1994) Algal biotechnological developments in East Africa: The case of *Eucheuma* farming in Tanzania. In Phang SM., Lee YK, Borowitzka MA, Whitton BA (eds), Algal Biotechnology in the Asia-Pacific Region, University of Malaya, Malaysia: 211–220.

Mtolera MSP (1996) Photosynthesis, growth and light-induced stress responses in the red alga *Eucheuma denticulatum*. PhD Thesis, Uppsala University, Sweden.

Ohno M, Largo DB, Ikumot T (1994) Growth rate, carrageenan yield and gel properties of cultured kappa-carrageenan producing red alga *Kappaphycus alvarezii* (Doty) Doty in the subtropical waters of Shikoku, Japan. Journal of Applied Phycology 6: 1–6.

Ohno M, Nang HQ, Hirase S (1996) Cultivation and carrageenan yield and quality of *Kappaphycus alvarezii* in the waters of Vietnam. Journal of Applied Phycology 8: 431–437.

Oyieke HA (1998) The seaweed resources of Kenya. In Critchley AT, Ohno M (eds), Seaweed Resources of the World. Japan International Cooperation Agency, Yokosuka: 385–388.

Parsons TR, Maita Y, Lalli CM (1984) A manual of chemical and biological methods of seawater analysis. Pergamon Press, Oxford, 173 pp.

Paula EJ, Pereira RTL (2003) Factors affecting growth rates of *Kappaphycus alvarezii* (Doty) Doty ex P. Silva (Rhodophyta, Solieriaceae) in sub-tropical waters of Sao Paulo State, Brazil. In Chapman, AR, Anderson ORJ, Vreeland VJ, Davison IR (eds), Proceedings of the 17th International Seaweed Symposium, Oxford University Press, Oxford: 381–388.

Prud'homme van Reine WF, Trono GC (eds) (2001) Plant resources of South-East Asia No 15. Cryptogams: Algae. Backhuys Publishers, Leiden, the Netherlands, 318 pp.

Russell DJ (1982) Introduction of *Eucheuma* to Fanning Atoll, Kiribati, for the purpose of mariculture. Micronesia 18: 35–44.

Silva PC, Basson PW, Moe Rl (1996) Catalogue of the benthic marine algae of the Indian Ocean. University of California Publications in Botany vol. 79: 1259 pp.

United Nation Environment Programme (1998) Eastern Africa Atlas of Coastal Resources. 1. Kenya, 119 pp.

Uyenco FR, Saniel LS, Jacinto GS (1981) The 'ice- ice' problem in seaweed farming. 10th International Seaweed Symposium. Walter de Gruyter, New York: 625–630.

Wheeler WN (1988) Algal productivity and hydrodynamics – A synthesis. Progress in Phycological Research 6: 23–58.

Yarish C, Wamukoya G (1990) Seaweeds of potential economic importance in Kenya: Field survey and future prospects. Hydrobiologia 204/205: 339–346.

Journal of Applied Phycology (2006) 18: 575–582
DOI: 10.1007/s10811-006-9063-5

Reproduction strategies of *Macrocystis pyrifera* (Phaeophyta) in Southern Chile: The importance of population dynamics

Alejandro H. Buschmann[1,*], Cristina Moreno[1], Julio A. Vásquez[2,3] & María C. Hernández-González[1]
[1]*i~mar, Universidad de Los Lagos, Camino Chinquihue km 6, Casilla 557, Puerto Montt, Chile;* [2]*Facultad de Ciencias del Mar, Universidad Católica del Norte, Coquimbo, Chile;* [3]*Centro de Estudios Avanzados de Zonas Áridas (CEAZA), Coquimbo, Chile*

*Author for correspondence: e-mail: abuschma@ulagos.cl

Key words: *Macrocystis pyrifera,* reproduction strategies, southern Chile, spore production, sporophylls

Abstract

Macrocystis pyrifera is an ecologically dominant species along the temperate Northern and Southern Pacific Coast of America, showing some similarities and differences at population and community level. In general, this kelp is reported to be reproductive all year round. Annual populations present in wave-protected areas of southern Chile suggest that the reproductive strategies of this population can be different. In this study we explore the reproductive strategies of annual *M. pyrifera* present in wave-protected areas and perennial populations encountered in exposed areas of southern Chile (41°S). Our results show that *M. pyrifera* present in wave-exposed locations has a reproductive strategy that is similar to populations in the northern hemisphere. These populations reproduce all year round and their strategy is to produce high numbers of sporophylls and ensure that most of them (over 90%) become sporogenous. On the other hand, the protected populations with an annual life cycle, produce more spores per area of sorus. In a few months, they are able to produce sufficient propagules to recolonize areas before the adult plants disappear in autumn.

Introduction

Macrocystis pyrifera (L.) C. Ag. is abundant along the Chilean coastline from Cape Horn up to Valparaíso (33°S), but has also been reported on the Peruvian coast (Hoffmann & Santelices, 1997). Abundant populations are located south of Concepción (37°S) down to Patagonia in protected bays open to the Pacific Ocean, as well as in channels and fjords, forming conspicuous belts along the coastline. In this southern region, fluctuation of environmental parameters, especially temperature, salinity and nutrients, are greater in the inner waters than in bays open to the Pacific Ocean (Buschmann et al., 2004). Also, the morphology of plants from these populations differs significantly. Plants present in the inner sea (the most wave-protected populations) have broader blades, smaller pneumatocysts, and more ribbed blades, among other characteristics suggesting that the environmental conditions have considerable individual and population consequences for this brown alga (Buschmann, 1992; Vásquez & Buschmann, 1997).

Previous results indicate that wave-exposed sites produce fewer perennial populations than wave-protected sites (e. g. Seymour et al., 1989; Harrold & Reed, 1985). Studies of the reproduction ecology of *Macrocystis pyrifera* in the North Pacific show that this species is reproductive all year round (Reed et al., 1996), as has also been established in protected areas of the Beagle Channel (Santelices & Ojeda, 1984). However, the wave-protected populations of *M. pyrifera* in the northern limit (41–44°S) of the archipelago area of southern Chile show a different trend (Buschmann, 1992; Vásquez & Buschmann, 1997). These are annual, raising several questions regarding factors that produce these unusual patterns and how these populations manage to recruit successfully year after year. In consequence, wave-protected and exposed kelp populations have important ecological differences that have not yet received attention for a more comprehensive understanding of *M. pyrifera* population dynamics.

This study outlines the annual population dynamics and reproductive patterns in areas with different water movement regimes that represent different reproductive strategies. We describe the reproductive strategy of *Macrocystis pyrifera* with respect to the size of the parent plants, indicating size of first reproduction and reproductive effort in relation to different (annual and perennial) population dynamics.

Material and methods

Study areas

This study was carried out in southern Chile (40–41°S) where different water movement regimes can be found as a consequence of the presence of channels, fjords and wave-protected bays (Figure 1). In this region a wave exposed locality (Bahía Mansa; 40°34′S, 73°44′W) and a protected locality (Metri; 41°36′S, 72°43′W) were identified on the basis of carbonate-block dissolution rates, as described by Buschmann et al. (2004). The study was carried out, within each locality, at two sites separated by <500 m to assure the representation of the variability of each locality. The exposed area, Bahía Mansa (Figure 1), is characterized by variable depths between 0 to 12 m, with a substratum mostly composed of compact rock and boulders and where carbonate dissolution rates varied from 0.6 to 0.9 g h^{-1}. At exposed localities, the giant kelp *Macrocystis pyrifera* populations are perennial, but show fluctuations in abundance due to increased water movement in winter (Westermeier & Möller, 1990). This kelp is the only subtidal canopy-forming species in the area, while the main understory species are *Ulva* sp. and *Sarcothalia crispata*. In wave-protected locations, such as Metri (Figure 1), *M. pyrifera* forests are present from the low intertidal down to 10 m depth and have typical annual population dynamic (Buschmann, 1992). Carbonate dissolution rates at Metri varied from 0.12 to 0.18 g h^{-1}, which is significantly lower ($P < 0.001$) than the exposed locality. The bottom is mainly granite-consolidated rock, with some boulder patches. The most abundant understorey species are the red alga *Sarcothalia crispata* and the green alga *Ulva* sp. At both Bahía Mansa and Metri the most conspicuous grazer

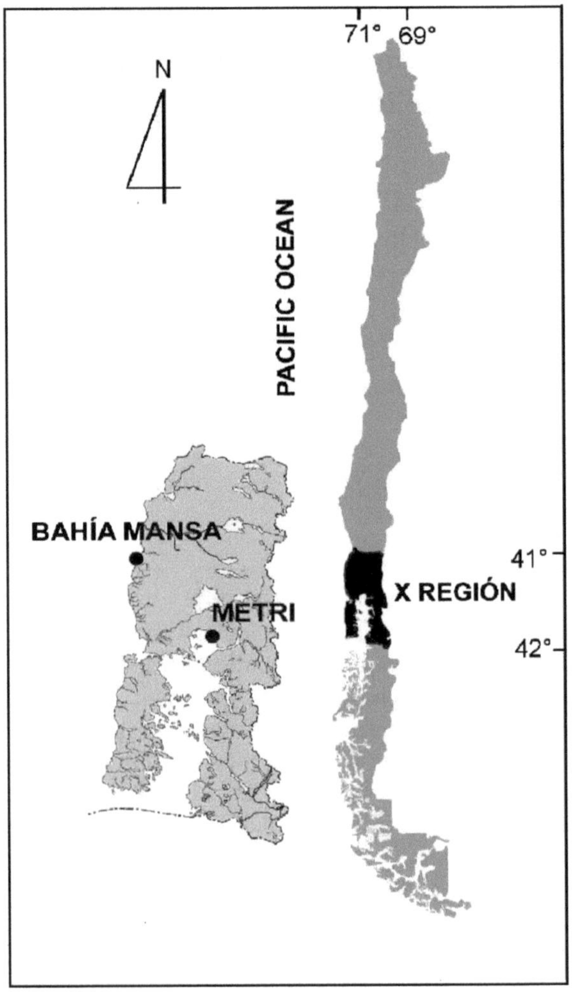

Figure 1. Map showing the study sites in southern Chile. Bahía Mansa is an exposed area and Metri is a protected area.

is the snail *Tegula atra*, but a lower number of sea urchins, chitons and limpets are also found. In general, it has been demonstrated that *Tegula* has no effect on the population dynamics of *Macrocystis*, but does have a moderate effect on the abundance of understory algal species (Moreno & Sutherland, 1982). The main difference between these areas is the absence in the protected areas of the kelp *Lessonia*, lower species diversity, and high abundance of the filter-feeding gastropod *Crepidula*.

Population and reproduction patterns

In each study area, eighteen 0.25 m^2 random samples (9 per location) were taken by scuba diving, to include a range of depths. Each location had a coastline length of at least 250 m. In each area, locations were duplicated to ensure better representation of the population dynamics of *M. pyrifera*. The diameter (±0.5 cm) of all holdfasts inside each quadrat was measured *in situ* with a plastic ruler, and the number of sporophylls on each plant was counted. A previous study demonstrated that the holdfast diameter correlates well with plant length (Buschmann, unpublished). This procedure was repeated monthly throughout one year and only interrupted twice due to dangerous diving conditions. This allowed us to describe the annual variation in plant size and reproductive tissue production for both populations (Metri & Bahía Mansa).

For each sampling date, the holdfasts with sporophylls ($n > 100$) were sorted by size, and a random number of 50 sporophylls from each location was collected monthly. In the laboratory, the area of each sporophyll was measured by digitizing the shape using an image analyzer. The sorus area (A) was measured by using the equation given by Reed et al. (1996) where $A = 2slw$, (s = total number of sporophylls containing sori; l = mean length of sori and w = mean width of the sori). These data were polled and allowed the calculation of the following variables for each plant: size (holdfast diameter in cm), plant density (no. m^{-2}), sporophyll abundance (no. plant $^{-1}$) and the reproduction frequency (%). Also, the ratio of sorus to sporophyll area and the total sorus area per plant, for two different size classes (small plants 0–6 cm and large plants 6–12 cm of holdfast diameter) were determined. Statistical analysis was by two-way ANOVA where the factors were plant size (holdfast diameter) and life history type (annual and perennial), using SYSTAT, after ensuring normality and homoscedasticity of the data.

Spore release

An additional reproductive variable, the number of spores per cm^2 of fertile tissue, was determined by cutting five independent 1 cm^2 discs from different sori every month. Each disc was rinsed with tap water, gently brushed and placed in a Petri dish filled with 10 mL of filtered (0.2 μm) and sterilized Provasoli culture medium (according to McLachlan, 1973). After 1.5 h, the disc was removed from each dish. Immediately, an aliquot was placed in a 1/10 mm deep Neubauer cell-counting chamber to determine the number of spores, under an inverted Nikon microscope. These data were used to calculate the number of spores per sorus area, which was later related to plant size to calculate the total release of spores per plant. Furthermore, the number of spores released per sorus area and the total number of spores released per plant for different sorus size classes was also determined. Using a one-way ANOVA, we performed pairwise comparisons between the numbers of spores released per fertile sorus.

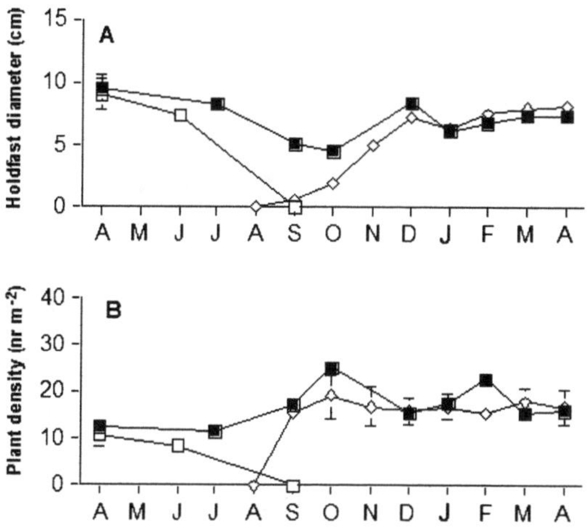

Figure 2. Mean (± 1 SE; $n = 18$) annual variation of the (A) plant size (holdfast diameter; cm) and the (B) plant density (Nr. m^{-2}) in Bahia Mansa (ν; exposed) and Metri (□ and ◇; protected). The □ represents the 2001 cohort and ◇ the 2002 *M. pyrifera* cohort. Absence of error bars indicates small variation of the data.

Results

Average plant size of *Macrocystis pyrifera* (as holdfast diameter) varies seasonally in both protected and exposed areas (Figures 2A). In protected areas, the size variation is greater because the 1-year old cohort disappears in September (Figures 2A). The protected populations recover again through massive recruitment during the following spring (September), reaching the highest holdfast diameter (similar to exposed populations), in summer. The exposed *M. pyrifera* populations have a rather constant population density (varying between 12 and 25 individuals per m^2), however, the protected population is lost in September and a new cohort starts in August-September (Figure 2B).

Due to this annual abundance pattern, the sporophylls disappear during winter (June to September), in Metri (Figure 3A). In contrast, the exposed population of *Macrocystis pyrifera* produce sporophylls throughout the year, although numbers of sporophyllic plants decreases strongly towards winter (Figure 3A). The percentage of plants carrying sporophylls shows a similar pattern to the above (Figure 3B). In exposed

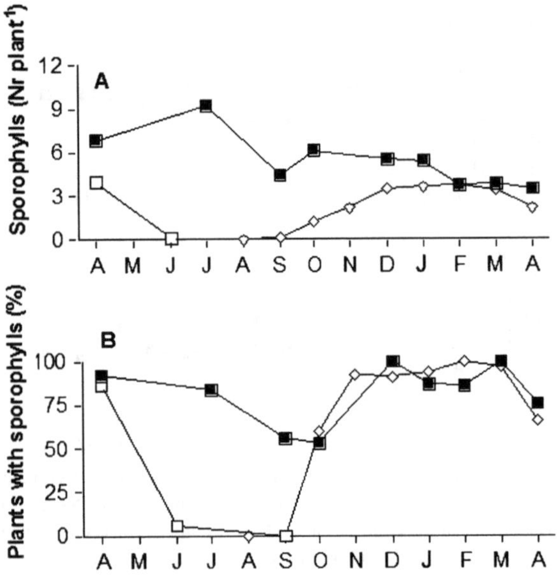

Figure 3. Mean (± 1 SE; n=18) annual variation of (A) sporophyll production and the (B) plants with sporophylls (%) in Bahia Mansa (ν; exposed) and Metri (□ and ◇; protected). The □ represents the 2001 cohort and ◇ the 2002 *M. pyrifera* cohort. Absence of error bars indicates small variation of the data.

M. pyrifera locations 50% of the plants carry sporophylls in October and the number increases to almost 100% during the rest of the year. In contrast, during the winter there are no plants in the protected locations and thus no sporophylls. In spring, Metri again shows a high (ca. 100%) number of plants with sporophylls (Figures 3B).

The average number of sporophylls per plant varies from 3 to 13 (Figure 4A). Smaller plants show a significantly ($F = 16.49$; $P \ll 0.001$) lower sporophyll number than the bigger plants at both sites. Also, the annual populations produce significantly ($F = 22.82$; $P \ll 0.001$) fewer sporophylls per plant than perennial ones (Figure 4A). The total sporophyll area per plant is significantly ($F = 5.76$; $P < 0.017$) higher in smaller plants, in sheltered conditions, but the significant ($F = 5.91$; $P < 0.015$) interaction between plant size and type of life history (annual and perennial) indicates that this difference is due mainly to the increased sporophyll area of plants collected from the annual population (Figure 4B). For the total sorus area per plant, the statistical analysis showed that the perennial populations produce significantly ($F = 35.64$; $P \ll 0.001$) more fertile tissue than the annual population from the protected locations (Figure 4C). However, the significant ($F = 11.17$; $P < 0.001$) interaction of plant size

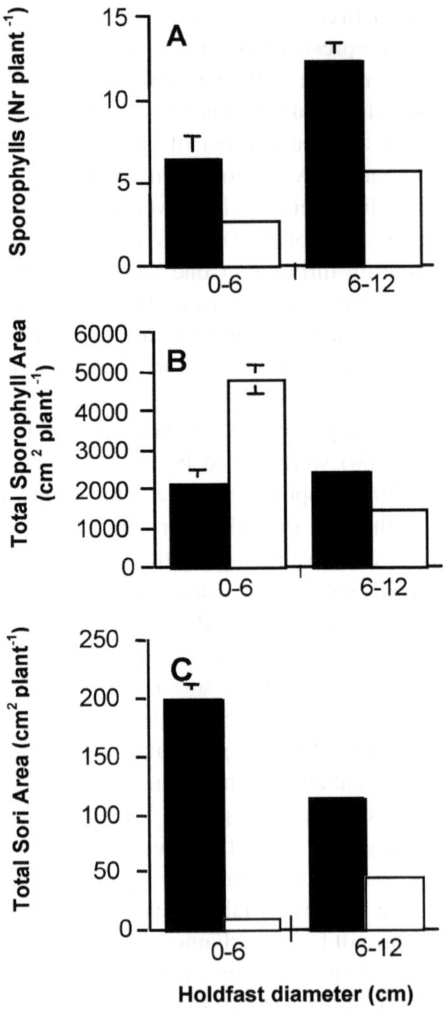

Figure 4. Mean (± 1 SE; $n \geq 10$) values of two reproduction variables: (A) number of sporophylls per plant; (B) total sporophyll area per plant and (C) total sorus area per plant in relation to two plant size categories, 0–6 cm and 6–12 cm holdfast diameter. White bars represent Metri (wave protected) and the black bars Bahía Mansa (wave exposed population). Absence of error bars indicates small variation of the data.

and type of life history indicates that the effect is due to the greater sorus area present in the exposed kelp populations (Figure 4C).

The annual release of spores for the Metri and Bahía Mansa populations was calculated at 760,000 and 480,000 spores cm^{-2}, respectively. Spore release of annual *Macrocystis pyrifera* is low in small plants and is equal to perennial populations for taller plants (Figure 5). However, a significant ($F = 330.4$; $P \ll 0.001$) difference exists between perennial and annual protected kelp populations, determined mainly by the

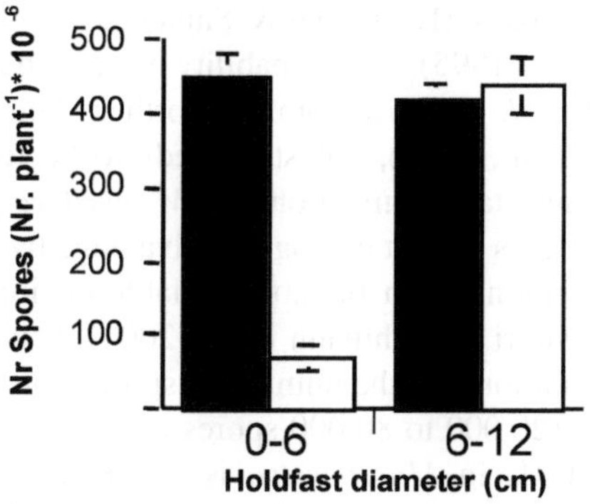

Figure 5. Mean (± 1 SE; $n = 15$) values of the number of spores produced per plant in relation to two plant size categories, 0–6 cm and 6–12 cm holdfast diameter. White bars represent Metri (wave protected with annual populations) and the black bars Bahía Mansa (wave exposed with perennial populations).

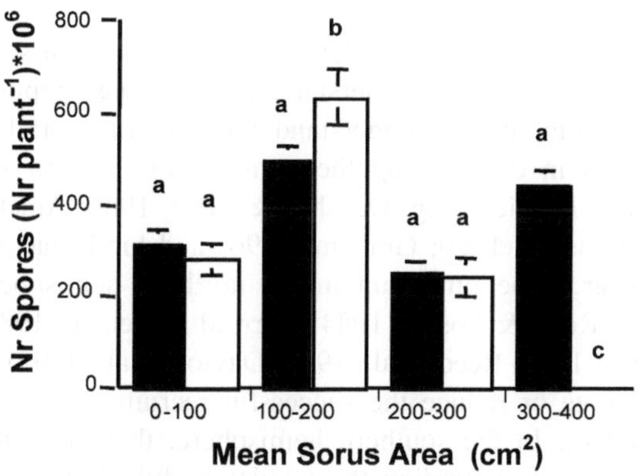

Figure 6. (A) Number of spores produced per plant in relation to sorus size. All data represent mean values (± 1 SE; $n = 15$). White bars refer to Metri (wave protected with annual populations) and the black bars to Bahía Mansa (wave exposed with perennial populations). The letters over the bars indicate statistical differences ($p < 0.05$) between exposed and protected populations.

significant ($F = 75.3$; $P \ll 0.001$) interaction of kelp size and life history strategy (Figure 5). The total spore release per plant for sori of the size class 100 to 200 cm2 from the annual kelp populations is also significantly ($P < 0.024$) higher than for the perennial populations (Figure 6). It should be emphasized that the perennial populations produce larger sori with spores, whereas the annual kelps release significantly more spores in smaller sori, especially in medium-sized sori (Figure 6).

Discussion

In our study, we demonstrate that wave-protected and exposed populations have different reproductive strategies. Wave protected areas with annual populations have a very distinctive strategy compared to perennial populations in exposed localities. These annual populations recruit in September (Figure 2A) and, within months, start to produce sporophylls and allocate biomass to fertile tissues. The spore production takes place in plants with smaller sori, and perhaps in younger plants than in the perennial populations. In contrast, perennial populations reproduce all year round, with a similar reproductive pattern to those described for the northern hemisphere (e.g. Reed et al., 1996).

It is clear that the potential reproductive effort can vary in space and time due to biological and abiotic factors. Because canopy density is related to water movement (e.g. Tegner & Dayton, 1987; Seymour et al., 1989; Graham et al., 1997) and grazing activity (e.g. Harrold & Pearse, 1987; Dayton, 1985; Dayton et al., 1992) causing a loss of the blades, the vegetative regrowth of the sporophyte is induced in spite of a reduction in the production of sporophylls (e.g. Graham, 2002). The removal of *Macrocystis pyrifera* canopy affects sporophyll production, since the experimental removal of 75% of the canopy resulted in a significant reduction in the sporophyll production (Reed, 1987). However, the expected responses between exposed and protected locations of *M. pyrifera* in southern Chile are different. The annual population in protected areas should not lose its canopy due to wave action or herbivory (Dayton, 1985; Vásquez & Buschmann, 1997), but the high summer mortality of this population seems to be related to higher temperature and low nutrient concentration (Buschmann, unpublished data; Buschmann et al., 2004; Muñoz et al., 2004). For these reasons the growth and reproduction of the kelp population in Metri is tightly coupled: when reproduction finishes in June recruitment and growth will start in late winter again. The perennial populations can lose part of their canopy, but still maintain a remaining sporophyll stock and reproduce all year round.

The contrasting life history strategies of *M. pyrifera* in protected and exposed sites described above, raises the question as to whether these differences have some genetic basis. There are no significant genetic differences between them in term of ITS1 and ITS2 sequences, indicating that we are dealing with one species (Coyer et al., 2001). On the other hand, ropes seeded from mother plants collected in exposed sites and transplanted for cultivation in protected environments did not survive (Gutierrez et al., 2006). This result suggests that not only phenotypical characters differ between these *M. pyrifera* populations, but that there are some intra-population genetic differences in this kelp in southern Chile.

Another general thought related to the reproduction strategy of kelps is that environmental conditions exert a much greater effect on the reproduction strategy of species like *Macrocystis*, which reproduce continuously, than on other strictly seasonal species (Reed et al., 1996). This suggests that short-lived species should reproduce during the complete growth period rather than risk delaying reproduction. Here we present data for an annual kelp that produces significantly fewer sporophylls, and spores, and less sorus tissue. These protected kelp populations present a paradox, by enhancing the risk of collapsing through reduction of their reproductive potential, as their recovery depends on a successful spore production and sporophytic recruitment. The strategic advantage of this protected population is related to an increased sporophyll area produced by a smaller number of sporophylls. However, this strategy alone cannot counterbalance the reduction of the reproduction potential of this kelp as the sorus area does not reach over 75% of the sporophyll area. The only reproductive strategy capable of increasing the reproductive success of these annual kelp populations seems to be an increase in the number of spores produced by middle-sized sori. Nevertheless, this strategy raises the question: why invest energy in producing large sporophylls with a high proportion of non-reproductive tissues?

This leads us to ask how these annual populations recover regularly year after year, even during a four to five month period without the presence of a seed bank? It has been suggested that the strategy followed by these protected populations involves massive spore production during summer and autumn (Buschmann et al., 2004). Annual comparisons show that spore release per adult plant of the Metri population reaches the same numbers as the exposed population (Figures 5 and 6). Thus, the higher release of spores per area unit of the annual population in Metri compensates for its prolonged absence of spore production. The annual population produces spores massively during summer and autumn, presenting a high number of spores per sorus area, which may produce sufficient gametophytes to ensure the recovery of populations during the following spring (Buschmann et al., 2004). This strategy suggests that the *M. pyrifera* propagules or more likely the resulting gametophytes have the capability of dormancy (Kinlan et al., 2003), and effectively create a seed bank (Hoffmann & Santelices, 1991; Santelices et al., 1995). This capability has also been suggested for *M. pyrifera* from the northern hemisphere (Ladah et al., 1999), but still needs to be tested. It is also important to mention that *M. pyrifera* plasticity constitutes a great ecological advantage by permitting the colonization of more variable environments such as Metri (Buschmann et al., 2004). It is important to mention that the number of spores released per sori area (20,000 to 80,000 spores cm^{-2}; Buschmann et al., 2004) in *Macrocystis*, is considerably lower than for *Laminaria* spp. (Chapman, 1984). This difference may be related to differences in the estimation method employed, as we estimated the release of spores instead of the total spore production counted microscopically in sections of the sorus (Chapman, 1984).

We need not only to determine the success of seed banks in order to understand these annual populations, but also to understand the role of mortality factors in determining the numbers of the new recruits. Physical (e.g. Deysher & Dean, 1986; Amsler & Neushul, 1990; Graham, 1996) and biotic factors appear to be important in the northern hemisphere (e.g. Reed & Foster, 1984; Harrold & Pearse, 1987; Reed, 1990; Reed et al., 1991; Dayton et al., 1992) to promote or reduce the successful recruitment of *M. pyrifera.* In the southern hemisphere, there is some controversy regarding the regulatory function of sea urchins and gastropods (see Dayton, 1985; Castilla & Moreno, 1982; Moreno & Sutherland, 1982). Whether the present distribution success of *Macrocystis* is related to over-fishing of sea urchins in these protected areas (Dayton, 1985) or failure of consumption, as claimed by Castilla and Moreno (1982), remains to be studied. The extent to which recruitment success is influenced by abiotic and biotic factors in the Chilean coast is another question that requires further attention.

Based on our results, we propose that our exposed and perennial *M. pyrifera* populations in the southern and southernmost part of the Chilean Coast (Santelices & Ojeda, 1984) show a reproductive strategy similar to that of the northern hemisphere populations, in contrast to protected and annual populations in southern Chile. The exposed populations reproduce all year round and their strategy is to produce high numbers of sporophylls to ensure sufficient reproductive tissue and to transform most of the sporophyll (over 90%) into fertile sorus tissue. On the other hand, the annual populations of protected sites invest in greater production of spores per sorus area, so that they are able to produce, in a few months, sufficient propagules to recolonize areas where the adult plants will disappear in the next months.

Acknowledgements

This study was financially supported by FONDECYT (Nr. 1010706) Chile. Furthermore, the help of Luis Filún, René Espinoza, Tomás Correa, Carlos García, Ulises Yagode, Patricio Ojeda and Verónica Muñoz is especially recognized. Comments by Daniel Varela, Daniel López and Robert Stead and two anonymous reviewers significantly improved this manuscript, as well as, the English review by Susan Angus.

References

Amsler CD, Neushul M (1990) Nutrient stimulation of spore settlement in the kelps *Pterygophora californica* and *Macrocystis pyrifera*. Mar. Biol. 107: 297–304.

Buschmann AH (1992) Algal communities of a wave-protected intertidal rocky shore in southern Chile. In: Seeliger U (ed.), Coastal Plant Communities of Latin America, Academic Press, Florida, pp. 91–104.

Buschmann AH, Vásquez JA, Osorio P, Reyes E, Filún L, Hernández-González MC, Vega A (2004) The effect of water movement, temperature and salinity on abundance and reproductive patterns of *Macrocystis* spp. (Phaeophyta) at different latitudes in Chile. Mar. Biol. 145: 849–862.

Castilla JC, Moreno CA (1982) Sea urchins and *Macrocystis pyrifera*: Experimental test of their ecological relations in southern Chile. In: Lawrence JM (ed.), International Echinoderm Conference, Tampa bay, A. A. Balkema Press, Rotterdam, pp. 257–263.

Chapman ARO (1984) Reproduction, recruitment and mortality in two species of *Laminaria* in southwest Nova Scotia. J. Exp. Mar. Biol. Ecol. 78: 99–109.

Coyer JA, Smith GJ, Andersen RA (2001) Evolution of *Macrocystis* spp. (Phaeophyceae) as determined by ITS1 and ITS2 sequences. J. Phycol. 37: 574–585.

Dayton PK (1985) The structure and regulation of some South American kelp communities. Ecol. Monogr. 55: 447–468.

Dayton PK, Tegner MJ, Parnell PE, Edwards PB (1992) Temporal and spatial patterns of disturbance and recovery in a kelp forest community. Ecol. Monogr. 62: 421–445.

Deysher LE, Dean TA (1986) *In situ* recruitment of sporophytes of the giant kelp, *Macrocystis pyrifera* (L.) C. A. Agardh: Effects of physical factors. J. Exp. Mar. Biol. Ecol. 103: 41–63.

Graham MH (1996) Effect of high irradiance on recruitment of the giant kelp *Macrocystis* (Phaeophyta) in shallow waters. J. Phycol. 32: 903–906.

Graham MH (2002) Prolonged reproductive consequences of short-term biomass loss in seaweeds. Mar. Biol. 140: 901–911.

Graham MH, Harrold C, Lisin S, Light K, Watanabe JM, Foster MS (1997) Population dynamics of giant kelp *Macrocystis pyrifera* along a wave exposure gradient. Mar. Ecol. Prog. Ser. 148: 269–279.

Harrold C, Reed DC (1985) Food availability, sea urchin grazing, and kelp forest community structure. Ecology 66: 1160–1169.

Harrold C, Pearse JS (1987) The ecological role of echinoderms in kelp forests. In: Jangoux M, Lawrence JM (eds.), Echinoderm Studies, Vol. 2 A. A, Balkema Press, Rotterdam, pp. 137–233.

Hoffmann A, Santelices B (1991) Banks of algal microscopic forms: Hypotheses and their function and comparison with seed banks. Mar. Ecol. Prog. Ser. 79: 185–194.

Hoffmann A, Santelices B (1997) Marine flora of central Chile, Ediciones Universidad Católica de Chile, Santiago, pp. 434.

Gutiérrez A, Correa T, Muñoz V, Santibañez A, Marcos R, Cáceres C, Buschmann AH (2006) Farming of the giant kelp *M. pyrifera* in southern Chile for development of novel food products. J. Appl. Phycol. (in press).

Ladah LB, Zertuche-González JA, Hernández-Carmona G (1999) Giant kelp (*Macrocystis pyrifera*, Phaeophyceae) recruitment near its southern limit in Baja California, after mass disappearance during ENSO 1997–1998. J. Phycol. 35: 1106–1112.

Kinlan BP, Graham MH, Sala E, Dayton P (2003) Arrested development of giant kelp (*Macrocystis pyrifera*, Phaeophyceae) embryonic sporophytes: A mechanism for delayed recruitment in perennial kelps?. J. Phycol. 39: 47–57.

McLachlan J (1973) Growth media – marine. In: Stein, J., (ed.), Handbook of Phycological Methods. Culture Methods and Growth Measurements Cambridge University Press, Cambridge, pp. 25–51.

Moreno CA, Sutherland JP (1982) Physical and biological processs in a *Macrocystis pyrifera* community near Valdivia, Chile. Oecologia 55: 1–6.

Muñoz V, Hernández-González MC, Buschmann, AH, Graham MH, Vásquez JA (2004) Variability in per capita oogonia and sporophyte production from giant kelp gametophytes (*Macrocystis pyrifera*, Phaeophyceae). Revista Chilena Historia Natural 77: 639–647.

Reed DC (1987) Factors affecting the production of sporophylls in the giant kelp *Macrocystis pyrifera* (L.) C. Ag. J. Exp. Mar. Biol. Ecol. 113: 61–69.

Reed DC (1990) The effect of variable settlement and early competition on patterns of kelp recruitment. Ecology 71: 776–787.

Reed DC, Foster M (1984) The effect of canopy shading on algal recruitment and growth in giant kelp (*Macrocystis pyrifera*) forest. Ecology 65: 937–948.

Reed DC, Neushul M, Ebeling AW (1991) Role of settlement density on gametophyte growth and reproduction in the kelps *Pterygophora californica* and *Macrocystis pyrifera* (Phaeophyceae). J. Phycol. 27: 361–366.

Reed DC, Ebeling AW, Anderson TW, Anghera M (1996) Differential reproductive responses to fluctuating resources in two seaweeds with different reproductive strategies. Ecology 77: 300–316.

Santelices B, Ojeda FP (1984) Population dynamic of coastal forests of *Macrocystis pyrifera* in Puerto Toro, Isla Navarino, Southern Chile. Mar. Ecol. Prog. Ser. 14: 175–183.

Santelices B, Hoffmann AJ, Aedo D, Bobadilla M, Otaíza R (1995) A bank of microscopic forms on disturbed boulders and stones in tide pools. Mar. Ecol. Prog. Ser. 129: 215–228.

Seymour RJ, Tegner MJ, Dayton PK, Parnell PE (1989) Storm wave induced mortality of giant kelp, *Macrocystis pyrifera,* in southern California. Estuar. Coast. Shelf Sci. 28: 277–292.

Tegner MJ, Dayton PK (1987) El Niño effects on southern California kelp forest communities. Adv. Ecol. Res. 17: 243–279.

Vásquez JA, Buschmann AH (1997). Herbivore-kelp interactions in Chilean subtidal communities: A review. Revist. Chilena Hist. Nat. 70: 41–52.

Westermeier R, Möller P (1990) Population dynamics of *Macrocystis pyrifera* (L.) C. Agardh in the rocky intertidal of southern Chile. Bot. Mar. 33: 363–367.

Journal of Applied Phycology (2006) 18: 583–589
DOI: 10.1007/s10811-006-9067-1

Group recruitment and early survival of *Mazzaella laminarioides*

B. Santelices* & D. Aedo
Departamento de Ecología, Facultad de Ciencias Biológicas, P. Universidad Católica de Chile, Casilla 193, Correo 22, Santiago, Chile

*Author for correspondence: e-mail: bsanteli@bio.puc.cl

Received 21 June 2003; accepted 6 January 2005

Key words: coalescence, recruitment, *Mazzaella*, Ca^{++}, spore abundance

Abstract

Several phycocolloid-producing Rhodophyta of significant economic importance are coalescing species, able to fuse with conspecifics during recruitment, reach larger sizes and increase their survival. In these species spores are needed to start cultivation (e.g. *Gigartina, Mazzaella*) or to increase the seed stocks, to renew senescent clones or to enlarge the base of genetic variation of vegetatively propagated species (e.g. *Chondrus, Gracilaria, Eucheuma*). This study uses *Mazzaella laminarioides* to evaluate some key features that influence recruitment success. Field measurements indicate that in any recruitment event a variable amount of the spores reaching a given place may form groups of 2 to over 100 coalescing spores, while field experiments support the idea that early recruitment success is a function of the number of coalescing spores forming the individual, as multisporic, coalescing recruits have higher survival rates than sporelings formed by one or a few spores. Therefore, group recruitment (spores settling and recruiting in close spatial proximity) appears as a prerequisite for sporeling coalescence and early recruitment success. In turn, laboratory experiments suggest that the frequency of group recruitment and coalescence increases with increasing spore abundance and with slight Ca^{++} additions to the culture medium. These last two factors could be handled by farmers to improve the success of spore inoculations of coalescing species.

Introduction

Several economically important red seaweeds are coalescing species, with the capacity to fuse and form genetically composite entities (e.g. *Chondracanthus, Chondrus, Eucheuma, Gigartina, Gracilaria, Mazzaella, Sarcothalia*). Fusion can occur among crustose basal portions of grown thalli (Santelices et al., 2003a), but occurs more frequently among spores, spore groups and sporelings (see Santelices, 2004, for a review). Laboratory results indicate that coalescence leads to larger sizes, reduced susceptibility to herbivory, enhanced competitive performance for space and higher survival (Maggs & Cheney, 1990; Santelices et al., 1996, 1999, 2003b).

Many of the above coalescing species are farmed for commercial purposes, with intensive use of spore inoculations. In species without commercial vegetative propagation (e.g. *Gigartina, Mazzaella, Sarcothalia*), spores are needed to propagate the crop and to initiate the farm (Buschmann et al., 1999). In species propagated by vegetative fragments (e.g. *Eucheuma, Gracilaria, Chondrus),* spores are needed to enlarge the base of genetic variation of the farmed stock (Santelices, 1992), to increase the seed stock (Azanza-Corrales et al., 1996) and to renew senescent clones (Alveal et al., 1995).

Among the above species, group recruitment (spores settling and recruiting in close spatial proximity) is a prerequisite for spore and sporeling coalescence, and therefore, a condition for higher survival during early recruitment. In spite of its biological and economic importance, however, no study seems to have addressed the question of how to increase group recruitment among these species. Available information (see Santelices, 1990, for a review) suggests that mode and periodicity of spore release, total spore abundance and dissolution rates of the mucilage layer maintaining the released spores in close proximity are important factors affecting group recruitment. Group recruitment is more likely to occur among species exhibiting massive spore shedding during comparatively short periods (see data in Boney, 1978; Ngan & Price, 1983) than among species releasing spores one by one over comparatively longer periods. After massive spore shedding, groups of spores may remain together, settling and recruiting as a group, although a few spores may break away from the main spore mass, drifting away and eventually settling in isolation from other spores. It would be expected, therefore, that increasing spore abundance will increase the frequency of group recruitment over solitary recruitment.

On the other hand, the chemical charges of the sulphated polysaccharides and glycoproteins of the mucilage around the spores (Pueschel, 1979; Chamberlain & Evans, 1981; Diannelidis & Kristen, 1988; Apple & Harling, 1995) probably influence the dissolution rate of mucilage while it is free-floating. As in other phycocolloids, adding positive ions to the culture medium (e.g. Ca^{++}) would probably promote gelification of the mucilage layer (Craigie, 1990), thereby delaying its dissolution rate and promoting group recruitment.

In this study we first compare the relative representation of solitary and group recruitment in a natural recruitment event of a coalescing species *(Mazzaella laminarioides).* Then we evaluate survivorship in the field during early recruitment of sporelings formed by a large number of spores in comparison to sporelings formed by one or a few spores. Finally, we measure the relationships between the relative frequency of group recruitment and increasing spore abundance, and Ca^{++} additions to the culture medium.

Materials and methods

Relative frequency of solitary and group recruitment at different levels of spore abundance

Relative frequency of solitary and group recruitment in a recruitment event were measured using 10 recruitment plates exposed in the middle level of two *Mazzaella laminarioides* belts on rocky outcrops in the Parque Pedro del Río Zañartu, Hualpén Peninsula, Concepción (36°48′S; 73°10′W) on September 7th, 2002. Plates were made of epoxy resin (Sea-Goin Poxy Putty), 5 cm. diam., 1 cm thick and with a coarse surface (see Brawley & Johnson 1991 for fabrication details). Average depth of the rugosity (measured as in Norton & Fetter 1981) was 783.35 ± 64.9 μm. They were attached to the substratum with stainless steel screws. After 72 h in the field, they were removed, transported to the laboratory in Santiago in glass containers containing 0.2 μm-filtered seawater and maintained at 8 ± 3°C. In the laboratory, the plates were placed in Petri dishes inside growth chambers and cultured under constant conditions of temperature (14°C $\pm$ 1 °C), photon flux density (40 μmol photons m^{-2} s^{-1}) and photoperiod (12:12) for 20 h. Previous experimental studies with this species (e.g. Hannach & Santelices, 1985; Santelices et al., 1999, 2003b) have shown that the above conditions are most suitable to transport the fertile blades from the field and to incubate spores in laboratory cultures.

A total of 24, 1.3 mm^2 circular microscopic fields were randomly marked on each of the 10 plates. Using a camera Cool Snap-Pro (Media Cybernetics) on a compound stereo-microscope Nikon Sm2-Udia, the image of each field was captured and stored in a computer. Images were then analysed using an Image Pro-plus 4.5 Program (Media Cybernetics). All images on all plates were captured within 20 h. The number of spores recruiting individually or in groups was quantified in each microscopic field and the data used to evaluate the relative frequency of solitary and group recruitment. Two or more spores settling at distances of 10 μm or less (cell wall to cell wall distance) were considered aggregated settlement.

The above spore recruitment data were also used to compare the frequency distribution of spore groups in plates with different values of spore abundance. Analysis of covariance (ANCOVA) was used (Snedecor & Cochran, 1967) to compare the percent contribution of either type of recruitment (solitary or in group) as a function of spore density.

Field survival of solitary and coalescing recruits

Experimental evaluation of differential survival in the field between sporelings built with different numbers of spores was done using thin rectangular (5 cm $\times$ 4 cm $\times$ 4 mm) ceramic plates that were previously detoxified by maintaining them for 48 h in running seawater. Spore groups formed by 1, 10, 50 ± 5 or 100 ± 10 spores were seeded on each plate in the laboratory, incubated for 5 days under the above controlled conditions and then transferred to the field. Germination rate of spores in each treatment was measured 2 days after seeding to assure that all the plates transferred from the laboratory to the field contained 100% germinated spores. Ten replicate plates per treatment were used. In the field they were randomly separated into two rocky outcrops in Maitencillo with abundant cover of *Mazzaella laminarioides* at mid-intertidal levels and separated by a sandy beach. The ceramic plates were placed in an aluminium framework that was attached to the rocky surface at the middle part of the *M. laminarioides* belt with stainless steel screws. Plates were left in the field for 30 d (October 14 to November 13, 2003) and transported back to the laboratory to measure survival rates of the different treatments. Data did not satisfy Cochran's test for homogeneity of variance, therefore a Kruskal-Wallis test was used for comparisons among treatments followed by the test of Nemeyi for multiple comparisons (Zar, 1996).

Effects of Ca^{++} additions on solitary and group recruitment

Fertile cystocarpic blades of *Mazzaella laminarioides* were collected in Caleta Maitencillo (32°39′S; 71°29′W) in September 2002. Experimental treatments included 4 Ca^{++} levels (1.0, 1.5, 2.0 and 3.0 times the Ca^{++} mean concentration in seawater). The 4 Ca^{++} treatments were obtained by adding 200 $\mu g\ g^{-1}$ (1.5×), 400 $\mu g\ g^{-1}$ (2×) and 800 $\mu g\ g^{-1}$ (3.0×) of Ca^{++} to normal seawater. Mean Ca^{++} concentration in seawater is 412 $\mu g\ g^{-1}$ (1.0×; De Boer 1981, Lobban & Harrison, 1994).

Following a single factor design with 4 levels, this experiment used 5 replicate multiplates (Bibby Sterilin Ltd., U.K.), each with 25 cellwells. Each level of factor was sub-replicated 4 times in each multiplate. Thus, 4 ml of the respective treatment solution was placed in each of 16 cellwells. Then 35 μl of a recently prepared solution of naturally released spores was placed in each cellwell. Average spore concentration in the solution was 9.328.000 ± 988.562 spores ml^{-1} measured in a Neubauer chamber (Bright-line, Hirschmann). The total time needed to complete spore deposition in a given plate was 2–3 min and treatments were assigned at random among the cellwells in a given plate. The whole plate was then agitated with the help of a Vortex (Termolyne Speed Control/Type 37600 mixer) at maximum speed for 5 seconds. The cultures were then incubated for 2 h under constant conditions of temperature (14 ± 1°C), photoperiod (12:12) and photon flux density (35–40 μmol photons $m^{-2}\ s^{-1}$). The number of spores settling singly and in aggregation was then counted in 3 ocular fields within each experimental cellwell using an inverted microscope (Nikon Eclipse TE 300).

After counting single and aggregated recruitment, the seawater of each cellwell was discharged to eliminate unattached spores and replaced by SWM-3 culture medium. Plates were then incubated under the above controlled conditions for 48 h and the cellwells were then re-examined to measure germination rates.

Data on total frequency of aggregated recruitment and on germination rates were tested for homogeneity of variance using Cochran's test and for normality using Shapiro-Wilk's test. Since data satisfied both requirements, results were then compared using a Model I one way analysis of variance (ANOVA). Since significant differences among main factors were found, ANOVA was followed by an *a posteriori* Tukey test (Snedecor & Cochrane, 1967; Sokal & Rohlf, 1969).

Results

Frequency of solitary and group recruitment at different levels of spore abundance

The 240 microscopic fields examined in the experimental plate comprised a total surface of 312 mm^2 and included 5297 spores; 55.9 ± 24.29% of the spores recruited without contacting neighboring spores, while the remaining 44.1 ± 20.10% recruited forming groups of 2 to over 100 spores (Figure 1). About 75% of the groups were formed by 2–10 spores, the remainder being formed by groups of 20 to over 100 spores.

The total number of spores, as well as the relative importance of solitary and group recruitment varied from one plate to another. Plotting the number of sporelings as a function of spore abundance (Figure 2) indicates that with increasing spore abundance the relative importance of group recruitment increases significantly, while the importance of single recruitment decreases. ANCOVA analysis indicate significant differences in the slope of both curves (F=142.1197; $p < 0.0001$).

Field survival of solitary and coalescing recruits

Average survival rates after 30 days of field exposure of the solitary and coalescing recruits was 18.75%. Lowest survival values occurred in sporelings formed by 1 and 10 spores (5% in both cases, Figure 3) while highest survival rates were observed in sporelings formed by 50 (30% survival) and 100 (35% survival) spores.

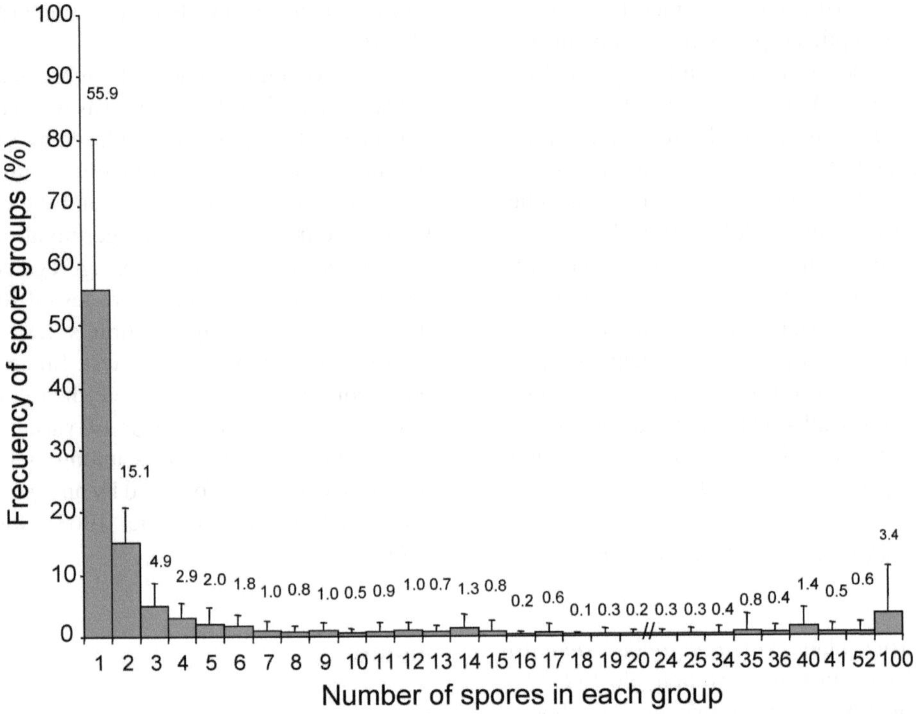

Figure 1. Average frequency of individual spores and spore groups recruited on 10 epoxy-putty plates maintained for 72 h in the field. Bars are SE.

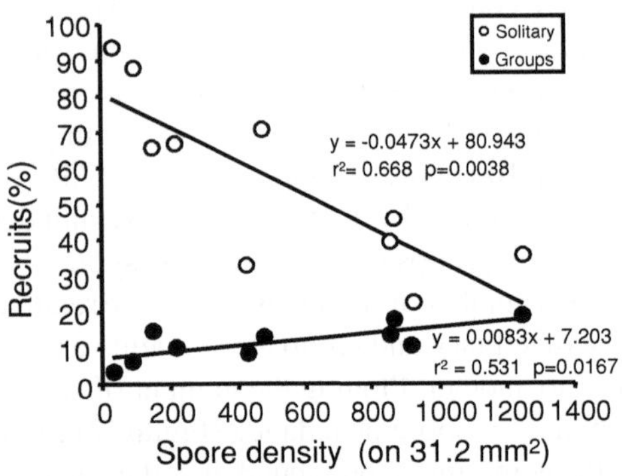

Figure 2. Variation in the relative importance (%) of single (empty circles) and aggregated (filled circles) recruitment as a function of spore abundance. Each pair of values corresponds to recruits (solitary recruitment or in group) counted in 24 fields within each of 10 plates exposed for 72 h to natural recruitment.

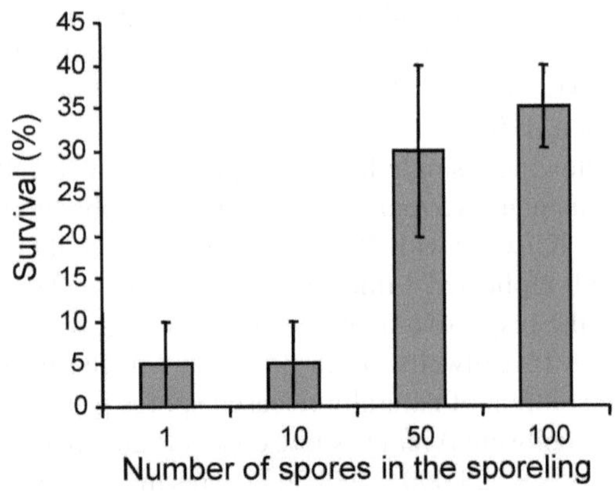

Figure 3. Average sporeling survival of *Mazzaella laminarioides* after 30 days of field exposure. Sporelings were built in the laboratory with different number of spores (1 to 100) and then transplanted to the field. Bars are range values.

Differences between treatments with low (1 and 10) and high (50 and 100) spore numbers were statistically significant (Kruskal-Wallis test, $H = 9.966154$, $p = 0.0189$; Nemenyi test, $p = 0.04$ for comparisons between 1 or 10 spores with 50 spores and $p = 0.019$ for comparisons between 1 or 10 spores with 100 spores).

Effects of Ca^{++} additions on group recruitment

Ca^{++} addition to the culture medium modified the percentage of spores of *Mazzaella laminarioides* that settled in groups. The highest values (Figure 4) were obtained at Ca^{++} concentrations of 600 $\mu g\ g^{-1}$ (1.5×)

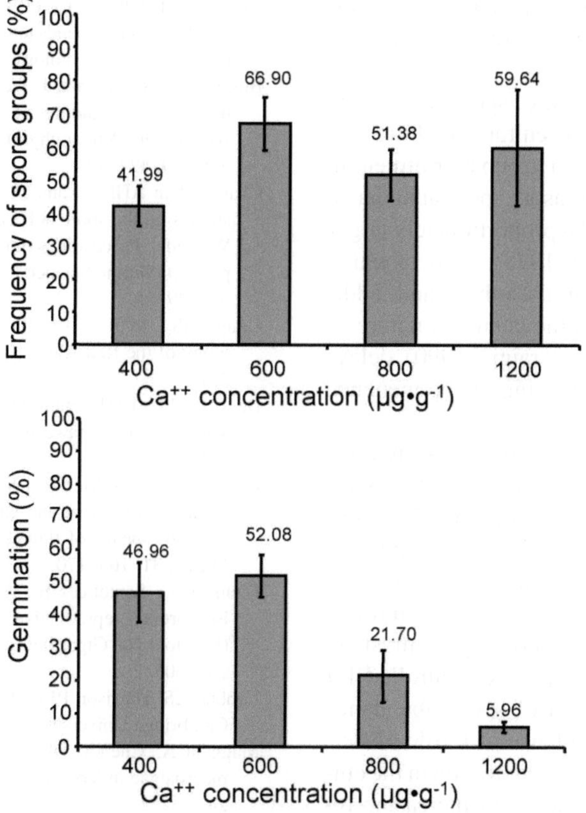

Figure 4. Changes in the relative frequency of spore groups and in the germination rates of spores of Mazzaella laminarioides as functions of several concentrations of Ca^{++} in the culture medium. Bars are ± SE.

and 1200 $\mu g\ g^{-1}$ (3.0×). Only in the first case (0.68 L^{-1}) were the differences with normal seawater significant (ANOVA, $F = 50.975$; $p = 0.0001$; Tukey test $p < 0.001$). On the other hand, germination data indicated low germination values under the two highest Ca^{++} concentrations (ANOVA, $F = 227, 757$; $p < 0.000001$; Tukey test $p < 0.0001$ in both cases) while germination values under concentrations of 600 $\mu g\ g^{-1}$ were not affected compared to normal seawater. Thus, concentrations of 600 $\mu g\ g^{-1}$ (1.5×) of Ca^{++} in the water increase spore aggregation without affecting germination.

Discussion

Our results suggest that spore abundance and the effects of some abiotic factors, such as Ca^{++} ions, may modify the relative importance of group recruitment and coalescence in the recruitment of *Mazzaella laminarioides.* In turn, our results also show that group recruitment and coalescence increase the probabilities of field survival during early recruitment.

Previous studies (e.g. Maggs & Cheney, 1990; Santelices et al., 1999; Morley et al., 2003) have already shown that spores of coalescing seaweeds may recruit singly or by forming groups. Our present results indicate that in the case of *Mazzaella laminarioides* the relative frequency of group recruitment was 44.1% of the total number of spores recruited and that the maximum number of spores that coalesce forming a single sporeling, can be well over a hundred.

Previous laboratory studies (Santelices et al., 1999; Morley et al., 2003) have also shown that multisporic germlings of coalescing seaweeds germinated and survived better than unisporic germlings during early recruitment. Our present results, gathered in the field, support the idea of higher survival rates of multisporic recruits. Even though we did not experimentally examine the factors determining this selective mortality, field observations during the experimental period indicate that dislodgement by water movement and sand abrasion and grazing affected solitary sporelings more than multisporic recruits.

As expected from the literature, spore abundance and Ca^{++} concentration in the culture medium can modify the relative importance of group recruitment in a given recruitment event. Increasing spore abundance probably involves liberation of a proportionately larger number of spore masses (Boney, 1978), many of which settle and recruit in groups. On the other hand, addition of ions probably modifies the chemical nature of the mucilage around the spores (Craigie, 1990), delaying its dissolution rate and increasing the permanence time of free-floating spore groups.

Since the above factors may modify the relative importance of group recruitment in the field, this response is expected to vary in space and time. For example, since spore abundance decreases exponentially with increasing distance from the spore sources (Norton, 1992), the importance of group recruitment is likely to follow a similar pattern. By contrast, in nursery facilities, these factors can be controlled for farming purposes. Manipulation of spore abundance through regulation of the abundance of fertile blades, and slight increments of Ca^{++} concentration in the culture medium should be most useful for increasing recruitment success of massive spore inoculations of coalescing species.

Acknowledgments

Our gratitude to M. Hormazábal for help with some of the field studies and to R. Finke for improving the grammar of earlier drafts. This study was supported by grant FONDECYT 1020855 to BS.

References

Alveal K, Romo H, Werlinger C (1995) Cultivo de Gracilaria a partir de esporas. In Alveal K, Ferrario ME, Oliveira EC, Sar E (eds.), Manual de Métodos Ficológicos. Universidad de Concepción, Concepción, Chile: pp. 599–609.

Apple ME, Harlin MM (1995) Inhibition of tetraspore adhesion in *Champia parvula* (Rhodophyta). Phycologia 34(5): 417–423.

Azanza-Corrales R, Aliaza TT, Montano NE (1996) Recruitment of *Eucheuma* and *Kappaphycus* in a farm in Tawi-Tawi, Philippines. Hydrobiologia 326/327: 235–244.

Boney AD (1978) The liberation and dispersal of carpospores of the red alga *Rhodymenia pertusa* (Postels et Rupr.) J. Ag. Journal of Experimental Marine Biology and Ecology 32: 1–6.

Brawley SH, Johnson LE (1991) Survival of fucoid embryos in the intertidal zone depends upon developmental stage and microhabitat. Journal of Phycology 27: 179–186.

Buschmann AH, Correa J, Westermeier R (1999) Recent advances in the understanding of the biological basis for *Gigartina skottsbergii* (Rhodophyta) cultivation in Chile. Hydrobiologia 398/399: 427–434.

Chamberlain AHL, Evans LV (1981) Chemical and histochemical studies on the spore adhesive of *Ceramium*. In Fogg GE, Jones WE (eds), Proceedings of the Eight International Seaweed Symposium. Marine Science Laboratories, Menai Bridge, Anglesey: pp. 539–542.

Craigie JS (1990) Cell walls. In Cole KM, Sheath RG (eds.), Biology of the Red Algae. Cambridge University Press: pp. 221–257.

De Boer JA (1981) Nutrients. In Lobban CS, Wynne MJ (eds.), The Biology of Seaweeds. Blackwell Scientific Oxford: pp. 356–391.

Diannelidis BE, Kristen U (1988) Comparative histochemical studies of the reproductive and gametophytic tissue of the marine red algae by means of fluorescent and light microscopy. Botanica Marina 31: 163–170.

Hannach G, Santelices B (1985) Ecological differences between isomorphic reproductive phases of two species of *Iridaea* (Rhodophyta: Gigartinales). Marine Ecology Progress Series 22: 291–303.

Lobban CS, Harrison PJ (1994) Seaweed ecology and physiology. Cambridge University Press, USA: 366 pp.

Maggs CA, Cheney DP (1990) Competition studies of marine macroalgae in laboratory culture. Journal of Phycology 26: 18–24.

Morley TL, Bolton JJ, Anderson RJ (2003) Phase dominance and reproductive characteristics in two co-occurring Rhodophyta from the west coast of South Africa. In Chapman ARO, Anderson RJ, Vreeland VJ, Davison IR (eds.), Proceedings of the 17th International Seaweed Symposium. Oxford University Press. Oxford and New York: pp. 365–371.

Ngan Y, Price IR (1983) Periodicity of the spore discharge in tropical Florideophyceae (Rhodophyta). British Phycological Journal 18: 83–95.

Norton T (1992) Dispersal by macroalgae. British Phycological Journal 27: 293–301.

Norton TA, Fetter R (1981) The settlement of *Sargassum muticum* propagules in stationary and flowing water. Journal of the Marine Biological Association of the U.K. 61: 929–940.

Pueschel CM (1979) Ultrastructure of the tetrasporogenesis in *Palmaria palmata* (Rhodophyta). Journal of Phycology 15: 409–424.

Santelices B (1990) Patterns of reproduction, dispersal and recruitment in the seaweed. Oceanography and Marine Biology: An Annual Review 28: 177–276.

Santelices B (1992) Strain selection of clonal seaweeds. In Round FE, Chapman DJ (eds.), Progress in Pycological Research. Biopress Ltd. England: pp. 85–116.

Santelices B (2004) A comparison of ecological responses among aclonal (unitary), clonal and coalescing seaweeds. Journal of Experimental Marine Biology and Ecology 300: 31–64.

Santelices B, Correa JA, Meneses I, Aedo D, Varela D (1996) Sporeling coalescence and intraclonal variation in *Gracilaria chilensis* (Gracilariales, Rhodophyta). Journal of Phycology 32: 313–322.

Santelices B, Correa JA, Aedo D, Flores V, Hormazábal M, Sánchez P (1999) Convergent biological processes in coalescing Rhodophyta. Journal of Phycology 35: 1127–1149.

Santelices B, Aedo A, Hormazábal M, Flores V (2003a) Field testing of inter- and intraspecific coalescence among middle intertidal red algae. Marine Ecology Progress Series 250: 91–103.

Santelices B, Correa JA, Hormazábal M, Flores V (2003b) Contact responses between spores and sporelings of different species, karyological phases and cystocarps of coalescing Rhodophyta. Marine Biology 143: 381–392.

Snedecor GW,. Cochrane WG (1967) Statistical Methods, 6th edn. Iowa State University Press, Ames, 593 pp.

Sokal RR, Rohlf FJ (1969) Biometry. The principles and practice of statistic in biological research. Freeman, San Francisco.

Zar JH (1996) Biostatistical analysis. Prentice Hall, 931 pp.

Journal of Applied Phycology (2006) 18: 591–598
DOI: 10.1007/s10811-006-9060-8

Distribution and recent reduction of *Gelidium* beds in Toyama Bay, Japan

D. Fujita[1], T. Ishikawa[2], S. Kodama[3], Y. Kato[4] & M. Notoya[1]
[1]*Tokyo University of Marine Science and Technology, Minato, Tokyo, 108-8477 Japan;* [2]*Institute of General Science for Environment, Metocean Environment Inc.,Ohigawa, Shizuoka, 21-0212 Japan;* [3]*Fushiki-Toyama Port Office, Hokuriku Regional Development Bureau, Ministry of Land Infrastructure and Transport, Toyama, 934-0001 Japan;* [4]*Fushiki Port Office Management, Toyama Prefecture, Takaoka, 933-0104 Japan*

*Author for correspondence: e-mail: d-fujita@s.kaiyodai.ac.jp

Key words: *Gelidium*, monitoring, sedimentation, stagnation of water, yield

Abstract

The distribution and recent reduction of *Gelidium* beds, i.e. mat-like beds dominated by the agarophyte *G. elegans* Kützing in Toyama Bay (Sea of Japan), in which 95% of the coastline is protected artificially, are reported. *Gelidium* beds were common in shallow waters (usually <10 m deep); most of the large beds (>1 ha) were restricted to the inner coasts of the bay. In calm and eutrophic areas, however, *G. elegans* was heavily colonized by epiphytes. In the last decade, two beds were buried *in situ* and beds in their vicinity were damaged by the stagnation of coastal water and/or sedimentation by silts which accompanied land reclamation. At the other two beds monitored since 1988, *Gelidium* declined a few times but most prominently in 1998, when episodic long summer rain was recorded. This is the first report, not only on the current status of *Gelidium* beds other than for the central Pacific Coast of Honshu in Japan, but also concerning reduction of the beds caused by both anthropogenic and natural events.

Introduction

Gelidium beds have been one of the most important seaweed beds in Japan because *Gelidium* (and species of related genera) have been the major materials for commercial extraction of agar (Akatsuka, 1986). Recently, however, less attention has been paid to *Gelidium* beds than before because domestic agar yields have declined drastically. The major reason for the reduction is the lower price of dried agarophytes, resulting from an increase in imports from other parts of the world and the decrease in the use of agar among a variety of polysaccharides (Fujita, 2004). Although data in the Nature Conservation Bureau (1994) showed that decrease of *Gelidium* beds was the most serious among eight types of domestic seaweed beds, little is known about the current status of *Gelidium* beds other than for the central Pacific (e.g. Yanase et al., 1982, Tokyo Metropolitan Fisheries Experiment Station, 1996).

Among prefectures of Honshu Island facing the Sea of Japan, Toyama Prefecture has ranked first, providing 89% of the annual *Gelidium* yield for more than a decade because the yields in 11 other prefectures have declined more rapidly (Fujita, 2004). In this prefecture, *Gelidium* has been cropped in Toyama Bay by hand or harvested using a stick with iron claws or triangular rakes called 'Manga' dropped from boats. These *Gelidium* fisheries have been summer jobs for diver-fishermen and labourers working for set-net fisheries (Fujita, 1994). Furthermore, *Gelidium* beds have been re-evaluated as nurseries for fish and shellfish, particularly for horned turban, a commercial gastropod *Turbo cornutus* (Fujita et al., 1990; Fujita & Kamono, 1998). However, *Gelidium* yield has also decreased along the prefectural coast (Fujita & Shozen, 1999, and revised in Figure 1) and little has been known about *Gelidium* beds up to now. In the present paper, we describe the distribution, current status and recent reduction of *Gelidium* beds in Toyama Bay.

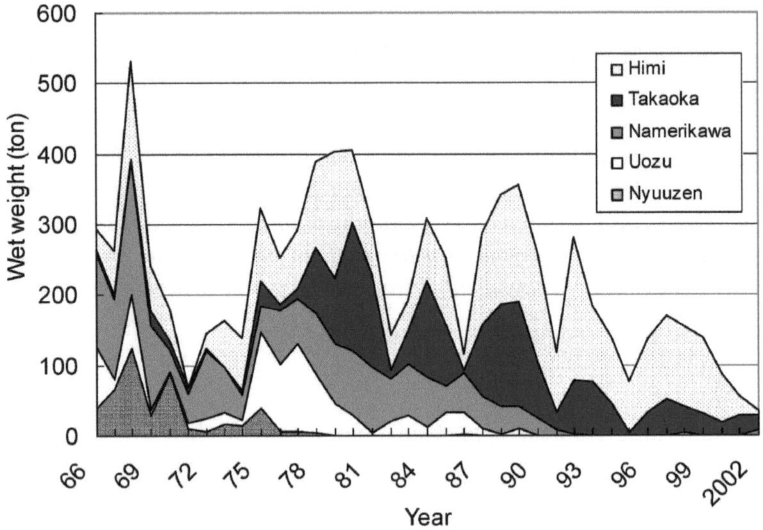

Figure 1. Yearly changes in the *Gelidium* yields (raw) by towns in Toyama Bay between 1966 and 2002 (revised from Fujita & Shozen (1999)).

Materials and methods

Toyama Bay (largely from 36°50′N to 37°N and 137°E to 137°40′E) is located on the central Sea of Japan coast of Honshu Island, Japan. The prefectural coastline of the bay is ca. 100 km long and shared by nine towns. The inner coasts of the bay, particularly the coast of Himi and east coast from Namerikawa to Uozu, are highly sheltered by the Noto Peninsula and Ikujibana Cape (Figure 2). The notable characteristics of the coast are the small tidal range (<0.5 m), eutrophication and low salinity (15 to 30 PSU in surface seawater) caused by inflow of many rivers as well as a high rate (95%) of artificial protection (Nature Conservation Bureau, 1998). Hard substrata comprise cobbles and rocks in the inner and outer coasts of the bay, respectively. The detailed environments have been described elsewhere (Fujita & Shozen, 1999, 2002 and papers cited in these reports).

The current status of *Gelidium* beds was surveyed by SCUBA diving during the Comprehensive Environmental Survey of Fishery Grounds in Toyama Bay made in 2001 and some additional visits. Distributions of seaweed beds were determined by taking aerographs (1/5000 in scale) of the whole prefectural coast. Areas of *Gelidium* and other seaweed beds were estimated using a planimeter after their outlines were traced on the contoured map. Here we described the detailed information on *Gelidium* beds that was not given in the report (Fujita & Shozen, 2002). The status of beds in the past was known from previous papers cited in Fujita and Shozen (1999, 2002) and unpublished photographic data.

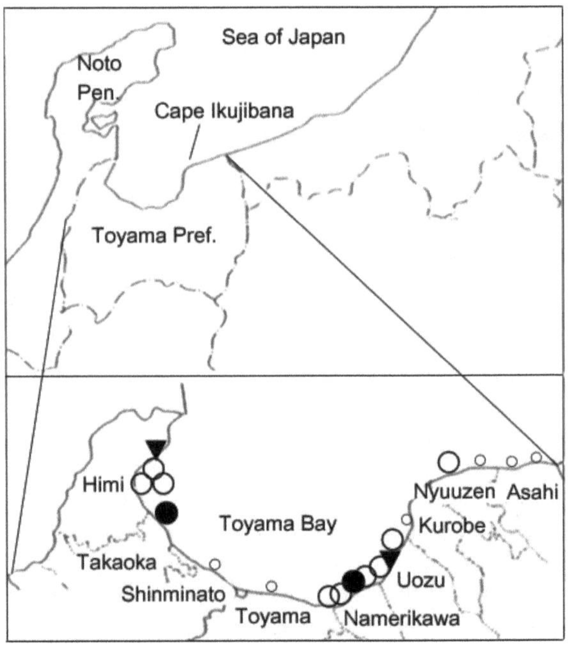

Figure 2. Map showing the location of Toyama Bay and the occurrence of *Gelidium* beds. Large (○) and small (○) open circles indicate large (>1 ha) and smaller *Gelidium elegans*-dominated red-mat like beds, respectively. Solid circles (●) = monitored *Gelidium* beds; Triangles (▼) = *Gelidium* beds declined after 1990.

Long-term monitoring began in 1988 at two *Gelidium* beds off Namerikawa (36°46′N 137°21′E) and Takaoka (36°48′N 137°3′E) on the east and west sides of the bay, respectively (Figure 2). At Namerikawa, observations were made bimonthly along one monitoring line located in the central part of the bed that extended from the shore to 11 m deep in 1988. The lower limit of *Gelidium* was marked on the cobbles using a sand bag or a stainless steel chain when the limit moved inshore. At other monitoring stations, *Gelidium* growing on six marked boulders (0.5 to 0.7 m in height) and thirty artificial concrete reefs (0.5 to 1.5 m in height) set on natural cobble beds (9 m in depth) has been monitored by taking photographs since 1993. From October 10th to December 21st in 2000, open-ended plastic bottles (300 ml in volume) were fixed at depths of 7 m and 11 m (i.e., present and past *Gelidium* beds, respectively) to trap sediments and compare the rate of sedimentation between the two depths. The bottles had enough shoulder to avoid re-suspension of sediments. On the last day, the bottles were capped at the sea bed and brought back to the laboratory. After all of the sediments settled, the volumes of sediments were estimated to calculate the rates of sedimentation. On December 21st in 2000, sediments covering cobbles were also carefully collected using plastic syringes (120 ml in volume) and observed with a light microscope in the laboratory.

At Takaoka, observations were made four times a year in a *Gelidium* bed (where the brown algae *Sargassum* and *Ecklonia* also coexist) which had extended from the shore to 6 m deep in 1988. Two meshed quadrats (5 m × 5 m), tied down by ropes, were fixed at depths of 4 m and 6 m in the bed and SCUBA divers recorded the percentage cover of each algal component.

The dominant and commercial gelidiacean species in Toyama Bay is *Gelidum elegans* Kützing, which is a perennial and lives up to three years (Fujita, 1994). The other common species are *Pterocladiella tenuis* (Okamura) Shimada, Horiguchi *et* Masuda and *Gelidium vagum* Okamura; the former species often grows with *G. elegans* but is rarely harvested together with it. In the present paper, the term *Geldium* bed refers to a *G. elegans*-dominated red mat-like bed, whether it is large (>1 ha) or small, and even when it is associated with epiphytic and/or non-epiphytic algae.

Results and discussion

Distribution of Gelidium *beds*

The vertical distribution (as a maximum depth) of and estimated area of *Gelidum* beds in Toyama Bay are shown in Figure 3. The total area of *Gelidium* beds was 227 ha comprising 33% of total seaweed beds occurring on hard substrata. *Gelidium* beds were common in the bay; large beds were found in five *Gelidium*-yielding areas: Nyuuzen, Uozu, Namerikawa, Takaoka and Himi (Figure 1). In the other four towns, only small beds were found because of the abundance of *Sargassum* spp and *Ecklonia kurome* Okamura on subtidal rocky bottoms (Asahi), dominance of unstable gravels on steep slopes (Kurobe) and presence of soft bottoms (Toyama and Shinminato).

In Nyuuzen, where the harvest of *Gelidium* stopped in 1979, one large *Gelidium* bed was present between the shore and 10 m in depth. This was the only large *Gelidium* bed located in the outer coast of the bay. In this bed, however, other algae such as *Gracilaria textorii* (Suringar) Hariot, *Martensia fragilaris* Harvey, *Sargassum* spp and *Ecklonia stolonifera* Okamura coexisted and replaced *Gelidium* in deeper beds (10–22 m in depth). In addition, thalli of *G. elegans* were heavily covered with epiphytes such as *Acrosorium venulosum* (Zanardini) Kylin and *Hypoglossum nipponicum* Yamada. Although the coast is highly protected by concrete, water motion is still active because of high wave heights. In Uozu, where the harvest of *Gelidium* ceased in 1998, large and small *Gelidium* beds were present in shallow waters from shore to 10 m in depth; dense beds were restricted to between 1 m to 5 m in depth. However, reduction of *Gelidium* is most prominent in the east coasts of Toyama Bay and the thalli of *G. elegans* were heavily colonized by the sessile diatom, *Arachnoidiscus ornatus* Ehrenberg, and macroepiphytes such as *Colpomenia sinuosa* (Martens ex Roth) Derbès et Solier. The most likely cause is the frequent construction of concrete structures, including breakwaters, and land-reclamation that allowed stagnation of coastal water and sedimentation of silts. In the newest case, not only were two hectares of *Gelidium* beds buried *in situ* by land reclamation (1995–2002) but *Gelidium* in the vicinity was also reduced. The extensive sedimentation was first reported in 1995; the sticky sand-dwelling diatom *Amphitetras antediluviana* Eherenberg was found to bind sand granules (Fujita et al., 1996). A recent notable event in the silt-covered area was the appearance of blue-green and unknown green algae (Fujita & Shozen, 2002), which had never been found before 2002.

In Namerikawa, where the harvest stopped in 1992 but reopened until 2004, both large and small *Gelidium* beds were found. Most beds were restricted

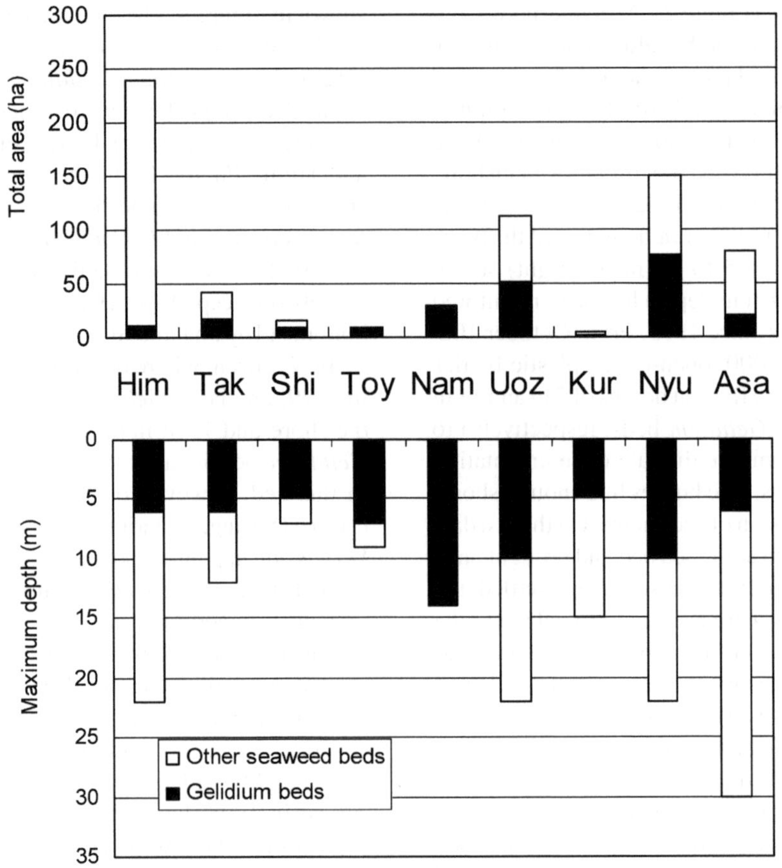

Figure 3. Maximum depth and total area of *Gelidium* and other seaweed beds of 9 coastal towns on Toyama Bay. See Figure 2 for the abbreviated town names.

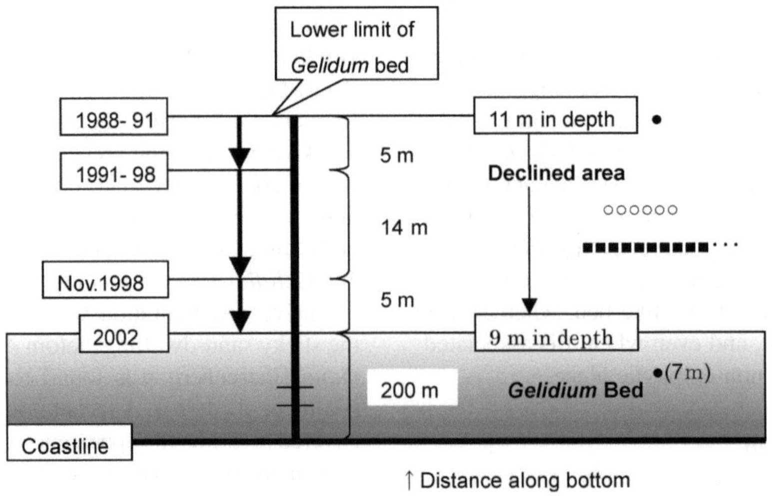

Figure 4. Diagram of reduction of *Gelidium* bed at Namerikawa. Closed circles = sampling points of accumulated silts; Open circles = marked cobbles; Squares = concrete blocks for stock enhancement of abalone and horned turban shell. (See text for explanation).

to shallow waters (<10 m in depth) but one extended down to 14 m in depth. *Gelidium* beds off Namerikawa are unique because they never co-existed with other types of seaweed beds (Figure 3) probably because of low water transparency. Thalli of *G. elegans* were heavily colonized by epiphytes including *C. sinuosa*, various unknown red algae and *A. discus* from shallow to deep water, respectively. A small *Gelidium* bed was also found just off a small river mouth, suggesting some tolerance to low salinity (ca. 30 PSU).

In Takaoka, where the harvest has continued up to 2004, one large *Gelidium* bed was present on rocky bottoms down to 6 m in depth and gave way to *Sargassum* and *Ecklonia* beds extending down to 12 m. The northern and southern neighbourhoods have developed as a fishing port and an exporting port, respectively, accompanied with long offshore and onshore breakwaters. Thalli of *G. elegans* were heavily colonized by *A. ornatus.*

In Himi, where the harvest has also continued up to 2004, large *Gelidium* beds were present around rocks

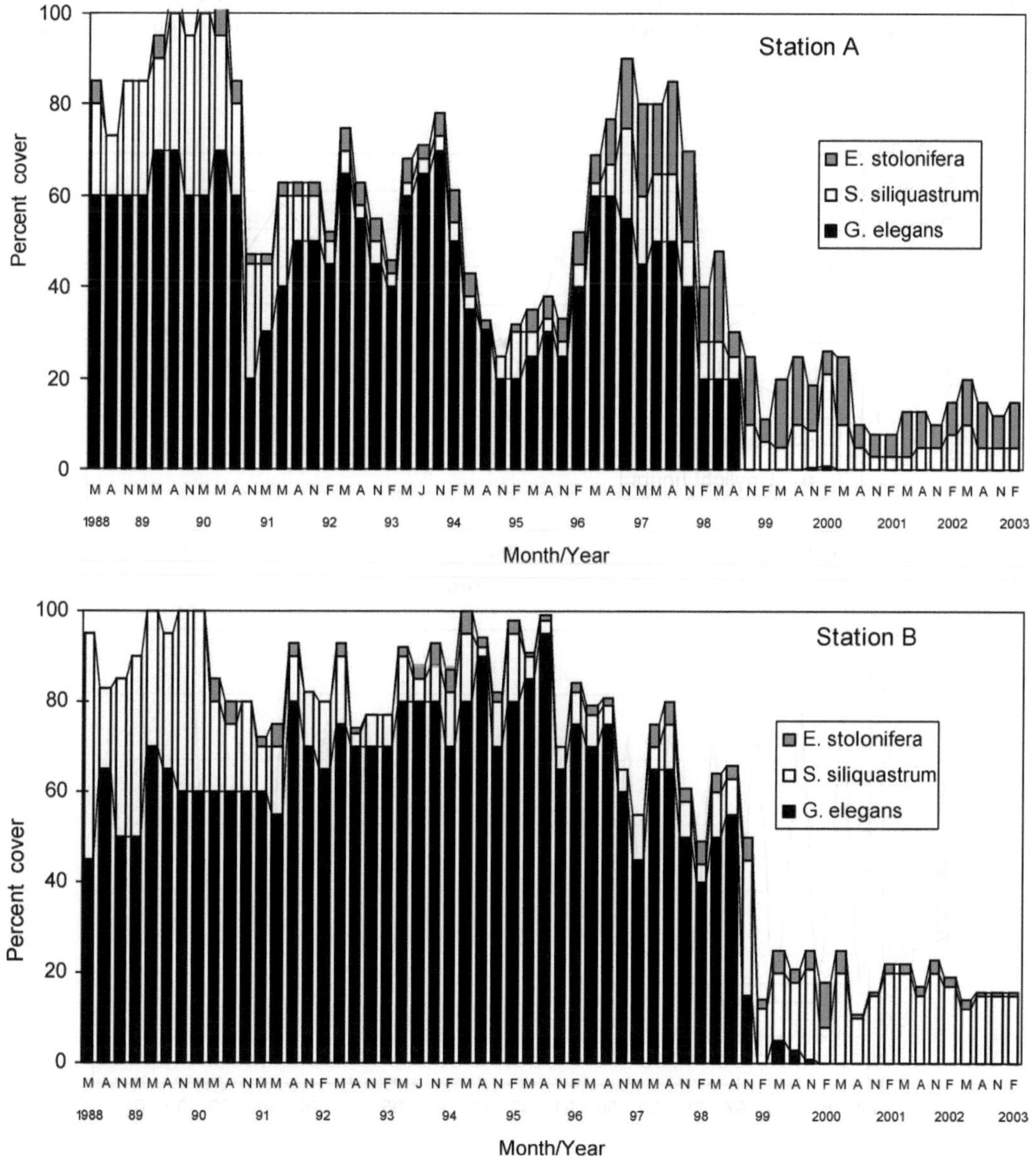

Figure 5. Seasonal changes in coverage of seaweed at two fixed quadrats 5 × 5 m placed at depths of 6 m (A) and 4 m (B) in a *Gelidium* bed off Takaoka. Note the abrupt reduction in 1998.

and isolated rocks, including submerged rocks just off the mouth of a small river (suggesting some tolerance to low salinity again: ca. 28 PSU). All of these beds were restricted to shallow areas (<6 m in depth), and *Sargassum* and *Ecklonia* beds were dominant in deeper waters (down to 22 m in depth). Thalli of *G. elegans* were heavily colonized by *A. ornatus.* At least until 1993, *Gelidium* beds (still colonized by *A. ornatus*) were also present 2 km north of the present bed; these beds (<5 m in depth) have been completely replaced by *C. sinuosa* and *Sargassum* spp. The most likely cause is again the reclamation of land (1993–1998); not only were two hectares of *Gelidium* beds buried *in situ* but *Gelidium* completely disappeared from the vicinity.

Monitoring in marked Gelidium *beds*

Results of monitoring in marked *Gelidium* beds at Namerikawa and Takaoka are shown in Figures 4 and 5, respectively. At Namerikawa, the lower limit of a *Gelidium* bed moved inshore by 5 m in 1991, by 14 m in 1998 and by 5 m in 2002 on the monitoring line (Figure 4). In total, the lower limit moved by 24 m in bottom distance or by 2 m (from 9 m to 11 m) in depth, resulting in the reduction of bed by 1 ha in estimated area. No recovery, i.e., the offshore movement of these lower limits, was detected during 16 years. In addition, all *G. elegans* colonizing concrete artificial reefs and marked boulders (see Figure 4 for the location)

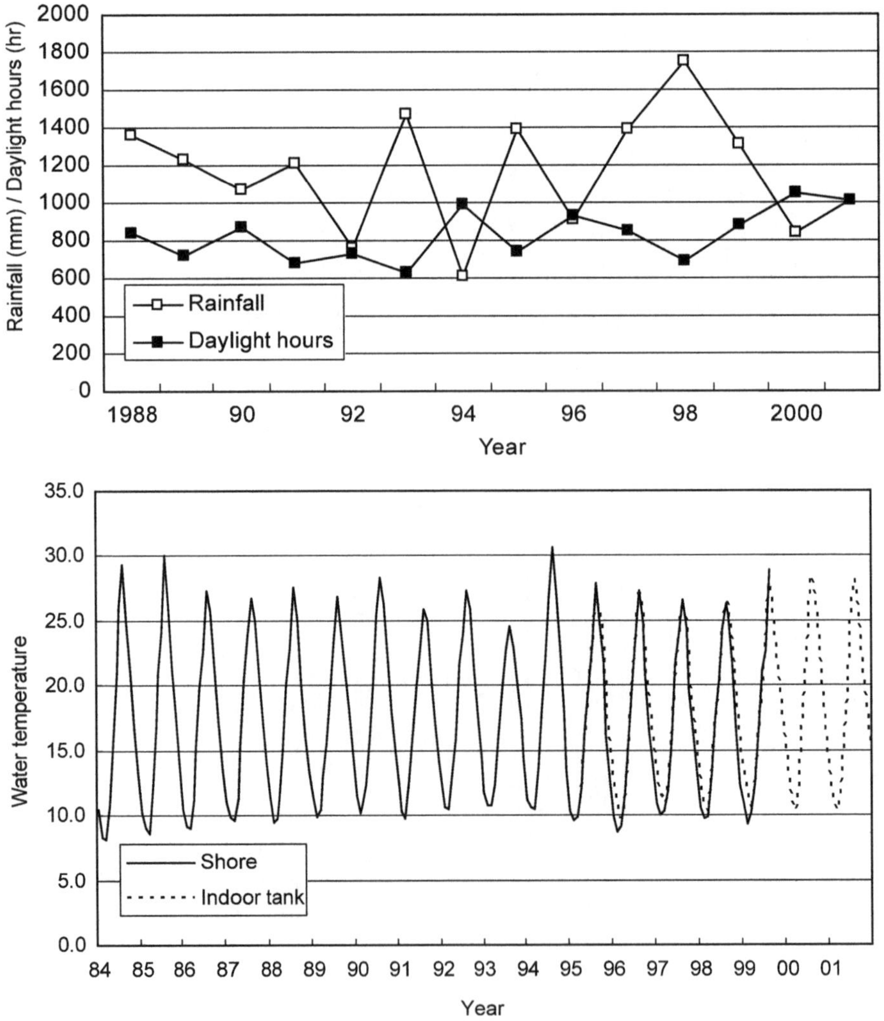

Figure 6. Yearly changes in rainfall and daylight hours at Uozu (Data: Meteorological Agency, Japan) between April to September (above) and water temperature at Toyama Prefectural Aquaculture Center at Himi (below).

also completely disappeared during summer of 1998. In bottles placed on bottoms for 72 days, sedimentation rates were 5.3 cm^3 cm^{-2} and 3.1 cm^3 cm^{-2} at 11 m and at 7 m depth, respectively. In addition, germlings of *G. elegans* were found in silt samples collected from the surface of cobbles (11 m in depth), suggesting that silts inhibit the attachment of *G. elegans* spores and germlings to the hard substrata (not shown).

At Takaoka, coverage of *G. elegans,* within two quadrats, fluctuated a few times at station A (6 m in depth) and less at station B (4 m in depth), respectively (Figure 5). However, at both sites, coverage of *G. elegans* was abruptly reduced between August and November 1998, followed by a small recovery in 1999 only at station B. After the reduction of *Gelidium elegans*, *Sargassum siliquastrum* (Mertens ex Turner) C. Agardh and *Ecklonia stolonifera* survived with low cover values (<25%). Substrata inside the quadrats and outside were extensively covered with silts after 1998.

At these monitoring sites, the *Gelidium* bed declined in deeper areas abruptly in 1998, when episodic long summer rain was recorded (Figure 6). This is the only year since 1971 when the Meteorological Agency did not announce the end of the rainy season (usually in July to early August). The long lasting rainy season must have caused the increase of turbidity and sedimentation of silts; decrease in salinity could be neglected because of the presence of *Gelidium* beds just off river mouths. Other factors, such as grazing by herbivores, water temperature and daylight hours were also examined, but none of them was unlikely to induce unrecoverable reduction of *Gelidium* beds during the monitoring periods. For example, mass mortality of the most important herbivore, the sea urchin *Strongylocentrotus nudus* (A. Aggasiz) was recorded at the monitored bed in Namerikawa in the episodic hot summer of 1994, but this never brought about the recovery of *Gelidium* (Fujita, 1998). In addition, no serious influence was recognized in the episodic highest and lowest summer water temperatures (recorded in 1994 and 1993, respectively) and in the lowest record (1993) of daylight hours during the growth period of *G. elegans*. The slight reduction of daylight hours in 1998 may result from the extensive cloud cover causing the long rain. As the monitoring sites were free from being harvested, over-harvesting was not a factor.

Conclusions

In monitoring *Gelidium* beds in central Pacific coasts of Honshu Island (e.g., Yanase et al., 1982 in Izu Peninsula, Tokyo Metropolitan Fisheries Experiment Station 1996 in Izu Islands), destructive sampling was done before every harvest season to forecast the crop of the year. In contrast, detecting fluctuation has been the target in our non-destructive monitoring. Our data reveal not only the current status of *Gelidium* beds in Japan other than for the central Pacific Coast of Honshu but also the recent reduction of *Gelidium* beds. Reduction in *Gelidium* beds was also recorded around the Izu Islands, following volcanic eruption which buried hard substrata from the sedimentation of ashes (Tokyo Metropolitan Fisheries Experiment Station 1986). Some of the reductions we observed must be anthropological, caused by stagnation of seawater and sedimentation of silts resulting from recent coastal protection and reclamation of lands. However, natural events may also affect the fate of *Gelidium* beds. At least in two monitoring sites, an episodic, long lasting rainy season is the most likely event. Dominance of the sessile diatom *A. ornatus* on *G. elegans* thalli growing in inner (sheltered and further protected) coasts of the bay may be another evidence of stagnation of coastal water, as suggested by Yamada (1967), and may be a sign of reduction of *Gelidium* beds in the stagnating water. Such spoiled thalli may be more sensitive to an extreme natural event. We should remind ourselves of the Japanese proverb, 'Yowarime ni tatarime' meaning 'misfortunes come after weakening'. Therefore, we should minimize the enhancement and enlargement of 'yowarime' (weakened status) before 'tatarime' (misfortune) comes.

References

Akatsuka I (1986) Japanese Gelidiales (Rhodophyta), especially *Gelidium*. Oceanography Marine Biology Annual Review 24: 171–263.

Fujita D (1994) In: Fujita D, Watanabe S, Hamada J (eds.), Algae of Toyama: Fisheries and enhancement of *Gelidium*. Toyama Prefectural Research Institute: pp 12–13.

Fujita D (1998) Strongylocentrotid sea urchin-dominated barren grounds on the Sea of Japan coast of northern Japan. In: Mooi R.,Telford, M. (eds.) Echinoderms: San Francisco. A. A. Balkema: pp. 659–664.

Fujita D (2004) Tengusarui (Gelidiaceae). In: Ohno M (ed.), Biology and technology of economic seaweeds. Uchida Rokakuho Publishing pp. 201–225.

Fujita D, Kamono H (1998) Horned Turban in Toyama. Toyama Prefectural Fisheries Research Institute.

Fujita D, Okada H, Sakata K (1990) The importance of some marine algae inhabiting fishing-port breakwater vertical surface as natural food for juvenile horned turban *Turbo* (*Batillus*) *cornutus*. Bulletin of Toyama Prefectural Fisheries Research Institute 2: 41–51.

Fujita D, Okamoto Y, Mayama S (1996) A sand-dwelling diatom found in the sand overlaying the shallow water stones off Uozu City, Toyama Prefecture. Toyama Prefectural Fisheries Research Institute 8: 25–29.

Fujita D, Shozen K (1999) Environments of fisheries grounds in eastern areas (Kurobe, Nyuuzen and Takaoka) of Toyama Prefecture. Toyama Prefectural Fisheries Research Institute.

Fujita D, Shozen K (2002) Environments of fisheries grounds in Toyama Prefecture (2001) Toyama Prefectural Fisheries Research Institute.

Nature Conservation Bureau, 1994. The report of the marine biotic environment survey in the 4th national survey on the natural environment. Vol. 2. Algal and sea-grass beds.

Nature Conservation Bureau 1998. Data book of the 5th National Survey on the Natural Environment.

Tokyo Metropolitan Fisheries Experiment Station, 1996. Report of the survey on the fishing ground damaged by the 1986 eruption of Izu-Ohshima Island. Report of the Tokyo Metropolitan Fisheries Experiment Station 209: 1–56.

Yamada N (1967) Studies on the manuring for *Gelidium*. Bulletin of Izu branch of shizuoka prefectural fisheries experiment station 32: 1–96.

Yanase R, Sasaki T, Nonaka M (1982) An assessment on the *Gelidium* yield forecasting off Izu Peninsula. Bulletin of Shizuoka Prefectural Fisheries Experiment Station 16: 43–50.

Journal of Applied Phycology (2006) 18: 599–610
DOI: 10.1007/s10811-006-9064-4

Overgrazing of kelp beds along the coast of Norway

Knut Sivertsen
Finnmark University College, N-9509 Alta, Norway

e-mail: knuts@hifm.no

Key words: barren grounds, grazing, kelp beds, life history, population dynamics, sea urchins, *Strongylocentrotus droebachiensis*

Abstract

The aim of this study was to better understand the down-grazing of kelp beds by sea urchins (*Strongylocentrotus droebachiensis*) along the coast of Norway. Barren grounds were first observed in sheltered areas along the coast of the counties of Trøndelag, Nordland and Troms in 1974. In the 1980s, the barren grounds spread to areas more heavily exposed to waves. In the 1990s, the kelp beds were re-established in some localities in southern Trøndelag, initially in wave-exposed areas. In the northernmost parts of Norway, i.e. the counties of Troms and Finnmark, the barren ground areas may still be increasing. Crabs (*Cancer pagurus*) and common eiders are the most common predators on urchins. Predation on sea urchins in kelp beds is probably not among the factors that limit the sea urchin populations. Along the coast of Nordland and further north, sea urchins are infected by nematodes, resulting in a low, but significant increase in their mortality. No re-growth of kelp beds has been found in the most infected areas. In the late 1960s and the early 1970s, a high occurrence of echinoderm larvae was observed in deeper waters. This was a period with cold water, which may have caused high recruitment of sea urchins. The bet-hedging life strategy of sea urchins may account for the sudden increase in the size of the populations. In the present paper I propose the hypothesis that higher individual growth rates and higher mortality rates in the south than in the north may explain the decrease in the populations, which may in turn account for the re-growth of kelp in the southern areas.

Introduction

Sea urchin populations are known to increase to such a high level that they overgraze the kelp beds in many temperate areas around the world, with barren grounds as a result. Several theories and critical reviews have been presented to explain why the densities of sea urchins have increased so dramatically (e.g. Lawrence, 1975; Dayton, 1985; Elner & Vadas, 1990; Steneck et al., 2002). One frequently proposed explanation is low predation on sea urchins. High recruitment of sea urchin larvae in combination with favourable hydrographic conditions is another hypothesis (Forman, 1977; Ebert, 1983; Hart & Scheibling, 1988; Wing et al., 1995). Parasites in the sea urchins may also cause large oscillations in the size of the populations (Scheibling & Hennigar, 1997).

Re-growth of kelp beds in barren grounds dominated by sea urchins has been observed in several areas around the world, e.g. California and Nova Scotia (Hawkins & Hartnoll, 1983; Scheibling & Hennigar, 1997, respectively). Will variation in the factors that increase sea urchin populations (e.g. high recruitment, low predation, or low parasite infection) also cause a reduction in the populations? Will other factors such as abnormal weather events and changes in hydrographical conditions affect the populations (Parsons & Lear, 2001; Steneck et al., 2002), or may variation in the parameters related to population dynamics explain the oscillations (Ebert, 1985)?

The cycles from kelp beds to barren grounds and back again to kelp beds may be explained with reference to the life history of the sea urchins, which is characterised by bet-hedging strategies (Ebert, 1982, 1985).

The bet-hedging theory postulates that an extended reproductive life-span, a high rate of adult survival, and low annual reproductive effort are adaptations to compensate for the low and highly variable survival rate of first-year juveniles (Stearns, 1976, Roff, 1992).

The proliferation of barren grounds along the coast of Norway, from Nordmøre and further north to the Russian border, has been reported since the early 1980s (Skadsheim et al., 1995; Sivertsen, 1997a). Barren grounds have previously been observed in the areas around Stavanger by von Düben (1847) and around Tromsø by Döderlein (1900). Re-growth of kelp beds in barren ground areas, observed from the late 1980s, have been recorded in some overgrazed areas in the south, i.e. in Nordmøre, Trøndelag and southern Nordland (Hagen, 1995; Skadsheim et al., 1995; Christie et al., 1995; Sivertsen, 1997a).

The aim of this paper is to get a better understanding of the down-grazing and re-growth of kelp beds by the sea urchin *Strongylocentrotus droebachiensis* (O. F. Müller), in view of both its regional and local distribution on the coast of Norway. The focus will be on predation, parasitism, recruitment, population dynamics and the life history of the sea urchins.

Materials and methods

The area investigated stretches from west of Lindesnes (58°N, the southernmost point of Norway) to the North Cape (71°N, the northernmost point of Norway) and eastwards to the Russian border. This coastline consists of large fjords and an archipelago of hundreds of thousands of large and small islands, some of them at a distance of more than 50 km from the mainland. The archipelago is here divided into three zones. The outer archipelago consists of areas where the swells frequently reach the shores. In the inner archipelago the sea water is influenced by fjord water. In between these zones is the middle archipelago, which is affected by swells and fjord waters to a lesser extent or not at all. This division is suitable for describing the pattern of barren grounds and of sea urchin density and size frequency (Sivertsen, 1997a; 2003). The Norwegian Coastal Current runs northwards along the entire coast.

Laminaria hyperborea (Gunn.) Foslie. dominates the kelp beds in wave-exposed areas, while *L. saccharina* (L.) Lamour. is most common in sheltered areas. *L. saccharina*, with its prostrate stipes and lamina may be easier for sea urchins to graze than *L. hyperborea*, with its stiff erect stipes. During the over-grazing process the kelp and the undergrowth are first grazed. In wave-exposed areas juvenile *Laminaria* sp. were first grazed down, which inhibited their re-growth and left only remainders of canopy individuals, and gradually the canopy kelp disappeared (Sivertsen, 1997a).

An Echinoderm larvae index (EI) was used to estimate the occurrence of Echinoderm larvae, based on unpublished zooplankton samples from 1969–1983 from The Institute of Marine Research (IMR) in Bergen. Plankton hauls were taken in six localities. Three southern localities were at 59°N, 61°N and 63°N respectively, and three northern localities at 68°N, 68°N and 71°N respectively (Figure 1). Two plankton net hauls were performed about twice per month, a shallow one from 50 m depth to the surface and a deep one from 300 m to the surface in each locality. A short-cut method was used to identify dominating plankton species in the samples. The names and stages of the first 100 individuals in sub-samples were identified microscopically, and then the number (N) of Echinoderm larvae out of 100 zooplankton individuals was counted. To estimate the number of Echinoderm larvae in a whole sample, the volume of the sub-sample, or the number and the volume of each identified species, is required (Hjort & Ruud, 1927; Wiborg, 1962). These estimates were not made here. Instead an EI (Echinoderm Index) was made. The Echinoderm larvae were not identified to each species, which brings an element of uncertainty into the data. The net volume (when large individuals such as medusae and large euphausiides were removed) in mL (V) of each sample was measured. The EI used was $EI = N*V$. The average EI from samples taken in March, April and May (zero indices included) was used. Deep and shallow hauls were separated. At the same time temperature and salinity were measured close to the localities where the zooplankton samples were taken. The hydrographic data are stored, and mean values of temperature for March, April and May have been estimated by The Norwegian Oceanographic Datacenter at IMR.

Many species prey on sea urchins. Hooper (1980) lists e.g. *Tealia*, other anemones, small sea stars, *Solaster*, *Leptasterias*, crabs (*Hyas*, *Canser*), lobsters, cod, flounders, wolf-fish and sea-birds as predators on sea urchins in the Newfoundland waters. These groups also occur on the coast of Norway. Predation on sea urchins was estimated from counts of sea birds and landings of fished stocks. Between 1983 and 1986 the sea-bird abundance was estimated in Trøndelag and Helgeland, 63°N–66°N (Follestad et al., 1986), both of which are large areas dominated by barren grounds

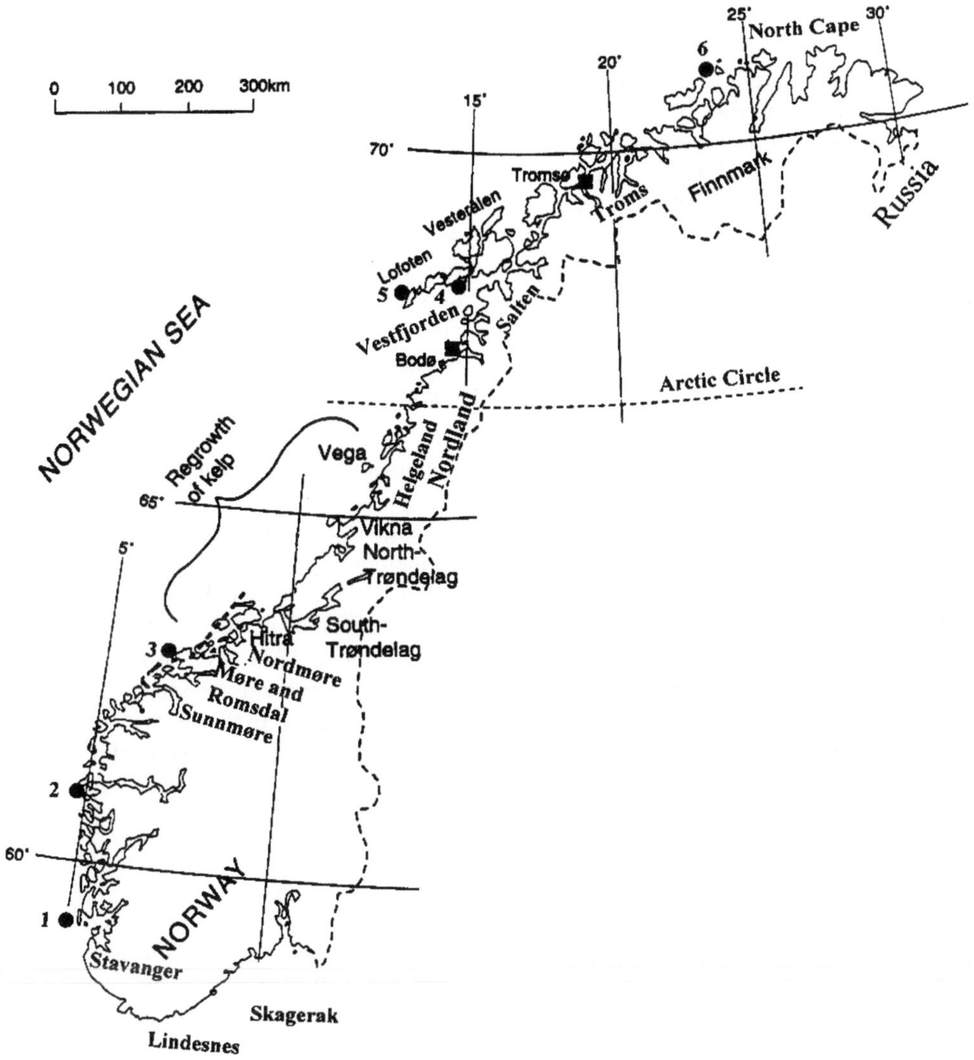

Figure 1. Map of the coast of Norway with reference to regions and localities. Broken line outside Nordmøre indicates southern border of barren grounds.

(Sivertsen, 1997a). These counts are of sea-bird numbers 10–20 years ago. Estimates of fishable stocks have not been made for this area. Landing statistics were therefore examined, assuming that the landings reflect variations in the size of the stocks (Fishery statistics, 1961–1973).

The consumption of sea urchins has been estimated in relation to the size of the predator populations, the percentage of sea urchins in their diet, the number of days per year of feeding on sea urchins, and also the amount of food needed by the individual species assumed to prey most on sea urchins in the Trøndelag and Helgeland areas (presumably eight species altogether). Wolf-fish, plaice, lobsters and crabs are assumed to be the most important fishable stocks consuming sea urchins. To estimate the size of fishable stocks, the mean landings in the areas in Trøndelag and Helgeland for the years 1961–74 were used (Fishery statistics, 1961–1973). The landings for plaice, wolf-fish, lobsters and crabs are multiplied by five, assuming a mortality of 0.2 for all of them. There are, however, obvious problems in using fish statistics for this purpose, since fishing gear, fishing intensity, prices, and management may influence landings as well as stock abundance. Plaice and wolf-fish feed in shallow areas half the year, and in this period 10% of their consumption is assumed to consist of sea urchins. Miller (1985) assumes that lobsters consume an annual amount of sea urchins that corresponds to their own weight, and crabs one and a half times their own weight. These estimates are also applied here. It is assumed that 20–25% of the diet of common eiders and king eiders are sea urchins (Bustnes & Lønne, 1996) and that they feed 250 and 150 days a year, respectively (Mehlum & Gabrielsen, 1995). Sea urchins also constitute 5% of the food for gulls, and gulls feed every day throughout the year (Karl Birger Strand, personal communication). The physiological estimates for need of food for each species are shown in Sivertsen (1997b).

The pattern of barren grounds along the coast of Norway is based on the surveys in the 1980s and the 1990s by Sivertsen (1997a) and Skadsheim et al. (1995). While making these surveys in the early 1980s, we also interviewed fishermen, asking them when they had first observed barren grounds. In addition, we also investigated the occurrence of kelp beds and barren grounds at Hitra in South-Trøndelag (63°33′N) during summer time every second or third year since 1980. Bottom areas from low water tide to 5–7 m depth were investigated by using a small boat, going along the shoreline. Dominant kelp species, and densities and size distribution of sea urchins were recorded using a sea telescope (a tube with a transparent clear glass in the lower end). Sea urchin occurrence and the borders between kelp beds and barren grounds were registered.

Results

Distribution of barren grounds

From about 1974, barren grounds were seen by fishermen at Hitra (63°30′N), Vikna, Vega, Bodøand Troms (70°N). Surveys in 1980 and later showed that barren ground areas were dominated by large densities of sea urchins, *S. droebachiensis*, but *Echinus esculentus* Linné occurred frequently during the overgrazing process (Sivertsen, 1997a). Stretches of barren grounds now occur from Nordmøre (63°N) and further north along the entire coast to the Russian border (Figure 1). Along the entire stretch of this coastline, barren grounds occurred more frequently in the middle and inner archipelago than in the outer archipelago and were rare or absent in the fjords. In general, barren grounds occurred in sheltered and moderately exposed areas, but not in areas heavily exposed to waves (Sivertsen, 1997a, 2003).

Predation

Common eiders and crabs are estimated to be the dominant predators on sea urchins along the coast of Trøndelag and Helgeland (Table 1). A total of 43,200,000 kg were consumed in the 2400 km^2 area, corresponding to 19 g m^{-2} y^{-1} or one individual of 46 mm in diameter m^{-2} y^{-1}. Wolf-fish, plaice, herring gulls, great black-backed gulls and common eiders are common all along the coast. Crabs and lobster are mainly found from Helgeland (66°N) and southwards, while king eiders are most common in the north.

Echinoderm larva recruitment

EI ranged between 0.5 and 21, reaching its maximum at temperatures below 4 °C, while at higher surface temperatures EI exceeded 3.5 only twice (Figure 2). High EIs were found in deep waters in the northern area from 1969 to 1974. This was in the period prior to the first observations of sea urchins dominating barren ground areas. In the south (59°N–63°N), where no barren grounds were found, EI did not exceed 6 in the period before 1974, neither in samples from deep nor shallow water.

Re-growth of kelp beds

Re-growth of kelp in previously barren ground areas dominated by sea urchin has been observed in the southern part of the overgrazed areas in Nordmøre,

Table 1. Predation on sea urchins on the coast of Trøndelag and Helgeland, an area of 2 400 km^2. The consumption is estimated from the mean value of catches of fish and Crustacean populations from 1961–1973. Birds counted in 1983–1986.

Species	Consumption of sea urchins (1000 kg y^{-1})
Common eiders *(Somateria mollissima)*	23 500
Crab *(Cancer pagurus)*	11700
Herring gulls *(Larus argentatus)*	2 650
Blackbacked gulls *(L. marinus)*	2 250
Plaice *(Pleuronestes platessa)*	1 215
King eiders *(S. spectibilis)*	940
Wolf-fish *(Anarchichas lupus)*	820
Lobster *(Homarus vulgaris)*	125
Sum	43 200

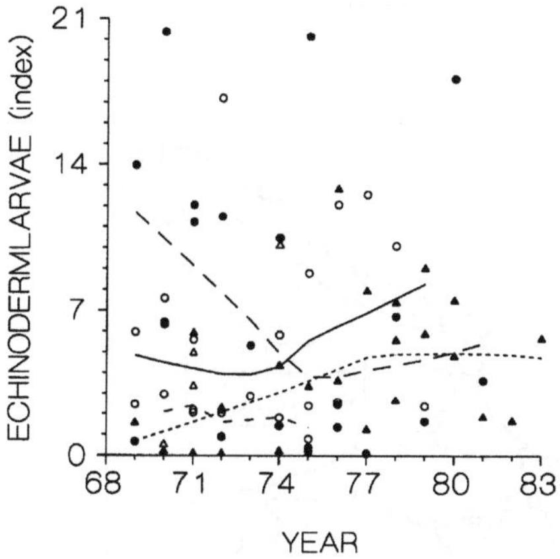

Figure 2. The Echinoderm larvae index (EI) is shown as a function of year of sampling (see text). Samples were taken by vertical plankton hauls in three locations (see Figure 1) at Vestlandet (1, 2, 3) and three locations (4, 5, 6) in northern Norway north of the Arctic Circle. The samples were pooled and divided into four groups to present EI in: 1) North-Norway, depth 50–0 m (empty circles and solid line). 2) North-Norway, depth 300-0 m (filled circles and large broken line). 3) Western Norway, depth 50–0 m (empty triangles and small heavy broken line). 4) Western Norway, depth 300-0 m (filled triangles and very small broken line).

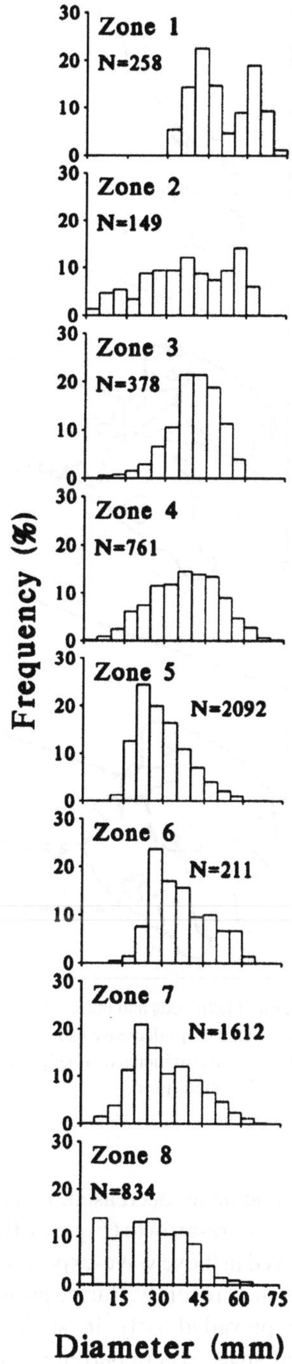

Figure 3. The size distribution of sea urchins (*S. droebachiensis*) on barren grounds along the coast of Norway. Zones refer to areas 1: Møre and Romsdal, 2: South Trøndelag, 3: North Trøndelag, 4: Helgeland, 5: Salten, 6: Lofoten and Vesterålen, 7: South Troms, and 8: North Troms. (From Sivertsen 1997).

Trøndelag, and occasionally in the southern part of Nordland (Skadsheim, 1995; Stien et al., 1995; Hagen, 1995; Sivertsen, 1997a), but not further north (Sivertsen 1997a, 2003). Investigations in the early 1980s showed that higher densities of *S. droebachiensis* were found more frequently in barren ground areas south of the Arctic Circle (52.2 ± 6.9 individuals m^{-2}) than north of it (26.1 ± 2.5 individuals m^{-2}) (Sivertsen, 1997a). New investigations in 1992 showed a decrease in densities to about 20 individuals m^{-2} in the area south of the Arctic Circle (Skadsheim et al., 1995) to even lower densities than in the north. No changes in densities from the early 1980s to 1992 were observed in the north (Skadsheim et al., 1995; Sivertsen, 1997a). The size distribution of sea urchins in barren grounds, pooled from eight or more localities from eight geographical zones in the south-north gradient, showed that large sea urchins dominated in the south (Figure 3). The mean size gradually decreased northwards, where individuals of 20–30 mm in diameter dominated in the northern-most area.

In an area north of Hitra (63°33′N) in Trøndelag, which was investigated every second or third year since 1980, the extent of barren grounds increased through the 1980s, reaching a maximum around 1987 (Figures 4A). The barren grounds expanded gradually outwards to more wave-exposed localities. After 1987

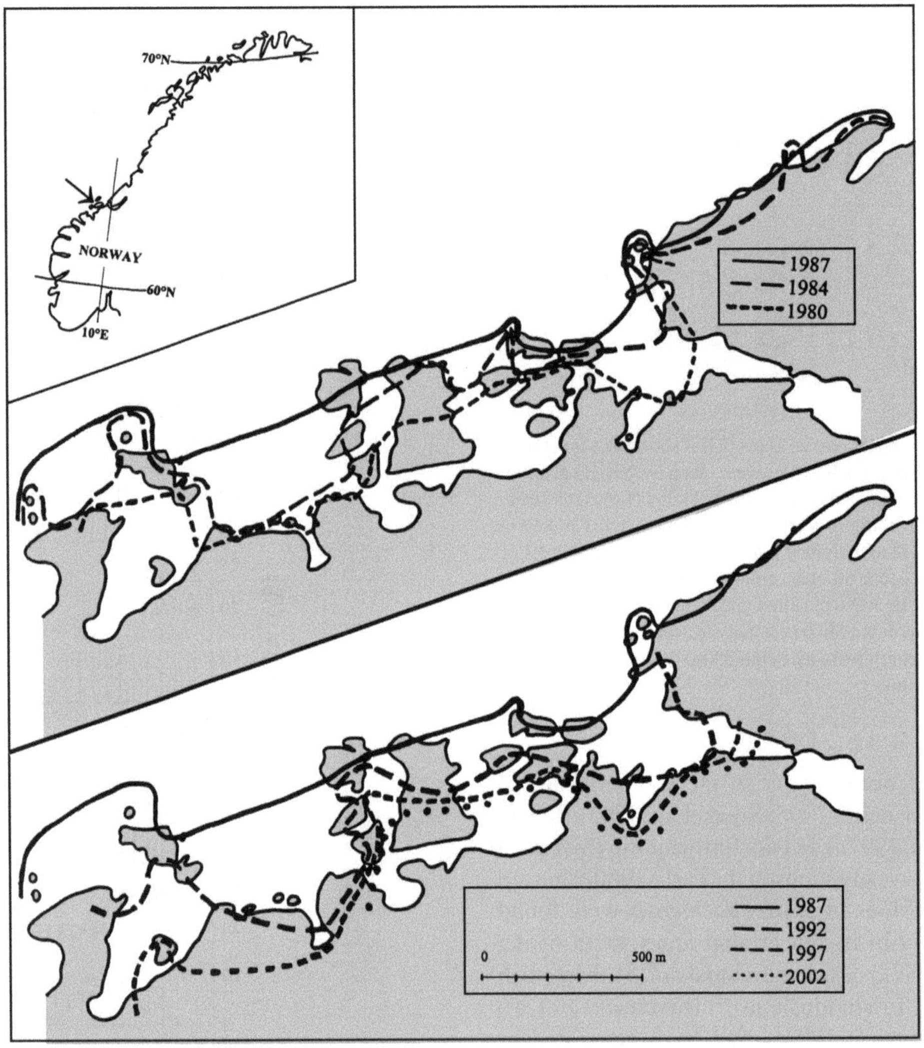

Figure 4. The extent of kelp beds and barren grounds in an area at Hitra, South-Trøndelag. Lines show the border between kelp beds and barren grounds in different years. Top: decrease of kelp beds, and increase of barren grounds from 1980 (small broken lines) 1984 (large broken lines) and 1987 (solid lines). Bottom: increase of kelp beds and decrease of barren grounds from 1987 solid line), 1992 (large broken line), 1997 (small broken line) and 2002 (dotted lines).

the barren ground areas decreased, and kelp beds underwent a period of re-growth (Figures 4B). Re-growth was first observed in those wave-exposed areas that had become barren most recently. Barren grounds gradually retreated and prevailed only in sheltered areas. In some sheltered areas, where barren grounds were observed in the 1980s, areas of them still remain. One locality that was found barren in 1974 is still barren 30 years later. In localities with re-growth of kelp, *Saccorhoza polyschides* (Lightf.) Batt. often dominates together with *L. hyperborea*, but also *L. saccharina*, *Desmarestia aculeata* (L.) Lamour. *Halidrys siliquosa* (L.) Lyngb. and annual brown algae frequently occur.

Discussion

The fluctuations in densities and size distribution in sea urchin populations along the coast of Norway may be explained with reference to population dynamics. My hypothesis is that the bet-hedging strategy is a characteristic of sea urchin populations along the coast of Norway, and that predation and parasitism may have only minor influences.

Predation

An estimate of predation on sea urchins was made in order to establish if low predation could account for the increase in sea urchin populations (Table 1). The eight species assumed to prey on sea urchins most effectively in the area Trøndelag and Helgeland (2400 km^2), have a mean consumption of 43 200 000 kg y^{-1} or 19 g m^{-2} y^{-1}. A production of 19 g m^{-2} y^{-1} is needed to compensate for this mortality. The mean annual mortality rate (Z) of *S. droebachiensis* in kelp beds was estimated as 0.47, which corresponds to a 40 % mortality of standing stock of sea urchin per year (Sivertsen & Hopkins 1995). If $Z = P/B$, where $P =$ productivity of *S. droebachiensis* and B=mean biomass of *S. droebachiensis* (Pitcher & Hart, 1982), and assuming that sea urchin mortality is caused exclusively by predation, the above equation gives $B = 41$ g m^{-2}. Biomasses were estimated from individual densities and mean individual weight from localities described in Sivertsen (1997a). The mean sea urchin biomass was 400 g m^{-2} in the area north of Bodø where there was no sign of overgrazing. During the overgrazing process, the mean sea urchin density was about 1500 g m^{-2} both south and north of Bodø (Sivertsen, 1997b). These localities did not show any sign of overgrazing. The recorded biomass of sea urchins was 10–40 times larger than what is needed to compensate for the predation. From these estimates it may be concluded that the predation on sea urchin populations is too low to support claims that these populations are predator-limited, at least in the last years prior to the appearance of barren grounds. It is possible that predation affects only sparse populations of sea urchins. There is no indication that there are any keystone predators on the urchin populations along the coast of Norway, as is proposed in Nova Scotia (review in Elner & Vadas, 1990). However, Jackson et al. (2001) have used historical data to show that in the West Atlantic there have been time-lags of decades or even centuries between the onset of over-fishing and consequent changes in the ecological communities.

It has been proposed that extensive fishing of cod (*Gadus morhua* L.) and other predatory fish in spawning aggregations in Maine and Nova Scotia in the Northwest Atlantic in the 1930s and later was accompanied by a rapid decline in the numbers and the body size of coastal cod in the Gulf of Main, and this coincided with their extirpation from the coastal zones (Steneck et al., 2002). Dominant fish predators in the coastal zone have been replaced by small, commercially less-important species, and large predatory finfish have remained functionally absent from the coastal regions of the western North Atlantic (Steneck et al., 2002). The extirpation of coastal cod and other fishes by the 1940s in the Gulf of Maine has resulted in the functional loss of apex predators, which has fundamentally altered the coastal food web, as lobster, crabs and sea urchins have become more abundant (Steneck et al., 2002). This change was observed in the 1960s in the Gulf of Main and a decade later in Nova Scotia (Steneck et al., 2002). They conclude that after the loss of apex predators the high density of sea urchins caused the denudation of the coastal zone.

In a large research program on the sea-ranching of cod on the coast of Norway (PUSH), during the 1990s, the diet of cod was investigated (Svåsand et al., 1998). Three of the regions investigated were in North Norway where large areas were overgrazed and dominated by sea urchins. Cod at the age of two years or younger, and smaller than 30 cm, mainly fed on benthic invertebrates. According to Svåsand et al. (1998), Echinoderms and Mollusca are not important as food for cod. As cod grow, first krill and later fish become a more important part of their diet. The Norwegian Coastal cod is managed separately from the Norwegian Arctic cod. The quota for harvest is 40,000,000 kg a year, and 30–40,000,000 kg have been harvested yearly since the 1950s. Cod has probably little or no direct influence on the occurrence of sea urchins on the coast of Norway.

Recruitment

The first observations of barren grounds were made in several localities along the coast of Norway at about the same time, in 1974. High densities of sea urchins may be caused by high recruitment of sea urchin larvae that have spread with the Norwegian Coastal Current. A high occurrence of Echinoderm larvae was found in North Norway in the late 1960s and early 1970s. These larvae were not identified to species. However, *S. droebachiensis* spawns in March, and its larvae have a pelagic stage for 6–8 weeks (Emlet et al., 1987). Ophiuroids spawn in late April, and the sea cucumber *Cucumaria frondosa* spawns at the end of March, and the larvae settle in May in the Tromsø area (Falk-Petersen, 1982). No large increases in these Echinoderm species (other than sea urchins) have been reported on the coast of Norway. However, it cannot be definitely concluded that the observed larvae were sea urchin larvae.

If environmental factors affect recruitment, growth and mortality of Echinoderm larvae, then it might be possible to explain temporal variations in recruitment in the benthic populations in terms of large-scale oceanographic events such as temperature anomalies (Hart & Scheibling, 1988). Ebert (1983) has reviewed the episodic nature of annual recruitment of echinoids, and considers the most important factors to be upwelling, the abundance of planktonic and benthic predators, and transport by water currents. He also reviews evidence for correlations between temperature and recruitment anomalies. His results indicate that strong high-temperature abnormalities correlate with high recruitment for some sea urchins, but that recruitment of *S. purpuratus* off southern California may be inversely related to temperature: in this area recruitment is most abundant after the coldest winters. Forman (1977) observed unusually large populations of *S. droebachiensis* in locations in the Strait of Georgia, British Columbia. These populations consisted mainly of a single cohort that settled in 1969. He suggests that the spring temperature normally is marginal for the development of *S. droebachiensis* larvae, and that 10 °C is an upper limit for larval development in that area. He concludes that record low spring temperatures in 1969 probably led to intensive recruitment that year. Wing et al. (1995) monitored settlement of sea urchins (*Strongylocentrotus* spp.) and crabs (*Cancer* spp.) and concurrent physical variables in northern California. Winds favourable to upwelling led to lower temperatures, higher salinities, and lower subsurface pressure, while periods of relaxation from upwelling typically caused a lagged reversal of each of these trends. Sea urchins settled primarily during an event of unusual relaxation which possibly involved remote physical forcing. Sea urchin and crab settlement were negatively correlated. Barren grounds were first reported from St. Margaret's Bay, Nova Scotia, in 1968, but the overgrazing probably began a few years earlier (Breen & Mann, 1976a,b). Hart and Scheibling (1988) investigated deviations from long-term monthly mean spring temperature in Halifax Harbor, Nova Scotia, Canada, in the period 1952 to 1986 and found a single large positive deviation of 3.5 °C higher than normal in June 1960. This was eight years before the first report of destructive grazing, and Hart and Scheibling (1988) propose that high urchin recruitment in 1960 as a result of high temperature caused the subsequent high population densities of sea urchins.

Braarud and Nygaard (1980) studied the phytoplankton communities in coastal waters off the coast from Møre and Romsdal to Vesterålen in the years 1968–1971. In the Vestfjord – Vesterålen area they found a mixed spring diatom community during the four-year period. Here they found remains of the same plankton association that they found on the coast of Møre and Romsdal and Trøndelag one month earlier, suggesting that Helgeland is a transition area for the two regions. This shows that offshore transport of plankton takes place, and in some years this transport is sufficiently strong to be identified far from its source. Braarud and Nygaard (1980) suggest that these "offshore coastal areas" may become far more extensively influenced by such long-distance transport than the shore regions from Møre to Vesterålen. Variability in the meteorological and hydrographical situations may lead to different effects from year to year. This transport shows that plankton communities can spread over long distances with the Norwegian Coastal Current.

Fishermen were interviewed randomly as we met them during our research, but the interviews were not performed in a systematic way. Doubt about larval species and the haphazard way in which the interviews on observations of barren grounds were made obviously weaken the claim that barren grounds appeared simultaneously along the coast of Norway. Therefore it seems reasonable to question the importance of these two factors for the increase in the size of sea urchin populations causing the overgrazing of kelp beds on the coast of Norway.

In 1970 the temperature both at the surface and in deep areas along the Coast of Norway was 1–2 °C lower than the normal annual mean value, and the Atlantic water had a cold period from 1965 to 1970, followed by a warmer period (Asplin & Dahl, 2003). The North Atlantic Oscillation index (NAO) is used to explain large-scale oscillations of many important species in the North Atlantic (Parsons & Lear, 2001). High NAO indices benefit some species while low indices benefit others. The NAO index was at an extreme low in the 1960s; it changed to positive after 1970 and reached its positive extreme during the late 1980s (Parsons & Lear, 2001). It may be possible to explain temporal variations in recruitment in terms of abnormal oceanographic events, as proposed by Hart and Scheibling (1988), or of long-term oceanographic events such as temperature variation, as explained by the NAO index. Larvae in areas along the northern coast can be spread over long distances by the Norwegian Coastal Current. High larval recruitment along the coast of Norway may have occurred in periods of cold water in the late

1960s. Fast individual growth of sea urchins to adult size may have occurred in the warm period in the early 1970s. From 1974 on, this may have led to a rapid increase in the sea urchin population, which became high enough to overgraze the kelp beds, resulting in barren grounds.

Re-growth of kelp beds

The patterns of density, the size distribution of sea urchins, and the re-growth of kelp in previously barren ground areas may yield important knowledge about the sea urchin populations and the dynamics between sea urchins and kelp beds. The densities and size distribution of the populations vary along the coast. North of 67°N small individuals were more abundant than larger ones, and the densities were relatively constant over time. This may explain the stability in populations with good recruitment. South of the Arctic Circle large individuals were more abundant than smaller ones, and the densities of urchins decreased significantly since the early 1980s (Sivertsen, 1997a) to 1992 (Skadsheim et al., 1995). This may indicate unstable populations with low recruitment, and the sea urchin populations may die out as the individuals become old and die. These results may explain differences in the population dynamics of sea urchins from the south to the north. Differences in mean annual temperature may explain these changes. At 63°N, the mean annual water temperature is 2.5 °C higher than at 71°N. As a consequence of physiological processes in sea urchins, the higher temperature in the south may account for the higher somatic growth rates and higher mortality rates of the sea urchins in the south than in the north (e.s. Pauly, 1982; Roff, 1992). Thus the duration of the life cycle for the sea urchin may be shorter in the south than in the north. Stability in the sea urchin populations in the southern areas may not be attained until the first few cohorts have gone through their life cycle and died. When the old individuals die, the densities may decrease to a level at which the re-growth of kelp will start.

Growth rates of sea urchins have been measured only in the areas from Bodø to Tromsø (Sivertsen & Hopkins, 1995), a distance which is too short to predict anything about the variation along the coast as a whole. Therefore growth parameters for cod and prawn (*Pandalus borealis* Krøyer) are here used for comparison. Growth rates of cod have been investigated in Vestlandet (60°N) and Troms (70°N) in the PUSH program (Svåsand, 1998). Mean individual length at the same age were about 30% shorter in Troms than in Vestlandet. Rasmussen (1967) found that female prawns normally spawn at the age of 2.5 y in Skagerak when the temperature is 6°C. In the period 1963 to 1966 the bottom temperature in Skagerak decreased by two degrees, and the females did not spawn until the age of 3.5 to 4.5 y. Hopkins and Nilssen (1990) found prawns spawning at the age of 4 at 4–5°C and at the age of five at 2°C in Troms. The mean carapace length increased 1–1.5 mM in the spawning stock in warm waters. If sea urchins follow the same patterns of growth rates and population dynamics parameters as cod and prawn, it would indicate that they grow faster and develop to maturity at a younger age in Nordmøre and Trøndelag than in areas further north.

The local pattern of overgrazing and subsequent regrowth of kelp beds shows that barren grounds first appeared in sheltered areas and gradually extended to moderately wave-exposed areas. This may be explained in view of the following factors: sea urchin recruitment, kelp bed productivity, wave activity, and adaptation to the areas most preferred by the sea urchins. Sea urchins are probably best adapted to sheltered areas. Here they are less disturbed by wave activity. Here also the bottom substrate is probably most suitable for juveniles to settle in and survive. Kelp beds have the lowest productivity in sheltered areas, increasing gradually to the highest level in heavily wave-exposed areas. *L. saccharina* usually dominates the kelp beds in sheltered areas. This kelp is prostrate and therefore more readily grazed by the sea urchins than *L. hyperborea.* Sea urchins may be abundantly recruited by one or a few cohorts both in sheltered and wave-exposed areas. As the sea urchins become large in test size, they may overgraze the kelp beds, but as kelp beds are more productive, or the sea urchins more disturbed by increased exposure to waves, the sea urchins need more time to graze down the kelp beds. Possibly, they graze on the juvenile kelp, thereby inhibiting the recruitment of kelp (Sivertsen, 1997a). In localities with heavy wave activity, fewer sea urchins may survive to maintain a high density long enough to graze down the kelp beds. Juvenile sea urchins live cryptically. They may find shelter in the kelp haptera or hide in crevices; they are rarely found in such places, though. In barren grounds the small individuals are very rare. or, usually, not found at all. It is more probable that juveniles settle and survive on substrata of gravel, dead shells or loose coralline algae. As they become 15–20 mM in diameter they move upwards to rocky and stony bottoms and join the adult populations (Sivertsen & Hopkins, 1995). As kelp beds disappear in a locality, the effects of the surge increase at the bottom. In moderately wave-exposed areas the juveniles may have difficulties in recruiting on the barren grounds. The recruitment in these areas may be too low to sustain the sea urchin populations, and then the grazing pressure can decrease to a level where re-growth can start, while in sheltered localities the recruitment is sufficient to maintain a stable population.

Parasites

Hagen (1987) and Stien et al. (1995) propose the hypothesis that infection by the nematode *Echinomermella matsi* in sea urchins increases the mortality to a level where re-growth of kelp beds can start. However, no support has been found for this hypothesis to date (Hagen, 1987; Christie et al., 1995; Sivertsen, 1996). Re-growth of kelp was found in the Trøndelag areas where nematodes were *not* found. Low sea urchin densities were found in the Bodøareas, where the occurrence of nematodes was highest, the prevalence ranging from 40 % to 88 %, but here there were no observations of re-growth of kelp (Hagen, 1987; Sivertsen, 1996). A sudden decrease to a fourth in the density of sea urchins in less than 6 months was found in an area in Helgeland, where re-growth of kelp was observed. (Stien et al., 1995). About 60% of the sea urchins were infected by nematodes in both samples, and the occurrence of nematode-infected individuals tended to increase rather than decrease. Here, there was no evidence that the nematodes were a factor causing a decrease in the sea urchin population; the prevalence of nematodes was unchanged after the decrease in sea the urchin density (Stien et al., 1995).

Population dynamics

The cycles from kelp beds to barren grounds and back again to kelp beds may be explained with reference to the population dynamics and the life history of the sea urchins, which is characterised by bet-hedging strategies (Ebert, 1982, 1985). The annual growth rate of a population, as presented by Schaffer (1974) and Schaffer and Gadgil (1975), is:

$$= cB + p \qquad (1)$$

where B is the total number of new-born individuals, c is the juvenile survival rate to the age of first reproduction, and p is the adult survival rate. Growth and mortality factors change with temperature variation. As temperature increases, growth and mortality rates also increase (Roff, 1992). Increased mortality leads to shorter life length, corresponding to a decrease of p in (1). To compensate for a low p, then c or B, or both, may increase to keep the size of the population constant.

Ebert (1982) proposes that sea urchins may have a bet-hedging life history. This postulates that an extended reproductive life-span, a high rate of adult survival, and low annual reproductive effort are an adaptation to the low and highly variable survival rates of first-year juveniles (Stearns, 1976, Roff, 1992). According to the bet-hedging theory, longevity is a response that reflects variations in juvenile survival (Murphy, 1968; Schaffer, 1974; Ebert, 1982). If c is very small, then one way of having 1 is for p to be large. Large p means long life. Very high but constant juvenile mortality may act as a constraint on longevity in this, in different species, this may reflect the presence of the barrier at different positions in the species' adapted space. These are the design constraints (Schaffer, 1974; Stearns, 1977; Ebert, 1982). Long-lived species, operating under the rule of bet-hedging populations, will usually be declining (births $<$ deaths), but there will be occasional recruitment episodes that increase stability in the population (Ebert, 1982).

The bet-hedging life history model may explain the cycles in the size of sea urchin populations found on the coast of Norway. The low predation found on sea urchins (Table 1) indicates that if the populations are not predator-controlled, there may be other factors that limit the increase in the size of the population. An event that triggers a very successful recruitment may lead to a sharp increase in the juvenile sea urchin population, which in turn will lead to a large increase in the adult population. This is what may have occurred in the late 1960s to early 1970s along the coast of Norway. High occurrence of Echinoderm larvae (although whether it was sea urchin larvae is unproved) was found in the coastal waters from 1969 to 1973, i.e. some years prior to the first observations of the emergence of barren grounds along the coast. This may be an indication (albeit weakly supported in our material) that the sea urchin populations had increased, with barren grounds as a result.

Re-growth of kelp beds has occurred in the southern parts of the overgrazed areas since the late 1980s. In the south we may find shorter life length, as a consequence of warmer water, and the short life expectancy may prevent the size of the population from being sustained at a high level. In Trøndelag where a decrease in the sea urchin populations was observed in some localities, the sea urchins compensate with higher recruitment or lower mortality only in those biotopes that they are best adapted to. In sheltered areas they still appear, while they have disappeared in marginal biotopes, such as wave-exposed areas.

The mean predation pressure on the sea urchins over the last 13 years prior to the appearance of barren ground was found to be 10–40 times lower than the mean live weight of sea urchin found in kelp beds and therefore too low to hold down the sea urchin populations. In some localities high parasite infection has been found to decrease the density and the mean size of the sea urchins significantly, but the decrease in the grazing pressure has not encouraged re-growth of kelp. Re-growth is usually found in areas where parasites have not been observed. Growth and mortality rates in a population are generally higher in warm than in cold environments. Variation in the water temperatures may consequently be a plausible explanation of the natural regulation of sea urchin populations; the decrease in the populations and the re-growth of kelp in Nordmøre, Trøndelag and Helgeland, which are the southernmost overgrazed areas, may be a result of higher water temperatures.

A sudden occurrence of high recruitment may be the most important factor for increase in the sea urchin populations. My hypothesis is that a sudden high recruitment in a population, followed by reduced recruitment with reduction of the population, is a result of the bet-hedging life strategy of sea urchins. This may explain the cyclical variation between kelp beds and barren grounds. The variations in population parameters discussed here are of a general character. The parameters should be estimated for each sea urchin species, in a temperature gradient that covers the whole range of distribution of the species in question.

Kelp beds have a primary productivity similar to cultivated areas ashore. The kelp production contributes to the benthic food chain in shallow areas. The fact that half of the kelp beds investigated in North Norway were grazed down indicates that in this area the food chain capacity has been reduced dramatically. Kelp beds are also important as habitats for many invertebrates and as nursery areas for fish. To test the hypothesis of bet-hedging we need studies on the recruitment of sea urchin larvae and on population parameters such as growth and mortality, covering the whole range of distribution of the species.

References

Asplin L,Dahl E (2003) Havets miljø2003, Fisken og havet. Særnummer 2: 2003. 122pp

Braarud T, Nygaard I (1980) Phytoplankton observations in offshore Norwegian coastal waters between 62°N and 69°N. 2. Diatom societies from Møre to Vesterålen, March – April 1968–1971. Sarsia 65: 93–114.

Breen PA, Mann KH (1976a) Changing lobster abundance and the destruction of kelp beds by sea urchins. Mar. Biol. 34: 137–142.

Breen PA, Mann KH (1976b) Destructive grazing of kelp by sea urchins in eastern Canada. J. Fish. Res. Bd. Canada 33: 1278–1283.

Bustnes JO, Lønne OJ (1996) Sea ducks as predators on sea urchins in a northern kelp forest. In Skjoldal HR, Hopkins CCE, Erikstad KE, Leinaas HP (eds) Ecology in fjords and coastal waters, Elsevier, Amsterdam, pp 599–608.

Christie H, Leinaas HP, Skadsheim A (1995) Local patterns in mortality of the green sea urchin, *Strongylocentrotus droebachiensis*, at the Norwegian coast. In Skjoldal HR, Hopkins CCE, Erikstad KE, Leinaas HP (eds), Ecology of fjords and coastal waters, Elsevier, Amsterdam, 573–584.

Dayton PK (1985) Ecology of kelp communities. Ann. Rev. Ecol. Syst. 16: 215–245.

Düben GVJ von (1847) In Forhandlinger ved de skandinaviske naturforskeres fjerde møte i Christiania den 11–18 Juli 1844. Christiania (Oslo): 250–255.

Döderlein L von (1900) II Die Echinodermen. In Wissenschaftliche Meeresuntersuchungen. Kommission zur wissenschaftlichen Untersuchungen der deutsche Meer in Kiel und der Biologische Astalt auf Helgoland. 195–249.

Ebert TA (1982) Longevity, life history, and relative body wall size in sea urchins. Ecol. Monogr. 52: 353–394.

Ebert TA (1983) Recruitment in echinoderms. In Jangoux M, Lawrence JM (eds) Echinoderm studies I, Balkema, Rotterdam, pp. 169–201.

Ebert TA (1985) Sensitivity of fitness to macroparameter changes: An analysis of survuvorship and individual growth in sea urchin life histories. Oecologia 65: 461–467.

Elner RW, Vadas RL (1990) Inference in ecology: The sea urchin phenomenon in the northwestern Atlantic. Am. Nat. 136: 108–125.

Emlet RB, McEdward LR, Strathmann RR (1987) Echinoderm larval ecology viewed from the egg. In Jangoux M, Lawrence JM (eds), Echinoderm studies, Balkema, Rotterdam, pp 55–136.

Falk-Petersen I-B (1982) Ecological investigations on zooplankton community of Balsfjord, northern Norway. Planktonic eggs and larvae from March 1978 to February 1980. Sarsia 67: 69–78.

Fishery statistics, (1961–1973) Central bureau of statistics, Oslo. Annual reports.

Follestad A, Larsen BH, Nygaard T (1986) Sjøfuglundersøkelser langs kysten av Søróg Nord-Trøndelag og sørlige del av Nordland 1983–1986. Direktoratet for naturforvaltning, Trondheim, Viltra pp 41: 1–113. (In Norwegian).

Foreman RE (1977) Benthic community modification and recovery following intensive grazing by *Strongylocentrotus droebachiensis*. Helgol. Wiss. Meeresunters. 30: 468–484.

Hagen NT (1987) Sea urchin outbreaks and nematode epizootics in Vestfjorden, northern Norway. Sarsia 72: 213–229.

Hagen NT (1995) Recurrent destructive grazing of successionally immature kelp forests by green sea urchins in Vestfjorden, north Norway. Mar. Ecol. Prog. Ser. 123: 95–106.

Hart MW, Scheibling RE (1988) Heat waves, baby bloom, and the destruction of kelp beds by sea urchins. Mar. Biol. 99: 167–176.

Hawkins SJ, Hartnoll RG (1983) Grazing of intertidal algae by marine invertebrates. Oceanogr. Mar. Biol. Ann. Rev. 21: 195–286.

Hjort J, Ruud JT (1927) A method for the analysis and comparison of plankton samples. Nytt. Mag. Nat. 65: 137–144.

Hooper R (1980) Observations on algal-grazer interactions in Newfoundland and Labrador. In Pringle, JD, Sharp GJ, Caddy JF (eds), Proceedings of the workshop on the relationship between sea urchin grazing and commercial plant/animal harvesting. Canadian Technical Report of Fisheries and Aquatic Science 954: 120–124.

Hopkins CCE, Nilssen EM (1990) Population biology of the deep-water prawn (*Pandalus borealis*) in Balsfjord, northern Norway: I. Abundance, mortality, and growth, 1979–1983. J. Explor. Mer 47: 148–166.

Jackson JBC, Kirby MX, Berger WH, Bjorndal KA, Botsford LW, Bourque BJ, Brandbury RH, Cooke R, Erlandson J, Estes JA, Hughes TP, Kidwell S, Lange CB, Lenihan HS, Pandolfi JM, Peterson CH, Steneck RS, Tegner MJ, Warner RR (2001) Historical overfishing and the resent collapse of coastal ecosystems. Science 293: 629–638.

Lawrence JM (1975) On the relationship between marine plants and sea urchins. Oceanogr. Mar. Biol. Ann. Rev. 13: 213–286.

Mehlum F, Gabrielsen GW (1995) Energy expenditure and food consumption by sea bird populations in the Barents sea region. In Skjoldal HR, Hopkins CCE, Erikstad KE, Leinaas HP (eds), Ecology in fjords and coastal waters, Elsevier, Amsterdam, pp 457–470.

Miller RJ (1985) Seaweeds, sea urchins, and lobsters: A reappraisal. Can. J. Fish. Aquat. Sci. 42: 2061–2072.

Murphy GI (1968) Pattern of life history and the environment. Am. Nat. 102: 391–403.

Parsons LS and Lear WH (2001) Climate variability and marine ecosystem impacts: A North Atlantic perspective. Prog. Oceanogr. 49: 167–188.

Pauly D (1982) Studying single-species dynamics in a tropical multispecies context. In Pauly D, Murphy GI (eds), Theory and management of tropical fisheries. ICLARM Conference Proceedings 9: 33–70.

Pitcher TJ, Hart PJB (1982) Fisheries ecology, Chapman and Hall, London, pp 1–414.

Rasmussen B (1967) Temperaturforhold og rekefiske i Skagerak 1962–66. Fiskets Gang 47: 842–847.

Roff DA (1992) The evolution of the life histories. Theory and analysis, Chapmann and Hall, London, New York, pp 1–535.

Schaffer MW (1974) Optimal reproductive effort in fluctuating environments. Am. Nat. 108: 783–790.

Schaffer WM, Gadgil M (1975) Selection for optimum life history in plants. In Cody ML, Diamond JM (eds), Ecology and evolution of communities, Belknap Press, Cambridge, Massachusetts, USA. pp 142–157.

Scheibling RE, Hennigar AW (1997) Recurrent outbreaks of disease in sea urchins *Strongylocentrotus droebachiensis* in Nova Scotia: Evidence for a link with large-scale meteorologic and oceanographic events. Mar. Ecol. Prog. Ser. 152: 155–165.

Sivertsen K (1996) Incidence, occurrence and distribution of the nematode *Echinomermella matsi* in its echinoid host, *Strongylocentrotus droebachiensis,* in northern Norway. Mar. Biol. 126: 703–714.

Sivertsen K (1997a) Geographic and environmental factors affecting the distribution of kelp beds and barren grounds and changes in biota associated with kelp reduction at sites along the Norwegian coast. Can. J. Fish. Aquat. Sci. 54: 2872–2887.

Sivertsen K (1997b) Dynamics of sea urchins and kelp during overgrazing of kelp forests along the Norwegian coast. Thesis. Norwegian College of Fishery Science. University of Tromsø, Norway. 127 pp.

Sivertsen K (2003) Harvestable sea urchin (*Strongylocentrotus droebachiensis*) resources along the Norwegian coast. Proceedings. International Conference on Sea-urchin Fisheries and Aquaculture, Puerto Varas, Chile, 25–27 March 2003. pp. 1–7.

Sivertsen K, Hopkins CCE (1995) Demography of the echinoid *Strongylocentrotus droebachiensis* related to biotope in northern Norway. In Skjoldal HR, Hopkins CCE, Erikstad KE, Leinaas HP (eds), Ecology of fjords and coastal waters, Elsevier, Amsterdam, pp 573–584.

Skadsheim A, Christie H, Leinaas HP (1995) Population reductions of *Strongylocentrotus droebachiensis* (Echinodermata) in Norway and the distribution of its endoparasite *Echinomermella matsi* (Nematoda). Mar. Ecol. Prog. Ser. 119: 199–209.

Stearns SC (1976) Life-history tactics: A review of ideas. Q. Rev. Biol. 51: 3–47.

Stearns CS (1977) The evolution of life history traits: A critique of the theory and a review of the data. Ann. Rev. Ecol. Syst. 8: 145–171.

Steneck RS, Graham MH, Bourque BJ, Corbett D, Erlandson JM, Estes JA, Tegner MJ (2002) Kelp forests ecosystems: Biodiversity, stability, resilience and future. Env. Cons. 29: 436–459.

Stien A, Halvorsen O, Leinaas HP (1995) No evidence of *Echinomermella matsi* (Nematoda) as a factor in a local mass mortality of *Strongylocentrotus droebachiensis* (Echinodea). In Skjoldal HR, Hopkins CCE, Erikstad KE, Leinaas HP (eds) Ecology of fjords and coastal waters, Elsevier, Amsterdam, pp 585–592.

Svåsand T, Kristiansen TS, Pedersen T, Salvanes AGB, Engelsen R, Nødtvedt M (1998) Havbeite med torsk-artsrapport. Norges forskningsråd. Oslo.1–78.

Wiborg KF (1962) Estimation of numbers in the laboratory. Rapp. et proc.-verb 153: 74–77.

Wing SR, Largier JL, Botsford LW, Quinn JF (1995) Settlement and transport of bentic invertebrates in an intermitted upwelling region. Limnol. Oceanogr. 40: 316–329.

Journal of Applied Phycology (2006) 18: 611–617
DOI: 10.1007/s10811-006-9062-6

Seasonal occurrences of epiphytic algae on the commercially cultivated red alga *Kappaphycus alvarezii* (Solieriaceae, Gigartinales, Rhodophyta)

Charles S. Vairappan
Institute for Tropical Biology and Conservation, Universiti Malaysia Sabah, Kota Kinabalu, 88999, Sabah, Malaysia

e-mail: csv@ums.edu.my; fax: +6088-320291

Key words: Kappaphycus alvarezii, epiphyte, seasonal, *Neosiphonia savatieri*

Abstract

Common problems faced in farming of the red algal genus *Kappaphycus/Eucheuma* are "ice-ice disease" and the occurrence of epiphytes. Considerable work has been documented on "ice-ice disease" and it's mode of infection but limited information is available on the emergence of epiphytes. The present study addresses the phenomenon of epiphyte infection, its prevalence in commercially cultivated red alga, *Kappaphycus alvarezii,* and their variability associated with seasonality. Cultured seaweed became susceptible to epiphytes in the dry seasons (1) between March – June and (2) September – November. Findings revealed *Neosiphonia savatieri* (Hariot) M. S. Kim *et* I. K. Lee, as the dominant infecting epiphyte, representing up to 80–85% of the epiphyte present during peak seasons. Besides *N. savatieri*, *Neosiphonia apiculata*, *Ceramium* sp., *Acanthophora* sp. and *Centroceras* sp. were observed in smaller quantities. SEM (Scanning Electron Microscope) micrographs revealed the epiphyte's attachment to the host. Further histological study showed the extent of penetration of epiphytes into the host's cortex tissues and condition of its surrounding tissues. The outbreak of epiphytic filamentous red algae also correlated with drastic changes in seawater temperature and salinity during March – June and September – November.

Introduction

In Malaysia, seaweed farming was introduced in the late 1970s and commercial cultivation started few years after, with the culture of *Gracilaria* in Peninsula Malaysia and *Eucheuma* in North Borneo. Today, after almost 35 years, the red algal genera *Kappaphycus* and *Eucheuma* are intensively cultivated in two locations; Semporna (east coast of Borneo), and Kudat (north Borneo). Concomitant with the increase in farm size and intensification of culture practices is the rise in seaweed diseases, particularly; "ice-ice disease" and invasion of epiphytes (filamentous red algae). Epiphytic invasion is not a new phenomenon and has been known to exist since the dawn of farming practices (Doty & Alvarez, 1975, 1981). However, little is known of their causative agents, seasonality, mode of action and factors causing outbreaks (Parker, 1974; Collen et al., 1995; Fletcher, 1995; Ask, 1999, Ask & Azanza, 2002; Critchley et al., 2004). Knowledge of these aspects is vital, since epiphytic algae invade regularly, at times affecting the marketability of the harvested seaweeds. Epiphyte outbreaks have also shown to weaken the seaweed, making it susceptible to bacterial attack (unpublished data).

Hence, the present study was initiated to gather substantial information and in-depth understanding of the causative agent and some indication of the factors affecting outbreaks. Dynamics of epiphyte outbreaks were monitored in relation to abiotic seawater factors such as temperature and salinity. In addition, histology and Scanning electron microscopy (SEM) investigations were carried out to determine the epiphyte's attachment and intrusion into the host.

Materials and methods

Sampling location

Infected seaweed specimens were collected from a seaweed farm in Teluk Lung, Pulau Balambangan, Kudat, Sabah (04°32′43″N, 116°58′30″E). Sampling was conducted fortnightly during the epiphyte outbreak. Specimens were transported in a cold ice-chest (4 ∼ 6 °C) and fixed in 10% formalin in seawater at the laboratory. Sections and epiphytes were removed from the host by hand under stereo microscope (Stemi-2000 CS, Carl Zeiss, Germany) using a razor blade and pith stick and stained with 0.5% (w/v) cotton blue in lactic acid/phenol/glycerol/water [1:1:1:1 (v/v)] solution and mounted in 50% glycerol/seawater microscope slides. Some of the isolated epiphytes were immediately mounted in seawater on microscope slides and observed. Epiphytes were counted randomly in an area of 1 cm × 1 cm under a stereo microscope at 2 ∼ 5 times magnification. A total of 20 individuals were observed during each observation and data presented in Figure 1 represent average numbers of epiphytes for 12 months observation.

Histological study

Sections (0.5 cm in length) of the infected algal thallus were fixed for 24 h in 4% glutaraldehyde in 0.1 M cacodylate buffer (salinity; 30 ppt). Fixed specimens were then dehydrated in the following series of ethanol; 70, 80, 90, 95 and 99%, each for 1 h. Dehydrated specimens were then cleared twice with xylene: each step lasted for 1 h. Finally, specimens were impregnated with paraffin (twice), each for 80 min, before blocking. Sections were cut at 15 μm thickness and stained in haematoxilin/eosin or pirodic acid Schiff's reagent prior to viewing.

Scanning electron microscope study

Algal thalli with epiphytes were cut ca. 1.0 cm in length and fixed for 24 h in 4% glutaraldehyde in 0.1 M cacodylate buffer before post-fixation in 1% OsO_4 at 4 °C for 2 h, followed by dehydration with a graded acetone series and finally, critical point drying. Dehydrated specimens were mounted on stubs and coated with a 10–30 nm layer of gold before observation with a Leica Cambridge S360 electron microscope.

Physical water parameters

Seawater temperature and salinity was recorded on a multi-probe water quality checker (YSI meter, Model 85–25) three times daily at 6 am, 12 pm and 6 pm. Measurements were taken at a maximum depth of 0.5 m below the seawater surface. Data are presented as averages of those measured, for each two-week interval.

Results and discussion

Seasonal variation of epiphytes

Occurrence of epiphytes was monitored from January to December 2003. The presence of epiphytes was first observed at the end of February and vegetative plants emerged 3–4 weeks later and persisted until the end of June. A second invasion was shorter: the first sign was seen in early September and the epiphyte disappeared by late November. A total of 5 epiphyte species were

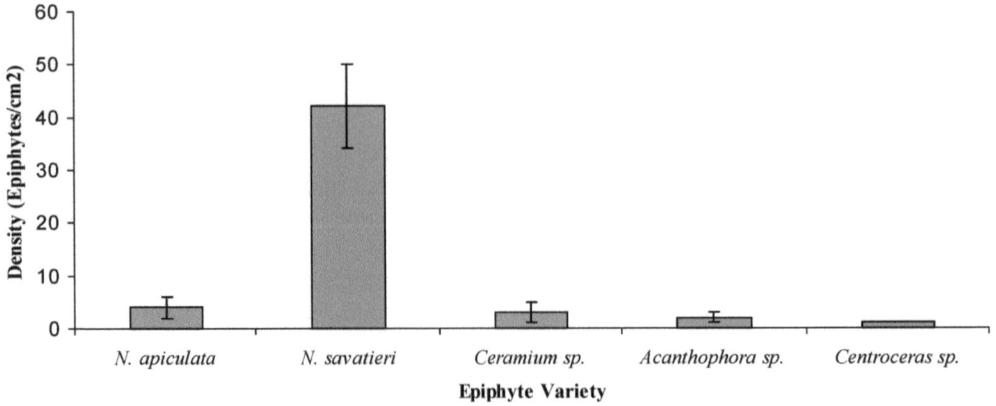

Figure 1. Average species composition of epiphytes isolated from the surface of the cultivated red alga, *Kappaphycus alvarezii*, during outbreaks.

isolated in both outbreaks, identified and their average distribution during the outbreak is shown in Figure 1. Species abundance was: *Neosiphonia savatieri* > *Neosiphonia apiculata* > *Ceramium* sp. > *Acanthophora* sp. > *Centroceras* sp. This trend was consistent at the study site for almost two years, from 2003 ~ 2004. Detailed observations of the dominant epiphyte were made and the findings are illustrated below.

Description of Neosiphonia savatieri

The first emergence of an epiphytic infection on *K. alvarezii* was observed in late February with the appearance of tiny black spots on surface cuticle cells layer (Figure 2A). These black spots then became rough and the vegetative epiphyte surfaced after 3–4 weeks (Figure 2B). Tissue cross-sections of the black spots revealed the presence of tetrasporelings embedded between the outer cortex cells (Figure 2C). By the end of March, the vegetative state of the epiphytes could be seen as shown in Figure 2D. Epiphytes were observed as solitary plants growing on the algal surface with rhizoids penetrating into the tissue of the cortical cell layers. In the peak season, the dominant epiphytes, *N. savatieri*, were seen to grow close to each other at a maximum density of 40–48 epiphytes/cm^{-2}.

N. savatieri is a plant with vertical axes and a height of 4–20 mm. Its basal attachment system is composed of a primary rhizoid with one or several secondary rhizoids. The latter were cut off from their pericentral cells at lower segments of the axis. These rhizoids penetrate into tissue of the basiphyte *via* cortical cells and at times are seen to penetrate down to the inner cortical

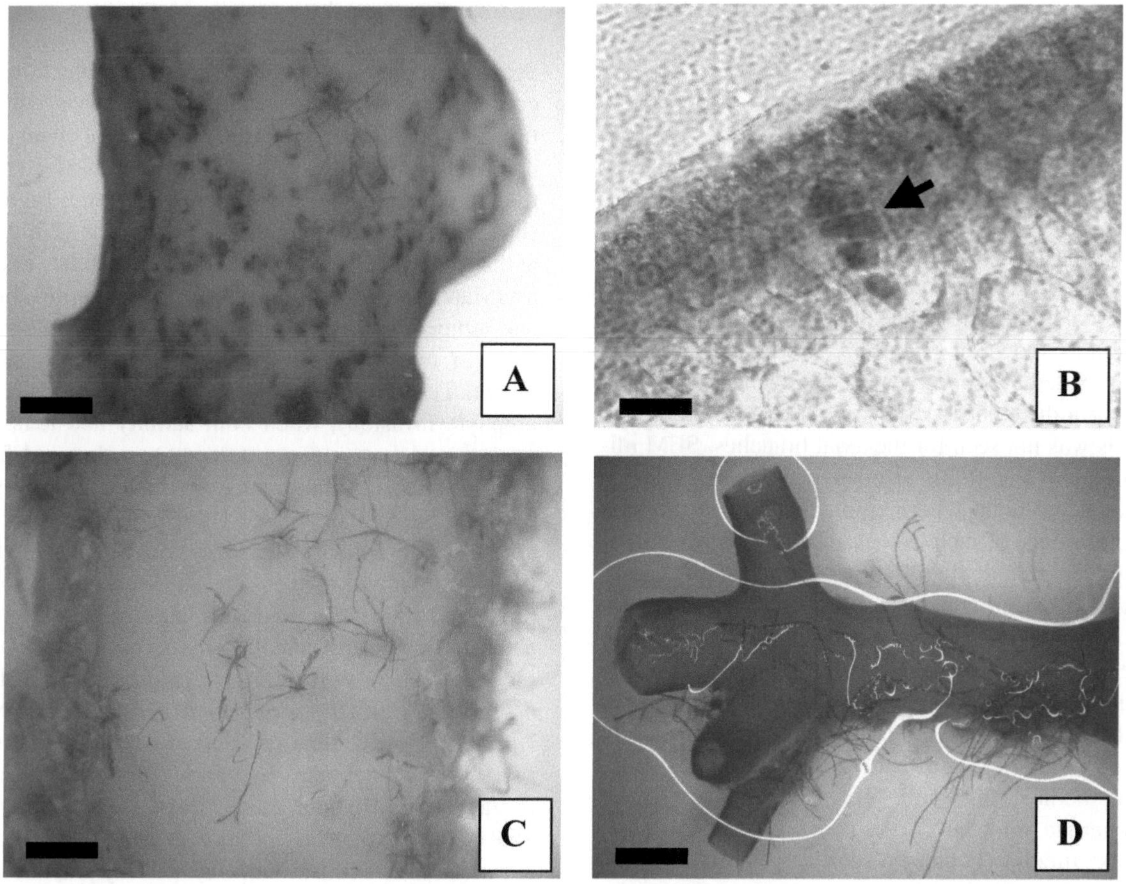

Figure 2. Kappaphycus alvarezii infected with *Neosiphonia savatieri*. (A) Host plant with early stage of epiphytes imbedded as tiny black spots in outer and inner cortex cell layers in late February (scale bar = 300 μm); (B) Tetrasporeling implanted between the outer and inner cortex cell layers (scale bar = 50 μm); (C) Host plant with epiphyte germlings by end of March (scale bar = 300 μm); (D) Host plant with mature epiphyte (scale bar = 300 μm).

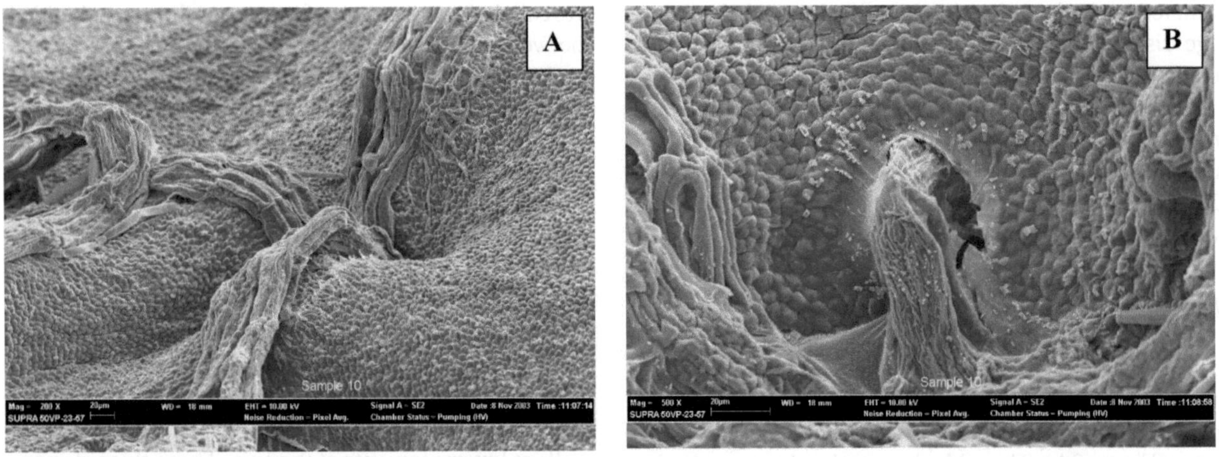

Figure 3. Scanning Electron Microscopy micrographs showing epiphyte's attachment to the host plant; (A) Emergence of epiphyte branching from a common location (200 ×); (B) Areas around the point of attachment have lesions/openings, providing ideal opportunity for grazers and microbial attack (500×).

cell layers as described by Kim and Lee (1999), Masuda et al. (2001), and Hollenberg (1968).

Epiphytic attachment phases

N. savatieri attached itself to the host *via* a basal attachment using its primary rhizoid or/and secondary rhizoids. Close examination of its attachment location revealed the emergence of 2–4 branches from one location as shown in Figure 3A. Horizontal branches that exceeded 8–10 mm were seen attached to the host again at a different locations *via* secondary rhizoids, but this was not seen for the axial branches. SEM micrographs in Figure 3B also showed slight lesions or cracks at the point where epiphytes entered the host plant. Presence of such lesions and cracks weakens the host plants, making them vulnerable to thallus breakage and bacterial attack. Histological study was carried out on tissue cross-sections at locations where the epiphyte was seen to penetrate the host, and the findings are shown in Figures 4A and 4B. Figures 4A shows an epiphyte with its main axis horizontally aligned to the host and with two of its rhizoids penetrating the host. In Figure 4B, two rhizoids are shown to penetrate into the cortex cell layers of the host; (1) first rhizoid **(I)** can be seen to penetrate deeply, to the inner cortex cells, where it is surrounded by outer cortex cells and inner cortex cells, (2) the second rhizoid only penetrates to the outer cortex cell layers **(II)**.

Seasonality of epiphytic invasion

Seasonal variation in epiphyte species composition was consistent for two consecutive year of crop cultivation (2003–2004). Correlation between the changes in epiphyte composition and physical abiotic factors at the culture site showed an interesting trend. Figure 5 shows the fluctuation in abiotic factors for 12 months at the study site. Emergence of epiphytes in late February/March coincides with drastic increases in salinity and temperature, e.g. seawater temperature increased from 27 to 31 °C, and salinity increased from 28 to 34 ppt. The opposite occurred in September (second epiphyte invasion), when both salinity and temperature were decreasing, e.g. temperature dropped from 30 to 25 °C, and salinity decreased from 29 to 27 ppt, between September to late November. Hence, there could be a possible correlation between the fluctuations in the abiotic factors and the emergence of epiphytes. Drastic changes in the abiotic factors could act as a triggering mechanism or cue for the epiphytes to infect *K. alvarezii*, as described by Mtolera et al. (1996) for *Eucheuma denticulatum* (N.L. Burman) F.S. Collins *et* Hervey. A similar situation has been reported in various bacterial diseases in Japanese kelp and *K. alvarezii* (Glenn & Doty, 1990; Vairappan, 2001; Largo et al., 1995a,b). Besides seawater temperature and salinity fluctuations, other physical factors such as seawater nutrient levels and photoperiod could also play an important role in epiphyte germination and outbreak. However, both

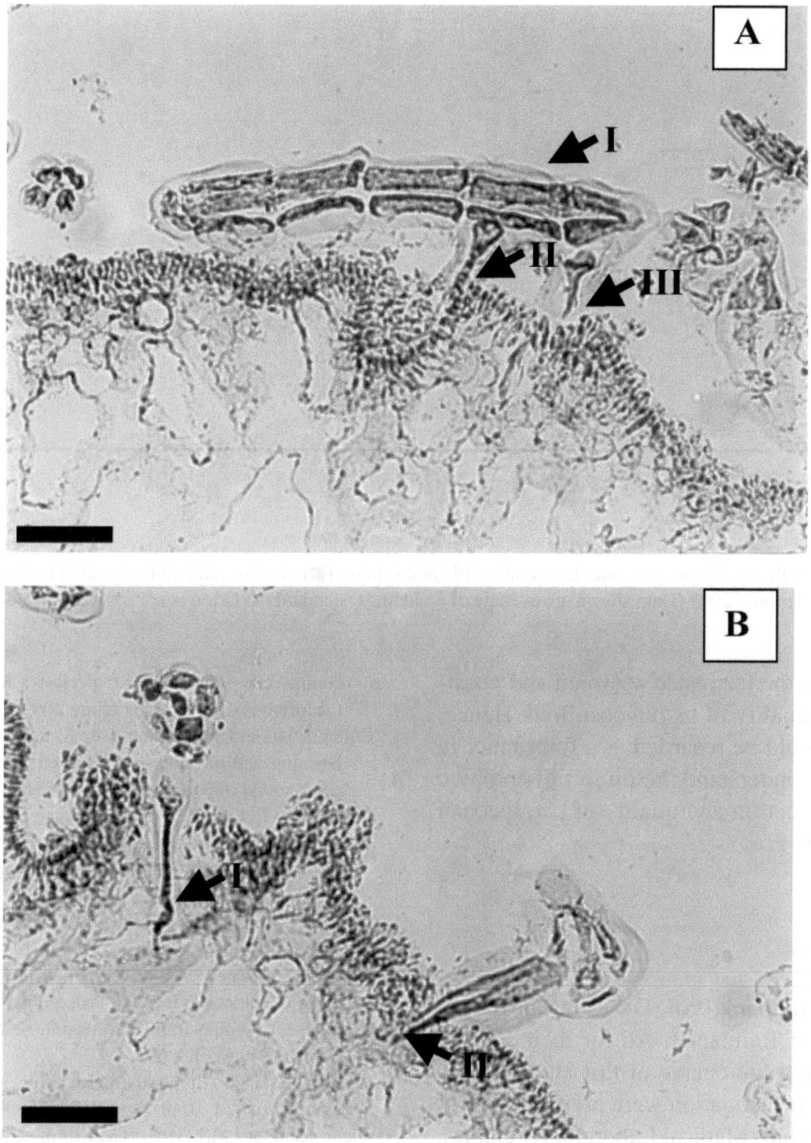

Figure 4. Histological section showing attachment of the epiphyte to the host plant; (A) Cross section of epiphyte's first-order branch (I) and two rhizoids attached to the host (II & III); (B) Slides showing penetration of rhizoids to the inner cortex cell layers (I) and outer cortex cell layers (II).

these factors could not possibly play any important role in this case since the culture site was far from anthropogenic sources of nutrients and there were no significant difference in photoperiod at the study site.

Conclusion

In the South East Asian region, where most commercial red algae are cultivated for carrageenan, occurrence of epiphytes is not given due importance. This could be due to lack of understanding or because cultivation is often a practice in remote areas, by poor farmers. However, the fact remains that epiphytic growth is on the rise and it has always been seen as a serious constraint to commercial seaweed cultivation, particularly in the tropics (Wheeler et al., 1981; Zemke-White & Ohno, 1999). Emergence of an epiphytic outbreak is a complex problem and the extent of the outbreak often depends on the quality of the cultivated strain, abiotic parameters of the culture site and seasonal weather fluctuations. However, considerable effort is warranted to better understand this problem due to its importance in

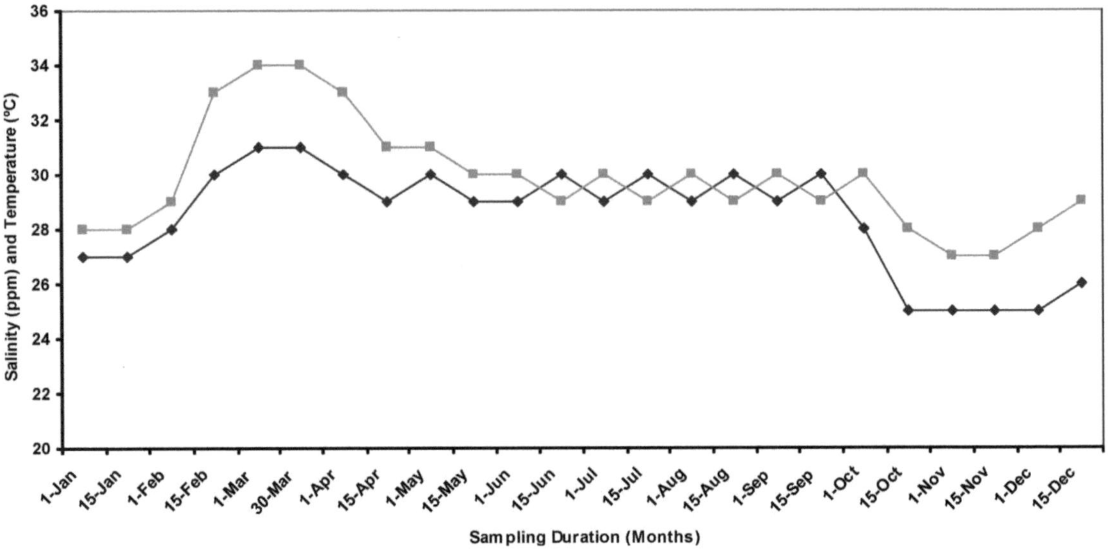

Figure 5. Fluctuations of abiotic factors, seawater temperature (♦) and salinity (■), at seaweed culture location in Teluk Lung, Pulau Banggi, Kudat, during the study period – 2003 (Data shown are average of 3 readings, standard deviation is not shown to avoid data congestion).

the marketability of the harvested seaweed and possible impacts on the quality of its phycocolloid. Hence, the present study could be regarded as a forerunner in our attempt to fully understand the impact of epiphyte on the growth, production and quality of carrageenan produced by *K. alvarezii.*

Acknowledgements

Author would like to thank Prof. Dr. Ridzwan B. Abdul Rahman and Mr. Ramlan B. Ali for their support and assistance during the course of this study. Slides for the histological investigation were prepared by Mr. Kandasamy S. (Gribbles Medical Laboratory, Penang). The author would also like to thank Prof. Dr. Michio Masuda for his assistance in identifying the epiphytes and facilities made available during the author's visit to the Graduate School of Biological Sciences, Hokkaido University, Sapporo, Japan. Funding to present this paper at ISS2004 was provided by Universiti Malaysia Sabah and Fischer Scientific (M) Ltd. and is gratefully acknowledged.

References

Ask E (1999) *Cottonii* and *spinosum* cultivation handbook. FMC Food Ingredients Division, Philadelphia.

Ask EI, Azanza RV (2002) Advances in cultivation technology of commercial eucheumatoid species: a review with suggestions for future research. Aquaculture 206: 257–277.

Collen J, Mtolera M, Abrahamsson K, Semesi A, Pedersen M (1995) Farming and physiology of the red algae *Eucheuma*: Growing commercial importance in East Africa. Ambio 24: 497–501.

Critchley AT, Largo D, Wee W, Bleicher Lhonneur G, Hurtado AQ, Schubert J (2004) A preliminary summary on *Kappaphycus* farming and the impact of epiphytes. Jpn. J. Phycol. 52: 231–232.

Doty MS, Alvarez VB (1975) Status, problems, advances and economics of *Eucheuma* farms. J. Mar. Technol. Soc. 9: 30–35.

Doty MS, Alvarez VB (1981) *Eucheuma* farm productivity. In Fogg GE, Jones WE (eds), Proceedings of the 8th International Seaweed Symposium, The Marine Science Laboratory, Menai Bridge, Wales, 688–691.

Fletcher RL (1995) Epiphytism and fouling in *Gracilaria* cultivation: An overview. J. Appl. Phycol. 7: 325–333.

Glenn EP, Doty MS (1990) Growth of seaweeds *Kappaphycus alvarezii, K. striatum* and *Eucheuma denticulatum* as affected by environment in Hawaii. Aquaculture 84: 245–255.

Hollenberg GJ (1968) An account of the species of *Polysiphonia* of the central and western tropical Pacific Ocean. Pac. Sci. 22: 56–98.

Kim MS, Lee IK (1999) *Neosiphonia flavimarina* gen. et sp. nov. with a taxonomic reassessment of the genus *Polysiphonia* (*Rhodomelaceae, Rhodophyta*). Phycol. Res. 47: 271–281.

Largo DB, Fukami F, Nishijima T, Olino M (1995a) Laboratory-induced development of the ice-ice disease of the farmed red algae *Kappaphycus alvarezii* and *Eucheuma denticulatum* (Solieriaceae, Gigartinales, Rhodophyta). J. Appl. Phycol. 7: 539–543.

Largo DB, Fukami F, Nishijima T (1995b) Occasional bacteria promoting ice-ice disease in the carrageenan-producing red algae *Kappaphycus alvarezii* and *Eucheuma denticulatum* (Solieriaceae, Gigartinales, Rhodophyta). J. Appl. Phycol. 7: 545–554.

Masuda M, Abe T, Kawaguchi S, Phang SM (2001) Taxonomic notes on marine algae from Malaysia. VI. Five species of *Ceramiales* (*Rhodophyceae*). Bot. Mar. 44: 467–477.

Mtolera MSP, Collen J, Pedersen M, Ekdahl A, Abrahamsson K, Semesi AK (1996) Stress-induced production of volatile halogenated organic compounds in *Eucheuma denticulatum* (Rhodophyta) caused by elevated pH and high light intensities. Europ. J. Phycol. 3: 89–95.

Parker HS (1974) The culture of the red algal genus *Eucheuma* in the Philippines. Aquaculture 3: 425–439.

Vairappan CS (2001) Study on chemical defense mechanism of seaweeds against marine microbes. Graduate School of Environmental Earth Science, Hokkaido University PhD. Dissertation. 180 pp.

Wheeler WN, Neushul M, Harger BWW (1981) Development of a coastal marine farm and its associated problems. In Levring T (ed.). Proceedings of the Tenth International Seaweed Symposium, Walter de Gruyter, Berlin, 631–636.

Zemke-White L, Ohno M (1999) World seaweed utilization: An end of-century summary. J. Appl. Phycol. 11: 369–376.

Journal of Applied Phycology (2006) 18: 619–627
DOI: 10.1007s/10811-006-9059-1

The role of encrusting coralline algae in the diets of selected intertidal herbivores

Gavin W. Maneveldt*, Deborah Wilby, Michelle Potgieter & Martin G.J. Hendricks
Department of Biodiversity and Conservation Biology, University of the Western Cape, P. Bag X17, Bellville 7535, South Africa

*Author for correspondence: e-mail: gmaneveldt@uwc.ac.za

Key words: encrusting coralline algae, diet, grazers, herbivory, organic content, rocky shore

Abstract

Kalk Bay, South Africa, has a typical south coast zonation pattern with a band of seaweed dominating the mid-eulittoral and between two molluscan-herbivore dominated upper and lower eulittoral zones. Encrusting coralline algae were very obvious features of these zones. The most abundant herbivores in the upper eulittoral were the limpet, *Cymbula oculus* (10.4 ± 1.6 individuals m^{-2}; 201.65 ± 32.68 $g.m^{-2}$) and the false limpet, *Siphonaria capensis* (97.07 ± 19.92 individuals m^{-2}; 77.93 16.02 $g.m^{-2}$). The territorial gardening limpet, *Scutellastra cochlear*, dominated the lower eulittoral zone, achieving very high densities (545.27 ± 84.35 m^{-2}) and biomass (4630.17 ± 556.13 $g.m^{-2}$), and excluded all other herbivores and most seaweeds, except for its garden alga and the encrusting coralline alga, *Spongites yendoi* ($35.93 \pm 2.26\%$ cover). In the upper eulittoral zone, encrusting coralline algae were only present in the guts of the chiton *Acanthochiton garnoti* ($30.5 \pm 1.33\%$) and the limpet *C. oculus* ($2.9 \pm 0.34\%$). The lower eulittoral zone limpet, *Scutellastra cochlear* also had a large percentage of encrusting coralline algae in its gut with limpets lacking gardens having higher ($45.1 \pm 1.68\%$) proportions of coralline algae in their guts than those with gardens ($25.6 \pm 0.8\%$). Encrusting coralline algae had high organic contents, similar to those of other encrusting and turf-forming algae, but higher organic contents than foliose algae. Radula structure, grazing frequencies as a percentage of the area grazed (upper eulittoral $73.25 \pm 3.60\%$ d^{-1}; lower eulittoral $46.0 \pm 3.29\%$ d^{-1}), and algal organic content provided evidence to support the dietary habits of the above herbivores. The data show that many intertidal molluscs are actively consuming encrusting coralline algae and that these seaweeds should be seen as an important food source.

Introduction

In South Africa there is a high level of seaweed endemism (Lüning, 1990; Stegenga et al., 1997), and large numbers of herbivores. These herbivores are important in the intertidal zone as they control the abundance and distribution of algae through their grazing activities (Branch, 1975, 1985). Much of the research on herbivore-algal interactions in South Africa has focused on grazing interactions involving fleshy seaweed (Branch, 1971, 1985). Encrusting coralline algae have been cited as important food sources for many intertidal herbivores (Steneck, 1982, 1985; Steneck & Watling, 1982; Paine, 1984; Steneck et al., 1991; Fujita, 1992; Littler et al., 1995; Raffaelli & Hawkins, 1996; Littler & Littler, 2003) and a few South African studies have shifted their research focus toward herbivore-algal interactions involving coralline algae (Keats et al., 1994b; Maneveldt, 1995).

Encrusting coralline algae are important occupiers of space in rocky marine intertidal environments (Adey & McIntyre, 1973; Paine, 1984; Steneck, 1982, 1985, 1986; Dethier et al., 1991; Steneck et al., 1991; Keats & Maneveldt, 1994; Keats et al., 1994a, b) and often become abundant in areas of intense herbivory (Adey & McIntyre, 1973; Steneck, 1983; Breitburg, 1984; Sousa & Connell, 1992; Dethier, 1994; Steneck & Dethier, 1994). Despite their ubiquity, they are a poorly known group of seaweeds (Keats et al., 1994a). Nonetheless, they are a very obvious feature of the South African

Table 1. Taxa surveyed during the study and their main features

Species	Main feature	Zone occupied	Sub study
Spongites yendoi (Foslie) Chamberlain	Encrusting coralline red alga	U & L	* ! #
Leptophytum foveatum[1] Chamberlain & Keats	Encrusting coralline red alga	L	*
Leptophytum ferox (Foslie) Chamberlain & Keats	Encrusting coralline red alga	U	* ! #
Hildenbrandia lecanellierii Hariot	Encrusting fleshy red alga	U	* !
Ralfsia verrucosa (Areschoug) J. Agardh	Encrusting fleshy brown alga	U & L	* (in 'other') !
Gelidium micropterum Kuetzing	Turfy limpet-garden red alga	L	* !
Gelidium pristoides (Turner) Kuetzing	Turfy red alga	U & L	* !
Porphyra capensis Kuetzing	Foliose red alga	U	* (in 'other') !
Gigartina polycarpa (Kuetzing) Setchell & Gardner	Foliose red alga	L	* !
Sarcothalia stiriata (Turner) Leister	Foliose red alga	L	* (in 'other') !
Ulva capensis Areschoug	Foliose green alga	U & L	* !
Enteromorpha intestinalis (Linnaeus) Link	Foliose green alga	U	* (in 'other') !
Cymbula oculus (Born)	Limpet – herbivore	U	@ + $
Siphonaria capensis Quoy & Gaimard	False limpet – herbivore	U	@ +
Scutellastra cochlear (Born)	Gardening limpet – herbivore	L	@ + $
Acanthochiton garnoti[2] (Blainville)	Chiton – herbivore	U	+ $
Oxystele variegata (Anton)	Winkle – herbivore	U	@ +
Tetraclita serrata Darwin	Barnacle	U	@
Octomeris angulosa Sowerby	Barnacle	U	@

U = Upper eulittoral zone; L = Lower eulittoral zone.

* – Algal cover; ! – Organic content; @ – Density & Biomass; # – Grazing frequency; + – Gut contents; $ – Radula anatomy.

1 – *L. foveatum* is extremely thin and nearly impossible to remove for organic content.

2 – *A. garnoti* was not observed on the exposed substratum during low tide but is known to graze coralline algae.

rocky intertidal with a few species occurring in high abundance (Stegenga et al., 1997).

The rocky intertidal of the South African south and west coasts has been divided into four zones: the *Littorina* zone (supralittoral fringe); the upper balanoid zone (upper eulittoral zone); the lower balanoid zone (mid-eulittoral zone); and the *cochlear* zone (lower eulittoral zone) (Branch & Branch, 1988). Within the intertidal, a band of seaweed dominates the mid-eulittoral zone and is sandwiched between a herbivore-dominated upper and lower eulittoral zones. In the upper eulittoral zone, a number of molluscan grazers are very abundant, feeding on a broad range of available seaweeds (Branch, 1971). Algal diversity generally increases down the shore. In the lower eulittoral zone, however, this diversity is abruptly reduced because a dense band of territorial, gardening limpets *Scutellastra cochlear*, exclude not only other herbivores, but all seaweed except for their gardens of fine red, turfy algae (*Gelidium micropterum* or *Herposiphonia heringii* – see Table 1 for authorities) and for two encrusting coralline red algae (*Spongites yendoi* and *Leptophytum foveatum*) (Branch, 1975; 1976; Branch & Griffiths, 1988; Keats et al., 1994b). The degree to which many of these grazers feed on a specific seaweed is relatively unknown and furthermore, the role of encrusting coralline algae, which form such a conspicuous feature of the intertidal, has not been addressed. In this study we asked: (1) which algae are readily available to grazers and what are their nutritional qualities? (2) are any grazers incorporating coralline algae in their diets? and if so (3) how often are coralline surfaces being grazed?

Materials and methods

Study site

The study site, Kalk Bay (34°8′S, 18°27′E), is situated within False Bay, South Africa. This site has a typical south coast zonation pattern (Branch & Branch, 1988). A number of molluscan grazers (the true limpet, *Cymbula oculus*, the false limpet, *Siphonaria capensis* and the winkle, *Oxystele variegata*) and barnacles (*Tetraclita serrata* and *Octomeris angulosa*) occur abundantly within the upper eulittoral zone. The territorial, gardening limpet, *Scutellastra cochlear* dominates the lower eulittoral zone.

Algal availability and nutritional quality

Algal availability and nutritional quality were estimated by determining: (1) natural algal cover; and (2) relative organic content.

Algal cover was determined along three randomly chosen transect lines running down the shore. Five quadrats (each $0.3 \times 0.3 = 0.09\,m^2$), placed at regular intervals along each transect were used to estimate algal cover. To determine cover of encrusting algae, that of the upright fleshy algae was estimated first and then they were removed to obtain estimates of the encrusting algae underneath. Using this method, summed cover estimates for both encrusting and upright algae may exceed 100%.

Organic content is often represented as a proportion of dry weight. Algae, however, vary greatly in their water content, which has a diluting effect and no-doubt affects the nutritional value to herbivores. We therefore assessed the organic content of algae on a 'per-bite basis' by representing organic content as a proportion of wet weight, i.e. ash-free dry weight as a proportion of wet weight (AFDW:WW). Samples for organic content analysis comprised algae found throughout the intertidal including those that were not abundant (Table 1). Algae were first hydrated in seawater for 30 min prior to examination and then blotted dry to remove excess water. Samples were ashed by burning in a furnace at 450 °C for 16 h.

Dietary incorporation of coralline algae

The incorporation of coralline algae into the diet was assessed via four determinations: (1) the densities and biomass of commonly observed invertebrates; (2) a comparison of the grazing frequencies experienced within each of the two herbivore dominated zones; (3) an analysis of gut contents of common molluscan herbivores; and (4) examination of the radula of those molluscs shown to graze corallines.

Invertebrate densities and biomass were determined along three randomly chosen transect lines running down the shore. Five quadrats ($0.5 \times 0.5 = 0.25\,m^2$) placed at regular intervals along each transect were used to calculate densities. Animals from these quadrats were collected and brought back to the lab for biomass determinations.

Grazing frequency was measured in the field by recording the number of graphite dots removed (*sensu* Steneck et al., 1991) from the surfaces of the two dominant encrusting coralline algae found within both the upper and lower eulittoral zones ($n = 25$ crusts per species per zone). A series of 4×4 dots, 2 mm in diameter and 10 mm apart (area per treatment: 3×3 cm $= 9\,cm^2$), was used to estimate the relative grazing frequencies experienced by encrusting coralline algae. The number of dots removed by limpets was recorded daily for two days and expressed as grazing frequency per day. Ungrazed dots persist for several days on the coralline surface.

For gut contents analysis, the foregut of 20 individuals from each of the molluscan herbivore species observed (see Table I) was dissected. The foregut was then extruded into a 5 ml vial containing 1 ml distilled water, agitated, and then emptied into an ice cube tray with 20×20 mm compartments. This method was effective at distributing the gut contents evenly in the ice cube tray compartments. Using a stereo microscope equipped with an eyepiece quadrat divided into 100 squares, the percentage of coralline algae was estimated; coralline algal cells are easily discernable from those of fleshy algae. Coralline 'chalk' from the posterior part of the gut was not included as this has been shown to produce inaccurate measures of actual consumption because 'indigestible' material is known to persist for a long time in the posterior part of the gut.

The radulae of herbivores found to contain coralline algae in their guts were then removed and a representative radula from each of these herbivore species was examined and photographed under a scanning electron microscope (SEM). Using X-ray analysis, the radulae were also scanned to determine the concentrations and relative amounts of hardening agents in the functional teeth.

Statistical analyses

Unless otherwise stated, all data are presented as means ± standard errors. Grazing frequencies in the upper and lower eulittoral zones were compared using a Mann-Whitney nonparametric test (Mann-Whitney, 1947).

Results

Algal availability and nutritional quality

Foliose algal cover was greater in the lower (39.67 ± 4.54%, $n = 15$) than in the upper (14.54 ± 4.65%, $n = 15$) eulittoral zone. In both zones the most abundant alga was an encrusting alga, occupying similar proportions of the substratum. The fleshy red alga

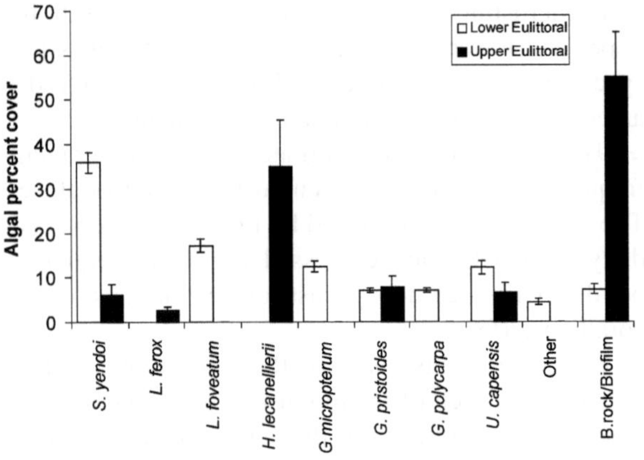

Figure 1. Natural algal cover within upper and lower eulittoral zones ($n = 15$ quadrats per zone) at Kalk Bay, South Africa.

Hildenbrandia lecanellierii dominated the upper eulittoral, while the coralline red alga *Spongites yendoi* dominated the lower eulittoral zone (Figure 1). In the upper eulittoral zone, a large percentage of the primary substratum was free of macroscopic algae as denoted by the "bare rock/biofilm" component.

Organic content (as a proportion of the algal wet weight) was highly variable among the different algae tested. Foliose algae (*Gigartina polycarpa, Sarcothalia stiriata, Ulva capensis* and *Enteromorpha intestinalis*), with the exception of *Porphyra capensis*, had much lower organic content than both encrusting (*S. yendoi*; *Leptophytum ferox, Ralfsia verrucosa* and *H. lecanellierii*) and turfy algae (*Gelidium micropterum* and *G. pristoides*) (Figure 2).

Dietary incorporation of coralline algae

A greater number of different invertebrate species was observed in the upper than the lower eulittoral zone (Table 2). The invertebrate densities and biomass in the upper eulittoral zone were highly variable, with barnacles (*T. serrata*) characteristically occurring in high numbers and attaining a high biomass. Although the false limpet *Siphonaria capensis* occurred in high numbers, this tiny limpet in no way matched the biomass of the larger true limpet, *Cymbula oculus*. The lower eulittoral zone contained only one species of limpet, *S. cochlear,* which attained very high densities and biomass (Table 2).

Grazing frequency (as a percentage of dots grazed d^{-1}) on encrusting coralline algae was significantly higher in the upper than in the lower eulittoral zone ($U = 82.0$, $Z = 4.47$, $P < 0.01$, $n1 = 25$, $n2 = 25$) (Figure 3).

Of the herbivores examined (Table 1), only three were found to contain coralline algae in their guts (Figure 4). Both the limpet *Scutellastra cochlear* and the chiton *Acanthochiton garnoti* contained a substantial percentage of coralline algae in their guts and were probably both actively grazing these algae. *Scutellastra cochlear* individuals without gardens,

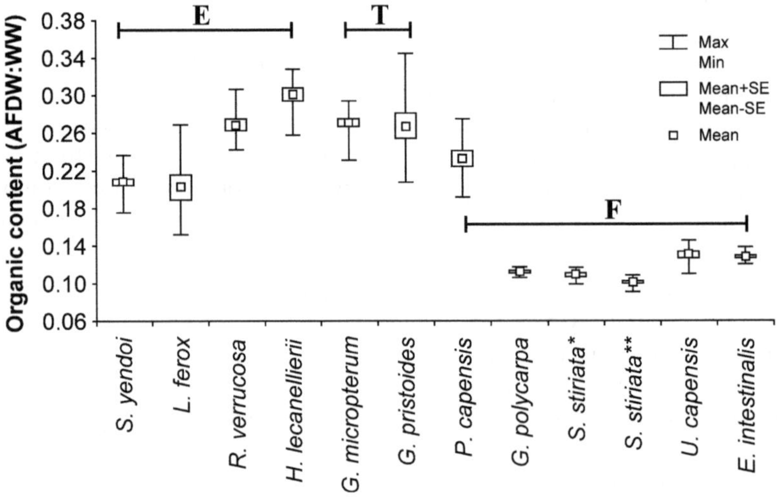

Figure 2. Organic content as a proportion of the wet weight of common intertidal algae ($n = 10$). E = encrusting algae; T = turfy algae; F = foliose algae; $*$ = *S. stiriata* sporophyte; $**$ = *S. stiriata* gametophyte.

Table 2. Densities (number m^{-2}) and biomass (gm^{-2}) of common invertebrates within the upper and lower eulittoral zones at Kalk Bay, South Africa.

Species/zone	Density	Biomass
Upper Eulittoral zone:		
C. oculus	10.4 ± 1.6	201.65 ± 32.68
S. capensis	97.07 ± 19.92	77.93 ± 16.02
T. serrata	62.13 ± 15.82	428.31 ± 111.62
O. angulosa	2.4 ± 2.13	14.16 ± 12.57
O. variegata	19.73 ± 2.93	28.76 ± 4.31
A. garnoti	Not observed on the exposed substratum during low tide	
Lower Eulittoral zone:		
S. cochlear	545.27 ± 84.35	4630.17 ± 556.13

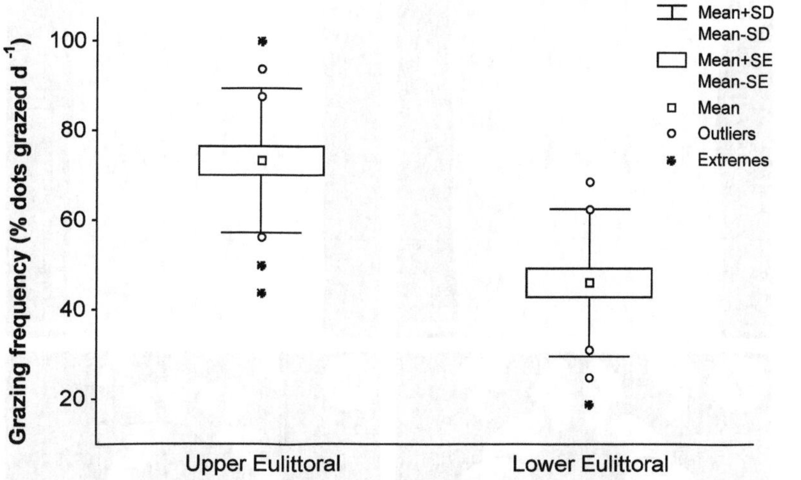

Figure 3. Grazing frequencies (% dots grazed d^{-1}) experienced by encrusting coralline algae from upper and lower eulittoral zones ($n = 25$ crusts per species per zone). Outliers and extreme values are included to show the distribution of the data. These values were incorporated in the analysis of the data and still produced highly significant values.

however, had much greater proportions of coralline algae in their guts than those individuals with gardens (Figure 4). By contrast, the limpet *C. oculus* had a small percentage of coralline algae in its gut.

The radulae of the three herbivores found to contain coralline algae in their guts (Figures 5 and 6) differed to varying degrees in their design, tooth structure and elemental concentrations of their functional teeth (see Steneck and Watling, 1982 for a description of different moluscan radulae). Limpets (*S. cochlear* and *C. oculus*) have a docoglossan radula, while chitons (*A. garnoti*) have a polyplacophoran radula that is more effective at grazing hard substrata. Although the design of the radulae of the two limpets are similar (3 marginals: 3 laterals: 1 central: 3 laterals: 3 marginals), they differed with respect to the shape of their teeth (Figure 5). *Scutellastra cochlear*'s teeth have rounded distal ends

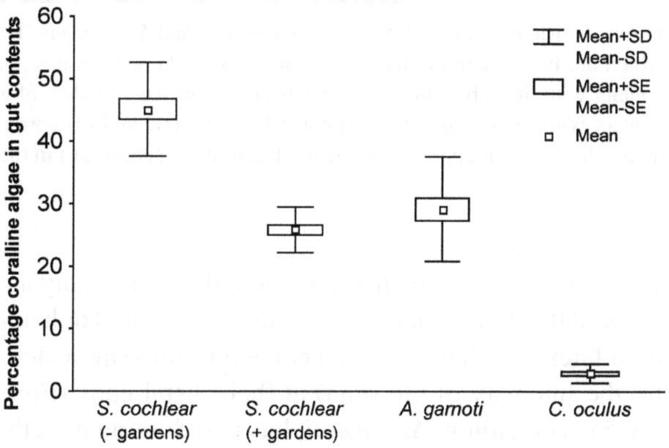

Figure 4. Percentage coralline algae in the guts of herbivores found to contain the algae ($n = 20$).

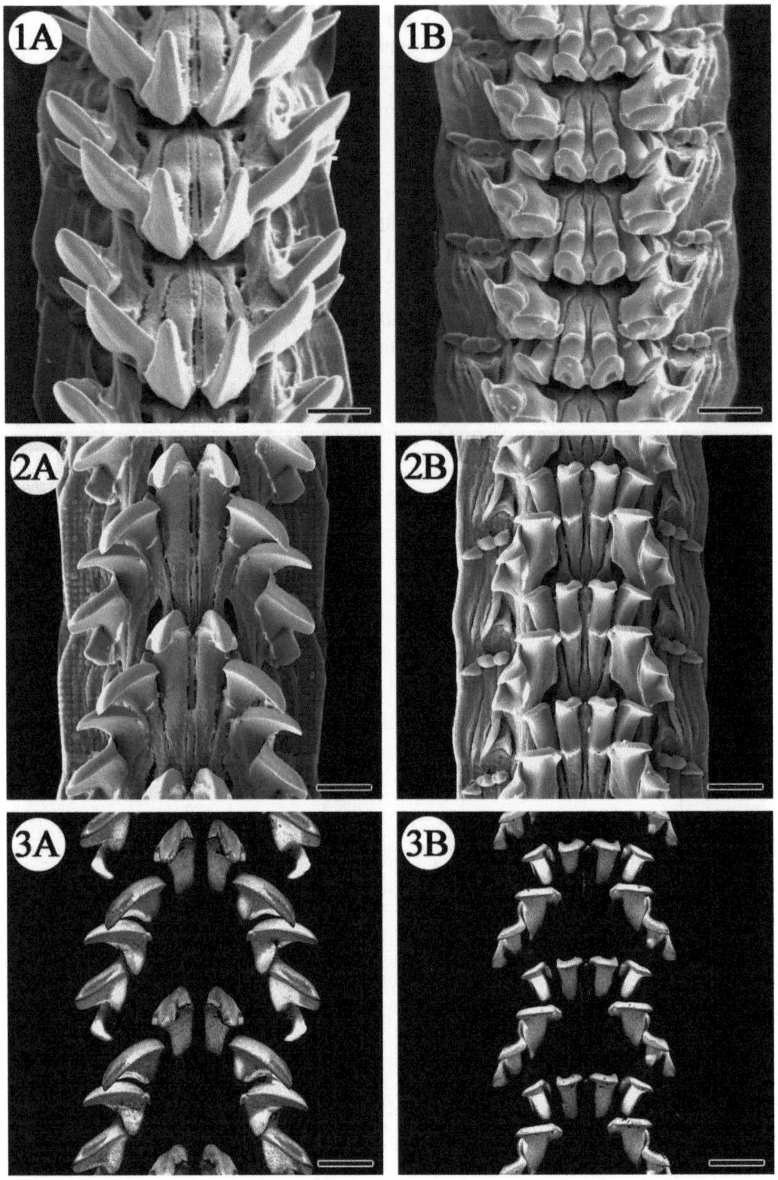

Figure 5. Radulae of the limpets *C. oculus* (A) and *S. cochlear* (B) in: (1) reverse SEM view (Note the shapes of the distal ends of the teeth which are better seen in this reverse view. *Cymbula oculcus* has pointed, almost rake-like teeth with narrow ends that have a reduced area of contact with the substratum. *Scutellastra cochlear* has rounded, blunt and shovel-like teeth with relatively broad ends that increase the area of contact with the substratum); (2) normal SEM view; and (3) corresponding X-ray imaging of the relative amounts of iron (visible as lighter areas; the lighter it appears, the greater the iron concentration) in the functional teeth (all scale bars = 100 μm).

while those of *C. oculus* are pointed. X-ray analysis showed that both limpets have iron concentrated only in their lateral teeth with greater concentrations (indicated by the intensity of the scan) at their distal ends (Figure 5). The chiton, *A. garnoti*, has two dominant teeth bearing three cusps (Figure 6). Chitons only use four marginal teeth during grazing, the two dominant teeth serving to excavate the substratum (Fretter & Graham, 1994).

Discussion

Both the upper and lower eulittoral zones at Kalk Bay have relatively high densities and biomass of herbivores

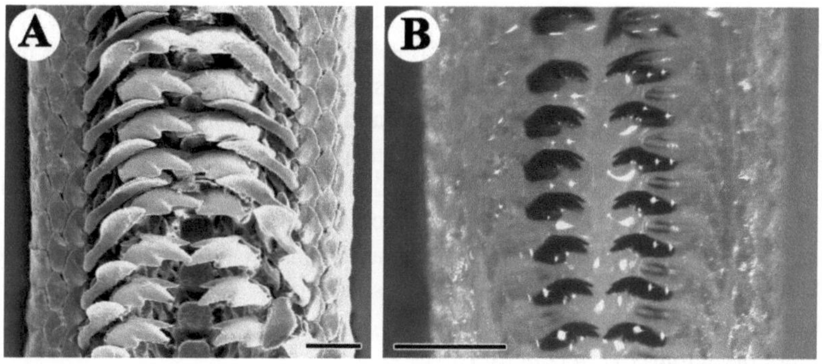

Figure 6. Radula of *A. garnoti* in SEM (A, scale bar = 200 μm) and light microscope (B, scale bar = 500 μm) view. Note the two dominant lateral teeth in each row which are completely impregnated with iron as shown by the dark colouration (B).

that have reduced the algal cover and diversity to but a few species (Branch, 1971, 1975, 1980, 1985). Encrusting algae are the only group of algae that are able to survive the high grazing pressures that occur within both zones. This is not surprising, as many studies have shown that encrusting algae, and corallines in particular, thrive under, and even require, intense herbivory (Paine & Vadas, 1969; Steneck, 1982, 1983; Breitburg, 1984; Sousa & Connell, 1992; Dethier, 1994; Steneck & Dethier, 1994).

The Kalk Bay intertidal is capable of supporting high herbivore densities and biomass despite the low algal cover. The dominant algae within both zones are high in organic content. Within the lower eulittoral, the territorial limpet, *Scutellastra cochlear*, appears to rely heavily on a diet of encrusting coralline algae despite many limpets having their own algal gardens. Branch (1980) stated that production and energy content of *Spongites yendoi* (as *Lithothamnion*) is so low that it cannot alone support the energy needs of *S. cochlear*. Branch (1980) further stated that the gardens of fine red algae are vital for the densely packed *S. cochlear* and although they often form only a small fringe around each animal, their production and energy contents are high. Our results show that *S. yendoi* is similarly high in organic content when compared to the garden alga, *G. micropterum*, and this explains why *S. cochlear* lacking gardens, consume proportionately more coralline algae than those limpets with gardens. It may be that the coralline and the garden alga are fulfilling different requirements and that *S. cochlear* with gardens are maintaining a mixed diet of algae. This, however, needs to be evaluated through further studies. Recent work by M. E. Hay and Q. E. Kappel (pers. comm.) has, however, also shown that organic content is high in many corallines relative to many fleshy algae.

Within the upper eulittoral zone, the large numbers of grazers may be supplementing their diets with the abundant encrusting, fleshy red alga, *H. lecanellierii* which is also high in organic content. Grazers typically consume a variety of plant matter and within the upper eulittoral zone, where algae are low in abundance, this probably comprises a mixture of predominantly microscopic algae, algal spores and larvae. An examination of the grazing frequencies experienced by encrusting corallines clearly demonstrates the high grazing pressures exerted within the upper eulittoral zone.

In the upper eulittoral, only the chiton *A. garnoti* and the limpet *C. oculus* were found to contain coralline algae in their guts. Chitons are known to be formidable grazers, many of them including large proportions of coralline algae in their diets (Steneck & Watling, 1982; Littler et al., 1995; Littler & Littler, 2003). Similarly, *A. garnoti* appears to be actively grazing the surfaces of encrusting coralline algae. From the structure of the radula, it is evident that *A. garnoti* is capable of excavating coralline algae and this ability probably plays an important role in its dietary choice. Although not as effective as chitons, some limpets are capable of excavating the surfaces of coralline algae (Steneck, 1982; 1983, 1985; Steneck & Watling, 1982). However, this does not explain the exceptionally low proportions of coralline algae in the gut of *C. oculus* despite its having a radula very similar to *S. cochlear* (Figures 5).

X-ray analysis of the radulae of the limpets *S. cochlear* and *C. oculus* revealed very little difference in the relative amounts of iron in their functional teeth (Figure 5) and so the shape of the functional teeth may provide the answers to the limpets' excavation potential. Unlike *S. cochlear*'s teeth that are rounded, blunt and shovel-like with relatively broad ends that increase the area of contact with the substratum, the teeth of *C. oculus* are pointed and almost rake-like, with narrow ends that have a reduced area of contact with the substratum (Figure 5). Radulae with shovel-like, blunt teeth are known to be deeper excavators of encrusting coralline algae than those with rake-like, pointed teeth (Steneck & Watling, 1982) and *C. oculus* is probably not as effective at excavating encrusting coralline algae as *S. cochlear*. Also, gut content analysis of *C. oculus* by Branch (1971) revealed a high proportion of encrusting fleshy algae, particularly *H. lecanellierii* (as a black paste) and *R. verrucosa*, both algae that we have shown to be high in organic content. Like *S. cochlear*, *C. oculus* may preferentially be consuming a mixture of algae but, unlike *S. cochlear*, may only be acquiring encrusting coralline algae in its diet through its foraging habit. A number of other grazers are known to perform best when foraging and maintaining mixed diets of intertidal algae, preferentially consuming mixtures of algae even when each of these foods is readily available (Kitting, 1980; Hagele & Rowell-Rahier, 1999; Cruz-Rivera & Hay, 2001; 2003). It may well be this grazing habit that reduces interspecific competition amongst foraging grazers but at the same time leads to high grazing frequencies.

Conclusion

It was previously suggested that fleshy algae are more nutritious than encrusting coralline algae (Branch, 1980). Our results have shown that this is not entirely true. The two corallines sampled (*L. ferox* and *S. yendoi*) are equally high in organic content compared to many other encrusting and turfy algae and even higher compared to many foliose algae. We have shown that encrusting coralline algae make up a large proportion of the diets of many intertidal grazers, some consuming more coralline algae than others. Herbivores capable of doing so may be incorporating much encrusting coralline alga in their diets, as these are generally high in organic content, and they may supplement them with other seaweeds that may or may not be readily available. Certainly, encrusting coralline algae play an important role in the diets of many intertidal herbivores.

Acknowledgements

We thank the University of the Western Cape for providing funding, research equipment and space, the South African National Research Foundation for research grants to GWM and the International Ocean Institute of Southern Africa for study bursaries to DW and MP. Basil Julies and Gerald Malgas provided valuable assistance with the operation of the SEM and the X-ray analysis of the various herbivore radulae. A special thanks to Gordon Harkins and Verno Gordon for assistance with statistical analyses and field work respectively.

References

Adey WH, McIntyre IG (1973) Crustose coralline algae: A re-evaluation in the geological sciences. Geological Society of America Bulletin 84: 883–904.

Branch GM (1971) The ecology of *Patella* Linnaeus from the Cape Peninsula, South Africa. I. Zonation, movements and feeding. Zoologica Africana 6: 1–38.

Branch GM (1975) Mechanisms reducing intraspecific competition in *Patella* spp.: Migration, differentiation and territorial behaviour. Journal of Animal Ecology 44: 575–600.

Branch GM (1976) Interspecific competition experienced by South African *Patella* species. Journal of Animal Ecology 45: 507–529.

Branch GM (1980) Territoriality in limpets: Manipulation experiments and energy budgets. Journal of the Malacological Society of Australasia 4: 245–246.

Branch G (1985) Limpets: Their role in littoral and sublittoral community dynamics. In PG Moore & R Seed (eds.), The Ecology of Rocky Coasts. Hodder & Stoughton Educational: 97–116.

Branch GM, Branch M (1988) The Living Shores of Southern Africa. Struik, Cape Town, 272 pp.

Branch GM, Griffiths L (1988) The Benguela Ecosystem part V. The coastal zone. Oceanography and Marine Biology. An Annual Review 26: 395–486.

Breitburg DL (1984) Residual effects of grazing: Inhibition of competitor recruitment by encrusting coralline algae. Ecology 65: 1136–1143.

Cruz-Rivera E, Hay ME (2001) Macroalgal traits and the feeding and fitness of an herbivorous amphipod: The roles of selectivity, mixing and compensation. Marine Ecology Progress Series 218: 249–266.

Cruz-Rivera E, Hay ME (2003) Prey nutritional quality interacts with chemical defenses to affect consumer feeding and fitness. Ecological Monographs: 73: 483–506.

Dethier MN (1994) The ecology of intertidal algal crusts: Variation within a functional group. Journal of Experimental Marine Biology and Ecology 177: 37–71.

Dethier MN, Paull KM, Woodbury MM (1991) Distribution and thickness patterns in subtidal encrusting algae from Washington. Botanica Marina 34: 201–210.

Fretter V, Graham A (1994) British Prosobranch Molluscs. Their functional anatomy and ecology. Ray Society, London, 820 pp.

Fujita D (1992) Grazing on the crustose coralline alga *Lithophyllum yessoense* by the sea urchin *Strongylocentrotus nudus* and the limpet *Acmaea pallida*. Benthos Research 42: 49–54.

Hagele BF, Rowell-Rahier M (1999) Dietary mixing in three generalist herbivores: nutrient complementation or toxin dilution? Oecologia 119: 521–533.

Keats DW, Maneveldt G (1994) *Leptophytum foveatum* Chamberlain & Keats (Rhodophyta, Corallinales) retaliates against competitive overgrowth by other encrusting algae. Journal of Experimental Marine Biology and Ecology 175: 243–251.

Keats DW, Matthews I, Maneveldt G (1994a) Competitive relationships and coexistence in a guild of crustose algae in the eulittoral zone, Cape Province, South Africa. South African Journal of Botany 60: 108–113.

Keats DW, Wilton P, Maneveldt G (1994b) Ecological significance of deep-layer sloughing in the eulittoral zone coralline alga, *Spongites yendoi* (Foslie) Chamberlain (Corallinaceae, Rhodophyta) in South Africa. Journal of Experimental Marine Biology and Ecology 175: 145–154.

Kitting CL (1980) Herbivore-plant interactions of individual limpets maintaining a mixed diet of intertidal marine algae. Ecological Monographs 50: 527–550.

Littler DS, Littler MM (2003) South Pacific Reef Plants: A divers' guide to the plant life of South Pacific Coral Reefs. Offshore Graphics Inc., Washington, 331 pp.

Littler MM, Littler DS, Taylor PR (1995) Selective herbivore increases biomass of its prey: A chiton-coralline reef-building association. Ecology 76: 1666–1681.

Lüning K (1990) Seaweeds. Their Environment, Biogeography and Ecophysiology. Wiley-Interscience, New York, 527 pp.

Mann HB, Whitney DR (1947) On the test of whether one or two random variables is stochastically larger than the other. The Annals of Mathematics and Statistics 18: 50–60.

Maneveldt G (1995) Geographical studies on the interaction between the pear limpet, *Patella cochlear*, and the encrusting coralline alga, *Spongites yendoi*. Unpublished MSc thesis. University of the Western Cape, Bellville, South Africa, 154 pp.

Paine RT (1984) Ecological determinism in the competition for space. Ecology 65: 1339–1348.

Paine RT, Vadas RL (1969) The effects of grazing by the sea urchins, *Strongylocentrotus* spp. on benthic algal populations. Limnological Oceanography 14: 710–719.

Raffaelli D, Hawkins S (1996) Intertidal Ecology. Chapman and Hall,London, 356 pp.

Sousa WP, Connell JH (1992) Grazing and succession in marine algae. In John, D. M., HawkinsS. J. , & PriceJ. H. (eds), Plant-Animal Interactions in the Marine Benthos. Systematics Association Special Volume Number 46. The Systematics Association. Clarendon Press, Oxford: pp. 425–441.

Stegenga H, Bolton JJ, Anderson RJ (1997) Seaweeds of the South African West Coast. Contributions from the Bolus Herbarium, University of Cape Town 18: 655 pp.

Steneck RS, (1982) A limpet-coralline alga association: Adaptations and defences between a selective herbivore and its prey. Ecology 63: 507–522.

Steneck RS, (1983) Escalating herbivory and resulting adaptive trends in calcareous crusts. Paleobiology 9: 44–61.

Steneck RS (1985) Adaptations of crustose coralline algae to herbivory: Patterns in space and time. In Toomy, D. F., & Nitecki M. H. (eds), Paleoalgology: Contemporary Research and Applications. Springer-Verlag, Berlin: pp. 352–366.

Steneck RS (1986) The ecology of coralline algal crusts: Convergent patterns and adaptive strategies. Annual Review of Ecological Systematics 17: 273–303.

Steneck RS, Dethier MN (1994) A functional group approach to the structure of algal-dominated communities. Oikos 69: 476–498.

Steneck RS, Watling L (1982) Feeding capabilities and limitations of herbivorous molluscs: A functional group approach. Marine Biology 68: 299–319.

Steneck RS, Hacker SD, Dethier MN (1991) Mechanisms of competitive dominance between crustose coralline algae: An herbivore-mediated competitive reversal. Ecology 72: 938–950.

Journal of Applied Phycology (2006) 18: 629–636
DOI: 10.1007/s10811-006-9071-5

Phenology of *Sargassum spp.* in Tung Ping Chau Marine Park, Hong Kong SAR, China

Put O. Ang, Jr.
Environmental Science Programme, Department of Biology, The Chinese University of Hong Kong, Shatin, N.T. Hong Kong SAR, China
e-mail: put-ang@cuhk.edu.hk

Key words: *Sargassum*, seaweeds, Hong Kong, phenology, growth, plant length, reproduction

Abstract

Eight species of the brown alga *Sargassum* have been recorded from Tung Ping Chau Marine Park in Hong Kong and the phenology of four of these was monitored from 1996 to 2000. All four species followed a typical growth cycle of *Sargassum* species reported elsewhere but with some annual variations. For *S. hemiphyllum*, the maximum mean (±SD) plant length, ranging from 38.5 ± 10.5 to 61.9 ± 19.9 cm, was recorded in January to March. The peak reproductive season was also mainly in February to March with up to 89% of the plants being reproductive. Some plants, however, remained reproductive until May. For *S. henslowianum*, the maximum mean plant length, ranging from 45.5 ± 25.5 to 77 ± 24.8 cm, was recorded mainly in November to January. The peak reproductive season was in November to February with up to 100% of the plants being reproductive. For *S. siliquastrum*, the maximum mean plant length, ranging from 48.2 ± 29.9 to 63.4 ± 22.1 cm, was also recorded mainly in January. The reproductive plants were found mainly between late December and mid February with up to 98% of the plants being reproductive. For *S. patens*, the maximum mean plant length ranging from 87.6 ± 62.4 to 118.7 ± 41.3 cm was recorded in January to March. Reproduction of this species was not monitored. Changes in water temperature over seasons were likely to be critical in affecting the phenological patterns of these species.

Introduction

Hong Kong is located on the southern coast of China. About 200 species of marine macrobenthic algae have been reported from this area (Ang, 2005). These algae are mainly distributed on the eastern shore and are characterized by both temperate as well as tropical species. Hong Kong algae were reported to exhibit a strong seasonality in their abundance (Hodgkiss, 1984). Most algae grow during the colder months from October to May, and disappear completely during the hot summer of June to September. This pattern is particularly true for the intertidal species. Information on subtidal algae, however, is very limited or almost non-existent. Until recently, there have been no detailed studies on the phenology of any algal populations within Hong Kong waters. This is partly because many of the areas where algal species are found are not readily accessible, and partly because the interest in algal studies has been limited.

On tropical and subtropical coasts exposed to strong or moderately strong waves, brown algae of the genus *Sargassum* are usually dominant in terms of algal cover and standing crop. The coast around Hong Kong is no exception. This study is an initial attempt to examine the phenology of *Sargassum* populations in Hong Kong water in greater detail, to contribute to our general understanding of the Hong Kong marine environment. Many of the coastal areas in Hong Kong where *Sargassum* spp. are dominant are currently threatened by marine pollution and coastal reclamation projects. An understanding of the dynamics of these algal species is essential in providing baseline information to assess environmental impacts caused by coastal developments.

Most studies undertaken elsewhere have indicated that *Sargassum* populations have a seasonal cycle of growth, reproduction, senescence and die-back (e.g. De Wreede, 1976; Ang, 1985; Largo & Ohno, 1992; Kendrick 1993). The timing of die-back, however, appears to vary according to species and locality. Tide has been suggested to be a critical factor in structuring the phenological patterns of two *Sargassum* populations from the Philippines (Ang, 1985). *Sargassum* died back at the time when lowest low tide of the year exposed the plants for a prolonged period. It is, however, not clear if the influence of tide plays a similar critical role in other *Sargassum* populations. Many *Sargassum* populations are found in the subtidal, i.e. they are never exposed to air. Questions remain as to whether these plants also exhibit seasonal die-back, just like their intertidal counterparts.

Materials and methods

This study was carried out in Long Lok Shui, on the western side of the island of Tung Ping Chau (114°26′E, 22°33′N), on the northeastern part of New Territories, Hong Kong (Figure 1). This island and the surrounding sea area are now a marine park. The study site is characterized by the presence of outcrops of sedimentary rocks, mainly shale, to which the marine algae are attached. Most of the algae are confined within the intertidal and shallow subtidal areas, extending to a depth of 10 m below Chart Datum. At least eight species of *Sargassum*, *S. angustifolium* (Turn.) Ag., *S. enerve* Ag., *S. fusiforme* (Harv.) Setch., *S. glaucescens* J. Ag., *S. hemiphyllum* (Turn.) Ag., *S. henslowianum* C. Ag., *S. patens* Ag. and *S. siliquastrum* (Turn.) Ag., may be found in the low intertidal to shallow subtidal, with the latter four species, *S. hemiphyllum, S. henslowianum, S. patens* and *S. siliquastrum,* being the most dominant.

The phenology of these four dominant *Sargassum* species was monitored. Biweekly to monthly visits to the site were conducted over the period of three and a half years, from November 1996 to June 2000, to study *S. hemiphyllum, S. henslowianum* and *S. siliquastrum* and from November 1997 to May 2000 to study *S. patens*. On each visit, 100 individuals of each of these four *Sargassum* species were haphazardly selected and measured. The maximum length was measured as the distance between the holdfast and the longest branch. For each individual, the presence or absence of reproductive structures (receptacles) was also noted. The peak reproductive season was considered to be the time when the highest percentage of the plants bore receptacles. All measurements were done *in situ* by snorkelling or SCUBA diving.

Results

Over the three and a half year sampling period, *S. hemiphyllum* attained a maximum mean plant length (± SD), ranging from 38.5 ± 10.5 to 61.9 ± 19.9 cm, mainly in January to March (Figure 2A). The longest mean plant length was recorded in February 28 1999 and the longest plant recorded at this same time was 109 cm in length. The peak reproductive season was recorded also mainly in February to March with up to 89% of the plants being reproductive (Figure 2B). Some individuals may remain reproductive until May. The plants died back very quickly from March to May after releasing their gametes. New laterals started to emerge from the holdfast and grew very slowly from May to September, before starting their rapid growth phase in November. There is evidence to indicate that some individuals survived only for one year.

Based on monitoring of tagged individuals, *Sargassum henslowianum* was shown to be a perennial species (Ang, unpublished). Individuals of this species attained their maximum mean plant length (± SD), ranging from 45.5 ± 25.5 to 78.36 ± 27.63 cm, mainly in November to January (Figure 3A). The longest mean plant length was recorded in Nov 28 1998 with the longest plant of 169.9 cm also recorded then. The peak reproductive season was in November to February, with up to 100% of the plants being reproductive (Figure 3B). Die-back occurred from February to April, and new laterals from the holdfast grew slowly from May to September. Rapid growth took place from September onward. There were some annual variations in the time the plants attained their maximum size and their peak reproductive period. Individuals of this species attained their maximum length in January 1997 and January 1998, whereas they attained their maximum length much earlier in November 1998 and November 1999 in the subsequent annual growth cycles. The rapid growth period in both 1998 and 1999 appeared to be shorter by two months than that in 1997 (and presumably also in 1996). This variation was not obvious in other species. The reproductive period was also earlier and longer in 1999. There were apparently two peaks of reproduction and some plants were found to be reproductive in May to September 1999. There was also a shift in the onset of reproductive period with some plants becoming reproductive earlier in October

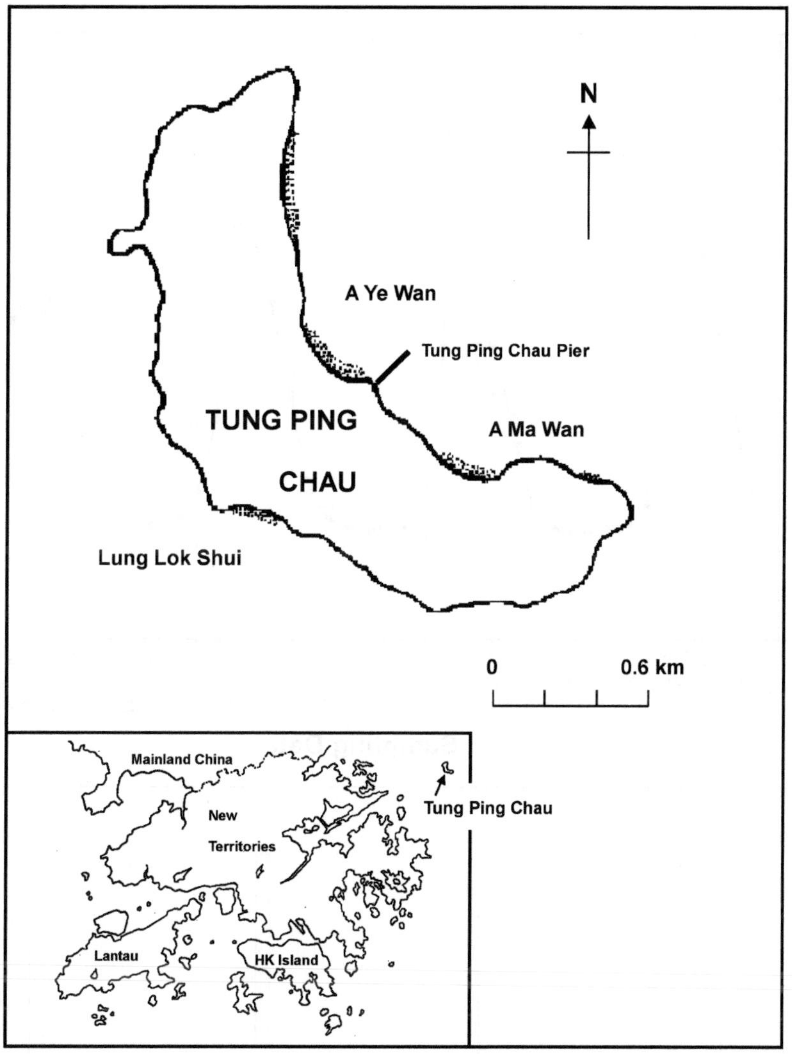

Figure 1. Map of Tung Ping Chau Marine Park showing the location of the study site, Lung Lok Shui. Insert: map of Hong Kong showing the location of Tung Ping Chau.

1998 and 1999 as compared with December in 1996 and late November in 1997.

Sargassum siliquastrum is also a perennial species (Chan, 2002). Individuals of this species attained their maximum mean plant length (± SD), ranging from 48.2 ± 29.9 to 63.4 ± 22.1 cm, mainly in January (Figure 4A). The longest mean plant length was recorded in January 4 1998 and the longest plant measured was 114.2 cm in length. There was a second, but much smaller peak of mean plant length recorded in June 1997, but this was not obvious in July 1998 nor was it observed in 1999. The reproductive plants were found mainly between late December and mid February with up to 98% of the plants being reproductive (Figure 4B). The onset of the reproductive period was apparently earlier in 1999. A few individuals (<2%) were found to have their receptacles in May 2000. This appears to be an exception rather than the rule as no reproductive plants were observed within the same period in the previous three years.

For *Sargassum patens*, also a perennial species, the maximum mean plant length (± SD), ranging from 87.6 ± 62.4 to 118.7 ± 41.3 cm, was recorded in January to March (Figure 5). Individuals of this species were the longest among the four species of *Sargassum* examined. The longest individual was 204 cm in length, recorded on March 29 1998. The longest mean plant length (±SD) was also recorded on March 29 1998.

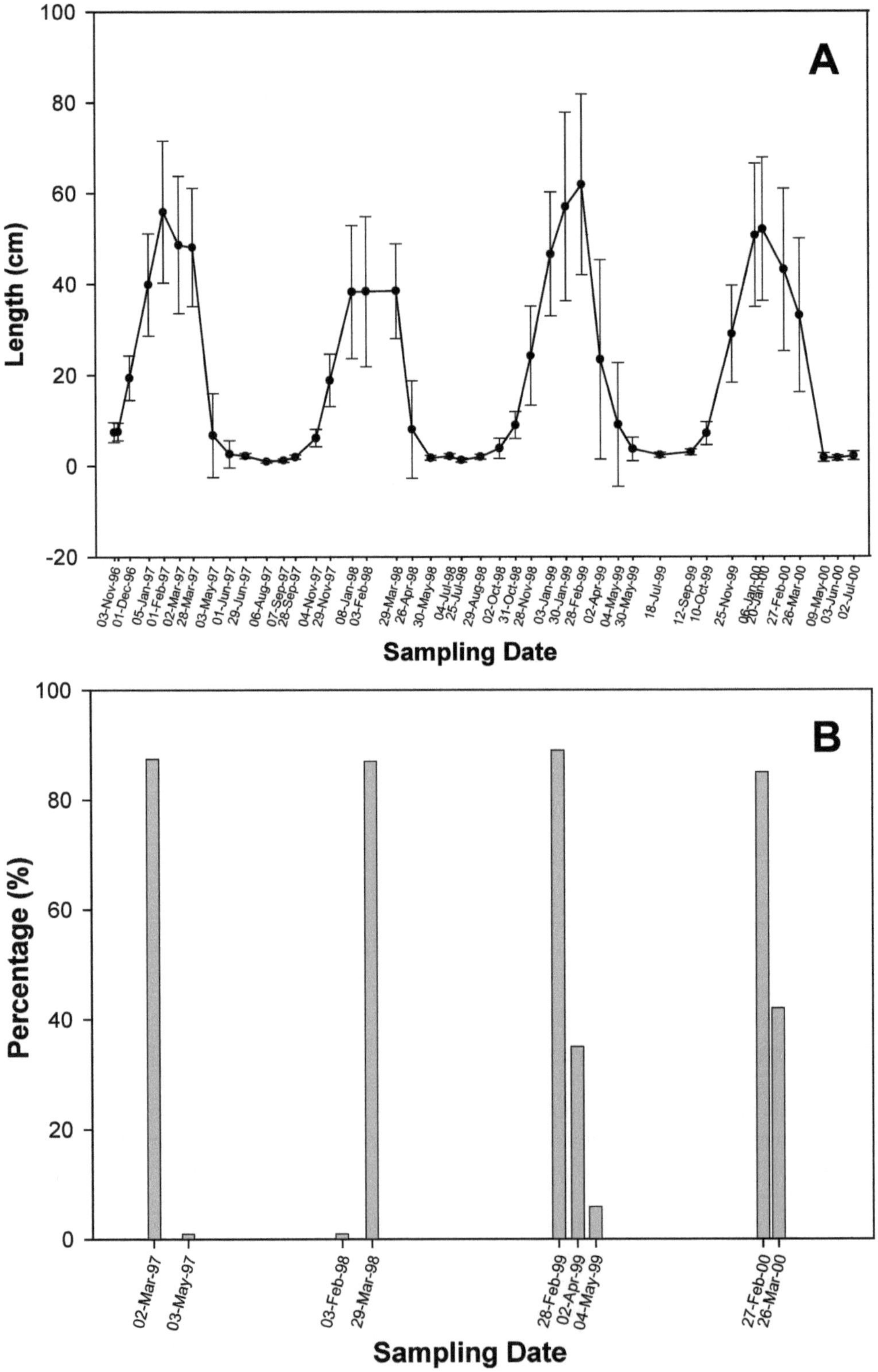

Figure 2. (A) Mean length (±SD) of 100 individuals in the *Sargassum hemiphyllum* population at different sampling dates over the study period. (B) Percentage of individuals in the population of *Sargassum hemiphyllum* that became reproductive at different times.

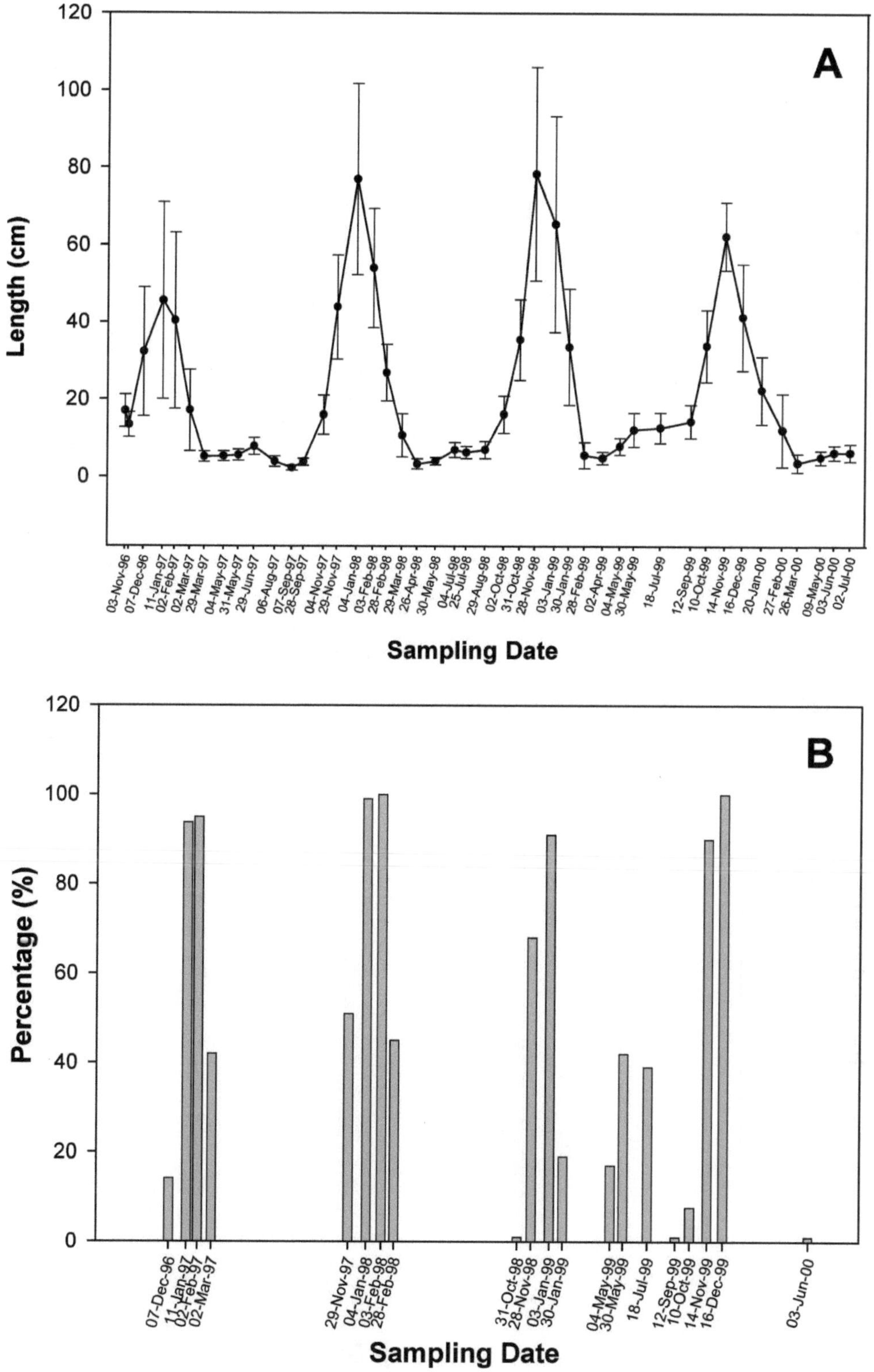

Figure 3. (A) Mean length (±SD) of 100 individuals in the *Sargassum henslowianum* population at different sampling dates over the study period. (B) Percentage of individuals in the population of *Sargassum henslowianum* that became reproductive at different times.

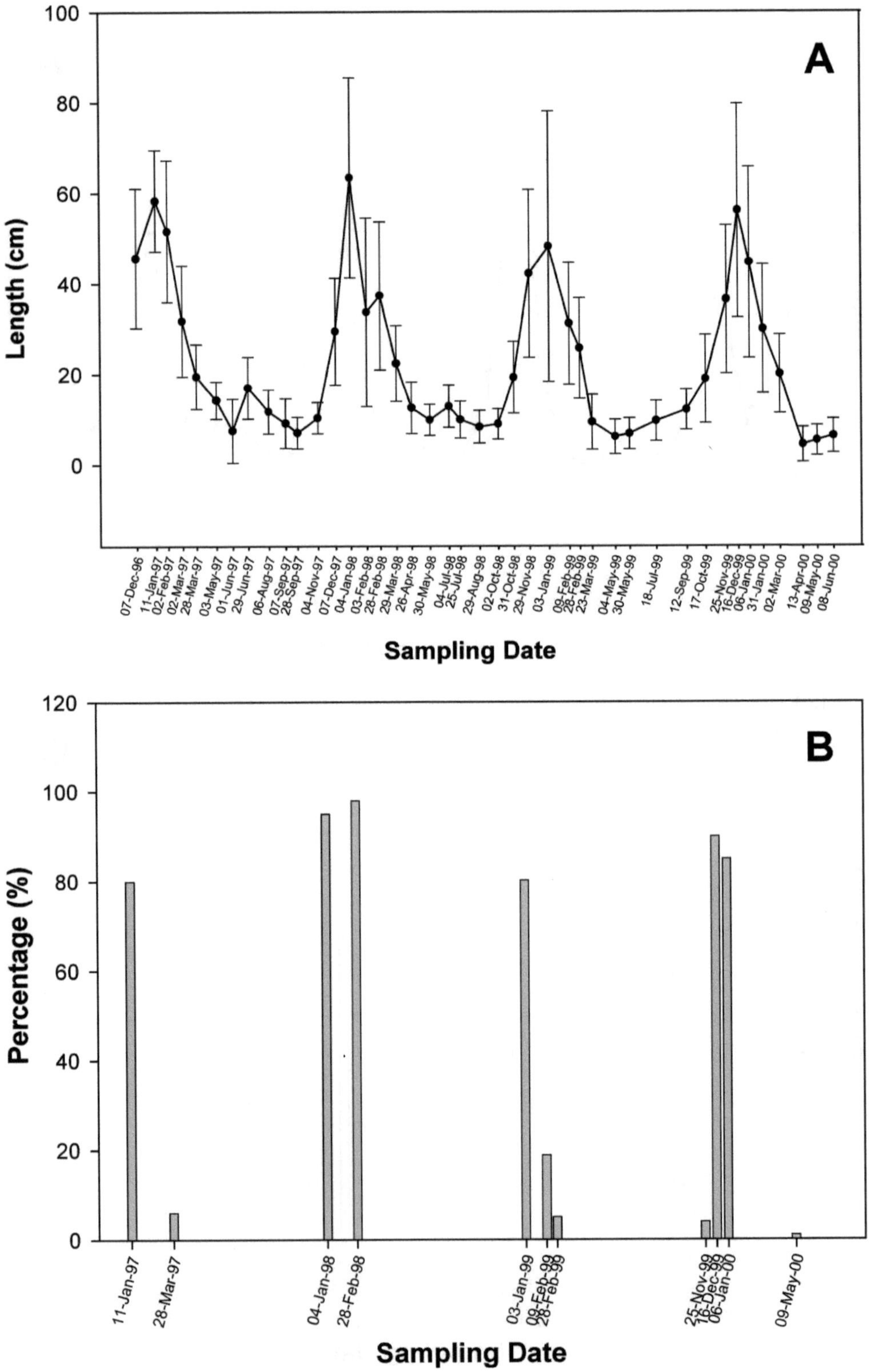

Figure 4. (A) Mean length (±SD) of 100 individuals in the *Sargassum siliquastrum* population at different sampling dates over the study period. (B) Percentage of individuals in the population of *Sargassum siliquastrum* that became reproductive at different times.

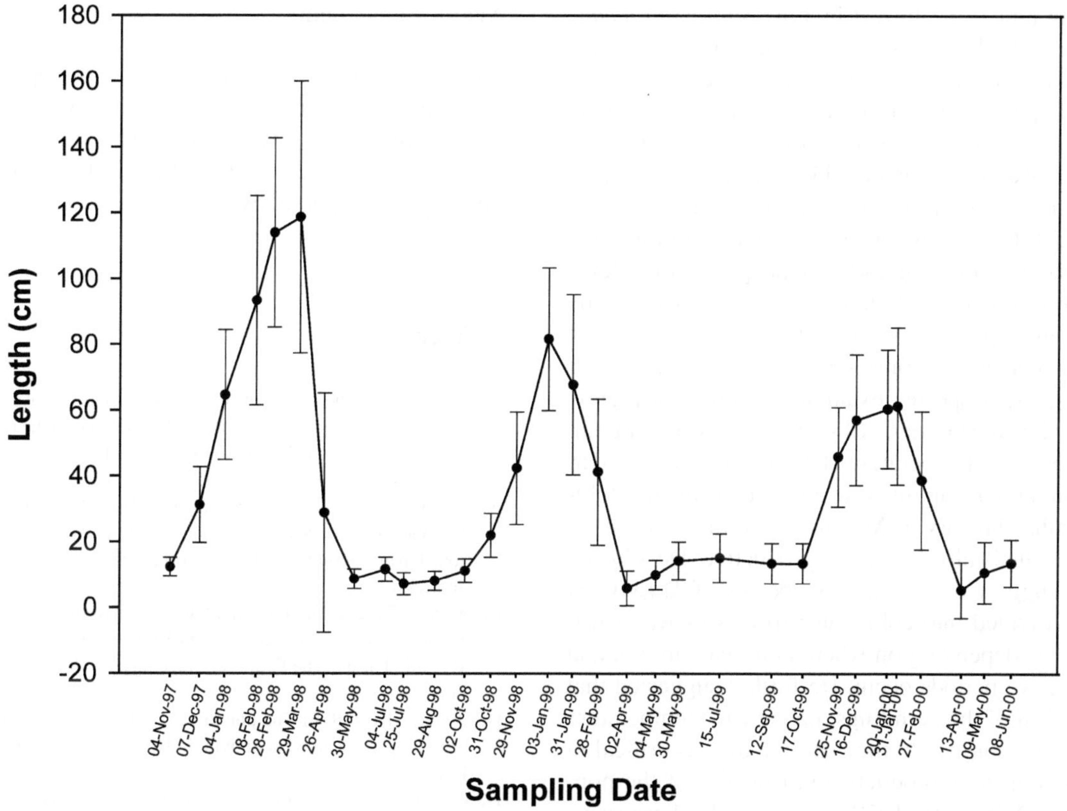

Figure 5. Mean length (±SD) of 100 individuals in the *Sargassum patens* population at different sampling dates over the study period.

Reproduction of this species was not monitored but it was likely to be around January to March as die-back commenced rapidly in March to April in 1998 or even earlier in 1999 and 2000. Plants were slow in growth from May to October and, rapid growth took place from October to December. There was an apparent drop in the size of the plants from 1998 to 2000. Around 70% of the individuals measured in March 1998 were longer than 1 m, while only 5% of the individuals were more than 1 m in length in January 2000. The longest plant measured in January 2000 was only 132.2 cm in length.

Discussion

All four species examined in this study followed a typical growth cycle of *Sargassum* species as reported elsewhere (e.g. McCourt, 1984; Ang, 1985; Huang et al., 1990; Largo & Ohno, 1992; Kendrick, 1993; Wong & Phang, 2004). This cycle is characterized by the presence of a slow growth phase, a rapid growth phase, and a reproductive phase that is followed by senescence and die back. Some annual variations, however, were observed between and within species. This is especially so with respect to the timing of rapid growth, the time when maximum plant size is attained and the timing and duration of reproductive period.

Sargassum hemiphyllum exhibited rapid growth and attained maximum size and reproduction later in the season than the other three species. These differences in growth and reproductive seasonality may be related to its spatial location along the tidal gradient. *Sargassum hemiphyllum* is present mainly in the lower intertidal and shallow subtidal zones. It occurs at the highest tide level of the *Sargassum* species examined in this study. *Sargassum henslowianum* is found mainly in the shallow subtidal, followed by *S. siliquastrum* and *S. patens* which are found in slightly deeper waters than *S. henslowianum.* The distribution range of *S. siliquastrum* extends to 10 m below Chart Datum, but the phenological patterns observed here are mainly those of individuals found in less than 3 m depth.

The tidal cycle in Hong Kong is mainly mixed semidiurnal. Daytime lowest low tide starts to occur in spring and is the lowest in July. *Sargassum hemiphyllum* could be exposed at extreme low tide in mid-spring to summer but the other three species are true subtidal species and are not exposed to air even during the lowest tide in July. Die-back of *S. hemiphyllum* plants in March to May could be correlated with their emergence in the spring low tide. However, the other three subtidal *Sargassum* species started to die back before March, or even earlier. Changes in tidal level are therefore unlikely to have significant effect on the phenological patterns observed for these species.

Notwithstanding some inter-annual variations, all the *Sargassum* species examined in this study grew slowest during summer. All grew fast during autumn, when seawater temperature started to become colder. Maximum plant lengths were attained during the coldest months (January to March), when the majority of the individuals also became reproductive. De Wreede (1976) suggested that various species of *Sargassum* may be divided into cold water species or warm water species depending on when their peak growth and reproductive periods occur. He further suggested that the survival of the germlings, i.e., at what temperature range or season would they survive best, is critical in determining the reproductive seasonality of the populations (DeWreede, 1978). Some of the *Sargassum* species found in HongKong, e.g. *S. siliquastrum* and *S. patens,* are also found in temperate regions such as Japan and Korea. In Japan, *S. patens* attains its maximum sizes and becomes reproductive in late spring (Taniguchi & Yamada, 1978). Hong Kong winter water temperatures, at 14 to 16 °C, are comparable to the spring or summer temperature in many temperate regions. The shift in peak growth and reproductive periods from winter in subtropical areas like Hong Kong to spring and summer in temperate regions like Japan suggests temperature to be one of the most critical factors in affecting the phenological patterns of *Sargassum* spp. observed in Hong Kong and elsewhere.

The present study provides one of the longest terms (three and a half years) of phenological data on *Sargassum* ever collected. Although some inter-annual variations are observed, in general the phenological pattern among species is consistent over the years. Differences at the micro-level, e.g. between same month in different years, need further evaluation. Further assessment will be carried out to evaluate differences in phenological patterns observed among *Sargassum* species with respect to temporal changes in the environmental parameters as well as spatial changes along a latitudinal gradient.

Acknowledgements

The author wishes to acknowledge with appreciation the help provided by Tso Ka Yip, Kong Sau Lai, Chan Wai Yi and Tam Tze Wai in the field. Comments from two anonymous reviewers were most helpful. This project was supported by a RGC Earmarked Research Grant.

References

Ang, PO (1985) Phenology of *Sargassum siliquosum* Ag J. and *S. paniculatum* J. Ag. (Sargassaceae, Phaeophyta) in the reef flat of Balibago (Calatagan, Philippines). Proceedings of the 5th International Coral Reef Congress 5: 51–57.

Ang, PO (2005) Studies on marine algae in Hong Kong. In Critchley AT, Ohno M (eds), Seaweed Resources of the World. Second Edition, Japan International Cooperation Agency, Yokosuka, Japan (in press).

Chan, WY (2002) Phenology and the Cost of Reproduction of *Sargassum siliquastrum* (Turn.) Ag. in Tung Ping Chau, Hong Kong. M. Phil. Thesis. The Chinese University of Hong Kong, Hong Kong.

De Wreede, RE (1976) The phenology of three species of *Sargassum* (Sargassaceae, Phaeophyta) in Hawaii. Phycologia 15: 175–183.

De Wreede, RE (1978) Growth in varying culture conditions of embryos of three Hawaiian species of *Sargassum* (Phaeophyta, Sargassaceae). Phycologia 17: 23–31.

Hodgkiss, IJ (1984) Seasonal patterns of intertidal algal distribution in Hong Kong. Asian Marine Biology 1: 49–57.

Huang Z, Li C, Zheng D, Lin S, Zheng C, Wang J, Yan S, Lin N (1990) Population dynamics of *Sargassum* from the west nearshore waters of Daya Bay. Collection of Papers on Marine Ecology in Daya Bay II. China Ocean Press, Beijing: 305–314.

Kendrick GA (1993) *Sargassum* beds at Rottnest Island: Species composition and abundance. In Wells FE, Walker DI, Kirkman H., Lethbridge R (eds), Proceedings of the 5th International Marine Biology Workshop: The Marine Flora and Fauna of Rottnest Island, Western Australia, Western Australian Museum, Perth. Vol. 2, pp. 455–472.

Largo DB, Ohno M (1992) Phenology of two species of brown seaweeds, *Sargassum myriocystum* J. Agardh and *Sargassum siliquosum* J. Agardh (Sargassaceae, Fucales) in Liloan, Cebu, in Central Philippines. Bull. Mar. Sci. Fish. Kochi University 1: 17–27.

McCourt RM (1984) Seasonal patterns of abundance, distributions, and phenology in relation to growth strategies of three *Sargassum* species. J. Exp. Mar. Biol. Ecol. 74: 141–156.

Taniguchi K, Yamada Y (1978) Ecological study on *Sargassum patens* C. Agardh and *S. serratifolium* C. Agardh in the sublittoral zone at Iida Bay of Noto Peninsula in the Japan Sea. Bull. Jpn. Sea Regional Fish. Res. Lab. 29: 239–253.

Wong CL, Phang SM (2004) Biomass production of two *Sargassum* species at Cape Rachado, Malaysia. Hydrobiologia 512: 79–88.

Journal of Applied Phycology (2006) 18: 637–641
DOI: 10.1007/s10811-006-9076-0

Biogeography of Alaskan seaweeds

Sandra C. Lindstrom
Department of Botany, #3529-6270 University Blvd., University of British Columbia, Vancouver, BC, Canada V6T 1Z4

e-mail: sandracl@interchange.ubc.ca

Key words: Alaska, biogeography, macroalgae, seaweeds

Abstract

A recent survey of seaweed specimens collected in Alaska over the past two centuries, together with the application of molecular techniques to recent collections, has revealed a surprisingly diverse flora given the history of glaciation, large areas of unsuitable habitat, and otherwise harsh environmental conditions. The number of recognized species has increased from 376 in 1977 to about 550 today. Species show a variety of biogeographic patterns: species that occur primarily to the south and have their northern limit in Alaska, species that occur primarily to the west and have their eastern limit in Alaska, species that are primarily Atlantic but extend through the Arctic to Alaska, and a number of endemics. Within these broad distribution patterns are more localized patterns often involving disjunctions. These disjunctions, the occurrence of endemic species, patterns of genotype distributions, and the overall richness of the seaweed flora support the idea that marine refugia must have existed in Alaska during Pleistocene glaciations.

Introduction

Situated at the convergence of the Asian and North American continents and at the confluence of the Pacific and Arctic Oceans, Alaska has a coastline of 75,000 km, longer than all the rest of the United States. Its morphologically diverse coastal areas support a rich assortment of macroalgae and other marine life, despite having been largely glaciated during Pleistocene Ice Ages.

Collection of seaweeds in Alaska began with the Russian expeditions in the eighteenth century, and a number of well-known northeast Pacific Ocean species are from types collected during these early Russian explorations in Alaska: *Alaria fistulosa* Postels et Ruprecht, *Ahnfeltia fastigiata* (Postels et Ruprecht) Makienko, *Bossiella cretacea* (Postels et Ruprecht) H.W. Johansen, *Calliarthron tuberculosum* (Postels et Ruprecht) E.Y. Dawson, *Endocladia muricata* (Endlicher) J. Agardh, *Osmundea spectabilis* (Postels et Ruprecht) Nam and *Soranthera ulvoidea* Postels et Ruprecht (1840).

Late nineteenth century collections were made by the Vega Expedition (Kjellman, 1889), the Harriman Alaska Expedition (Saunders, 1901) and the University of California Botanical Expedition to Alaska (Setchell & Gardner, 1903). Two expeditions in 1913, the U.S. Bureau of Soils Kelp Expedition (Frye, 1915; Rigg, 1915) and the Canadian Arctic Expedition (Collins, 1927), added to knowledge of the seaweed flora. Interest revived in the 1960s, with a number of expeditions organized in relation to the underground nuclear tests carried out at Amchitka Island (Lebednik et al., 1971; Wynne, 1970a,b, 1971a,b, 1980a,b, 1981), the Great Alaska Earthquake (Johansen, 1971) and by the University of British Columbia (Druehl, 1968, 1970; Widdowson, 1971; Scagel et al., 1989). More recent ecological and environmental studies have added significantly to knowledge of the flora, although many collections remain unexamined and, in some instances, even unprocessed.

Lindstrom (1977) provided a summary of knowledge of the seaweeds of Alaska and included a brief history of collecting efforts up to that time. The present effort, to inventory existing collections of seaweeds, was in response to a recognized need for an updated summary and as the first phase of a Seaweed Flora of Alaska. In this paper I summarize some of the conclusions that can be drawn from these collections.

Materials and methods

Seaweed specimens collected in Alaska were examined, species identifications were updated or corrected, and collection data were entered into a Microsoft Access database. Specimens came from the following herbaria (abbreviations after Holmgren et al., 1990): ALA, ALAJ, CANA, DUKE, HSC, L, LD, LE, MASS, MIL, SAP, SAPA, TEX, TNS, UC, UPS, as well as a number of private herbaria. Previously entered specimen data for UBC collections were appended to the database, and only specimens that appeared anomalous or questionable were examined.

Results and discussion

The Alaska Seaweed Database housed at the University of British Columbia currently consists of 22,413 records of seaweed specimens collected in Alaska (the database continues to be updated as more specimens are examined). Nearly 1000 names of specific and infraspecific taxa have been used for Alaskan seaweeds, of which approximately 550 are currently recognized as distinct species.

Among these 550 species, 321 that also occur in more southerly parts of North America (California, Oregon, Washington and/or British Columbia) reach their northwestern distribution limit in Alaska (Table 1). These data indicate that some areas of Alaska have a larger loss of species than others, and these trends are upheld even when the numbers are normalized to the total number of species occurring in the area or to the number of specimens examined from the area. The greatest proportion of southern species reach their northern limit in Southeast Alaska, followed by Prince William Sound, the Bering Sea, and the Western Aleutian Islands. Data for the latter two areas is somewhat artificial since some of the species indicated as dropping out in these areas actually extend further west to Siberia. The Kodiak Archipelago and the Eastern Aleutian Islands represent other significant distribution endpoints.

A smaller percentage of species (ca 200 of 550) occur primarily to the west and have their eastern distribution limit in Alaska. Among these are *Dumontia simplex* Cotton, *Neohypophyllum middendorfii* (Ruprecht) M.J. Wynne, *Pleonosporium kobayashii* Okamura, and *Porphyra pseudolinearis* Ueda, all of which show disjunctions within Alaska, between the northwestern Gulf of Alaska and the cold inside waters of northern Southeast Alaska. Table 2 lists species that fall into this category but represent new records for Alaska (i.e., not previously recorded in Lindstrom, 1977, or Scagel et al., 1989).

Table 1. Number of "southern" species that reach their northern distribution limit in each area

From To	California	Oregon	Washington	British Columbia	Total	As % of local species
Southeast	60	2	9	2	73	73/368 = 19.8%
Yakutat	9	0	0	1	10	10/182 = 5.6%
Prince William Sound	34	3	3	1	41	41/273 = 15.0%
Kenai Fjords	9	0	3	1	13	13/142 = 9.2%
Cook Inlet	5	1	0	0	6	6/160 = 3.8%
Kodiak Archipelago	19	1	7	2	29	29/210 = 13.8%
Alaska Peninsula	8	2	1	0	11	11/166 = 6.6%
Eastern Aleutian Is.	23	2	3	3	29	29/183 = 15.8%
Central Aleutian Is.	6	1	0	0	7	7/117 = 6.0%
Western Aleutian Is.	29	2	3	3	37	37/170 = 21.8%
Pribilof Islands	20	0	2	0	12	12/97 = 12.4%
Bering Sea	19	2	11	6	38	38/117 = 32.5%
Arctic Ocean	13	1	1	0	15	15/70 = 21.4%
Total	244	17	43	17	321	

Table 2. New records of species occurring in Alaska, previously known from more westerly locations

Species	Alaska distribution	Western distribution
Ahnfeltia plicata (Hudson) Fries	Bering Sea coast and Arctic Ocean	Kamchatka[1]
Beringia castanea Perestenko	Western Aleutian Islands	Southeast Kamchatka, Commander Islands[2]
Callophyllis beringensis Perestenko	St. Lawrence Island	Northern Kamchatka[1]
Callophyllis platyna Perestenko	St. Lawrence Island	St. Lawrence Bay, Russia[2]
Callophyllis radula Perestenko	Amchitka Island	Southeast Kamchatka[1]
Chordaria chordaeformis (Kjellman) Kawai et S.H. Kim	St. Lawrence Island[3], Cook Inlet	Hokkaido, Kamchatka[3]
Cirrulicarpus ruprechtiana (E.S. Zinova) Perestenko	St. Lawrence Island, Pribilof Islands, Central & Western Aleutian Islands	Kurile Islands (Urup, Paramushir), Commander Islands[2]
Congregatocarpus pacificus (Yamada) Mikami	Pribilof Islands, Western Aleutian Islands	Japan Sea, Hokkaido, Sakhalin, Kurile Islands[2], Commander Islands[4]
Kallymeniopsis lacera (Postels et Ruprecht) Perestenko	Bering Sea to Southeast Alaska	Okhotsk Sea, Southeast Kamchatka, middle & northern Kurile Islands, Commander Islands[2]
Kallymeniopsis verrucosa Zinova et Gussarova	Pribilof Islands, Eastern Aleutian Islands	Sakhalin, Kurile Islands[2]
Laminariocolax tomentosoides (Farlow) Kylin	Barren Islands (Kenai Peninsula)	Southeast Kamchatka[1]
Lukinia dissecta Perestenko	Pribilof Islands, Western Aleutian Islands	Sakhalin, Commander Islands[2]
Neoabbottiella araneosa (Perestenko) S.C. Lindstrom	St. Matthew Island (Bering Sea), Kodiak Island	Japan Sea, Sakhalin Island, middle Kurile Islands, Paramushir Island, Commander, Islands[2]
Phyllariella ochotensis Yu.E. Petrov et Vozzhinskaya	Pribilof Islands	Okhotsk Sea[5]
Polysiphonia morrowii Harvey	Kodiak Island area	Yellow Sea, Japan Sea,Okhotsk Sea, Honshu, Hokkaido, Sakhalin, Kurile Islands[2], Commander Islands[4]
Rhodomela sibirica Zinova et K.L. Vinogradova	Arctic Ocean, Chukchi Sea, St. Lawrence Island	Okhotsk Sea, Kamchatka, East Siberian Sea[2]
Velatocarpus pustulosus (Postels et Ruprecht) Perestenko	Pribilof Islands, Eastern Aleutian Islands	Japan Sea, Okhotsk Sea, Sakhalin, Southeast Kamchatka, middle and northern Kurile Islands[2], Commander Islands[4]

[1]Klochkova (1998).
[2]Perestenko (1994)
[3]Kim and Kawai (2002).
[4]Selivanova and Zhigadlova (1997).
[5]Petrov and Vozzhinskaja (1966).

About twenty species are primarily Atlantic but extend through the Arctic Ocean to Alaska. Among these are *Phyllophora truncata* (Pallas) Zinova and *Fimbrifolium dichotomum* (Lepechin) G.I. Hansen. *Rbc*L gene sequences of Alaskan specimens of these species show near-identity or identity, respectively, to their Atlantic conspecifics (Lindstrom, 2001; pers. obs.). *Phyllophora truncata* has thus far been recorded from only the Arctic coast, the northern Bering Sea, and the cold inside waters of northern Southeast Alaska. *Fimbrifolium dichotomum* has been collected in the Bering Sea and the Barren Islands in the northwest Gulf of Alaska.

The number of endemics is uncertain. The recently described *Dumontia alaskana* V. Tai, S.C. Lindstrom et G.W. Saunders and *Porphyra aestivalis* S.C. Lindstrom et Fredericq, currently known only from Alaska, are probably more widespread and have yet to be reported from elsewhere. Other species reported to be endemic include *Alaria crispa* Kjellman, *Asterocolax hypophyllophila* M.J. Wynne, *Compsonema tenue* Setchell et N.L. Gardner, *Neohalosacciocolax aleutica* I.K. Lee, *Odonthalia uniseriata* Masuda et K.A. Miller, *Orculifilum denticulatum* S.C. Lindstrom, *Pleonosporium pedicellatum* S.C. Lindstrom, M.J. Wynne, et N.I. Calvin,

Punctaria lobtata (Saunders) Setchell et N.L. Gardner, *Rhodophyllis aleutica* M.J. Wynne, *Tokidaea chilkatensis* S.C. Lindstrom et M.J. Wynne, *Zinovaea acanthocarpa* M.J. Wynne, and at least two undescribed foliose red algae from the Aleutian Islands (R.E. Norris, K.A. Miller, pers. comm.). Some of these species need to be re-examined using modern biosystematic methods to ensure their distinctness from congeners.

Within these broad distribution patterns are more localized patterns often involving disjunctions. A pattern in which predominantly Arctic and western Pacific species are restricted to the cold inside waters of northern Southeast Alaska was previously described by Lindstrom et al. (1986). The much larger database on which the present study is based supports the restriction of many of these and other species to sites that remain cold year-round. The present study also found a complementary pattern in which more southerly species occur as disjuncts in other areas of Alaska. *Phaeostrophion irregulare* Setchell et N.L. Gardner, *Ahnfeltiopsis gigartinoides* (J. Agardh) P.C. Silva et DeCew, *Callithamnion biseriatum* Kylin, *Hollenbergia subulata* (Harvey) E.M. Wollaston, *Kallymeniopsis oblongifructa* (Setchell) G.I. Hansen, *Mazzaella sanguinea* (Setchell et Gardner) Hommersand, and *Pikea californica* Harvey have disjunct populations in the northern Gulf of Alaska but have yet to be recorded from Southeast Alaska.

Disjunctions, occurrence of endemic species, patterns of genotype distributions (e.g., Lindstrom et al., 1997), and the overall richness of the seaweed flora all lend support to the idea that marine refugia must have existed in Alaska during Pleistocene glaciations. Ice Age refugia would have provided space for species that are now restricted to cold inside waters, areas that were recolonized after the ice receded. But refugia would have also provided space for species that still occur in outer coastal areas but are disjunct from more southerly populations. Present-day seaweed distributions indicate that one or more refugia must have been located in the northeastern Gulf of Alaska.

The Alaska Seaweed Database is not currently online, although plans are underway to make it available on the worldwide web. A subset of data collected as part of this project, the Algal Types Database, is accessible at: http://herbarium.botany.ubc.ca/herbarium_data/algaltypes_web/default.htm.

Acknowledgments

The research reported herein was supported by grants from the U.S. National Science Foundation (DEB-9870215) and from the Natural Sciences and Engineering Research Council of Canada.

References

Collins FS (1927) Marine algae from Bering Strait and Arctic Ocean collected by the Canadian Arctic Expedition, 1913–1916. Report of the Canadian Arctic Expedition, 1913–1918, Vol. IV: Botany, 3–16.

Druehl LD (1968) Taxonomy and distribution of northeast Pacific species of *Laminaria*. Can. J. Bot. 46: 539–547.

Druehl LD (1970) The pattern of Laminariales distribution in the northeast Pacific. Phycologia 9: 237–247.

Frye TC (1915) The kelp beds of Southeast Alaska. U.S.D.A. Report No. 100, Pt. IV: 60–104.

Holmgren PK, Holmgren NH, Barnett LC (1990) Index Herbariorum. Part I: The Herbaria of the World, 8th edition. New York Botanical Gardens, New York. (available online: http://www.nybg.org/bsci/ih/ih.html)

Johansen HW (1971) Effects of elevation changes on benthic algae in Prince William Sound. In The Great Alaska Earthquake of 1964: Biology. National Academy of Sciences, Washington 35–68.

Kim S-H, Kawai H (2002) Taxonomic revision of *Chordaria flagelliformis* (Chordariales, Phaeophyceae) including novel use of the intragenic spacer region of rDNA for phylogenetic analysis. Phycologia 41: 328–339.

Kjellman FR (1889) Om Beringhafvets algflora. Kongl. Svensk. Vetenskaps-Akad. Handl. 23(8): 1–58, pl. 1–7.

Klochkova NG (1998) An annotated bibliography of marine macroalgae on northwest coast of the Bering Sea and the southeast Kamchatka: The first revision of flora. Algae 13: 375–418.

Lebednik PA, Weinmann FC, Norris RE (1971) Spatial and seasonal distributions of marine algal communities at Amchitka Island, Alaska. Bioscience 21: 656–660.

Lindstrom SC (1977) An annotated bibliography of the benthic marine algae of Alaska. Alaska Department of Fish & Game, Juneau, Technical Data Report No. 31.

Lindstrom SC (2001) The Bering Strait connection: Dispersal and speciation in boreal macroalgae. J. Biogeogr. 28: 248–251.

Lindstrom SC, Calvin NI, Ellis RJ (1986) Benthic marine algae of the Juneau, Alaska area. Contribution to Natural Sciences, British Columbia Provincial Museum, Number 6, 10 pp.

Lindstrom SC, Olsen JL, Stam WT (1997) Postglacial recolonization and the biogeography of *Palmaria mollis* (Rhodophyta) along the Northeast Pacific coast. Can. J. Bot. 75: 1887–1896.

Perestenko LP (1994) Red algae of the far-eastern seas of Russia. Komarov Botanical Institute, Russian Academy of Sciences, St. Petersburg, 331 pp. [in Russian].

Petrov YE, Vozzhinskaja VB (1966) De genere ac specie novis Laminarialium e Mari Ochotensi notula. Novitates Systematicae Plantarum Non Vascularium 4: 100–102. [in Russian].

Postels A, Ruprecht FJ (1840) Illustrationes algarum in itinere circa-orbem. . .exsecuto in oceano Pacifico imprimis septentrionali ad littora rossica Asiatico-Americana collectarum. Pp. iv + 22 + [2], 40 pls. St. Petersburg.

Rigg GB (1915) The kelp beds of western Alaska. U.S.D.A. Report No. 100, Part V: 105–122, 40 pl.

Saunders DA (1901) The algae of the expedition. Proc. Wash. Acad. Sci. 3: 391–487.

Scagel RF, Gabrielson PW, Garbary DJ, Golden L, Hawkes MW, Lindstrom SC, Oliveira JC, Widdowson TB (1989) A synopsis of the benthic marine algae of British Columbia, Southeast Alaska, Washington and Oregon. University of British Columbia, Vancouver, Phycological Contribution No. 3, 532 pp. Reprinted 1993.

Selivanova ON, Zhigadlova GG (1997) Marine algae of the Commander Islands. Preliminary remarks on the revision of the flora. III. Rhodophyta. Bot. Mar. 40: 15–24.

Setchell WA, Gardner NL (1903) Algae of northwestern America. Univ. Calif. Publ. Bot. 1: 165–418, pl. 17–27.

Widdowson TB (1971) A taxonomic revision of the genus *Alaria* Greville. Syesis 4: 11–49.

Wynne MJ (1970a) Marine algae of Amchitka Island (Aleutian Islands). I. Delesseriaceae. Syesis 3: 95–144.

Wynne MJ (1970b) Marine algae of Amchitka Island (Aleutian Islands). II. Bonnemaisoniaceae. Pac. Sci. 24: 433- -438.

Wynne MJ (1971a) Concerning the Phaeophycean genera *Analipus* and *Heterochordaria*. Phycologia 10: 169–175.

Wynne MJ (1971b) The genus *Porphyra* at Amchitka Island, Aleutians. Proc. Int. Seaweed Symp. 7: 100–104.

Wynne MJ (1980a) *Boreothamnion* (Ceramiaceae, Ceramiales), a new red algal genus from Alaska. Contrib. Univ. Mich. Herb. 14: 209–219.

Wynne MJ (1980b) *Beringiella* (Rhodomelaceae, Ceramiales), a new red algal genus from Alaska. Contrib. Univ. Mich. Herb. 14: 221–229.

Wynne MJ (1981) A new species of *Rhodophyllis* (Gigartinales, Rhodophyta) from Amchitka Island, the Aleutians. Proc. Int. Seaweed Symp. 8: 528–534.

Journal of Applied Phycology (2006) 18: 643–651
DOI: 10.1007/s10811-006-9066-2

Systematics and genetic variation in commercial *Kappaphycus* and *Eucheuma* (Solieriaceae, Rhodophyta)

Giuseppe C. Zuccarello[1,*], Alan T. Critchley[2], Jennifer Smith[3], Volker Sieber[4], Genevieve Bleicher Lhonneur[2] & John A. West[5]
[1]*School of Biological Sciences, Victoria University of Wellington, P.O. Box 600, Wellington 6001, New Zealand;* [2]*Degussa Texturant Systems France SAS, Baupte, France;* [3]*Department of Botany, University of Hawaii Manoa, Honolulu, HI 96822, USA;* [4]*Degussa Food Ingredients GmbH, Freising, Germany;* [5]*School of Botany, University of Melbourne, Victoria 3010, Australia*

*Author for correspondence: e-mail: joe.zuccarello@vuw.ac.nz

Key words: *cox2-3* spacer, *Eucheuma*, genetic variation, *Kappaphycus*, RuBisCo spacer, "sacol"

Abstract

The systematics and taxonomy of *Kappaphycus* and *Eucheuma* (Solieriaceae) is confused and difficult due to morphological plasticity, lack of adequate characters to identify species and commercial names of convenience. These taxa are geographically widely dispersed through cultivation. Commercial, wild and herbarium sources were analysed; molecular markers provided insights into taxonomy and genetic variation, and where sources of genetic variation may be located. The mitochondrial *cox*2-3 and plastidal RuBisCo spacers were sequenced. There is a clear genetic distinction between *K. alvarezii* ("cottonii") and *K. striatum* ("sacol") samples. *Kappaphycus alvarezii* from Hawaii and some samples from Africa are also genetically distinct. Our data also show that all currently cultivated *K. alvarezii* from all over the world have a similar mitochondrial haplotype. Within *Eucheuma denticulatum* ("spinosum") most African samples are again genetically distinct. Our data also suggest that currently cultivated *E. denticulatum* may have been "domesticated" several times, whereas this is not evident for the cultivated *K. alvarezii*. The present markers used do not distinguish all the morpho-types known in cultivation (e.g. var. *tambalang*, "giant" type) but do suggest that these markers may be useful to assess introductions and species identification in samples.

Introduction

The seaweeds most commonly cultivated for the carrageenan industry belong to the genera *Kappaphycus* Doty and *Eucheuma* J. Agardh. These crops are almost entirely farmed and are usually referred to by the commercial names "cottonii", "sacol" and "spinosum". The formal taxonomy of these taxa has for a long time been in confusion due to misapplication of commercial and scientific names, the known general paucity of adequate morphological characters and the morphological plasticity of seaweeds. Much of the taxonomic confusion was addressed by the pioneering work of Maxwell Doty (Doty, 1985, 1988; Doty & Norris, 1985). Even in the detailed work of Doty, variability in the presence or absence of diagnostic morphological characters within taxa was noted, especially in non-ideal specimens (i.e. non-reproductive specimens and specimens lacking typical attachment structures) and this was addressed by the caveat that descriptive paragraphs must carry the preamble "there is a tendency…" (Doty, 1988, p. 166).

Doty (1988), however, formally recognized certain species of *Eucheuma* as *Kappaphycus*, mostly based on their production of κ-carrageenan, and this generic circumscription has been supported, for the most part, by molecular studies (Fredericq et al., 1999; Aguilan et al., 2003). Nevertheless, questions remain as to the taxonomic identity of commercially produced strains. *Kappaphycus alvarezii* (Doty) Doty *ex* P. Silva is the most-grown commercial κ-carrageenan producer and many varieties and local strains are known (www.surialink.com). One of the commercially used 'strains' of *Kappaphycus* is the so-called "sacol" variety, but its scientific name is still unresolved. While it was originally considered to be *K. striatum* (Schmitz) Doty *ex* P. Silva (Trono, 1997), recent molecular investigations suggested that it could be a form of *K. cottonii* (Weber-van Bosse) Doty *ex* P. Silva (Aguilan et al., 2003). *Kappaphycus cottonii* is morphologically quite distinct from either *K. alvarezii* or *K. striatum* as it mostly forms prostrate branches. Culture studies have shown that many of the characters used to separate *Kappaphycus* species (e.g. habit, decumbent versus dichotomous) are found to segregate in tetraspore progeny (de Paula et al., 1999) from single plants, and it is likely that the identification of individual specimens based on morphology is unreliable.

Molecular markers have proven useful in not only elucidating red algal systematics but also in discovering genetic variation within red algal species. Commonly used intraspecific markers are the nuclear-encoded internal transcribed spacers of the ribosomal cistrons (ITS 1 and 2, e.g. Marston & Villalard-Bohnsack, 2002), the plastid-encoded RuBisCo spacer (e.g. Zuccarello et al., 2002) and the mitochondrial-encoded *cox2-3* spacer (Zuccarello & West, 2003), although these markers have their limitations, such as uniparental inheritance and limited variation (i.e. they do not reflect all the genetic variation found within groups). Studies using the RuBisCo spacer have shown that even single base pair changes could indicate reproductively isolated cryptic species (Zuccarello & West, 2003), while there is more variation within species at the *cox2-3* spacer region, due to its higher mutation rate (Zuccarello & West, 2002).

This work aimed to: (1) determine the levels of genetic variation in commercially grown species of *Kappaphycus* and *Eucheuma*; (2) clarify some of the taxonomic confusion in commercial strains and wild strains of *Kappaphycus* and *Eucheuma*; (3) determine which geographic regions contain samples with ecologically superior genotypes or with genetic variation that is potentially useful to the industry.

Materials and methods

Samples were for the most part obtained from commercial supplies. Thalli were selected from the corners of representative bales delivered from suppliers for industrial extraction of carrageenan. Samples were placed in silica gel until DNA extraction. Although drying and storage methods may have differed, nearly all samples were adequate for DNA extraction and amplification. Often exact provenance (specific region, farm) of the samples was unknown. Other samples were collected (Hawaii and Indonesia) and dried immediately in silica gel. Hawaiian samples collected at Kane'ohe Bay spanned the range of morphologies at this site where material is introduced and invasive, and contained many cystocarpic or tetrasporic specimens. Other samples were removed from herbarium sheets and processed as below. All samples used are listed in Table 1.

DNA extractions were performed using a modified CTAB extraction procedure. Dried samples (approx. 1 cm tip) were pulverized using a shaking mill (Retsch, type MM200) and then placed in a microfuge tube containing 500 μL of CTAB extraction buffer (2% CTAB, 0.1 M Tris-HCl (pH 8.0), 1.4 M NaCl, 20 mM EDTA, 1% PEG 8000) plus 50μg RNAse A and 80 g Proteinase K (Promega, Madison, USA). Samples were then placed at 60°C for 30 min, mixing occasionally. Two extractions using an equal volume of chloroform:isoamyl alcohol (24:1), mixing, and spinning for 10 min at 12,000 g were preformed. DNA was precipitated with an equal volume of 100% isopropanol, the tube inverted and placed at room temperature for 30 min. The sample was spun for 30 min at 12,000 g and decanted and then washed in 70% ethanol, air-dried and 50 μL of 0.1 X TE buffer was added.

Amplification of the plastid-encoded RuBisCo spacer followed procedures outlined in Zuccarello et al. (1999b). Amplification of the mitochondrial-encoded *cox2-3* spacer and sequencing followed procedures outlined in Zuccarello et al. (1999a). Sequences were aligned by eye. Phylogenetic relationships were inferred with PAUP*4.0b10 (Swofford, 2002). Data sets from different genomic regions were tested for incongruence using the partition homogeneity test (PHT) (Farris et al., 1994) as implemented in PAUP* (1000 replicates, 5 random additions, 100 trees per addition saved). A combined dataset was subjected to maximum-parsimony (MP) analysis, using the heuristic search option, 500 random sequence additions, 100 trees per addition saved, TBR branch swapping, unordered and unweighted characters, and gaps treated as missing data. The program Modeltest version 3.06 (Posada & Crandall, 1998) was used to find the model of sequence evolution that best fits each data set by a hierarchical likelihood ratio test ($\alpha = 0.01$). When the best sequence evolution model had been determined, maximum-likelihood searches were performed in PAUP* using the estimated parameters (substitution model, gamma distribution, proportion of invariable

Table 1. Samples used in molecular analysis

Code	*Sample information*
1	1 PIL (Dublin) Zamboanga del Norte, Philippines
2	*E. isiforme*
3	'cottonii', commercial Venezuela
4	'cottonii', 3096, Philippines
5	'cottonii', Panama
6	'cottonii', 3005, Indonesia
7	*K. striatum*, Madagascar, wild collected
8	'denticulatum', 2189, Madagascar
13	'spinosum', 3000, Indonesia
14	'spinosum', 3127, Philippines
15	'spinosum', 3036, Tanzania
16	Madagascar, '*striatum*', cultivated
17	Bongae, Tani Tauf. (Dublin)
18	Calaguas, Philippines
19	Zamboanga, Philippines (Sacol)
20	Jolo, Philippines (Socal) E. Dublin
21	Mindoro, Philippines (Lopez)
22	'cottonii', 3054, Tanzania, (*K. alvarezii*)
23	'cottonii', Jolo, Philippines (Dublin)
24	Zamboanga, Philippines (Marcel)
32	*E. spinosum*, L992029702, Snellius II, 15/9/84, NE Sumba, Indonesia
33	*E. spinosum*, L991332595, Snellius II, 14/9/84, NE Sumba, Indonesia
34	*E. johnstonii* S&G, L94034716, California (?) W.van Bosse
35	*E. isiforme*, L01791 Bahia Honda State Park, Florida, USA
37	*E. isiforme*, L987276949, Summerland Key, Florida, USA
44	*E. denticulatum*, L992 274 234, Bone Tambung Is., S.W. Sulawesi, Indonesia
45	*E. denticulatum*, L992274274, Langkai Is., S.W. Sulawesi, Indonesia
46	*E. denticulatum*, L370570, Dar es Salaam, Tanzania,
48	*K. striatum*, L993114217, Kudingareng Keke Is. S.W. Sulawesi, Indonesia
49	*K. striatum*, L992274420, Langkai Is., S.W. Sulawesi, Indonesia
50	"cottonii" 03 240, Oct 2003, Vietnam
51	*K. alvarezii*, R: Oct 2003, Colombia
53	'cottonii' (best)-Solomon Islands (1), Nov 2003
54	*K. alvarezii*, 4264–2, (John West)
55	'cottonii' 03 241, Oct 2003, Vietnam, large
56	*E. denticulatum*, Apr 2003, AH-RU3–3, Philippines.
57	SS1, Reef 44, Kane'ohe Bay, Hawaii
58	GS1, Reef 44, Kane'ohe Bay, Hawaii
59	C2A, Kane'ohe Bay, Hawaii
60	*E. odontophorum* var. *mauritianum*, HEC 14606, Cotton Bay, Rodrigues, Mauritius

(*Continued on next page*)

Table 1. (*Continued*)

61	*Eucheuma spinosum*, HEC 11474, Pemba Is., Tondooni, Tanzania
62	'cottonii' Parara, Philippines, Feb 2003
63	'cottonii', Calaguas, Philippines, May 2003
64	*Betaphycus philippinensis*, Type locality, HG-PH267, DapDap, Bulusan, Sorsogon, Luzon, Philippines
65	*E. platycladum*, HEC 11311, Mbudya Is. (near Dar es Salaam), Tanzania
66	*Eucheuma* sp. HG 6933, Tudor Creek, Mombasa, Kenya,
68	C2B, Kane'ohe Bay, Hawaii
69	C3A, Kane'ohe Bay, Hawaii
70	C3B, Kane'ohe Bay, Hawaii
71	O1, Hawaii, store bought
72	GS2, Reef 44, Kane'ohe Bay, Hawaii
73	SS2, Reef 44, Kane'ohe Bay, Hawaii
74	GS3, Reef 43, Kane'ohe Bay, Hawaii
75	GS4, Reef 43, Kane'ohe Bay, Hawaii
76	GS5, Coconut Is., Kane'ohe Bay, Hawaii
77	SS3, Reef 44, Kane'ohe Bay, Hawaii
78	SS4, Reef 29, Kane'ohe Bay, Hawaii
79	SS5, Reef 29, Kane'ohe Bay, Hawaii
80	C2C, Kane'ohe Bay, Hawaii
81	C2D, Kane'ohe Bay, Hawaii
82	C3C, Kane'ohe Bay, Hawaii
83	C3D, Kane'ohe Bay, Hawaii
84	GS6, Coconut Is., Kane'ohe Bay, Hawaii
85	SS6, Reef 29, Kane'ohe Bay, Hawaii
86	'*Kappaphycus*', Colombia, new cultivation trial-Raul, Jan 2004
87	Panama, recent commercial supply-fine form
88	Panama, recent commercial supply-large form
89	'cottonii' Flower-type (Sacol) (Marcel sample)-13/2/04, Jolo, Philippines
90	'cottonii' Giant, (Marcel sample)-13/2/04, Jolo, Philippines
91	'cottonii' 'tambalang', (Marcel sample)-13/2/04, Jolo, Philippines
92	Kane'ohe Bay, Hawaii 'cottonii-like', 6/2/04, Prostrate large branch
93	'*E. denticulatum*', Kane'ohe Bay, Hawaii, 6/2/04
94	'*E. denticulatum*-like', Kane'ohe Bay, Hawaii, 6/2/04, Prostrate lateral branch B/
95	Philippines, sample A2, Feb 04
96	Philippines, sample A1, Feb 04
97	Philippines, "C", Feb 04
99	Kane'ohe Bay, Hawaii 6/2/04, B/, lateral branch
100	Philippines, Mr Dublin material. 2 local names-Adic-Adic; Maka-Purdoy
101	Philippines, Mr Dublin material. 3 local names-Pataka; Flower, sacol
102	'Eucheuma tip', Panama (commercial supply)

(*Continued on next page*)

Table 1. Continued

103	'Eucheuma tip'-2, Panama (commercial supply)
104	Kane'ohe Bay, Hawaii 6/2/04, lateral branch a/
105	Philippines, sample B2, Feb 04
106	'cottonii', Vietnam, Mar 2004
107	'cottonii', Vietnam, Mar 2004, -2
108	*K. cottonii*, HG-PH125, Dumaluan Beach, Panglao, Bohol. Philippines
109	*E. spinosum*, Tanzania, HEC11401, Pemba Is., Tanzania
110	*Eucheuma* sp., HEC11501, Kunduchi, Dar es Salaam, Tanzania
111	*E. platycladum*, HEC9452, Chale Is., Kenya
113	*E. denticulatum* Bali A, Nusa Lembongan-Ceningan Channel, Indonesia
114	*E. denticulatum* Bali B, wild, Tanjung Aan, Lombok, Indonesia
115	*E. denticulatum* Bali C, Tanjung Aan, Lombok, Indonesia
116	*E. denticulatum* Bali D, Tanjungan, Nusa Lembongan, Indonesia
117	*Eucheuma* sp. L BER03-464, F30, Maratua Is., Indonesia, wild sample
118	*Betaphycus philippinensis* 4, Dancalan, Bulusan, Sorsogun, Luzon, Philippines
119	*K. striatum* 6, western side Hilutangal Is. Cebu, Philippines
120	*K. alvarezii* Bali E, Nusa Lembongan-Cennigan Channel, Indonesia
121	*K. alvarezii* Bali F, Brown, Gunung Payon, Nusa Dua, Bali, Indonesia
122	*K. alvarezii* Bali G, Gerupak Lombok, Indonesia
123	*K. alvarezii* Bali H, Tanjungan, Nusa Lembongan, Indonesia
124	*K. alvarezii* Bali I, Tanjung Aan, Lombok, Indonesia
125	*K. alvarezii* Bali J, Red, Gunung Payon, Nusa Dua, Bali, Indonesia
126	*K. alvarezii*, BZ1 (brown strain) sporophyte (Edison de Paula)
127	*K. alvarezii*, BZ2 (brown strain) sporophyte (Edison de Paula)
128	*K. alvarezii*, BZ3 (brown strain) female (Edison de Paula)
129	*K. alvarezii* BZ5, sporophyte (Edison de Paula)
130	*E. striatum* 5, Zanzibar, Tanzania
131	*Kappaphycus striatum* 2, Little Santa Cruz Is. Mindanao, Philippines
132	*K. cottonii* 3, western side Hilutangal Is. Cebu, Philippines
133	*Kappaphycus alvarezii*, Bali K, sumba-type, Gunung Payon, Nusa Dua, Bali
134	*E. denticulatum* 8, Zanzibar, Tanzania
135	*Betaphycus speciosum* 1, Ravin Reef, Rottenest Is. WA, Australia
137	*K. striatum* BZ4, green strain (Edison de Paula)

Samples given names derived from suppliers if available. (L = National Herbarium Netherlands Leiden; HEC = Herbarium Eric Coppejans, HG = Herbarium Gent).

sites, transition–transversion ratio). Maximum likelihood (ML) was used to construct the most likely tree from the data set. Support for individual internal branches was determined by bootstrap analysis (Felsenstein, 1985), as implemented in PAUP*. For bootstrap analysis, 1000 bootstrap data sets were generated from resampled data (5 random sequence additions, 1,000,000 rearrangements per replicate). Haplotype networks (gene genealogies) were calculated using TCS 1.13, (Clement et al., 2000) that produces an estimation of gene genealogies for DNA sequences.

Sequences are deposited in Genbank (*cox2-3* spacer: AY687417–39; RuBisCo spacer: AY687400–16).

Results

Molecular phylogeny

The results from the PHT test indicated that the mitochondrial and plastid data sets are not significantly different from each other and could be combined ($p = 0.689$). This combined data set contained only samples from which both sequences were available and had 83 taxa and 668 characters, 193 of which were potentially informative. Maximum-parsimony produced 461 trees of 461 steps ($CI = 0.6573$) (Figure 1). Maximum likelihood (estimated evolution model = HKY85; Ti/Tv ratio = 6.0729, proportion of invariable sites = 0.515, gamma parameter = 2.9212; $-\ ln\ L$ = 3007.31052) produced a topology identical to the MP tree.

The tree topology shows that the *Kappaphycus* samples are distinct from the *Eucheuma* samples. One sample from the Philippines identified as *K. cottonii* (108) collected in the wild is a sister sample to the mainly cultivated species *K. alvarezii* and *K. striatum.* Within *Kappaphycus* there is a clear distinction between the two species *K. alvarezii* and *K. striatum* (the "sacol" strain). Within *K. striatum* the two wild-collected samples (48, 117) are distinct from the cultivated samples. Within the *K. alvarezii* clade three main groups are found: (1) A sample from Africa (plus other African samples as seen in the haplotype network, see below); (2) samples from Hawaii; and (3) cultivated samples from around the world.

Samples of *Eucheuma denticulatum* are also divided into three groups, though the composition of the three groups is not as clear as for *K. alvarezii*. The first group contains samples from Africa, including a sample identified as *E. odontophorum* var. *mauritianum* (60). A

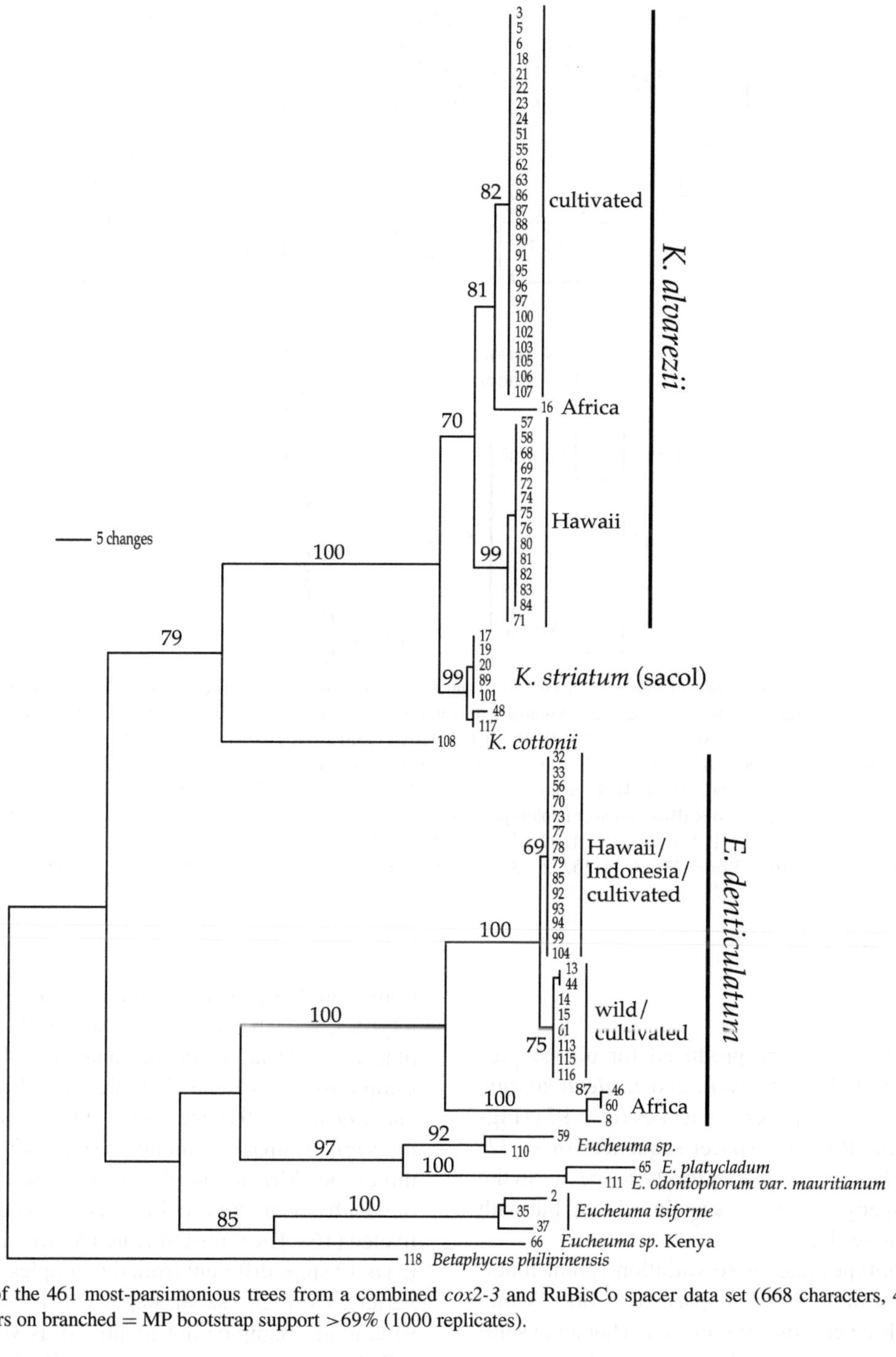

Figure 1. One of the 461 most-parsimonious trees from a combined *cox2-3* and RuBisCo spacer data set (668 characters, 461 steps, CI = 0.6573), Numbers on branched = MP bootstrap support >69% (1000 replicates).

second group contains the samples from Hawaii, plus some wild samples from Indonesia (32, 33) and a cultivated sample from the Philippines (56); the final group contains wild samples from Indonesia (44) and Tanzania (61) plus cultivated samples from the Philippines, Indonesia and Tanzania (15).

Other samples of *Eucheuma* were sequenced, with many of these only identified to genus. Three samples of *E. isiforme* (C. Agardh) J. Agardh grouped together. There is also a supported grouping of a sample from Tanzania (110) and a sample collected from Hawaii (59).

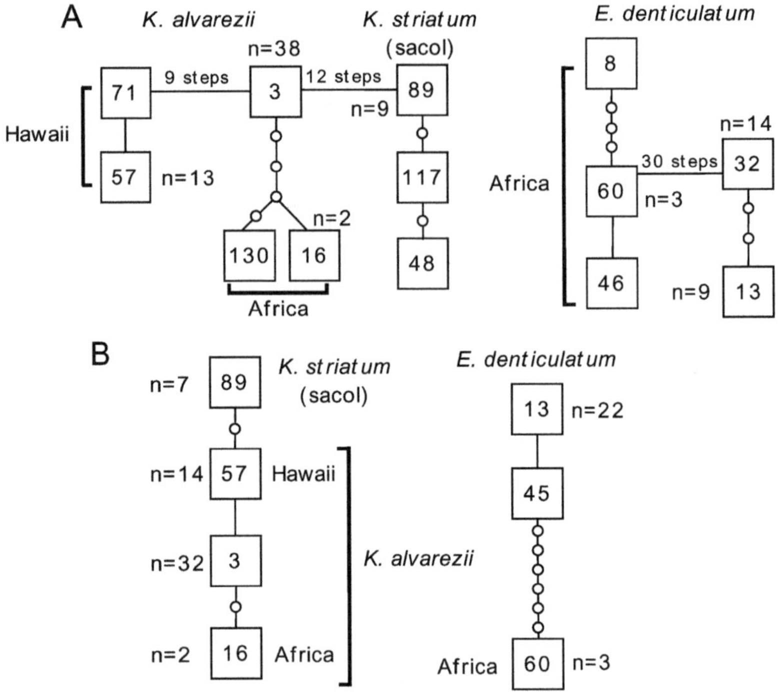

Figure 2. Haplotype networks of samples of *Kappaphycus alvarezii* and *K. striatum* and *Eucheuma denticulatum*. n = number of samples. Line indicates a point mutation, empty circle = intermediate hypothetical haplotype. (A) *cox2-3* spacer haplotypes: 71 = 71; 57 (n = 13; 57, 58, 68, 69, 72, 74, 75, 76, 80, 81, 82, 83, 84); 89 (n = 9; 17, 19, 20, 89, 101, 131, 132, 133, 137); 117 = 117; 48 = 48; 3 (n = 38; 3, 5, 6, 18, 21, 22, 23, 24, 51, 54, 55, 62, 63, 86, 87, 88, 90, 91, 95, 96, 97, 100, 102, 103, 105, 106, 107, 119, 120, 121, 122, 123, 124, 125, 126, 127); 16 (n = 2; 7, 16); 130 = 130; 8 = 8; 60 (n = 3; 60, 109, 134); 46 = 46; 32 (n = 14; 32, 33, 56, 70, 73, 77, 78, 79, 85, 92, 93, 94, 99, 104); 13 (n = 9; 13, 14, 15, 44, 61, 113, 114, 115, 116). (B) RuBisCo spacer haplotypes: 89 (n = 7; 17, 19, 20, 48, 89, 101, 117); 57 (n = 14; 57, 58, 68, 71, 72, 74, 75, 76, 80, 81, 82, 83, 84); 3 (n = 32; 1, 3, 4, 5, 6, 18, 21, 22, 23, 24, 49, 50, 51, 53, 54, 55, 62, 63, 86, 87, 88, 90, 91, 95, 96, 97, 100, 102, 103, 105, 106, 107); 16 (n = 2; 7, 16); 13 (n = 22; 13, 14, 15, 32, 33, 44, 56, 61, 70, 73, 77, 78, 79, 85, 92, 93, 94, 99, 104, 113 115, 116); 45 = 45; 60 (n = 3; 8, 46, 60).

Haplotype network

Haplotype networks were produced for all samples of *K. alvarezii* and *K. striatum*, and *E. denticulatum* with both the *cox2-3* spacer sequences ($n = 93$) (Figure 2A) and the RuBisCo spacer sequences ($n = 82$) (Figure 2B). The grouping seen for these species in the overall phylogeny (Figure 1) is again evident, but with increased sample sizes.

More haplotypes and more variation (point mutations) are seen in the *cox2–3* spacer network. Within *K. alvarezii* three groups are seen. The Hawaiian samples which contain two haplotypes: the single store-bought sample (71) (bought fresh in Honolulu and presumably collected from a location on Oahu); and the samples from Kane'ohe Bay (haplotype 57, $n = 13$), a location known to contain an invasive population of *K. alvarezii* (Conklin & Smith, 2004; Smith et al., 2002). This Hawaiian group is separated by 9 mutations from the cultivated samples of *K. alvarezii* (haplotype 3, $n = 38$), comprising samples from the Philippines, Indonesia, Vietnam, Panama, Colombia, and a sample from Tanzania (22), plus samples identified as var. *tambalang* (91) and 'giant' (90). This cultivated *K. alvarezii* group is 4–5 mutation steps different from the three other African samples, all identified as *K. striatum* of which two are wild-collected (130, 7) and one is cultivated (16). The main cultivated *K. alvarezii* haplotype (3) is 12 steps different from the samples of the "sacol" variety. The main "sacol" haplotype (haplotype 89, n = 9) contains samples not identified as variety "sacol" (17, 133), plus wild samples identified as *K. cottonii* (132) and *K. striatum* (131). The two divergent haplotypes of *K. striatum* (117, 48) are both wild-collected plants from Indonesia. There are five *cox2–3* spacer haplotypes of our sampled *E. denticulatum*. Again three groups are seen, corresponding to: (1) samples from

Hawaii ($n = 14$); (2) cultivated and wild samples (haplotype 13, $n = 9$) from Indonesia, Tanzania and the Philippines; and (3) three haplotypes of samples from Madagascar, Mauritius and Tanzania (8, 46, 60).

The RuBisCo spacer haplotypes (Figure 2B) form groups congruent with the *cox2-3* spacer haplotype, though less variation is found in this genetic region plus the sample size is slightly smaller. For example, only one RuBisCo spacer haplotype is found for all the "sacol" samples versus three for the *cox2-3* spacer, and only one haplotype for the *K. alvarezii* samples from Hawaii versus two for the *cox2–3* spacer. For *E. denticulatum* the same RuBisCo spacer haplotype occurs in all Hawaiian and non-African or cultivated samples (haplotype 13, $n = 22$). African samples are quite distinct from the other *E. denticulatum* samples.

Discussion

Our data show that the genetic regions used are useful for resolving inter- and intra-specific relationships. The utility of these short genetic markers (approximately 300–350 base pairs) in resolving intra- and inter-specific relationships that are supported by longer regions has been documented (Zuccarello et al., 2002) and means that sample sizes can be drastically increased for equal time and money. Shorter segments are also more easily amplified in less well-preserved material (i.e. herbarium specimens, commercially dried samples, etc.).

Our molecular phylogeny clearly indicates that the two genera *Eucheuma* and *Kappaphycus* are distinct, but that *Eucheuma* could be paraphyletic (no bootstrap support for its distinction from *Kappaphycus*). Our choice of *Betaphycus* as an outgroup was based on a mid-point rooted tree that indicated that the *Betaphycus* sequences were the most divergent. Analyses with *E. isiforme* as an outgroup, a suggested sister taxon to the remaining *Eucheuma/Kappaphycus/Betaphycus* species (Fredericq et al., 1999), placed *Betaphycus philippinensis* Doty (4) as a sister taxon to *Kappaphycus* and therefore made *Eucheuma* paraphyletic, although it did not change the relationships within *Kappaphycus* or the remaining *Eucheuma* samples (data not shown). More research into the higher level relationships of these genera is needed, that should include wide-spread geographic and taxonomic sampling and use of molecular data from types (Hughey et al., 2001).

Kappaphycus

A lone sample identified morphologically as *K. cottonii* (103) is a highly divergent sister taxon to the remaining *Kappaphycus* samples. The true identity of this sample will require clarification. Our data clearly indicate that *K. alvarezii* can be divided into several supported evolutionary lineages. Firstly, there is a lineage that contains mostly, but not exclusively, samples grown as the "sacol" variety. Of note is that two wild collections of this variety from Indonesia are distinct from the cultivated samples, based on the *cox2–3* spacer. The taxonomic status of the "sacol" variety of *Kappaphycus* is problematic. It is clear that the morphological variability is such that species can only be "approximately" identified on the sole basis of morphology. Previous molecular studies suggested that the "sacol" variety was *K. cottonii*. This was based on only one sample identified as *K. cottonii* and one of the "sacol" variety using the *rbc*L gene (Aguilan et al., 2003). Incorporation of the terminal end of the *rbc*L gene of the Aguilan et al. (2003) samples into our RuBisCo spacer data set shows that these samples (AF481500 and AF481499) are part of our clade of the "sacol" variety. Our data indicate that the "sacol" variety is probably a distinct *Kappaphycus* species, although its proper name will depend on more sampling of wild collections and continued morphological examination. We suggest the use of the epithet *K. striatum* for this variety for now. The only other samples that could be considered as *K. striatum* are the cultured and wild samples from Africa (Tanzania and Madagascar) that have unique haplotypes (Figure 2) but are nested within the *K. alvarezii* samples (Figure 1).

Another lineage that forms a distinct evolutionary group is the set of samples from Hawaii. This was surprising as these samples were introduced to Hawaii, presumably from the Philippines, but are distinct from all other *Kappaphycus* samples. The unique genotypes, as expressed in unique haplotypes, of these samples may explain their invasive nature in Hawaii (Conklin & Smith, 2005). The final lineage contains cultured samples of *K. alvarezii* from around the world. Obviously these organellar markers do not reflect all the genetic variation existing in these samples as they can be morphologically quite variable and even contain genetically stable variants (var. *tambalang*). It appears that "cultivated" *K. alvarezii* is already found throughout the world in cultivation.

The RuBisCo spacer haplotypes again reveal the geographic/taxonomic patterns seen in the *cox2-3* haplotype data and this is reflected in the combined phylogeny, although the level of variation is lower compared to the mitochondrial marker. The RuBisCo spacer is known to be more conserved and in other red algae more closely indicates reproductively isolated or partially isolated groups (Zuccarello et al., 2002; Zuccarello & West, 2003) or morphologically identifiable species (Brodie et al., 1996) with only 1–2 bp changes. It is therefore likely that all four groups of *Kappaphycus* studied are reproductively isolated from each other (African, cultivated, Hawaiian, "sacol" samples). Due to the rarity of reproduction in "cultivated" *Kappaphycus* this may be difficult to test but important if crop improvement by hybrid formation is pursued.

Eucheuma

The phylogeny, including single samples from herbaria, indicates that many 'species' of *Eucheuma* exist. In some cases we thought it prudent not to try a preliminary identification and left the samples as unidentified. It is clear that a complete monograph, including a molecular data set, of *Eucheuma* is desperately needed. Our sampling of *E. denticulatum* is much lower than for *Kappaphycus*, partially reflecting its lower commercial production (www.surialink.com). Again three groups are seen although they do not clearly reflect geographic/commercial patterns as seen in *K. alvarezii*. An African group consists of wild and cultivated samples. A mainly Hawaiian group also contains wild samples from Indonesia (32, 33) and a cultivated sample from the Philippines (56), plus a group containing both wild and cultivated samples from Indonesia, the Philippines and Tanzania. It is likely that the cultivation history of *E. denticulatum* is different from that for *K. alvarezii* with the species brought into cultivation several times in various locations. *E. denticulatum* is known to have been cultivated in larger volumes than at present, due to greater market demands for κ-carrageenan from *Kappaphycus* species. Our haplotype data again show that the *cox2–3* spacer is more variable in this species than the RuBisCo spacer. The RuBisCo spacer is not able to distinguish the predominantly Hawaiian *cox2-3* haplotype from the predominantly cultivated haplotype.

Our results indicate many interesting aspects of the evolutionary history of the genera *Kappaphycus* and *Eucheuma*. These data also highlight the limitations in our taxonomic knowledge and the need for more variable markers that can lead to a better analysis of genotypes that correlate with economically and ecologically important variables (colloid quality and content, growth rate, warm water susceptibility, disease resistance, invasive potential) in these commercially important seaweeds.

Acknowledgments

We thank Eric Coppejans, Mr Expeditio Dublin, Max Hommersand, Anne Hurtado, Frederik Leliaert, Edison de Paula and Marcel Carrageenan for samples. Funding from COSTAI (Collaboration in Science and Technology between Australia and Indonesia) was provided to JAW for the collections in Bali and Lombok.

References

Aguilan JT, Broom JE, Hemmingson JA, Dayrit FM, Montano MNE, Dancel MCA, Ninonuevo MR, Furneaux RH (2003). Structural analysis of carrageenan from farmed varieties of Philippine seaweed. Bot. Mar. 46: 179–92.

Brodie J, Hayes PK, Barker GL, Irvine LM (1996) Molecular and morphological characters distinguishing two *Porphyra* species (Rhodophyta, Bangiophycidae). Europ. J. Phycol. 31: 303–08.

Clement M, Posada D, Crandall KA (2000) TCS: A computer program to estimate gene genealogies. Mol. Ecol. 9: 1657–59.

Conklin EJ, Smith JE (2005) Abondance and spread of the invasive red algae *Kappaphycus* spp., in Kane'ohe Bay, Hawai'i and an experimental assessment of management options. Biological Invasions, 1029–39.

Doty MS (1985) *Eucheuma alvarezii sp. nov.* (Gigartinales, Rhodophyta) from Malaysia. In Abbott, I. A. (ed.), Taxonomy of Economic Seaweeds. California Sea Grant College Program, La Jolla, California, 37–45.

Doty MS (1988) Prodomus Ad systematica eucheumatioderum: A tribe of commercial seaweeds related to *Eucheuma* (Solieiriaceae, Gigartinales). In Abbott, I. A. (ed.), Taxonomy of Economic Seaweeds. California Sea Grant Program, La Jolla, California, 159–207.

Doty MS, Norris JN (1985) *Eucheuma* species (Solieriaceae, Rhodophyta) that are major sources of carrageenan. In Abbott, I. A. (ed.), Taxonomy of Economic Seaweeds. California Sea Grant College Program, La Jolla, California, 47–61.

de Paula EJ, Lima Pereira RT, Ohno M (1999) Strain selection in *Kappaphycus alvarezii* (Solieriaceae, Rhodophyta) using tetraspore progeny. J. Appl. Phycol. 11: 111–21.

Farris JS, Källersjö M, Kluge AG, Bult C (1994) Testing for significance of incongruence. Cladistics 10: 315–19.

Felsenstein J (1985) Confidence intervals on phylogenies: An approach using the bootstrap. Evolution 39: 783–91.

Fredericq S, Freshwater DW, Hommersand MH (1999) Observations on the phylogenetic systematics and biogeography of the Solieriaceae (Gigartinales, Rhodophyta) inferred from *rbc*L sequences and morphological evidence. Hydrobiologia 399: 25–38.

Marston M, Villalard-Bohnsack M (2002) Genetic variability and potential sources of *Grateloupia doryphora* (Halymeniaceae, Rhodophyta), an invasive species in Rhode Island waters (USA). J. Phycol. 38: 649–58.

Hughey JR, Silva PC, Hommersand MH (2001) Solving taxonomic and nomenclatural problems in Pacific Gigartinaceae (Rhodophyta) using DNA from type material. J. Phycol. 37: 1091–109.

Posada D, Crandall KA (1998) Modeltest: Testing the model of DNA substitution. Bioinformatics 14: 817–18.

Smith JE, Hunter CL, Smith CM (2002) Distribution and reproductive characteristics of nonindigenous and invasive marine algae in the Hawaiian Islands. Pac. Sci. 56: 299–315.

Swofford DL (2002) PAUP*. Phylogenetic analysis using parsimony (*and Other Methods). Version 4. Sinauer Associates, Sunderland, Massachusetts.

Trono GC (1997) Field guide: Atlas of the seaweed resources of the philippines. National Bookstore, Inc., Manila, Philippines, 306.

Zuccarello GC, Burger G, West JA, King RJ (1999a) A mitochondrial marker for red algal intraspecific relationships: variable within populations and maternally inherited. Mol. Ecol. 8: 1443–1448.

Zuccarello GC, West JA, Kamiya M, King RJ (1999b) A rapid method to score plastid haplotypes in red seaweeds and its use in determining parental inheritance of plastids in the red alga *Bostrychia* (Ceramiales). Hydrobiologia 401: 207–214.

Zuccarello GC, West JA (2002) Phylogeography of the *Bostrychia calliptera/B. pinnata* complex (Rhodomelaceae, Rhodophyta) and divergence rates based on nuclear, mitochondrial and plastid DNA markers. Phycologia 41: 49–60.

Zuccarello GC, Sandercock B, West JA (2002) Diversity within red algal species: Variation in world-wide samples of *Spyridia filamentosa* (Ceramiaceae) and *Murrayella periclados* (Rhodomelaceae) using DNA markers and breeding studies. Europ. J. Phycol. 37: 403–17.

Zuccarello GC, West JA (2003) Multiple cryptic species: Molecular diversity and reproductive isolation in the *Bostrychia radicans/B. moritziana* complex (Rhodomelaceae, Rhodophyta) with focus on North American isolates. J. Phycol. 39: 948–59.

Journal of Applied Phycology (2006) 18: 653–661
DOI: 10.1007/s10811-006-9068-0

Phylogenetic diversity of New Zealand Gelidiales as revealed by *rbc*L sequence data

W.A. Nelson[1,*], T.J. Farr[1] & J.E.S. Broom[2]
[1]*National Institute of Water and Atmospheric Research Ltd (NIWA), Private Bag 14901, Wellington, New Zealand;* [2]*Department of Biochemistry, University of Otago, PO Box 56, Dunedin, New Zealand*

*Author for correspondence: e-mail: w.nelson@niwa.co.nz

Key words: Gelidiales, New Zealand, *Capreolia, Pterocladia,* phylogeny, *rbc*L sequence data

Abstract

Diversity and phylogenetic relationships of New Zealand representatives of the red algal order Gelidiales have been examined using *rbc*L sequence data. Extensive field collections have been made from throughout the New Zealand region. Six genera have been reported previously from New Zealand (*Capreolia, Gelidium, Pterocladia, Pterocladiella, Pterocladiastrum, Ptilophora*). This research has revealed species with very restricted local distributions, as well as the discovery of several undescribed, cryptic taxa. The common and widespread *Gelidium caulacantheum* is confirmed to be more closely related to *Capreolia* than to other species of *Gelidium*. The generic concept of *Capreolia*, based on life history characters, will need to be modified to accommodate additional species possessing "*Gelidium*" life histories. A species endemic to New Zealand, *Gelidium ceramoides*, has been found to differ significantly from all other members of the Gelidiales and requires reclassification in another genus and order. Examination of field collections and herbarium specimens in addition to molecular sequence data have led us to conclude that specimens previously placed in the genera *Ptilophora* and *Pterocladiastrum* belong within *Pterocladia lucida*.

Introduction

The red algal order Gelidiales contains a number of commercially valuable agarophyte genera and is thus economically important. In New Zealand there has been an agar industry for ca. 60 years primarily based on the use of *Pterocladia lucida* (Turner) J. Agardh (Schiel & Nelson, 1990). Interest in potential new sources of valuable polysaccharides has stimulated research on the Gelidiales in New Zealand.

Definition of genera and species in the Gelidiales has been considered problematic for a very long time, in part because of the morphological variability that appears to be a feature of many of the taxa. Although there has been considerable effort expended in the search for reliable taxonomic characters in *Gelidium* (e.g. Stewart, 1976; Rodriguez & Santelices, 1987; Santelices, 1990) to date there has been little success. Over the past 15 years research on the Gelidiales has seen a focus on regional taxonomic studies employing traditional morphological and anatomical techniques (e.g. Santelices, 1994; Lee & Kim, 1995). Molecular sequence data are providing fresh insights into the phylogenetic relationships within the Gelidiales as well as an improved understanding of generic boundaries (Freshwater et al., 1995; Bailey & Freshwater, 1997; Freshwater & Bailey, 1998; Patwary et al., 1998; Shimada et al., 1999; Tronchin et al., 2002, 2003).

Chapman (1969) recorded two genera of Gelidiales from the New Zealand region, recognising 15 taxa within the genus *Gelidium* (including nine species with three varieties, three forms and one ecad) and three species of *Pterocladia.* Chapman also listed three species as 'species excludendae'. Adams (1994) recorded four genera, six species of *Gelidium,* three species of *Pterocladia,* and one species each of *Ptilophora* and *Pterocladiastrum,* although she expressed doubt about the recognition of two of these species (*Pterocladia lindaueri* Fan and *Pterocladiastrum robustum* Akatsuka). Guiry and Womersley (1993) erected the genus *Capreolia* for a mat-forming species found in the mid-intertidal zone on both exposed and sheltered coasts from south-eastern Australia and New Zealand, placing material that had been previously identified as *Gelidium pusillum* (Stackhouse) Le Jolis into the monotypic *C. implexa* Guiry and Womersley. Santelices and Hommersand (1997) established the genus *Pterocladiella* for species previously assigned to *Pterocladia* that possessed cystocarps with unequally developed locules. Nelson et al. (1994) and Nelson and Farr (2003) examined the endemic species *Gelidium allanii* and *G. longipes* respectively.

Over a period of 6 years, collections of Gelidiales have been made throughout the New Zealand region from the Three Kings Islands (34°S 172°E) to Stewart Island (47°S 168°E) and including the Chatham Islands (44°S 167°W). In this paper we report on the results of molecular sequencing techniques applied to a selection of these collections. The value of *rbc*L sequence data for developing hypotheses about phylogenetic relationships of genera within this order has been established (Freshwater et al., 1995; Shimada et al., 1999; Tronchin et al., 2002, 2003) although a more variable marker may be required to elucidate relationships at the species level. Our objective in this study was to survey diversity and examine current understandings of New Zealand representatives of the order.

Materials and methods

Field material was sorted and treated in three ways. Samples were preserved in 3–5% formalin/seawater for anatomical and morphological examination, material for extraction of molecular sequence data was dried in silica gel, and fresh material was pressed as herbarium sheets to serve as voucher specimens (Table 1) which are lodged in the herbarium of the Museum of New Zealand Te Papa Tongarewa (WELT, Holmgren et al., 1990).

Molecular biology methods: DNA extraction and PCR amplification

DNA was extracted using the Chelex method of Goff and Moon (1993) or the CTAB extraction method

Table 1. GenBank accession numbers, collection information and voucher numbers for samples from the New Zealand region sequenced in the course of this study (Vouchers marked * differ in the collection date from the sample sequenced)

Sample	GenBank accession no	Location/Strain	Date	Collectors	WELT no.
Capreolia implexa Guiry et Womersley	AY648012	Hooper Pt, Spirits, Bay, North I.	15 Nov 1996	Nelson	A26812
Capreolia implexa	AY648009	Castlepoint, Wairarapa, North I.	31 Mar 1998	Nelson	A22316*
Capreolia implexa	AY648010	Evans, Bay, Wellington, North I.	13 Nov 1997	Nelson & Knight	A26811
Capreolia implexa	AY648008	Cable Bay, Nelson, South I.	8 May 1997	Nelson	A26821
Capreolia implexa	AY648011	Gentle Annie Pt, Westland, South I.	9 Mar 2000	Nelson & Russell	A26818
Capreolia implexa	AY648013	Ringaringa, Stewart I.	9 Oct 1998	Nelson & Broom	A26817
Gelidium caulacantheum J. Agardh	AY648017	Puheke, Northland, North I.	8 May 2001	Nelson & Farr	A25776
Gelidium caulacantheum	AY648015	Matauri Bay, Northland, North I.	28 Oct 2003	Nelson	A26819
Gelidium caulacantheum	AY648016	Piha Beach, Auckland, North I.	5 Apr 2000	Nelson & Farr	A26816
Gelidium caulacantheum	AY648020	Tauranga Harbour, North I.	31 July 2003	Nelson	A26824
Gelidium caulacantheum	AY648014	Castlepoint, Wairarapa, North I.	31 Mar 1998	Nelson	A22317*
Gelidium caulacantheum	AY648018	Puponga, South I.	28 Sep 1996	Nelson	A26823
Gelidium caulacantheum	AY648019	Queen Charlotte Sound, South I.	30 Dec 2001	Broom	A26822
Gelidium sp. "Northland"	AY648024	Te Ngaire, Northland, North I.	9 May 2001	Nelson & Farr	A25789
Gelidium sp. "Fiordland"	AY648023	Fiordland, South I.	13 Oct 2000	Wing & Goebel	A25798
Gelidium microphyllum (Crosby Smith) Kylin	AY648022	Ringaringa, Stewart I.	9 Oct 1998	Nelson	A26813
Gelidium longipes J. Agardh	AY648021	Northland, North I.	7 May 2001	Nelson	A25764
Pterocladia lucida (Turner) J. Agardh	AY648025	Maketu, North I	31 July 2003	Nelson	A26820

of Hillis et al. (1996). PCR products spanning the *rbc*L-*rbc*S region were amplified using primers from Freshwater and Rueness (1994); for sequencing of *Gelidium longipes* samples, primer F-492 (cgt atg gat aaa ttt ggt cg) was replaced by primer F-492ga (cgt atg gat aa**g** ttt gg**a** c). This primer matches the *G. longipes rbc*L sequence, and was much more effective for sequencing PCR products derived from this taxon. It differs from F-492 by two nucleotide substitutions (indicated in bold) and in the removal of the terminal guanine residue. In some instances a nested amplification strategy was used to amplify the target. Amplification using primers F-57 and R-*rbc*S-start was followed by amplification using F-57 with R-752, and F-492 R-*rbc*S-start, diluting the original amplification product 1:100. This yielded two overlapping products spanning the *rbc*L gene and associated spacer. Amplifications were performed in a Stratagene Robocyler (Stratagene Corporation, La Jolla, Ca) with parameters as follows: an initial denaturation of 30 s at 94 °C was followed by 35 cycles of 25 s at 94 °C, 45 s at 45 °C, 1 min 30 s at 72 °C and a final extension for 10 min at 72 °C. Amplification products were purified by PEG precipitation and sequenced on an ABI 377 automated sequencer according to standard methods. Sequences were aligned with 36 Gelidiales *rbc*L sequences downloaded from GenBank (Table 2) using Se-Al v2.0a11 (Rambaut, 1996). Where *rbc*L/*rbc*S spacer sequences were available, these were also included in the alignment. Three taxa, from the Bonnemaisoniales (*Delisea pulchra* (Grev.) Mont.) and Ceramiales (*Laingia hookeri* (Lyall ex Harv.) Kylin and *Hypoglossum hypoglossoides* (Stackh.) Collins & Herv.), were chosen as outgroup taxa.

Phylogenetic analysis

Maximum parsimony (MP) and neighbour joining (NJ) phylogenetic trees were constructed using PAUP*4.0b10 (Swofford, 2002). MP trees were obtained by a heuristic search strategy (100 replicates of random-order stepwise sequence addition followed by tree-bisection-reconnection (TBR) branch swapping) with gaps treated as unknown bases. A less computer intensive search strategy of 3 replicates of random-order sequence addition followed by subtree-pruning-regrafting (SPR) branch swapping returned an identical consensus tree, and this search strategy was used to find MP trees for 100 bootstrap replicates. Modeltest 3.06 (Posada & Crandall, 1998) was used to identify an appropriate model of sequence evolution for the

Table 2. Published samples used in this study. Taxa are in the order that they appear in Figure 1

Sample	Origin	GenBank accession no.
Delisea pulchra		U26812
Hypoglossum hypoglossoides		AF257368
Laingia hookeri		AF257371
Pterocladia lucida	Owhiro Bay NZ	U01048
Pterocladia lucida	SA Australia	AF305808
Pterocladia lucida	Lord Howe Is	AY352423
Pterocladiella melanoidea	Spain	U01046
Pterocladiella bartlettii	Costa Rica	AF305806
Pterocladiella bartlettii	TX USA	AF305807
Pterocladiella caloglossoides	Australia	AY352422
Pterocladiella beachiae	Costa Rica	AF305811
Pterocladiella caerulescens	Hawaii	AF305805
Pterocladiella capillacea	Japan	AB023849
Pterocladiella capillacea	Owhiro Bay NZ	U24156
Pterocladiella capillacea	Venezuela	U01893
Pterocladiella capillacea	CA USA	U01896
Pterocladiella capillacea	Ireland	U01891
Pterocladiella capillacea	SA Australia	U01898
Gelidiella acerosa	Hawaii	L22457
Gelidiella ligulata		AB017678
Ptilophora pectinata	NSW Aus	AY344043
Ptilophora subcostata	Japan	U16835
Ptilophora scalarimosa	Philippines	AF305804
Ptilophora hildebrandtii	South Africa	AF522359
Ptilophora mediterranea	Greece	AF522360
Ptilophora pinnatifida	South Africa	AF522361
Ptilophora leliaertii	South Africa	U16834
Ptilophora helenae	South Africa	AY344045
Ptilophora rhodoptera	South Africa	AF522365
Ptilophora coppejansii	South Africa	AF522366
Ptilophora diversifolia	South Africa	AF305803
Gelidium divaricatum	Japan	U16828
Gelidium caulacantheum	Porirua NZ	U00103
Capreolia implexa	Australia	L22456
Gelidium caulacantheum	Piha NZ	U01043
Gelidium sp.	NSW Australia 1	AY352420
Gelidium sp.	NSW Australia 2	AY352418
Gelidium asperum	Vic Australia	AY350782
Gelidium australe	SA Australia	AY350783
Gelidium micropterum	SA	U00446
Onikusa pristoides	South Africa	U01044
Gelidium microdonticum	Costa Rica	AF305799
Gelidium isabelae	South Africa	AY350778
Gelidium sesquipedale	Spain	L22071
Gelidium sp.	South Africa	AY350775

(*Continued on next page*)

Table 2. (*Continued*)

Sample	Origin	GenBank accession no.
Gelidium pulchellum	Ireland	U01969
Gelidium latifolium	France	U01818
Gelidium latifolium	Ireland	U01821
Gelidium floridanum	FL USA	U00107
Gelidium allanii	Doubtless Bay NZ	L22458
Gelidium sp.	Lord Howe Is	AY350777
Gelidium tenuifolium	Japan	AB030628
Gelidium elegans	Japan	AB030623
Gelidium pacificum	Japan	AB030627
Gelidium omanense	Oman 1	AY346460
Gelidium pluma	Hawaii USA	AF522367
Gelidium omanense	Oman 3	AY346462
Acanthopeltis japonica	Japan	AB017673
Gelidium vagum	Japan	AB017680
Gelidium rex	Chile	AF305801
Onikusa japonicum	Japan	AB017676
Gelidium chilense	Chile	AF305800
Gelidium capense	South Africa	L22461
Gelidium pusillum	CA USA	U00984
Gelidium coulteri	CA USA	U00105
Gelidium pusillum	Japan	AB017679
Gelidium crinale	Spain	AF308792
Gelidium pusillum	NC USA	U00981
Gelidium pusillum	Puerto Rico	U00983
Gelidium crinale	Canary Is	AF308793
Gelidium crinale	WA Australia	AY350780
Gelidium crinale	NSW Australia	AY350781

dataset. The model selected was the General Time Reversible model allowing for invariant sites and with rate heterogeneity modeled by four gamma-distributed rate classes (GTR + I + Γ). Parameters estimated by Modeltest were used to calculate genetic distances and build a neighbour joining (NJ) tree; support for clades under NJ was estimated by 1000 bootstrap replicates.

Bayesian trees were constructed using MrBayes v3.01 (Ronquist & Huelsenbeck, 2003) to run four Metropolis-coupled MCMC chains (one cold and three incrementally heated, temperature parameter = 0.2). Three independent MrBayes analyses were run under the GTR + I+ Γ model of sequence evolution, each for 1,000,000 generations. Model parameters were treated as unknown and were estimated in each analysis. Random starting trees were used and the trees were sampled every 100 generations. Appropriate burn-in values were determined by inspection of plots of log-likelihood against generation time for each run, and trees obtained before this value were discarded. The remaining trees were used to calculate 50% majority rule consensus trees in PAUP*4.0b10, in which each clade posterior probability value is represented by the proportion of trees containing that clade.

The maximum likelihood tree was estimated under an heuristic search strategy starting from one of the trees found in the third Bayesian analysis after burn-in, and proceeding via TBR branch swapping for 70000 iterations, using the model of sequence evolution identified by Modeltest.

Results

Sequences generated in the course of this study were deposited in GenBank under Accession nos AY648008-25 (Table 1). The phylogenetic matrix consisted of 1491 characters, of which 514 were parsimony informative, 852 constant and 125 parsimony-uninformative. Relatively few *rbc*L/*rbc*S spacer sequences were available, and since these contributed few parsimony-informative characters (3 out of 514) their inclusion had little effect on the analysis. Maximum parsimony analysis identified 84 trees of length 2838. Trees obtained from the three Bayesian analyses converged to similar -ln likelihoods and consensus trees of the three independent analyses had an identical topology. For simplicity only the results of the third run are shown here. All four methods of analysis produced trees with very similar structures. Figure 1 shows the maximum likelihood phylogram with support values from the other analyses overlaid.

The analyses presented here provide strong support for clades that equate to the genera currently recognised in this order, and New Zealand taxa are resolved in clades that correspond to the genera *Pterocladia, Pterocladiella, Capreolia* and *Gelidium.*

Pterocladia: The four *Pterocladia* samples shown on the tree form a well supported clade within which the two New Zealand samples group together as do the two Australian samples.

Pterocladia lucida (Turner) J. Agardh 1852: 483
NZ distribution: Three Kings, North, northern South, and Chatham Is.
Basionym and type locality: *Fucus lucidus* Turner 1819: 98, pl.238; BM; southern Australia.
Synonyms: *Pterocladiastrum robustum* Akatsuka 1986: 57, *Ptilophora pectinata* (A. & E.S.Gepp) R.E. Norris sensu Adams 1994: 154.

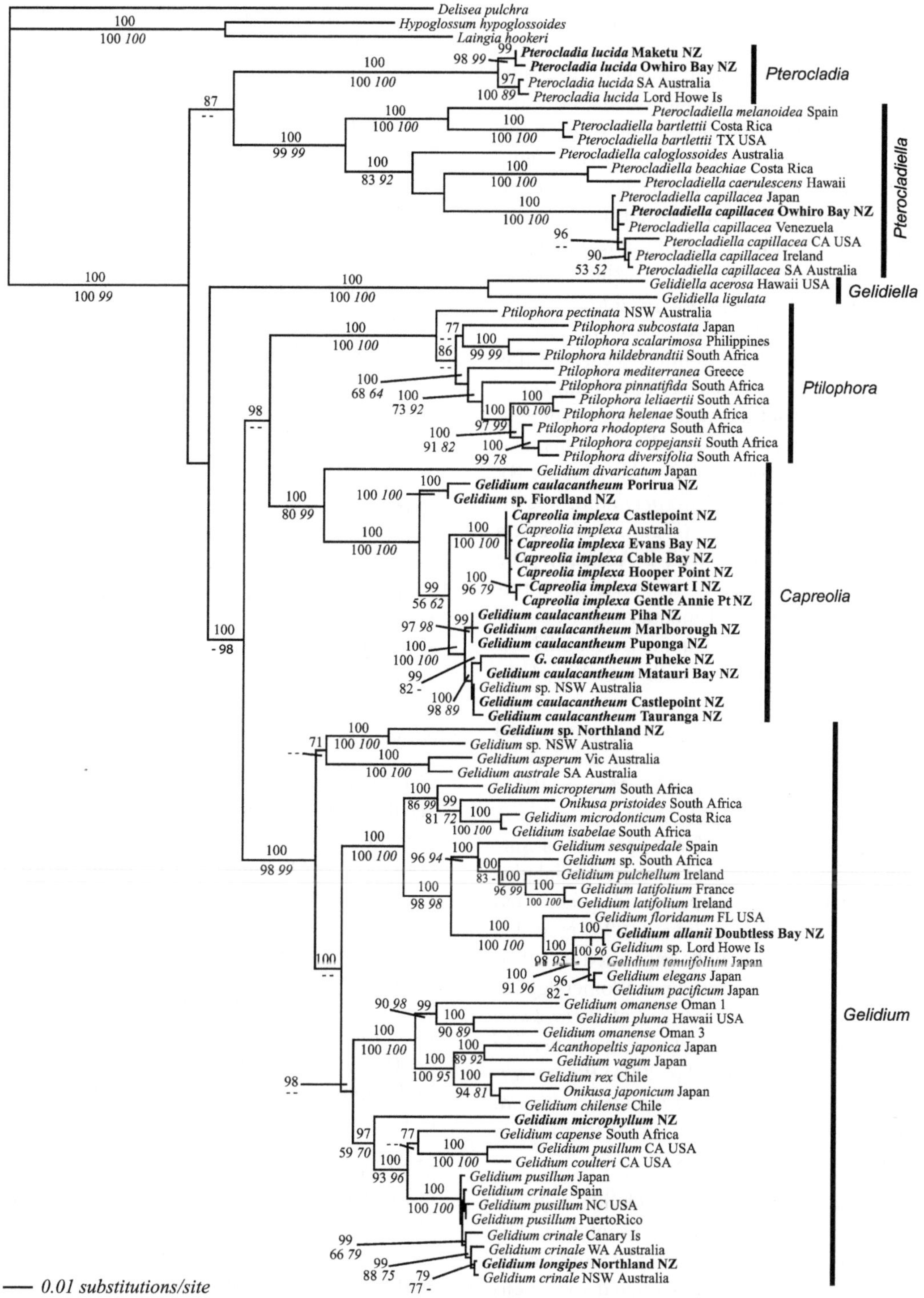

Figure 1. Maximum likelihood phylogram (–ln likelihood 16022.19) found by PAUP*4.0b10. Support values are shown on each node: Bayesian posterior probability values above the line, and MP and NJ bootstrap values below, NJ in italics. Support values are shown only on those nodes supported by >70% on at least one analysis. New Zealand taxa are in bold type.

Remarks: In New Zealand *P. lucida* is morphologically very variable: Moore (1945) discussed five growth forms although she considered all belonged to a single variable species. V.W. Lindauer in unpublished manuscripts had referred previously to one of the growth forms as *P. lucida* var. *sublittoralis*. Adams (1994) transferred the specimens with this distinctive growth form, all collected from subtidal locations, to *Ptilophora pectinata,* noting that only tetrasporangial material had been seen in the New Zealand collections. The basionym of *Ptilophora pectinata* is *Pterocladia lucida* f. *pectinata* A. & E.S. Gepp (1906), based on material from New South Wales, and this form was raised to species status as *Pterocladia pectinata* by Lucas (1931). Subtidal material consistent with the morphological concept employed by Adams, and with the illustration in Norris (1987: 247, Figure 8), and initially identified as *Ptilophora pectinata,* was included in our sequencing study (Figure 1: *P. lucida* Maketu, Table 1: voucher = WELT A26820). The sequence data, however, placed this material corresponding to "*sublittoralis*" of Lindauer and *P. pectinata* of Adams, with *Pterocladia lucida* from both New Zealand and Australia, and not with the *Ptilophora* clade.

Akatsuka (1986) established a new genus and species, *Pterocladiastrum robustum*, for the growth forms of *Pterocladia lucida* named by Moore (1945) as "Robust" and "Poor Knights". The new taxon was established on the basis of the "pattern of surface cells coupled with distinctive tetrasporangial pinnules", characters that have not been widely accepted as reliable at a generic level. Adams (1994) recorded this monotypic genus although expressed reservations about its distinctiveness, and Womersley & Guiry (1994) in a taxonomic treatment of *Pterocladia lucida* in southern Australia rejected *P. robustum*. In AlgaeBase (Guiry & Nic Dhonncha, 2004) *Pterocladiastrum* is listed as a synonym of *Ptilophora pectinata*, which we show above to be a synonym of *Pterocladia lucida* for New Zealand specimens.

Pterocladiella: This genus has a worldwide distribution, with a single New Zealand representative. The sequence data place the New Zealand sample of *P. capillacea* within a clade of closely related specimens sourced from throughout the world (Figure 1), a result that is in agreement with the findings of Freshwater et al. (1995) and Shimada et al. (2000).

Pterocladiella capillacea (Gmelin) Santelices et Hommersand 1997: 117–118.
NZ distribution: Kermadec, Three Kings, North, northern South, and Chatham Is.
Basionym and type locality: *Fucus capillaceus* Gmelin 1768: 146, pl. 15, Figure 1; Lectotype (original illustration); Mediterranean.
Synonyms: *Pterocladia lindaueri* Fan 1961: 335.

Remarks: Fan (1961) distinguished *Pterocladia lindaueri,* described from a collection from the northern North I., on the basis of unequal locules in the cystocarps. Cystocarps of unequal size also occur in *Pterocladiella capillacea,* and material from the type locality of *P. lindaueri* is indistinguishable from *P. capillacea*. Santelices and Hommersand (1997) concluded that material described by Fan was likely to have been material of *P. capillacea*. These species are considered here to be synonymous.

Capreolia: A clade containing *Capreolia implexa* is clearly differentiated in the phylogenetic analyses. Samples of taxa in this clade included in this study were sourced from a wide geographic spread (northern North Island to Stewart Island) in order to assess diversity within New Zealand. The results suggest that there are at least three species of *Capreolia* in New Zealand, including two or three taxa currently grouped under *G. caulacantheum.*

Gelidium caulacantheum J. Agardh 1876: 548.
NZ distribution: North, South, and Chatham Is.
Type and type locality: LD; Berggren collections from both the Bay of Islands and Tauranga, North I.

Remarks: Within New Zealand, a number of names have been applied to small, intertidal species commonly found in turfs (e.g. *Gelidium subulifolium* (Harv.) V.J.Chapm., *G. subuliferum* Laing), as well as two forms (f. *laxiforme*, f. *fasciculatum*) and one variety (var. *pygmaeum*) of *G. caulacantheum* established by Chapman (1969). At present these are all grouped as *G. caulacantheum* sensu Adams (1994).

Capreolia implexa Guiry et Womersley 1993: 267.
NZ distribution: Three Kings, North, South, Stewart, and Chatham Is.
Type and type locality: MEL; Sandringham, Port Phillip, Victoria, Australia.

Remarks: Sequence data place New Zealand and Australian material of *C. implexa* within the same clade. The name *G. pusillum* (Stackhouse) LeJolis was previously applied to *C. implexa* in New Zealand (Adams, 1994).

Gelidium: Four samples sequenced here are grouped in the *Gelidium* clade.

"*Gelidium* Northland" is known from a single, sterile collection and is thus unable to be characterised and described at present. The sequence data indicate that this species is clearly distinct from all other New Zealand taxa.

Gelidium allanii V.J. Chapm. 1969: 98.
NZ distribution: northern North I.
Type and type locality: AKU in AK; Waitata Rocks, Bay of Islands.

Remarks: Molecular sequence analyses place this species within the *Gelidium* clade, well separated from other New Zealand Gelidiales. Previous research on *Gelidium allanii* (Nelson et al., 1994) revealed that this species, which is found on both the east and west coasts of the northern North I., has a restricted distribution and possesses an unusual pyruvulated agar.

Gelidium microphyllum (Crosby Smith) Kylin 1934: 56.
NZ distribution: southern North, South, Stewart, Chatham, Snares, Antipodes, Auckland, and Campbell Is.
Basionym, type and type locality: *Nitophyllum microphyllum* Crosby Smith in Laing 1902: 344; CHR; Green Island Beach, Dunedin.

Remarks: The placement of *G. microphyllum* within the *Gelidium* clade is well supported. This species is restricted to cold waters around the New Zealand mainland and subantarctic islands. Although specimens similar to *G. microphyllum* have been found at the Kermadec Islands, they are unlikely to be conspecific as the Kermadecs (29–32°S) are tropical to subtropical.

Gelidium longipes J. Agardh 1876: 547.
NZ distribution: northern North I.
Type and type locality: LD; Berggren, Bay of Islands.

Remarks: Material from the type locality of *G. longipes* grouped in a clade containing specimens identified as both *G. pusillum* and as *G. crinale*, both European species reported to have cosmopolitan distributions but for which species concepts remain poorly understood (Freshwater & Rueness, 1994). Although this endemic taxon has a very restricted distribution in New Zealand (Nelson & Farr, 2003), the relationship with material from other parts of the world is clearly seen by the results presented here. Clarification of the species concepts of European *Gelidium* and a further examination of species–level diversity in this group are required before nomenclatural change is undertaken.

Excluded taxon: Levring (1949) described *Gelidium ceramoides* from material collected from Kaka Point, Otago, and subsequently it has been found infrequently from only Otago and Stewart I. (Nelson & Phillips, 2001). Preliminary microscopic examination and culture studies clearly establish that this species is not a member of the Gelidiales: material collected from the type locality was sequenced and found to be unrelated to members of the Gelidiales, and more closely related to Gigartinales sensu lato. This species is the subject of a further study.

Discussion

Molecular sequence data have enabled an examination of the phylogenetic relationships of New Zealand species of Gelidiales. New Zealand taxa are found across the order, belonging to well resolved clades corresponding to four genera. Interpretation of morphological variability in *Pterocladia lucida* in New Zealand has been problematic: this study has clarified the identity of material recently transferred to *Ptilophora pectinata* as more correctly placed in *Pterocladia lucida*. The genus *Capreolia* is clearly differentiated in the phylogenetic analyses, and the results of this study indicate that there are likely to be at least three species of *Capreolia* in New Zealand. The relationship between *Gelidium caulacantheum* and *Capreolia implexa* has been recognised previously (Bailey & Freshwater, 1997; Freshwater et al., 1995, Freshwater & Bailey, 1998). *Capreolia* was established by Guiry and Womersley (1993) on the basis of its biphasic life history in which fertilised carpogonia develop directly into tetrasporophytes. Further research is required on this group to clarify species boundaries and to reconsider the criteria used to distinguish the genus, particularly life history characters. Fertile material of *G. caulacantheum* indicates that at least some of the material resolved in this clade does not have the same life history as *C. implexa*. The sequence of the Japanese *G. divaricatum* is resolved as a sister taxon to the *Capreolia* clade. The generic placement of this species warrants attention.

Within the *Gelidium* clade in our analyses there are four New Zealand species that are distinguished by *rbc*L sequence data: *G. allanii, G. longipes, G. microphyllum* and *G.* "Northland". From examination of herbarium collections, it is clear that there are additional undescribed taxa in the New Zealand region, currently represented by fragmentary and incomplete specimens. In particular it is likely that there will be greater diversity in this order when fuller collections, including samples purpose-collected for molecular sequencing, can be made from the northern islands of the New Zealand archipelago (Kermadec Islands) and from the subantarctic islands (Antipodes, Bounty, Auckland and Campbell Is).

In the data presented here *Pterocladia* is an Australasian genus, and the *Capreolia* clade reveals an Australasian radiation. The clade equating to the genus *Gelidium* contains four phylogenetically distinct New Zealand taxa. Increased taxon sampling from throughout the southern hemisphere may illuminate further relationships within this clade.

Acknowledgements

We thank our colleagues Kate Neill for assistance in the field and Jennifer Dalen for discussions on the manuscript. This research was funded by the New Zealand Foundation for Research, Science and Technology MNZX0202.

References

Adams NM (1994) Seaweeds of New Zealand. Canterbury University Press, Christchurch 360 pp.

Agardh JG (1852) Species genera et ordines algarum. Vol. 2, part 2, Gleerup, Lund, pp. 335–720.

Agardh JG (1876) Species genera et ordines algarum. Volume 3, part 1 – Epicrisis systematiis Floridearum. Weigel, Leipzig. pp. viii + 1–724.

Akatsuka I (1986) *Pterocladiastrum*, a new genus segregated from *Pterocladia* (Gelidiales, Rhodophyta). Bot. Mar. 29: 51–58.

Bailey JC, Freshwater DW (1997) Molecular systematics of the Gelidiales: Inferences from separate and combined analyses of plastid *rbc*L and nuclear SSU gene sequences. Europ. J. Phycol. 32: 343–352.

Chapman VJ (1969) The marine algae of New Zealand. Part III: Rhodophyceae. Issue 1: Bangiophycidae and Florideophycidae (Nemalionales, Bonnemaisoniales, Gelidiales). Lehre, Cramer.

Fan KC (1961) Morphological studies of the Gelidiales. Univ. Calif. Publ. Bot. 32: 315–368.

Freshwater DW, Bailey JC (1998) A multigene phylogeny of the Gelidiales including nuclear large-subunit *r*RNA sequence data. J. Appl. Phycol. 10: 229–236.

Freshwater DW, Fredericq S, Hommersand MH (1995) A molecular phylogeny of the Gelidiales (Rhodophyta) based on analysis of plastid *rbc*L nucleotide sequences. J. Phycol. 31: 616–632.

Freshwater DW, Rueness J (1994) Phylogenetic relationships of some European *Gelidium* (Gelidiales, Rhodophyta) species, based on *rbc*L nucleotide sequence analysis. Phycologia 33: 187–194.

Gepp A, Gepp ES (1906) Some marine algae from New South Wales. J. Bot. 44: 249–261, pl. 481.

Gmelin SG (1768) *Historia Fucorum*. 4 pls., 239 pp., 35 pls. St. Petersburg.

Goff LJ, Moon DA (1993) PCR amplification of nuclear and plastid genes from algal herbarium specimens and algal spores. J. Phycol. 29: 381–384.

Guiry MD, Nic Dhonncha E (2004) *AlgaeBase version 2.1*. Worldwide electronic publication, National University of Ireland, Galway. http://www.algaebase.org; searched on 15 June 2004.

Guiry MD, Womersley HBS (1993) *Capreolia implexa* gen. et sp. nov. (Gelidiales, Rhodophyta) in Australia and New Zealand: An intertidal mat-forming alga with an unusual life history. Phycologia 32: 266–277.

Hillis DM, Mable BK, Larson A, Davis SK, Zimmer EA (1996) Nucleic Acids IV: Sequencing and Cloning. In: Hillis DM, Moritz C, Mable BK (eds.), Molecular Systematics, 2nd ed, Sinauer Associates, Sunderland Massachusetts, pp. 321–81.

Holmgren PK, Holmgren NH, Barnett LC (1990). Index Herbariorum, Part 1: The Herbaria of the World. New York Botanical Garden, New York.

Kylin H (1934) Bemerkungen über einige Florideen aus Neu-Seeland. Kung. Fysiograf. Sällsk. Lund. Förhandl. 4: 55–62.

Laing RM (1902) Revised list of New Zealand seaweeds, Part II. Transactions and Proceedings of the N.Z. Institute 34: 327–359.

Lee H-B, Kim J-I (1995) Notes on Gelidiales species from Korea. In: I. A. Abbott (ed.) Taxonomy of Economic Seaweeds, with reference to some Pacific species. Vol. V. California Sea Grant College, University of California, La Jolla, California, pp. 161–174.

Levring T (1949) Six new marine algae from New Zealand. Trans. Proc. Roy. Soc. NZ 77: 394–397.

Lucas AHS (1931) Notes on Australian marine algae. VI. Proc. Linn. Soc. N.S.W. 56: 407–411, 23–37.

Moore LB (1945) The genus *Pterocladia* in New Zealand. Trans. Roy. Soc. NZ 74: 332–342, 45–49.

Nelson WA, Knight GA, Falshaw R, Furneaux RH, Falshaw A, Lynds SM (1994) Characterisation of the enigmatic, endemic red alga *Gelidium allanii* (Gelidiales) from northern New Zealand – morphology, distribution, agar chemistry. J. Appl. Phycol. 6: 497–507.

Nelson WA, Farr TJ (2003) Field and morphological observations of *Gelidium longipes* J.Agardh (Gelidiales, Rhodophyta), a rare endemic red alga from northern New Zealand. NZ J. Bot. 41: 707–713.

Nelson WA, Phillips LE (2001) Locating the type specimens of New Zealand marine algae described by Levring. NZ J. Bot. 39: 349–353.

Norris RE (1987) A re-evaluation of *Ptilophora* Kützing and *Beckerella* Kylin (Gelidiales, Rhodophyceae) with a review of South African species. Bot. Mar. 30: 243–258.

Patwary MH, Sensen CW, MacKay RM, van der Meer JP (1998) Nucleotide sequences of small-subunit and internal transcribed spacer regions of nuclear rRNA genes support the autonomy of some genera of the Gelidiales (Rhodophyta). J. Phycol. 34: 299–305.

Posada D, Crandall K (1998) MODELTEST: Testing the model of DNA substitution. Bioinformatics 14: 817–818.

Rambaut A (1996) Se-Al: Sequence Alignment Editor. Available at http://evolve.zoo.ox.ac.uk/.

Rodriguez D, Santelices B (1987) Patterns of apical structure in the genera *Gelidium* and *Pterocladia* (Gelidiaceae, Rhodophyta). Hydrobiologia 151/152: 199–203.

Ronquist F, Huelsenbeck JP (2003) MrBayes 3: Bayesian phylogenetic inference under mixed models. Bioinformatics 19: 1572–1574.

Santelices B (1990) New and old problems in the taxonomy of the Gelidiales (Rhodophyta). Hydrobiologia 204/205: 125–135.

Santelices B (1994) A reassessment of the taxonomic status of *Gelidium amansii* (Lamouroux) Lamouroux. In: IA Abbott, (ed.) Taxonomy of Economic Seaweeds, with reference to some Pacific species. Volume IV. California Sea Grant College, University of California, La Jolla, California, pp. 37–53.

Santelices B, Hommersand MH (1997) *Pterocladiella*, a new genus in the Gelidiaceae (Gelidiales, Rhodophyta). Phycologia 36: 114–119.

Schiel DR, Nelson WA (1990) The harvesting of macroalgae in New Zealand. Hydrobiologia 204/205: 25–33.

Shimada S, Horiguchi T, Masuda M (1999) Phylogenetic affinities of genera *Acanthopeltis* and *Yatabella* (Gelidiales, Rhodophyta) inferred from molecular analyses. Phycologia 38: 528–540.

Shimada S, Horiguchi T, Masuda M (2000) Confirmation of the status of three *Pterocladia* species (Gelidiales, Rhodophyta) described by K. Okamura. Phycologia 39: 10–18.

Stewart JG (1976) Gelidiaceae. In: Abbott IA, Hollenberg GJ (eds.), Marine Algae of California. Stanford University Press, California, pp. 340–352.

Swofford DL (2002) PAUP*. Phylogenetic Analysis Using Parsimony (*and Other Methods). Version 4., Sinauer Associates, Sunderland, Massachusetts.

Tronchin EM, Freshwater DW, Bolton JJ, Anderson RJ (2002) A reassessment and reclassification of species in the genera Onikusa Akatsuka and *Suhria* J.Agardh ex Endlicher (Gelidiales, Rhodophyta) based on molecular and morphological data. Bot. Mar. 45: 548–558.

Tronchin EM, Freshwater DW, Bolton JJ (2003) A re-evaluation of the genera *Beckerella* and *Ptilophora* (Gelidiales, Rhodophyta) based on molecular and morphological data. Phycologia 42: 80–89.

Turner D (1819) Fuci sive Plantarum Fucorum Generi a Botanicis Ascriptarum Icones Descriptiones et Historia. Volume 4, pp. 1–153, pl. 197–258. London.

Womersley HBS, Guiry MD (1994) Gelidiales. In: HBS Womersley, The marine benthic flora of Southern Australia. Part IIIA, ABRS, Canberra, pp. 118–142.

Journal of Applied Phycology (2006) 18: 663–669
DOI: 10.1007/s10811-006-9069-z

Taxonomic considerations of a foliose *Grateloupia* species from the Straits of Messina

R.J. Wilkes[1], M. Morabito[2] & G.M. Gargiulo[2]*
[1]*Irish Seaweed Centre, Martin Ryan Institute, National University of Ireland, Galway, Ireland;* [2]*Department of Botanical Sciences, University of Messina, Salita Sperone, 31, S. Agata, 98166 Messina, Italy*

*Author for correspondence: e-mail: ggargiulo@unime.it

Key words: *Grateloupia*, Halymeniales, *rbc*L gene, molecular systematics, Mediterranean Sea, Rhodophyta.

Abstract

Grateloupia turuturu Yamada is the currently accepted name for the invasive red alga that is present on coasts of the North Atlantic. Previously considered as *G. doryphora* (Montagne) M.A. Howe, populations of this invasive species were examined and their taxonomic position revised using molecular and morphological techniques. It was also thought that similar invasive populations in the Mediterranean should be identified as *G. turuturu*. This investigation used *rbc*L based molecular analyses to clarify the taxonomic position of *Grateloupia "doryphora"* from the Straits of Messina. Our results indicate that this population is neither *G. doryphora* nor *G. turuturu*. It was placed separately in all analyses and grouped consistently with other *Grateloupia* species from the Pacific. On the basis of molecular data from this and previous investigations, it is evident that the status of the foliose Atlantic and Mediterranean entities is still unclear and a re-evaluation of the old names connected to them should be undertaken.

Introduction

Grateloupia turuturu Yamada is the currently accepted name for the invasive red alga, previously considered as *Grateloupia doryphora* (Montagne) M.A. Howe, present on coasts of the UK, France and north eastern America (Gavio & Fredericq, 2002). Until recently, both Atlantic and Mediterranean foliose *Grateloupia* was referred to as *G. doryphora* based on the papers of Ardré and Gayral (1961) and Dawson et al. (1964). Ardré and Gayral (1961) placed in synonymy some foliose Atlantic and Pacific entities under the name *G. lanceola* (J. Agardh) J. Agardh *emend.* Ardré et Gayral, originally described as *Halymenia lanceola* J. Agardh (1841: 19) for the Atlantic coasts of Morocco and South Spain. *Grateloupia doryphora* was not included in this due to an absence of data (Ardré & Gayral, 1961). Subsequently, Dawson et al. (1964) associated *G. doryphora* to the *G. lanceola*-complex, keeping the former name as the first validly published. Since then reports for different localities, both Mediterranean and Atlantic, such as the records from Morocco and Senegal (Gayral, 1958; Ardré & Gayral, 1961) or those from southern Spanish coasts (Pérez-Cirera et al., 1989) of *G. lanceola*, have been considered *G. doryphora.*

Recent investigations have compared some invasive foliose *Grateloupia* populations with specimens from the type localities of *G. doryphora* (Peru) and of *G. turuturu* (Japan), and, based on *rbc*L gene sequences and anatomical observations, these were shown to be conspecific to *G. turuturu* (Gavio & Fredericq, 2002). This result was previously hypothesised by Verlaque (2001) on the basis of phytogeographical considerations. *Grateloupia "doryphora"* specimens from the Mediterranean Sea were also supposed to be *G. turuturu* (Verlaque, 2001; Gavio & Fredericq, 2002), but molecular investigations had not been carried out to test this hypothesis.

Grateloupia "doryphora" was first reported from the Straits of Messina, Mediterranean Sea (De Masi & Gargiulo, 1982) but specimens of a foliose *Grateloupia* species, identified as *G. cuneifolia* J. Agardh *ex* Kützing in Giaccone's herbarium (CAT), date back to 1969 (Giaccone, 1969). Subsequently, a similar foliose species was recorded in Thau lagoon (Riouall et al., 1985) and in Venice lagoon (Gargiulo et al., 1992; Tolomio, 1993). In this paper comparative *rbc*L molecular analyses have been undertaken in order to clarify the taxonomic position of *Grateloupia "doryphora"* from the Straits of Messina.

Materials and methods

Samples of *Grateloupia* sp. were collected from both sides of the Straits of Messina, on rocks and concrete blocks near the water line. Samples used in the present study, with voucher numbers and collection information, are: Gra002 (Villa S. Giovanni, Reggio Calabria, Italy, 16/02/2000); Gra004 (Villa S. Giovanni, Reggio Calabria, Italy, 16/02/2000); Gra015 (Villa S. Giovanni, Reggio Calabria, Italy, 17/03/2004); Gra016 (Villa S. Giovanni, Reggio Calabria, Italy, 17/03/2004); Gra017 (Villa S. Giovanni, Reggio Calabria, Italy, 17/03/2004); Gra018 (Torre Faro, Messina, Italy, 22/03/2004).

DNA was isolated from freshly collected thalli (within 48 h) and from silica gel and herbarium preserved samples, with a modified CTAB protocol (Doyle & Doyle, 1987). Voucher specimens were preserved in 4% formalin in seawater, dried in silica gel, pressed as herbarium sheets and deposited in the Phycological Herbarium of the Department of Botanical Sciences of the University of Messina (MS). Anatomical observations were made on hand sections of the thalli. Micrographs were taken by a Diaplan Leica microscope equipped with a camera. In order to prevent errors in sorting of samples, each DNA isolation was performed from a single individual, a fragment of which was kept as voucher formalin preserved and/or pressed for further inspections.

The *rbc*L gene was PCR amplified using primers listed in Wang et al. (2000). Sequencing was performed by an external company (MWG Biotech AG, Ebersberg, Germany). Six separate thalli from the samples of the Straits of Messina were sequenced but as they differed by only 1bp only one was used for the analyses (Gra017). Sequences were aligned manually using GeneDoc 2.6.002 (Nicholas & Nicholas, 1997). No insertions or deletions were found, making the alignment unambiguous. Previously published sequences were obtained from Genbank and added to the dataset to give a 1253-bp sequence alignment for analyses (Table 1). Trees were rooted with *Sebdenia monardiana* (Montagne) Berthold, which is the generitype of the family Sebdeniaceae, a sister group to the Halymeniaceae (Saunders & Kraft, 1996).

The data were analysed for maximum parsimony (MP), neighbour joining (NJ), using a Kimura 2-parameter distance matrix as input, and for maximum likelihood (ML) using PAUP* 4b10 (Swofford, 2002). MP analysis was performed with 50 random sequence additions. Modeltest version 3.06 (Posada & Crandall, 1998) was used to determine the parameters for the ML analyses, and specified a Transition model with a proportion of invariable sites and a gamma distribution (TIM+I+G). The rate matrix was specified as [A−C] = 1.0, [A-G] = 4.4776, [A−T] = 0.6298, [C−G] = 0.6298, [C−T] = 11.0935 and [G−T] = 1.0, with the base frequencies at A = 0.3164, C = 0.1435, G = 0.2059, and $T = 0.3342$. The proportion of invariable sites was set at 0.4904 with a gamma distribution of 0.6904. The robustness of each analysis was tested by bootstrapping the dataset 1000 times for MP and NJ, and 100 times for ML analysis (Felsenstein, 1985).

Results

Thalli of *Grateloupia* sp. from the Straits of Messina are solitary or gregarious, purplish-red to yellowish-red. Fronds are lanceolate or irregularly lobate, with one or more blades connected to a short stipe, arising from a discoid holdfast (Figure 1). They are up to 70 cm long and 15 cm large, and 500–2000 μm thick. The consistency of the frond ranges from leathery to somewhat gelatinous.Transverse sections show a compact outer cortex of 4–6 cells periclinally oriented, with the most external ones elongated (6–8 × 1.5–2.5 μm), and the others roundish (5–6 μm in diameter), an inner cortex of 2–3 irregularly stellate cells (8–10 × 16–18 μm), anticlinally oriented (Figure 2). The medulla is loose, with filaments mainly anticlinally oriented (1.5–2.5 × 25–30 μm). The cortex pattern is very constant both at different levels of the blade and in variously aged thalli.

The auxiliary ampullae are branched up to secondary order (Figure 3).

The alignment of *rbc*L gene sequences contained 263 parsimony informative sites with 869 invariable positions. The single maximum-likelihood tree with bootstrap values from each analysis overlaid on the branches is shown in Figure 4. NJ and MP trees had topologies similar to ML and are not shown. All the *Grateloupia* samples were resolved in a single large

Table 1. Samples used for *rbc*L molecular analyses

Samples	Collection data	References	GenBank accession numbers
G. acuminata	Japan	Kawaguchi et al. (2001)	AB055480
G. angusta	Japan	Wang et al. (2001)	AB061378
G. asiatica	China	Kawaguchi et al. (2001)	AB055484
G. chiangii	Japan	Wang et al. (2001)	AB061385
G. cornea	Japan	Wang et al. (2001)	AB061381
G. crispata	Japan	Wang et al. (2001)	AB061383
G. divaricata	Japan	Wang et al. (2000)	AB038609
G. doryphora	Falklands	Gavio and Fredericq (2002)	AF488827
G. doryphora	Peru	Gavio and Fredericq (2002)	AF488817
G. elata	Japan	Kawaguchi et al. (2001)	AB061388
G. elliptica	Japan	Wang et al. (2000)	AB038605
G. filicina	Quercianella, Italy	Kawaguchi et al. (2001)	AB055470
G. filicina	Quercianella, Italy	Kawaguchi et al. (2001)	AB055471
G. imbricata	Japan	Wang et al. (2000)	AB038607
G. kurogii	Japan	Wang et al. (2000)	AB038606
G. lanceolata	Japan	Kawaguchi et al. (2001)	AB055478
G. livida	Japan	Wang et al. (2000)	AB038610
G. lyalli	USA	Fredericq et al. (1996)	U04217
G. patens	Japan	Kawaguchi et al. (2001)	AB061390
G. ramossissima	Japan	Kawaguchi et al. (2001)	AB061393
G. schmitziana	Japan	Kawaguchi et al. (2001)	AB061394
G. sparsa	Japan	Kawaguchi et al. (2001)	AB055473
G. turuturu	Japan	Wang et al. (2001)	AB038611
G. turuturu	Japan	Gavio and Fredericq (2002)	AF488820
G. turuturu	Langstone Harbour, UK	Present study	AY654891
Grateloupia sp.	Villa S. Giovanni, Italy	Present study	AY651060
Halymenia dilatata	Japan	Wang et al. (2000)	AB038604
H. floresia	Malaysia	Wang et al. (2000)	AB038603
Sebdenia monardiana	Italy	Fredericq et al. (1996)	U21600

clade with good support (bootstrap proportion value = 86–95%). The arrangement of the taxa within this clade was the same in each analysis. The Straits of Messina sample was placed in a strongly supported subclade with 100% bootstrap support. This entity allied strongly with *G. lanceolata* (Okamura) Kawaguchi in each analysis (100%). A second subclade was resolved with good bootstrap support of 89–99%. The remaining species were placed in small groups with *G. turuturu* (98–100%) allied with *G. sparsa* (Okamura) Chiang (98–100%) and a cluster with *G. doryphora* samples (100%). Although consistent in each of the analyses, the relationships between *Grateloupia filicina* (J.V. Lamouroux) C. Agardh, *G. ramosissima* Okamura and these smaller groupings was not well supported.

Discussion

Grateloupia doryphora had been reported as an introduced species on both the western and eastern coasts of North Atlantic (Farnham, 1973; Cabioch et al., 1997; Villalard-Bohnsack & Harlin, 1997; Maggs & Stegenga, 1999), and also in the Mediterranean Sea (De Masi & Gargiulo, 1982; Riouall et al., 1985; Tolomio, 1993). Different authors had suggested that its introduction might be due to oyster transfer from Japan (Farnham, 1980; Riouall et al., 1985; Ribera & Boudouresque, 1995), but Verlaque (2001) noted that *G. doryphora* was not reported either in Japan or in Korea and suggested that the introduced species should be better compared to *G. turuturu*, a species endemic to these countries. Among the Mediterranean

Figure 1–3. *Grateloupia* sp. from Villa S. Giovanni, Italy. (1) Herbarium specimen of a tetrasporangial thallus. (2) Cross section of a female gametophyte, 5 cm below the apex. (3) Cross section of a female gametophyte showing an auxiliary cell ampulla.

localities from which *G. "doryphora"* had been reported, Thau lagoon and Venice are indeed oyster farming sites. Gavio and Fredericq (2002) investigated several North Atlantic populations of *G. "doryphora"* with the use of both molecular and morphological tools, demonstrating that they were conspecific with *G. turuturu* from Japan. Although not included in their analyses, they hypothesised on the basis of available RAPDs and ITS studies (Marston & Villalard-Bohnsack, 1999) that the Mediterranean populations were also *G. turuturu* (Gavio & Fredericq, 2002). However, even if the hypothesis regarding this introduction by oyster transfer might be valid for Venice and Thau lagoon, it does not seem convincing for the Straits of Messina because of the absence of oyster farming.

The *rbc*L analysis presented here clearly shows that the foliose *Grateloupia* species present in the Straits of Messina is a separate entity, distinct from either *G. turuturu* or *G. doryphora*. While the *rbc*L sequence of this entity does not match that of any other *Grateloupia* species included in the analyses, it was placed robustly in a subclade of Japanese samples. Within this grouping it was most closely related to *Grateloupia lanceolata* but is clearly distinct, with a sequence divergence of 1.27%, well within the reported interspecific limits for different *Grateloupia* species (e.g. Gavio & Fredericq, 2002; Kawaguchi et al., 2002). The latter species was originally described as *Aeodes lanceolata* Okamura (1934), later transferred to the genus *Pachymeniopsis* (Yamada in Kawabata, 1954) and recently included in *Grateloupia* (Kawaguchi, 1997). *G. lanceolata* and *Grateloupia* sp. from the Straits of Messina share a similar external morphology, as well as a *Grateloupia*-type auxiliary ampulla, but they differ somewhat in their anatomy, because the latter has a thinner cortex. No *rbc*L data are available for the populations present in Venice and Thau lagoon, and no direct comparison was possible. However, Marston & Villalard-Bohnsack (2002), on the basis of RAPDs, ITS and *cox* data, showed that Thau lagoon specimens, as *G. doryphora*, were conspecific with other North Atlantic populations presently considered to be *G. turuturu* (Gavio & Fredericq, 2002). Therefore, it seems reasonable to expect that Thau lagoon samples are conspecific with *G. turuturu*, even if this hypothesis needs testing.

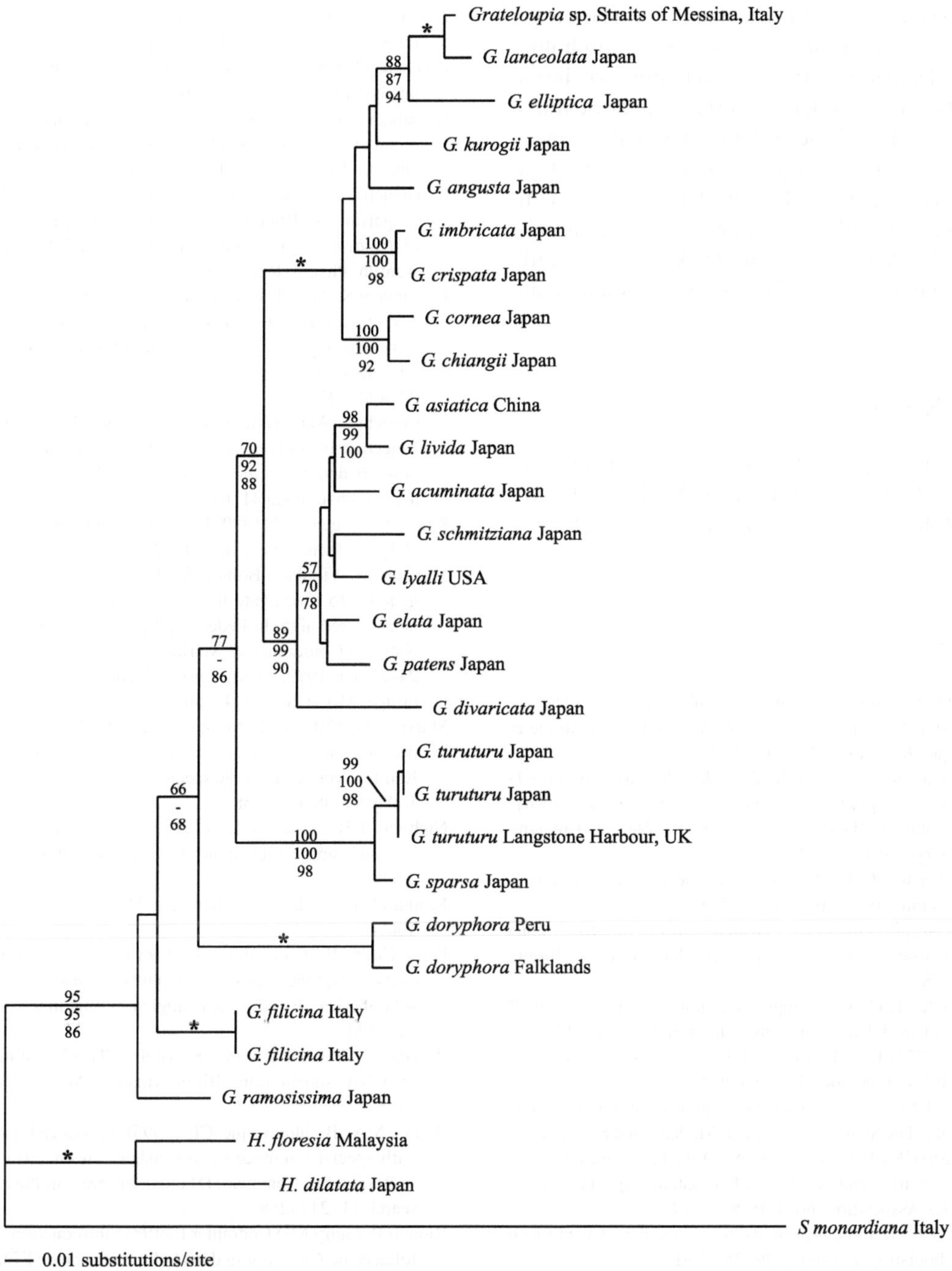

Figure 4. ML tree with bootstrap values inferred from MP, NJ and ML analyses overlaid on the branches. Branches with 100% support in all analyses are marked with an asterisk.

Conclusion

At present we cannot give a definite taxonomic attribution to the population from the Straits of Messina, but it is likely that at least two entities, sharing the same foliose habit, are present in the Mediterranean Sea. It is also not possible, on the basis of the information available, to know whether the population under study is an introduction or not. Its presence in the Straits has been known for certain since the 1960s (Giaccone, 1969) and it is possible to hypothesize that it could be a relict of Indo-Pacific origin. On the basis of both morphological and molecular data from this and previous investigations (Gavio & Fredericq, 2002), it is evident that the status of some foliose Atlantic and Mediterranean *Grateloupia* species is not yet clear. Further investigations together with a re-evaluation of the old names connected to them should be undertaken, with special consideration afforded to *G. lanceola* (J. Agardh) J. Agardh from the Atlantic coasts of Morocco and South Spain.

Acknowledgements

This study was supported by a grant from the University of Messina, Italy (PRA2003) to G.M. Gargiulo. We thank Dr Robert Fletcher for providing material.

References

Agardh JG (1841) In historiam algarum symbolae. Linnaea 15: 19.

Ardré F , Gayral P (1961) Quelques *Grateloupia* de l'Atlantique et du Pacifique. Rev. Algol. N. S. 6: 38–48.

Cabioch J, Castric-Fey A, L'hardy-Halos M, Rio A (1997) *Grateloupia doryphora* et *Grateloupia filicina* var. *luxurians* (Rhodophyta, Halymeniaceae) sur les côtes de Bretagne (France). Cryptogam. Algol. 18: 117–137.

Dawson EY, Acleto OC, Foldvik N (1964) The seaweeds of Peru. J. Cramer, Weinheim. 111 pp + Pl. LXXXI.

De Masi F, Gargiulo GM (1982) *Grateloupia doryphora* (Mont.) Howe (Rhodophyta, Crytonemiales) en Méditerranée. Allionia 25: 105–108.

Doyle JJ, Doyle JL (1987) A rapid isolation procedure for small quantities of fresh leaf tissue. Phytochem. Bull. 19: 11–15.

Farnham WF (1973) The addition of a foliose species of *Grateloupia* to the British marine flora. Br. Phycol. J. 8: 208–209.

Farnham WF (1980) Studies on aliens in the marine flora of southern England. The Shore Environment, Methods and Ecosystems, Portsmouth (UK). In Price, JH, Irvine, DEG, Farnham, WF (eds), The shore environment. Volume 2: Ecosystems. Special Volume Systematics Association, no. 17B: 875–914

Felsenstein J (1985) Confidence limits in phylogenies: An approach using the bootstrap. Evolution 39: 783–791.

Fredericq S, Hommersand M, Freshwater DW (1996) The molecular systematics of some agar- and carrageenan-containing marine red algae based on *rbc*L sequence analysis. Hydrobiologia 326/327: 125–135.

Gargiulo GM, De Masi F, Tripodi G (1992) *Sargassum muticum* (Yendo) Fensholt (Phaeophyta, Fucales) is spreading in the lagoon of Venice (Northern Adriatic Sea). Gior. Bot. Ital. 126: 259.

Gavio B, Fredericq S (2002) *Grateloupia turuturu* (Halymeniaceae, Rhodophyta) is the correct name of the non-native species in the Atlantic known as *Grateloupia doryphora*. Europ. J. Phycol. 37: 349–360.

Gayral P (1958) Algues de la côte atlantique marocaine. Collection La Nature au Maroc, Rabat. 524 pp.

Giaccone G (1969) Raccolte di fitobenthos sulla banchina continentale italiana. Gior. Bot. Ital. 103: 485–514.

Kawabata S (1954) On the structure of the frond, and the reproductive organ of *Pachymeniopsis lanceolata* Yamada (*Aeodes lanceolata* Okam.). Bull. Jpn. Soc. Phycol. 2: 11–15.

Kawaguchi S (1997) Taxonomic notes on the Halymeniaceae (Gigartinales, Rhodophyta) from Japan. III. Synonymization of *Pachymeniopsis* Yamada in Kawabata with *Grateloupia* C. Agardh. Phycol. Res. 45: 9–21.

Kawaguchi S, Wang HW, Horiguchi T, Sartoni G, Masuda M (2001) A comparative study of the red alga *Grateloupia filicina* (Halymeniaceae) from the northwestern Pacific and Mediterranean with the description of *Grateloupia asiatica*, sp. nov. J. Phycol. 37: 433–442.

Kawaguchi S, Wang HW, Horiguchi T, Lewis JA, Masuda M (2002) Rejection of *Sinkoraena* and transfer of some species of *Carpopeltis* and *Sinkoraena* to *Polyopes* (Rhodophyta, Halymeniaceae). Phycologia 41: 619–635.

Maggs C, Stegenga H (1999) Red algal exotics on North Sea coasts. Helgol. Meeresunters. 52: 243–258.

Marston M, Villalard–Bohnsack M (1999) The use of molecular genetics to investigate the geographic origin and vector of an invasive red alga. In Pederson, J (ed.), Proceedings of the First National Conference of Marine Bioinvasions, Cambridge, MA, 24–27 Jan 1999. Massachusetts Institute of Technology, Cambridge, MA (USA): 244–250.

Marston M, Villalard-Bohnsack M (2002) Genetic variability and potential sources of *Grateloupia doryphora* (Halymeniaceae, Rhodophyta), an invasive species in Rhode Island waters (USA). J. Phycol. 38: 649–658.

Nicholas KB, Nicholas HBJ, (1997) GeneDoc: a tool for editing and annotating multiple sequence alignments. Version 2.6.002.

Kamura K (1934)Icones of Japanese Algae VII (4, 5). K. Okamura, Tokyo.

Pérez-Cirera JL, Cremades J, Barbara I (1989) *Grateloupia lanceolata* (Cryptonemiales, Rhodophyta) en las costa de la Peninsula Ibérica: Estudio morfologico y anatomico. Lazaroa 11: 123–134.

Posada D, Crandall KA (1998) MODELTEST: testing the model of DNA substitution. Bioinformatics Application Note 14: 1998.

Ribera MA, Boudouresque CF (1995) Introduced marine plants, with special reference to macroalgae: mechanism and impact. In Round, FE, Chapman, DJ (eds), Progress in Phycological Research 11: 217–268.

Riouall R, Guiry MD Codomier L (1985) Introduction d'une espèce foliacée de *Grateloupia* dans la flore marine de l'Étang de Thau (Hérault, France). Cryptogam. Algol. 6: 91–98.

Saunders GW, Kraft GT (1996) Small-subunit rRNA gene sequences from representatives of selected families of the Gigartinales and Rhodymeniales (Rhodophyta). 2. Recognition of the Halymeniales ord. nov. Can. J. Bot. 74: 694–707.

Swofford DL (2002) PAUP*. Phylogenetic Analisis Using PArsimony (*and Other Methods). Version 4.0b10. Sinauer Associates, Sunderland, Massachusetts.

Tolomio C (1993) Prima segnalazione di *Grateloupia doryphora* (Mont.) Howe (Rhodophyceae) nella Laguna di Venezia. Lavori della Società Veneziana di Scienze Naturali 18: 215–220.

Verlaque M (2001) Checklist of the macroalgae of Thau Lagoon (Herault, France), a hot spot of marine species introduction in Europe. Oceanol. Acta 24: 29–49.

Villalard-Bohnsack M, Harlin M (1997) The appearance of *Grateloupia doryphora* (Halymeniaceae, Rhodophyta) on the northeast coast of North America. Phycologia 36: 324–328.

Wang HW, Kawaguchi S, Horiguchi T, Masuda M (2000) Reinstatement of *Grateloupia catenata* (Rhodophyta, Halymeniaceae) on the basis of morphology and *rbc*L sequences. Phycologia 39: 228–237.

Wang HW, Kawaguchi S, Horiguchi T, Masuda M (2001) morphological and molecular assessment of the genus *Prionitis* J. Agardh (Halymeniaceae, Rhodophyta). Phycol. Res. 49: 251–261.

Journal of Applied Phycology (2006) 18: 671–678
DOI: 10.1007/s10811-006-9072-4

Observations on flattened species of *Gracilaria* (Gracilariaceae, Rhodophyta) from Taiwan

Showe-Mei Lin
Department of Natural Science Education, National Taitung University, Taitung 950, Taiwan

e-mail: smlin@nttu.edu.tw

Key words: taxonomy, Taiwan, Rhodophyta, Gracilariaceae, *Gracilaria punctata*, *G. vieillardii*, *G. spinulosa*, *Hydropuntia*

Abstract

Four flattened *Gracilaria* species have been reported from Taiwan: *G. spinulosa, G. vieillardii, G. textorii* and *G. punctata*, identified based on branching pattern, the presence or absence of spines, and characters that often vary seasonally. *Gracilaria spinulosa* was originally described from the type locality, Tainan. Species with toothed margins are usually referred to *G. "vieillardii"*; those with smooth margins to *G. "textorii"*, and those with smooth margins and dark spots scattered over the blade to *G. "punctata"*. Molecular analyses show that specimens with marginal teeth cluster in three different groups: a *G. "vieillardii"* clade, a *G. spinulosa* clade, and a clade sister to *G. spinulosa*. An undescribed species comprises the third clade, which is distinguished by its relatively large gonimoblast cells and weakly developed tubular nutritive cells. The three clades can be separated by the character of the tubular nutritive cells, the size of gonimoblast cells and certain vegetative features. Plants with entire margins form a single clade characterized by cystocarps with basal tubular nutritive cells and their absence in the cystocarp cavity. They are nested in the *Hydropuntia* complex and are referred to as *Gracilaria "punctata"* here. The records of *G. textorii* and *G. punctata* from Taiwan require reinvestigation in comparison with the Japanese species.

Introduction

Species in the red algal genus *Gracilaria* (C. Agardh) Greville 1830 in the family Gracilariaceae include some economically important agarophytes. Historically, there were eighteen species recorded from Taiwan (Chiang, 1985; Lewis & Norris, 1987; Huang, 1999); among them, four flattened species, *G. spinulosa* (Okamura) Chang et Xia 1976, *G. vieillardii* Silva in Silva, Meñez et Moe 1987, *G. textorii* (Suringar) De Toni 1895, and *G. punctata* (Okamura) Yamada 1941. These were identified based on branching pattern and the presence or absence of marginal spines. *G. spinulosa* was originally described from Tainan, Taiwan, based on *Rhodymenia spinulosa* Okamura 1934. Species with toothed margins were usually referred to *G. "vieillardii"* (Chiang, 1985), those with smooth margins to *G. "textorii"* (Huang, 1999), and those with smooth margins and dark spots scattered over the blade to *G. "punctata"* (Yamada, 1941; Ohmi, 1958). Recent collections around the coasts of Taiwan have permitted a new interpretation of the Taiwan species. In this study, the vegetative and reproductive morphology of the four flattened species are described in detail and their taxonomic status is discussed based on *rbc*L sequence analyses.

Material and methods

Collections were made either by SCUBA or snorkel. Treatment of the algal samples, sectioning and staining techniques used in the morphological studies, DNA sequencing procedures and phylogenetic analyses are as described in Lin et al. (2004). Voucher specimens are deposited in the Herbarium of the National Taitung University, Taiwan. Collection information and new *rbc*L sequences generated in this study and those available from GenBank are shown in Table 1.

Table 1. List of species used in *rbc*L analysis and accession numbers in GenBank. The number after the accession number is the percentage of the gene sequenced

Species	Collection information/GenBank accession number
Gracilaria "punctata"	Sail Rock, Kenting National Park, southern Taiwan, coll. S.M. Lin, 1.x.2002. AY737447, 98%
Gracilaria "punctata"	Lungkeng, Kenting National Park, southern Taiwan, coll. S.M. Lin, 2.iv.2001. AY737446, 96.1%
Gracilaria "punctata"	Five Caves, Orchid Island, Taiwan, coll. S.M. Lin, 17.iv.2003. AY737448, 98.4%
Gracilaria "vieillardii"	Houwan, Kenting National Park, southern Taiwan, coll. S.M. Lin, 24×.2001. AY737436, 99.5%
Gracilaria "vieillardii"	Five Caves, Orchid Island, coll. S.M. Lin, 17.iv.2002. AY737437, 98.3%
Gracilaria beckeri (J. Agardh) Papenfuss	AY049377*, 96.3%
Gracilaria bursa-pastoris (Gmelin) Silva	AY049376*, 91.6%
Gracilaria capensis Schmitz ex Mazza	AY049378*, 96.5%
Gracilaria flabelliforme (P. et H. Crouan) Fredericq et Gurgel	AY049343*, 98.8%
Gracilaria hayi Gurgel, Fredericq et J. N. Norris	AY049319*, 95.6%
Gracilaria multipartite (Clement) Harvey	AY049322*, 98.6%
Gracilaria occidentalis (Børgesen) Bodard	AY049322*, 98.6%
Gracilaria smithsoniensis Gurgel, Fredericq et J. N. Norris	AY049321*, 97.3%
Gracilaria sp.	Yeliu, northern Taiwan, coll. Allen Liu, 26.vii.2002. AY737438, 98.1%
Gracilaria sp.	Sail Rock, Kenting National Park, southern Taiwan, coll. S.M. Lin, 14.iii.2002. AY737439, 96.8%
Gracilaria sp.	Keelung, northern Taiwan, coll. S.M. Lin, 1.v.2002. AY737440, 98.1%
Gracilaria spinulosa	Wind Blow Sand, Kenting National Park, southern Taiwan, coll. S.M. Lin, 21.vii.2001. AY737441, 95.7%
Gracilaria spinulosa	Ya Din, Tainan, western Taiwan, coll. D.T. Lin, 26.v.2003. AY737442, 98.3%
Gracilaria spinulosa	Lin Ping, Pingtung County, southwest Taiwan, coll. Y.S. Huang, 10.ii.2002. AY737443, 98.4%
Gracilaria spinulosa	Little Yeliu, Taitung, eastern Taiwan, coll. S.M. Lin& F.K. Huang, 18.iii.2003. AY737444, 98.3%
Gracilaria textorii (Suringar) De Toni	AY049325*, 97.5%
Gracilaria venezuelensis Taylor	AF539603*, 95.4%
Gracilaria vieillardii	Bulusan, N. Philippines, coll. Allen Liu, 18.ii2003. AY737445, 98.4%
Gracilaria yoneshigueana Gurgel, Fredericq et J. N. Norris	AY049372*, 93.4%
Gracilariopsis bailiniae Zhang et Xia	AY049411* regarded as *Gracilariopsis heteroclada* 91.1%
Gracilariopsis lemaneiformis (Bory de Saint-Vincent) E.Y. Dawson, Acleto et Foldvik	AY049415*, 97.6%
Hydropuntia caudata (J. Agardh) Gurgel et Fredericq	AY049358*, 76.4%
Hydropuntia cornea (J. Agardh) Wynne	AY049338*, 98.8%
Hydropuntia crassissima (P. et H. Crouan) Wynne	AY049351*, 98%
Hydropuntia eucheumatoides (Harvey) Gurgel et Fredericq	AY049389*, 93.3%
Hydropuntia urvillei Montagne	AY049402*, 97.4%
Hydropuntia usneoides (C. Agardh) Gurgel et Fredericq	AY049346*, 98%

*Refers to Gurgel and Fredericq (2004).

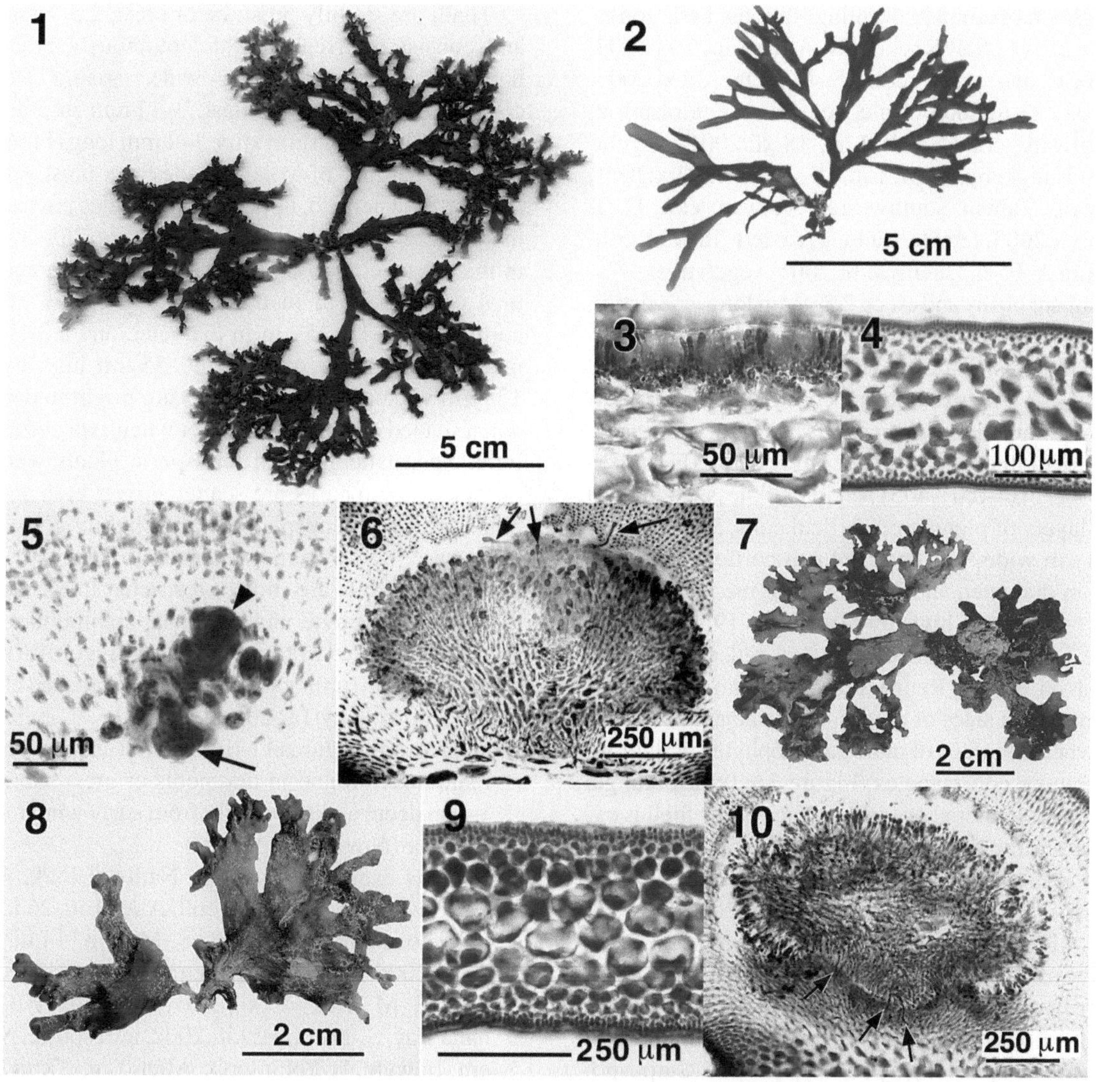

Figure 1–10. (1–6)*Gracilaria spinulosa* (Figures 1 and 6, Tainan, southwestern Taiwan, Figures 2–5, Sail Rock). (1) A typical cystocarpic specimen showing dense branches in upper parts of the thallus. (2) A typical male plant showing loose branches with more or less entire margins. (3) Cross-section of spermatangial conceptacles. (4) Cross-section of vegetative thallus. (5) Transverse section of a young, multinucleate fusion cell (arrowhead) and the supporting cell (arrow). (6) Transverse section of a mature cystocarp showing the gonimoblasts and some tubular nutritive cells (arrows) arisen from the upper parts of gonimoblasts. *Gracilaria "vieillardii"* (Kenting National Park, southern Taiwan: (7), Sail Rock, (8–10), Houwan) (7) Female specimens with marginal spines. (8) Female plant with marginal lobes. (9) Cross-section through a vegetative thallus. (10) Transverse section of a mature cystocarp showing the gonimoblasts and some tubular nutritive cells penetrating the cystocarp floor.

Results

Observations

Gracilaria spinulosa (Okamura) Chang et Xia 1976, 11: 148, Figure 42. (Figures 1–6).

Synonym: *Rhodymenia spinulosa* Okamura 1934, 7(4): 33, pl. 318, Figures 1–6.

Distribution in Taiwan: This alga is distributed in eastern (Taitung) and southwestern (Tainan, type locality) to southern (Kenting) Taiwan.

Habitat and seasonality: Plants grew all year round and were attached to rocky coral reefs or man-made concrete substrata in intertidal to subtidal zones, 0–2 m deep, sometimes in association with *G. "vieillardii"*.

Specimens examined: Kenting National Park, southern Taiwan: (1) Sail Rock, coll. Allen Liu, 26.i.2004, tetrasporic and male; coll. S.-M. Lin, 28.xi.2001, 17.i.2002, 27.iii.2004, male, female and tetrasporic; (2) Houwan, coll. S.-M. Lin, 18.xii.2001, vegetative; (3) Lungkeng, S.-M. Lin, 19.vii.2001, 03.x.2001, tetrasporic. Tainan, southwestern Taiwan: coll. D.-T. Lin, 16.vii.2003, female. Taitung, eastern Taiwan: coll. S.-M. Lin & F.-K. Huang, 8.iii.2003, vegetative.

Thalli are bushy and erect, 2.8–9 cm long, consisting of loose to dense, irregularly dichotomously branched, flattened blades, 2–7 mm wide, arising from a discoid holdfast, 2–3 mm in diameter, occasionally with a short stipe 1–2.5 mm long (Figures 1, 2). The margins of blades are toothed (Figure 1) or rarely entire (Figure 2). Blades are 110–160 (-405) μm in thickness, composed of 1–2 layers of pigmented cortical cells, 5–8 μm long by 5–6 μm wide, 1–2 layers of sub-cortical cells, 30–65 μm in diameter, and 2–3 layers of medullary cells, 75–90 (−120) μm high by 90–120 (−195) μm wide (Figure 4). Blades are rose red to dark red, occasionally greenish in colour. Reproductive structures are scattered over both sides of blades. Spermatangia are scattered over the surface of male gametophytes in shallow, *textorii* -type conceptacles (Figure 3). Tetrasporangia are cruciate, scattered over the surface of the thallus except the basal part. Cystocarps are hemispherical and slightly constricted at the base, 1.1–1.4 mm in diameter (Figure 6), scattered over the mid to upper part of the thallus. Gonimoblast initials (not shown) are cut off from the multinucleate fusion cell (Figure 5). Tubular nutritive cells are present in the cystocarp cavity and in the floor of the cystocarp and inner gonimoblast cells are 55–28 μm long by 25–35 μm wide. Carposporangia are uninucleate, borne in branched chains, 15–20 μm wide by 15–23 μm long.

Gracilaria "vieillardii" Silva in Silva, Meñez et Moe 1987: 44 (Figures 7–10)

Distribution: Eastern (Taitung, Green Island, Orchid Island) to southern (Kenting) Taiwan.

Habitat and seasonality: Plants grew from early winter (October) to spring (April) and were attached intertidally to rocky coral reefs in tide pools, usually in association with *G. spinulosa*.

Specimens examined: Kenting National Park, southern Taiwan: 1) Sail Rock, coll. Allen Liu, 26.i.2004, female; coll. S.-M. Lin, 27.iii.2004, vegetative; 2) Houwan, coll. S.-M. Lin, 18.xii.2001, female. Eastern Taiwan: 1) Taitung, coll. F.-K. Huang, 12.i.2002, vegetative; 2) Orchid Island, coll. S.-M. Lin, 17.iv.2002, vegetative.

Thalli are slightly prostrate or erect, 2.5–6 cm long and consist of irregularly dichotomously branched, flattened blades, 3–12 mm wide, arising from a conspicuous, discoid holdfast, 3–11 mm in diameter, occasionally with a short stipe 1–4 mm long (Figures 7, 8). The margins of young blades are mostly entire (Figure 8); when old, upper parts of blades possess fine marginal spines (Figure 7). Blades are 255–395 μm in thickness, composed of 2 layers of pigmented cortical cells, 6–7 μm in diameter, 2–3 layers of sub-cortical cells, 15–35 μm in diameter, and a 3-layered medulla composed of cells, 30–55 μm high by 50–125 μm wide (Figure 9). Blades are bright to dark red when shaded or greenish in color when exposed to sunlight. Spermatangial and tetrasporic plants were not observed. Cystocarps are scattered over both sides of blades and are hemispherical and slightly constricted at the base, 1.5–1.9 mm in diameter (Figure 10). Tubular nutritive cells are mostly restricted to the base of the carposporophyte and the inner gonimoblast cells are interconnected to form a network (Figure 10). Carposporangia are uninucleate, borne in branched chains, 12–15 μm wide by 15–20 μm long.

Gracilaria sp. (Figures 11–16) Habitat and seasonality: Plants were found in tide pools or grew subtidally, 1–2 m deep, and seasonally from early winter to late summer (November-August).

Specimens examined: Kenting National Park, southern Taiwan: 1) Sail Rock, coll. Allen Liu, 26.i.2004, tetrasporic and male; coll. S.-M. Lin, 14.iii.2002, 27.iii.2004, female and tetrasporic; 2) Houwan, coll. S.-M. Lin, 13.xii.2001, vegetative; 3) Banana Bay, S.-M. Lin, 29.iii.2002, tetrasporic. Northern Taiwan: 1) Yeliu, coll. Allen Liu, 27.vii.2002, female; 2) Keelung, coll. S.-M. Lin, 1.v.2002, vegetative.

Thalli are erect, 4–8 cm long, and consist of irregularly dichotomously branched, flattened blades, 5–13 mm wide, arising from a stipe, 5–18 mm long, with a discoid holdfast, 2–4 mm in diameter (Figures 11, 12). The blades are associated with marginal spines or lobes (Figure 11), and also have numerous, tiny lobes or bladelets, 1–3 mm long by 1–4 mm wide arising from their surfaces (Figure 12). Blades are 250–500 μm thick, composed of 1–2 layers of pigmented cortical cells, 5–6 μm in diameter, 1–2 layers of sub-cortical cells, 12–20 μm in diameter, and one layer of medullary cells, 75–100 μm high by 150–200 μm wide (Figure 13). Blades are rose to dark red, occasionally greenish in colour. Reproductive structures are scattered over both sides of blades. Spermatangia are

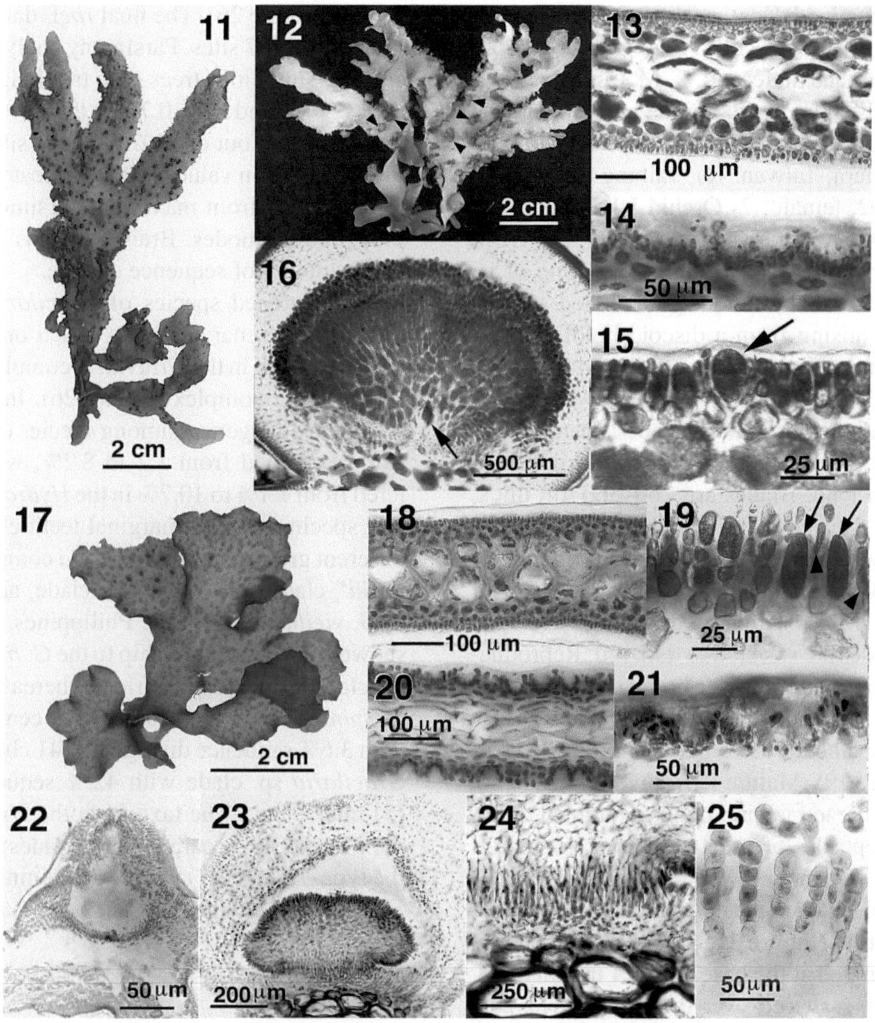

Figure 11–25. Gracilaria sp. (Sail Rock, Kenting National Park, southern Taiwan) (11) Female thallus. (12) Fresh plant showing numerous bladelets (arrowheads) born on the thallus surface. (13) Cross section of vegetative thallus. (14) Cross-section of male gametophyte showing shallow *textorii*-type spermatangial conceptacles. (15) Cross-section of tetrasporophyte showing a mature tetrasporangium (arrow) and some young ones. (16) Transverse section of a mature cystocarp showing the gonimoblasts and the remaining fusion cell (arrow).*Gracilaria "punctata"* ((17) Taitung, eastern Taiwan, (18–25), Sail Rock) (17) Cystocarpic plant. (18) Cross-section of vegetative thallus. (19) Transverse section of immature tetrasporangia. (20) Transverse section of immature spermatangial conceptacles. (21) Transverse section of mature, P*olycavernosa*-type spermatangial conceptacles. (22) Transverse section of a very young cystocarp showing the pre-existing cavity. (23) Transverse section of mature cystocarp showing the gonimoblasts and basal tubular nutritive cells. (24) Same section as in (23) showing close up of basal tubular nutritive cells. (25) Close up of uninucleate carposporangia in unbranched chains.

scattered over the surface of male gametophytes in shallow, *textorii* -type conceptacles (Figure 14). Tetrasporangia are superficial and cruciately divided (Figure 15). Cystocarps are hemispherical and slightly constricted at the base, 1.6–2.2 mm in diameter with the inner gonimoblast cells 75–115 μm long by 40–60 μm wide (Figure 16). Tubular nutritive cells are seldom found in mature cystocarps, and carposporangia are uninucleate, 18–25 μm wide by 20–33 μm long and borne in branched chains.

Gracilaria "punctata" (Okamura) Yamada 1941: 203 (Figures 17–25)

Synonym: *Rhodymenia punctata* Okamura 1929: 13, pl. 258, Figures 1—6.

Distribution: Eastern (Taitung) to southern (Kenting) Taiwan.

Habitat and seasonality: Plants grew seasonally from winter (December) to summer in tide pools, or subtidally in sandy substrata or on rocky coral reef, 1–2 m deep.

Specimens examined: Kenting National Park, southern Taiwan: 1) Sail Rock, coll. Allen Liu, 26.i.2004, tetrasporic, female and male; coll. S.-M. Lin, 10.i.2002, 14.iii.2002, 27.iii.2004, male, female and tetrasporic; 2) Lungkeng, coll. S.-M. Lin, 2.iv.2001, 22.vii.2001, tetrasporic. Eastern Taiwan: 1) Taitung coll. F.-K. Huang, 12.i.2002, female; 2) Orchid Island, coll. S.-M. Lin, 17.iv.2002, vegetative.

Thalli are erect, 4.5–8.7 cm high, consisting of irregularly to pseudodichotomously branched blades, 7–28 mm wide, arising from a discoid holdfast, 2.5–5 mm in diameter (Figure 17), usually with a stipe 5–15 mm long. The margins of blades are wavy, ruffled or entire and the surfaces of blades contain scattered brown to dark red spots, sometimes with colorless hairs, 15–40 μm long. Blades are 260–560 μm thick, composed of 1–2 layers of pigmented cortical cells, 6–8 μm long by 5–7 μm wide, 1–2 layers of subcortical cells, 12–30 μm in diameter, and 1-2-3 layer of medullary cells, 40–140 μm high by 80–210 μm wide (Figure 18). Blades are rose red to dark red. Reproductive structures are scattered over both sides of blades. Tetrasporocytes are scattered over the thallus surface embedded in nemathecia, surrounded by elongated cortical cells (Figure 19). Mature tetrasporangia were not found in any of the tetrasporic plants examined. Spermatangial conceptacles are scattered over the thallus surface and are cup shaped when young (Figure 20), but become confluent as in the *Polycavernosa*-type conceptacle at maturity (Figure 21). Cystocarps are hemispherical, constricted at the base, 1.5–2.1 mm in diameter (Figure 23), scattered over the thallus. The cavity of the cystocarp is formed before the gonimoblast initials are cut off from the multinucleate fusion cell (Figure 22). Tubular nutritive cells are absent in the cystocarp cavity (Figure 23) and are restricted to the base of the carposporophyte where they penetrate the floor of the cystocarp (Figure 24). Carposporangia are uninucleate, 15–20 μm wide by 20–28 μm long, and are borne in straight chains (Figure 25).

Molecular analysis

The *rbc*L sequences of *G. spinulosa*, *G. "vieillardii"*, *Gracilaria* sp., *Gracilaria "punctata"* from Taiwan and *G. vieillardii* from the Philippines were newly generated, and a set of 17 additional representative taxa belonging to the genera *Gracilaria* and *Hydropuntia* were selected for analysis, together with two species of *Gracilariopsis* which served as the outgroup (see Table 1, Figure 26). The final *rbc*L data matrix was restricted to 1407 sites. Parsimony analysis revealed two most parsimonious trees with tree length of 925 steps, CI = 0.5459 and RI = 0.7077; there were 303 informative characters out of 1407 included sites (22%). Bootstrap proportion values (1000 replicates) and decay indices derived from maximum parsimony analysis are shown on the nodes. Branch lengths are proportional to the amount of sequence change.

The flattened species of *Gracilaria* from Taiwan formed four separate clades based on *rbc*L sequence analyses: three in the *Gracilaria* complex and one in the *Hydropuntia* complex (Figure 26). Interspecific *rbc*L sequence divergences among species of the *Gracilaria* complex varied from 1.2 to 8.2%, whereas they differed from 1.1% to 10.7% in the *Hydropuntia* complex. The specimens with marginal teeth clustered in three different groups in the *Gracilaria* complex: a *G. "vieillardii"* clade, a *G. spinulosa* clade, and a clade sister to *G. vieillardii* from the Philippines. *G. "vieillardii"* showed a close relationship to the *G. beckeri*/*G. capensis* clade from South Africa, whereas the position of *G. spinulosa* was unresolved between *G. "vieillardii"* with 3.6% sequence divergence (41 characters) and the *Gracilaria* sp. clade with 4.2% sequence divergence (51 characters). The taxa from the *Hydropuntia* complex formed two paraphyletic clades: one containing the type, *H. urvillei* and one containing *G. "punctata"* from Taiwan along with some other Caribbean species.

Discussion

The phylogenetic relationships among the species of the family Gracilariaceae have been extensively studied using DNA sequencing in recent years (Bird et al., 1994; Bellorin et al., 2002; Gurgel & Fredericq, 2004). Liao and Hommersand (2003) recently examined the types of genera that have been proposed historically for the species of *Gracilaria* and recognized nine types, based mainly on the formation of spermatangial conceptacles and cystocarp development. They provided a description of each type and assigned species presently placed in *Gracilaria* to each of the groups. Gurgel and Fredericq (2004) identified nine distinct evolutionary lineages in *Gracilaria sensu lato* based on *rbc*L sequence analyses and classified the species into two genera: *Gracilaria sensu stricto* and *Hydropuntia*, emphasizing both the manner of formation of the spermatangial conceptacles and the cystocarp features.

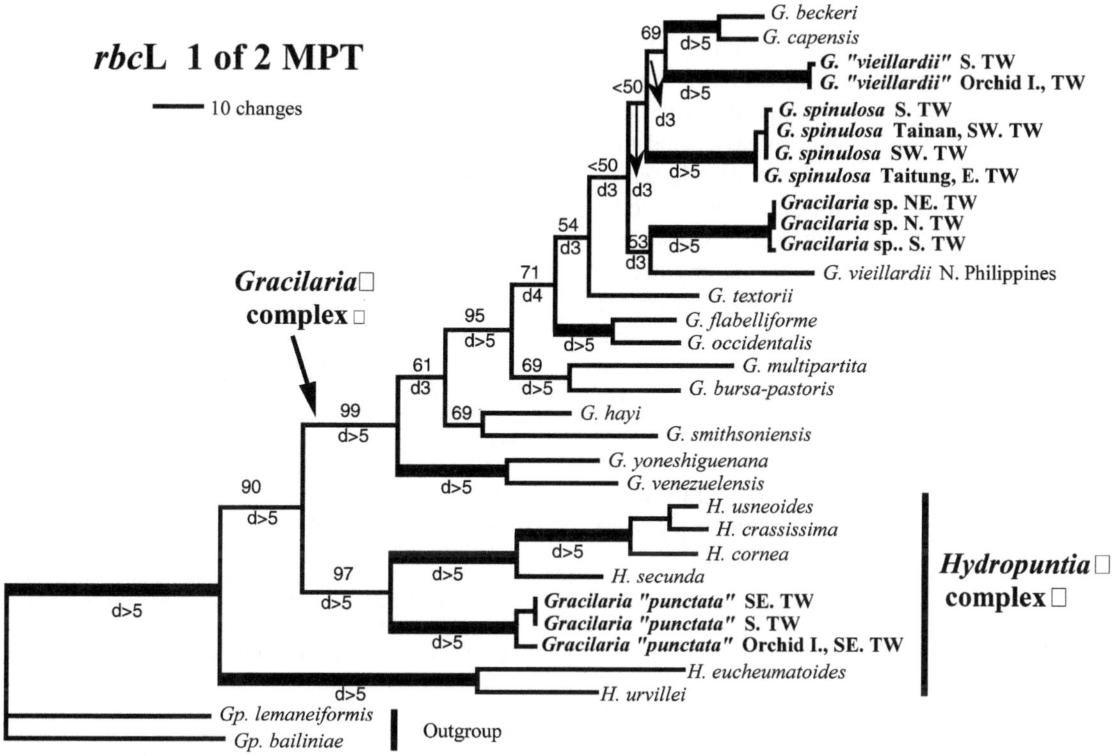

Figure 26. One of two most parsimonious trees from analysis of the *rbc*L sequence data. Bootstrap proportion values are shown above nodes and thick bold branches correspond to 100% support; decay indices are shown below nodes. Branch lengths are proportional to the amount of sequence change.

Molecular analyses in this study showed that the four flattened species of *Gracilaria* from Taiwan fall into four distinct clades: three (*G. spinulosa*, *G. "vieillardii"*, *Gracilaria* sp.) in *Gracilaria sensu stricto* and one (*G. "punctata"*) in *Hydropuntia*, as circumscribed by Gurgel and Fredericq (2004) (see Figure 26). Although *G. spinulosa*, *G. "vieillardii"* and *Gracilaria* sp. formed three distinct clades, they are closely related to *G. vieillardii* from the Philippines and *G. beckeri* and *G. capensis* from South Africa. All of these species consistently formed a single clade sister to *G. textorii* from Japan in the analyses, although with a low bootstrap support. *Gracilaria* sp. is an undescribed species characterized by a thallus surface bearing numerous lobes or bladelets (see Figures 11 and 12), by relatively large inner gonimoblast cells, and by a lack tubular nutritive cells in the cavity of cystocarp compared to *G. spinulosa*. This species will be published in a separate paper. *Gracilaria "vieillardii"* and *G. spinulosa* are morphologically similar to each other, but the former can be separated from the latter by its slightly prostrate habit, smaller and less branched thallus, relatively wider blades, and tubular nutritive cells that are mainly restricted to the floor of the cystocarp. *Gracilaria vieillardii* has been widely used as a name for the flattened species with marginal spines throughout the Western Pacific Ocean (Yamamoto, 1978; Chiang, 1985; Withell et al., 1994). *Gracilaria "vieillardii"* from Taiwan remains provisional because of the absence of a detailed description of *G. vieillardii* from the type locality, New Caledonia.

Okamura (1929) described a new species, *Rhodymenia punctata*, from Tosa, Japan in the absence of information about female reproductive structures. Yamada (1941) examined some female and tetrasporic specimens of a *Gracilaria* species collected from Tairi and Garanbi, Taiwan, and compared them with the type specimen of *R. punctata*. He came to the conclusion that the Taiwan species was the same as *R. punctata* and transferred the species to *Gracilaria*, as *G. punctata* (Okamura) Yamada without, however, providing any description or figures of the internal structure of the cystocarp. Ohmi (1958) reported that tubular nutritive cells were present and linked the outer gonimoblast cells to the outer pericarp and that the spermatangia were in shallow, solitary or confluent conceptacles in

G. punctata. Ohmi's observations were based on material collected in Taiwan. Later, Yamamoto (1978) investigated material from Komesu, Okinawa Prefecture, Japan and reported that the tubular nutritive cells were few in number in the cystocarp cavity and that the tetrasporangia were embedded in a slightly raised nemathecium surrounded by elongated cortical cells. *Gracilaria "punctata"* from Taiwan differs from Yamamoto's (1978) description only in the absence of tubular nutritive cells in the cavity. The material of *Gracilaria punctata* from Taiwan examined by Ohmi (1958) may be related to *Gracilaria* sp. examined in this study. None of the flattened species of *Gracilaria* from Taiwan sequenced in this study matches *G. textorii* from Japan (See Figure 26). This suggests that the records of *G. textorii* and *G. punctata* from Taiwan require reinvestigation and comparison with the Japanese species.

Acknowledgements

This study was mainly supported by National Science Council research grants 92-2621-B-143-001 and 92-3114-B-143-001 to SML. SML thanks her collecting partners F.-K. Huang and Allen Liu for assisting with fieldwork and D.-T. Lin for sending the material of *G. spinulosa* from Tainan. Dr. F. Gurgel provided the additional *rbc*L sequences of the Gracilariaceae is highly acknowledged. Special thanks go to Dr. Max H. Hommersand for his critical comments and suggestions, which have improved the quality of this article.

References

Bellorin AM, Oliveira MC, Oliveira EC (2002) Phylogeny and systematics of the marine algal family Gracilariaceae (Gracilariales, Rhodophyta) based on small subunit rDNA and ITS sequences of Atlantic and Pacific species. Journal of Phycology 38: 551–563.

Bird CJ, Ragan MA, Critchley AT, Rice EL, Gutell RR (1994) Molecular relationships among the Gracilariaceae (Rhodophyta): Further observations on some undetermined species. European Journal of Phycology 29: 195–202.

Chang CF, Xia BM (1976) Studies on Chinese species of *Gracilaria.* Studia Marina Sinica 11: 91–163.

Chiang Y-M (1985) *Gracilaria* from Taiwan: key, list and distribution of the species. In Abbott IA, Norris JN (eds), Taxonomy of Economic Seaweeds with References to Some Pacific and Caribbean Species, California Sea Grant College Program, La Jolla, CA, 81–83.

De Toni GB (1895) Phyceae japonicae novae addita enumeratione algarum in ditione maritima Japoniae hucusque collectarum. Memorie del Reale Istituto Veneto di Scienze, Lettere ed Arti 25: 1–78.

Greville RK (1830) Algae Britannicae.... Maclachlan & Stewart, Edinburgh.

Gurgel CFD, Fredericq S (2004) Systematics of the Gracilariaceae (Gracilariales, Rhodophyta): A critical assessment based on *rbc*L sequence analysis. Journal of Phycology 40: 138–159.

Huang S-F (1999) Floristic studies on the benthic marine algae of northeastern Taiwan. Taiwania 44: 217–298.

Lewis JE, Norris JN (1987) A History and Annotated Account of the Benthic Marine Algae of Taiwan, Smithsonian Institution Press, Washington, D.C.

Liao LM, Hommersand MH (2003) A morphological study and taxonomic reassessment of the generitype species in the Gracilariaceae. Journal of Phycology 39: 1207–1232.

Lin S-M, Hommersand MH, Fredericq S (2004) Two new species of *Martensia* (Delesseriaceae, Rhodophyta) from Kenting National Park, southern Taiwan. Phycologia 43: 13–25.

Ohmi H (1958) The species of *Gracilaria* and *Gracilariopsis* from Japan and adjacent waters. Memoirs of the Faculty of Fisheries, Hokkaido University 6: 1–66.

Okamura K (1929) Icones of Japanese algae. Vol. VI, Tokyo.

Okamura K (1934) Icones of Japanese algae. Vol. VII, Tokyo.

Silva PC, Meñez EG, Moe RL (1987) Catalog of the benthic marine algae of the Philippines. Smithsonian Contributions to Marine Sciences 27: 1–179.

Yamada Y (1941) Notes on some Japanese algae IX. Scientific Papers of the Institute of Algological Research, Hokkaido Imperial University 2: 195–215.

Yamamoto H (1978) Systematic and anatomical study of the genus *Gracilaria* in Japan. Memoirs of the Faculty of Fisheries, Hokkaido University 25: 97–152.

Withell AF, Millar AJK, Kraft GT (1994) Taxonomic studies of the genus *Gracilaria* (Gracilariales, Rhodophyta) from Australia. Australian Systematic Botany 7: 281–352.

Journal of Applied Phycology (2006) 18: 679–697
DOI: 10.1007/s10811-006-9073-3

Phylogenetic re-evaluation of the *Laurencia* complex (Rhodophyta) with a description of *L. succulenta* sp. nov. from Korea

K.W. Nam
Department of Marine Biology, Pukyong National University, Busan 608–737, Korea
e-mail: kwnam@pknu.ac.kr

Key words: *Laurencia* complex, *L. succulenta* sp. nov., Korea, *Palisada* (Yamada) stat. nov., phylogeny, cladistics

Abstract

Laurencia succulenta sp. nov. (Rhodophyta) is described from Korea. This species exhibits vegetative and reproductive structures typical of the genus, but is distinct from similar species in its epiphytic habitat and the fleshy, robust, thick and subcompressed thalli with basically distichous branching. In addition, it is readily distinguished from the most similar species, such as *L. nipponica* Yamada and *L. okamurae* Yamada, by the cystocarps with a somewhat protuberant ostiole. In a phylogenetic analysis of 47 species of the *Laurencia* Lamouroux complex from various localities around the world based on 49 morphological characters, four major clades (*Laurencia*, *Chondrophycus palisadus* (Yamada) Nam group, *C. cartilagineus* (Yamada) Garbary et Harper group and *Osmundea* Stackhouse assemblage), each of which forms a monophyletic group, were recognized. Among these, the *Laurencia* clade is basal to the overall assemblage, and is defined by the vegetative axis with four rather than two pericentral cells. The *Osmundea* clade is supported by autapomorphic characters for the genus, features associated with spermatangial formation of the filament type and tetrasporangial production from epidermal cells. By contrast, *Chondrophycus*, a genus characterized by a combination of features (vegetative axis with two pericentral cells, trichoblast-type spermatangial development and tetrasporangial production from pericentral cells), is paraphyletic, and the species were separated into two well-supported clades, the *C. palisadus* group and *C. cartilagineus* group. These clades are distinguished from each other by the position of the first pericentral cell relative to the trichoblast, the presence or absence of fertility at the second pericentral cells and number of sterile pericentral cells in the tetrasporangial axis, the pattern of formation of spermatangial branches on trichoblasts, post-fertilization feature associated with the formation time of the auxiliary cell, and, probably, the number of pericentral cells in the procarp-bearing segment. Of these features, the side position of the first pericentral cell in the latter group (a synapomorphy for the *C. cartilagineus* group plus *Osmundea*) suggests that the *C. cartilagineus* group is more closely related to *Osmundea* than to the *C. palisadus* group. This cladistic analysis indicates that *Chondrophycus* is not monophyletic, suggesting that the *C. palisadus* group should be separated from *Chondrophycus* at the genus level. Based on this result, *Palisada* (Yamada) stat. nov. is proposed for the group, together with an emendation of the generic delineation of *Chondrophycus*, and relevant nomenclatural changes for several *Chondrophycus* species are also included. In addition, *Corynecladia* J. Agardh is reinstated for the type species *L. clavata* Sonder.

Introduction

The *Laurencia* complex Lamouroux (Rhodophyta) has been separated into three genera: *Laurencia*, *Chondrophycus* (Tokida et Saito) Garbary et Harper and *Osmundea* Stackhouse, based on vegetative and reproductive structures (Garbary and Harper, 1998; Nam, 1999). These genera share the typical and superficial rhodomelacean morphology with apical cells sunk in apical pits of branchlets, a recognizable axial cell row only near the apical cell and an extensive cortex (Kylin, 1956). However, *Osmundea* is readily distinguished from *Laurencia* and *Chondrophycus* in having filament-type rather than trichoblast-type spermatangial development and tetrasporangial production from random epidermal cells rather than particular pericentral cells (Nam et al., 1994; Nam, 1999). *Laurencia* is delimited from *Chondrophycus* by the vegetative axial segment feature of four instead of two pericentral cells (Nam, 1999). Of these genera, *Laurencia* has a worldwide distribution, although the majority of species are found in the Southern Hemisphere (McDermid, 1988). By contrast, *Osmundea* shows a disjunct distribution, occurring in Pacific North America, Brazil, the Mediterranean Sea, Atlantic Europe, India, Australia (?) and probably North Africa (Harper and Garbary, 1997; Serio et al., 1999; Nam et al., 2000; Yoneshigue-Valentin et al., 2003; Furnari et al., 2004), and currently fifteen species are recognized in those areas (Nam et al., 1994, 2000). *Chondrophycus* has been reported mainly in the Pacific (Garbary and Harper, 1998; Nam, 1999). This genus is the most diverse, and seems to be actively evolving within the complex, considering the number of species and infrageneric categories which have been established based on morphological features with phylogenetic significance (Nam, 1999).

Recently, a morphological phylogenetic analysis of this complex was reported (Garbary and Harper, 1998). However, some features with phylogenetic importance were not considered in the analysis. Subsequently, the vegetative and reproductive structures of *L. clavata* Sonder, which may phylogenetically link the *Laurencia* complex and *Chondria* C. Agardh, were clarified (Nam and Choi, 2001). Nam et al. (2000) and McIvor et al. (2002) also carried out phylogenetic analyses, but their work was focused largely on *Osmundea*.

In this study, the Korean *L. succulenta* sp. nov. is described, and, in order to reevaluate phylogenetic relationships among the assemblages within the *Laurencia* complex, a cladistic analysis of 47 species including *L. clavata* from various localities around the world was performed based on 49 morphological characters including the additional elaborate features (such as the position of the first pericentral cell relative to trichoblast, the fertility or sterility of the second pericentral cell and number of sterile pericentral cells in tetrasporangial axis, and the formation pattern of spermatangial branches on trichoblasts), which are of major phylogenetic significance (Nam, 1999; Nam and Choi, 2000) but were not considered in the previous works (Garbary and Harper, 1998; Nam et al., 2000; McIvor et al., 2002).

Materials and methods

Data for the new species were obtained from liquid-preserved and herbarium specimens collected from Korea. Methods for microscopic examination of anatomical features are the same as those given in Nam and Saito (1990, 1991b). For phylogenetic analysis, 47 species including the type species of each genus of the *Laurencia* complex, *Laurencia obtusa* (Hudson) Lamouroux, *Chondrophycus cartilagineus* (Yamada) Garbary et Harper and *Osmundea osmunda* (S.G. Gmelin) Nam et Maggs were selected from various localities around the world (Figure 1) (Table 1). *Chondria* C. Agardh based on *C. tenuissima* (Goodenough et Woodward) C. Agardh (type species) and *C. dasyphylla* (Woodward) C. Agardh (Gordon-Mills, 1987) was adopted as the outgroup. 49 morphological characters were used in the analysis (Table 2). Cladistic analysis was carried out using PAUP* v.4.0b10 for the Macintosh (Swofford, 2002) by the maximum parsimony method and optimum trees were searched for the heuristic algorithm with the options Simple additions, TBR swapping, COLLAPSE (max), MULTREES and Steepest descent (No). Character optimization on trees was made by ACCTRAN.

Results and discussion

Laurencia succulenta sp. nov. (Figures 2–6)

Type locality: Sungsanpo (33°27′N, 126°56′E), Korea

Distribution: Korea (Sungsanpo, Ulleungdo, Tongbaeksum in Busan)

Representative specimens examined: N860105♂♀ ⊕ (holotype) and N860107 ⊕ (paratype) deposited in the Herbarium of the Department of Marine Biology, Pukyong National University, Korea. Busan (6.vi.1986, N860106 sterile); Ulleungdo (9. ix.1986, N960230 sterile), Sungsanpo (21.v.1985, N861116♂).

Etymology: The epithet is derived from the comparatively fleshy thalli.

Habitat: Epiphytic on coarse algae near lower tidal zone

Discription: *Thalli epiphytii, caespites laxos facientes, subcompressi, fusci, carnosi, subcartilaginei, affixa ad substratum per ramos stoloniformes accessorios numerosos; axes erecti unus vel pli ramulis repentibus basalibus exorientes; ramificatio disticha; cellulae epidermales leviter procurrentes vel non procurrentes*

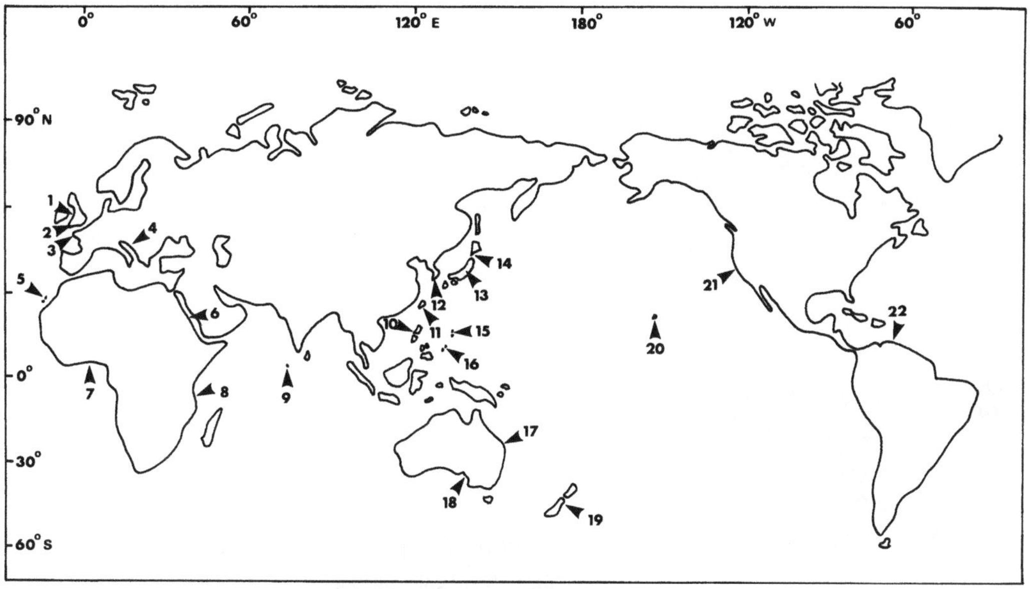

Figure 1. Map showing the localities of the examined specimens of the selected species for this study. 1. England (Anglesey); 2. England (Devon); 3. France; 4. Yugoslavia; 5. Canary Isl.; 6. Egypt; 7. Ghana; 8. Tanzania; 9. Maldive Isls.; 10. Philippines; 11. Taiwan; 12. Korea; 13. Japan (Niigata); 14. Japan (Hakodate); 15. Guam; 16. Palau; 17. Queensland; 18. South Australia; 19. New Zealand; 20. Hawaii; 21. California; 22. Venezuela.

prope apicem ramuli, in sectione transversali ramuli pro vallo nec radians elongatae nec dispositae, fovea-connetiva inter eas facientes; crassitudines lenticulares absentes in parietibus cellularum medullosarum; ramuli antherideales cum depressionibus apicalibus cupulatis; trichoblastus antheridialis exorientes cellua axialibus, constans ex ramis fertilibus et sterilibus; rami fertiles desinens in cellulam sterilem grandem singularem, cellula terminalis; spermatangium cum nucleo apicali; segmentum fertile trichoblasti feminei cellulas pericentrales quinque instructum; cystocarpuim subconicum, ostiolumrostre cum; initium tetrasporangii cellula pericentrali oruindum, in cursu abaxiali abscissum; segmentum axiale omne plerumque cellulam pericentralem fertilem duas efferens, eas accessorias non efferens; cellula pericentralis tertia et quarta semper fertilis; cellulae tegentes duae presporangii pro cellulis indivisis remanentes, ab latitudine in aspectu paginae stichidii dispositae; tetrasporangia tetraedrice divisa et in latibus lateralibus stichidii dispersa.

Thalli 5–8 cm high, epiphytic, forming loose clumps, fleshy, robust and thick, subcompressed, brown in color, subcartilaginous, attached to substratum by numerous accessory stoloniferous branches; one or more erect axes arising from basal creeping branches, 1–2 mm in diameter, usually percurrent, sometimes with adventitious branchlets but in lower parts often denuded or sparsely branched; branching basically distichous; all branches slightly attenuated toward the tip; ultimate branchlets cylindrical to clavate, truncate or rounded at the apices, 0.3–0.6 mm in diameter (Figure 2). Vegetative axis with four pericentral cells (Figure 3). Epidermal cells slightly projecting near branchlet apex, without palisade-like arrangement, $11–13 \times 11–14\,\mu m$ in size, with secondary pit connections; lenticular thickenings usually absent in medullary cells but thickened cell walls often observed; secondary cortication extensively developing. Male branchlets turbinate, 1.4–1.6 mm in diameter, with cup-shaped spermatangial pits, $470–670 \times 750–1080\,\mu m$, spermatangial trichoblast arising from axial cell, consisting of fertile and sterile branches; the fertile branches ended in a single large sterile cell, $28–36 \times 22–29\,\mu m$, spermatangia with apical nucleus, $11–15 \times 8–10\,\mu m$ (Figure 4). Procarps with five pericentral cells, consisting of four-celled carpogonial branch with trichogyne and two groups of sterile cells; with typical post fertilization process of *Laurencia*; cystocarps subconical, $1100–1200 \times 1100–1200\,\mu m$, with somewhat protuberant ostiole (Figure 5). Tetrasporangial branchlets clavate, more or less constricted at the base, usually simple, 0.8–1.0 mm in diameter; tetrasporangial axis with tetrasporangia abaxially produced from the third and fourth pericentral cell, without additional fertile

Figure 2. Laurencia succulenta sp. nov. (A) Habit of female plant on other algae. (B) Details of female branch. (C) Habit of male plant. (D) Details of male branch. (E) Habit of tetrasporangial plant.

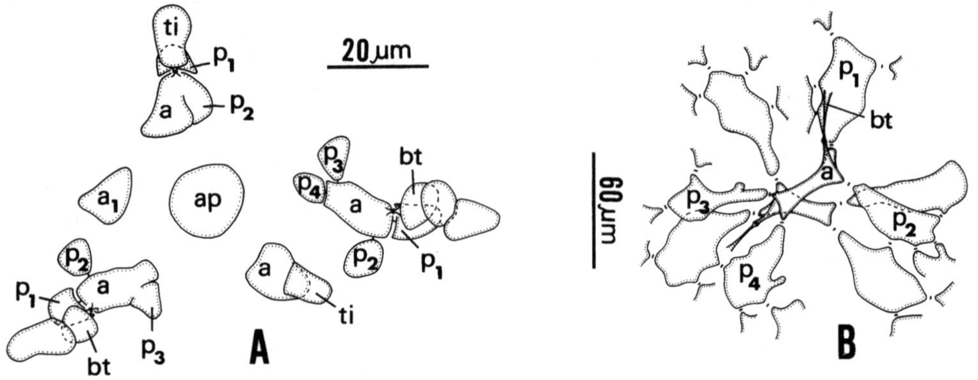

Figure 3. Laurencia succulenta sp. nov. (A) Successive axial segments, showing formation sequence of trichoblast and pericentral cells. (B) Two superimposed vegetative axis with four pericentral cells. a, axial cell; ap, apical cell; bt, basal cell of trichoblast; p, pericentral cell; ti, trichoblast initial. The numbers indicate formation sequence.

Table 1. The selected species for this cladistic analysis.

Species	Major references
Chondria C. Agardh	Gordon-Mills (1987); Gordon-Mills and Womersley (1987)
Laurencia clavata Sonder	Saito and Womersley (1974); Nam and Choi (2001)
L. obtusa (Hudson) Lamouroux	Saito (1967, 1982); Nam et al. (1994)
L. intricata Lamouroux	Yamada (1931); Saito (1967); Nam (1990)
L. nipponica Yamada	Yamada (1931); Saito (1967); Nam et al. (1991)
L. okamurae Yamada	Yamada (1931); Saito (1967); Nam (1990)
L. intercalaris Nam	Nam (1994)
L. venusta Yamada	Yamada (1931); Saito (1967); Nam et al. (2000a)
L. succulenta Nam	This study
L. pinnata Yamada	Yamada (1931); Saito (1967); Nam (2004)
L. elata (C. Agardh) Harvey	Saito and Womersley (1974); Nam and Choi (2001)
L. brongniartii J. Agardh	Saito and Womersley (1974); Nam and Sohn (1994); Abe et al. (1998)
L. similis Nam et Saito	Nam (1990); Nam and Saito (1991); Masuda et al. (1997a)
L. botryoides (Turner) Gaillon	Saito and Womersley (1974); Nam (1990)
L. crustiformans McDermid	McDermid (1989)
Chondrophycus cartilagineus (Yamada) Garbary et Harper	Yamada (1931); Saito (1967); Nam and Saito (1990)
C. undulatus (Yamada) Garbary et Harper	Yamada (1931); Saito (1967); Nam (1999)
C. dotyi (Saito) Nam	Saito 1969b; Nam (1999)
C. succisus (Cribb) Nam	Saito (1969b); Cribb (1983); Nam (1999); Furnari et al. (2004)
C. carolinensis (Saito) Nam	Saito (1969b); Nam (1999)
C. kangjaewonii (Nam et Sohn) Garbary et Harper	Nam and Sohn (1994)
C. translucidus (Fujii et Cordeiro-Marino) Garbary et Harper	Fujii and Cordeiro-Marino (1996)
C. palisadus (Yamada) Nam	Masuda et al. (1998a); Nam (1999)
C. intermedius (Yamada) Garbary et Harper	Yamada (1931); Saito (1967); Nam and Saito (1995)
C. capituliformis (Yamada) Garbary et Harper	Yamada (1931); Saito (1967); Nam and Saito (1995)
C. tumidus (Saito et Womersley) Garbary et Harper	Saito and Womersley (1974); Nam and Saito (1995)
C. perforatus (Bory) Nam	Masuda et al. (1998b); Nam (1999)
C. dinhii (Masuda et Kogame) Nam	Masuda and Kogame (1998); Nam (1999)
C. papillosus (C. Agardh) Garbary et Harper	Nam and Saito (1991a); Masuda et al. (1997a); Nam (1999)
C. maris-rubri (Nam et Saito) Garbary et Harper	Nam and Saito (1995); Nam (1999)
C. parvipapillatus (Tseng) Garbary et Harper	Masuda et al. (1997b); Nam (1999)
C. iridescens (Wynne et Ballantine) Garbary et Harper	Wynne and Ballantine (1991); Nam (1999)
C. gemmiferus (Harvey) Garbary et Harper	Fujii et al. (1996); Nam (1999)
C. poiteaui (Lamouroux) Nam	Fujii et al. (1996); Nam (1999)
Osmundea osmunda (S.G. Gmelin) Nam et Maggs	Maggs and Hommersand (1993); Nam et al. (1994); Nam and Choi (2000)
O. crispa (Hollenberg) Nam	Smith and Hollenberg (1943); Saito (1969); Nam and Choi (1999)
O. splendens (Hollenberg) Nam	Smith and Hollenberg (1943); Saito (1969); Abbott and Hollenberg (1976); McIvor et al. (2002)
O. pinnatifida (Hudson) Stackhouse	Saito 1982; Maggs and Hommersand (1993); Nam et al., (1994)
O. hybrida (A.P. de Candolle) Nam	Saito 1982; Maggs and Hommersand (1993); Nam and Saito (1994)
O. pelagiensis (Cormaci, Furnari et Serio) Furnari	Cormaci et al. (1994)
O. spectabilis (Postels et Ruprecht) Nam	Saito (1969); Abbott and Hollenberg (1976); Nam et al. (1994)
O. pelagosae (Schiffner) Nam	Furnari and Serio (1993b)
O. verlaquei (Cormaci, Furnari et Serio) Furnari	Cormaci et al. (1994)
O. truncata (Kützing) Nam et Maggs	Maggs and Hommersand (1993); Furnari and Serio (1993a); Nam et al. (1994)
O. oederi (Gunnerus) McIvor, Maggs, Guiry et Hommersand	Nam et al. (2000b); McIvor et al. (2002)
O. maggsiana Serio, Cormaci et Furnari	Serio et al. (1999)
O. lata (Howe et Taylor) Yoneshigue-Valentin, Fujii et Gurgel	Yoneshigue-Valentin et al. (2003)

Table 2. Morphological characters used in the cladistic analysis (NA, non-applicable characters).

Habit
1. Holdfast: discoid (0); crust (1); tangled (2); stolonous (3)
2. Axis compression: none (0); none/slight (1); slight/strong (2); strong (3)
3. Branch basal constriction: present (markedly)* (0); absent (1)
4. Main leading branch: present (0); absent (1)
5. Branching: radial (0); distichous (1)
6. Size of mature thallus: 1–5 cm (0); 6–10 cm (1); 11–20 cm (2); >20 cm (3)
7. Axis diameter: <1.5 mm (0); 1.5–2.5 mm (1); > 2.5 mm (2)
8. Texture: soft (0); somewhat rigid (1); rigid (2); very rigid (3)
9. Iridescence: absent (0); present (1)

Vegetative structure
10. Pericentral cells in axis: 5 (0); 4 (1); 2 (2)
11. Position of the first pericentral cell relative to trichoblast: underneath (0); beside (1)
12. Axial cell row: fully extended (0); somewhat extended (1); limited to apex (2)
13. *Corps en cerise*: absent (0); present (1)
14. Secondary pit connections: present (0); absent (1)
15. Presence of secondary pit connection: almost always (0); frequently (1); rarely (2); absent (3)
16. Lenticular thickenings: always present (0); may be present (1); absent (2)
17. Epidermal cell projection: absent (0); occasionally slightly present (1); present (2)
18. Shape of epidermal cell projection: plate (0); dome (1); conical (2)
19. Palisade structure or radial elongation of epidermal cell in TS: absent (0); present (1)
20. Regeneration of trichoblast branches: absent (0); present (1)
21. Starch grains in medullary cells: present (0); absent (1)

Male reproductive structure
22. Spermatangial development: trichoblast type (0); filament type (1)
23. Spermatangial pit: absent (0); cup-shaped (1); pocket-shaped (2)
24. Spermatangial axial cell row: recognizable (0); unrecognizable(1)
25. Spermatangial nucleus: apical (0); central (1)
26. Spermatangia length: < 7–10 μm(0); 10–15 μm (1); > 15 μm(2)
27. Origin of spermatangial branch: trichoblast (0); apical and epidermal cell (1)
28. Spermatangial branch development in trichoblasts: partly (0); intercalary (1); NA (2)
29. Branching of spermatangial branch: subdichotomous to dichotomous (0); alternate (1)
30. Terminal sterile cells: single (0); clusters (1)
31. Shape of terminal sterile cells: spherical (0); ovoid to ellipsoidal (1); pyriform (2); elongate (3)
32. Length of terminal sterile cells: < 25 μm (0); 25–35 μm (1); > 35 μm (2)
33. Sterile branches in spermatangial branch: present (0)**; partly present (1)***; absent (2)
34. Arrangement of spermatangial receptacle: determinate (0); indeterminate (1)

Female reproductive structure
35. Pericentral cell number in procarp segment: 5 (0); 6 (1); 4 (2)
36. Auxiliary cell timing: normal (0); delayed (1)
37. Protuberance of ostiole in cystocarp: absent (0); present (1)
38. Cystocarp shape: ovoid (0); ovoid to subconical (1); conical (2)
39. Cystocarp diameter: <800 μm (0); 800–1000 μm (1); > 1000 μm(2)
40. Carposporangial diameter: <50 μm (0); 50–80 μm (1); > 80 μm (2)

Tetrasporangia
41. Tetrasporangial origin: pericentral cells (0); epidermal cells (1)
42. Tetrasporangial initial formation: adaxial (0); abaxial (1); lateral (2)
43. Tetrasporangial arrangement: parallel (0); right angle (1)
44. Cover cell arrangement: horizontal (0); longitudinal (1)
45. Cover cell division: absent (0); present (1)
46. Sterile pericentral cell number in axis: 2–3 (0); 1 (1); NA (2)
47. Additional tetrasporangial pericentral cell: absent (0); present (1); NA (2)
48. Fertility in the second pericentral cell: absent (0); present (1); NA (2)
49. Diameter of tetrasporangia: to ca 100 μm(0); 100–150 μm (1); >150 μm (2)

*Abrupt or internode-like constriction derived from having only one or a few layers of cells in axial segments particularly of young branchlets (not gradual constriction seen in clavate branchlets).
**Formation of spermatangial branches from one of two laterals on suprabasal cell of trichoblasts (the other one remains sterile).
***Formation of spermatangial branches from two laterals on suprabasal cell of trichoblasts, but partly remain sterile.

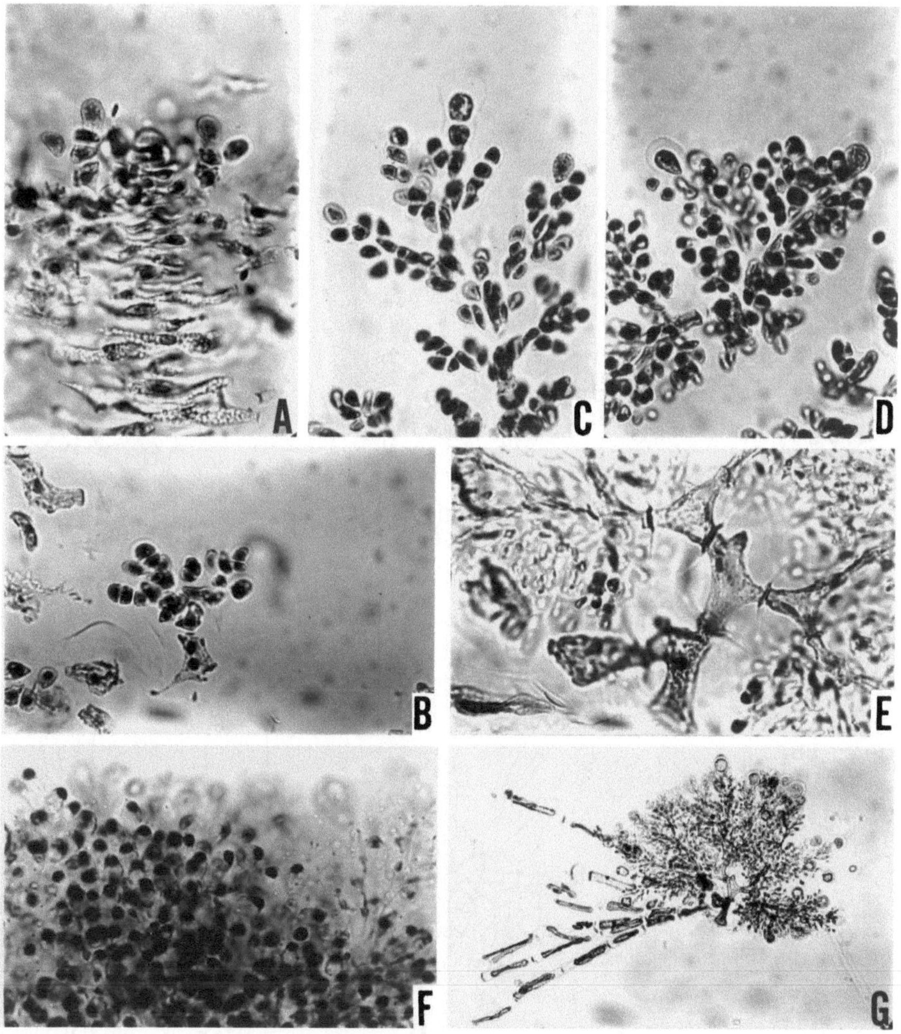

Figure 4. Laurencia succulenta sp. nov. (A–B) Young male trichoblast derived from axial cell near apical cell. (C–D) Young and mature spermatangial branches with terminal sterile cells and spermatangia. (E) Basal part of spermatangial branches. (F) Mature spermatangia with apical nucleus. (G) Trichoblast with two laterals (sterile and spermatangial branches) on its suprabasal cell (×85). (A – D, ×340).

pericentral cells; tetrasporangia with two presporangial cover cells undivided and arranged transversely in surface view of stichidium, with tetrahedral division and parallel arrangement to stichidial axis, 110–140 μm in diameter (Figure 6).

Laurencia succulenta is superficially most similar to *L. nipponica* and *L. okamurae*. However, it differs from both species in its possession of subcompressed thalli with basically distichous branching. It is also readily distinguished from these species by the presence of protuberant ostioles in the cystocarps, a feature very useful for distinguishing among similar species of the *Laurencia* complex (Nam and Saito, 1990, 1995; Masuda and Kogame, 1998; Nam et al., 2000). *L. succulenta* has subconical cystocarps with somewhat rostrate ostioles, whereas the latter two species have ovoid cystocarps with nonprotuberant ostioles. The colour and spermatangial features are also useful for distinguishing *L. succulenta* from *L. nipponica* and *L. okamurae*. In *L. succulenta* thalli are brown, and the spermatangia possess apical nuclei, whereas, *L. nipponica* features spermatangia with central nuclei (Saito, 1967; Nam et al., 1991) and *L. okamurae* thalli are usually purplish-green (Saito, 1967; Nam, 1990), rather than brown. *L. succulenta* is distinct from other similar species in being epiphytic on other coarse algae, and is characterized by the fleshy, robust, thick and subcompressed thalli with distichous branching.

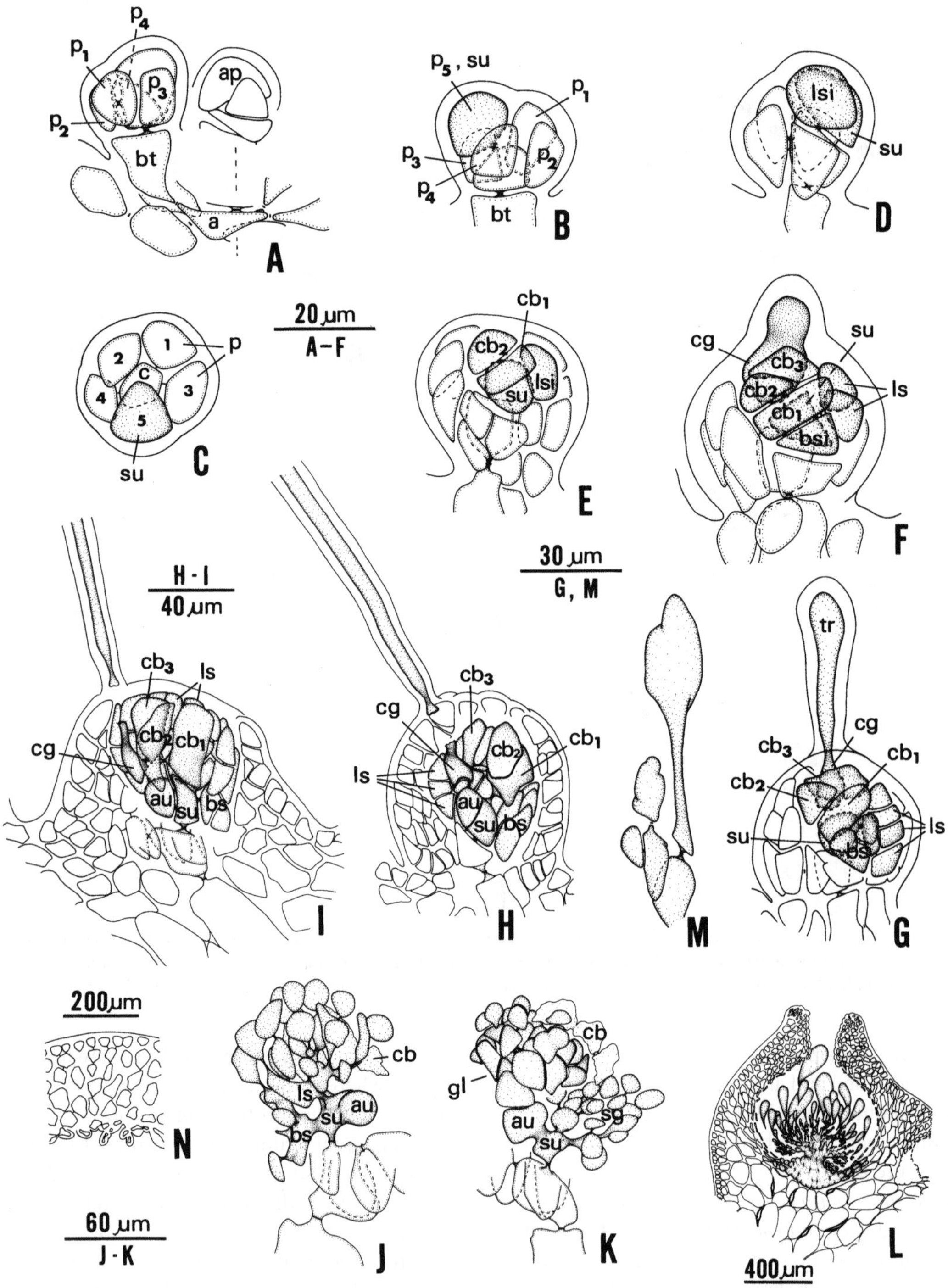

Figure 5. Laurencia succulenta sp. nov. (A-G) Development of young procarp. (C) Procarp-bearing segment with five pericentral cells. (H-K) Procarp after fertilization, showing auxiliary and sterile cell development from supporting cell. (L) Cystocarp with somewhat protuberant ostiole. (M) Young carposporangia. (N) Pericarp in TS. a, axial cell; ap, apical cell; au, auxiliary cell; bs, basal sterile group; bt, basal cell of trichoblast; c, central cell of procarp-bearing segment; cb, carpogonial branch; cg, carpogonium; p, pericentral cell; gl, gonimoblast; su, supporting cell; tr, trichogyne. The numbers indicate formation sequence.

Table 3. Data matrix of the morphological characters used in this analysis (?, missing data)

Taxa	Character 1	2	3	4	5	6	7	8	9	10	11	12	13	14	15	16	17	18	19	20	21	22	23	24	25	26	27	28	29	30	31	32	33	34	35	36	37	38	39	40	41	42	43	44	45	46	47	48	49
Chondria	0	0	0	0	0	3	0	0	?	0	0	0	0	0	0	0	0	0	0	0	0	0	0	0	0	0	0	0	0	0	0	0	0	0	0	0	0	0	0	0	0	0	0	0	0	0	0	0	1
L. clavata	0	0	0	1	0	2	2	1	0	1	0	1	?	0	0	0	0	0	0	0	0	0	1	0	0	1	0	0	0	0	1	2	0	0	?	?	?	?	?	?	?	?	?	?	?	?	?	?	?
L. obtusa	0	0	1	0	0	2	1	0	0	1	0	2	1	0	0	2	1	1	0	0	1	0	1	0	0	0	0	0	0	0	1	0	0	0	0	0	0	0	1	0	0	1	0	0	0	0	0	0	0
L. intricata	2	0	1	1	0	1	0	0	0	1	0	2	1	0	0	2	1	1	0	0	1	0	1	0	0	1	0	0	0	0	1	0	0	0	0	0	0	1	1	1	0	1	0	0	0	0	0	0	2
L. nipponica	3	0	1	0	0	3	2	1	0	1	0	2	1	0	0	1	1	1	0	0	1	0	1	0	1	0	0	0	0	0	1	0	0	0	0	0	0	0	0	0	0	1	0	0	0	0	0	0	1
L. okamurae	3	0	1	0	0	3	1	1	0	1	0	2	1	0	0	1	1	1	0	0	1	0	1	0	0	0	0	0	0	0	1	0	0	0	0	0	0	1	0	1	0	1	0	0	0	0	0	0	0
L. intercalaris	3	0	1	0	0	2	0	1	?	1	0	2	?	0	0	0	1	1	0	1	1	0	1	0	0	0	0	1	0	0	1	0	0	0	?	?	?	?	?	?	?	?	?	?	?	?	?	?	?
L. venusta	2	1	1	1	0	1	0	1	0	1	0	2	1	0	0	1	1	1	0	0	1	0	1	0	0	1	0	0	0	0	1	1	0	0	0	0	0	0	0	0	0	1	0	0	0	0	0	0	0
L. succulenta	3	2	1	0	1	1	0	1	?	1	0	2	?	0	0	1	1	1	0	0	1	0	1	0	0	1	0	0	0	0	1	1	0	0	0	0	0	1	2	2	0	1	0	0	0	0	0	0	1
L. pinnata	0	3	1	1	1	1	1	0	0	1	0	2	1	0	0	2	1	1	0	0	1	0	1	0	0	0	0	0	0	0	1	1	0	0	0	0	0	0	2	2	0	1	0	0	0	0	0	0	2
L. elata	3	2	1	0	1	3	1	1	?	1	0	2	?	0	0	0	0	0	0	0	1	0	1	0	0	1	0	0	0	0	1	0	0	0	0	0	0	0	2	0	0	1	0	0	0	0	0	0	2
L. brongniartii	0	3	1	0	1	1	2	0	?	1	0	2	1	0	0	0	1	1	0	0	1	0	1	0	0	0	0	0	0	0	1	1	0	0	0	0	?	?	?	?	0	1	0	0	0	0	0	0	1
L. similis	0	0	1	0	0	2	1	2	?	1	0	2	1	0	0	2	0	0	0	?	1	0	1	0	0	0	0	0	0	0	1	0	0	0	?	?	0	0	0	?	0	1	1	0	0	0	0	0	1
L. botryoides	3	1	1	0	1	2	1	1	?	1	0	2	?	0	0	1	0	0	0	0	1	0	1	0	0	1	0	0	0	0	1	0	0	0	0	0	0	0	2	0	0	1	1	0	0	0	0	0	2
L. crustiformans	1	0	1	0	0	0	0	2	1	?	?	2	1	0	0	2	0	0	0	?	1	?	?	?	?	?	?	?	?	?	?	?	?	?	?	?	0	1	0	?	0	1	1	?	?	?	?	?	0
C. cartilagineus	0	1	1	1	0	2	1	2	0	2	1	2	0	1	3	2	2	1	0	0	1	0	1	0	0	2	0	0	0	0	2	2	1	0	0	1	1	2	0	1	0	1	1	0	0	0	1	0	2
C. undulatus	0	3	1	1	1	1	2	2	0	2	1	2	0	1	3	2	0	0	0	0	1	0	1	0	0	1	0	0	0	0	0	2	1	0	0	1	0	1	2	1	0	1	1	0	0	0	1	0	2
C. dotyi	0	1	1	0	0	0	1	2	?	2	?	2	?	1	3	2	2	1	0	?	1	?	?	?	?	?	?	?	?	?	?	?	?	?	?	?	?	?	?	?	0	1	1	0	?	?	?	?	?
C. succisus	0	3	1	?	1	0	1	2	?	2	?	2	?	1	3	2	0	0	0	?	1	?	?	?	?	?	?	?	?	?	?	?	?	?	?	?	?	?	?	?	0	1	1	0	?	?	?	?	?
C. carolinensis	0	1	1	0	0	0	?	2	?	2	?	2	?	1	3	2	2	1	0	?	1	?	1	?	?	?	?	?	?	?	?	?	?	?	?	?	?	?	?	?	0	1	1	?	?	?	?	?	?
C. kangjaewonii	0	3	1	1	1	1	2	2	0	2	1	2	0	1	3	2	0	0	0	0	1	0	1	0	0	2	0	0	0	0	0	1	1	0	0	1	0	?	1	2	0	1	0	0	0	0	1	0	2
C. translucidus	3	1	1	0	0	2	?	0	?	2	1	2	0	1	3	2	0	0	0	?	1	0	1	0	0	1	0	0	0	0	1	2	?	?	0	?	1	2	1	1	0	1	0	0	0	0	1	0	1
C. palisadus	0	0	1	0	0	2	1	3	?	2	0	2	0	1	3	2	0	0	1	?	1	0	1	0	0	1	0	0	0	0	1	0	0	0	2	?	1	2	0	?	0	1	1	0	0	1	1	1	1
C. intermedius	3	0	1	0	0	2	2	3	0	2	0	2	0	1	3	2	0	0	1	0	1	0	1	0	0	1	0	0	0	0	1	1	0	0	2	0	0	1	1	1	0	1	1	0	0	1	1	1	2
C. capituliformis	0	0	1	0	0	2	1	3	0	2	0	2	0	1	3	2	0	0	1	0	1	0	1	0	0	1	0	0	0	0	1	0	0	0	2	0	1	2	0	1	0	1	1	0	0	1	1	1	2
C. tumidus	1	0	1	0	0	2	1	3	?	2	0	2	?	1	3	2	0	0	1	0	1	0	1	0	0	2	0	0	0	0	2	2	1	0	2	0	0	1	2	1	0	1	1	0	1	1	1	1	2
C. perforatus	0	0	1	0	0	0	0	3	?	2	?	2	?	1	3	2	0	0	1	?	1	0	?	?	?	?	?	0	?	0	?	?	?	?	?	?	?	?	?	?	0	1	1	0	0	1	1	1	0
C. dinhii	0	0	1	0	0	1	1	3	?	2	?	2	?	1	3	2	0	0	1	?	1	0	?	?	?	?	?	?	?	?	?	?	?	?	2	?	1	2	1	?	0	1	1	0	0	1	1	1	1
C. papillosus	0	0	1	0	0	1	0	3	?	2	0	2	0	1	3	2	0	0	0	?	1	0	1	?	0	0	0	0	0	0	1	1	0	0	?	?	1	2	1	?	0	1	1	0	0	1	1	1	0
C. maris-rubri	0	0	1	0	0	1	0	3	?	2	0	2	?	1	3	2	0	0	0	?	1	?	?	?	?	?	?	?	?	?	?	?	?	?	?	?	?	?	?	?	0	1	1	0	0	1	1	1	2
C. parvipapillatus	0	3	1	1	1	0	2	3	?	2	0	2	1	0	1	2	2	2	1	0	1	0	1	?	0	0	0	0	0	0	1	0	0	0	?	?	?	?	?	?	0	1	1	0	0	1	1	1	1
C. iridescens	3	0	1	1	0	0	0	2	1	2	0	2	?	0	0	2	2	2	0	?	1	?	?	?	?	?	?	?	?	?	?	?	?	?	?	?	?	?	?	?	0	1	1	0	0	1	1	1	0
C. gemmiferus	0	0	1	1	0	2	1	1	?	2	?	2	?	0	0	2	2	2	0	?	1	0	1	0	0	1	0	0	0	0	1	1	?	0	0	?	1	2	0	0	0	1	1	0	0	1	1	1	0
C. poiteaui	0	0	1	0	0	2	1	1	?	2	?	2	0	0	0	2	0	0	0	?	1	0	1	0	0	1	0	0	0	0	1	1	?	0	0	?	1	2	0	0	0	1	1	0	0	1	1	1	1
O. osmunda	0	3	1	0	1	2	2	1	0	2	1	2	0	1	3	1	0	0	0	0	1	1	2	1	0	2	1	2	1	1	1	1	2	1	0	0	0	0	2	1	1	2	0	1	0	2	2	2	1
O. crispa	2	1	1	0	0	1	0	0	0	2	1	2	0	1	3	2	0	0	0	0	1	1	2	1	0	0	1	2	1	1	1	0	2	1	0	0	0	1	2	0	1	2	0	1	0	2	2	2	1
O. splendens	0	3	1	0	1	1	2	1	?	2	?	2	?	1	3	2	?	?	0	?	1	1	2	1	?	?	1	2	?	?	?	?	2	?	?	?	1	0	?	?	1	2	0	1	0	2	2	2	?
O. pinnatifida	3	3	1	0	1	1	1	1	0	2	1	2	0	1	3	1	0	0	0	0	1	1	2	1	0	2	1	2	1	1	1	1	2	1	0	0	0	0	2	1	1	2	0	1	0	2	2	2	1
O. hybrida	0	0	1	0	0	2	0	1	0	2	1	2	0	1	3	2	0	0	0	0	1	1	1	1	0	1	1	2	1	1	1	2	2	1	1	0	0	1	2	2	1	2	0	1	0	2	2	2	1
O. pelagiensis	1	2	1	0	1	1	1	?	?	2	?	2	?	1	3	2	0	0	1	?	1	1	1	1	?	?	1	2	1	?	?	?	2	1	?	?	0	0	0	?	1	2	0	1	?	2	2	2	?
O. spectablilis	0	3	1	0	1	3	2	1	?	2	1	2	?	0	2	2	1	1	0	0	1	1	2	1	0	0	1	2	1	1	1	1	2	1	0	0	1	1	2	2	1	2	0	1	0	2	2	2	2
O. pelagosae	0	2	1	0	1	2	1	?	0	2	?	2	?	0	?	0	?	?	?	?	1	1	?	1	0	2	1	2	?	0	3	2	2	1	?	?	0	0	0	?	1	2	0	1	?	2	2	2	?
O. verlaquei	1	2	1	0	1	0	1	?	?	2	1	2	?	0	0	2	0	0	0	?	1	1	1	1	0	?	1	2	1	0	3	2	2	?	?	?	?	?	?	?	1	2	0	1	?	2	2	2	1
O. truncata	0	3	1	0	1	1	0	1	0	2	1	2	0	0	1	1	0	0	0	?	1	1	1	1	0	0	1	2	1	0	0	?	2	1	0	0	1	2	2	1	1	2	0	1	0	2	2	2	1
O. oederi	0	2	1	0	1	2	0	1	?	2	1	2	0	0	1	2	0	0	0	0	1	1	1	1	0	1	1	2	1	1	1	2	2	1	0	0	0	0	1	1	1	2	0	1	0	2	2	2	1
O. maggsiana	0	0	1	0	0	0	2	2	?	2	?	2	?	1	3	2	0	0	1	?	1	1	1	1	0	1	1	2	1	0	1	2	2	?	?	?	?	?	?	?	1	2	0	1	?	2	2	2	?
O. lata	0	3	1	0	1	2	2	0	?	2	?	2	?	0	?	2	?	?	0	?	1	1	1	1	0	0	1	2	1	1	1	1	2	1	?	?	0	1	1	0	1	2	0	1	0	2	2	2	2

In cladistic analyses, maximum parsimonious trees (MP) with treelength of 350, consistency index of 0.674, retention index of 0.927 and rescaled consistency index (RC) of 0.625 were found. Whereas in previous analyses (Garbary and Harper, 1998), only three clades were found, here, four major clades (*Laurencia*, *Chondrophycus palisadus* group, *C. cartilagineus* group and *Osmundea* assemblage) each formed a monophyletic group (Figures 7, 8). The groups were characterized by the number and position of pericentral cells in axial segments and by the origin of tetrasporangia and spermatangial branches. Of the four clades, the first one, which is basal to the overall assemblage, corresponds to *Laurencia*. This clade is primarily defined by the vegetative axis having four pericentral cells rather than two as in the remainder of the assemblage. The *Osmundea* clade is supported by synapomorphic characters for the species of the genus, the features associated with spermatangial formation of filament type and tetrasporangial production from epidermal cells (Nam et al., 1994). By contrast, *Chondrophycus*, which up to now has

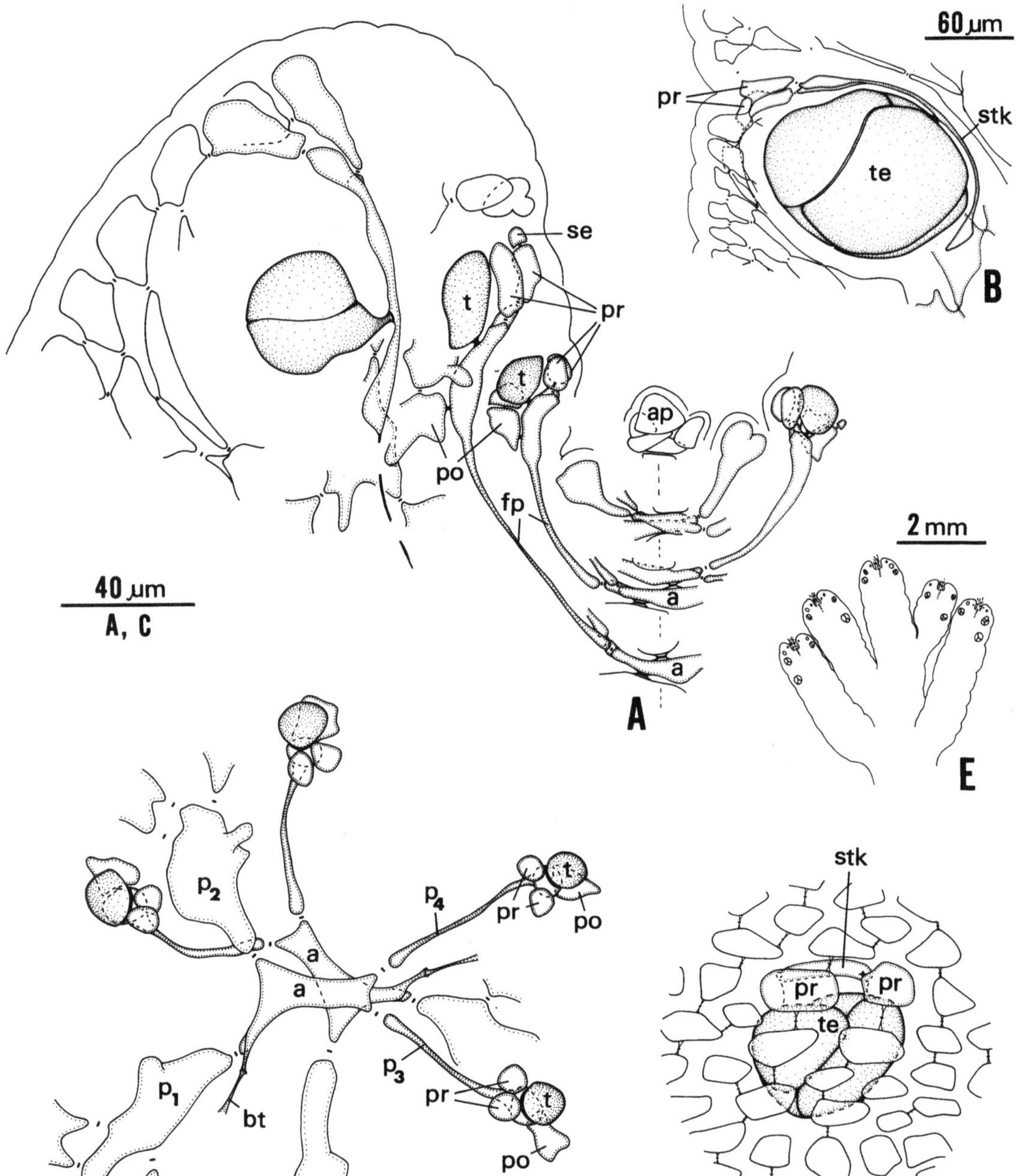

Figure 6. Laurencia succulenta sp. nov. (A) Development of tetrasporangia. (B) Mature tetrasporangium. (C) Two superimposed tetrasporangial axis with two fertile pericentral cells. (D) Two presporangial cover cells with transverse arrangement in surface view. (E) Stichidia with parallel arrangement of tetrasporangia. a, axial cell; ap, apical cell; bt, basal cell of trichoblast; fp, fertile pericentral cell; p, pericentral cell; se, conjuctor cell for secondary pit connection; po & pr, post- and presporangial cover cell; stk, stalk cell; t, tetrasporangium initial; te, tetrasporangium. The numbers indicate formation sequence.

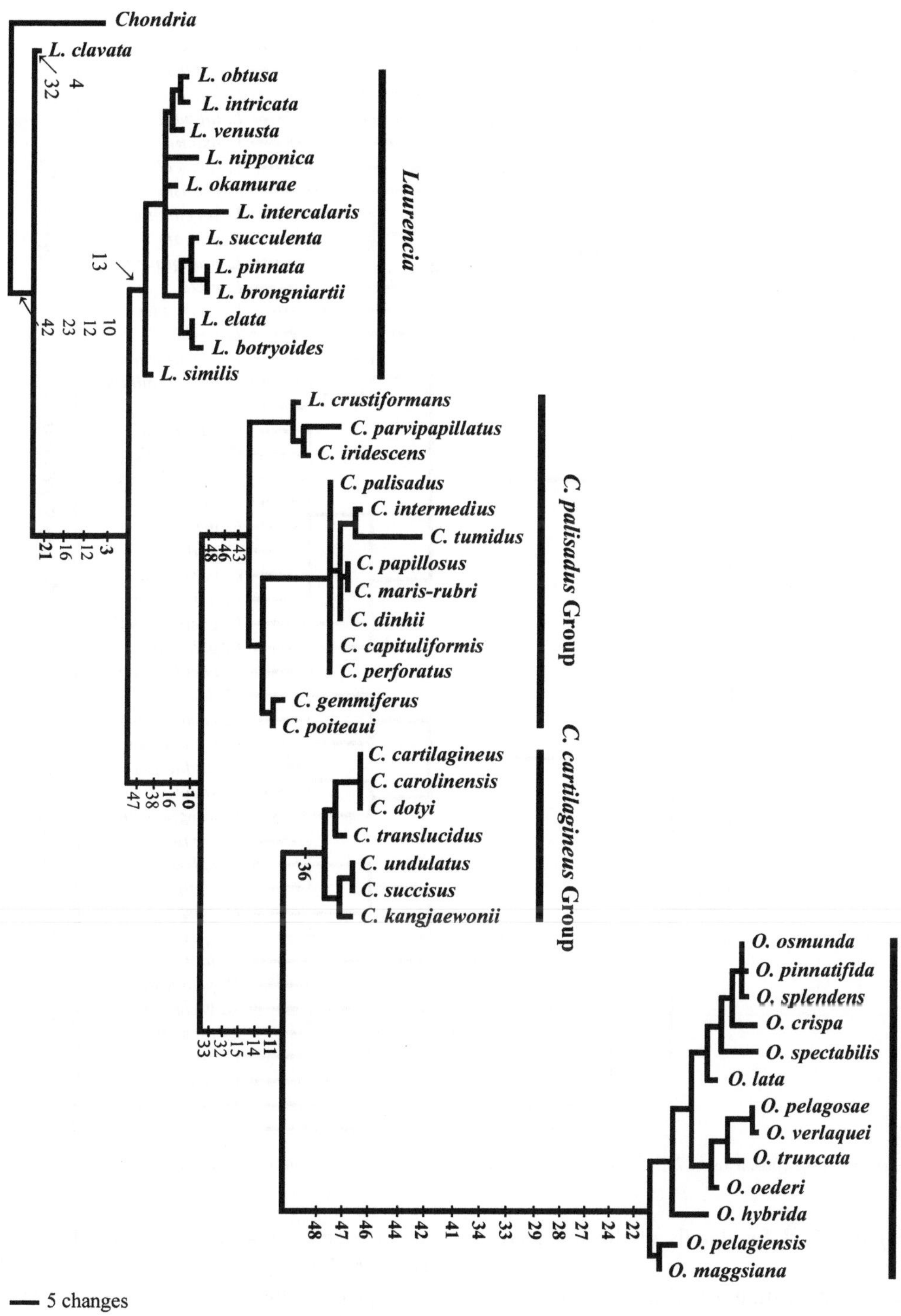

Figure 7. Most parsimonious tree from heuristic search for optimum tree. Numbers below internodes indicate the major characters that have changed state (Tables 2, 3). Bold numbers indicate synapomorphies for that clade. Apomorphies for terminal and some subterminal clades were not indicated.

been characterized by the combination of vegetative axis with two pericentral cells, trichoblast-type spermatangial development and tetrasporangial production from pericentral cells (Nam, 1999) is now paraphyletic. The species are separated into two well-supported clades, the *C. palisadus* and *C. cartilagineus*

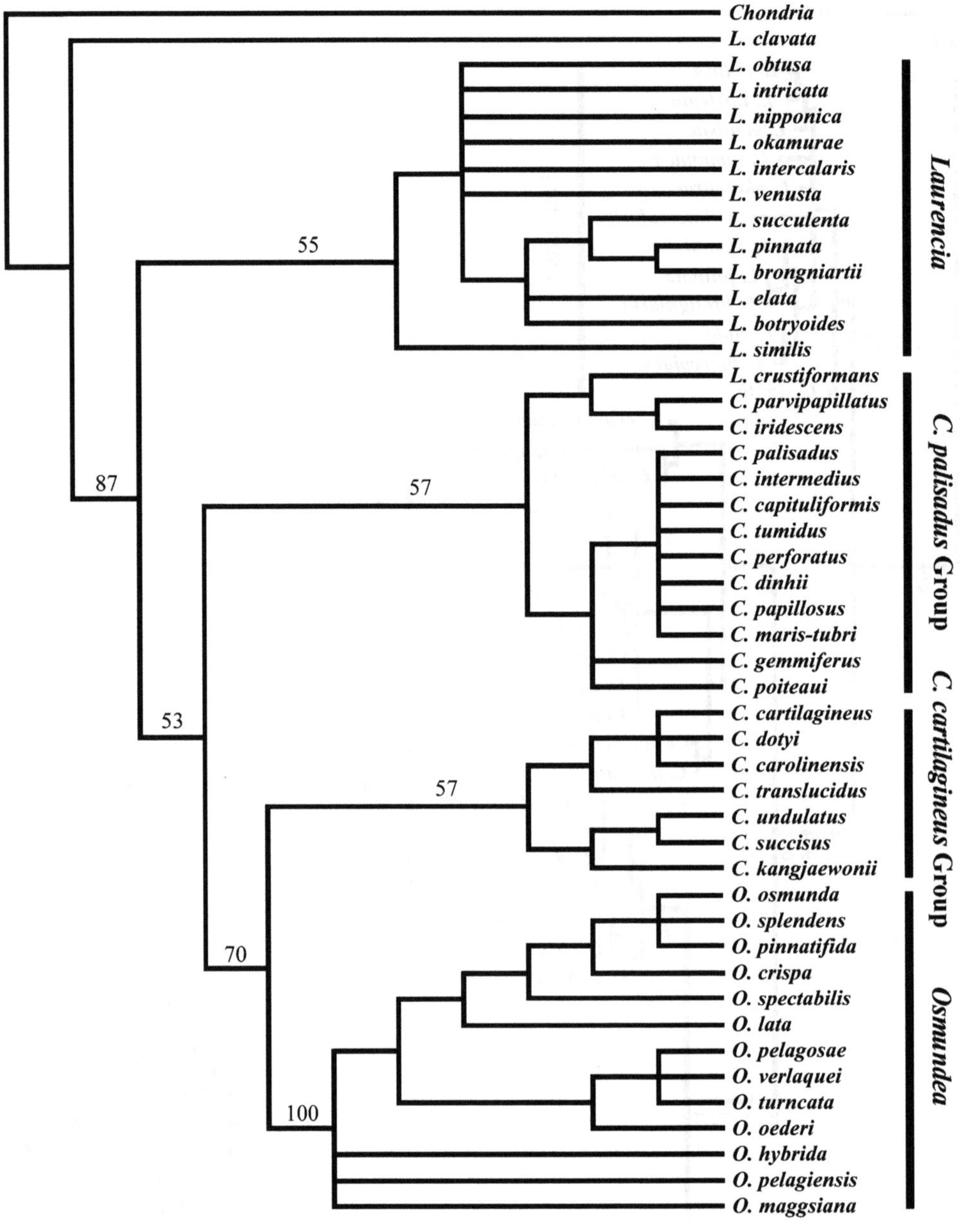

Figure 8. Strict consensus of 27500 most parsimonious (MP) trees. Numbers on branches show bootstrap values.

groups (see Figures 7, 8 for details of the included species).

The two groups are distinguished from each other by the following features: position of the first pericentral cell relative to the trichoblast, absence or presence of fertility of the second pericentral cells, number of sterile pericentral cells in the tetrasporangial axis, and a post-fertilization feature associated with formation time of the auxiliary cell. In addition, the formation pattern of spermatangial branches on trichoblasts and probably the number of pericentral cells in procarp-bearing segments of female trichoblasts can be adopted for the characterization of each group. In the *C. palisadus* group, the first pericentral cell is produced underneath the basal cell of the trichoblast, then the second one is formed at some distance from the cell, as in *Laurencia* (Tables 2, 3). The second pericentral cell in the tetrasporangial axis is always fertile, the resulting axis has one sterile pericentral cell, and auxiliary cells in the post fertilization process show a normal developmental pattern after presumed fertilization. With the exception of *C. tumidus* (Nam and Saito, 1995; Nam, 1999), spermatangial branches are produced from one of two laterals on the suprabasal cell of trichoblasts, as in *Laurencia*. Procarp-bearing segments have four pericentral cells rather than five, except in *C. gemmiferus* and *C. poiteaui* (Fujii et al., 1996). In contrast, in the *C. cartilagineus* group, the first pericentral cell is produced on the side of the trichoblast basal cell, then the second one on the other side, as in *Osmundea*, and in the tetrasporangial axis the second pericentral cell is never fertile, resulting in an axis with two sterile pericentral cells. Spermatangial branches are produced from two laterals on suprabasal cell of trichoblasts, but remain partly sterile. There are five procarpic pericentral cells, and there is apparently a delayed formation pattern of auxiliary cells, although more data are required.

Of these features, the lateral position of the first pericentral cell is synapomorphic for the *C. cartilagineus* group and the *Osmundea* clade, while the position underneath the first pericentral cell, together with the formation of spermatangial branches from one lateral, is symplesiomorphic for the *Laurencia* clade and the *C. palisadus* group. Garbary and Harper (1998) noted an evolutionary reduction in the number of vegetative pericentral cells. Figure 7 is also congruent with the evolutionary trend of the character. Probably, with the reduction of the pericentral cells from four to two, the formation position of the first pericentral cell would be changed from underneath the trichoblast basal cell, the position of which in the axial segment determines the formation sequence of pericentral cells (Nam, 1999), to the side. In the parasitic *Janczewskia* Solms-Laubach, the same evolutionary pattern is found for this feature (Nam et al., unpublished data). However, any possible evolutionary advantage conferred by the positional change in this cell is unknown. In the spermatangial feature associated with branch formation, as seen in the *C. palisadus* group, *Laurencia* shows the formation of spermatangial branches from one of two laterals on the suprabasal cells of the trichoblasts (the other lateral remains sterile), which appears to be an ancestral condition. However, *Osmundea* lacks this kind of sterile branches in spermatangial filaments, even though their origin differs (Nam et al., 1994). This distribution of the character state probably suggests a progressive reduction of sterile branches in spermatangial trichoblasts/filaments. It is difficult to understand how evolutionary pressure could exist for the reduction of sterile branches in spermatangial trichoblasts/filaments across the whole lineage. However, this could be interpreted in relation to the shape of the spermatangial pit. As reported previously (Nam et al., 1994), two typical shapes of spermatangial pits, cup- and urn-shaped, are found in the *Laurencia* complex, although a variation in the shape has been reported in the Mediterranean species, *O. pelagosae* (Furnari and Serio, 1993b; McIvor et al., 2002). On the other hand, typically in trichoblast-type spermatangial structures, while the adaxial-side lateral one of two on the suprabasal cell of the trichoblasts develops into spermatangial branches, the abaxial side one remains sterile (Nam, 1990). As noted by Nam (1994), the sterile branches may play a functional role to protect spermatangial branches inside the pit, similar to paraphyses for protection of sporangia in Phaeophyceae (Bold and Wynne, 1978). It is assumed that the urn-shaped spermatangial pit with an ostiole-like upper opening is more protective of the spermatangial trichoblasts/filaments than is the open cup-shaped one. This may have led to the evolutionary loss of the sterile branches in *Osmundea* clade with urn-shaped pit structure. In terms of mass production of spermatangia, the reduction of sterile branches in spermatangial trichoblasts/filaments can be more easily understood. In trichoblast-type spermatangial development with a recognizable axial cell row at the base of the pit (indicating continuous growth of pit), spermatangial branches successively divide in a dichotomous pattern with comparatively high frequency, and the spermatangial pit continuously grows (Nam et al., 1994), probably leading to more production of spermatangia per pit. The cup-shaped pit would permit this mass production strategy. In contrast, in filament types without a recognizable axial cell row at the base of pit (meaning the pit cease to grow as the apical cell also develops into a spermatangial filament), spermatangial branches are unbranched, or sparsely or rarely branched in an alternate manner with lower frequency, and the pit stops further growth (Nam et al., 1994), resulting in (relatively) lower less spermatangial production. This production would benefit from more protective urn-shaped pits, which need no additional protection for spermatangial filaments. Hence, loss of sterile branches may occur. Consequently, the reduction of sterile branches in spermatangial trichoblasts/filaments appears to be closely related with reproductive strategy. The number of sterile pericentral cells in the tetrasporangial axis associated with fertility of the second pericentral cell also has phylogenetic significance. As mentioned above, the species of the *C. palisadus* group have one sterile pericentral cell (the first one) in the axis with the second of the two pericentral cells being fertile, while the remainder of the assemblage has two. In the polysiphonous structure, it appears that the number of sterile pericentral cells in the fertile axis is related to the structural stability of the thallus. Rigid thalli, being a usual feature in the group (Nam, 1999), may enable a reduction of the sterile cells in the tetrasporangial axis, thereby augmenting tetrasporangial production. Epidermal palisade structures, which are usually found in the group (Yamada, 1931; Saito, 1967; Saito and Womersley, 1974; Masuda and Kogame, 1998; Masuda et al., 1998), also seem to contribute to construction of rigid thalli, as noted by Nam (1999).

The *C. cartilagineus* clade is strongly supported by the developmental feature of spermatangial branches coupled with the post-fertilization features, being synapomorphies among the species of the clade. In contrast, the *C. palisadus* clade is characterized by autapomorphic tetrasporangial features. The procarpic feature, with four pericentral cells, may also be shared only by this clade, even though exception of *C. gemmiferus* and *C. poiteaui* has been reported (Fujii et al., 1996). A critical examination of this character in some species is required.

The results of this cladistic analysis strongly supports previous findings that *Laurencia* and *Osmundea sensu* Nam et al. (1994), Garbary and Harper (1998) and Nam (1999) are both monophyletic (Garbary and Harper, 1998; Nam et al., 2000; McIvor et al., 2002), but that *Chondrophycus* is not (see Figures 7, 8). The *C. cartilagineus* group is more closely related to the *Osmundea* clade rather than to the *C. palisadus* group (see Figures 7, 8), suggesting the possibility of separation of the *C. palisadus* group from *Chondrophycus* at the genus level.

Recently, Nam (1999) proposed four subgenera within *Chondrophycus* based on the presence or absence of secondary pit connections between epidermal cells, tetrasporangial arrangement type and number of sterile pericentral cells in tetrasporangial axis. The species included in the *C. cartilagineus* group (including the type of the genus) correspond to the subgenera *Chondrophycus*+*Kangjaewonia* Nam, and relationships within the group are well-defined. The group is divided into two subclades (Figures 7, 8). The *C. kangjaewonii*+*C. undulatus*+*C. succisus* clade is characterized by distichous branching with compressed thalli, whereas, the other clade, which includes *C. translucidus, C. cartilagineus, C. dotyi* and *C. carolinensis,* has radial branching. The character of tetrasporangial arrangement type (associated with the stichidial growth), which has been adopted as a distinguishing feature between the two subgenera (Nam, 1999) and was emphasized previously (Saito, 1967), shows homoplasy, because *C. kangjaewonii* and *C. translucidus,* both with a parallel arrangement (Nam and Sohn, 1994; Fujii and Cordeiro-Marino, 1996) are separated from each other (Figures 7, 8).

In contrast, the species of the *C. palisadus* group are referable to subgenera *Palisadi* (Yamada) Nam+*Yuzurua* Nam, but are divided into three subclades (Figures 7, 8). The *L. crustiformans*+*C. parvipapillatus*+*C. iridescens* clade is a sister group to the remaining species, and can be characterized by the presence of *corps en cerise*, although the character is unknown in *C. iridescens.* The *C. gemmiferus*+*C. poiteaui* clade also forms a successive sister clade to the remaining cluster of species (Figure 7). These species retain ancestral state in the epidermal secondary pit connections and procarpic features. The last assemblage is very well-defined based on the synapomorphic absence of secondary pit connections and procarps with four pericentral cells. *C. tumidus* is very distinct. Two presporangial cover cells, with divisions into several small cells, are autapomorphic for this species (Nam and Saito, 1995). The tumid thallus with extremely short branchlets and very elongated spermatangia are also unique for *C. tumidus* (Saito and Womersley, 1974; Nam, 1990; Nam and Saito, 1995). The formation pattern of spermatangial branches on trichoblasts is also noteworthy for this species. The spermatangial branches in *C. tumidus* show the same developmental pattern as those of *C. cartilagineus* group rather than *C. palisadus* group. The feature in *C. tumidus* seems to be independently evolved within the group.

In the *Laurencia* clade, relationships between subclades are comparatively well-resolved. *L. similis*, which is suggested as an intermediate between subgenera *Laurencia* and *Chondrophycus sensu* Saito (1967) (Nam and Saito, 1991), is apparently a sister clade to *Laurencia*, and *L. obtusa, L. intricata, L. venusta, L. okamurae, L. nipponica* and *L. intercalaris* form a subclade, which is supported by radial branching with terete or subterete thalli, although the latter three species show polytomy (Figure 7). In contrast, the last clade containing *L. succulenta, L. pinnata, L. brongniartii, L. elata* and *L. botryoides* is characterized by distichous branching with sub- or compressed thalli. The character of tetrasporangial arrangement type is also homoplasic in *Laurencia*, because *L. botryoides* with right-angle type (Saito and Womersley, 1974) is closely related to *L. succulenta, L. pinnata, L. brongniartii* and *L. elata* with parallel arrangement type (Saito, 1967; Saito and Womersley, 1974; Nam and Choi, 2001; Nam, 2004), rather than to *L. similis* with the same arrangement type (Nam and Saito, 1991b) (Figure 7). Spermatangia with central nuclei are autapomorphic for *L. nipponica* (Nam et al., 1991). In contrast, the features of trichoblast regeneration and intercalary formation of spermatangia in trichoblasts are autapomorphic for *L. intercalaris.* The budding-like regeneration of sterile laterals in spermatangial trichoblasts (Nam, 1994), which are deciduous, is particularly interesting, probably suggesting that male trichoblasts, in trichoblastic spermatangial development, require functionally sterile branches, which may protect spermatangial branches inside of opened cup-shaped pits, as mentioned above.

The species assigned to *Osmundea* form a clearly defined monophyletic group (Figures 7, 8), but relationships between subclades are not well-resolved. *O. maggsiana* and *O. pelagiensis,* which are characterized by epidermal palisade structure, absence of secondary pit connections and cup-shaped spermatangial pits, form a sister clade to the remaining species cluster, and *O. hybrida* is a successive sister clade supported by procarpic feature with six pericentral cells. The remaining assemblage is divided into two clades: one consisting of *O. oederi, O. truncata, O. pelagosae,* and *O. verlaquei* and the other containing *O. lata, O. spectabilis, O. crispa, O. osmunda, O. pinnatifida,* and *O. splendens.* The major features that supports the clades are epidermal secondary pit connections (presence or absence and occurrence frequency) and spermatangial pit shape (cup or urn shape). As pointed out by Nam et al. (2000) and McIvor et al. (2002), the former character seems to have more important phylogenetic significance.

Laurencia clavata, which is an outlier in the entire assemblage, forms an apparent sister group to the *Laurencia* complex (Figures 7, 8). As reported by Nam and Choi (2001), *L. clavata* exhibits a marked constriction at the base of the branches as in *Chondria,* but the constriction is likely to be internodal, as having only a few layers of cells in axial segments is unique to this species. The species also has a unique verticillate branching pattern, without a main leading branch. In addition, many starch grains are accumulated in medullary and subcortical cells, although not in the swollen or bulb-like cells of epidermal, subcortical or secondary cortex origin, as in some species of *Chondria* (Gordon-Mills and Womersley, 1987). Unlike in the *Laurencia* complex, apical cells also show comparatively less oblique division, with the resulting axial cell row extending somewhat below the distal branchlets. The axial cell rows are clearly recognizable throughout the branchlets particularly at a young stage. These features, which are distinctly different from the *Laurencia* complex, might have led to the previous assignment of *L. clavata* to *Chondria* (Harvey, 1862) or *Corynecladia* (J. Agardh, 1876) (as lectotype, see Garbary and Harper, 1998: 193). However, it differs from *Chondria* in having spermatangial structures without specialized plates, axial cell rows unrecognizable throughout mature thalli, apical cells without clear transverse divisions in branchlet apex and axis with four pericentral cells rather than five. Recently, Womersley (2003: 460) reported tetrasporangial features of *L. clavata.* According to his study, tetrasporangia seem to be abaxially produced from outer cortical cells, although a critical examination of their origin is needed. This feature, in combination with the trichoblast-type spermatangial development, is also very distinct for *L. clavata. L. clavata* appears to be in a phylogenetic position linking the *Laurencia* complex and *Chondria.* This analysis also strongly suggests it regardless of tetrasporangial origin in the species, leading to the conclusion that *Corynecladia* should be reinstated for the type species *L. clavata.* It is interesting that *L. crustiformans* is placed in the *C. palisadus* group rather than the *Laurencia* clade. Probably, this species should be removed from *Laurencia,* but prior to taxonomic change a confirmation of the generic features in the species is required.

Conclusions

L. succulenta sp. nov. is described from Korea. Based on this cladistic analysis, *Palisada* (Yamada) stat. nov. is proposed for the *C. palisadus* group, together with an emendation of the generic delineation of *Chondrophycus,* and relevant nomenclatural changes for several *Chondrophycus* species are included (Table 4). Also, the genus *Corynecladia* is reinstated for *L. clavata.*

Palisada (Yamada, 1931) stat. nov.

Apical cell always sunk in apical pit of branchlet; central axis recognizable only near apical cell; forming extensive cortex; vegetative axial segments with two pericentral cells; first pericentral cell with underneath position of trichoblast; spermatangial development

Table 4. New combinations in *Palisada* mainly from the Pacific.

Palisada robusta **nom. nov.** (renamed by tautonym)
Holotype: SAP 13883. Basionym: *Laurencia palisada* Yamada (1931: 196, figures c, d, pl. 4, figure a). Synonym: *Chondrophycus palisadus* (Yamada) Nam (1999: 463).
Palisada intermedia **(Yamada) comb. nov.**
Holotype: SAP 13876. Basionym: *Laurencia intermedia* Yamada (1931: 191, pl. 1, figure c, pl. 2). Synonym: *Chondrophycus intermedius* (Yamada) Garbary et Harper (1998: 195).
Palisada capituliformis **(Yamada) comb. nov.**
Holotype: SAP 13880. Basionym: *Laurencia capituliformis* Yamada (1931: 217, pl. 14). Synonym: *Chondrophycus capituliformis* (Yamada) Garbary et Harper (1998: 194).
Palisada tumida **(Saito et Womersley) comb. nov.**
Holotype: ADU A42777. Basionym: *Laurencia tumida* Saito et Womersley (1974: 846, figures 5E, F, 26, 27). Synonym: *Chondrophycus tumidus* (Saito et Womersley) Garbary et Harper (1998: 195).
Palisada iridescens **(Wynne et Ballantine) comb. nov.**
Holotype: Wynne 8278 (=Ballatine 2675). Basionym: *Laurencia iridescens* Wynne et Ballantine (1991: 395–397, figures 1–11). Synonym: *Chondrophycus iridescens* (Wynne et Ballantine) Garbary et Harper (1998: 195).
Palisada maris-rubri **(Nam et Saito) comb. nov.**
Holotype: N002(=PHYT227). Basionym: *Laurencia maris-rubri* Nam et Saito (1995: 162, figures 22–29). Synonym: *Chondrophycus maris-rubri* (Nam et Saito) Garbary et Harper (1998: 195).
Palisada papillosa **(C, Agardh) comb. nov.**
Holotype: Herb. Forsskal. No. 886. Basionym: *Chondria papillosa* C. Agardh (1822: 344). Synonym: *Chondrophycus papillosus* (C. Agardh) Garbary et Harper (1998: 195).
Palisada parvipapillata **(Tseng) comb. nov.**
Holotype: Tseng 2813. Basionym: *Laurencia parvipapillata* Tseng (1943: 204–205, pl. IV). Synonym: *Chondrophycus parvipapillatus* (Tseng) Garbary et Harper (1998: 195).
Palisada gemmifera **(Harvey) comb. nov.**
Holotype: TCD. Basionym: *Laurencia gemmifera* Harvey (1853: 73–74, pl. XVIII, B). Synonym: *Chondrophycus gemmiferus* (Harvey) Garbary et Harper (1998: 194).
Palisada cruciata **(Harvey) comb. nov.**
Type: TCD (Trav. set 209), Herb. Harvey (Saito and Womersley, 1974). Basionym: *Laurencia cruciata* Harvey (1855: 544). Synonym: *Chondrophycus cruciatus* (Harvey) Nam (1999: 463).
Palisada yamadana **(M.A. Howe) comb. nov.**
Holotype: No. ?, New York Bot. Gard. and United St. Nat. Museum. Basionym: *Laurencia yamadana* M.A. Howe (1934: 37, figure 4). Synonym: *Chondrophycus yamadanus* (M.A. Howe) Nam (1999: 463).
Palisada flagellifera **(J. Agardh) comb. nov.**
Type: No. 36604–36606, Herb. J. Agardh LD (Yamada 1931: 197). Basionym: *Laurencia flagellifera* J. Agardh (1852: 747). Synonym: *Chondrophycus flagelliferus* (J. Agardh) Nam (1999: 463).
Palisada surculigera **(C.K. Tseng) comb. nov.**
Holotype: Tseng 293, Herb. Tseng. Basionym: *Laurencia surculigera* C.K. Tseng (1943: 192, pl. 1, figures 4, 5). Synonym: *Chondrophycus surculigerus* (C.K. Tseng) Nam (1999: 463).
Palisada longicaulis **(C.K. Tseng) comb. nov.**
Holotype: Tseng 2579, Herb. Tseng. Basionym: *Laurencia longicaulis* C.K. Tseng (1943: 194, pl. 2, figures 1, 2). Synonym: *Chondrophycus longicaulis* (C.K. Tseng) Nam (1999: 463).
Palisada jejuna **(C.K. Tseng) comb. nov.**
Holotype: Tseng 2565, Herb. Tseng. Basionym: *Laurencia jejuna* C.K. Tseng (1943: 189, pl. 1, figures 1–3). Synonym: *Chondrophycus jejunus* (C.K. Tseng) Nam (1999: 463).
Palisada concreta **(Cribb) comb. nov.**
Holotype: BRIU 877.18, Herb. Dept. Bot. Univ. Queensland, Saint Lucia, Australia. Basionym: *Laurencia concreta* Cribb (1983: 116). Synonym: *Chondrophycus concretus* (Cribb) Nam (1999: 463).
Palisada dinhii **(Masuda et Kogame) comb. nov.**
Holotype: SAP (062606♀). Basionym: *Laurencia dinhii* Masuda et Kogame (1998: 205, figures 8–26). Synonym: *Chondrophycus dinhii* (Masuda et Kogame) Nam (1999: 463).
Palisada poiteaui **(Lamouroux) comb. nov.**
Type: CN. Basionym: *Fucus poiteaui* Lamouroux (1805: 63. pl. xxxi: figures 2, 3). Synonym: *Laurencia poiteaui* (Lamouroux) M.A. Howe (1918: 518). Synonym: *Chondrophycus poiteaui* (Lamouroux) Nam (1999: 463).
Palisada perforata **(Bory) comb. nov.**
Type: No. ?, Herb. Thuret in Paris (Yamada 1931: 193). Basionym: *Fucus perforatus* Bory (1803: 503, pl. 5, figure 1). Synonym: *Laurencia perforata* (Bory) Montagne (1840: 155). Synonym: *Chondrophycus perforatus* (Bory) Nam (1999: 463).

with trichoblast type (Nam et al., 1994); spermatangial branches produced from one of two laterals on suprabasal cell of trichoblasts; procarp-bearing segments with four or five pericentral cells: auxiliary cells normally formed after fertilization; tetrasporangial production from particular pericentral cells; tetrasporangial axis with one sterile pericentral cell and the second pericentral cell being fertile.

Type species: *Palisada robusta* nom. nov. (renamed by tautonym)

Etymology: The epithet is derived from the robust thallus.

Holotype: SAP (13883) (Masuda et al., 1998)

Basionym: *Laurencia palisada*, Yamada (1931: 196) designated as lectotype of the section *Palisadae* Yamada (1931) by Masuda et al. (1998)

Synonym: *Chondrophycus palisadus* (Yamada) Nam (1999: 463)

Chondrophycus (Tokida et Saito) Garbary et Harper (1998) gen. emend.

Apical cell always sunk in apical pit of branchlet; central axis recognizable only near apical cell; forming extensive cortex; vegetative axial segments with two pericentral cells; first pericentral cell with side position of trichoblast; spermatangial development with trichoblast type; spermatangial branches produced from two laterals on suprabasal cell of trichoblasts, but remain partly sterile; procarp-bearing segments with five pericentral cells; formation of auxiliary cell somewhat delayed after fertilization; tetrasporangial production from particular pericentral cells; tetrasporangial axis with two sterile pericentral cells including the second pericentral cell.

Type species: *Chondrophycus cartilagineus* (Yamada) Garbary et Harper (1998: 194)

Holotype: SAP (?; Yendo's specimen)

Basionym: *Laurencia cartilaginea* Yamada (1931: 230)

Acknowledgments

I am grateful to Professor Y. Saito (retired from Hokkaido University, Japan) for providing the *Laurencia* complex specimens from various localities around the world, and anonymous reviewers for improving and criticizing the manuscript. Thanks are also due to Miss S-J. Lee and Mr. P.J. Kang for their assistance in preparing the manuscript. This study was supported by a grant (no. 2000-1-20200-003-3) from the KOSEF.

References

Abe T, Masuda M, Kawaguchi S, Kamura S (1998) Taxonomic notes on *Laurencia brongniartii* (Rhodomelaceae, Rhodophyta). Phycological Research 46: 231–237.

Abbott IA, Hollenberg GJ (1976) Marine algae of California. Stanford Univ. Press.

Agardh C (1822) Species Algarum. Vol. 1, Pt. 2, Lund.

Agardh JA (1852) Species genera et ordines algarum. Vol. 2, Part 2, Lund.

Agardh JG (1876) Species genera et ordines algarum. Vol. 3. Lipsia [Leipzig].

Bold HC, Wynne MJ (1978) Introduction to the algae. Structure and reproduction. Prentice-Hall, INC., Englewood Cliffs, New Jersey.

Bory de Saint-Vincent JB (1803) Essai sur les îles Fortunées.

Cormaci M, Furnari G, Serio D (1994) Two new species of *Laurencia* (Ceramiales, Rhodophyta) from the Mediterranean Sea: *Laurencia pelagiensis* sp. nov. and *Laurencia verlaquei* sp. nov. Japanese Journal of Phycology 42: 365–375.

Cribb AB (1983) Marine Algae of the Southern Great Barrier Reef-Rhodophyta. Australian Coral Reef Society, Watson Ferguson and Co., Brisbane.

Fujii MT, Cordeiro-Marino M (1996) *Laurencia translucida* sp. nov. (Ceramiales, Rhodophyta) from Brazil. Phycologia 35: 542–546.

Fujii MT, Collado-Vides L, Cordeiro-Marino M (1996) Morphological studies of *Laurencia gemmifera* and *Laurencia poiteaui* (Rhodomelaceae, Rhodophyta) from the Nichupte Lagoon System, Quintana Roo, Mexico. Botanica Marina 39: 317–326.

Furnari G, Serio D (1993a) The distinction of *Laurencia truncata* (Ceramiales, Rhodophyta) in the Mediterranean Sea from *Laurencia pinnatifida*. Phycologia 32: 367–372.

Furnari G, Serio D (1993b) The reproductive structures of the Mediterranean alga *Laurencia pelagosae* (Ceramiales, Rhodophyta). European Journal of Phycology 28: 141–143.

Furnari G, Serio D, Cormaci M (2004) Revision of *Laurencia pedicularioides* (Ceramiales, Rhodophyta). Taxon 53: 453–460.

Garbary DJ, Harper JT (1998) A phylogenetic analysis of the *Laurencia* complex (Rhodomelaceae) of the red algae. Cryptogamie Algologie 19: 185–200.

Gmelin SG (1768) Historia Fucorum. Academia Scientarum, St Petersburg.

Gordon-Mills EM (1987) Morphology and taxonomy of *Chondria tenuissima* and *Chondria dasyphylla* (Rhodomelaceae, Rhodophyta) from European waters. British Journal of Phycology 22: 237–255.

Gordon-Millis EM, Womersley HBS (1987) The genus *Chondria* C. Agardh (Rhodomelaceae, Rhodophyta) in southern Australia. Australian Journal of Botany 35: 477–565.

Harper JT, Garbary DJ (1997) Marine algae of Northern Senegal: The flora and its biogeography. Botanica Marina 40: 129–138.

Harvey WH (1855) Some account of the marine botany of the colony of Western Australia. Transactions of the Royal Irish Academy 22: 525–566.

Harvey WH (1862) Phycologica Australica. Vol. 4, London.

Howe MA (1918) Class 3, Algae. In Britton NL (ed), Flora of Bermuda (Illustrated). New York: pp 489–540.

Howe MA (1934) Hawaiian algae collected by Dr. Paul C. Galtsoff. Journal of Washington Academic Science 24: 32–42.

Hudson V (1778) Flora anglica. Editio altera. London.

Kylin H (1956) Die Gattungen der Rhodophyceen. Gleerups. Lund.

Lamouroux M (1805) Dissertations sur plusieurs espèces de Fucus. Agen.

Lamouroux JV (1813) Essai sur les genres de la famille des Thallasiophyte non articulées. Annales du Musée d'Histoire Naturelle de Paris 20: 21–47, 115–139, 267–293.

Maggs CA, Hommersand MH (1993) Seaweeds of the British Isles. Volume 1. Rhodophyta, Part 3A. Ceramiales. Natural History Museum, London.

Masuda M, Kogame K (1998) A taxonomic study of the genus *Laurencia* (Ceramiales, Rhodophyta) from Vietnam. V. *Laurencia concreta* Cribb and *L. dinhii* sp. nov. Cryptogamie Algologie 19: 201–212.

Masuda M, Kawaguchi S, Phang SM (1997a) Taxonomic notes on *Laurencia similis* and *L. papillosa* (Ceramiales, Rhodophyta) from the Western Pacific. Botanica Marina 40: 229–239.

Masuda M, Kogame K, Abe Y, Kamura S (1997b) Taxonomic notes on *Laurencia parvipapillata* (Ceramiales, Rhodophyta) from the western Pacific. Cryptogamie Algologie 18: 319–329.

Masuda M, Kogame K, Abe Y, Kamura S (1998a) A morphological study of *Laurencia palisada* (Rhodomelaceae, Rhodophyta). Botanica Marina 41: 133–140.

Masuda M, Kogame K, Arisawa S, Suzuki M (1998b) Morphology and halogenated secondary metabolites of three Gran Canarian species of *Laurencia* (Ceramiales, Rhodophyta). Botanica Marina 41: 265–277.

McDermid KJ (1988) Section V. *Laurencia* (Rhodophyta, Rhodomelaceae). In Abbott I.A. (ed.), Taxonomy of economic seaweeds. With reference to some Pacific and Caribbean species Volume II. Publication of California Sea Grant College Program: 264 pp.

McDermid K (1989) *Laurencia crustiformans* sp. nov. (Ceramiales, Rhodophyta) from the Hawaiian Islands. Phycologia 28: 352–359.

McIvor L, Maggs CA, Guiry MD, Hommersand MH (2002) Phylogenetic analysis of the geographically disjunct genus *Osmundea* Stackhouse (Rhodomelaceae, Rhodophyta). Constancea 83 pp.

Montagne C (1840) Histoire naturelle des iles Canaries. Phytographia canariensis sectio ultima.

Nam KW (1990) Morphotaxonomic studies of genus *Laurencia* (Rhodomelaceae, Rhodophyta). Unpublished PhD Thesis. Hokkaido University, Japan.

Nam KW (1994) Vegetative and male reproductive anatomy of *Laurencia intercalaris* sp. nov. (Rhodomelaceae, Rhodophyta) in Korea. Journal of Plant Biology 37: 421–428.

Nam KW (1999) Morphology of *Chondrophycus undulata* and *C. parvipapillata* and its implications for the taxonomy of the *Laurencia* (Ceramiales, Rhodophyta) complex. European Journal of Phycology 34: 455–468.

Nam KW (2004) *Laurencia* complex: Morphology of *Laurencia pinnata* (Rhodophyta), with a key to Korean species. In Abbott IA, McDermid KJ (eds), Taxonomy of economic seaweeds. Vol. 9. Taxonomy of economic seaweeds. With reference to some Pacific and other locations Volume IX. Publication of Hawaii Sea Grant College Program: 255–271.

Nam KW, Choi HG (1999) Developmental morphology of *Osmundea crispa* (Ceramiales, Rhodophyta) from California. Journal of Fisheries Science and Technology 2: 25–31.

Nam KW, Choi HG (2000) A detailed morphological study of the type species of *Osmundea* (Rhodomelaceae, Rhodophyta). Botanica Marina. 43: 291–297.

Nam KW, Choi HG (2001) Morphology of *Laurencia clavata* and *L. elata* (Ceramiales, Rhodophyta) in relation to generic circumscription in the *Laurencia* complex. European Journal of Phycology 36: 285–294.

Nam KW, Saito Y (1990) Morphology of *Laurencia cartilaginea* Yamada (Rhodomelaceae, Rhodophyta). Bulletin of the Faculty of Fisheries, Hokkaido University 41: 107–120.

Nam KW, Saito Y (1991a) Anatomical characteristics of *Laurencia papillosa* (Rhodomelaceae, Rhodophyta) from Guam and Palau. Micronesica 24: 87–94.

Nam KW, Saito Y (1991b) *Laurencia similis* (Ceramiales, Rhodophyta), a new species from Queensland, Australia. British Phycological Journal 26: 375–382.

Nam KW, Saito Y (1994) A re-examination of *Laurencia hybrida* (Ceramiales, Rhodophyta) from the British Isles: vegetative and reproductive morphology. Phycologia 33: 34–41.

Nam KW, Saito Y (1995) Vegetative and reproductive anatomy of some *Laurencia* (Ceramiales, Rhodophyta) species with a description of *L. maris-rubri* sp. nov. from the Red Sea. Phycologia 34: 157–165.

Nam KW, Sohn CH (1994) *Laurencia kangjaewonii* sp. nov. (Ceramiales, Rhodophyta) from Korea. Phycologia 33: 397–403.

Nam KW, Saito Y, Sohn CH (1994) Vegetative structure and reproduction of *Laurencia nipponica* Yamada (Rhodomelaceae, Rhodophyta). Korean Journal of Phycology 6: 1–12.

Nam KW, Choi HG, Lee S, Park EJ, Kang KH, Kim YS (2000a) Vegetative and reproductive development of *Laurencia venusta* (Ceramiales, Rhodophyta). Cryptogamie Algologie 21: 97–110.

Nam KW, Maggs CA, Garbary DJ (1994) Resurrection of the genus *Osmundea* with an emendation of the generic delineation of *Laurencia* (Ceramiales, Rhodophyta). Phycologia 33: 384–395.

Nam KW, Maggs CA, McIvor L, Stanhope MJ (2000b) Taxonomy and phylogeny of *Osmundea* (Rhodomelaceae, Rhodophyta) in Atlantic Europe. Journal of Phycology 36: 759–772.

Saito Y (1967) Studies on Japanese species of *Laurencia,* with special reference to their comparative morphology. Memoirs of the Faculty of Fisheries, Hokkaido University 15: 1–81.

Saito Y (1969a) On morphological distinctions of some species of Pacific North American *Laurencia*. Phycologia 8: 85–90.

Saito Y (1969b) The algal genus *Laurencia* from the Hawaiian Islands, the Philippine Island and adjacent areas. Pacific Science 23: 148–160.

Saito Y (1982) Morphology and infrageneric position of three British species of *Laurencia* (Ceramiales, Rhodophyta). Phycologia 21: 299–306.

Saito Y, Womersley HBS (1974) The Southern Australian Species of *Laurencia* (Ceramiales: Rhodophyta). Australian Journal of Botany 22: 815–874.

Serio D, Cormaci M, Furnari G (1999) *Osmundea maggsiana* sp. nov. (Ceramiales, Rhodophyta) from the Mediterranean Sea. Phycologia 38: 277–282.

Smith GM, Hollenberg GJ (1943) On some Rhodophyceae from the Monterey Peninsula, California. American Journal of Botany 30: 211–222.

Sonder OG (1853) Plantae Muellerianae. Algae. Linnaea 25: 657–709.

Stackhouse J (1809) Tentamen marino-cryptogamicum. Mémoires de la Société Impériale des Naturalistes de Moscou 2: 50–97.

Swofford DL (2002) PAUP*: Phylogenetic analysis using parsimony (and other methods), version 4.0 beta. Sinauer Associates, Sunderland, Massachusetts.

Tseng CK (1943) Marine algae of Hong Kong. IV. The genus *Laurencia*. Papers of the Michigan Academy of Science, Arts and Letters 28: 185–208.

Womersley HBS (2003) The marine benthic flora of southern Australia. Part IIID. Australian Biological Resources Study, Canberra and the State Herbarium of South Australia. Pirion, Canberra.

Wynne MJ, Ballantine DL (1991) *Laurencia iridescens* sp. nov. (Rhodomelaceae, Ceramiales) from the Caribbean Sea. Phycologia 30: 394–401.

Yamada Y (1931) Notes on *Laurencia,* with special reference to the Japanese species. University of California Publications in Botany 16: 185–310.

Yoneshigue-Valentin Y, Fujii MT, Gurgel CFD (2003) *Osmundea lata* (M. Howe & W.R. Taylor) comb. nov. (Ceramiales, Rhodophyta) from the Brazilian south-eastern continental shelf. Phycologia 42: 301–307.

MIX
Papier aus verantwortungsvollen Quellen
Paper from responsible sources
FSC® C105338

If you have any concerns about our products,
you can contact us on
ProductSafety@springernature.com

In case Publisher is established outside the EU,
the EU authorized representative is:
Springer Nature Customer Service Center GmbH
Europaplatz 3, 69115 Heidelberg, Germany

Printed by Libri Plureos GmbH
in Hamburg, Germany